*Society of Exploration Geophysicists*

# SEISMIC PHYSICAL MODELING

*Edited by:*

*Daniel A. Ebrom*

*John A. McDonald*

*Series Editor:*

*Franklyn K. Levin*

*Geophysics reprint series*
*No. 15*

Seismic physical modeling / edited by Daniel A. Ebrom, John A. McDonald.
p. cm.—(Geophysics reprint series; no. 15)
Includes bibliographical references.
ISBN 1-56080-072-0
1. Seismic waves—Simulation methods. 2. Geological modeling.
I. Ebrom, Daniel A., 1957- . II. McDonald, John A. (John Andrew), 1931- . III. Series.
QE538.5.S42 1994
622'.1592'0228—dc20 94-34763
CIP

ISBN 0-931830-00-1 (Series)
ISBN 1-56080-072-0 (Volume)

Library of Congress Catalog Number

Society of Exploration Geophysicists
P. O. Box 702740
Tulsa, Oklahoma 74170-2740

Published in 1994

Printed in the United States of America

## Contents

# Preface

The most important criterion for selection of the articles in this volume is the significance of the *results* of the particular physical modeling experiment, rather than the cleverness of the experimental technique. Our hope is that each of the papers enlightens a facet of *wave behavior.* The scientists who performed these experiments were (like industrial geophysicists) more concerned with getting *answers* than doing experiments just for the sake of experimentation. Papers primarily concerned with instrumentation issues have been relegated to the Annotated Bibliography.

Selection of articles for the volume was also guided by several general guidelines (or biases, if you prefer). First of all, as befits a book published by an association of applied geophysicists, we have chosen those articles which emphasize exploration-relevant seismic modeling, rather than crustal and deep-earth seismic modeling. Second, we have not included any articles on borehole acoustic physical modeling, although we have pointed the interested reader to some relevant papers in the Annotated Bibliography. Third, we have not included any non-seismic physical modeling, i.e., no electrical or electromagnetic physical modeling. (Perhaps some enterprising electrical physical modelers will publish their own reprint volume!) Last, we emphasized publications written in (or translated into) English. This last criterion necessarily eliminated a number of fine papers. Again, some of these papers are referenced in the Annotated Bibliography.

The structure of this book is roughly chronological. In each chapter, the papers are listed in order of publication date. After the overviews (Chapter 1), the reader is introduced to the earliest physical modeling papers (Chapter 2). These papers are, much as the early seismic interpreters were, primarily concerned with the *kinematics* (timing) of wave propagation. In the next section (Chapter 3), physical model data are used to validate wave theoretical predictions, primarily by use of *amplitude* measurements. The advent of digital computers immediately suggested the possibility of testing seismic processing algorithms using physical model data (Chapter 4). The real earth contains many complicated rock geometries, and these are discussed in Chapter 5. In the concluding section, we address the question of where physical modeling is heading and its relevance to current research problems (Chapter 6).

Although a second edition of this reprint book is not planned, we encourage readers to inform the editors of significant physical modeling papers that have been overlooked in the current volume.

**Acknowlegments**

In selecting and editing these articles, we have been fortunate to be aided by a number of scholars, whose help we very much appreciate. A list of these colleagues follows:

| | |
|---|---|
| Al Balch | Doug McCowan |
| Jim Brown | Larry Myer |
| James Brune | Jack Oliver |
| Art Cheng | Sheila Peacock |
| Neville Cook | Patrick Rasolofosaon |
| Brian Evans | Jazz Rathore |
| Tony Gangi | Eike Rietsch |
| Gerry Gardner | Joe Rosenbaum |
| Tom Goforth | Mike Schoenberg |
| Paul Golden | K. K. Sekharan |
| Neil Goulty | Bob Sheriff |
| Gene Herrin | John Sherwood |
| Fred Hilterman | Rob Stewart |
| Julie Hood | Bob Tatham |
| Chaur-Jian Hsu | Sven Treitel |
| Jiri Jech | Norm Uren |
| Sid Kaufman | L. Waniek |
| Don Lawton | Ed White |
| Frank Levin | Paul Wuenschel |


NOTE: Within the editorial portions of this reprint book, references to papers are occasionally made for comparative purposes. If the referenced paper is not reprinted in this volume, then it will be found in the annotated bibliography.

# Chapter 1.
# Overview

In a seismic physical model experiment, the desired result, the seismic reflection response for a specific geological model, is determined through the process of measuring the reflected elastic or acoustic wavefield over a scaled model.

Physical modeling is a form of analog computing. Analog computers, once widespread, have become less common following the introduction of digital computers. Although analog computers are generally quite specialized and lack the flexibility of digital computers, they have some appealing features.

Advantages of analog computers include:

- no simplifying mathematical assumptions,
- no approximations to mathematical functions,
- no round-off errors, and
- *a priori* understanding of the mathematics of the modeled phenomenon is not the limiting factor in accuracy of results.

In point of fact, virtually all physical modeling done today would be better termed "hybrid analog/digital." Digital computers are used to automatically position source and receiver transducers, initiate the source pulse, digitize the received signal, and write the digitized data onto magnetic storage media.

Of course, every tool has its disadvantages. The principal disadvantages of analog computing in general, and physical modeling in particular, are the errors to be expected in any experiment. These include:

- bandwidth limited by available transducers,
- dynamic range limited by available electronics,
- precision in placement of sources and receivers limited by accuracy of available placement technologies,
- precision in placement of interfaces in physical models limited by available construction techniques, and
- parameter ranges for velocity and density limited to those materials that can actually be found or fabricated.

For piezoelectric transducers, the specific limitations are:

- strong resonance at one frequency (i.e., restricted bandwidth),
- source and receiver dimensions large compared to a wavelength (in the field, individual sources and receivers are generally small compared to a wavelength), and
- pronounced directionality.

Some geophysicists have had a philosophical reservation about physical modeling. The underlying assumption is that the scaled physical mechanisms are identical to field physical mechanisms. Stated more simply, we are assuming that waves propagate in model materials in the same way that they propagate in real-earth materials. While this may seem a strong assumption, many physical modeling experiments show that wave propagation in physical models is well explained by infinitesimal strain elastic-wave theory. As this same mathematical theory has been shown to apply to field seismic data, we may complete the syllogism to conclude that physical model data and field seismic data involve the same mechanisms.

However, not all earth materials are equally easy to reproduce in physical model experiments. A notoriously difficult case is that of the weathering layer, which generally has both very low Q and very high Poisson's ratio. Although the presence of the weathering layer has strong effects on land seismic data quality (often beneficial by attenuating ground roll), virtually all physical models lack a realistic weathering layer.

## Scaling and scale factors

Probably nothing has generated quite as much confusion amongst the users of physical model data as the twin topics of "scaling" and "scale factors."

First of all, we must decide which **physical quantities** will be preserved in a scale model experiment. Typically, the quantity that we wish to preserve is the kinematics, i.e., the relative traveltimes of specific events. For kinematics, physical model experimental design is relatively straightforward. There is a cardinal rule:

"The **ratio** of geological feature size to wavelength must be the **same** in both the field and the model."

This seemingly innocuous rule, if faithfully obeyed, will eliminate most uncertainties in physical modeling for kinematics.

As a practical example, suppose we are modeling the seismic response of an anticline with 100 m relief and 300 m width (Figure 1). If a seismic pulse has a velocity of 3000 m/s, then the wavelength at 30 Hz is 100 m. Thus, the ratio of the anticlinal relief to the wavelength is 100 m/100 m, or 1.0. The ratio of the anticline's width to the wavelength is 300 m/100 m, or 3.0. Our physical model will have to preserve these two ratios.

Let us now consider the physical model. For ease of construction, the model might be constructed from (RTV) silicone rubber, with a *P*-wave velocity of about 1000 m/s. Commonly available source transducers have central frequencies at about 100 kHz. Hence, the central wavelength in the physical model is about 1 cm. To preserve the ratios of our anticline's relief and width to the wavelength, our scaled anticline must have relief of 1 cm and width of 3 cm (Figure 2). Thus, our **length** scale factor is 10 000: 1; that is, we have scaled 100 m in the field to 1 cm in the model.

Chosen independently of the length scale factor is the **time** scale factor. The time scale factor is the ratio of the time digitization interval in the field to the time digitization interval in the model. A common field time digitization interval is 1 ms. If we chose the model time digitization interval at 0.3 $\mu$s, then we would have a time scale factor of 3333: 1. One result of this choice is that we have a **velocity** scale factor of 3.0. (The velocity scale factor is the ratio of the length scale factor to the time scale factor.) Thus, a model velocity of 1000 m/s scales to 3000 m/s in the field.

Many modeling materials have velocities much lower than true rock velocities. By using appropriate ratios of length and time scale factors, the scaled velocities of the modeling materials can be matched to the true field velocities.

A few comments about scale factors are in order. First, the choice of a different length scale factor is equivalent to a "stretch" or a "squeeze" of the horizontal (distance) axis when displaying a seismic section. Thus, the length scale factor is, in a very real sense, arbitrary. Second, the choice of a different time scale factor is equivalent to a "stretch" or "squeeze" of the vertical (time) axis when displaying a seismic section. The time scale factor is also arbitrary.

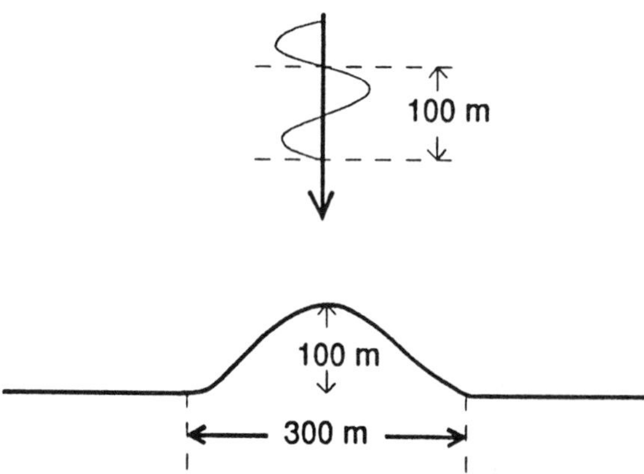


FIG. 1. A possible field scenario, with an anticline 100 m high and 300 m wide, and a dominant seismic wavelength of 100 m. The **ratio** of anticlinal **relief** to seismic **wavelength** is 100 m/100 m, or **1.0.** The **ratio** of anticlinal **width** to seismic **wavelength** is 300 m/ 100 m or **3.0.** These two ratios must be preserved in the physical model.

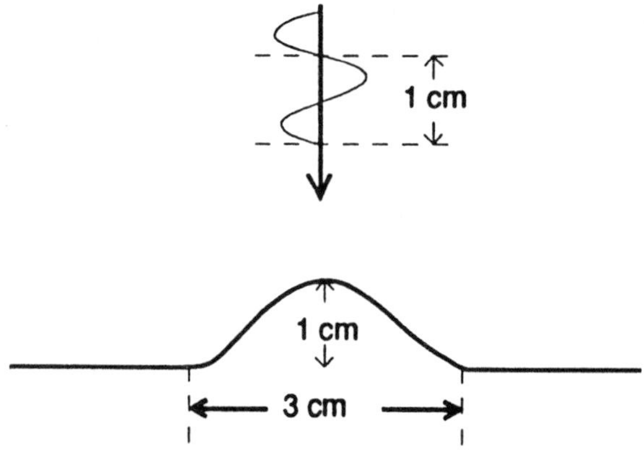


FIG. 2. A physical modeling scenario equivalent to Figure 1, with an anticline 1 cm high and 3 cm wide, and a dominant seismic wavelength of 1 cm. The **ratio** of anticlinal **relief** to seismic **wavelength** is 1 cm/1 cm, or **1.0.** The **ratio** of anticlinal **width** to seismic **wavelength** is 3 cm/1 cm or **3.0.**

What is not arbitrary? The requirements of minimum temporal and spatial sampling (the Nyquist criteria) must be obeyed regardless of scale factor choice. These criteria are best evaluated using the actual (unscaled) dimensions of the model experiment. For example, a wavelet with a central frequency of 100 000 Hz must be sampled with a time digitization interval no larger than 5 μs to satisfy the Nyquist criterion and avoid aliasing. Similar arguments apply to the trace spacing (CMP interval) on the model to avoid spatial aliasing.

Note that in the example above, nothing was mentioned regarding the **amplitudes** of the recorded events. To faithfully preserve amplitudes, a host of contributing factors must be measured and accounted for. These include, but are not limited to:

- source and receiver transducer directionality,
- source and receiver transducer coupling to the physical model,
- attenuative properties of modeling materials,
- geometrical spreading (if the model is not 3-D), and
- impedance contrasts (i.e., reflection coefficients) in the model.

## The place of physical modeling

Physical modeling competes in the research marketplace against other research techniques. As alternative research techniques have developed, physical modelers have moved on, like frontier pioneers, to those problems for which no other method is yet suitable.

When first introduced in the 1920s, the only competitor to physical modeling was hand calculation of ray-tracing or analytic numerical models. The introduction of electronic computing devices in the early 1960s represented the first major challenge to physical modeling as a tool for research into wave phenomena.

Primarily because of its flexibility and relatively low cost, numerical modeling on digital computers has undergone an enormous expansion in usage. Virtually any research group can now purchase a workstation capable of executing a variety of forward-modeling programs. Research into the seismic response of 2-D acoustic (or isotropic elastic) geologic models can be handled more easily and cheaply numerically rather than physically.

For the acoustic 3-D geological model response, physical modeling is still probably the preferred method. Chen and McMechan (1993) have demonstrated they can calculate 3-D shot gathers which are quite similar to shot gathers produced by physical modeling. Given likely decreases in the cost of computing, numerical modeling will probably replace physical modeling for the acoustic 3-D case in the next 10-20 years.

Areas where physical models are likely to remain most valuable include research into:

- the isotropic elastic 3-D geological model response,
- the anisotropic 2-D or 3-D geological model response, and
- studies into wave mechanisms, e.g., wave propagation in media containing microcracks, layering, fractures, etc.

The papers in chapter 6 of this volume were chosen to give some feel for those research areas in which physical modeling will continue to play a significant role.

Aside from supporting research into seismic wave behavior, physical modeling can provide technical support for some more applied endeavors. Nonresearch uses for physical modeling include:

- providing inexpensive data for evaluating proposed geological models related to specific oil and gas prospects,
- providing inexpensive data for educational purposes, and
- providing a check on the output of numerical models.

### Overview articles

A number of good general introductions to physical modeling have been written. We will reproduce here only three of them.

+Riznichenko (1965) examines the philosophical position of physical modeling, and defends its utility in the face of ever more powerful digital computers.

+Ivakin (1965) gives a short introduction to 2-D models and electrical modeling of seismic waves.

+O'Brien (1971) actually goes through much of the "nuts and bolts" of both 2-D and 3-D physical modeling.

Other articles that will give a good general grounding include chapter 6 of White (1965), chapter 1 of McDonald et al. (1983), Pant and Greenhalgh (1988) and Menke and Dubendorff (1989).

*Symposium on Seismic Models, Castle of Liblice near Prague, 9–12 November 1965*

# SEISMIC MODELLING, DEVELOPMENT AND OUTLOOK

YU. V. RIZNICHENKO

*Institute of Physics of the Earth, Acad. Sci. USSR, Moscow*)*

Seismic modelling is not in itself a goal but a means. It is one of the aids for studying such basic problems of seismology as the origin of earthquakes, the generation of seismic waves in the foci of earthquakes and by means of explosions, the relation of the wave pattern of seismic waves and the structure of the Earth, the effect of tremors on the Earth's surface and on structures on the Earth's surface during the passage of seismic waves.

All these problems have been more or less modelled but the greatest interest has been concentrated on the problem of the propagation of seismic waves in connection with the structure of the medium; here, too, the greatest attention will be paid to this problem. But in conclusion a number of other problems will also be dealt with.

Twenty years ago almost all quantitative deductions in the physics of elastic waves, relating to general seismology and to seismic prospecting, were based merely on kinematic considerations. The dynamic theory of elasticity was at that time mostly detached from the practical problems of general and applied geophysics. It was clearly felt that the kinematic methods of the interpretation of seismic data had to be supplemented by dynamic methods and that the gap between dynamic theory and actual practice in seismic observations had to be bridged.

One of the possibilities for bringing theory nearer to seismic experiment was the application of the general theory to the solution of concrete geophysical problems. This was later done to a great extent in the Soviet Union – in the work of groups of theoreticians headed by G. I. Petrašeň and A. S. Aleksejev. Another possibility, or aid, was seismic modelling. At that time, however, it was only in its beginnings. It had to be brought into the world, and it was.

Today, in 1965, it is just 20 years since systematic research into the modelling of seismic wave phenomena was first begun in a small laboratory of the Institute of Theoretical Geophysics (Acad. Sci. USSR) ([1], see also [2]). That was in the difficult but happy year when the war ended. A year of new hopes and beginnings. Our hopes proved right. Today seismic modelling is widespread in the Soviet Union and also in other progressive countries, and this international symposium is clear proof of this. Another proof is the number of papers published. In the Soviet Union alone, there have already been more than a hundred. The situation is similar in other countries.

## Stages of Development

I shall not deal here with a description of the prehistory of today's seismic modelling This has already been done [3]. I shall confine myself merely to the post-war development of the problems.

*) Address: B. Gruzinskaja 10, Moskva G-242, USSR.

Reprinted from Studia Geophysica et Geodaetica, 1966, Volume 10, 243-253 with permission of Plenum Press

The first attempts made by me and B. N. Ivakin to model seismic waves were done on a two-dimensional discrete model represented by a rubber net with brass weights in the nodes, which resembled a two-layered continuous medium. The movements of the weights were recorded with a loop oscillograph by means of photoelectric equipment. The first papers on this appeared in 1949–1950 [2, 4] when this stage of work ended. Modelling by means of nets was interrupted.

This was because soon after beginning the experiments with net models it became clear that on an arbitrary model of this type, either mechanical or electrical, containing a limited number of discrete elements, it is very difficult, if not impossible, to obtain a sufficiently detailed wave pattern. From net models we obtain a wave pattern as through a very coarse raster. It is true that this difficulty is not of a fundamental but of a technical character and it is possible that a return will be made to net models when real possibilities arise for making them sufficiently detailed, two-dimensional and even three-dimensional. The one-dimensional type, to which some authors later returned (particularly in the case of electrical analog simulation [5, 6]) has, of course, only limited significance.

The possibility of modelling by means of nets thus proved to be very limited at this stage of development. We were also not satisfied by the possibility of the primarily qualitative and demonstrative character which was provided by the earlier methods of producing elastic waves in agar-agar or analogous media having the character of gravitational waves on the surface of a liquid (Tindal, see e.g. [7, 8]) or various optical [9, 10] and stationary ultrasonic methods [11].

Earlier seismologists and others dealing with seismic prospecting remember the remarkable photographs of wave fronts once obtained by Frank Rieber et al. [9] in the United States by the shadow method for stroboscopic illumination or by O. v. Schmidt [10] in Germany by means of the schlieren method. Their effective photographs clearly showed the general pattern of wave propagation and permitted a characterization of their kinematics but provided little information on the dynamics of the wave process, i.e. often that which interested us most. There are undoubtedly possibilities for further supplementing this, such as the determination of the dynamic parameters of waves by means of photometry, particularly in polarized light, since at that time such methods were little known. As regards the use of ultrasonic apparatuses with a stationary regime, it is obvious that such a regime does not correspond to the pulse character of earthquake problems or to problems of seismic prospecting by means of explosions.

Already in 1947–48 we turned our attention to a fundamentally different method and modelling technique — to the pulse ultrasonic method. The basis for working out this method in our laboratory was provided in 1934 by S. Y. Sokolov's discovery of a pulse ultrasonic defectoscope [12]. The technique of this new type of modelling was similar to radar, which was intensively developed during the last years of the war.

The pulse ultrasonic method proved successful and it was soon used to perform research into a number of three-dimensional solid-liquid models with layered structures as well as structures with vertical steps or "faults". The first results obtained with this method were published in 1951 [3] and they were followed by others [13, 14]. From the very first experiments the questions of similarity were more or less discussed and were later worked out in much greater detail [15, 16] etc.

In our first papers dealing with this method we tried out electromagnetic, spark and piezoelectric transmitters and piezoelectric receivers. In later papers only piezoelectric transducers were as a rule used. However, the spark resembles the seismic effect of an explosion better and this can find application in modelling series of special problems, particularly in explaining the formation of waves in the region near to the source. Examples of the use of a spark in modelling in later periods are known ([17] etc.) but the advantages of such a source are still far from being exhausted.

Two years after the first Soviet publication on seismic modelling by means of ultrasonic pulses [3] the first American reports appeared: Howes et al. [17] and also Northwood and Anderson [18]. Reports on similar work then began to arrive from other countries, from Japan, France etc.

As I have already said, the first models investigated by means of ultrasonics were three-dimensional and solid-liquid and most frequently consisted of thin rigid plates submerged in the liquid. An important stage in the development of methods was the use of two-dimensional solid models — plates thin in comparison with the wave-length used. Such a model was tried out for the first time in the Soviet Union in 1952 [13] and two years later in the United States [19].

This method had the advantage not only of a technical character (the two-dimensional model is easier to produce than the three-dimensional one and the damping of the waves with distance is smaller in it) but also of a fundamental character. It is important that two-dimensional models are solid models, in the same way as the modelled natural medium — the Earth — and not solid-liquid (only in problems of sea seismics is the medium solid-liquid [20] etc.; there is also another possible exception — the Earth's crust in volcanic regions). The wave processes in a solid-liquid system distinctly differ from processes in a solid medium. The conditions on the interface are particularly different. In solid media the contact between layers is usually rigid, but between a solid and a liquid it is sliding. In connection with this, layered media in a solid-liquid have some dependences, important from the geophysical aspect, which are even of opposite significance to the dependences obtained in solid media. This relates, for example, to the dependence of the damping of a guided and head wave on the thickness in rigid thin plates with higher velocity [21].

In the field of two-dimensional modelling an important step forward was made in 1956 by Oliver [22] who proposed making bimorphic models, i.e. models composed of two layers, possibly with a variable thickness, glued together along their surfaces. This makes it possible to vary the density and elastic properties of the bimorphic sheet, and in particular to make its generalized parameters smoothly variable with depth (e.g. as in the case of transit layers or Gutenberg's low-velocity channel in the Earth's upper mantle). At the same time, perforated and other continuous discrete two-dimensional models began to be elaborated in the Soviet Union in 1948 with the same aim — to vary the parameters of the medium [23, 24].

Methods of three-dimensional modelling were also developed. A large step forward in this direction was made by the Czechoslovak seismologists Vaněk et al. [25] in 1964, when they elaborated a fine method for the production of a three-dimensional horizontally layered medium with practically continuously varying properties as a function of depth. This was done on the basis of gelatine materials. Unfortunately, such materials have the properties of a liquid in the ultrasonic frequency range used; this disadvantage for modelling seismic processes has already been discussed. The progressive use of higher frequencies of the order of Mc/s in comparison with the frequencies of the order of hundreds of kc/s used earlier has also appeared in Czechoslovak research. This permits an increase in the resolving power of the experiments. Earlier similar attempts to increase the frequencies in modelling are clear in some American papers [26] and later in Soviet ones [21].

Shamina [21, 27] in the Soviet Union has recently (1964—65) succeeded in elaborating an effective method of producing solid three-dimensional micromodels, especially for research into transit layers and wave-guides with a non-sharp boundary in a frequency range of the order of Mc/s. This was done on the basic of epoxide resins with powdered minerals or other fillers.

I have not mentioned here all the improvements in modelling methods and have not dealt at all with questions of apparatus. It is not possible to deal with such a wide subject as seismic modelling is rapidly becoming. Work is going on and I hope that at this symposium I shall hear much that is new and important.

## Mutual Relations

I shall now turn my attention to the mutual relation between seismic modelling and theoretical work and field observations.

It is natural that those problems were modelled which most interested the individual seismologists. At present, the range of problems modelled is almost as wide as the range of interests of seismologists. It is simply impossible to name them all here and it is even barely possible to give a simple classification of them. Instead, I shall confine myself to a few examples, perhaps rather dramatic ones.

In the Soviet Union over the last two decades we have been most interested in body waves, particularly longitudinal, but also shear and converted ones (in the United States and recently also in Germany, West Berlin [28], interest has been concentrated on surface waves). This has also played a role in modelling. From the beginning Soviet seismologists have clearly been oriented on problems of wave propagation in layered media with sharp boundaries of homogeneous layers: thick, thin and with intermediate thickness in comparison with the wave-length. This was closely connected with the development of the correlation method of refracted waves (CMRW) and with the method of deep seismic sounding (DSS) and to a smaller extent also with problems of the method of reflected waves (MRW), particularly with the method of controlled directional reception (CDR). But problems have also been modelled that are related to all above-mentioned spheres of experimental seismics and also to a number of problems of general seismology.

It was soon found in modelling that an important role is played by reflections, beginning beyond the critical point, in the formation of a wave pattern in media with sharp boundaries. It was more than once found when describing the results of modelling that reflections in such media are usually far more intensive than the main refracted waves. On the other hand, in field measurements by means of CMRW and DSS only small attention was previously unjustifiably paid to reflections beyond the critical point and most often they were simply not interpreted at all or in other cases were ascribed to refracted waves. Unfortunately, workers in modelling have not yet succeeded in directing the attention of the broad seismic public to this problem. A great contribution towards learning the objective reality in this direction has been made by theoreticians (A. S. Aleksejev et al.). As they made rapid progress forward in the field of the dynamics of waves in such media, theoreticians at this time had no close contact with workers in modelling and possibly also regarded their results disparagingly.

It is true, of course, that no experiment, even modelling, is so "pure" as theory itself. But in theory there always exists the danger of excessively simplifying the situation. From the objective point of view it requires the perfect formulation of the conditions of the problem in order to obtain a clear solution. It is necessary to have a feeling for reality and proportion, to understand the limited validity of one's assumptions and conclusions in relation with the surrounding world. As is known, however, the supply of such feeling is in reality distributed very unevenly between experimenters and theoreticians.

Theoreticians were at first occupied with the sharp and at the same time plane interfaces and did not appreciate the possible influence of unsharp and uneven second-order boundaries, which might increase the importance of refracted waves compared with reflected ones (this had already been known to seismologists for a long time, as is borne out by the many remarks in the literature). In practice, this led to all intensive waves observed in nature being explained by ardent theoreticians as reflected. This category included particularly the $P^*$ wave known in general seismology and in DSS, which was earlier ascribed to a wave refracted on the boundary of the "basalt layer", on the so-called Conrad discontinuity in the continental Earth's crust. Instead they indiscriminately began to ascribe it to reflections beyond the critical point on the Mohorovičić discontinuity. In connection with this the Earth's crust was described as one-layered with "granitic" velocities of wave propagation. Analogous extremist tendencies appeared also in evaluating the results of the interpretation of some prospecting work using the CMRW method.

Some investigators dealing with the field measurements blindly believed the theoreticians.

However, the majority defended the realistic possibility of the existence of strong refracted waves for many cases. Those dealing with modelling were at one with the workers in the field and sometimes they were even the same people. Modelling experiments were aimed at working out methods, the modelling of transit layers and uneven boundaries for which there really existed experimentally determined strong refracted waves ([29] etc.).

But not even the theoreticians were satisfied. They themselves worked out a method of calculating refracted waves in media with continuously variable properties, the so-called "refracted" waves in "gradient" media, and were themselves convinced of their great intensity. The scales at that time were weighted on the opposite side. Again extremists were found who would have declared all "geophysical" media to be gradient and all observed waves refracted. The experimenters again had to defend the stratification of the Earth's crust, this time from attacks from the other side. Today it seems that almost everyone has understood that nature is more varied than any extreme scheme. We meet with extremes in it but more often with situations lying between extreme cases.

It should be noted that the combined efforts of Soviet geophysicists succeeded in defending continental "basalt". Moreover, on the basis of all-round evaluations of all data obtained by work in the field, theoretical work and modelling, supporters of a one-layered and "gradient" Earth's crust were forced into the defensive: today it is thought that the Earth's crust is composed of many layers and that in its lower part it contains a thicker set of layers with higher "basalt" velocities than was thought earlier [30]. As regards the $P^*$ wave, it must be interpreted today as a complex interference wave group containing both waves reflected beyond the critical point and refracted waves; the importance of both may vary greatly in different parts of the Earth, ranging even to practically extreme cases.

Another fascinating problem of geophysics, which has particularly attracted the attention of workers in seismic modelling, is that of the existence of a wave-guide with low velocities in the Earth's crust and particularly in the upper mantle of the Earth. There is as yet no mathematical solution of this problem. In such a situation the role of modelling is particularly important. As a consequence of its twenty years of evolution, seismic modelling has at present great possibilities for studying this problem.

At this symposium we shall hear a number of reports on this theme and it is premature to sum up in any way now.

## Perspectives

I now raise the following, not entirely superfluous question. What would happen if theoreticians suddenly came forward tomorrow with a complete and mathematically exact solution of the problem of elastic wave propagation in a solid medium containing wave-guides with arbitrary boundaries – sharp or not sharp? What if, by means of powerful electronic computers they work out in detail within two or three days, or even in a few minutes, a solution of this problem in a thousand variants? Will it not put an end to our work and attempts to model this problem? And could not such a situation occur with any problem in seismic modelling? Will not seismic modelling then become a superfluous waste of strength? Would it not be better to concentrate our energy along some other lines: on the one hand, on the development of theory and, on the other hand, on field observations and laboratory experiments not connected with modelling?

Such opinions exist amongst geophysicists, especially among those with an analytical way of thinking whom we call theoreticians, in contrast to people with a predominantly symbolically imaginative way of thinking who usually include experimenters and those who deal with physical modelling. Regardless of the ever greater application of seismic modelling, theoreticians still energetically speak of its "crisis" and sometimes even of its complete failure.

I should like to make it clear that such opinions are the consequence of theoreticians not quite understanding what present-day seismic modelling actually means for experimenters, what are its main tasks and thus perspectives. I should like to make it clear that in reality we are faced not by crisis and failure but by a new, higher stage in the evolution of seismic modelling, caused by the general development of seismology, both by the successes of its theory, supported by the great possibilities of present-day computing techniques, and also by the ever greater flow of information from field and laboratory investigations.

Many theoreticians regard seismic modelling as merely a sort of variation of computing technique – the use of specialized analog computers. They think along the following lines: in principle, an arbitrary analog computer can be replaced by a universal digital one which is undoubtedly nothing more than an aid to theory. What is thus the sense, the raison d'être, of any special seismic modelling? It is thus clear that it can simply be replaced by theory and its computing technique, preferably fast digital computers.

I shall show that such considerations are erroneous for two main reasons:

*Firstly*, if we even allow that seismic modelling is merely a kind of computing technique, although this is not the case, such considerations are erroneous in that they over-estimate the real possibilities of present-day theory in solving seismological problems, that they make out the wanted or the possible to be really existing. An arbitrary problem can only be solved on a digital computer if its full algorithm is known (including that of the adaptive computers). In other words, it is necessary that the problem has its general theoretical solution even if this is only in a certain concrete situation. For many important problems in seismology this is still lacking. Such a problem today is that of the wave-guide. It is thus all to the good that it can be solved by means of seismic modelling, and this is being done. There are many other problems in the dynamic theory of elasticity where theory still has far to go before solving them. These are the problems of the propagation of an elastic pulse in the general case of a three-dimensional inhomogeneous perfectly elastic medium. This case is very close to that of a layered-block model of the Earth's crust and upper mantle, which is now looming forth on the basis of the field investigations into earthquakes and explosions [30, 31]. Even less clear today are the theoretical problems of wave propagation in imperfectly elastic media. And very little has been done on theoretical questions concerning processes in the foci of earthquakes. All this is food for seismic modelling as an analog computer.

It is probable that with the development of the real possibilities of theoretical calculation the range of problems which there is sense in modelling will become smaller in some directions. However, in other directions this range of possibilities will grow even more rapidly, due to the new investigations and new interests of seismology, and will extend into the region of problems which theory has not yet been able to master and which in some cases has not even begun to master.

*Secondly*, from the point of view of the experimenter, this does not comprise the main shortcoming of the above pessimistic considerations regarding the future of modelling. Models having the role of a computer do not exhaust the whole range of modelling, its significance and tasks. Indeed, a large part of work already done on seismic modelling relates to automodelling, i.e. to imitating the physical process on a laboratory scale. And this changes the whole situation.

It is not the physical process itself that can be studied by means of a computer but only an idealized conception of it – its mathematical model. If this model is not known, the computer is unable to say anything about the given process. With automodelling the case is quite different. It is true that there, too, it is necessary to know something about the process so as to be able to consider details, but not nearly so much as is necessary for the complete theoretical solution up to using a computer.

A theoretician may object and ask how can one speak of similarities when the exact formulation of the state equations and equations of motion and also of the initial and boundary conditions remains unknown? Formally, this is correct and if all this is given then the model really does become a regular computer.

But herein lies the core of the matter, that the experimenter regards the problem from a different aspect. He looks on automodelling as the contact between the researcher and a special case of nature itself. Nature is always more varied than any kind of mathematical abstraction expressing a theoretical concept. The experimenter is prepared to reject the complete and rigorous logic of the considerations and to preserve it only where he regards it as important. The same is the case with physical intuition which theoreticians so disdain although during research work they themselves make great use of it. The criterion for a correct choice of what is fundamental is for the experimenter a comparison of the results with reality, both with the model and in nature. When modelling is regarded from this aspect, an exact knowledge and complete formulation of all the properties of the model and nature is no longer necessary. Herein is the core of the method. Only this gives the experimenter a free hand, allows him to begin the research which cannot be carried out and completed without it.

If successful, the physicist discovers the basic laws governing the phenomen and gives the mathematician the necessary formulation of all the basic conditions and laws without which the latter cannot get by. If unsuccessful, the physicist begins his experiments and approximate considerations anew.

Who is more in the right in this argument regarding the significance and problems of seismic modelling? Such a question is nothing new and it is also not specific for our field. It is part of the more general problem of the ratio of physical and mathematical methods of thinking in the process of learning the objective truth. Each side has its strong and weak points. Each side sometimes sins against the absolute correctness of the methods of studying nature. The theoretician does so at the very beginning, when he replaces reality by its pale, idealized, in some way different picture even though it is strictly logical. The experimenter avoids such an error at the very beginning but errs during his not quite correct considerations. For this reason both argue and regardless of their disputes cannot get on without one another.

The best is if both – experimenter and theoretician – are united in one person. This makes it possible to work tête-à-tête. However, this hardly ever occurs so that nothing remains but that the experimenter and theoretician should meet more frequently, try to understand one another and work together. This will lead to a much more rapid rapprochement and to objective truth.

All these universally known rules are fully valid in seismic modelling too.

In conclusion I shall try to name the main trends of work in seismic modelling for the immediate future. Taking into consideration the general interests of seismology, the present real possibilities of theory and the perspectives for field investigations and also the possibilities of modelling, especially automodelling, the trends appear as follows:

1. The study of elastic wave processes in inhomogeneous media, where a large role is played by diffraction effects which have not yet been sufficiently analyzed theoretically; the solution of the problem of non-horizontally layered media by means of modelling, including problems of the type of a layered block Earth's crust and upper mantle.

2. The study of wave processes in media with fundamentally different properties to those of a perfectly elastic medium.

3. The study of processes taking place in the immediate neighbourhood of the source of seismic waves of the explosive type and a source of the shear type (earthquake focus) and possibly also of other types.

4. The study of the creep of solids accompanied by the formation of dislocations and the release of elastic energy (seismic creep of rock masses [32]). The modelling of groups of earthquakes (seismic regime) for various methods of loading materials of different properties. This research is aimed at helping to work out methods of earthquake prognoses.

5. The modelling of processes connected with the seismic movements of the Earth's surface layers and with their effects on structures, as a function of the local geological conditions. These problems are also connected with microseismic zoning and seismic engineering.

I make no claim to have completely exhausted the problems. Part of the work in seismic modelling will obviously continue along traditional lines.

For the realization of such research it is indisputably necessary to improve and work out new instruments and methods.

The next stage of development of seismic modelling will thus be characterized not by the narrowing down but by the further extension of the range of problems, by the improvement of methods and increase in the importance of modelling in seismology.

Received 9. 11. 1965 *Reviewer: J. Vaněk*

*References*

[1] Ю. В. Ризниченко: Изучение законов распространения сейсмических волн в реальных геологических средах и на моделях. Вестник АН СССР, 1946, 4.

[2] Ю. В. Ризниченко: О распространении сейсмических волн в дискретных и гетерогенных средах. Изв. АН СССР, сер. геогр. и геофиз., № 2 (1949), 115.

[3] Ю. В. Ризниченко, Б. Н. Ивакин, В. Р. Бугров: Моделирование сейсмических волн. Изв. АН СССР, сер. геофиз., № 5 (1951), 1.

[4] Б. Н. Иванкин: Упругие волны в одномерных и двумерных сеточных моделях непрерывных сред. Труды Геофиз. инст. АН СССР, № 9 (1950), 84.

[5] Б. Н. Ивакин: О моделировании поглощения сейсмических волн. Изв. АН СССР, сер. геофиз., № 7 (1958), 818.

[6] Б. Н. Ивакин: Расчет и моделирование поглощения сейсмических волн. Изв. АН СССР, сер. геофиз., № 11 (1958), 1288.

[7] A. B. Wood: A Textbook of Sound. London 1946.

[8] E. Oddone: Un modello inteso a mostrare la propagazione entro il globi delle onde sismiche longitudinali. Geofis. pura e appl., I (1939), 2.

[9] F. Rieber: Visual Presentation of Elastic Wave Patterns under Various Structural Conditions. Geophysics, 1 (1936), 196.

[10] O. v. Schmidt: Über Kopfwellen in der Seismik. Zs. f. Geophys., 15 (1939), 141. (not published)

[11] С. И. Кречмер, С. Н. Ржевкин: Исследование волновых процессов по методу моделей с применением ультразвуковых волн. Усп. физ. наук, 18 (1937), 1.

[12] С. Я. Соколов: Авторское свидетельство на использование импульсного У-3 метода, № 48894, класс 21. д. 30 от 31 окт. 1934 г.

[13] Ю. В. Ризниченко, Б. Н. Ивакин, В. Р. Бугров: Моделирование сейсмических волн при помощи ультразвуковых импульсов. Изв. АН СССР, сер. геофиз., № 3 (1952), 58.

[14] Ю. В. Ризниченко, Б. Н. Ивакин, В. Р. Бугров: Импульсный ультразвуковой сейсмокоп. Изв. АН СССР, сер. геофиз., № 1 (1953), 26.

[15] С. И. Чубарова: Изучение распространения сейсмических волн методом моделирования. Канд. дисс., фонды МГУ, 1954 (not published).

[16] Б. Н. Ивакин: Подобие упругих волновых явлений. Изв. АН СССР, сер. геофиз., 1956, I — № 11, 1269; II — № 12, 1384.

[17] E. T. Howes, L. H. Tejada-Flores, R. Lee: Seismic Model Study. Journ. Acoust. Soc. Am., 25 (1953), 915.

[18] T. D. Northwood, D. V. Anderson: Model Seismology. Bull. Seism. Soc. Am., 43 (1953), 239.

[19] J. Oliver, F. Press, M. Ewing: Two-dimensional Model Seismology. Geophysics, 19 (1954), 202.

[20] Б. А. Осипова: Моделирование морской сейсмической реверберации при помощи ультразвуковых импульсов. Уч. зап. Азерб. унив., сер. геол.-геогр., № 5 (1959).

[21] О. Г. Шамина: Затухание головных волн от тонких слоев при жестком и скользящем контакте. Изв. АН СССР, сер. геофиз., № 3 (1965), 11.

[22] J. Oliver: Body Waves in Layered Seismic Models. Earthq. Notes, 27 (1956), 29.

[23] Б. Н. Ивакин: Методы управления плотностью и упругостью среды при двумерном моделировании сейсмических волн. Изв. АН СССР, сер. геофиз., № 8 (1960), 1149.

[24] П. Г. Гильберштейн, И. И. Гурвич: О применении дырчатых материалов для двумерного моделирования. Изв. ВУЗ, Геология и разведка, № 1 (1960), 139.

[25] J. Vaněk, L. Waniek, Z. Pros, K. Klíma: Three-dimensional Seismic Models with Continuously Variable Velocity. Nature, 202 (1964), 995.

[26] L. Knopoff: Small Three-dimensional Seismic Models. Trans. Am. Geophys. Un., 36 (1955), 1029.

[27] О. Г. Шамина: Методика трехмерного моделирования волноводной мантии на твердых средах. Изв. АН СССР, Физ. Земли, № 7 (1965), 102.

[28] J. Steinbeck: Modellseismische Untersuchungen von Rayleighwellen unter besonderer Berücksichtigung einer Deckschicht von variabler Mächtigkeit. Geophys. Abh. d. Inst. f. Met. u. Geophys. d. fr. Univ. Berlin, 1965.

[29] В. И. Уломов: Изучение глубинного строения земной коры юговостока Средней Азии по данным сейсмологии. Канд. дисс., Фонды Инст. геол. и геофиз., АН УзбССР, Ташкент 1964 (not published).

[30] I. P. Kosminskaya, Yu. V. Riznichenko: Seismic Studies of the Earth's Crust in Eurasia. Res. in Geophys., Vol. 2., MIT Press, USA, 1964, 81.

[31] Ю. В. Ризниченко, И. П. Косминская: О природе слоистости земной коры и верхней мантии. ДАН СССР, 153 (1963), 323.

[32] Ю. В. Ризниченко: О сейсмическом течении горных масс. В кн. „Динамика земной коры", Изд. Наука, Москва 1965, 56.

Резюме

## СЕЙСМИЧЕСКОЕ МОДЕЛИРОВАНИЕ, ЭТАПЫ И ПЕРСПЕКТИВЫ

Ю. В. Ризниченко

*Институт физики Земли АН СССР, Москва*

Описаны этапы развития сейсмического моделирования, взаимоотношения между сейсмическим моделированием, теоретическими разработками и полевыми наблюдениями и сформулированы перспективные задачи сейсмического моделирования.

Поступило 9. 11. 1965

*Discussion*

Červený: All the problems mentioned in connection with modelling the wave processes are important but for many of them the wave pattern will be very complicated. Therefore I think that it might be useful to build some simple model for investigating the behaviour of individual waves and comparing it with theory. This would enable us also to extend the region of problems which can be solved theoretically.

Riznichenko: I agree that the comparison of theory with a physical model can be very useful, especially in the case when only approximative methods can be applied.

Vaněk: Seismic models based on gels cannot be denoted as liquid ones. The elastic properties of gels are similar to those of rubbers, their Poisson ratio being near to 0·5.

Riznichenko: If the Poisson ratio is near to 0·5, then $v_S/v_P \to 0$; therefore such bodies are similar to liquids in the sense that practically no shear elasticity and no shear waves exist in the frequency range used.

Vaněk: Can you give some details on modelling the process of solid flow to study the seismic regime?

Riznichenko: Details are given in my paper: Сейсмическое течение горных масс, Динамика земной коры, изд. Наука, Москва 1965.

# METHODS OF SEISMIC MODELLING

B. N. Ivakin

*Institute of Physics of the Earth, Acad. Sci. USSR, Moscow*)*

## 1. INTRODUCTION

The modelling of seismic wave processes in the laboratory is based on the invariancy of dimensionless equations of motion and conditions guaranteeing uniqueness (initial and boundary conditions) at transition from one scale to another [1]. Since on the whole it is possible to convert the equations of motion and conditions of uniqueness to the dimensionless form, we thus obtain directly the criteria of similarity for wave processes. The satisfaction of these criteria for modelling permits the model wave processes to be converted exactly to the wave processes in nature in which we are interested.

Two basic groups of models are obtained from the similarity criteria found for wave processes [1]: 1) the parameters of the models (velocities of wave propagation, density) are equal to the corresponding parameters in nature; 2) the more general case when the parameters of the models are in some way changed with respect to the corresponding parameters in nature.

The second group of models makes it possible, for example, to model wave processes in the mantle and core of the Earth where the velocities of propagation are higher than in all known materials on the Earth's surface.

However, in many cases the similarity criteria have been incompletely fulfilled which is due to difficulties in finding modelling materials with the appropriate parameters. Wave processes are often studied on models without any kind of relation to real seismic conditions.

In the initial stages of modelling liquid models with solid inserts were often used. However, the liquid medium often greatly differed from the seismic in that shear waves did not propagate in it. At present this approximate modelling is convenient in the first phases when learning modelling methods (particularly of the very complicated wave processes) where a simpler wave pattern and stability of recording guarantees a correct interpretation and treatment of the material obtained.

Since the occurrence of solid modelling materials with the necessary parameters is on the whole limited, solid two-dimensional and three-dimensional models with so-called variable properties are at present used [2–6]. In the case of two-dimensional models, first proposed in [7] and made in the form of elastic plates, their thickness being much smaller than the length of the waves used, the following methods of changing the elastic properties of the plates are employed: 1. the plates are glued to one another – bimorph models [5, 6]; 2. a net of openings (or protuberances) is made – perforated models [2, 3]; 3. certain parts of the plate are heated – thermal modelling [4]; 4. plates are made with variable thickness [2].

*) Address: B. Gruzinskaja 10, Moskva G-242, USSR.

Reprinted from Studia Geophysica et Geodaetica, 1966, Volume 10, 253-258 with permission of Plenum Press

With three-dimensional solid models, which have so far been rarely used due to their size on the frequencies used (around 100 kHz) the artificial creation of the heterogeneity of the model, for example, can be used for changing the properties [8, 9] (which is to a certain extent analogical to perforated modelling in the two-dimensional case) or the local heating of the model.

Let us now analyze in detail the above-mentioned modelling methods.

## 2. TWO-DIMENSIONAL MODELLING

### 2.1 Two-dimensional models

Two-dimensional modelling on thin plates was proposed some time ago, in 1952 [7], but still no theory exists although quite a lot has been done along these lines from the experimental point of view. The kinematic properties of waves propagating in a plate [10] (which also led to the proposal of two-dimensional modelling) and the dispersion curves of the velocity for a plate [11] have been known from theory for a long time. The questions of the damping of different waves with distance as a function of the ratio $\lambda/d$, where $\lambda$ is the wave-length and $d$ the thickness of the plate, have not been explained theoretically. However, this question is very important for the choice of thickness of the model plate. The theoretical results concerning some properties in the behaviour of a thin plate with free edges appeared recently [12]. That paper also gives the corresponding analytical limitations. It is shown that the given anomalies lead to the dispersion of Rayleigh waves in a homogeneous two-dimensional model but this was not observed experimentally.

Due to the lack of theoretical papers dealing with the dynamics of waves in a plate, it is naturally necessary to turn to experiment. The experimental bases of two-dimensional modelling and the choice of thickness of the sheet-model are studied in papers [13–15] concerning the wave properties of plates of finite thickness. Paper [14] deals with the study of the velocities of waves in a plate in cases when the waves propagate with a "plate" and "body" velocity and paper [15] studies the dynamic characteristics of longitudinal waves in plates of different thickness. Mention should also be made of the paper by Oliver et al. [16] which also gives some theoretical calculations and experiments concerning the bases of two-dimensional modelling.

It follows from the above and from our work that the use of two-dimensional models in the shape of plates for studying wave processes is sufficiently justified. Approximately from a value $\lambda/d = 10$ the plate behaves sufficiently exactly as a two-dimensional region with certain effective values of the velocities and damping of the compressional and shear waves.

### 2.2 Bimorph models

Bimorph modelling, which was proposed and tried out in various converted reflected and refracted waves by Oliver [5], is being successfully developed in the

Soviet Union by Riznichenko et al. [6]. Since there exists no exact dynamic theory for bimorph models, special experiments were proposed for studying so-called generalized waves [6]. These experiments made it possible to derive the most suitable conditions for bimorph models. The very important coefficients of absorption and propagation velocity for homogeneous bimorph models were obtained from the experiments. The measured velocities, for example, agree well with the values of the velocities calculated according to Riznichenko's approximate formula. It was found that the possible effect of the inversion of symmetrical vibrations to anti-symmetrical, which is mentioned for the first time in relation to bimorph modelling in [2], does not cause further considerable absorption of compressional waves. A theoretical paper [17] recently appeared which deals with the study of engineering equations of the vibratory motion of plates with a layered structure. The paper has a direct relation to some of the basic questions of bimorph and generally polymorph models.

Starting out from the experimental results, our attention is attracted by the fact that the generalized ("bimorph") wave sets in ever more gradually with distance. This effect of the bimorph model (complicated for theory) must be studied in detail experimentally, particularly after the passage of a wave through the boundary. In general, it should be noted that the occurrence of a generalized wave in a bimorph model is a much less natural process than, for example, in perforated models where the process of forming the wave to a certain extent recalls the same process in heterogeneous seismic media.

### 2.3 Perforated models

Perforated models, developed in the Soviet Union [2, 3], cannot be mastered theoretically. For example, the calculation for a homogeneous elastic plate with equilateral square or triangular network of openings is very difficult even for the quasi-classical case when the wave-length is much larger than the distance of the openings. So far only approximate theoretical calculations of the velocity of propagation have been made [2]. In view of these difficulties, the necessary research into the dynamic properties of waves in perforated plates was carried out experimentally. In these experiments, when the distance of the openings was 5 mm and the length of the compressional waves 50–30 mm, the dependence of the velocity and absorption of compressional, shear and Rayleigh waves was studied as a function of the hole density or porosity of the plate. It was found that the velocity in such a perforated plate can be decreased by as much as 50% of the initial value by increasing the hole density. It was also found that in a sufficiently wide interval of hole density the triangular net ensures practically the isotropy of the perforated plate with respect to the velocities of propagation; as regards the isotropy in the propagation of wave energy, this interval is somewhat narrower. During research into the process of wave formation in a perforated plate it was found that the distance at which the wave occurs is sufficiently small and is smaller than half the wave-length in the plate.

As regards the cause of the decrease in velocity in a perforated plate or, in general, in a heterogeneous medium, this phenomenon is connected with the interference of two waves existing in the effective wave which propagates through the perforated plate [19].

## 2.4 Thermal modelling

The thermal method of controlling the elastic properties of two-dimensional models was proposed in [4]. The alloy of paraffin and polyethylene they used provided sufficient changes in the velocities of the compressional and shear waves at a change in temperature of the material. This method was used to create a wave-guide with lower velocity and comparison with a homogeneous model showed its influence on the propagation of elastic waves. The method is relatively simple and has the advantage that different cases of the structure of an inhomogeneous medium can be created with one model. Unfortunately, the alloy has a relatively high absorption and therefore the future success of this method depends on finding suitable materials with the necessary properties.

## 2.5 Models with variable thickness

A change in thickness of the two-dimensional model causes a change in the effective density which is theoretically proved in [2]. This method of changing the density of a two-dimensional model is a good complement to the preceding methods since it permits both the velocity of propagation and the density to be varied on models independently, which is important for the exact fulfilment of the similarity conditions.

## 2.6 Discussion

The above modelling methods have their disadvantages. For example, with the perforated model it is difficult to make a fine enough network of openings compared with the wave-length. A shortcoming is the anisotropy that is produced, particularly when the density of openings is large. With bimorph models there are difficulties with the sufficiently firm glueing of the second layer and controlling the thickness of the layers. It is a disadvantage that the interference generalized wave is produced slowly and in some cases there even exist two waves – apart from the generalized wave there exists a wave in the layer with higher velocity. In the case of two-dimensional models of variable thickness there are still no theoretical or experimental works defining the allowed limits of a discontinuous-sudden change in thickness of the two-dimensional model. Despite these difficulties and shortcomings in modelling methods there exist many papers on seismic modelling in which such methods are used and which cannot be dealt with in this short report.

A survey of papers in modelling is given for example in [18]. Note, however, that the theoretical bases of the given methods lag behind experimental work in modelling.

## 3. MISCELLANEOUS METHODS OF MODELLING

Due to the large size of the models at working frequencies of around 100 kHz that are usually used, it is necessary to go over to a frequency of the order of 1 MHz. Such micromodelling permits the use of solid models having an acceptable size under laboratory conditions and in time will necessarily predominate in all work concerning seismic modelling. Here however, there are still many, mainly technical, difficulties related in particular to a sufficiently powerful ultrasonic source with a uniform directional characteristic and sensitive reception corresponding to reception under seismic conditions.

It should be stressed here that in practical seismic modelling it will be necessary to make greater use of the known method of rendering waves visible. Wave images, hitherto only qualitative, can substantially help in studying complicated seismic phenomena.

It is also necessary to pay attention to the possibilities of modelling the absorption of seismic waves by means of electric networks. The method is described in [20, 21]. Although little is yet known about the absorption properties of a real seismic medium, particularly as regards the frequency dependence of such properties, there exist several theories of wave absorption (afterworking, internal friction etc.), which to a certain extent describe the phenomenon in nature. In the above-mentioned papers the possibilities of modelling the absorption of seismic waves by means of uni-dimensional electric networks are studied. Such networks, it is found, are described by the same equations of motion and their fabrication and work with them are very simple. By means of a seismoscope an electric pulse is supplied directly to the electric model and by means of a cable and input amplifier the changes in shape and damping of the wave propagating in the given model of the medium are observed on the C.R.T. of the seismoscope and photographed. Such electric models, and particularly their mechanical analogs, are very clear and greatly help to understand the phenomenological aspect of the effect of absorption; they permit improvements in the analogy between the absorption model and the real absorbing medium. In a similar experiment, for example, an entirely new mechanism of absorption was discovered which was connected not with imperfect elasticity but with imperfect inertia of the mass of the absorbing medium [22]. This phenomen was then discovered experimentally.

### *References*


[1] Б. Н. Ивакин: Подобие упругих волновых явлений. Изв. АН СССР, сер. геофиз., № 11 (1956), 1269; № 12, 1384.

[2] Б. Н. Ивакин: Методы управления плотностью и упругостью при двумерном моделировании сейсмических волн. Изв. АН СССР, сер. геофиз., Но 8 (1960), 1149.

[3] П. Г. Гильберштейн, И. И. Гурвич: О применении дырчатых материалов для двумерного сейсмического моделирования. Изв. ВУЗ, Геология и разведка, № 1 (1960), 139.

[4] Л. Н. Рыкунов, В. В. Хорошева, В. В. Седов: Двумерная модель волновода с нерезкими границами. Изв. АН СССР, сер. геофиз., № 11 (1960), 1601.

[5] J. Oliver: Body Waves in Layered Seismic Models. Earthquake Notes, *27* (1956), 29.

[6] Ю. В. Ризниченко, О. Г. Шамина, Р. В. Ханутина: Упругие волны с обобщенной скоростью в двумерных биморфных моделях. Изв. АН СССР, сер. геофиз., № 4 (1961), 497.

[7] Ю. В. Ризниченко, Б. Н. Ивакин, В. Р. Бугров: Моделирование сейсмических волн при помощи ультразвуковых импульсов. Изв. АН СССР, сер. геофиз., № 3 (1952), 58.

[8] С. Ф. Больших, В. П. Горбатова, Л. Н. Давыдова: Изучение кинематических и динамических характеристик отраженных и головных волн на моделях слоистых сред. Прикл. геофиз., *30* (1961), 25.

[9] О. Г. Шамина: Методика трехмерного моделирования волноводной мантии на твердых средах. Изв. АН СССР, сер. геофиз., № 7 (1965), 102.

[10] Л. С. Лейбензон: Курс теории упругости. Гостехиздат, М.-Л. 1947.

[11] H. Lamb: On Waves in an Elastic Plate. Proc. Roy. Soc. London, *A 93*, (1917), 114.

[12] Г. И. Петрашень, Л. А. Молотков: О некоторых проблемах динамической теории упругости в случае сред, содержащих тонкие слои. Вестник ЛГУ, № 22 (1958), 137.

[13] Ю. В. Ризниченко, О. Г. Шамина: Об упругих волнах в твердой слоистой среде по исследованиям на двумерных моделях. Изв. АН СССР, сер. геофиз., № 7 (1957), 855.

[14] О. Г. Шамина, О. И. Силаева: Распространение упругих импульсов в слоях конечной мощности со свободными границами. Изв. АН СССР, сер. геофиз., № 3 (1958), 302.

[15] О. Г. Шамина: Изучение динамических характеристик продольных волн в слоях различной толщины. Изв. АН СССР, сер. геофиз., № 8 (1960), 1135.

[16] J. Oliver, F. Press, M. Ewing: Two-dimensional Model Seismology, Geophysics, *19* (1954), 202.

[17] Л. А. Молотков: Об инженерных уравнениях колебаний пластин, имеющих слоистую структуру. Сб. Вопросы динамич. теории распространения сейсм. волн, V (1961).

[18] Б. Н. Ивакин: Развитие методов и обзор результатов моделирования сейсмических волн. Изд. „Наука". Геоакустика, Москва 1966.

[19] Б. Н. Ивакин: О причине ниже средних скоростей распространения в тонконеоднородных средах. Изд. „Наука", Геоакустика, Москва 1966.

[20] Б. Н. Ивакин: О моделировании поглощения сейсмических волн. Изв. АН СССР, сер. геофиз., № 7 (1958), 818.

[21] Б. Н. Ивакин: Расчет и моделирование поглощения сейсмических волн. Изв. АН СССР, сер. геофиз., № 11 (1958), 1288.

[22] Б. Н. Ивакин: Упругие среды с неидеальной инерционностью и их модели. Изв. АН СССР, сер. геофиз., № 2 (1959), 210.

# Model seismology

P N S O'BRIEN AND M P SYMES

BP Research Centre, Sunbury-on-Thames, Middlesex

## Contents



**Abstract.** There are four main types of seismic wave. Compressional (P) and shear (S) waves which travel through the body of the Earth, and Rayleigh and Love waves which are bound to its surface. Their wavelengths are measured in kilometres. In model seismology laboratory experiments are made on elastic waves in which the wavelengths involved are usually measured in centimetres. These experiments are made on highly idealized models of the Earth or, more generally, of a small part of it, in which the Earth's interior is represented by a small number of elastic layers. Metals and plastics are the materials most commonly used.

Since the Earth's subsurface structures are often two-dimensional, seismic wave propagation in the Earth may be modelled by elastic wave propagation in a plate. Perhaps 50% of model seismic experiments have been made with plates. Some problems in seismology concern the propagation of essentially plane waves in a direction normal to a series of plane-parallel layers of rock. These may be modelled by one-dimensional elastic wave propagation along a bar or within a fluid-filled tube.

The source and detector are normally piezoelectric transducers, usually in the form of bars, discs or cantilevers. Their proper construction and the effect of the reaction of the model upon them is one of the major problems of model seismology.

Reprinted from Reports on Progess in Physics, Volume 34, 697-718, 753-764 with permission of American Institute of Physics.

In many situations the theory and practice of their design are in good agreement but a sound theoretical treatment is lacking for shear wave transducers and for piezoelectric discs used in the radial mode. Also, more work needs to be done on the radiation impedance of transducers operating on both two- and three-dimensional models.

In order to obtain wavelengths of a few centimetres frequencies have to be in the ultrasonic range and the detector output is usually displayed on a CRO. For ease of presentation the source is pulsed repetitively at a rate of a few tens of times per second. In addition to providing a 'stationary' trace on the CRO screen this enables a waveform translator to be used so that the signal may be recorded on a low frequency recorder such as an $XY$ plotter or a magnetic tape recorder. Such a magnetic tape record may be made suitable for direct entry into a digital computer for further processing of the signal.

Model seismic experiments provide a link between the highly idealized models of the theoreticians and the complex records of the observational seismologist. They are useful in both earthquake seismology and in seismic prospecting.

Examples are given which show body waves and surface waves. Also some examples are given of the effect of velocity layering and of buried structure on the reflection, refraction and diffraction of elastic (seismic) waves.

This review was completed in July 1971.

## List of symbols

$a$ radius of a circular disc transducer

$A$ area, usually of the active face of a transducer

$c$ elastic wave velocity

$c_{\mathrm{d}}$ velocity of dilatational (P) waves in an extensive body (bulk velocity)

$c_{\mathrm{p}}$ limiting value of low frequency extensional waves in a plate (plate velocity)

$c_{\mathrm{s}}$ velocity of rotational (S) waves in an extensive body

$c_{\mathrm{Y}}$ limiting value of low frequency extensional waves in a bar or rod (Young's modulus velocity)

$c_{\mathrm{d}}^{\mathrm{D}}$ bulk velocity in piezoelectric plate under conditions of constant electric displacement

$c_{\mathrm{Y}}^{\mathrm{E}}$ Young's modulus velocity in a piezoelectric rod under conditions of constant electric field

$c_{33}^{\mathrm{D}}$ an elastic stiffness in a piezoelectric material under conditions of constant electric displacement

$C_0$ electrical capacitance of piezoelectric transducer; its precise value depends upon the mechanical and electrical boundary conditions

$C_{\mathrm{m}}$ 'mechanical' capacitance (compliance) in the equivalent circuit for a piezoelectric transducer; its value also depends upon the mechanical and electrical boundary conditions

$d_{31}$ a piezoelectric constant relating elastic strain to electric field or electric displacement to elastic stress

$D_1$ electric displacement in the $x_1$ direction

$e_{33}^{\mathrm{T}}$ a dielectric constant of a piezoelectric material under conditions of constant stress

$e_{33}^{\mathrm{S}}$ a dielectric constant of a piezoelectric material under conditions of constant strain

$E_1$ electric field in the $x_1$ direction

$f$ frequency

$f_{\mathrm{n}}$ fraction of material n in a composite plate

| | |
|---|---|
| $F$ | mechanical force, usually associated with a propagating elastic wave |
| $h$ | thickness of a layer bonding together two plates or rods |
| $h_{33}$ | a piezoelectric coefficient relating stress to electric displacement, or electric field to strain |
| $k, k_0$ | elastic bulk modulus |
| $k_t$ | electromechanical thickness coupling factor for a piezoelectric material |
| $k_{31}$ | an electromechanical coupling factor for a piezoelectric material |
| $L$ | transducer length |
| $L_m$ | 'mechanical' inductance (mass) in the equivalent circuit for a piezoelectric transducer; its value depends upon the mechanical and electrical boundary conditions |
| $m$ | mass per unit length of a rod or bar |
| $m$ | $c_s/c_p$ |
| $N$ | transformer ratio in an equivalent circuit of a piezoelectric transducer |
| P wave | irrotational (dilatational) wave, the faster of the two possible types of wave in an extensive isotropic solid |
| $P(\theta)$ | transducer directivity for P waves |
| $r$ | source–detector distance |
| $r_{12}$ | reflection coefficient for a plane wave incident from medium 1 onto medium 2 |
| $R$ | reflection coefficient for a plane wave incident onto a layer |
| $R_m$ | 'mechanical' resistance in the equivalent circuit for a piezoelectric transducer |
| $s_{11}^E$ | an elastic compliance in a piezoelectric material under conditions of constant electric field |
| $S_1$ | elastic strain in the $x_1$ direction |
| S wave | equivoluminal (shear) wave, the slower of the two possible types of wave in an extensive isotropic solid |
| $S(\theta)$ | transducer directivity for S waves |
| $t$ | thickness of a plate |
| $t$ | time |
| $t_{12}$ | transmission coefficient for a plane wave incident from medium 1 onto medium 2 |
| $T_1$ | elastic stress in $x_1$ direction across a surface whose normal is also in the $x_1$ direction |
| $T$ | transmission coefficient for a plane wave passing through a layer |
| $u$ | component of particle displacement in an elastic wave, either in the $x_1$ direction or in the plane of a plate |
| $\dot{u}$ | time differential of $u$, that is a component of particle velocity |
| $V$ | voltage, usually either that applied to a source transducer or the open-circuit output of a detector |
| $w$ | component of particle displacement, either in the $x_3$ direction or perpendicular to the plane of a plate |
| $W$ | width of a transducer plate or bar |
| $x_1, x_2, x_3$ | Cartesian coordinates |
| $X$ | distance in a high velocity layer of the critically refracted ray |
| $Y$ | Young's modulus |
| $Z$ | mechanical impedance |
| $Z_0$ | acoustic impedance for a plane wave, $\rho c A$ |

$\alpha$ exponential time constant
$\beta^S_{33}$ a dielectric impermeability constant of a piezoelectric material under constant stress
$\theta$ ray angle
$\lambda, \lambda_0$ wavelength or one of Lamés' constants
$\mu, \mu_0$ rigidity, the other of Lamés' constants
$\rho$ density, either absolute or relative
$\sigma$ Poisson's ratio
$\sigma^E$ Poisson's ratio of a piezoelectric material under conditions of constant electric field
$\tau$ an interval of time
$\omega$ circular frequency, $2\pi f$

## 1. Introduction

Seismology is primarily the study of the propagation of elastic waves along the surface and through the body of the Earth. By placing a vibration detector (seismograph) on the surface of the Earth the seismologist can obtain a facsimile of the ground motion as the wave passes by. Such a facsimile is usually called a seismogram or seismic record, and the seismologist spends most of his working life trying to interpret such records in terms of the internal structure of the Earth. Academic seismologists are usually interested in the whole Earth and, because of this, need records produced by waves which have travelled through at least its major part. They therefore have to rely on large energy shocks such as those which produce earthquakes. On the other hand, prospecting seismologists are seldom interested in depths greater than about 10 km and can use low energy artificial shocks such as those produced by piledrivers or small charges of high explosive.

The efficacy of the seismic method results mainly from the fact that the Earth closely approximates a layered structure, both in the large, where it is essentially spherically symmetric, and in the small, where the layering is essentially plane-parallel. Where the Earth departs greatly from this geometry the seismic method is very difficult to apply. For this reason it is used hardly at all in prospecting for metallic minerals, which normally occur in nonlayered rocks, but is used extensively in prospecting for oil and gas which occur in the highly stratified sedimentary rocks.

There are two classes of seismic wave. One class consists of body waves; these can travel in all directions throughout the Earth and their energy decays approximately as the inverse square of the distance from their source; they have many affinities with light spreading out from a source of illumination. The other class consists of interface waves; these are bound to one of the interfaces between the Earth's layers and their energy loss is approximately directly proportional to the distance from their source; they have some slight resemblance to waves at sea. Because the seismologist is rarely able to place his seismograph at the boundary of one of the interior layers the interface waves observed are normally those bound to the Earth's surface and these are called, naturally enough, surface waves.

These two classes of wave can themselves be divided further into two subclasses. The body waves divide into those whose motion contains no rotation, they are usually called compressional waves, and those whose motion contains no volume changes, which are usually called shear waves. The compressional waves travel faster than the shear waves and are designated P, for primary, while the shear waves are designated S, for secondary. The surface waves divide into those whose particle motion is parallel to the layer boundary, which are called Love waves, and those whose motion contains a component perpendicular to the layer boundary, which are called Rayleigh waves. These waves are named after the distinguished scientists who predicted their existence.

Earthquake seismologists make frequent use of all the possible wave types. On the other hand the prospecting seismologists keep very strictly to P waves and use every trick of the trade to reduce or eliminate the S and surface waves.

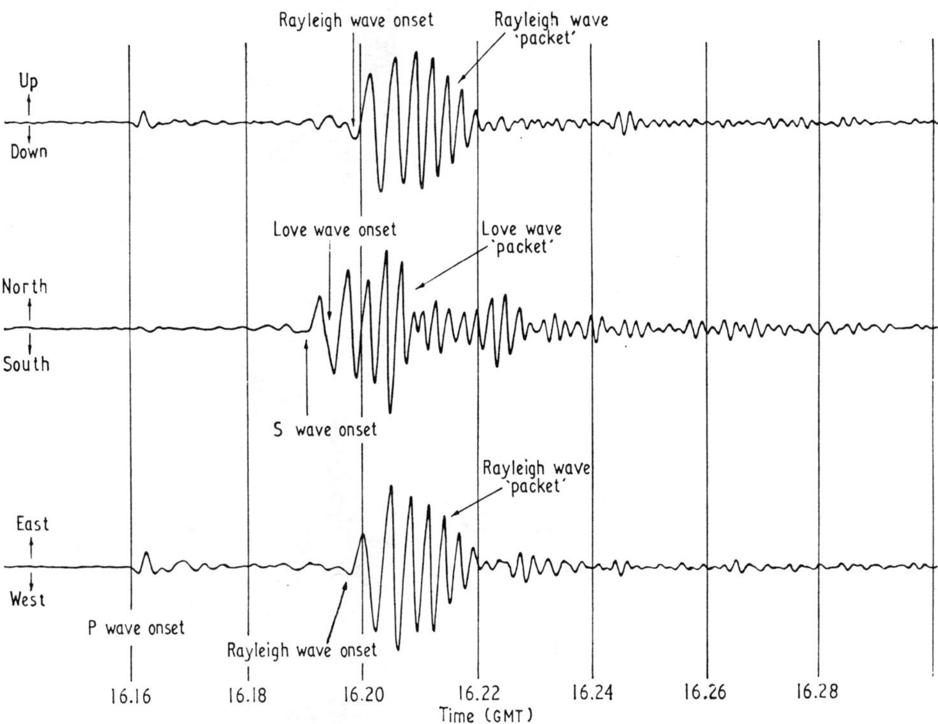


Figure 1. Three-component motion at the UKAEA seismograph station at Eskdalemuir, Scotland, due to an earthquake on the North Atlantic Ridge at 16h 12min 07·1s on 18 September 1970. The earthquake was exactly due west of the station so there is only slight N–S motion from the P and Rayleigh waves. For the same reason the Love waves give only a N–S motion. Due to the increase of velocity with depth the surface waves exhibit dispersion with the higher frequencies (shorter wavelengths) arriving the later (ie at the end of the wave 'packets'). The recording is by courtesy of Mr P D Marshall of the United Kingdom Atomic Energy Authority, Blacknest.

Figures 1 and 2 show two rather clear examples of seismic records, one of an earthquake and one of the vibration due to a small charge of high explosive. It may be seen that the earthquake record contains all types of waves, P, S, Rayleigh and Love, whereas the propecting record contains only P wave reflections.

Much of the advance in seismology during the past hundred years has been due to applied mathematicians working on problems of wave propagation in layered media. Of course, their numerical models are but highly idealized approximations to the real Earth and it has not always been easy to apply the results of their studies to better the interpretation of actual seismic records, particularly when attempts were made to use wave amplitudes as well as wave velocities. Also, the problems which the mathematicians were able to solve were not always those of most importance to the seismologist. Accordingly, scientists started to experiment with physical (analogue) models in the laboratory, partly to check directly the predictions of the mathematicians but mainly to act as direct aid in the interpretation of field (actual) records. This review will be concerned with such models.

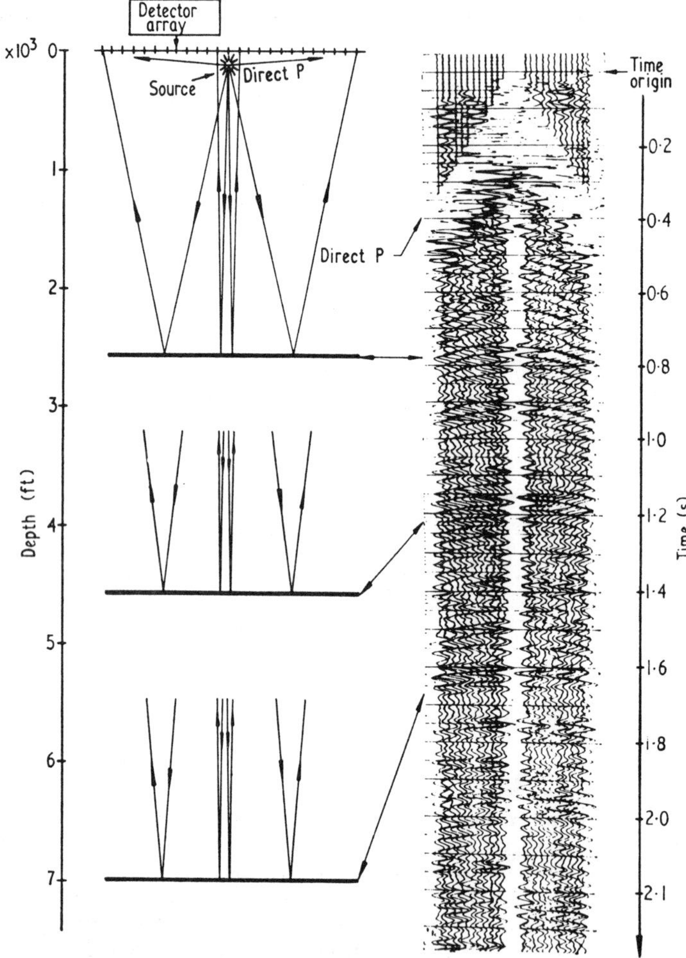


Figure 2. Seismic reflection record obtained while prospecting for oil. Each trace corresponds to the output of a detector array so connected as to reduce the amplitude of the horizontally travelling waves which are still further reduced by bandpass filtering. The reflection coefficients corresponding to the major reflections are around 0·05. There are many other reflectors than the three indicated.

It is sometimes convenient to divide seismology into four areas of investigation. These are, the seismic source, the transmission path between source and detector, the seismograph and its interaction with the ground, and the processing (filtering) of the recorded ground motion. Analogue models have been made in each of the first three areas but by far the greatest effort has been concerned with the transmission path. In this review we omit discussion of the other areas and so will make no mention, for instance, of models of the earthquake focus or of ground–seismograph coupling.

Probably the first published account of seismic model work was by Terada and Tsuboi (1927). Already they were expressing the main point of model seismology which was '... to compare the results of such experiments, on the one hand, with the outcome of theories and, on the other hand, with the facts of observations of actual earthquakes'. Also, they were already saying that the main benefit from such laboratory studies would be in the realm of wave propagation. Their transmission path consisted of layers of different strength gels of agar-agar (see also Tsuboi 1928) having elastic properties which resulted in a P wave velocity of about 10 $\mathrm{m\,s^{-1}}$ and a Poisson's ratio of about 0·45. The latter value is quite typical of those for porous near surface rocks so, in spite of its low velocity, the experiments gave results at least qualitatively similar to those observed in field studies. Their models were excited into steady-state vibration and the resulting surface motion was examined optically.

The next set of model experiments appear to have been those by Rieber (1936, 1937). Whereas Terada and Tsuboi were earthquake seismologists concerned with surface waves, Rieber was a prospecting seismologist concerned with reflected body (P) waves. Using a spark as a source of intense sound he took photographs of the wavefront as it travelled through the air and obtained striking pictures of it after reflection and diffraction from screens with curved surfaces and sharp edges. These pictures had a profound effect on prospecting seismology. Not only did they serve to emphasize that the amplitude of the reflected wave could vary dramatically with the curvature of the reflector and that diffractions from the buried edges of strata could help locate geological faults (see § 6.3) but they also served to hasten the application of frequency–wavenumber filtering (array processing) so that waves approaching the seismograph from different directions could be separately displayed on the final seismic record.

Another fine paper was that by Schmidt (1939). Using layered models of which at least part were transparent he employed schlieren techniques to record the fronts of waves which were reflected, refracted and diffracted at the various interfaces. In particular, he appears to have been the first to demonstrate the existence of the second order 'refracted' wave which comes into being when a curved wavefront propagating in a lower velocity medium is incident on a boundary with a higher velocity medium. Such a wave (known as a head wave in Germany, conical wave in France and lateral wave in the USSR) had been in use for many years in both earthquake and prospecting seismology but its mode of genesis was in considerable doubt until Schmidt's publication.

The early history of model seismology ended with Schmidt's paper and its recent history began in 1951 with the work of Kaufman and Roever (1951) and Riznichenko *et al* (1951). From that time on most experimenters used electromechanical transducers to convert transient motions in the model into voltages which were then amplified and displayed on a CRO. For some time sparks and minute explosive charges continued to be used as seismic sources but soon electromechanical transducers were used for both the source and detector and nowadays virtually all workers use a piezoelectric ceramic.

Although photoelastic and schlieren methods of recording the propagating wavefront were occasionally tried (eg Evans *et al* 1954, Thomson *et al* 1969) they have not been very popular, partly because the technique is more limiting in the materials which may be used but mainly because the results are not in the same form as field records and are difficult to interpret quantitatively. Nevertheless, one feels that the possibility of following the wavefronts as they travel through the body of the transmitting medium should give valuable insight into the relative importance of the various factors involved. Indeed, this has been well demonstrated by the production of moving films showing the genesis of reflections, refractions and diffractions.†

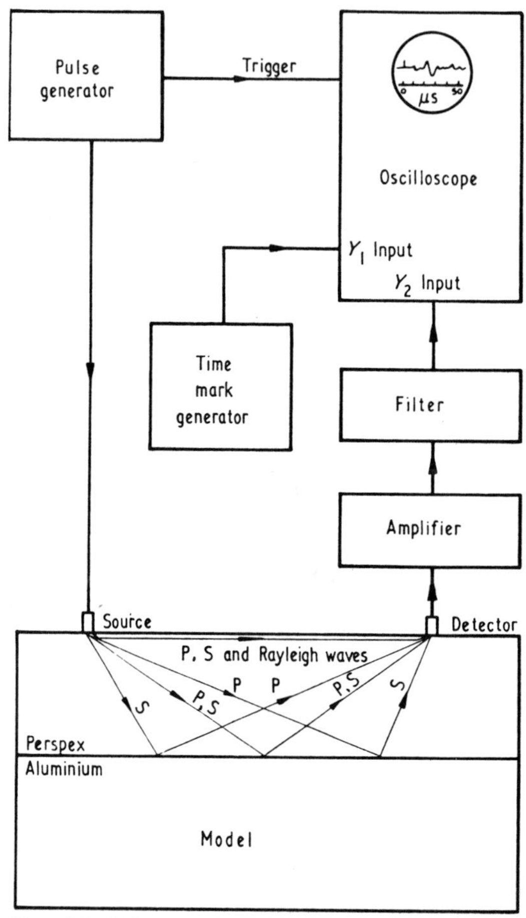


Figure 3. Simplified block diagram of a typical model seismic 'set-up'.

Figure 3 shows a simplified block diagram of a typical laboratory set-up such as was used throughout the 1950s and early 1960s and which, except for one vital difference in the method of recording, is still in use today. The ultrasonic source is driven by a voltage pulse to produce an ultrasonic pulse which usually lasts between 1 μs and 100 μs according to the size of the model. The voltage pulse is continuously repeated at a rate which is low enough to ensure that the signal at the receiver due to one pulse has died away before the succeeding pulse is generated. This repetition rate is usually around a few tens per second and so a persistent image is formed on the CRO screen which may be visually inspected or photographed for later analysis.

† A film *Schlieren Imaging of Acoustic Waves* is available (for loan or purchase) either from Automation Industries Inc, PO Box 950, Boulder, Colorado, USA or from Automation Industries UK, 108 Albert Street, Fleet, Hants, UK.

From the first there were two approaches to the CRO record. In one case the investigator might be most interested in the details of the amplitude and shape of a particular pulse at a particular place on the model. Accordingly, he arranges the trigger delay, timebase speed and $Y$ amplification so that this pulse fills the whole of the CRO screen. An example of this is shown in figure 4 where the observed and predicted Rayleigh wave pulses are compared for an impulsive source applied to the surface of a homogeneous halfspace (Lamb's problem). In the other case the investigator might be most interested in the time–distance relationships of one or

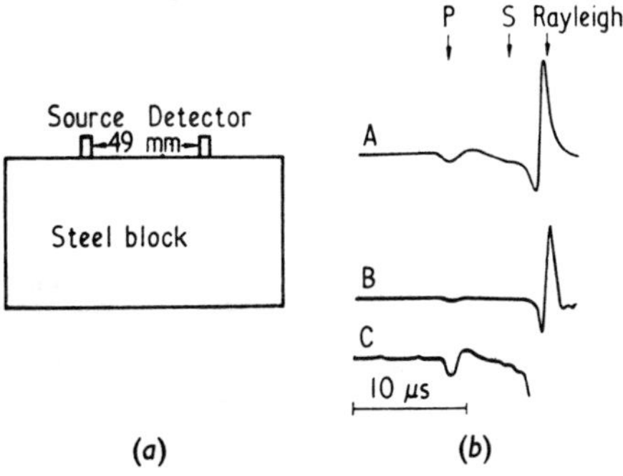


Figure 4. Lamb's problem. (*a*) experimental set-up. (*b*) theoretical and observed waveforms: A, approximate particle displacement for an impulse of pressure; B, observed waveform, normal gain; C, observed waveform, 10 times normal gain. (From Tatel 1954.)

more pulses and he then crams as many traces as possible onto one photograph so that a seismic 'section' is built up. Figure 5 shows such a section illustrating reflection and refraction from a buried high velocity layer.

During the 1950s the number of laboratories engaged in model seismology increased rather rapidly both in the oil industry and in academic institutions. The number of papers published continued to increase during the 1960s, the high point being 1966 when over seventy papers appeared, but at the same time there arose the feeling that the hopes for the usefulness of model seismology had not been fulfilled. This feeling was partly due to a rapid increase in the availability of digital computers which diverted many seismologists in Western Europe and North America from analogue model experiments but equally it arose because 'model' seismograms proved to be almost as complicated as real ones, with the result that they did not contribute very much to the better interpretation of field records. This was particularly the case for prospecting seismologists who, used to handling large quantities of data, found that their effort was much more usefully employed on improving methods of digital data processing than on laboratory experiments devoted to the better understanding of elastic wave propagation. Also, the indispensable aid to the geological interpretation of prospecting records is the seismic section. This consists of a large number of seismic traces placed side by side to enable visual correlation of the various reflected, refracted and diffracted pulses. Although, as has been illustrated in figure 5, such a display is possible with a CRO recorder, because of the limits on the spot size of the electron beam it was not easily possible to obtain sections with many tens of traces. Nor was it easily possible to record the ultrasonic traces in reproducible form—such as on magnetic tape—which would allow computer processing of the model results prior to display. For these reasons western seismologists considerably reduced their model seismology effort in the late 1950s, whilst those in the USSR and Eastern Europe continued with the technique and, until lately, provided the bulk of the published work on the subject.

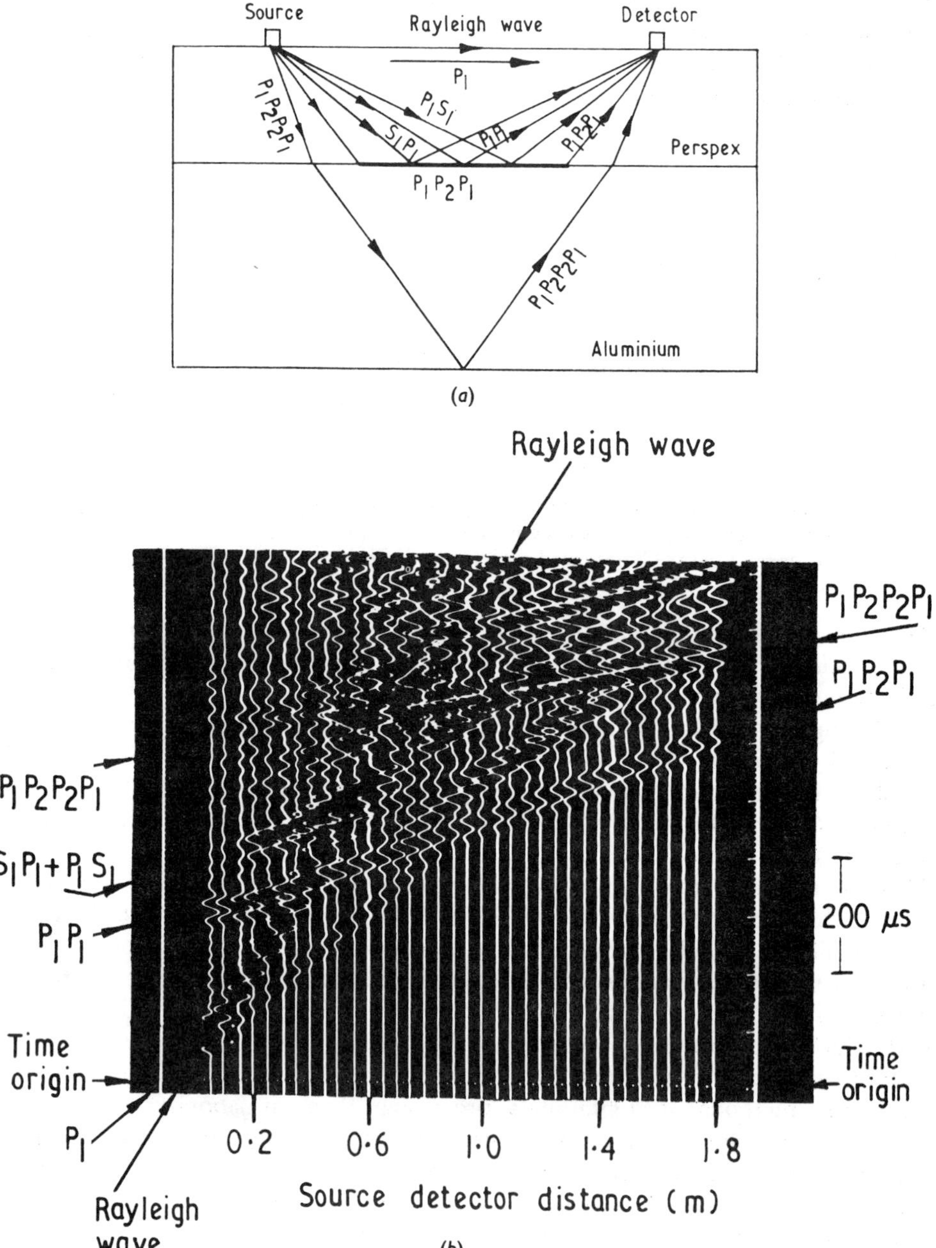


Figure 5. A wiggle-trace section obtained by photographing the face of a CRO. (*a*) the two-dimensional model. (*b*) the observed section.

In the early 1960s the waveform translator became commercially available. The action of this instrument will be more fully discussed in § 5.4 but its essential purpose is to sample the repetitive ultrasonic waveform as displayed on the CRO and output an exactly similar waveform with a timescale about 1000 times as long. This immediately made it possible to record the individual traces on an *XY* plotter or on magnetic or paper tape for direct entry into a computer, and to use normal prospecting equipment to display model sections. Figure 6 shows such a section produced

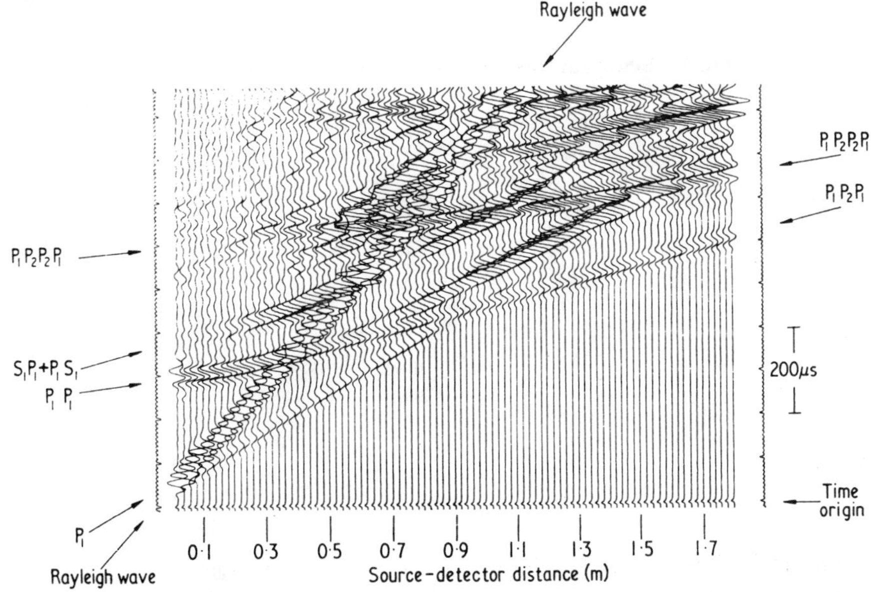


Figure 6. The same section as in figure 5(*b*) but recorded on an *XY* plotter. Note the increased number of traces and their much narrower width making for greater clarity, increased resolution and more accurate amplitude measurement.

from the same model as that used for figure 5 but recorded on an *XY* plotter, the increase in clarity is reasonably clear in these illustrations but is more obvious still on the original recordings.

## 2. Scaling factors

If the laboratory experiments were performed on exact scale models of the Earth the resulting records would, of course, be just as complex and difficult to interpret as the field seismogram. Sometimes this is what is required, the mode of operation being to alter the model in known ways so as to determine each of the contributing factors to the total complexity. Usually, however, there is a stage in the procedure in which those aspects of the field situation which are thought to be incidental to the problem are eliminated before construction of the model. For example, porous rocks, which in the real Earth are full of water, are never modelled as fluid–solid continua in the laboratory experiments. The correct decision on which factors should be omitted or inexactly scaled is of considerable importance. Unfortunately, it is a decision which has often been taken far too lightly.

The important parameters in seismic wave propagation are the linear dimensions of the transmission path, for example the depth and thickness of layers and the seismic pulse length, the P and S wave velocities, and the rock densities. Properties such as anelasticity, microinhomogeneity and stress amplitude are of minor or negligible importance and are seldom considered when building a model. In fact, many experiments have been made under extreme violation of proper scaling as, for example, those in which fluids have been used as part of the transmission path.

Taking mass, length and time as the fundamental quantities, each may be assigned an independent scaling factor in terms of which the scaling factors for all the other physical quantities may be obtained. Table 1 gives a short list of some

**Table 1. Scale factors for three-dimensional models**

| Quantity | Dimensions | Scale factor notation | Commonly used scale factor |
|---|---|---|---|
| Mass | M | $S_M$ | $10^{-15}$ |
| Length | L | $S_L$ | $10^{-5}$ |
| Time | T | $S_T$ | $10^{-5}$ |
| Density | $ML^{-3}$ | $S_M S_L^{-3}$ | 0·5–3 |
| Velocity | $LT^{-1}$ | $S_L S_T^{-1}$ | 1 |
| Strain | — | 1 | unscaled |
| Stress | $ML^{-1}T^{-2}$ | $S_M S_L^{-1} S_T^{-2}$ | unscaled |
| Elastic modulus | $ML^{-1}T^{-2}$ | $S_M S_L^{-1} S_T^{-2}$ | 0·5–3 |
| Viscosity | $ML^{-1}T^{-1}$ | $S_M S_L^{-1} S_T^{-1}$ | unscaled |
| Poisson's ratio | — | 1 | 1 |
| Quality factor, $Q$ | — | 1 | 0·2–100 |
| Wavelength | L | $S_L$ | $10^{-5}$ |
| Frequency | $T^{-1}$ | $S_T^{-1}$ | $10^5$ |

relevant physical quantities and their scaling factors. It is worth remembering that dimensionless quantities should be the same in the model as in the Earth. Poisson's ratio, certainly, and absorption loss, possibly, are two such factors. Also it is worth noting that if length and time are scaled by the same factor then the rocks themselves

**Table 2. Properties of some materials suitable for seismic models**

| Material | Density ($10^{-3}$ kg m$^{-3}$) | Velocity (m s$^{-1}$) Dilatational 3-D | 2-D | 1-D | Shear | Rayleigh | Poisson's ratio |
|---|---|---|---|---|---|---|---|
| Aluminium | 2·7 | 6400 | 5450 | 5150 | 3150 | 2950 | 0·34 |
| Araldite | 1·2 | 2650 | 2200 | 2050 | 1250 | 1170 | 0·36 |
| Brass | 8·5 | 4300 | 3750 | 3500 | 2100 | 1960 | 0·34 |
| Copper | 8·9 | 4700 | 3950 | 3700 | 2260 | 2120 | 0·35 |
| Dural | 2·8 | 6500 | 5400 | 5100 | 3150 | 2940 | 0·35 |
| Lead | 11·3 | 2300 | 1500 | 1650 | 900 | 850 | 0·41 |
| Paraffin wax | 0·9 | 2200 | 1450 | 1300 | 800 | 750 | 0·43 |
| Perspex | 1·2 | 2700 | 2300 | 2200 | 1350 | 1260 | 0·33 |
| Pitch | 1·3 | 2450 | 1800 | 1650 | 1000 | 940 | 0·40 |
| Plaster of Paris | 1·2 | 2800 | 2250 | 2100 | 1250 | 1170 | 0·37 |
| Polystyrene | 1·05 | 2350 | 2000 | 1850 | 1150 | 1070 | 0·35 |
| Polythene | 0·9 | 1800 | 950 | 800 | 500 | 475 | 0·46 |
| Polyvinyl chloride (rigid) | 1·35 | 2300 | 1800 | 1700 | 1050 | 990 | 0·37 |
| Steel | 7·8 | 5900 | 5450 | 5250 | 3260 | 3000 | 0·28 |
| Tin | 7·3 | 3300 | 2900 | 2750 | 1650 | 1550 | 0·33 |
| Tungsten | 19·3 | 5200 | 4750 | 4550 | 2850 | 2650 | 0·29 |

Values are approximate only and may not be mutually consistent. The 'two-dimensional' Rayleigh wave velocity is less than the three-dimensional value, usually by 2% to 5%.

may be considered for the laboratory experiment, although they may violate the scaling conditions due to lack of homogeneity or due to dimensional anelasticity (eg viscosity). In practice rocks are seldom used and more easily worked materials are chosen, a list of which is given in table 2.

Since it is impossible to model a section of the Earth in all its complexity the laboratory results can only be used in a semiquantitative manner. As a result many experiments have been performed with materials chosen more for their laboratory convenience than for their scaled properties. As mentioned above, the extreme example of this is the use of a fluid, usually water, as part of the transmission path. This has led to a large number of results which are not merely useless but positively misleading. Though exact scaling is seldom either possible or desirable sufficient consideration must be given to the problem to ensure that the model does relate to the Earth in some meaningful way.

One manner in which scaling is almost always violated is in the size of the source and receiver, which are normally much larger than proper scaling would allow. If these are important, for instance in multireceiver experiments where each individual detector acts as an appreciable scattering centre, then special calculations or experiments have to be made.

## 3. Types of model

Models can be most conveniently categorized as one-, two- or three-dimensional.

### 3.1. *One-dimensional models*

In many situations, both in earthquake and prospecting seismology, the source–detector distance is sufficiently great for the seismic wavefront to be considered plane. If such a wave travels through the earth so that its wavenormal (ray) is perpendicular to plane-parallel layering then the system becomes one-dimensional. Seismic reflection prospecting in sedimentary basins usually very closely approximates to this condition as do many earthquake body waves as they pass through the Earth's crust. In such a case it is not necessary to construct a three-dimensional model and propagation along a rod provides a close analogy to the field case. Also, since there is no mode conversion from P to S or vice versa when rays are perpendicular to the layering, a fluid-filled pipe is equally satisfactory.

In the case of a rod three types of wave motion are possible; extensional, rotational and flexural. Of these only the lowest mode rotational waves are without dispersion, although for extensional waves whose length is much greater than the diameter of the rod the dispersion is negligible. Since seismic body waves are without geometric dispersion it would seem that rod rotational waves would be most suitable for the model. However, it appears that only low frequency extensional waves have been used, possibly because they are somewhat easier to generate and detect than are rotational waves.

In order to ensure plane-wave propagation the wavelengths used must be many times as great as the rod or pipe diameter. In the case of the rod the velocity is the 'Young's modulus' velocity, that is $c_Y = (Y/\rho)^{1/2}$, while the velocity in the pipe is that of the contained fluid slightly modified by the rigidity of the pipe walls. In order to simulate the layering it is usual to alter the area of cross section either by machining the rod or by putting inserts in the pipe. This technique leaves unaltered the wave velocity but provides a variation in the mass per unit length. A change in mass per unit length from $m_1$ to $m_2$ will provide a reflection coefficient of $(m_2-m_1)/(m_2+m_1)$. By suitable machining of the rod or insertion of obstructions in the pipe any distribution of reflection coefficients may be obtained. In the Earth the reflection coefficients are usually caused by a variation in velocity rather than in cross-sectional density—the full formula for a one-dimensional reflection coefficient is $(m_2 c_2-m_1 c_1)/(m_2 c_2+m_1 c_1)$ where $c$ represents velocity—so that if precise modelling is required it is necessary to alter the distances between the model interfaces in the same proportion as the field velocities, so that the relative timing between reflections remains correct. Some experimenters have used composite rods made of materials of different velocities but the extra complication of butt-joining is seldom worthwhile. Yet another method is to alter the velocity of segments of the rod by altering their temperature. Berryman *et al* (1958) found that nylon was a suitable material in this regard, showing that its velocity decreased linearly with temperature by about a factor of two for a change in temperature of 70 °C. Perhaps the ultimate in such analogue models was made by Woods (1956) who used an air-filled pipe 100 m long with a loudspeaker as the source and a microphone as the detector.

Sherwood (1962) described a model in which electric waves were made to travel up and down a transmission line constructed of lumped components. Each layer in the Earth was represented by a capacitor, an inductance and, if necessary to allow for dissipation, a resistor. The capacitor was made continuously variable and calibrated in terms of the equivalent seismic velocity. With the component values chosen by Sherwood this ranged from about 1·5 km $s^{-1}$ to 7 km $s^{-1}$. The frequency band was from 25 kHz to 600 kHz and so conventional electronic and photographic equipment could be used for the source, detector and recorder. The great advantage of this type of model is that the transmission path may be altered essentially instantaneously by simply turning a number of knobs and the effect on the model seismogram monitored by a conventional cathode ray oscilloscope. Enough experience was obtained with this type of model to confirm that it was easy to use and of great educational benefit but, in spite of this, it was soon supplanted by the digital computer, whose ubiquity outweighed the disadvantage of an increased 'turn-round' time. Numerical modelling in one dimension is standard practice in prospecting but will not be pursued here as it does not fall within the province of this review.

Before leaving this section it is necessary to refer to the pioneer paper of Peterson *et al* (1955). It was they who first constructed an analogue model consisting of the hundreds of layers necessary to model accurately a section of sedimentary rock. In their system they first constructed a film strip whose transparency varied along its length in the same manner as the reflection coefficient varied with depth in the Earth's subsurface. This strip was then passed between a light source and a photocell in such a way that the output of the cell gave a voltage analogue of the reflection coefficients. This was then appropriately filtered to simulate the effect of the seismic source and detection system before recording with a conventional seismic oscillograph. The major drawback of this system is that it does not include the multiple reflections which occur within and between the various layers and which have subsequently been found to play such an important role in the formation of the field seismogram. Nevertheless, the paper introduced a new and valuable concept into the repertoire of the seismologist which has been of great interest and much utility.

3.2. *Two-dimensional models*

Although plane-parallel layering is often a good approximation to the outer layers of the Earth, seismologists are usually more interested in the less monotonous areas where, for instance, the strata may thicken in a given direction, exhibit curvature or terminate against a geological fault. A large number of these conditions are well approximated by two-dimensional structures and these may be modelled in the laboratory by propagation in an elastic plate. Also, even when the strata being modelled are essentially plane-parallel it is often the case that the field seismogram consists of waves which have propagated through the section at some angle to the layering and to model this it is also necessary to move from one to two dimensions. Further, one very important wave type—the 'refracted' or head wave—only exists when the propagating wavefronts are curved so that, once again, at least two dimensions are called for. Lastly, analytical solutions for two-dimensional problems may be more reliably extended to the three-dimensional geometry of field seismology after they have been confirmed in the laboratory by two-dimensional experiments.

Consider a homogeneous elastic plate of thickness $t$ placed in a vacuum. Let its major faces be parallel to the $xy$ plane of a set of Cartesian coordinates whose origin is in the centre of the plate. The plane $z = 0$ is therefore the median plane of the plate and the planes $z = \pm t/2$ represent its major faces. For the very low frequency waves in which the particle displacement in the plate is essentially independent of $z$, Oliver *et al* (1954) analysed the wave motion in terms of the concept of generalized plane stress (Love 1944). They showed that a P wave would travel in the plate with a velocity $c_{\text{p}}$ somewhat less than that for dilatational waves in an infinite solid $c_{\text{d}}$. This wave has its particle motion in the direction of propagation and is often called the longitudinal plate wave. They also showed that a rotational S wave would travel in the plate with the same velocity $c_{\text{s}}$ as S waves in an infinite solid, with its particle motion perpendicular to the direction of wave propagation. Finally, they demonstrated that the velocity of Rayleigh waves along the free edge of a plate was related to the plate wave dilatational and shear velocities in an identical manner to the relation between the halfspace Rayleigh wave velocity and the P and S wave velocities in an infinite medium. Therefore, in order to use a plate to model the Earth it is necessary to choose a material such that $c_{\text{p}} : c_{\text{s}}$ in the plate equals $c_{\text{d}} : c_{\text{s}}$ in the Earth. In other words, when modelling a three-dimensional geometry with one of two dimensions the nondimensional parameters such as Poisson's ratio or quality factor do not remain unchanged. Of course, it is not necessary to do the experiments in a vacuum, the acoustic impedance of air is so much less than that of any plate that its effect may be neglected.

More detailed analysis of plate wave propagation (eg Redwood 1960) shows that it is generally dispersive, with both P and S waves having a doubly infinite set of modes of propagation. Each set may be divided into two subsets, one of which contains waves whose motion is symmetrical with respect to the median plane of the plate while the other contains only those waves which have antisymmetric motion. These are illustrated in figure 7. The P and S waves investigated by Oliver *et al* (1954) are the fundamental symmetric modes for P and S wave propagation. For the S wave this is nondispersive and truly rotational but for the P wave the velocity is frequency dependent and, except in the zero frequency limit, its motion is not simply longitudinal.

This dispersion is a direct consequence of the plate geometry; it does not occur in an infinite medium and therefore when modelling propagation in the Earth by propagation in a plate it is necessary to minimize its effect. Further, since the antisymmetric modes are not required care must be taken to ensure that they are

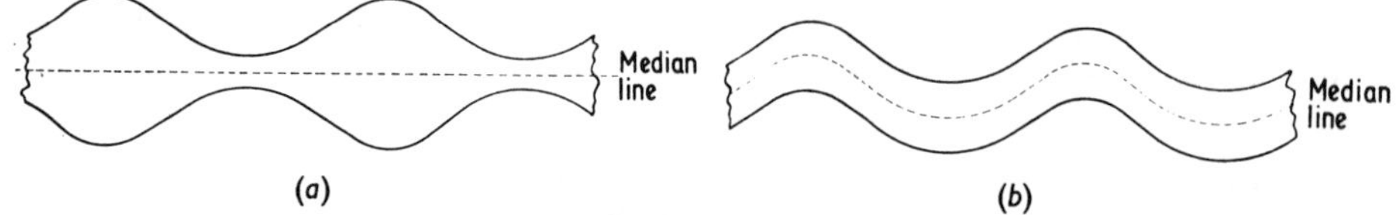


Figure 7. (*a*) symmetric and (*b*) antisymmetric plate (Lamb) waves. The motion is vastly exaggerated for purposes of illustration.

not produced in the laboratory model. Table 3 gives some values of 'longitudinal' plate wave velocity as a function of Poisson's ratio and the ratio of wavelength to plate thickness. How much dispersion is tolerable depends upon the problem under investigation. Most workers seem to use a rule of thumb value of 10:1 for wavelength-to-thickness ratio when doing qualitative work and to make exact calculations

**Table 3. Phase velocity, $c$, of plate waves as a function of Poisson's ratio, $\sigma$, wavelength, $\lambda$, and plate thickness, $t$**

| $\sigma$ | $100(c_p - c)/c$ (% dispersion) | $\pi t/\lambda$ |
|---|---|---|
| 0·20 | 0·01 | 0·0977 |
| | 0·10 | 0·3041 |
| | 1·00 | 0·8366 |
| 0·25 | 0·01 | 0·0734 |
| | 0·10 | 0·2300 |
| | 1·00 | 0·6714 |
| 0·30 | 0·01 | 0·0570 |
| | 0·10 | 0·1797 |
| | 1·00 | 0·5428 |
| 0·33 | 0·01 | 0·0490 |
| | 0·10 | 0·1543 |
| | 1·00 | 0·4732 |
| 0·35 | 0·01 | 0·0454 |
| | 0·10 | 0·1434 |
| | 1·00 | 0·4423 |
| 0·40 | 0·01 | 0·0367 |
| | 0·10 | 0·1160 |
| | 1·00 | 0·3626 |

$c_p$ is the limiting value for low frequency extensional waves in a plate.

of the dispersive effect in quantitative experiments. It should be emphasized that the deleterious effects of dispersion hardly ever arise in transit-time measurements but occur mainly when measuring the amplitude or shape of propagating pulses.

The need to keep the wavelength-to-thickness ratio fairly large imposes a rather irksome restriction on two-dimensional modelling. This is because many problems require that the transmission path should be at least some tens of wavelengths long.

In crustal studies, for instance, the model may have to simulate a shot-to-detector distance of 100 km with seismic wavelengths around 2 km. With a scaling factor of $10^5$ the model would have to be about 2 m long, so as to ensure that reflections from its boundaries arrive after all the waves of interest. With a wavelength-to-thickness ratio of 20 : 1 the plate would be 1 mm thick. Such a plate would be very floppy to handle and awkward to support and, while it obviously would be possible to use it, the care required would considerably lengthen the time necessary for the measurements.

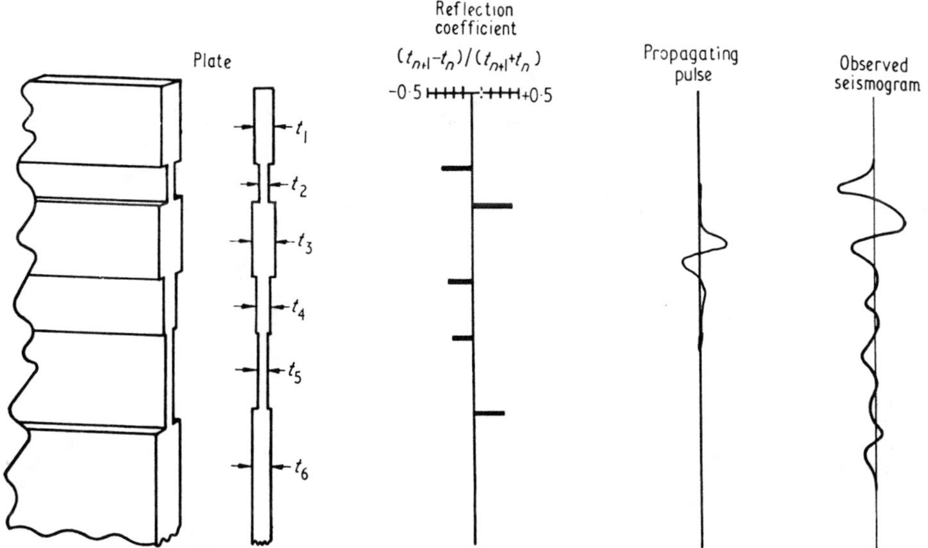


Figure 8. Reflection seismogram obtained from plate thickness variations. The observed amplitude is normalized to that of the first reflection and thereafter allows for attenuation due to enlargement of the wavefront and transmission loss across the reflecting boundaries. Transmission paths involving more than one reflection have been neglected.

The simplest two-dimensional model is obtained by making alterations in the plate thickness. In the same manner as with rods reflections take place wherever a propagating wave meets a variation in the density per unit length of the plate. Figure 8 shows a plate which has been made with a number of decreases and increases in its thickness. These correspond to negative and positive reflecting interfaces which give the simple reflection seismogram which is also shown in the figure. As in the one-dimensional case the reflection coefficient depends simply upon the contrast in density per unit length, the velocity remaining essentially constant, provided the plate is always sufficiently thin. Ideally, the thickness variations should be made symmetric about the median plane so as to prevent the excitation of the slowly propagating dispersive asymmetric modes. However, in order to keep the manufacture of the plate as simple as possible the machining is often done on one side only. A simpler method of altering the plate thickness is to stick pieces of adhesive tape on both sides of the plate. This apparently crude method, which was first suggested by Boucher (1964), is surprisingly effective and has the two great advantages that the same plate can be used over and over again and that each model can be made up and stripped down extremely rapidly. A worthwhile refinement is to use strips of lead foil rolled flat and stuck on to the plate with double-sided adhesive tape. Different reflection coefficients may be obtained by choosing different thicknesses of foil.

Probably, two-dimensional models have been most often made by taking two or more sheets of the same thickness but of differing materials, and bonding them edge to edge with an epoxy resin. This is illustrated in figure 9(*a*). By this means a transmission path may be built up consisting of a number of layers each with different values of velocity, density and Poisson's ratio. Because the range of usable materials is not sufficient and because it is sometimes required to make models with

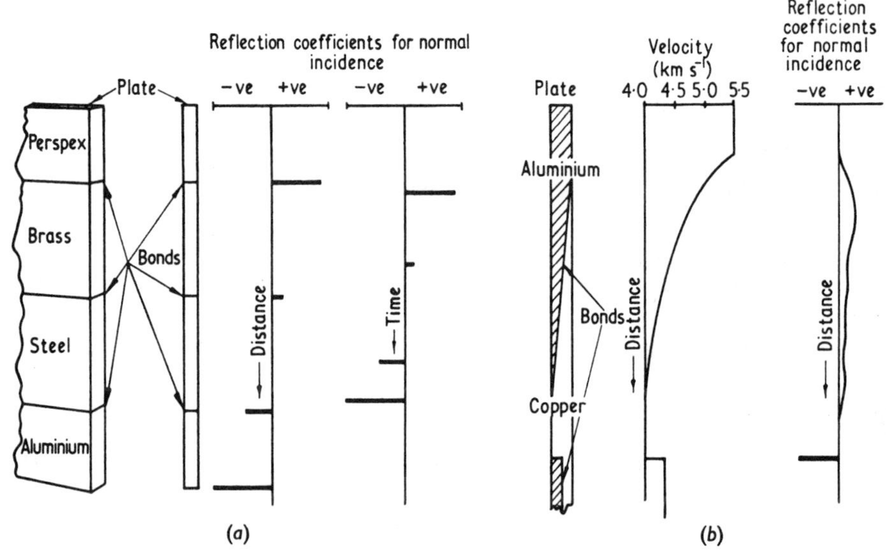


Figure 9. Layered models produced by plates bonded (*a*) edge to edge and (*b*) face to face. In the face to face model the reflection coefficient of the transition layer is frequency dependent.

continuously varying properties a certain amount of experimentation has been done with composite models. In these, two or more thin sheets with different properties are bonded together face to face. By altering the relative proportions of the individual sheets the density and velocity of the composite plate may be made to vary in a prescribed manner. Such a plate is illustrated in figure 9(*b*). This technique appears to have been first suggested by Oliver (1956) and subsequently to have been described and justified in more detail by Angona (1960), Healy and Press (1960) and Riznichenko *et al* (1961).

For very low frequencies the appropriate elastic constants approach those applicable to the static case. Using this principle the above authors showed that the velocity and density of a composite plate are given by $\bar{c}$ and $\bar{\rho}$ where

$$\bar{\rho} = \Sigma f_n \rho_n \quad \text{and} \quad \bar{c}^2 = (\Sigma f_n \rho_n c_n{}^2)/\bar{\rho}$$

and where $f_n$ is the fraction of material n, $\rho_n$ is its density and $c_n$ is the velocity of either the P wave or the S wave. These authors also compared the predicted value of $\bar{c}$ with measurements made on a variety of two-component plates and found any discrepancy between the two to be around 1%. This is well within the general experimental error which is due mainly to fabrication difficulties. Healy and Press also pointed out that the P wave dispersion appeared to be appreciably greater in a composite plate than in a homogeneous one. Consequently, in seismic modelling, composite plates have to be thinner than homogeneous ones. They attributed the increased dispersion to cross coupling between the symmetric and asymmetric modes, it being obviously impossible entirely to eliminate the latter due to the intrinsic asymmetry of the composite. Another difficulty is that the motion close to the source is a rapidly varying interference pattern between the motions in the individual sheets and it is not until the propagation distance has reached several wavelengths that the motion becomes stable. In one experiment Riznichenko *et al* (1961) reported that the source–receiver separation had to be 0·5 m before the propagating wave settled down to a stable waveform with a velocity given by the theoretical low frequency value. This distance is a large slice out of any model, so before committing oneself to composite plates it is desirable to perform preliminary experiments to determine the range of the 'near field'.

Another problem with composite plates comes from difficulties in obtaining a satisfactory bond. Amplitude measurements by Riznichenko *et al* (1961) gave attenuation coefficients which were many times those expected, especially for the shear wave. As they pointed out this additional attenuation was most probably due to a poor bond between the two sheets. In order to circumvent this, some workers have used electrolytically-clad materials such as those used in the manufacture of copper-bottomed aluminium saucepans.

Rykunov *et al* (1960) made plates of a paraffin–polyethylene mixture and then produced velocity gradients within them by varying the temperature throughout the plate. For this particular mixture they found that a practical range of temperatures was from 10 °C to 20 °C. In this range the absorption was essentially constant (whether per unit length or per wavelength they do not state) and both P and S wave velocities decreased with increase of temperature in the same proportion, so that Poisson's ratio remained essentially constant. At higher temperatures both the absorption and Poission's ratio increased rapidly. Schick and Schneider (1967) also used a paraffin plate and managed to obtain velocity variations of as much as 40% without appreciably altering either Poisson's ratio or the absorption. Their temperature range was from 20 °C to 30 °C. They modelled several possible velocity–depth distributions and demonstrated quite convincingly that whereas minor variations in the distribution made no measurable difference to the arrival time of the various pulses they had a very considerable effect on their amplitudes.

Not much work appears to have been done with heated plates. This is no doubt due to the fact that those materials which have a high velocity–temperature coefficient are also rather highly absorptive so that additional work is required to separate out the effect of the intrinsic absorption from the seismically relevant effect of the velocity distribution. Nevertheless, the technique is potentially useful. The continuously varying velocity–depth functions which may be produced by heating a plate should provide good models of the velocity variations in the deeper portions of the Earth, where the effect of increasing pressure and temperature would normally be expected to produce a continuously varying velocity.

Another method of varying the velocity and density through a plate is to perforate it with small holes. A great deal of experimental work and some theory (eg Ivakin 1960, Ivakin and Vasil'ev 1963, Ivakin and Aver'yanov 1963) has indicated that there should be at least ten holes per wavelength and that their diameter should be no more than about half their spacing. Also, their distribution should be on a triangular grid unless transverse isotropy is required in which case a square grid is necessary. By altering the porosity (proportion of void space) the velocity and density may be made to vary throughout the plate. In addition to lowering the effective elastic modulus, and hence the velocity, the holes introduce an appreciable attenuation over and above that due to the inherent anelasticity of the plate material. It has been reported (Ivakin and Vasil'ev 1963) that this attenuation is essentially independent of frequency in the bandwidth used. This seems surprising since one would expect the attenuation to be due to a scattering process and hence frequency dependent. Although the mean values of attenuation per wavelength are similar to those in porous rocks the fact that it decreases with increase in frequency (attenuation per unit distance constant) means that it does not model the field process very well, for in rocks the attenuation per wavelength is essentially independent of frequency. Increasing the porosity of the plate reduces the shear wave velocity more rapidly than it reduces that of the longitudinal waves, hence increasing the effective Poisson's ratio. This is exactly the type of velocity distribution exhibited by porous near surface rocks. They may, therefore, be well modelled by a perforated plate provided that its thickness is properly varied to model the density–depth relationship in the rock.

Ivakin (1960) provided a theoretical analysis of the variation with porosity of the P and S wave velocities of the plate. He assumed that the wavelength was long enough compared with the hole spacing to treat the problem as one in static elasticity. His paper contains a table of formulae from which the velocities and the mean density may be calculated and also a graph showing a comparison between calculated and measured values for the P wave velocity. The measured values were generally 5–10% less than those calculated; this difference he attributed mainly to the fact that the wavelengths used were only 6–10 times the hole spacing, with the result that static conditions were not approached sufficiently closely. His formula for the proportionate decrease in S wave velocity is independent of the Poisson's ratio of the material. Since the P wave velocity does depend upon the Poisson's ratio of the undrilled plate this is in qualitative agreement with the experimental result that the effective Poisson's ratio of a drilled plate increases with its porosity.

A more exact theory for deriving the static elastic constants of a three-dimensional material with spherical holes was developed by Mackenzie (1950) and applied by Sato (1952). In this the velocities of both the P and S waves depend upon the initial value of Poisson's ratio. However, the dependence is slight and there is not much difference between the two methods over the permissible range of porosity. The elastic constants as given by Sato are

$$\frac{k}{k_0} = 1 - \frac{1{\cdot}5(1-\rho)(1-\sigma_0)}{1-2\sigma_0}$$

$$\frac{\mu}{\mu_0} = 1 - \frac{15(1-\rho)(1-\sigma_0)}{7-5\sigma_0}$$

$$\frac{\lambda}{\lambda_0} = 1 - \frac{(1-\rho)(1-\sigma_0)}{\sigma_0}\left(\frac{0{\cdot}5(1+\sigma_0)}{1-2\sigma_0} - \frac{5(1-2\sigma_0)}{7-5\sigma_0}\right)$$

$$\frac{\sigma}{\sigma_0} = 1 - \frac{(1-\rho)(1-2\sigma_0)(1-\sigma_0^2)}{\sigma_0}\left(\frac{0{\cdot}5}{1-2\sigma_0} - \frac{5}{7-5\sigma_0}\right)$$

where $\rho$ is the density of the porous medium divided by the density of the solid material and the subscript 0 refers to the nonporous material. The theory is strictly applicable only for values of $\rho$ not too different from unity but seems quite good enough for design purposes for all feasible porosities, even in plates. In fact, Ivakin's (1960) measured velocities are significantly closer to those predicted using Mackenzie's theory than to those predicted by his own. It is preferable, therefore, to use the above formulae rather than to use those given by Ivakin.

We have outlined above the main types of two-dimensional models, in practice it is usually required to vary both the velocity and linear density of the plate and so a combination of techniques is often used. Most usually one uses a combination of butt-jointing plates of different velocities and of alteration of their thicknesses.

One major problem in butt-jointing is that of the adhesive layer between the dissimilar plates. This will be discussed in §3.4.

### 3.3. *Three-dimensional models*

The early experiments were with three-dimensional models (eg Kaufman and Roever 1951, Northwood and Anderson 1953). Many materials were used—the rocks themselves, metals, plastics, glass, concrete, pitch and various aqueous solutions and organic liquids. For quite a long time water was a favoured material for the uppermost layer of a model. This greatly facilitated the coupling of the transducers to the model and it was thought that the absence of shear waves in one layer would cause no essential difference to the P wave part of the seismogram. As mentioned in §1 this was sometimes not the case and nowadays very few laboratories use liquids in their models. However, the continuing search for a method of producing a velocity gradient in a three-dimensional model has led to one school doing a lot of work with multicomponent gels (Sobotova and Vanek 1966, Waniek 1966). Although gels have some rigidity it is so low that no measurements have been made concerning their shear wave propagation characteristics and a certain amount of doubt must always surround the reliability of the results obtained with such materials. Gels also have the disadvantage that they deteriorate rapidly under normal laboratory conditions and they are seldom usable for more than one or two weeks. Their single advantage is that it is possible to use them to build models with a wide range of continuous velocity–depth functions—although a total velocity range of 1·3–2·0 km s$^{-1}$ is about the maximum possible without introducing so much gelatine that the consequent absorption becomes intolerable.

A more promising method of producing a continuous velocity distribution was introduced by Shamina (1965a, 1966). She made up layers of polymerized epoxy resins loaded with varying amounts of very finely crushed quartz sand. Each layer was about 1 mm thick and the model was made up to its final dimensions of 200 mm × 100 mm × 80 mm by placing layers one on top of another. Succeeding layers were laid down before the previous one had finished polymerizing so that they fused into each other to form an essentially continuous velocity distribution. The sand was fine enough with respect to the 6–7 mm wavelengths used to cause negligible attenuation due to scattering. In fact, the greater the proportion of sand the less was the absorption, which at the greater concentrations reduced to values commensurate with that in steel.

For a while, cement–sand blocks were used. In making these it is necessary to take great care to ensure the homogeneity of the mix for it is difficult to prevent a residual uncontrollable velocity variation of 1 or 2% across the block. For propagation parallel to the gradient this is of no importance but for propagation perpendicular to it very large amplitude changes (as compared with those in a truly homogeneous block) may occur. Obviously, these may interfere with the true purposes of the experiment. A greater disadvantage of normal cement–sand mixes is that the relatively large grain sizes necessitate the use of large wavelengths (several centimetres) which implies models of the order of a cubic metre in size. This is too large and heavy to be usable except in extreme circumstances.

In order to keep three-dimensional models small and easy to handle and store it is necessary to work with wavelengths of a few millimetres. This requires using frequencies of around a megahertz. There is no particular difficulty in this except, perhaps, in feeding sufficient acoustic energy into the model.

Compared with two-dimensional models three-dimensional ones are not so easy to make, not so easy to store and not so easy to use. In particular, because the wavefront expands spherically rather than cylindrically attenuation rates are much higher, with the consequence that greater power needs to be fed into the three-dimensional model. This almost invariably means that 'off the shelf' power sources cannot be used and special ones have to be built. Also, it is not feasible to place detectors within the interior of a three-dimensional model. This may be desirable in order to record the motion at points close to boundaries and discontinuities where measurements may be particularly useful. The equivalent positions in two-dimensional models may easily be achieved by placing the detectors on the side of the plate.

There are, no doubt, a number of problems in which it is essential to use all three space dimensions, but these seem to be more concerned with making models of the seismic source than with models of the transmission path.

---

## 6. Seismic model results

### 6.1. *Proving the technique*

In order to check on their experimental technique most workers choose for their first problem one with a well known analytical solution. The problem most often chosen is that originally solved by Lamb (1904). In this most classic paper Lamb considered the propagation of vibrations over the surface of a semi-infinite solid and obtained their form and amplitude in the far field both for a vertically impulsive line source (two-dimensional) and for a vertically impulsive point source (three-dimensional). A typical model result is that obtained by Tatel (1954) which is illustrated in figure 4 where one of his experimental seismograms is compared with that predicted by Lamb. The agreement is good, in particular of the predominance of the Rayleigh wave and the long tails to the body wave onsets. Even the small change in gradient at the S wave onset is just perceptible, though the experimental procedure was not well designed for detecting it.

Lamb's problem is a convenient one on which to test the experimental technique because a semi-infinite solid—that is, one in which reflections from its boundaries occur too late to be of importance—is the simplest model to construct. However, any problem with a well established solution is satisfactory and, in fact, it is desirable to make two or three separate experiments to check transducer directivity, bonding characteristics, total impulse response, etc, before embarking on an experiment designed to produce new results.

6.2. *Checking the mathematics*

In explosion seismology, particularly as used in studying the Earth's crust, much use if made of head waves and wide-angle reflections. These and their associated ray paths are illustrated in figures 5 and 6. Although analytical expressions for the reflection and refraction coefficients of plane waves have been known for a long time (Knott 1899, Zoeppritz 1919) these are not strictly applicable to seismic waves which are produced by an essentially point source and therefore have curved wavefronts. In particular, the head wave does not exist at all in plane wave theory and the form and amplitude of the spherical wave reflected at angles greater than critical may depart appreciably from those for a plane wave.

Considerable theoretical work has been done on head waves, initially by Jeffreys (1926) who treated the case of two fluids and later by many investigators the most thorough of whom was probably Cagniard (1939). Extensions of head wave theory to include the case of a very thin high-speed† layer have been made by Rosenbaum (1965) and Donato (1965) with Donato also providing some model seismic results. These latter theoretical treatments were made consequent to the publications of the results of many model seismic experiments concerning the head wave generated by a thin layer (eg Lavergne 1961, Levin and Ingram 1962, Shamina 1965b). They account for the observations when the layer is very thin but for the more interesting case when its thickness is greater than about one half of a wavelength reliance has still to be placed upon the model results (eg Faizullin and Epinat'eva 1967, Siskind and Howell 1967).

The essential predictions for a head wave from a semi-infinite high-speed substratum are that, in the far field, its waveform should be the time integral of the incident pulse and its amplitude should decrease as $(rX^3)^{-1/2}$ where $r$ is the distance between the source and detector and $X$ is the distance travelled in the high-speed medium by the critically refracted ray. These predictions are for the three-dimensional case, for two dimensions the $r^{-1/2}$ is dropped. These predictions have been confirmed by model experiments made by O'Brien (1955) for the attenuation law and by Donato (1960) for the waveform.

Many theoretical papers on spherical waves reflected at angles greater than critical have been published but those by Cerveny (1966) and Muller (1968) are probably of most interest to seismologists, since they also consider the vital region where the head and reflected waves interfere with one another. There is great complexity in this region which is markedly frequency dependent. The major difference from the plane wave prediction is that the source–detector distances for minimum and maximum reflection amplitudes are significantly greater for a spherical wave. This difference decreases as the frequency increases disappearing at the high-frequency limit which is, of course, the condition for plane wave theory to hold. Model experiments giving very clear confirmation of the increase in range for the maximum reflection amplitude have been published by O'Brien (1963), Couroneau (1965), Behrens and Dresen (1969), Behrens *et al* (1969). Figure 41 shows the identical section to that in figure 6 except for the changes in recorder gain. Also shown on the figure is a comparison of the calculated amplitudes using *plane wave* reflection coefficients with the amplitudes actually measured. It may be seen that there are substantial differences.

The most impressive case of agreement between theory and experiment is that given by Roever *et al* (1959). They investigated the propagation of elastic waves

† It is, of course, the waves within the layer which have the high speed, not the layer itself!

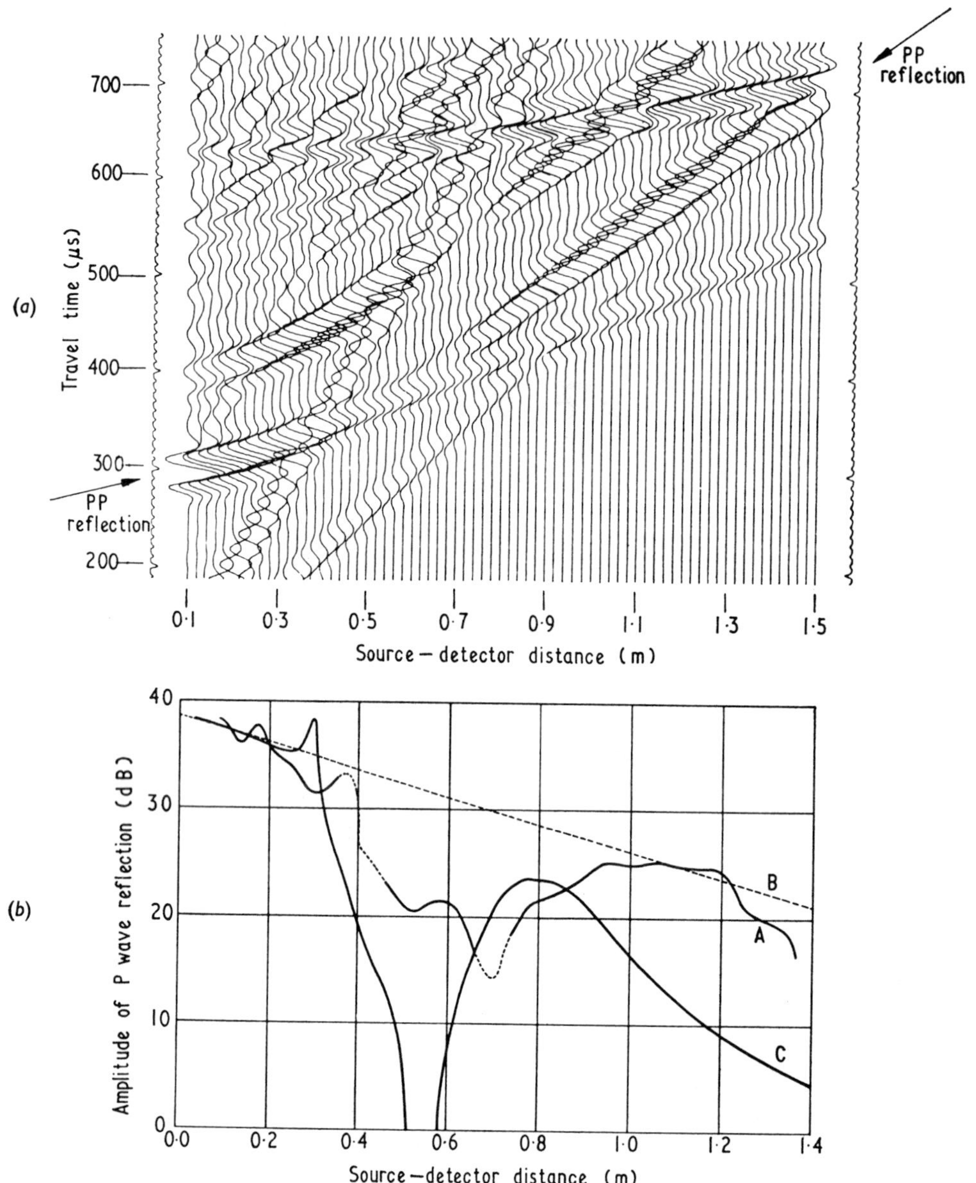


Figure 41. Comparison of measured reflection amplitudes with those calculated using plane wave reflection coefficients. (*a*) the observed section; the gain was increased by 1 dB every 80 mm. (*b*) a plot of the reflection amplitudes. A, observed peak-to-peak amplitudes, the broken section is where the measurements are uncertain due to interference with other events; B, a gradient of 12·5 dB m$^{-1}$; C, calculated values.

along a fluid–solid interface. The calculations were for a line source generating a Dirac impulse of pressure, though it was shown that in the far field the propagating waves should be essentially identical to those generated by a point source. The experiments were made with a spark source in water overlying either pitch, plaster of Paris or iron. The solid materials are chosen not only for experimental convenience but also to illustrate differences which occur due to differences in their mechanical properties. One of their examples is reproduced in figure 42, together with a recording of their source pulse. Roever *et al* give a detailed comparison of the experimental and theoretical results in their paper; it suffices to remark here that they cleared up many points concerning the nature of head and interface waves and provided the first experimental verification of the existence of Stoneley waves, whose existence has been predicted for some years (Stoneley 1924).

### 6.3. *Some problems in seismic prospecting*

Many of the analytical solutions for wave propagation in layered media are for plane-parallel layering. The prospecting seismologist normally becomes interested in a seismic section only when it shows departures from this, these being due mainly to either curvature of the reflecting interfaces or to their abrupt termination (faulting). When the curvature is large the consequent focusing and defocusing of the propagating wave may have very marked effects on the seismic section which may be well demonstrated with seismic models.

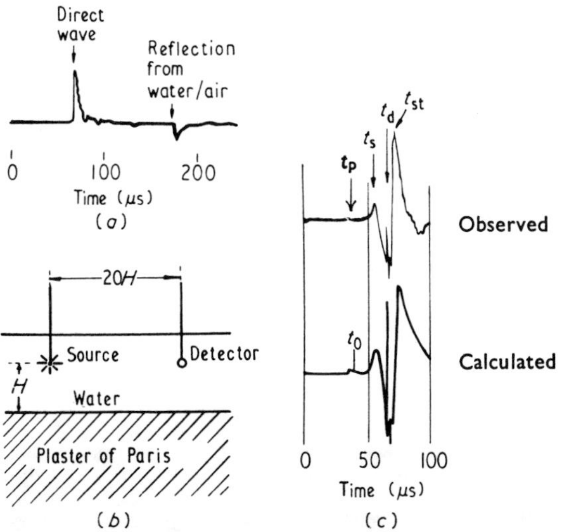


**Figure 42. Pressure response close to a water/plaster of Paris interface (from Roever *et al* 1959). (*a*) direct water-borne pulse 100 mm from the source. (*b*) diagram of the experiment. (*c*) observed and calculated waveforms for set-up in (*b*). $t_P$, arrival time of P head wave from the plaster; $t_0$, arrival time of first zero crossing of the head wave; $t_s$, arrival time of S head wave from the plaster; $t_d$, arrival time of direct water wave (the reflection from the water/plaster boundary occurs immediately after the direct wave and can just be resolved on the observed waveform); $t_{st}$, arrival time of the Stoneley wave.**

Figure 43 shows a reflection experiment carried out with a section containing two depressions of equal curvature but differing depths. The shallower boundary gives reflections whose time and amplitude may easily be correlated to give a curve which is fairly similar to the true depth section. The deeper boundary, however, gives a complex pattern of reflections which, if simply correlated, would imply the existence of a buried hill rather than the true situation, which is a buried valley. Sections like these are invaluable in building up one's appreciation of the effect of curvature. In cases where seismic sections observed in the field have been interpreted in terms of reflections of considerable curvature it may be desirable to make a model of the hypothetical depth section to ensure that it does produce reflections whose salient characteristics match those of the field seismograms.

When a propagating wave meets a 'corner' such as formed by the sudden termination of a horizontal layer or by the tip of a buried wedge, diffraction will take place. This problem is very difficult to tackle analytically and little of value to the seismologist has been published. Consequently, seismic models illustrating the properties of diffracted waves have been made right from the very early days of model seismology, with Rieber's (1936) results still being some of the best obtained. Some twenty to thirty papers on diffraction patterns in model seismology have been published of which perhaps those by Grannemann (1956), Kuhn (1960, 1961, 1962), Tsei-Wen (1963, 1965) and Ru-Teng (1966), are the most noteworthy. Figure 44 gives an example of simple diffraction patterns from a faulted reflecting layer and also shows a section over a continuous layer containing a flexure rather than a discontinuous fault. It may be seen that the two sections are quite similar. However,

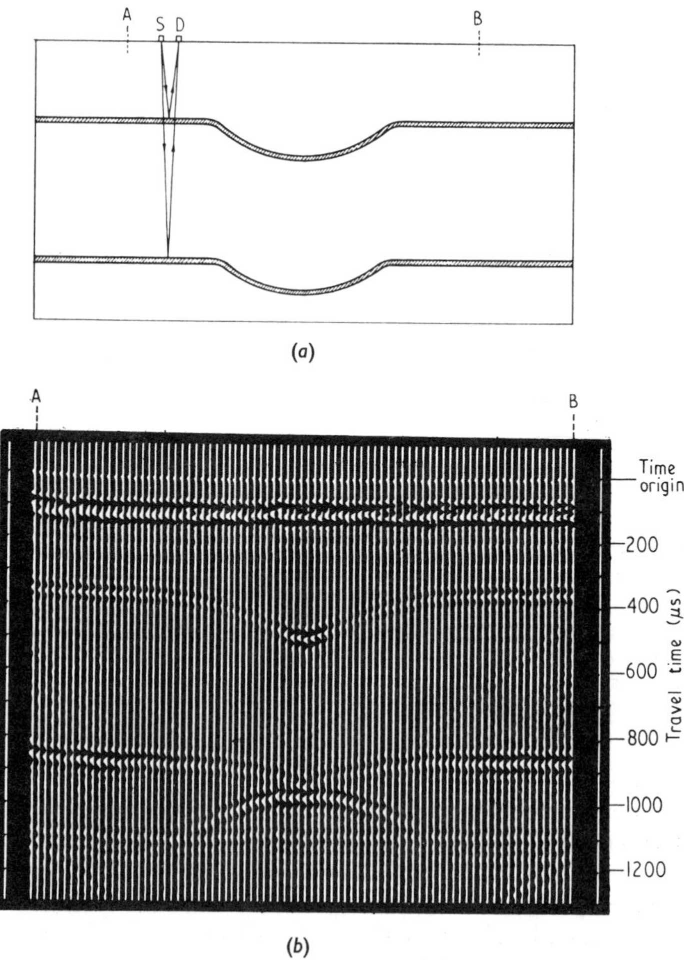


Figure 43. Section illustrating the effect of depth on the reflections received from a curved reflector. The source, S, and detector, D, were kept a fixed distance apart and moved as a pair along the top of the model. Each trace on the section corresponds to one position of the transducer pair.

the geological significance of the two situations may be very different and it is important to be able to discover from the seismic reflection section which one actually exists. Studies of model sections such as those illustrated in figure 44 may be quite useful in such cases.

The problem of the head wave arrival from a thin high-speed layer has already been mentioned in §6.2. While this is of some interest to seismologists the more important problem concerns the attenuating effect which the layer has on waves which pass through it at an obtuse angle and are returned to the surface from a deeper interface. This problem of the 'screening' or 'masking' high-speed layer does not appear to have been investigated theoretically but a number of studies have been made with models, notably those by Lavergne (1966) and Poley and Nooteboom (1966).

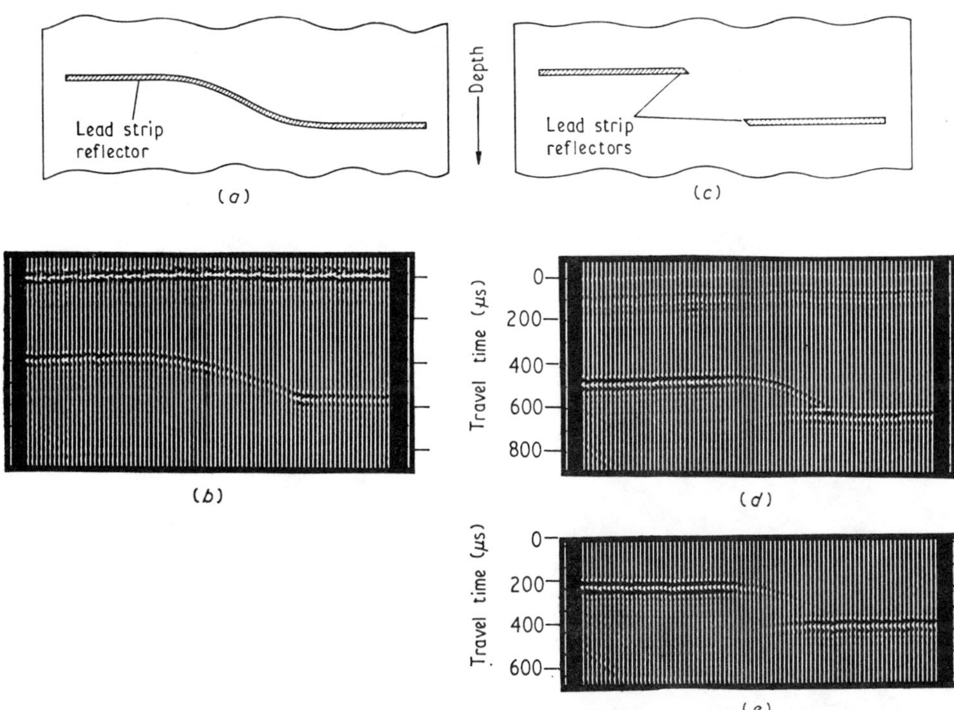


Figure 44. Sections illustrating the reflections obtained from a reflector containing either a flexure or a fault. (*a*) perspex plate model of a flexure. (*b*) section obtained from the flexure model using constant gain. (*c*) perspex plate model of a fault. (*d*) section obtained from the fault model using time-varying gain to keep the shallowest diffracted arrival at near constant amplitude. (*e*) section obtained from the fault model using constant gain.

## 6.4. *Some general seismological problems*

Due to continuously increasing temperature and pressure with depth it seems reasonable to suppose that, in addition to any acoustic discontinuities due to changes in material (rock) properties, the velocity of seismic waves also should vary continuously with depth. Indeed, this is amply confirmed by observation of the time–distance relationships of earthquake waves. It is, however, not yet possible to obtain from routine analysis of seismic records details of the velocity gradients over limited depth intervals. This is because the gradients which exist are so small that they have an insignificant effect on the travel times of seismic waves unless a very large depth interval is considered. They may, however, produce considerable short-range fluctuation in the amplitude–distance relationship, not unlike those produced by curvature of reflecting interfaces. With the improved instrumentation consequent upon seismic monitoring of underground nuclear explosions, observational seismologists are beginning to make a serious study of amplitude–distance relationships.

As an aid to the interpretation of such measurements many model studies have been made to illustrate the effects of various velocity–depth relationships. Much of this modelling has been done by numerical ray tracing using digital computers, but there are mathematical difficulties involved and a number of laboratory model studies have also been made, notably in the USSR and in Eastern Europe (Upper Mantle Committee 1966). As has already been mentioned in § 3.2 Schick and Schneider (1967) produced velocity gradients in a plate by producing a thermal gradient within it. Examples from their paper are shown in figure 45. With

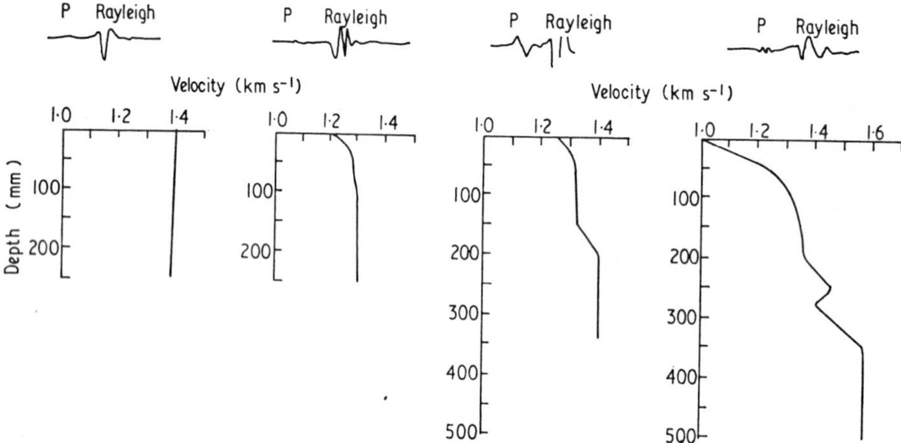


Figure 45. Refraction and Rayleigh wave arrivals for various velocity–depth functions obtained by heating a plate (from Schick and Schneider 1967).

source–detector distances corresponding to typical field cases they measured significant amplitude variations for changes in velocity gradient which produced changes in travel time of only one part in a thousand. Many other papers concerning the effects of velocity gradients have been published, among them being those by Ivakin and Aver'yanov (1963), Ivakin and Kapsan (1968) and Dresen (1971).

Shamina (1966) has investigated with three-dimensional models the effect on seismic wave amplitudes of the depth of the source when the transmission path contains a low-velocity zone (waveguide). Figure 46 shows a typical result from which it may be seen that the minimum amplitude is obtained when the source lies on the axis of the low-velocity zone. This seems quite reasonable, though the rather close similarity between the amplitude–depth and velocity–depth profiles is perhaps more surprising. In her paper Shamina compared her model results with amplitude–depth measurements made for earthquakes in the Hindu-Kush and deduced that a low-velocity zone existed with its axis at a depth of about 130 km.

Another problem of general seismological interest concerns the propagation of surface waves across a layer of varying thickness such as occurs in the region between a continent and an ocean. Some theoretical work has been done on the problem (eg Mal and Knopoff 1966) but the mathematics is difficult and so a small amount of model work has also been done. Much of the work has drastically simplified the problem by treating the propagation along a simple wedge with free boundaries (eg Gutdeutsch 1969, Lewis and Dally 1970) but Kuo and Thompson (1963) treated a more physically appropriate case with a two-layered model in which a thin surface layer of varying thickness represented the Earth's crust. The general conclusion so far is that provided measurements are made sufficiently far away from the tip of a wedge—whether free or covered—the phase velocity is controlled by the mean thickness close to the point of observation and does not depend upon the direction of arrival of the wavetrain. As well as being closer to a true model the work of Kuo and Thompson was also noteworthy for their method of introducing a train

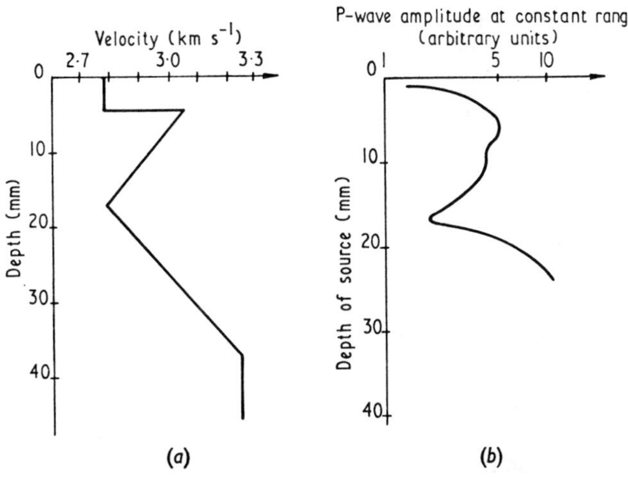


Figure 46. Transmission path containing a low-velocity layer (waveguide). (*a*) velocity–depth profile in the three-dimensional model. (*b*) surface amplitude as a function of source depth. (From Shamina 1966.)

of dispersed Rayleigh waves into their model. They did this by attaching to the model a 9 foot long rod of aluminium which they excited antisymmetrically at its free end. By the time the wave had reached the end of the rod attached to the model the initial 9 μs pulse had dispersed into a train of nearly sinusoidal oscillations with a duration of several hundreds of microseconds.

There are, of course, many other seismological problems which have been investigated by model seismic techniques. Many of these are too specific in application to be included in this review but some of more general interest concern reflection from a rough boundary (Rapaport 1961, Voskresensky 1962), the field close to a point source (Knopoff 1958), topographic scattering of surface and body waves (Tatel 1954, Tatel and Tuve 1955) and propagation in a wedge (Fuchs 1965).

## 7. Future work

Model seismology is flourishing. Not particularly in the number of papers which are published, which remains rather static at around a dozen or so a year, but in the increasing number of laboratories in which model experiments are performed and in the increasing diversity of techniques which are being used. This review has concentrated on the conventional systems using piezoelectric transducers but a renewed and active interest is being taken in schlieren and photoelastic experiments (Kozak and Waniek 1970, Lewis and Dally 1970), and holographic and similar experiments are also taking place (Kalra and Rodgers 1970) or are proposed (Winter and Wells 1971).

Probably it will become more routine to record on magnetic tape, so facilitating computer processing of the experimental results and building up a mixed analogue–digital system. Computer processing will increase the use of deconvolution and enable smaller models to be used. This will remove one of the main advantages of two-dimensional models and so facilitate a return to the more appropriate three-dimensional ones.

Whether or not it will come about we believe that the emphasis on the 'seismology' in model seismology should be removed so that the purpose becomes more related to elastic wave propagation in general. The technique is well suited to pedagogic and didactic purposes and could form the basis of many useful undergraduate experiments in elastic wave propagation, without any necessary reference to seismology.

But seismologists will remain the main users of the technique, investigating problems concerning the seismic source—whether earthquake or explosion—and the transmission path, particularly when it contains complicated structures. Numerical methods are paramount when it comes to investigating the effect of parameter variation in simple models, but physical experiments are needed when the numerical model departs too far from the physical actuality. There are any number of such cases.

More attention needs to be paid to transducer design and to the methods of coupling it to the medium. Although most of the problems to be investigated concern linear systems, so that once a seismogram is obtained for a particular source waveform that which would be produced by any other source may in principle be computed, it is obviously much to be preferred if the initial experiment is carried out with the source function of interest. For this, we need clear design criteria for surface and buried sources generating P or S waves in either two- or three-dimensional models.

Model seismology will no doubt continue to go in and out of fashion but, in the long term, it should find a secure place as an indispensable link between the idealized models of the applied mathematicians and the complex records of observational seismologists. In forging this link, it is to be hoped that the model 'seismologist' will keep one foot firmly in both camps.

## Acknowledgments

We are grateful to the Chairman and Board of Directors of the British Petroleum Company for permission to publish this article, to Mr P Seaman who produced many of the model sections used to illustrate this paper and to Dr M H Safar for discussions on the equivalent circuits of piezoelectric transducers.

## References

In addition to the references quoted in the text and listed below the following general references may be useful.


Berckhemer H and Waniek L 1970 *Bibliography on Seismic Modelling, Parts* 1, 2 *and* 3 (Prague: European Seismological Commission)

Brekhovskikh L M 1960 *Waves in Layered Media* (New York: Academic Press)

Brush Clevite 1966 *Piezoelectricity* (Southampton: Brush Clevite)

Ewing W M, Jardetzky W S and Press F 1957 *Elastic Waves in Layered Media* (New York: McGraw-Hill)

Upper Mantle Committee 1966 *Symposium on Seismic Models, Studia Geophys. Geod.* **10** 239–400b

White J E 1965 *Seismic Waves* (New York: McGraw-Hill)

---

Angona F A 1960 *Geophysics* **25** 468–82

Behrens J and Dresen L 1966 *Z. Geophys.* **32** 232–41

—— 1969 *Z. Geophys.* **35** 175–89

Behrens J, Dresen L and Hinz E 1969 *Z. Geophys.* **35** 43–68

Berckhemer H and Ansorge J 1963 *Geophys. Prospect.* **11** 459–70

Berlincourt D A, Curran D R and Jaffe H 1964 *Physical Acoustics* vol 1 pt A ed W P Mason (New York: Academic Press) pp169–270

Berryman L H, Goupillard P L and Waters K H 1958 *Geophysics* **23** 244–52

Bokanenko L I 1966 *Izv. Acad. Sci. USSR, Phys. of the Solid Earth* 41–6

Boucher F G 1964 US Patent 3154169

Bradfield G 1970 *Ultrasonics* **8** 177–89

Cagniard L 1939 *Reflexion et Refraction des Ondes Séismiques Progressives* (Paris: Gauthier-Villars) (1962 translated and revised by F A Flinn and C H Dix *Reflection and Refraction of Progressive Seismic Waves* (New York: McGraw-Hill))

Cerveny V 1966 *Studia Geophys. Geod.* **9** 259–69

Couroneau J 1965 *Geophys. Prospect.* **13** 405–32

Donato R J 1960 *Geophys. J.* **3** 270–1

—— 1965 *Geophys. Prospect.* **13** 387–404

Dresen L 1971 *Z. Geophys.* **37** 71–88

Evans J F, Hadley C F, Eisler J D and Silverman D 1954 *Geophysics* **19** 220–36

Faizullin I S and Epinat'eva A M 1967 *Izv. Acad. Sci. USSR, Phys. of the Solid Earth* 360–6

Filipczynski L 1963 *Proc. Vibration Problems* **4** No 1 17–25

Fuchs K 1965 *Z. Geophys.* **31** 51–89

Gangi A F and Disher D 1968 *Geophysics* **33** 88–104

Gilbershtejn P G and Gurvich I I 1963 *Bull. Acad. Sci. USSR Geophys. Ser.* 969–78

Grannemann W W 1956 *J. Ac. Soc. Am.* **28** 494–7

Groves I D Jr and Tims A C 1970 *J. Ac. Soc. Am.* **48** 725–8

Gutdeutsch R 1969 *Bull. Seismol. Soc. Am.* **59** 1645–52

Gutdeutsch R and Koenig M 1966 *Studia Geophys. Geod.* **10** 314–22

Healy J H and Press F 1960 *Geophysics* **25** 987–97

IRE 1961 *Proc. Inst. Radio Engrs* **49** 1161–9

Ivakin B N 1960 *Bull. Acad. Sci. USSR Geophys. Ser.* 761–71

Ivakin B N and Aver'yanov A G 1963 *Bull. Acad. Sci. USSR Geophys. Ser.* 546–57

Ivakin B N and Kapsan A D 1968 *Izv. Acad. Sci. USSR, Phys. of the Solid Earth* 225–30

Ivakin B N and Vasil'ev Yu V 1963 *Bull. Acad. Sci. USSR Geophys. Ser.* 149–56

Jeffreys H 1926 *Proc. Camb. Phil. Soc.* **23** 472–81

Kalra A K and Rodgers P W 1970 *Am. Petrol. Inst. Semi-a. Rep. Res. Project* 105 (obtainable only from American Petroleum Institute, New York)

Kaufman S and Roever W L 1951 *Proc. 3rd Wld Petrol. Cong.* (Leiden: Brill) section 1 pp537–45

de Klerk J 1971 *Ultrasonics* **9** 35–48

Knopoff L 1958 *J. Appl. Phys.* **29** 661–70

Knott C G 1899 *Phil. Mag.* **48** 64–97

Koenig M 1969 *Z. Geophys.* **35** 9–15
Kozak J and Waniek L 1970 *Z. Geophys.* **36** 175–92
Kuhn V V 1960 *Bull. Acad. Sci. USSR Geophys. Ser.* 434–42
—— 1961 *Bull. Acad. Sci. USSR Geophys. Ser.* 1136–47
—— 1962 *Bull. Acad. Sci. USSR Geophys. Ser.* 561–8
Kuo J T and Thompson G A 1963 *J. Geophys. Res.* **68** 6187–97
Lamb H 1904 *Phil. Trans. R. Soc.* A **203** 1–42
Langevin R A 1954 *J. Ac. Soc. Am.* **26** 421–7
Lavergne M 1961 *Geophys. Prospect.* **9** 60–73
—— 1966 *Geophys. Prospect.* **14** 504–27
Lavergne M and Chauveau J 1961 *Acustica* **11** 121–6
Levin F K and Ingram J D 1962 *Geophysics* **27** 753–65
Lewis D and Dally J W 1970 *J. Geophys. Res.* **75** 3387–98
Love A E H 1944 *A Treatise on the Mathematical Theory of Elasticity* 4th edn (London: Cambridge University Press) pp207–8
MacKenzie J K 1950 *Proc. Phys. Soc.* B **63** 1–11
Mal A K and Knopoff L 1966 *Bull. Seismol. Soc. Am.* **56** 455–66
Mason W P 1948 *Electromechanical Transducers and Wave Filters* 2nd edn (New York: Van Nostrand)
May J E Jr 1964 *Physical Acoustics* vol 1 pt A ed W P Mason (New York: Academic Press) pp417–500
Miller G F and Pursey H 1954 *Proc. R. Soc.* A **223** 521–41
—— 1955 *Proc. R. Soc.* A **233** 55–69
Muller G 1968 *Z. Geophys.* **34** 147–62
Northwood T D and Anderson D V 1953 *Bull. Seismol. Soc. Am.* **43** 239–45
O'Brien P N S 1955 *Geophysics* **20** 227–42
—— 1963 *Geophys. Prospect.* **11** 59–72
Oliver J 1956 *Earthq. Notes* **27** 29–31
Oliver J, Press F and Ewing M 1954 *Geophysics* **19** 202–19
Orwell R F J 1963 *Ultrasonics* **1** 49–52
Peterson R A, Fillipone W R and Coker F B 1955 *Geophysics* **20** 516–38
Poley J and Nooteboom J J 1966 *Geophys. Prospec.* **14** 184–203
Pursey H 1953 *Brit. J. Appl. Phys.* **4** 12–20
Rapaport M B 1961 *Bull. Acad. Sci. USSR, Geophys. Ser.* 118–24
Redwood M 1960 *Mechanical Waveguides* (London: Pergamon)
—— 1961 *J. Ac. Soc. Am.* **33** 527–36
—— 1962 *J. Ac. Soc. Am.* **34** 895–902
—— 1963 *Appl. Mater. Res.* **2** 76–84
—— 1964 *J. Ac. Soc. Am.* **36** 1872–80
Rieber F 1936 *Geophysics* **1** 196–218
—— 1937 *Geophysics* **2** 132–60
Riznichenko Yu V, Ivakin B N and Bugrov V R 1951 *Izv. Akad. Nauk SSSR Ser. Geofiz.* 1–30
Riznichenko Yu V, Shamina O G and Charutina R V 1961 *Bull. Acad. Sci. USSR, Geophys. Ser.* 231 321–34
Robinson E A 1967 *Statistical Communication and Detection* (London: Griffin)
Roever W L, Vining T F and Strick E 1959 *Phil. Trans. R. Soc.* A **251** 455–523
Romanenko E V 1957 *Sov. Phys.–Acoust.* 364–70
Rosenbaum J H 1965 *Geophysics* **30** 204–12
Ru-Teng L 1966 *Acta Geophys. Sin.* **15** 148–57
Rykunov L N, Khorsheva V V and Sedov V V 1960 *Bull. Acad. Sci. USSR Geophys. Ser.* 1069–71
Sato Y 1952 *Bull. Earthq. Res. Inst. Tokyo Univ.* **30** 179–90
Schick R and Schneider G 1967 *Proc. 9th Assembly Eur. Seismol. Commn.* (Copenhagen: Akademie Forlag) 413–22
Schmidt O 1939 *Z. Geophys.* **15** 141–8
Schwab F 1967 *Geophysics* **32** 819–26
Schwab F and Burridge R 1968 *Geophysics* **33** 473–80
Shamina O G 1965a *Izv. Acad. Sci. USSR, Phys. of the Solid Earth* 484–6
—— 1965b *Izv. Acad. Sci. USSR, Phys. of the Solid Earth* 148–53
—— 1966 *Studia Geophys. Geod.* **10** 341–50
Sherwood J W C 1958 *Proc. Phys. Soc.* B **71** 207–19
—— 1962 *Geophysics* **27** 19–34

Siskind D E and Howell B F Jr 1967 *Bull. Seismol. Soc. Am.* **57** 437–42
Sobotova C and Vanek J 1966 *Studia Geophys. Geod.* **10** 281–90
Stern E 1969 *Ultrasonics* **7** 227–33
Stoneley R 1924 *Proc. R. Soc.* A **106** 416–28
Tatel H E 1954 *J. Geophys. Res.* **59** 289–94
Tatel H E and Tuve M A 1955 *Geol. Soc. Am., Spec. Paper* 62 35–50
Terada T and Tsuboi Ch 1927 *Bull. Earthq. Res. Inst. Tokyo Univ.* **3** 55–65
Thompson K C, Ahrens T J and Toksöz M N 1969 *Geophysics* **34** 696–712
Toksöz M N and Schwab F 1964 *Geophysics* **29** 405–13
Tsei-Wen T 1963 *Bull. Acad. Sci. USSR, Geophys. Ser.* 985–92
—— 1965 *Izv. Acad. Sci. USSR, Phys. of the Solid Earth* 764–6
Tsuboi Ch 1928 *Bull. Earthq. Res. Inst. Tokyo Univ.* **4** 9–20
Upper Mantle Committee 1966 *Symp. Seism. Models, Studia Geophys. Geod.* **10** 239–400b
Voskresensky Yu V 1962 *Bull. Acad. Sci. USSR, Geophys. Ser.* 404–9
Waniek L 1966 *Studia Geophys. Geod.* **10** 273–81
Winter T G and Wells W 1971 *Research proposal to the Am. Petrol. Inst.* (obtainable only from American Petroleum Institute, New York)
Woods J P 1956 *Geophysics* **21** 261–76
Zoeppritz K 1919 *Göttingen Nachrichten* 66–84

# Chapter 2.
# Early Work

The earliest physical modeling paper in this volume is that of Terada and Tsuboi (1927), continued in Tsuboi (1928). Their experiments had a refreshingly pragmatic motivation. They wished to determine whether trenches built alongside physics laboratories might allow better isolation of the delicate instruments within from the vibrations caused by passing traffic. Their model in 2½ dimensions measured the attenuation of Rayleigh waves by a canal. The choice of agar-agar as the modeling medium (Poisson's ratio of nearly 0.5) is one striking difference from most later work, which generally used either true solids or true liquids. This choice of material was probably appropriate, given the high Poisson's ratio of the material to be modeled (soil).

Rieber (1936, 1937) had articles in both the first and the second volumes of *Geophysics*. He used a spark light source to capture an image of wavefronts "on the fly." The pre-World War II experimenters were hampered by lack of sophisticated electronics. The same war that led to major improvements in electronics also led to a decade-long hiatus in published physical modeling research.

After the second World War, physical modeling was soon revived in the Soviet Union. Detailed 3-D model studies were published in 1951 by Riznichenko, although earlier results had been published by Gamburtsev in 1947. The first plate (2-D) model seismology experiments appear to have been done by Riznichenko in 1952. The Russians were prolific in their pursuit of physical modeling, and created some fascinating innovations. One such innovation, plate perforation, allowed Ivakin (1963) to alter in a controlled fashion the density and Poisson's ratio of a model plate. The Russian effort appears to have been driven by investigations into the nature of earthquake seismograms, nuclear weapons test detection, and deep-earth structure.

Often overlooked are the contributions made by physical models in 1-D. Before the advent of digital computers, calculating the response of a series of flat beds to a vertically incident wavelet was not trivial. Woods (1956) used a hollow pipe with variable thickness inserts. The inserts varied the acoustic impedance of the pipe and served as reflectors for sound waves traveling down the pipe. Berryman et al. (1958) produced continuous velocity changes by heating a nylon rod (whose Young's modulus is quite sensitive to temperature). Bennett (1962) produced a model which was essentially a rod of variable diameter. The abrupt diameter changes, again, modeled acoustic impedance discontinuities.

Similar in philosophy to the preceding 1-D papers, but more flexible in operation, the Seismoline (Sherwood, 1962) was an *electric* analog computer that used variable capacitors and resistors to achieve impedance changes along an electric circuit to produce reflected electric pulses. (As an aside, it might be mentioned that Dix (1958) had shown how an analog *electrical* computer could be constructed to model Lamb's problem by calculating Cagniard's integrals.) Together, these modeling efforts were valuable in investigating the convolutional model of wave reflection in a vertically stratified medium.

Columbia University researchers can take credit for resurrecting 2-D physical modeling in the West, and for the introduction of the "sandwich" (or bimorph) approach to 2-D model construction. A few of the papers put out by this group include Oliver et al. (1954), Press et al. (1954), Press et al. (1957), and Healy et al. (1960).

A gradient in velocity is one of the most difficult features to incorporate into a physical model. In 2-D models, the thickness, composition percentages, or temperature can produce such a gradient. Using the sandwich technique, Healy and Press (1960) investigated Rayleigh wave dispersion in a two-layered earth. In 3-D models, gradients are more difficult to achieve. Sobotova and Vanek (1966) obtained a depth-dependent velocity gradient by the laminations of many gelatine layers of successively stiffer composition.

Oliver et al. (1954) summarized the advantages of a 2-D physical model versus a 3-D physical model. The points raised by Oliver reduce fundamentally to issues of money. As we have noted in the introduction, this is not a trivial matter, as intelligent research managers will put their dollars where they can get the greatest research return. The issue of cost meant that the majority of academic efforts in physical modeling in the 1950s were 2-D. The majority of 3-D work was conducted by, or underwritten by, oil companies.

Published Western post-war 3-D physical modeling resumed with the experiment of Shell researchers Kaufman and Roever (1951). Roever would continue these experiments with increasing experimental accuracy to the point that they could serve as an experimental test for the Cagniard-deHoop method of seismogram synthesis. This test of Cagniard-deHoop is included in the next chapter (Roever et al., 1959).

The reader may note in the experiments of the 1950s something of an overemphasis on detecting refracted events. We use the term ''overemphasis'' because refracted events have had a peripheral value in exploration geophysics since the end of the large scale refraction shoots of the 1930s. Why, then, this curious propensity for the refracted arrival? There are several answers. First, and probably most important, the limited dynamic range of the early systems made it difficult to record reflections in the presence of the Rayleigh wave. The refracted arrival generally appeared before the Rayleigh wave so that its waveform and amplitude could be measured with a fair degree of accuracy. Second, the headwave is not predictable on the basis of ray theory. Hence, the presence of the headwave is something of a ''smoking gun'' for validating analytic solutions of infinitesimal strain wave motion. Third, the headwave does not suffer from transducer directionality as do reflected arrivals, because the refracted arrival always comes in at the same angle. Fourth, the refracted events are of great importance to earthquake seismologists, and many geophysicists were (and still are) recruited from the ranks of earthquake seismologists.

A quick summary follows:

+ Terada and Tsuboi (1927) described their methodology for experiments with Rayleigh waves using agar-agar (gelatin) as the medium, an electromechanical source, and an optical detector. The authors make a point of specifying that their apparatus has dimensions of 5 to 10 wavelengths in length. They showed that the amplitude of particle motion of the Rayleigh wave diminished with depth, in qualitative agreement with theory.

+ Tsuboi (1928) used the experimental arrangement described in the preceding paper to show that a trench dug to a depth of a good fraction of Rayleigh wavelength will substantially reduce the amplitude of the propagating Rayleigh wave.

+ Rieber (1936) used an optical method to capture 2-D wavefronts as they progressed in time. These photographs showed the diffractions caused by truncating reflectors as well as reflections from curved beds. Wavefront images of this clarity could not be produced by any digital method until the introduction of high-order finite-difference numerical modeling in the 1970s.

+ Kaufman and Roever (1951) recorded the vertical particle motion as a wiggle trace. They observed direct body waves, refracted body waves, and surface waves in a 3-D homogeneous medium. This paper has some historical importance in its support for Lamb's theory. At that time, there was a belief by some in the nuclear test detection community that Lamb's theory did not strictly apply to wave propagation in real materials.

+ Howes et al. (1953) showed that the events from 3-D faulted models can be correlated with those same faults.

+ Oliver et al. (1954) laid out the theory of plate-wave modeling in this paper and experimentally demonstrated equations for wave dispersion. They included a lengthy defense of the relative merits of 2-D physical modeling relative to 3-D physical modeling.

+ Evans (1959) performed a 3-D shear-wave physical model experiment to demonstrate the clarity of SH profiling. He also showed that Love waves are generated by an SH source when there is a shallow low-velocity layer.

+ Angona (1960) investigated diffractions from terminating beds, as well as reflections from curved reflectors in 2-D plate models, much as Rieber did in 1936. The improvement here stems both from the ability to better manipulate the velocity field (Rieber was restricted to constant velocity) and to produce measurements corresponding to discrete surface sampling locations.

+ Healy and Press (1960) used a novel laminated construction method (alternatively termed ''sandwich'' or ''bimorph'' construction) to create 2-D physical models with elastic parameters that varied smoothly with increasing depth.

+ Levin and Ingram (1962) successfully matched a simple theory based on superposition of head waves and direct waves against amplitude measurements acquired from 2-D models with various refractor layer thicknesses.

# *Experimental Studies on Elastic Waves* (Part 1)

By

**Torahiko TERADA and Chûji TSUBOI**

Earthquake Research Institute

---

## 彈性波の實驗　（第一報）

所員　寺田寅彦

所員　坪井忠二

地震學上の諸問題で、實驗物理的の研究を要するものが多數にあるに拘らず、從來此の方面の研究が餘り進んで居ない樣に見える。地震波傳播に關する諸問題も其例である。此方面の理論的の諸問題は從來偶然有數な數理物理學者の興味をひいた爲、可也な發展をしたやうであるが、元來簡單な彈性方則に基礎を置いた理論を、地殼の如きものに應用した場合に、如何なる程度まで適用され得るかといふ事に就いては疑を挿む餘地がある。それで若し不完全な彈性を有する物質に就いて、純實驗的に波動傳播に關する諸現象を研究して、一方では此れを從來の理論と對照し、又一方では地震波の實測の結果と比較すれば、此れによつて地殼の彈性に關して何等かの新しい知識を得る事が出來る見込があるかと思はれる。

又應用地震學の方面でも、例へば地震による建物の搖動に關する諸問題、地表の微動に對する溝や堀の效果などでも、理論的には甚だ困難なものが、容易に實驗によつて解決せらるゝ場合が多からうと思ふ。

以上の樣な考から彈性波に關する實驗を始める事にしたのであるが、未だ始めたばかりで、豫備的の粗雜な方法により大體の見當をつけたに過ぎない。

以下に報告する實驗の諸考案は主に坪井君のものであり、其の遂行は同君と助手矢田君によつてなされたものである。それで以下の報告も坪井君の筆を煩はし、自分は唯茲で紹介の辭を述べるに止めた。　　寺田寅彦

此の實驗の目的に副ふものとして、第一に用ひた媒質は寒天である。先づ實驗の準備として、寒天の彈性の常數の大體の値を測つたが、其の結果、ヤングの彈性率は $2.0\times10^5$ c.g.s. ポアソンの比は 0·47 といふ値を得た。之から計算すると、毎秒數十回振動する源から生ずる彈性波の波長は、數センチメートル乃至十數センチメートルの桁になるので、實驗に使ふ寒天の容器は、之に適當する大きさに作つた。振動の源としては、鐵の球を寒天の中に埋めたものを使ひ、其の上に近付けた電磁石に交流を通して、その球を振動させる樣にした。併し、之では一定の振動數のものだけしか得られないので、寒天の中に埋めた眞鍮の棒を電氣モートルに取付けたエキセントリック、ホイールで動かす樣にし、電氣抵抗でモートルの廻轉數を加減する事によつて、種々の振動數の振動を得る工夫もした。

Reprinted from Bulletin of the Earthquake Research Institute, 1927, Volume 3, pages 55-65 with permission of University of Tokyo

かうして生じた振動を觀測するには、寒天の表面に小さい鏡を並べて之に光をあてて、鏡から反射して出來る像の振動する有樣を見る方法を採つた。寒天の容器の側面や底から波が反射するのを防ぐ爲には、其處に綿を充分厚く敷いて、之によつてダンプさせる方法を採り、尙實際、反射が此れによつて防がれると云ふ事も實驗で確めた。

始め注目したのは表面波であつて、深さが增すと共に、其の振幅は、理論が要求するのと殆んど同じ樣に減少して居るのを見た。寒天を切つて崖を作ると、其の緣で波が反射して、定常波が出來る事、併し、反射が不完全である爲に節にあたる所も相當の振幅を持つて居る事などを明にし、此の振幅から反射の良否の程度を定め得る事を示した。次に堀を掘つて、其の對岸の振動を減少させる事を調べ、波長の數分の一の幅や深さの堀によつても、對岸の振幅を著しく減少させる事が出來る事を示した。しかも、一定の大きさの堀は、或るきまつた波長の波をさへぎるのに都合がよい事、即對岸の振幅が、此の波長に對して極小になる事が明になつた。同時に、前に述べた堀の緣での反射の程度が、此の波長に對して極大になる事が解つたので、堀が波をさへぎるメカニズムが多少明にされた。それから、媒質に半ば埋まつて居る物體が、來る波に適應して如何に動くかと云ふ事も實驗したが、未だはつきりした結論は得られて居ない。

---

## 1. Introduction.

Among the diverse fields of researches in the domain of seismology, those quarters which may be effectively explored within the walls of a physical laboratory by the hands of experimental physicists, seem to have been unduly neglected. The cause of the neglect seems, in our opinion, to have been an accidental one, neither due to the essential futility of the attempt nor due to any practical difficulty of the method. Late Professor Kusakabe's research on the elastic properties of rocks may be mentioned as a beautiful example of the possibilities in this direction.

Problems of seismology awaiting thorough experimental studies are numerous. However, those connected with the propagation of elastic waves seem to form a category which is one of the most inviting and most promising. The problems of seismic waves have hitherto been the favourable topics of many eminent mathematical physicists and we owe, indeed, a lot of interesting acquirements of modern seismology to their labours. We may, however, scarcely assume that the mathematical theory of elasticity based on the most simple assumption on the elastic property of matter may be applied in all its consequence to the actual wave phenomena in the earth crust, especially because we are still in profound ignorance respecting the elastic properties and the structure even of its most superficial layer. In view of these considerations, it seems not quite superfluous to carry out a more or less systematic study on the various phenomena of elastic waves, from a purely experimental point of view, especially with a substance which may show marked deviations from the perfect elasticity. We may compare the results of such experiments, on one hand, with the outcomes of the theories and, on the other hand, with the facts of observations of actual earthquakes. By these means, it may be hoped, we may come across some clues for elucidating the actual properties and structures of the crust. Moreover, by means of various model experiments, we may solve comparatively easily a number of problems concerning the nature and mechanism of earthquake origins which may appear to surpass the power of mathematical analysis. Ample fields of useful researches are also in prospect in the domain of practical seismology.

Our present attempt is merely a preparation for the first step in the direction above pointed out. The experiments[1] which will be briefly described in the following are of a preliminary nature carried out for the purpose of sheer orientation.

T. Terada.

## 2. Physical Properties of Agar-agar.

As a suitable substance for studying the propagation of elastic waves, agar-agar was firstly chosen. Besides its deviation from the perfect elasticity, this substance has many favourable properties for the objects of the present experiment.

In order to determine the proper size of the vessel for containing this medium, its elastic constants[2] are needed. For this purpose, Young's modulus and Poisson's ratio were determined, both of which are easily measurable.

With the sample of agar-agar to be used in the determination, a test piece was made in the form of a circular cylinder. For this purpose, a solution of agar-agar was poured into a hollow circular cylinder of glass of a suitable size, placed upon a plane glass plate, and was there solidified. The solidified mass was taken out of the glass cylinder and placed on the brass

(1) The designs of the apparatuses and the methods of experiments as well as the discussions of the results are chiefly due to Tsuboi. The experiments were carried out by the able assistance of Mr. Y. Yada of the Institute.

(2) It is of course doubtful whether the usual word "elastic constant" may have any strict meaning for such a substance as agar-agar. Here, the word is used only in its approximate sense.

plate of the measuring apparatus. The arrangement for measuring the elastic constants is shown in Fig. 1, in which $S$ is a spherometer, $G$ a thin glass plate, $K$ the agar cylinder, $W$ weights, and $A$ aluminium plate.

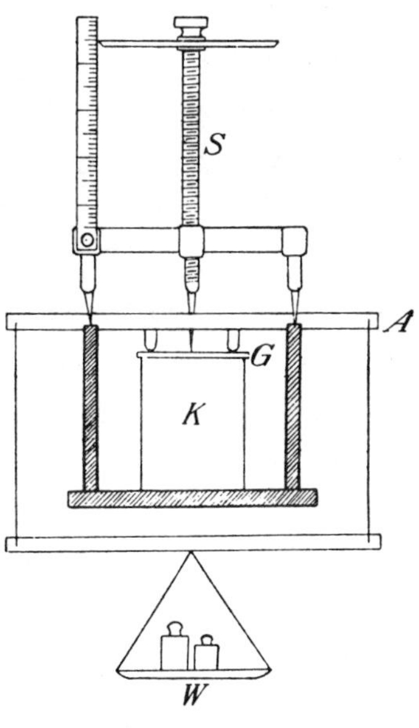

Fig. 1.

A suitable series of weights was placed on the aluminium pan below. The corresponding depression in height and the lateral expansion at about the middle height of the agar cylinder were measured by the spherometer and a micrometer screw gauge respectively. The micrometer screw gauge is not shown in the figure. From these two readings, Young's modulus and Poisson's ratio were calculated. Their values, for example, for a sample of agar-agar of 1.2 % are:

$$\text{Young's modulus} = E = 2.0 \times 10^5 \text{ c.g.s.}$$
$$\text{Poisson's ratio} = \sigma = 0.47.$$

These values, of course, depend on temperature,[1] but this will not matter much in the present degree of accuracy of our experiment. Then, from the ordinary formula, the velocities of elastic waves were estimated; the results are:

$$\text{Longitudinal Wave} \quad V_l = \sqrt{\frac{E}{\rho}\frac{1-\sigma}{(1+\sigma)(1-2\sigma)}} = 1096 \frac{\text{cm.}}{\text{sec.}}$$

$$\text{Transverse Wave} \quad V_t = \sqrt{\frac{E}{\rho}\frac{1}{2(1+\sigma)}} = 261 \frac{\text{cm.}}{\text{sec.}}$$

where $\rho$ is the density of the medium which is approximately 1.1 in this case.

The velocities of the elastic waves being of such magnitude as obtained above, their wave lengths will range from a few cm. to 10 cm., according as the frequency of the vibration is several ten times in a second. Thus a vessel 50 cm. in length will contain from 5 to 10 waves. The actual vessel used in the present experiment was 50 cm. in length, 25 cm. in depth, and 20 cm. in width.

(1) H. J. Poole, Trans. Faraday Soc., **22** (1926), 82.

In summer, agar is liable to putrefaction but this can be easily avoided by mixing a small quantity of mercury bichloride in the medium.

From the surface of the solidified agar, the constituent water constantly evaporates, thus causing slight changes in the values of the elastic constants and the density of the medium. Many trials have been made to keep away this evaporation. The surface of the medium was oiled, for example, but the attempt proved to be unsuccessful for the purpose. The effect of evaporation of the constituent water on the elastic constants and the density of the medium is, however, so small that it will matter little for the present purpose of our experiments.

### 3. Source of Vibration.

A small electromagnet fed by an alternating current of 50 cycles was used to excite the origin of the vibration in the medium. An iron sphere about 1 cm. in diameter which was imbedded at a certain depth from the surface of the medium, was set in vertical vibration under the influence of this electromagnet. The frequency of the vibration of the sphere is 50 or 100 times in a second according as it is magnetised or not respectively. By this method, however, a vibration of a definite frequency can only be produced. When the frequency of vibration is needed to be varied, as was actually the case, another method should be devised. For this purpose, an eccentric wheel driven by a *D.C.* shunt motor was used. A brass rod of about 1 cm. in diameter which was horizontally imbedded in the medium was set in vertical vibration by the eccentric wheel. (Fig. 2.) The lateral vibration of this source was avoided by means of a suitable guide. By adjusting the electric resistance, potentiometrically connected in the circuit of the motor as is shown in Fig. 2, the revolution of the motor could be varied at will from 0 to 2000 times per minutes. The constancy of the revolution velocity of this kind of electric motor proved to be a very favourable property to be used as a source of vibration.

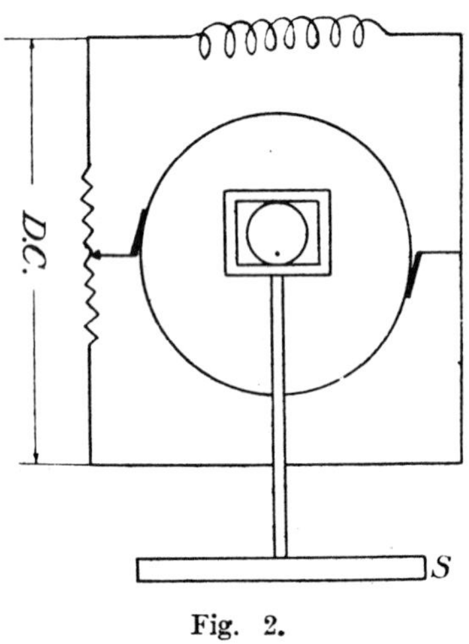


Fig. 2.

## 4. Method of Observation.

An optical method was used for observing the vibrations thus produced in the medium. The general scheme of arrangements is shown in Fig. 3, in which $S$ is a light source, $M$ reflecting mirrors, $m$ small mirrors, $K$ the agar

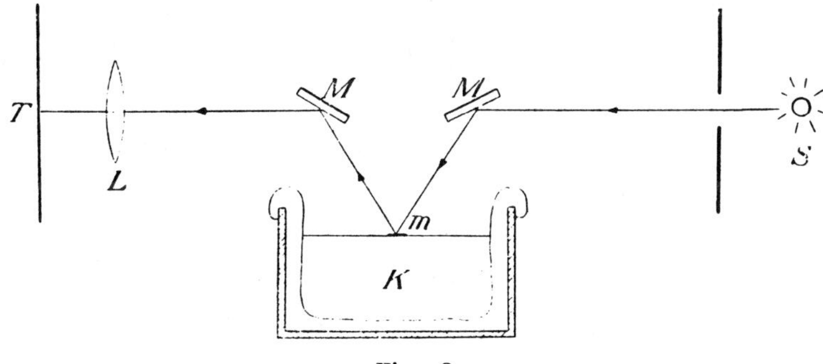


Fig. 3.

mass, $L$ a lens, and $T$ a frosted glass plate. A number of small mirrors $m$ were placed on the surface of the medium along a straight line and illuminated by light from a point source. Light reflected by these mirrors was forcussed by means of the lens $L$ on the screen $T$. The image due to each mirror vibrates according to the vibration of the medium. The mirrors placed on the surface of the medium were a few mm. square in their area and a few tenth of mm. in their thickness, so that it is almost at once evident that they follow the vibration of the medium faithfully. Instead of the plane mirrors, concave ones may conveniently be used, for they dispence of the use of a lens for forcussing the images. As the source of light, a tungsten pointolite of Tôkyô Electric Company was used.

After the vibration has attained the stationary state, the image by each mirror will repeatedly describe one and the same path. This path could be traced with a pencil upon a tracing paper. For the observation of transient phenomena, however, the screen was replaced by a photographic film wound around a cylindrical drum. By rotating the drum, the motion of the image of the mirror was recorded on the film. If, in Fig. 3, the elastic wave propagates perpendicularly to the plane of the paper, the rotation axis of the small mirrors would be parallel to the plane of this paper, so that the images vibrate perpendicularly to this plane. The amplitude of the vibration of the image by the mirror is proportional to its inclination.

## 5. Boundary Conditions.

It is of utmost importance in the present experiment to get free from the reflection of the waves at the sides and at the bottom of the vessel. For this purpose, pads of cotton wool were spread at the sides and at the bottom of the vessel to a sufficient thickness. A sheet of absorbent cotton was further spread upon the ordinary cotton. By these arrangements, the abrupt change in the density at the boundary of the medium was practically excluded.

That no sensible reflection of elastic wave takes place at the sides or at the bottom of the vessel was ascertained by the following experiment. Two mirrors were placed on the surface of the agar mass at different distances from the source of vibration, and the elastic wave was suddenly made to propagate from the source. If a sensible reflection of the wave is actually present, some different phase, other than the direct progressive wave from the source, must appear in the records of the vibrations of the images of these mirrors. But this was not the case. In the neighbourhood of the walls of the vessel, however, the motion of the medium is somewhat constrained, so that the mirrors should be placed at a sufficient distance from the walls of the vessel to study the unconstrained motion of the medium.

## 6. The Variation of the Amplitude of Vibration with Depth.

Whether the wave now under discussion is really a surface wave or not is the next question to be determined. For this purpose, the variation of the amplitude of vibration with depth was investigated. A vessel was specially constructed for this purpose, of which the side walls could be removed after the medium was solidified. After removing the side walls, pieces of very thin and short hair wire were stuck horizontally on the side face of the medium at every half cm. of depth. Their amplitudes of vibration were measured by a telescope fitted with a micrometer gauge. The results of the measurements are shown in Table 1 and graphically in Fig. 4.

TABLE 1.

| Depth (cm.) | 0.5 | 1.0 | 1.5 | 2.0 | 2.5 | 3.0 | 3.5 | 4.0 | 4.5 | 5.0 | 5.5 | 6.0 | 6.5 |
|---|---|---|---|---|---|---|---|---|---|---|---|---|---|
| Amplitude | 100 | 94.2 | 82.2 | 70.5 | 57.0 | 50.0 | 39.8 | 25.8 | 21.3 | 13.3 | 7.0 | 3.0 | 0.8 |

Fig. 4.

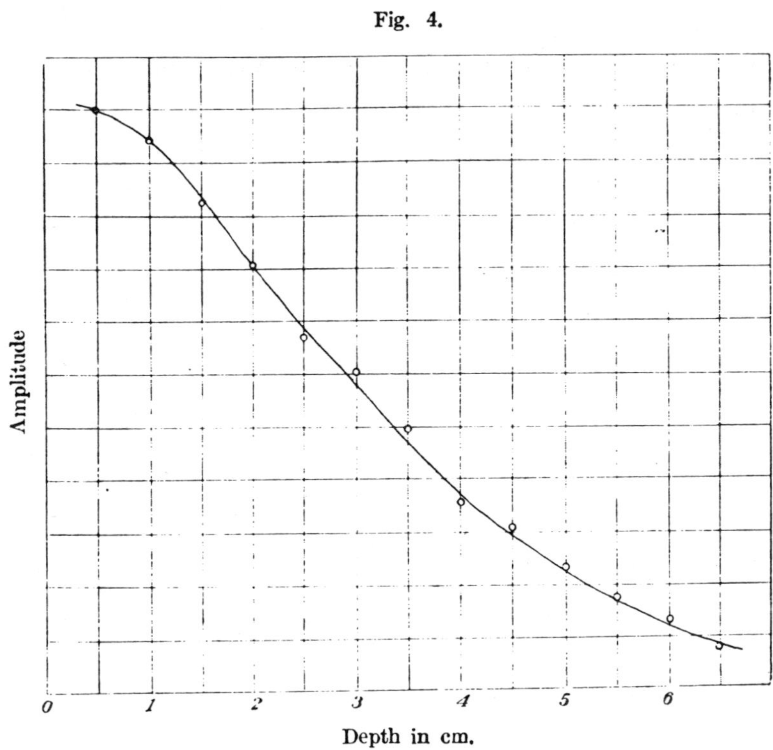


In the table, the amplitude of vibration at the depth of 0.5 cm. was taken to be 100, and the amplitudes at any depth were represented in their percentage values as compared with that at 0.5 cm. The relation between the depth and the amplitude of vibration above obtained is exactly what is expected if the wave generated is a surface one. The depth of the medium in this case was about 20 cm., whereas the amplitude of vibration at the depth of 10 cm. is practically zero. This fact is sufficient to prove that the bottom has no sensible influence on the propagation of elastic waves along the surface of the medium.

## 7. Experiments Regarding the Reflection of Waves at the Edge of a Cliff.

The source of the waves in this case was line-shaped, vibrating vertically by means of an eccentric wheel. The edge of the cliff was parallel to the source of vibration. Small mirrors were placed at every cm. from the source up to the edge of the cliff and their amplitudes of vibrations were measured.

By cbanging the frequency of the vibration, the amplitude distribution is consequently altered. The results shown in Fig. 5, are typical of the several cases examned. The distance between the source and the edge of the cliff in this case was 23 cm.

Ffg. 5.

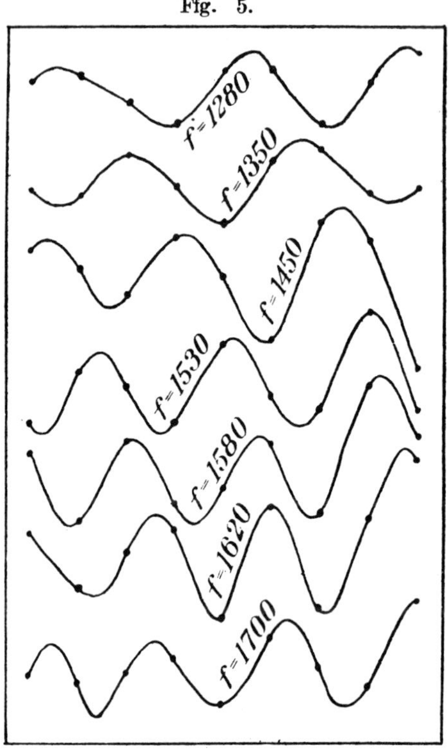


As is evident from the result, the stationary wave is taking place. Suppose the progressive wave to be rapresented by

$$y_1 = A \sin (x - vt)$$

and the retrogressive wave by

$$y_2 = B \sin (x + vt)$$

where $x$-axis is parallel to the direction of the propagation of the wave, $v$ the velocity of the wave, and $t$ the time. At the edge of the cliff, the wave is imperfectly reflected, so that $A = B + C$ where $C > 0$.

The resultant motion of the waves will be

$$y = C \sin (x - vt) + 2B \sin x \cos vt.$$

Therefore the amplitude of the resultant vibration will depend on the value of $x$ through the relation

$$y_0 = C + 2B |\sin x|$$

The nodes of vibration are not exactly at rest.

## 8. Experiments Regarding the Effect of a Canal as a Screen for Waves.

In connection with the experiments of the preceeding article, the effect of a canal as a screen for waves was investigated. This problem has some practical interest because a hint may be drawn from the results of the experiments regarding the possibilities of screening off the actual waves on ground due to earthquakes or traffic disturbances by a suitable canal.

A rectangular canal which was 5 cm. deep and 2 cm. wide was dug parallel to the direction of the line source so as to obstruct the propagation of the waves. In the opposite side of the canal, the amplitudes of vibration is greatly reduced.

The ratio $S$ of the mean amplitudes on both sides of the canal varies according to the frequency of vibration. The curve in full line of Fig. 6 shows the way in which this ratio varies with the frequency of vibration.

Fig. 6.

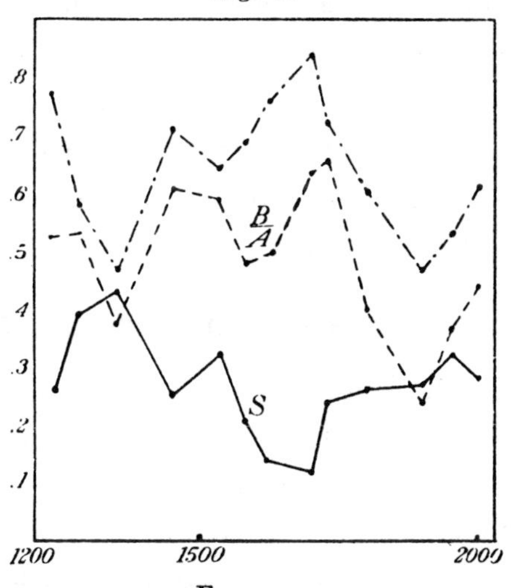


As can be seen from the figure, the ratio $S$ takes its minimum value when the frequency of vibration is 1700 per minutes. Below this frequency, the ratio decreases with the increasing frequency, and after this minimum value has been passed. however, the ratio increases with the frequency.

Like in the case of the cliff described in the preceeding article, the elastic wave is here also imperfectly reflected at the edge of the canal. The resultant amplitudes of the progressive and retrogressive waves are

$$y_0 = C + 2B|\sin x|$$

using the same notation as the preceeding article.

In the experiments, $C$ and $C+2B$ are directly measurable, so that the amplitude of the incident wave $A=B+C$ will at once be deduced.

The ratio $\frac{B}{A}$ will, at any rate, be a measure of the efficacy of the edge of the canal in reflecting the incident wave. The curve in broken line of Fig. 6 shows the ratio $\frac{B}{A}$ thus obtained as the function of the frequency of the vibration.

As can be seen from the figure, the ratio $\frac{B}{A}$ takes its maximum value at the same frequency at which the amplitude of vibration at the opposite side of the canal takes its minimum value.

The curve in chain line of Fig. 6 shows the relative value of the reflecting power $\frac{B}{A}$ as compared to the sum $S+\frac{B}{A}$. Some irregularities seen in the $\frac{B}{A}$ curve are here very much smoothed out.

A part of the energy of the incident wave is transmitted to the opposite side of the canal and another part is reflected backwards at the edge of the cliff. The proportion in which the energy of the incident wave is divided into these two parts depends on the relative magnitude of the canal with respect to the wave length, and there is a definite optimum width and depth of the canal for screening off a wave of a definite wave length.

How this optimum size of the canal is determined for a given wave length is not yet certain.

## 9. The Vibration of Wooden Block Half Imbedded in the Medium.

Rectangular pieces of wood of different sizes were half imbedded on the surface of the medium. Minute mirrors were stuck on these pieces. By observing the vibrations of the images of these mirrors, the mode was studied in which such an imbedded block behaves under the action of an incident wave. One of the record thus taken is reproduced in Fig. 7. The series of

Fig. 7.

small spots in the figure are time marks indicating every $\frac{1}{50}$ of a second. As can be seen from the figure, the vibration of the wooden pieces with its own rocking period predominates at the beginning of the vibration. After some time, this vibration continuously damps away, and the state of forced oscillation comes in turn. When the incident wave ends, the piece comes to rest after some vibrations with its own rocking period. These experiments are intimately connected with the practical problems regarding the rocking vibration of the pedestal of a seismograph or that of a rigid building imbedded in a weak ground in the case of an earthquake. The experiments are now going on to study what change in the mode of vibration takes place when the wave length is gradually changed relative to the magnitude of the wooden piece.

(*to be continued.*)

# *Experimental Studies on Elastic Waves.* (Part 2)

By

**Chûji TSUBOI.**

Earthquake Research Institute.

---

## 彈性波の實驗(第二報)

所員 坪井忠二

次に述べる彈性波の傳播に關する實驗は、此の彙報の第三號に述べた報告の續きであつて、其の後得られた實驗の結果をまとめたものである。實驗の方法は第一報に述べたものと大差無く、寒天を媒質として、其の表面を傳はる彈性波に就いて、種々の實驗をしたのである。

此の報告で取扱つて居る第一の問題は、彈性係數と密度とを異にする二つの層が重なつて居る時に、其の一番の表面を傳はる表面波の速度が、其の波長や、又層の厚さなどによつて、どんなに變化するかと云ふ事である。先づ固い寒天を容器の中に固めてから、其の上にやはらかい寒天を流して固めて層を作つた。表面には固まる時に、不規則な凹凸を生じるけれども、其處はさしみ庖丁で平に削り取つて仕舞ふ事にした。本文中の第一圖に示した樣に、寒天の一端は垂直に削り取つてあつて、振源から來る波が此處で反射して、表面に沿つて定常波が出來る樣にしてある。此の定常波の波長と、彈性波の源の振動數とを測り、夫等の値から表面波の傳播の速度を計算した。色々な層の厚さや波長に對する傳播の速度は本文中の第一表、第二表に示した通りで、波長を $\lambda$. 上層の厚さを $H$ とすると、速度は $\frac{\lambda}{H}$ の函數として與へられ、$\frac{\lambda}{H}$ が大きい程速度が早く、$\frac{\lambda}{H}$ が大きくなるに從ひ下の媒質のみ場合の表面波の速度に、$\frac{\lambda}{H}$ が 0 に近づくに從ひ、上の媒質のみの場合の表面波の速度に、漸近的に近付く事が明にされた。之等の事は、當然彈性論の立場から期待される所であり、本彙報の第三號で妹澤所員の發表された計算とも大體一致した結果なのであるが、元來寒天の樣に彈性的には不完全な物質に彈性論が其儘の形で適用されるとも考へられないし、又實際の地殼を傳はる所謂「地震波」に就いても同樣の事が云へると思はれるのであつて、種々のモデルに就いて行つた寒天の實驗と、實際の地震記象と、彈性論の結果とを比較研究する事に依つて、始めて地殼の構造に關する正當な手がかりが得られるものと考へられる。併し此の點に就いて、現在の粗雜な實驗から決定的の議論をなす事は差ひかへて置く事にした。

第二の問題は、堀を掘る事によつて、媒質の表面を傳はる振動が遮斷されるかと云ふ問題である。之は實際問題として精密な測定等を行つて居る研究所等を、汽車、電車、自働車や其他のものから來る振動から遮斷したいと云ふ時等に屢々議論に上る所である。堀の深さや幅が、來る波長の數分の一のものでも、其の對岸の振幅は堀が無い時に較べて著しく減じる事は、第一報で報告して置いた通りである。其の後實驗した所に依ると、堀の兩岸の振幅の平均の比は、堀を深くすればする程小さくなるが、其の減少の割合は、始めの中

Reprinted from Bulletin of the Earthquake Research Institute, 1928, Volume 4, pages 9-20 with permission of University of Tokyo

が急で後は次第にゆるくなるので、或る一定の深さになると、夫以上堀を深くしても大した效果が無いと云ふ事になつた。之は平均振幅に就いてであるが、對岸のきまつた點に就いて見ると、堀に近い點は淺い堀によつても其の振幅が非常に減るが、遠くの點では殆んど影響が無い。所が堀を段々深くして行くと、一旦非常に振幅が小さくなつた堀に近い點は、最早夫以上大した影響を受けないが、堀から遠い點は次第に其の振幅が小さくなつて行く。之等の結果を實際に適用しやうと云ふに就いては、實際の地面を傳はる表面波の振幅が深さと共に減少する割合が知られて居ないから、はつきりした事は云へないけれども、寒天に於けるよりは急な割合で減るのであらうと云ふ事は略想像される所である。若しさうだとすれば、問題になる樣な振動は、傳播速度が毎秒數百米、周期が一秒の十分のいくつと云ふ位であるから、深さが十米内外の堀で充分所要の目的が達せられるものと考へられる。尙堀に近い所は淺い堀でもかなり振幅が減少するといふ實驗の結果から見ると、實際の堀は問題になつてゐる建物になるべく近く作るのが有效であらう。

---

The experimental investigation described in the following pages forms a part of our scheme of researches into different fields of phenomena concerning the propagation of elastic waves. The first report of the experiments has appeared in the last number of this Bulletin.(1) Since that time, some new results have been acquired of which the present paper is a report.

The general arrangements of the experiments are essentially the same as those described in the previous paper, but some modifications have since been introduced here and there when they seemed to be necessary or desirable.

Briefly, a mass of agar-agar was solidified in a rectangular vessel with a dimension of 80 cm. in length, 25 cm. in width, and 30 cm. in depth. In the solidified mass, a circular brass rod with its diameter 1 cm. and its length 24 cm. was imbedded horizontally at a certain depth from the surface. The rod was connected by means of a thin vertical rod to an eccentric wheel which is driven by an electric motor. The brass rod thus set in vertical vibration served as a source of elastic waves. The reflection of the elastic waves either at the bottom and the sides of the vessel could be practically excluded by spreading pads of cotton wool on these sides to a sufficient thickness, by means of which the waves were continuously damped away without any sensible reflection.

The vessel for containing the medium was so constructed that its side walls can be removed after the agar mass was solidified. Fine powders of aluminium were then blown onto the side face of the mass, which served to

(1) T. Terada and C. Tsuboi, Bull. Earthq. Res. Inst., Vol. 3, (1927) 55.

indicate the path of the vibrational motion of each individual particle on the lateral surface of the medium. Unlike in the previous experiments, the modes of vibration of the medium thus revealed were observed directly through a telescope fitted with a micrometer gauge, by means of which the measurement were made of the dimensions of the trajectory of each individual particle of the medium.

It will be not out of place here to remark that when the surface of the mass is sufficiently plane, the path of each particle of the medium as indicated by the aluminium powder is observed to be an ellipse whose major axis is vertical and remains in a ratio approximately 10 : 7 to the minor axis. The ratio of the two axes observed in this case is almost exactly what is required by the theory of the Rayleigh wave. On the other hand, the decrease in the amplitudes of vibration with depth from the surface was also seen in the previous paper to be almost exactly what is required by the same theory. Thus there remains hardly any doubt that what we are observing is nothing else than the Rayleigh wave. So far as the writer is aware, the positive identification of the Rayleigh wave in the case of the actual earthquake has not yet been made. The cause seems to be generally attributed to the imperfection of the vertical seismograph to record the motion of the earth. In these connections, it seems not uninteresting that we have demonstrated the positive existence of the Rayleigh wave experimentally.

## 10. Dispersion of the Surface Waves along the Surface of a Stratified Layer.

A somewhat concentrated solution of agar-agar was solidified in the rectangular vessel. The surface of the solidified mass became corrugated into an irregular wavy form owing to its own shrinkage during the course of solidification. For cutting off these wavy parts of the surface, use was made of a flat long knife, called "Sasimi-Bôtyô" specially deviced for the use in Japanese cookery. After cutting off these wavy parts, a more dilute solution of agar-agar than the first was poured upon the latter to be solidified into the form of a horizontal superficial layer. These two masses stick to each other so firmly that there is no slipping along the boundary between them. The uppermost surface was also shaved into a horizontally plane surface by cutting off the superficial wavy parts by the knife. The source of vibration was imbedded in the lower medium. For the present, it will not be of a very important difference, however, whether the source of vibration is imbedded in the lower medium or in the upper, because what we are observing is the surface wave at a distance sufficiently large from the source compared with the wave-length of the elastic wave.

As is shown in fig. 1, one end of the agar mass was cut down vertically so as to form a sharp edge or cliff. At this edge, the progressive wave was reflected backwards and consequently a stationary wave was generated along the surface of the medium.

Fig. 1.

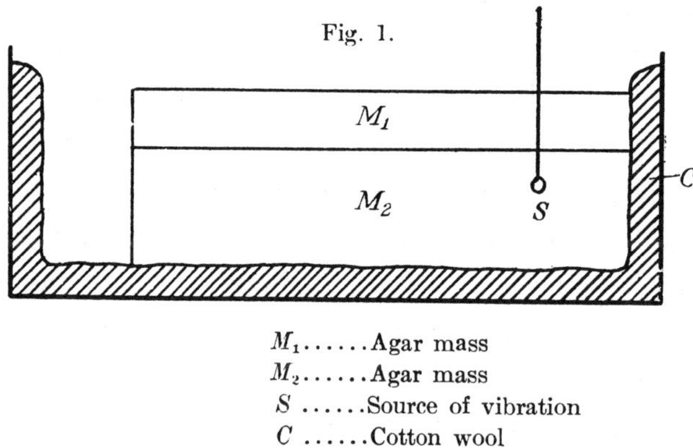


$M_1$......Agar mass
$M_2$......Agar mass
$S$ ......Source of vibration
$C$ ......Cotton wool

The nodes and loops of the stationary wave appeared very distinctly so that the wave-length of the stationary wave could be measured at once. The product of twice the wave-length into the frequency of the vibration will give the velocity of propagation of the elastic wave.

By these processes, the velocities of propagation of the elastic waves were determined for different wave-lengths for a given thickness of the upper layer. After a series of these measurements were carried out, the thickness of the layer was diminished by 1 cm. by cutting off the surface with the flat long knife. With thus reduced thickness of the layer, the velocities of the wave were again determined for different wave-lengths and so on.

It is evident that the wave-velocity depends on the wave-length $\lambda$ and the thickness of the superficial layer $H$ only through the quotient $\frac{\lambda}{H}$ if we neglect the effect of the surface tension of the medium.

A few examples of the results of the measurements are shown in the following tables.

TABLE I.

| Thickness of the upper layer $H$ (cm.) | Frequency of Vibration $f$ (per min.) | Wave Length $\lambda$ (cm.) | $\frac{\lambda}{H}$ | Velocity $v\left(\frac{\text{cm.}}{\text{sec.}}\right)$ |
|---|---|---|---|---|
| 5.5 | 1930 | 5.90 | 1.07 | 380 |
| 4.5 | 1950 | 5.86 | 1.30 | 381 |
| 3.5 | 1830 | 6.31 | 1.80 | 385 |
| 3.5 | 1720 | 7.30 | 2.06 | 413 |
| 2.5 | 1890 | 7.12 | 2.85 | 449 |
| 2.5 | 1720 | 8.70 | 3.48 | 499 |
| 1.5 | 2000 | 7.20 | 4.80 | 480 |

The upper medium......1.6% of agar-agar
The lower medium......2.4% of agar-agar

Fig. 2. The Velocity for Different $\frac{\lambda}{H}$

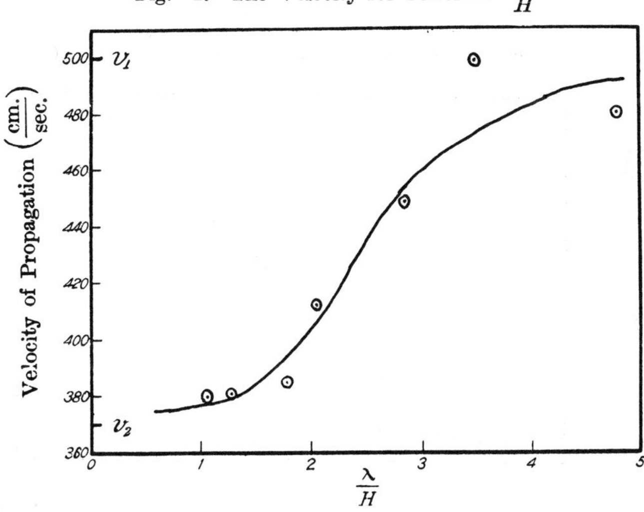

TABLE II.

| $H$ | $f$ | $\lambda$ | $\frac{\lambda}{H}$ | $v$ |
|---|---|---|---|---|
| 5.5 | 1910 | 5.50 | 1.00 | 350 |
| 5.5 | 1980 | 5.28 | 0.96 | 348 |
| 4.5 | 1620 | 7.05 | 1.57 | 381 |
| 4.5 | 1900 | 5.50 | 1.22 | 348 |
| 3.5 | 1900 | 5.83 | 1.67 | 369 |
| 3.5 | 1790 | 5.88 | 1.68 | 351 |
| 2.5 | 1720 | 7.50 | 3.00 | 430 |
| 2.5 | 1890 | 6.40 | 2.56 | 403 |
| 2.5 | 1920 | 5.78 | 2.31 | 370 |
| 1.5 | 2000 | 7.10 | 4.73 | 440 |

The upper medium......1.8% of agar-agar
The lower layer ........2.4% of agar-agar

Fig. 3. The Velocity for Different $\frac{\lambda}{H}$

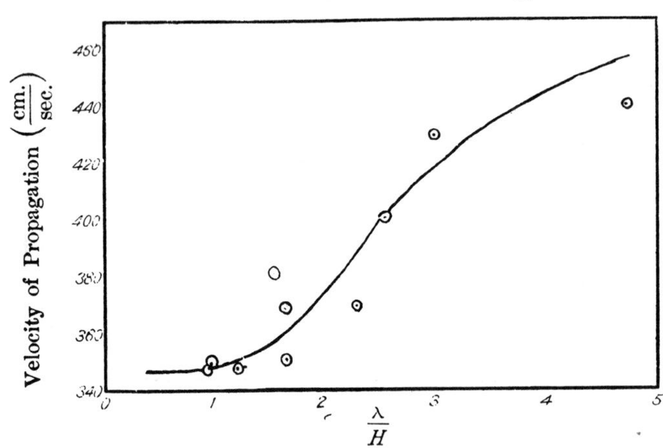

In the above results graphically represented in fig. 2 and 3, $v_1$ and $v_2$ are the velocities of the elastic waves in the case when the thickness of the upper layer tends to infinity or zero, or the wave-length to zero or infinity. In other words, $v_1$ and $v_2$ are the velocities of the Rayleigh wave on the surface of the single medium $M_1$ and $M_2$ respectively.

The medium of the lower layer in the two cases of the experiments was one and the same sample and it is seen in both cases that the velocities of the wave tend to one and the same value when the wave-length tends to infinity, as might be naturally expected.

When the thickness of the upper layer is not uniform along the path of the elastic wave but changes gradually as is shown in fig. 4, the velocity

Fig. 4.

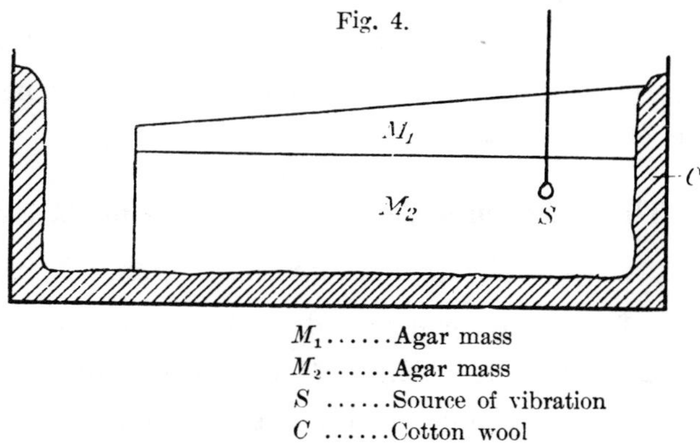


$M_1$......Agar mass
$M_2$......Agar mass
$S$ ......Source of vibration
$C$ ......Cotton wool

of propagation of the elastic wave is of different values from point to point. The wave-length of the stationary wave along the surface is the longer, the nearer is it to the edge. The elongation of the wave-length must mean so much increase in its velocity of propagation. An example of the measurements in such a case is summarised in the following table.

TABLE III.

| Mean Depth $H$(cm.) | 1.65 | 2.15 | 2.75 | 3.35 | 3.90 |
|---|---|---|---|---|---|
| Wave-length $\lambda$ (cm.) | 7.0 | 6.5 | 6.2 | 4.9 | 4.8 |
| $\lambda/H$ | 4.24 | 3.02 | 2.25 | 1.96 | 1.23 |
| Velocity $v$ (cm./sec.) | 432 | 401 | 382 | 302 | 296 |

Frequency of vibration......1850 per min.

The general features of the relation inferred by these experimental results are in qualitative agreement with the outcomes of the elastic theory recently developed by K. Sezawa(1) of our Institute. Any detailed discussions, however, regarding, for example, the effect of the imperfection of the elasticity of the medium cannot be drawn in the present state of our experiment.

On the other hand, B. Gutenberg(2) pointed out the dispersion phenomena in the actual earthquakes, which shows in its features a substantial coincidence with the present experimental results.

It may of course not be justified to identify the actual earthquake "waves" at once with such a mathematically pure waves as treated in the present experiments or in the elastic theory, yet we may hope to get some clue for elucidating the physical configuration of the earth crust, on one hand by paying special attentions to dispersion phenomena of the earthquake "waves", and on the other hand, carrying out more or less systematic experiments on various classes of models of the crust.

## 11. The Effect of a Canal as a Screen for Waves.

The preliminary discussions of the said effect were given in the previous paper where the facts were emphasised that a canal of which both the width and the depth are a few tenth of the wave-length was found to be sufficiently effective in reducing sensibly the amplitudes of vibration on its opposite side. The efficacy of a given canal in this respect depends on its relative size to the wave-length of the incident wave, and is greatest at a certain definite value of the latter. In the present article, it will be studied how the efficacy of a canal in this respect dapends upon its depth.

A number of small concave mirrors were arranged on the surface of the medium along a straight line perpendicular to the length of the canal as is shown in fig. 5.

A light from a tungsten pointolite was projected on all of the mirrors and was reflected from them to make small bright images of the light source on a frosted plane glass screen placed in a suitable position. After the source was set into vibration and the motion of the medium has attained its stationary state, each of those spots describe one and the same track repeatedly. Photographs

(1) K. Sezawa, Bull Earthq. Res. Inst., Vol. 3, (1927) 1.
(2) B. Guteuberg, Phys. Zeits., 25 (1924) 377.

Fig. 5.

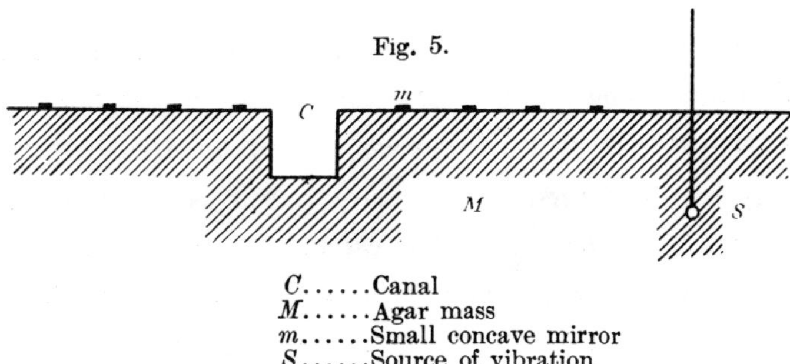


*C*......Canal
*M*......Agar mass
*m*......Small concave mirror
*S*......Source of vibration

were taken of these paths of moving images by an ordinary camera. The measurements of the amplitudes of vibration of the spots were made with thus taken images on the photographic plates. The ratio of the mean amplitudes of vibration of both sides of the canal is very much affected by the depth of the canal. A few examples of the results of measurements are shown in the following table and figure.

TABLE V.
Ratio of the Mean Amplitudes of Both Sides of the Canal

| Depth of the Canal | | 0 cm. | 2 | 4 | 6 | 8 |
|---|---|---|---|---|---|---|
| Experiment | 1 | 0.481 | 0.402 | 0.250 | 0.303 | 0.220 |
| | 2 | 0.435 | 0.447 | 0.220 | 0.242 | 0.202 |
| | 3 | 0.417 | 0.195 | 0.157 | 0.148 | 0.081 |
| | 4 | 0.474 | 0.277 | 0.278 | 0.219 | 0.199 |
| Mean | | 0.452 | 0.330 | 0.226 | 0.203 | 0.176 |

Width of the Canal......2 cm.

Fig. 6. The Decrease of the Ratio of Mean Amplitudes of Both Sides of the Canal with its Depth.

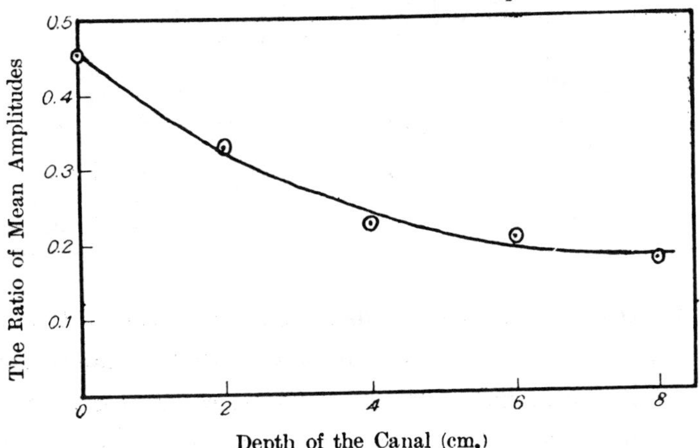

As can be seen from the above results, the ratio of the mean amplitudes decreases with increasing depth of the canal. But the decreases becomes comparatively slow after the depth has passed a certain value.

What has been observed in the above results is chiefly connected with the mean amplitudes on both sides of the canal. In the following, the modes will be studied in which the amplitudes at a point at a definite distance from the canal decrease with the increasing depth of the canal.

As seen with a telescope, the surface of the medium in motion appear, somewhat like the fig. 7. The thickness $d$ of the parts in half shadow indicates the amplitudes of vibration at the surface. The distribution of the amplitudes of vibration thus manifested along the surface of the medium were surveyed by a micrometer gauge mounted in the telescope.

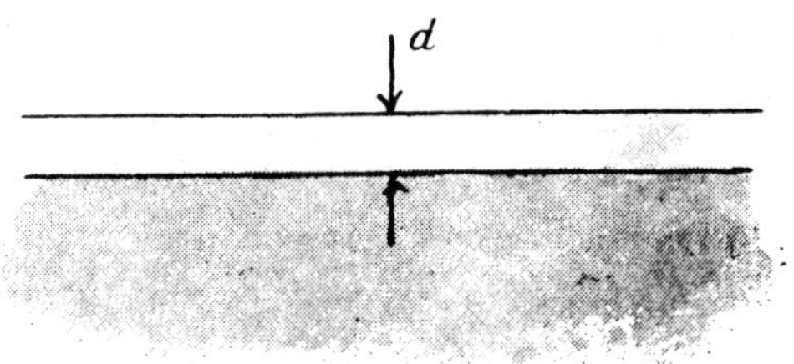


Fig. 7. The Surface of the Medium in Motion as seen with a Telescope.

In the following example, the modes can be seen in which different parts of the opposite side of the canal decrease in amplitude with increasing depth of the canal.

TABLE VI.

Amplitudes of Vibration (in arbitrary scale)

| Distance from the Canal \ Depth of the Canal | 0 | 2 | 4 | 6 | 8 |
|---|---|---|---|---|---|
| 2 cm. | 100 | 51 | 48 | 42 | 39 |
| 6 | 100 | 64 | 46 | 58 | 36 |
| 10 | 100 | 103 | 47 | 60 | 46 |

As can be seen from the results above, the amplitude in the point nearest to the canal is greatly decreased even by a very shallow canal while those in distant points are not so much affected. As the canal is gradually deepened, the amplitude in the point nearest to the canal which has once been greatly reduced suffers scarcely any further effect, while those at distant points are gradually reduced.

The general features of the relation implied in the above statements will be summarised in the following schematic figure.

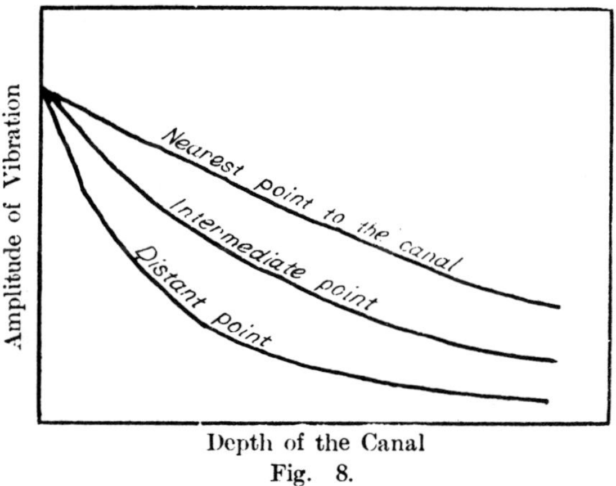


Fig. 8.

These features are quite analogous with those experienced in the familiar phenomena of sound shadow behind an obstacle. The canal in the present experiment may correspond to an obstacle for the sound wave. In the latter case, even a very narrow obstacle is sufficient to throw shadow just behind of it, but is incapable to produce any sensible effect upon the intensity of the sound at a distant point from itself.

It has not infrequently become the favorite topic among the physicists and the engineers in our country how will it be possible to screen from traffic and other artificial disturbances a research laboratory or an observatory of any kind constructed on the weak ground where different kinds of fine measurements are going on. If such disturbances are justified to be regarded as a kind of surface wave, the experimental results obtained above might afford some clue to the possibility of the attempt.

We have good reasons to expect that the amplitudes of vibration in the superficial soil decrease with depth more rapidly than in the agar-agar of the present model experiments. Therefore, if a canal is constructed on the superficial soil in the same relative scale of magnitude with the wave-length as in the present experiments, we may naturally expect its effect to be more efficient than in the present experimental case. The velocity of propagation of such disturbances being of an order of a few hundred meters per second and its period a few tenth of a second, the wave-length will be a few tens of a meter. Thus a canal whose width and depth are less than ten meters will be sufficient to produce a desired reduction in the amplitudes of vibration. And it might be added further that the nearer the canal is constructed to the laboratory in question, the more efficient is its effect.

In conclusion, the writer wishes to express his sincere thanks to Professor Torahiko Terada for the advices given and the interests taken throughout the course of the present experiment.

October, 1927.

*(to be continued)*

# VISUAL PRESENTATION OF ELASTIC WAVE PATTERNS UNDER VARIOUS STRUCTURAL CONDITIONS

FRANK RIEBER[1]

ABSTRACT

The usual form of reflection seismograph operates satisfactorily over simple structural conditions, but frequently fails to obtain part or all of the desired information when structures are steeply folded, faulted or otherwise complicated. The reasons for this are plainly evident if the paths of the waves in the earth can be visualized. This has been done by the use of a technic originally developed for acoustical measurements.

A miniature explosion radiates waves into various models of structure, where reflection and diffraction take place in the same manner as in the earth. The various moving waves are actually photographed in flight. A series of plates is presented, showing wave patterns in various types of structures, ranging from simple to complex. A new type of equipment and technic are briefly described, with which exploration may be carried into the more complex structural regions successfully.

The appearance of an ideal reflection record is well known to all of those familiar with seismograph work. Definite, well marked bands or patterns of vibrations, more or less parallel to each other are seen to traverse the record. These bands persist in amplitudes sufficient to be readily seen and marked, for a considerable length of record.

The good shooting conditions, permitting such records to be taken, occur chiefly in regions where strata are definite and well marked, and relatively flat lying.

The appearance of a poor or low grade record is, unfortunately, almost as well known, especially to those who have had occasion to attempt reflection shooting in regions of relatively steep folding or faulting. These poor records, while they contain vibrations of good amplitude persisting for a satisfactory distance down the strip, show very few patterns or "line-ups" which might be marked as reflections. Furthermore, it is frequently impossible to plot from them any consistent structural condition.

Records of this latter type are customarily marked "N. R.," presumably meaning "no reflections." However, a simple consideration of the space geometry of the reflected wave paths will show that, in very many cases, such confused records are due to the presence of too many, rather than too few, reflected waves.

Consider first the fact that such poor records are very frequently obtained in the vicinity of steep folding and faulting, where the

[1] Rieber Laboratory, 1007 Braxton Ave., Los Angeles, Calif.

Reprinted from Geophysics, 1, 196-218

rapidly changing attitude of the beds must necessarily result in simultaneous arrival of groups of reflections from a wide variety of different directions.

For example, take the well known case of shooting over a syncline. If we could by some means remove either side of the syncline and shoot for the other side alone, we might expect to get a high grade record showing a succession of relatively parallel bands from which the plotted results would correspond accurately to the dip of the beds in that side of the structure.

Ordinary shooting methods do not permit us to make such separations, however, and reflections from both sides of the syncline enter the system with equal freedom and appear with equal prominence on the record. The vibration patterns thus produced will fall in a criss-cross fashion on the record, giving a result that is difficult and often impossible to read.

As is well known, the clearest records are obtained in regions where all reflections arrive at the receiving system from substantially the same direction. Steeply folded or faulted regions contain very few places where this ideal condition prevails. Hence the predominance of poor records and the great difficulty of reflection exploration in such places.

The writer has devised a receiving system,[2] employing a sound track record and an optical analyzer. This system has been in field use for some time, in locations where structural conditions were sufficiently complex so that the usual type of visual record was unsatisfactory. In such places, the new system has done exactly what was expected of it, namely, broken down the complex group of arriving waves into its components. From these components it has been possible to map structure quite satisfactorily.

A majority of the identified individual waves recovered from the complex vibration by the analyzer, under such conditions, plot satisfactorily as strata, and permit the delineation of the structure. A smaller, but very definite number of the returned waves, however, cannot be plotted as strata without conflicting seriously with the majority evidence.

This minority of wave trains is quite real, and the places in the earth from which they are returned to the receiving system are quite

[2] "A New Reflection System with Controlled Directional Sensitivity," by F. Rieber, January 1936 issue of "Geophysics," Vol. 1, No. 1, p. 97. This system is the subject of several pending applications for patent.

definite and can be separately identified from several successive shooting or receiving positions at the surface.

The question then naturally arises as to what geologic conditions, other than stratification, capable of returning a wave to the surface, might exist in the structures under examination.

The possibility suggested itself that some of these waves were diffractions rather than reflections. Theoretically, such diffraction patterns could arise at any place where stratified material is broken or faulted, or suffers any abrupt change in its elastic properties, its density, or its power to absorb vibrations, or in short, any corner or edge where there is a change of mechanical impedance.

In other words, it seemed probable that, if we were to explain all of the waves which were analyzed out of the complex earth vibrations over steeply folded structure, it would be necessary to go a step beyond the ordinary geometric paths assumed for reflected waves where the angle of incidence equals the angle of reflection.

Accordingly, an attempt was made to reproduce, in model form, certain possible elements of structure, and to photograph the actual wave patterns produced when a compression wave impinged on the model.

The technic for doing this has occasionally been used in acoustical investigations. The writer is greatly indebted to Dr. Vern O. Knudsen and Prof. L. P. Delsasso, of the University of California at Los Angeles, for information on this earlier technic, and also to Mr. B. F. McNamee and Mr. Frank McCullough, who carried out the difficult and exacting work of constructing the actual apparatus[3] used in these investigations, made the models, and photographed the wave patterns here presented.

Briefly, the technic is one of shadow photography, no lens of any sort being employed. A bright electric spark, lasting only about one millionth of a second, is spaced about four feet from a photographic plate. The model of the structure to be investigated is placed part way between the spark and the plate, in such a position that its shadow will be photographed.

If no sound waves are present in the field between the light source and the plate, the shadow of this model will be the only thing shown in the photograph. The rest of the field will be uniformly exposed to

[3] An article by B. F. McNamee, describing in detail the mechanical and electrical features of this equipment, is to appear in a forthcoming issue of *Electronics* magazine.

the light from the spark and will correspondingly show uniform photographic density.

If an abrupt sound wave happens to be passing through the field, however, the light from the spark will be bent slightly at the places where it passes through the denser air of the wave front. This bent light will be superposed upon other illumination arriving directly on the plate from the spark, thereby causing a dark line. The part of the plate from which the bent light was diverged will show, correspondingly, as a lighter line.

Only a very abrupt wave front of sound may be pictured in this manner. Such a wave, fortunately, can be created by another electric spark, acting, in this instance, as an explosion rather than as a source of light. Such an explosion, or sound spark, is placed above the structure model, its location being indicated in the subsequent plates by the shadow of its electrode structure, as cast on the plate by the light spark.

In operation, this sound spark or explosion is repeated about five times per second, sending a succession of sound waves down to strike on the model. Following each sound spark at exactly the correct interval is a light spark, so timed as to catch the sound wave at some desired point in its progress, and flash an image of this wave against the photographic plate.

A ground glass can be substituted for the plate, and the actual shadow of the wave may be observed. For any adjustment of the apparatus the wave pattern is practically stationary. Turning a knob, which adjusts the very critical time interval between the sound spark and the light spark, results in causing the wave image to move forwards or backwards. The downgoing wave front may thus be advanced until it just strikes upon the structural element to be investigated, at which point reflected or diffracted waves will begin to appear. Advancing the downgoing wave still further will cause the returned waves to progress a corresponding distance back from the surface. Fig. 1 shows the arrangement in somewhat diagrammatic form. The electrical control circuits regulating the interval between the sound spark and the light spark are purposely omitted from this diagram. These circuits, however, as will be immediately recognized, are the most critical part of the equipment. They must control the time interval between the successive sparks to an accuracy of approximately one millionth of a second, and must permit the continued repetition of these spark groups with this identical time interval, regardless of such variations as may occur in the fluctuating regions of heated gas in the spark gaps themselves.

The photographic portion of the equipment is contained in a box about one foot high, one foot wide, and four feet long, at one end of which a plate holder and ground glass arrangement is mounted, similar to that on a camera. A shutter is also provided, interlocked with the electrical mechanism, by the use of which the plate may be exposed to the light of a single spark only.

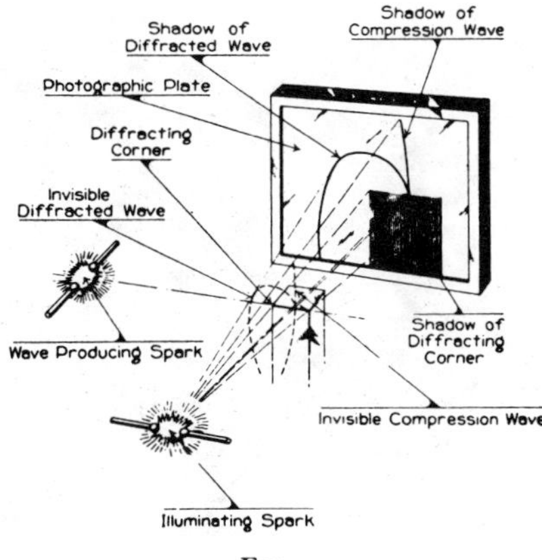


FIG. 1

Referring now to the models, the first of these is intended to represent a succession of planes of stratification, and to show how a downgoing wave, passing through this succession of planes, would send back from each plane a reflected component. It was accordingly necessary to find material through which a compression wave could pass without great loss, but from which a small portion of the wave would be reflected. Fine mesh copper screen was used for this purpose, and a model showing four successive stratifications, with a wave passing through all of them, is shown in Fig. 2. It will be observed that the downgoing wave does not lose appreciable energy in passing through the successive strata. It is also of interest to note that several of the reflected waves, in passing upwards through the strata are in part reflected downwards again.

Two other points of minor interest appear in Fig. 2. One is a second downgoing wave, appearing just below the location of the explosion. This is due to the reflection of the wave from the original explosion, where it struck the top of the box. Even though lined with felt for the purpose of absorbing such waves, this top sent back a sufficient impulse to be clearly recorded. The center of this down-going wave is missing, having been broken up on passing through the heated air in the spark. A second point of minor interest lies in the clearly shown diffraction pattern from the individual wires of the screen, which gives a definite texture to the background region in the vicinity of the explosion spark.

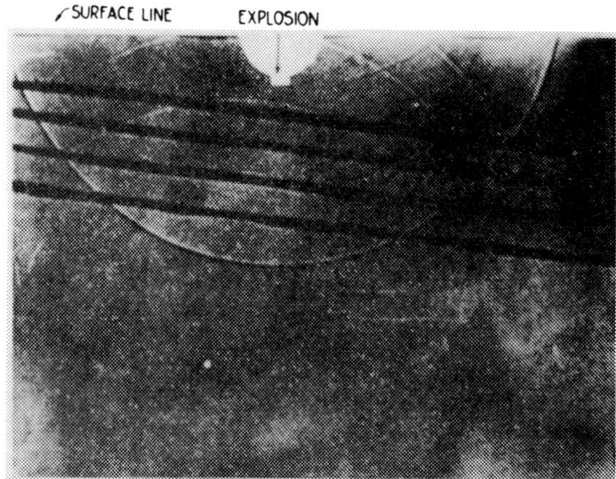


FIG. 2

With the exception of this first screen model, the subsequent plates are all from models employing a single structural element of material through which the original waves did not pass. This choice was made in the interests of simplifying the early part of these investigations by confining them to the wave patterns produced by one individual element. A succession of such elements would obviously produce a succession of later phenomena, following one another at definite time intervals.

Fig. 3 shows the wave pattern produced when a broken-off stratum is introduced into the field. Such a pattern would be expected for each successive bed in a region of normal faulting. The original or down-going wave is clearly seen as a large circle centering at the location of the explosion. Above the stratum, starting where the original wave meets the stratum, a reflected wave will be seen curving up towards the explosion, where it suddenly changes to a weaker or diffracted wave pattern of approximately circular form described around the upper corner of the broken-off stratum as a center. A second circular diffracted pattern will be seen described around the lower corner of the stratum as a center and extending until it strikes the lower surface

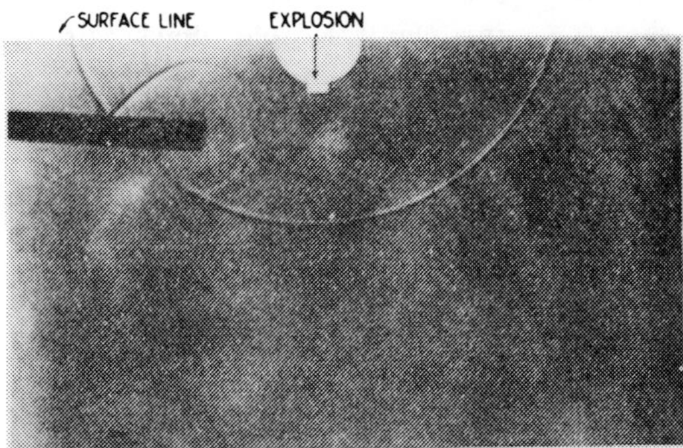


FIG. 3

of the stratum. The relative intensities of the reflected and diffracted waves as shown in this plate indicate that diffraction patterns are of sufficient magnitude to be useful in indicating the presence of faulting,

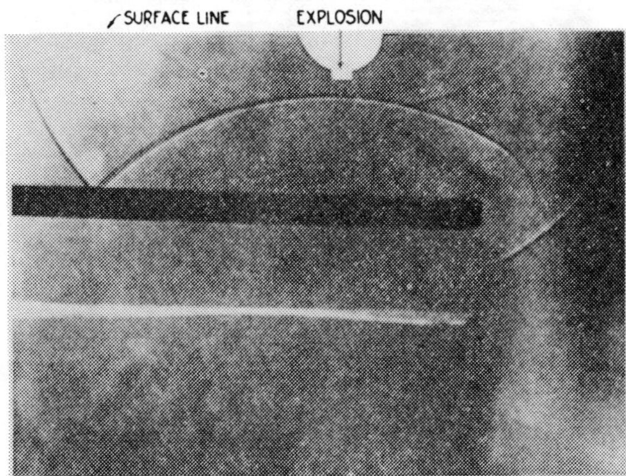


FIG. 4

provided appropriate means are used for separating and identifying diffracted waves in the presence of much stronger reflected waves from other directions. This problem will be gone into in more detail later.

Fig. 4 shows the same type of diffraction pattern from the same type of broken stratum, with the sole difference that the stratum extends past the explosion. The diffracted wave, while slightly weaker, can be very definitely seen here also.

Fig. 5 shows a normal fault located almost directly below the explosion, a slight but definitely noticeable diffraction pattern being visible here also. It seems rather evident that if diffraction patterns are to be used as a means of locating normal faulting this would be best accomplished by placing the explosion to one side of the fault, rather than directly over it, in order that the diffracted waves may arrive at the receiving system from a direction substantially different from that of the reflected waves.

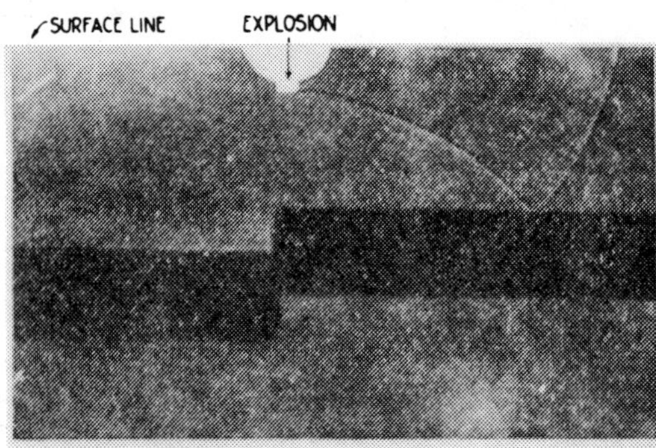


FIG. 5

Fig. 6 shows a normal fault to one side of the explosion. The reflected wave from the stratum on each side of the fault is clearly seen, as are also the two diffraction patterns due to the discontinuities in the upper and lower strata. This plate will be used as a basis for the diagram given later as Fig. 15, showing the problem of fault location in greater detail.

Fig. 7 shows another fault, with the two reflected waves and several diffracted waves clearly evident. The downgoing waves seen below the strata have not passed through them, as might be supposed, but have passed by them in front and behind the model, which did not extend to any great distance in the direction to and from the plane of the picture.

Fig. 8 shows another fault in which each of the wave patterns is due largely to reflection, and only a small portion to diffraction.

Fig. 9 shows the same model after the waves have progressed somewhat farther. It is clearly evident that waves from two different directions will shortly reach the surface slightly to the left of the explosion.

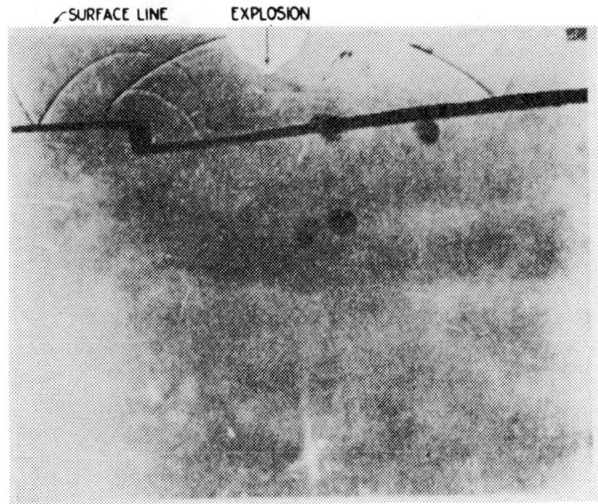


Fig. 6

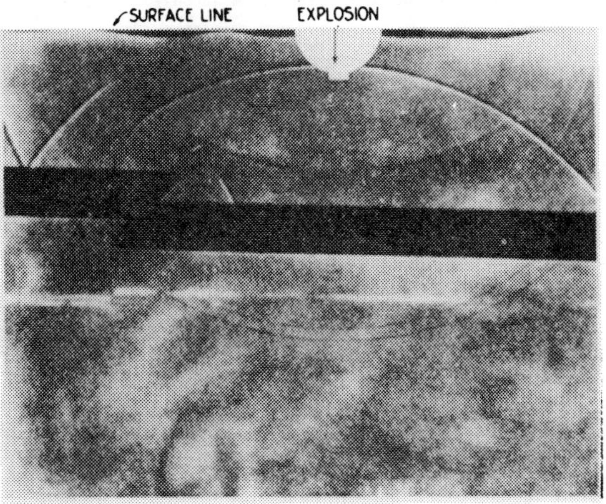


Fig. 7

If a receiving system were placed here, the identification of both of these arriving waves would be necessary if the structural conditions were to be properlp defined.

Fig. 10 shows a sharp fold which has returned some reflected and some diffracted energy.

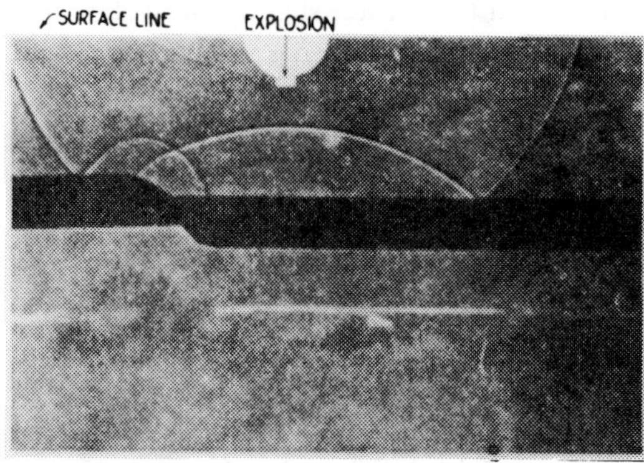


FIG. 8

Figs. 11 and 12 show two successive stages in the return of waves from a sharply folded syncline. Here again it is quite evident that a number of waves from this group will shortly reach the surface to the left of the explosion, more or less simultaneously, but from a number of different directions, thus presenting a problem in separation and identification.

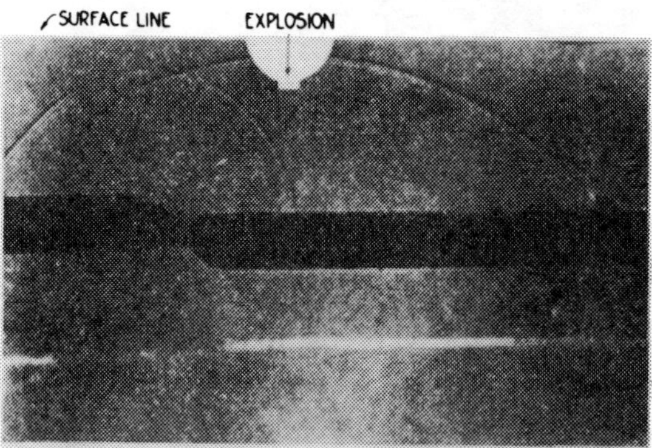


FIG. 9

Figs. 13 and 14 show two successive stages in the spread of the reflected and diffracted wave pattern from a sequence of sharp folds or wrinkles. Here, again, it is evident that a complex wave pattern containing more or less simultaneous arrivals from various directions

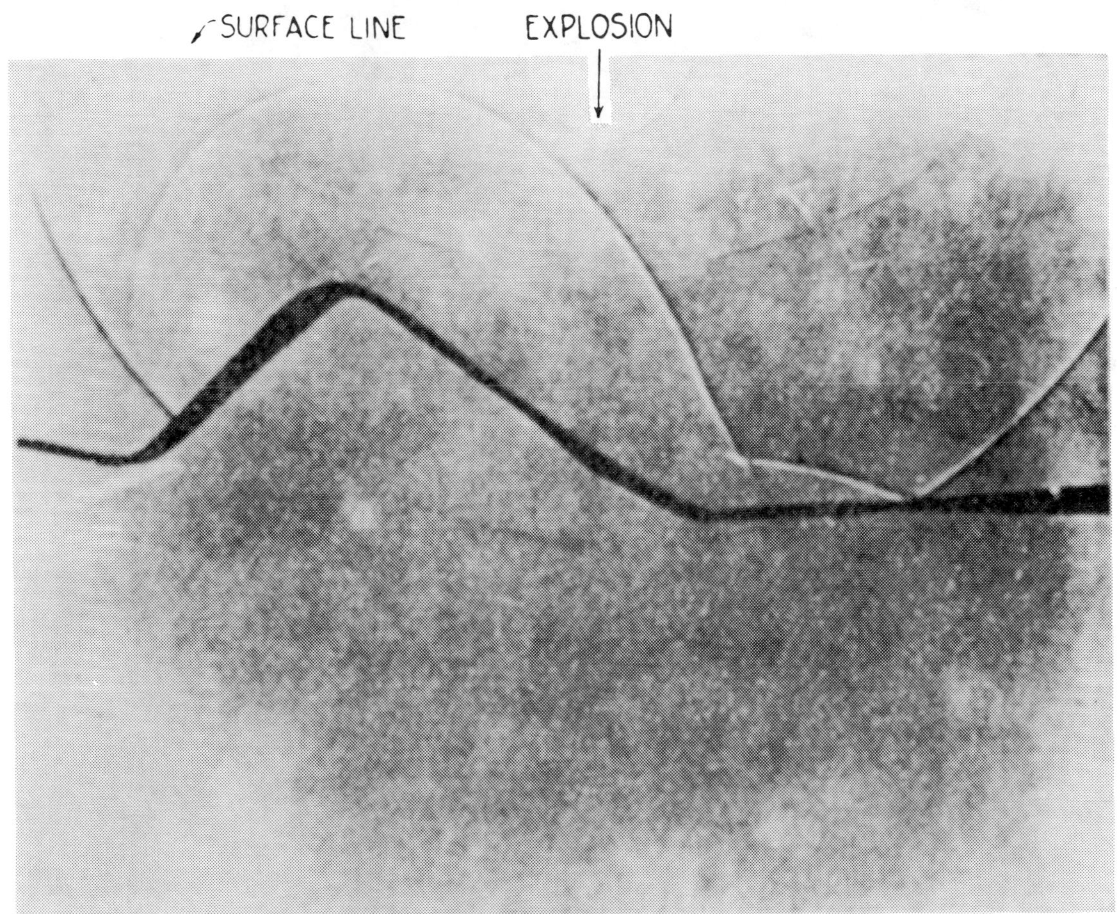


Fig. 10

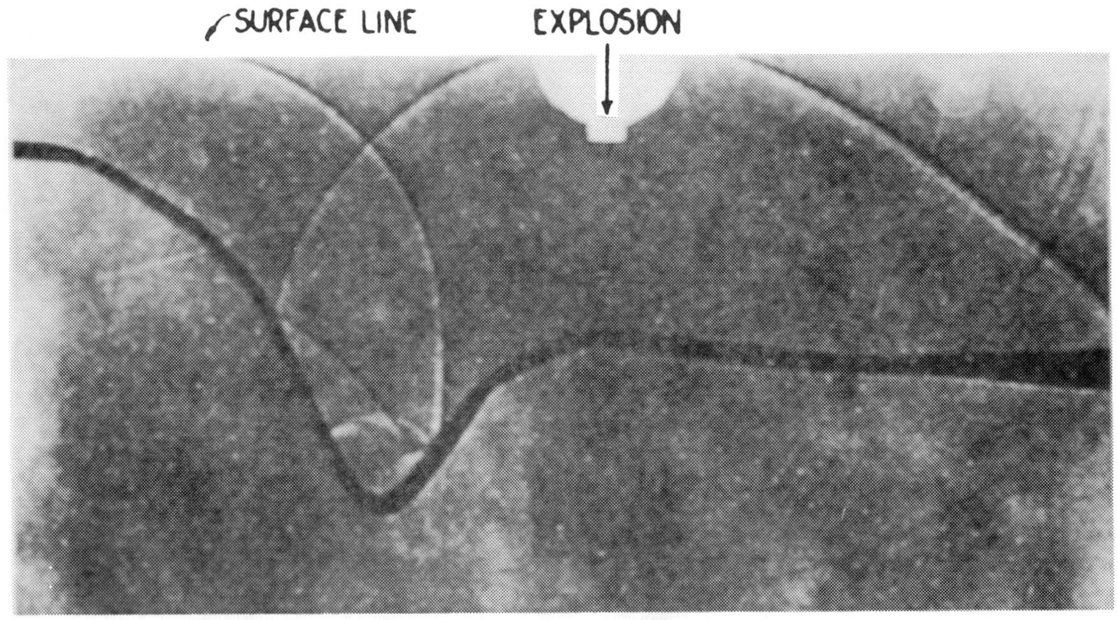


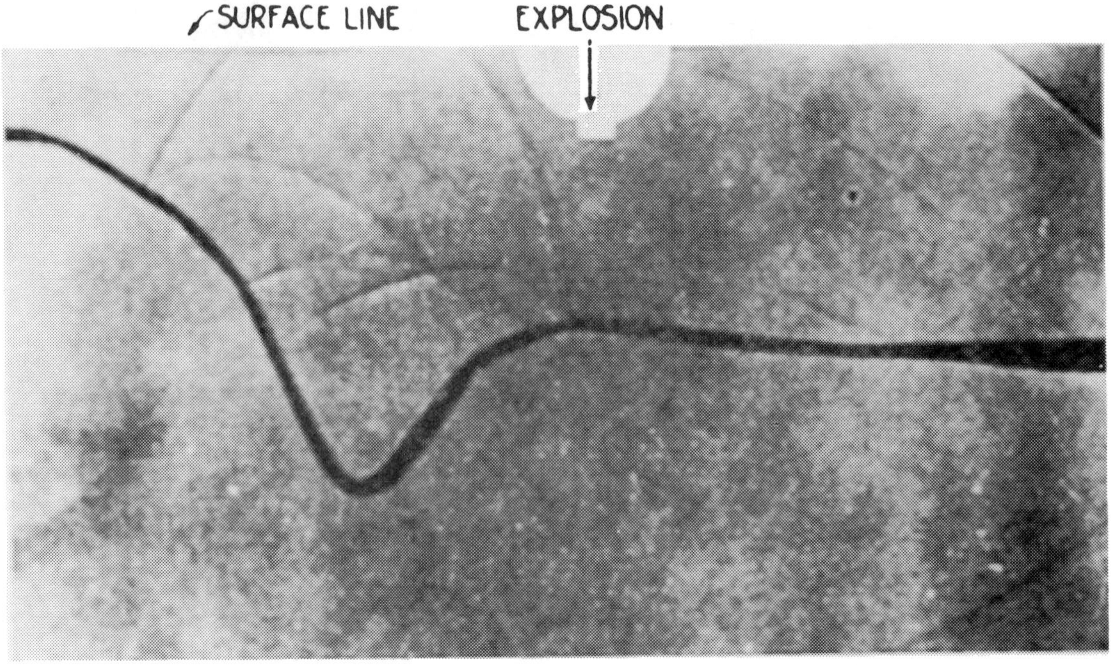


FIGS. 11 AND 12

will ultimately reach the surface. Two such waves, in fact, may be observed on Fig. 14, just reaching the surface to the left of the explosion.

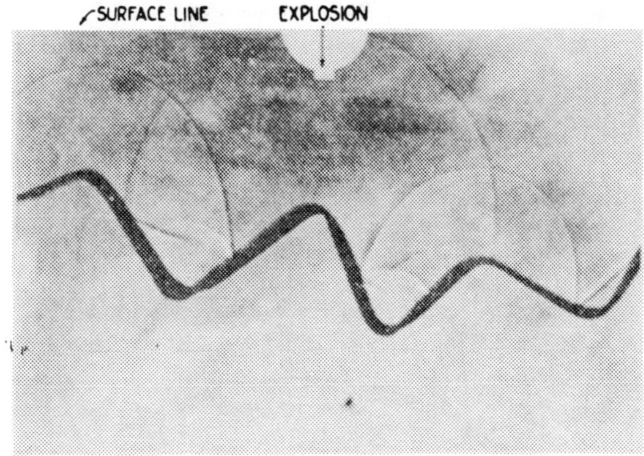


Fig. 13

From all of the foregoing two definite points may be emphasized, both of which have great importance in the development of a technic for reflection work in steeply folded or faulted regions.

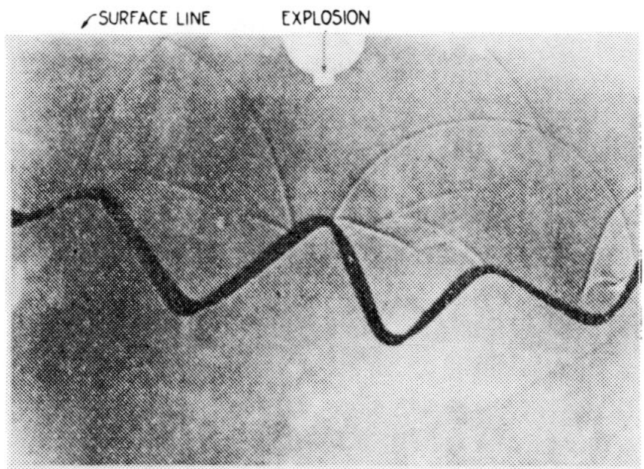


Fig. 14

First, not all returned waves may be treated as simple reflections whose places of origin are to be plotted by geometrical means as portions of strata. Diffracted waves, and waves returned by non-stratified causes can play a large and significant part in the total wave pattern received in such regions.

Second, such structural conditions must necessarily return many groups of waves from different directions, but with simultaneous or overlapping arrivals. Any attempt to map such regions must successfully receive and define the returned waves. If any wave is lost, the thing in the earth which originated it will also be lost from the picture. As a specific instance, there are many places where a succession of sedimentary beds exists, none of which give outstanding or distinctive reflections which permit them to be used as markers. If a normal fault occurs in such a region, we cannot hope to detect its

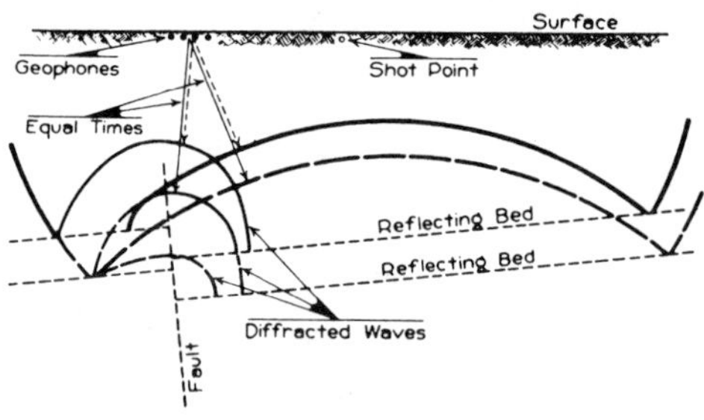


FIG. 15

presence satisfactorily by correlation, since the interval pattern between successive reflections on one side of the fault will very frequently resemble that on the other side to a sufficient degree, regardless of possible vertical displacements between the patterns, so that no definite proof of faulting may be obtained by correlation.

Likewise, normal faulting may often occur in such regions without producing a marked change in attitude between the beds on one side and the beds on the other side of the fault.

In such cases, the only means left by which such a fault might be detected would seem to lie in the detection and identification of diffracted waves originating from discontinuities in the successive strata where they reached the fault.

Fig. 15 shows such a region and is patterned after the model photographed on Fig. 6. The position of the receivers in Fig. 15 has been so chosen that reflected waves from two successive reflecting beds will arrive from one direction, while diffracted waves from two successive discontinuities in a fault will arrive from a definitely different direction. In order to separate and identify both of these sets of waves, the receiving apparatus must obviously be capable of angular discrimination—that is, it must be able to favor waves arriving from any chosen direction, and at the same time suppress waves arriving from other undesired directions.

Conditions where a number of different waves are traveling simultaneously in the same medium, but may be separated and identified, are common in the modern technical world. Telephone conversations, for example, are conducted through the same copper wire which serves for a number of other two-way conversations at the same time. And the ether carries, any hour of the day, a far greater miscellany of criss-cross vibrations, each of which may be separated from the total group and separately used.

It will be noted that, if we were to attach an oscillograph to a telephone wire which was carrying a number of conversations at one time, the wave record from the oscillograph would be undecipherable. Even though we knew the precise appearance of the waves, for, say, the vowel *a*, we would look in vain on our oscillograph records for such a group unless, by sheer accident, everyone else had happened to be silent while one conversing party said *a*.

Similarly, a radio set without tuning would receive from the ether so many interlocking sounds that the ear would be unable to identify any of them separately. However, the confusion in an oscillograph record connected to a multi-party telephone conversation does not indicate that these waves are inseparable. It merely indicates that they cannot be separated by an oscillograph, and that some other means must be used.

Correspondingly, if we find in the earth good evidence that wave groups are arriving simultaneously from various directions, and if a visual record of the commonly used type becomes so confused in such regions as to be unreadable, this does not necessarily mean that the individual waves in the earth cannot be separately identified. It merely means that we cannot conveniently separate them by the use of a simple visual record.

Where waves are traveling in a common medium, and we desire to separate them at a receiving point, it is necessary that the wave to be separated differ, in some one of its properties, from all of the other waves. In the case of the telephone conversations, each conversation is sent as a modulated carrier wave having a definite and very high frequency. Electrical tuning can, therefore, be used, and set to respond to any desired carrier wave, thereby releasing only that speech which has been consigned to that particular carrier. Similarly, the carrier waves in the ether have different, tunable, vibratory frequencies by which any one may be selected, and the subject or message consigned to that frequency may be separated and identified.

Wave groups returned from the earth, as a result of a downgoing wave from an explosion, do not differ materially in frequency. Hence, they cannot be tuned in or out individually. If they are to be separated at all, some other property than frequency must be used.

Fortunately, these arriving wave groups differ in one other property, which is precisely the one in which we are interested—namely, direction of arrival. Accordingly, if a directionally sensitive receiving means could be devised, it would be possible to separate and identify waves reaching the receiver with various directional properties.

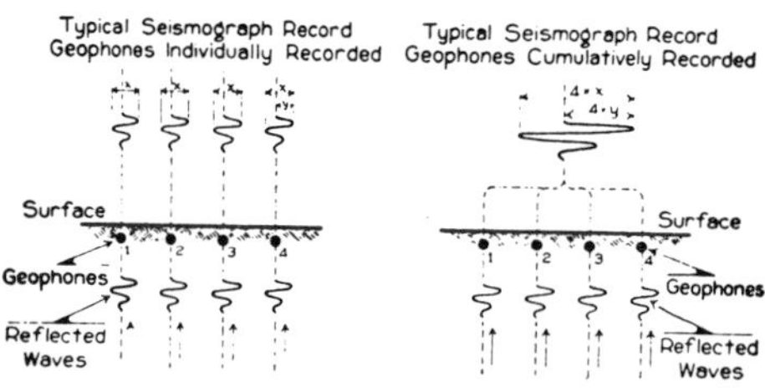


Fig. 16

A means for conferring directional properties on receiving systems has been known for some time, and consists in the so-called "grouped geophone" arrangement. This is shown diagrammatically on Fig. 16. On the left side of the plate, four individual geophones are shown, and it is assumed that they are being acted upon by a vertically arriving wave. For the purposes of the diagram, this arriving wave is shown as four individual arriving elements, one under each geophone. Further, and solely for the purposes of convenient diagramming, the waves are indicated as vibrations transverse to the direction of propagation. Above the geophones, waves are indicated whose amplitudes correspond to the electrical energy traveling from the geophone along the cable to the recording system.

To the right of the picture these four geophones are shown acted upon by a similar wave front, but having their outputs connected to a common circuit which correspondingly contains an electrical impulse with four times the energy of that produced by any individual geophone.

Such a group of geophones has distinct directional properties, as clearly shown in Fig. 17. To the left of the plate four such groups of geophones are shown, each group feeding into its own electrical circuit. When such an arrangement is used in the field, each group is made to produce a single trace on the ultimately recorded record. To the right of Fig. 17 is shown a single such group when acted on

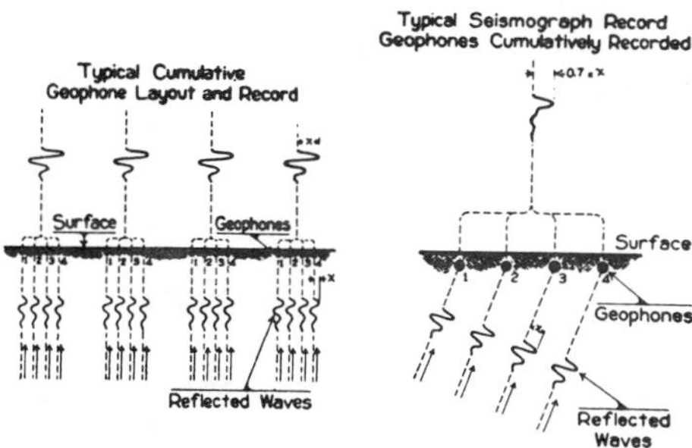


FIG. 17

by a wave which is arriving from a non-vertical direction. It will be readily seen that the wave elements do not reach the geophones 1, 2, 3 and 4 at identical times. Hence, they will not be added together to produce four times the energy, as was the case on the preceding plate. Instead, the outputs of the several geophones will conflict with each

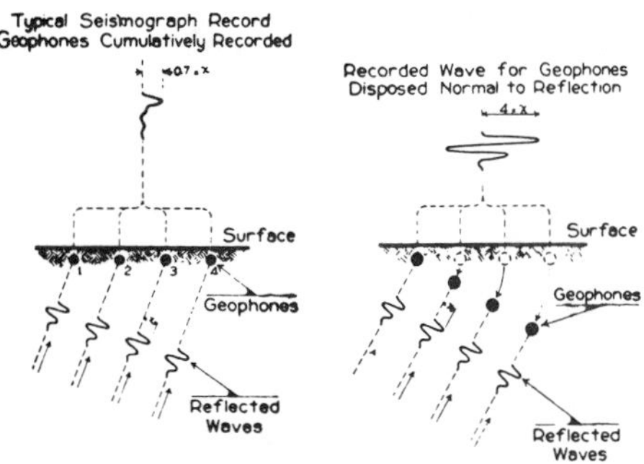


FIG. 18

other, and an electrical impulse somewhat of the form shown in the diagram will be sent out when this group of geophones is acted on by such a non-vertical wave.

Fig. 18 indicates one manner in which a group of four geophones could be made to give a good response to a non-vertical wave. To the left of the plate, for purposes of comparison, one of the preceding diagrams has been repeated, showing that the ordinary arrangement of four geophones connected in a group give relatively poor response to a non-vertical wave direction.

At the right of the plate it is shown that if these geophones could be moved from their former positions at the surface and planted instead at successively increasing depths, a position could be found for them at which they would once more receive the arriving wave elements simultaneously.

Under such conditions, their cumulative output would again contain four times the amount of energy available from each individual geophone.

As a practical matter, therefore, if we knew in advance the direction from which an arriving wave was expected, we could hope to plant a group of geophones at successively increasing depths, and connect them to any circuit, in such manner as to accentuate any reflected waves arriving from this predetermined direction. Under such conditions, however, there would be little point in making the experiment, since the answer would be already known.

If we were to attempt to conduct exploration in an unknown territory using this method, it would be necessary to plant each group of geophones in some definite relation of increasing depth, and then to fire a shot and record whatever waves might be arriving from this direction. Thereafter, all the geophones could be replanted to favor some new direction, and another shot fired to determine what reflections, if any, might be arriving from the new direction.

As a practical field procedure such a method is obviously ridiculous, on account of the large number of placements and shots required for each position. The diagram has been given merely to illustrate the fact that a group of geophones placed horizontally constitute a means of selectively receiving waves from one direction—but that, by placing the geophones other than horizontally, waves from any other chosen direction could be selectively received.

The system devised by the author involves the ability to examine any complex group of waves arriving from different directions and to select from this group and emphasize waves from any chosen direction, suppressing those from other directions. This is not done during the original recording, however, but subsequently by making the original record as a sound track film, which can be passed through an analyzer whose optical and electrical elements resolve the complex vibration into its various directional elements. This method permits the directions of all the various waves to be determined from a single shot in the field.

A difficulty commonly experienced when grouped geophones are used, in an attempt to confer directional properties on a system, lies in the fact that the low velocity or weathered surface layer of the earth may not have uniform thickness or properties under the successive geophones.

Fig. 19 illustrates such a condition. At the left of the plate a uniform surface layer is shown, through which the reflected group of waves may pass without altering the time of arrival of each wave element at the respective geophones. Under such conditions good cumulation of the outputs of the individual geophones will result and the equipment will operate satisfactorily as a means for selectively increasing the amplitude of vertically arriving waves.

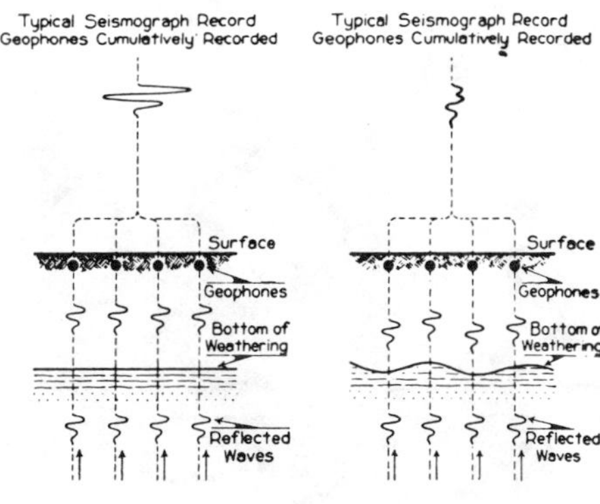


FIG. 19

On the right of Fig. 19 is shown the condition resulting from an irregular weathered layer. It will be observed that the elements of reflected waves arrive below the weathered layer at identical times, but that, after passing into the weathered layer, these wave elements reach the geophones at non-identical times and hence tend to interfere with each other and partially cancel the cumulative effect which would otherwise have been obtained. Under such conditions the cumulative output of the four geophones in the group will resemble that shown above them. Comparing this cumulation with the preferred one shown at the left of the plate, it will be seen that an irregular weathered layer may almost entirely destroy any advantage obtained by grouping geophones.

By observing individual times of arrival of reflected waves on these geophones, however, this irregularity in surface layer conditions may be detected and evaluated. Thereafter, in the author's equipment, such irregularities can be compensated for by appropriate adjustments of the analyzing apparatus. As a matter of possible interest, some of the equipment for the new system[4] is shown herewith. Fig. 20 pictures the recording truck for a ten track wave recording unit. Cable reels are protected as shown by a folding cover as a convenient means of

Fig. 20

avoiding moisture trouble and otherwise providing against theft or tampering. The water supply for film washing is contained in the tank over the cab. The ice box, visible through the opened operating compartment door, contains ice for controlling developer temperature when needed, and also serves as a seat for the operator. The operator's compartment is heat insulated, and means are provided for circulating the air in the compartment past the ice chamber when desired on account of hot weather.

Fig. 21 shows the interior of the operating compartment in the recording truck. At the upper left of the picture is a terminal and switch panel box where all battery circuits are controlled, fused,

[3] This new technic has been termed controlled directional selectivity, and is designated, for the sake of brevity, as the C. D. S. method.

switched and measured. A volt meter with a percentage scale and selector switches permits the measurement of any battery voltage at will with the same meter.

At the upper center of Fig. 21 are seen the ten amplifier units, and below them gain control and other adjustments, while individual volume level indicators for the ten channels are shown at the bottom of the instrument channel. At the right of the plate the film drying rack can be seen.

FIG. 21

At the lower left of the plate is the film developing and fixing tank with its water tight top closed down. This tank is heat insulated and thermally ballasted with a large amount of wash water which may be either heated or iced. Temperature control within 2° is thus readily possible, and this accuracy is more than sufficient to permit maintenance of good photographic quality in the sound tracks. Immediately to the right of the developer tank is a small illuminated glass slide on which film may be inspected. Farther to the right is the operator's desk, under the top of which record books and the like may be kept. To the right of the desk is the recorder proper, driven by a governed spring motor. It may be loaded by opening the hinged top, the operating compartment being light proof when the outer door is closed.

Fig. 22 shows the automatic optical analyzer which combines the records of the ten sound tracks on any one film to produce a sequence of analyzed components, each pertaining to a definite direction of arrival in space.

The magnetic drive for the analyzer pen is shown at the upper left of the illustration, and is mounted on tracks permitting it to occupy

Fig. 22

successively a number of predetermined positions, one for each successive repetition of the analyzing process. The large drum is rotated by motor contained within it, a paper strip serving to record the pen movements. Recorded waves can be clearly seen on this strip in the illustration. The film on which the analysis is being conducted passes from the small upper cylindrical container, shown in front of the drum, to a small lower cylinder, being driven by a sprocket contained within the analyzing mechanism. This sprocket automatically returns the film to the starting position at each revolution of the drum.

The optical elements of the analyzer are contained in the rectangular compartment in front of the main recording drum whose front is hinged so that it may be opened for inspection. The small control knobs on the top of the analyzer compartment serve to apply individual time corrections to compensate for irregular surface layers.

The amplifying and control equipment for the electrical circuits are shown at the left of the plate. The entire apparatus is driven from alternating current, the power supply source being mounted on a separate panel not shown in the illustration. This power supply source is

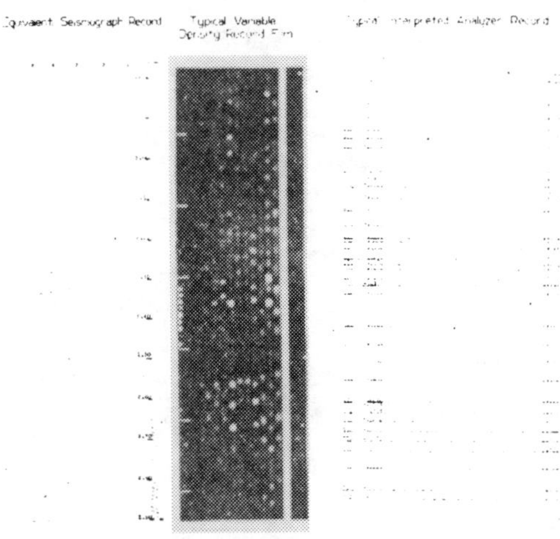

Fig. 23

compensated against line voltage changes, so that the amplitude of the pen record is at all times determined solely by the amplitude of the sound track record on the film.

Fig. 23 shows a variably density film, at the center of the plate, with a typical analyzer record as obtained from this film shown at the right of the plate. The transverse marks are put on by the person doing the interpreting, the designations on each line referring to the successive identified wave trains, giving for each a grade of dependability, and other quantitative factors from which the position in the structure of the point from which the wave was returned may be determined.

The central trace on the analyzer strip appears darker than the others. This is due to the fact that on the original strip the ink in the pen is changed for this trace from blue to red, for convenience in identifying the successive traces. For comparison, this same sound track film has been reproduced at the left, not as an analyzer strip, but as a succession of individual waves such as are customarily recorded by other forms of reflection equipment. Repeated comparisons made in this manner have shown that records may be analyzed successfully in cases where the ordinary oscillograph form of record could not have been handled visually at all.

Fig. 24 shows a short section of an analyzer record, on which two reflections have been marked, arriving almost simultaneously but from different angular directions. Below the record is a diagram show-

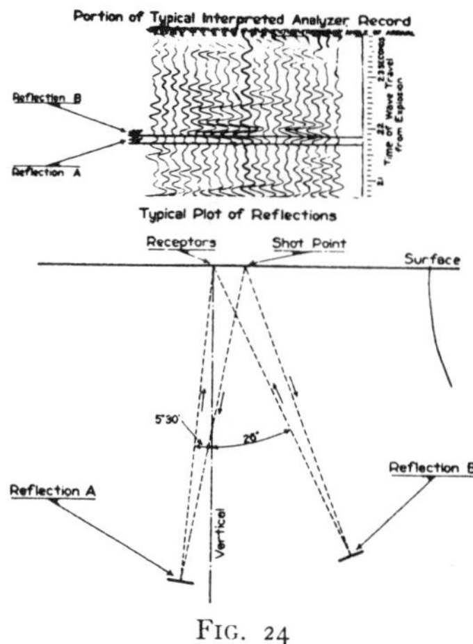


FIG. 24

ing the reconstructed section of structure in which the sources of these two reflections have been plotted.

In conclusion, the author desires to acknowledge his indebtedness to the members of his laboratory staff for their many valuable contributions to the development of the apparatus here described, and the technic for its use.

Particular credit is due to Mr. Curtis H. Johnson for working out the underlying mathematical factors involved in the analysis and plotting of complex structural conditions, and to Mr. B. F. McNamee and Mr. William T. Jochum for designing and supervising the construction of the recording and analyzing equipment. Acknowledgment is likewise made of the valuable services of Mr. E. M. Lane in developing a satisfactory and economical field operating technic for the new method.

# LABORATORY STUDIES OF TRANSIENT ELASTIC WAVES

BY

S. KAUFMAN * and W. L. ROEVER **

## Synopsis

An experimental procedure for model studies of the propagation of transient elastic waves is described, and some of the preliminary results are presented. These indicate that this method of approach to the complicated seismic problem may be a fruitful one. Although some workers, on the basis of earthquake and atomic bomb observations, have questioned the results of classical theory, evidence concerning the character of the surface-propagated Rayleigh wave has been obtained which is in agreement with theory.

## Résumé

La communication décrit une méthode expérimentale sur modèles réduits de la propagation d'ondes élastiques passagères et présente quelques-uns des résultats préliminaires. De ceux-ci il ressort que cette méthode d'entamer ce problème sismique compliqué pourrait bien être féconde. Quoique certains investigateurs aient mis en doute les résultats de la théorie classique à la suite d'observations faites lors de tremblements de terre et d'explosions de bombes atomiques, on a obtenu au sujet du caractère de l'onde de Rayleigh à propagation en surface, des données positives qui concordent avec la théorie.

## Introduction

The seismic method for determining the contours of subsurface formations in order to locate features favorable for the accumulation of oil is markedly successful over many portions of the earth. The method consists, in very general terms, of generating a seismic disturbance at a point on or near the surface of the earth, observing the resulting disturbances at a number of surface points in the vicinity of the initial disturbance, and determining, from selected features of the observed surface disturbances, the depths of the subsurface formations causing reflection or refraction of the energy emitted from the initial disturbance. As the subsurface formations are continuous over large regions, their features can be mapped by repeating the process at neighboring points along the surface.

Since the method has become such an excellent means for subsurface exploration, much effort has been expended on improvements in the instrumentation and the field techniques. Yet, although the state of the art is at a high level, there are still many questions to be answered, particularly those questions concerned with the propagation of seismic energy through the earth. For example, what is happening in areas where the method yields no results? Or in those areas where disturbances occur which cannot be associated with the subsurface formations? Also, is it known that all the possible information is being obtained even from areas which are now considered good areas? It is realized that the seismic method has been successfully utilized in spite of a high degree of ignorance as to many details of sound propagation through a complex medium, and that improvements would be expected with an increased knowledge of the processes involved. These problems are being attacked by experimentation in the field, a costly and cumbersome procedure which often leads to indefinite results and additional problems, and by theoretical approaches, which are beset with complicated and lengthy computations. However, in spite of the difficulties, the work continues because of the prospect of large rewards.

An approach to the problems which has been rather neglected is offered in laboratory studies of elastic wave propagation. Here a medium can be

* Senior Physicist, Exploration and Production Research Division, Shell Oil Co.

** Geologist, Exploration and Production Research Division, Shell Oil Co.

Reprinted from Proceedings of the Third World Petroleum Congress, 1951, pages 537-545 with permission from E. J. Brill.

selected which is far simpler than the earth in its acoustic behavior, and one would expect, therefore, that the results would be easier to interpret in terms of the fundamental processes involved. Key experiments could be performed, designed not only to check the theoretical work at various stages of its development, but also to determine, in part, the channels into which the theoretical work might be directed. The medium could be modified to simulate various surface or boundary conditions, and observations could be made which might aid in the interpretation of the complex effects noted in the field data; and the laboratory conditions could readily be changed to test variations of current field techniques as well as to test different procedures. Furthermore, the possibility of scaling of the elastic medium with respect to certain conditions of the earth exists, and such data would be extremely valuable in providing ideas, or corroboration of ideas, on the effect of earth conditions on the observed data.

With these considerations in mind a program of laboratory work has been undertaken designed to study the propagation of elastic waves under those conditions of interest to the field geophysicist, and it is the purpose of this report to indicate the method of approach and to present some of the preliminary results. The problem is quite complex, and the possibilities of experimentation are so numerous that, it is felt, such a presentation might result in valuable suggestions and perhaps encourage others to engage in a similar activity.

## Apparatus

It was considered desirable to have the laboratory procedure similar in basic details to the procedures used for seismic exploration in the field, i.e., the work would be directed towards a study of the behavior of transients in an elastic medium. A spark discharge, whose circuit details are schematically shown in Figure 1, is used as the source of energy. This could be fired in the air above the elastic material, or imbedded within the elastic material. The switch shown is actually an electronically operated one which, when closed, discharges the 0.5 mfd capacitor through the low-voltage winding of the ignition coil. The high-voltage pulse from the secondary of the coil causes ionization between the central electrode and the outer electrodes, decreasing the impedance between the two outer electrodes, and allowing the main capacitor to discharge across the gap between the two outer electrodes. The total impedance across the main capacitor is very low during the discharge. A trigger pulse, obtained by tapping across a portion of the discharge line, is shown in Figure 2a. The electrical discharge is seen to reach its maximum current value within less than one microsecond, and to have a time constant of about ten microseconds. Figure 2b shows the pressure pulse created in water with the spark 5 mm above the surface and a pressure-sensitive detector placed directly beneath the spark and 6 cm below the surface of the water. (It was necessary to retouch the originals of Figures 2a and 2b since the writing speed was too fast to photograph well.)

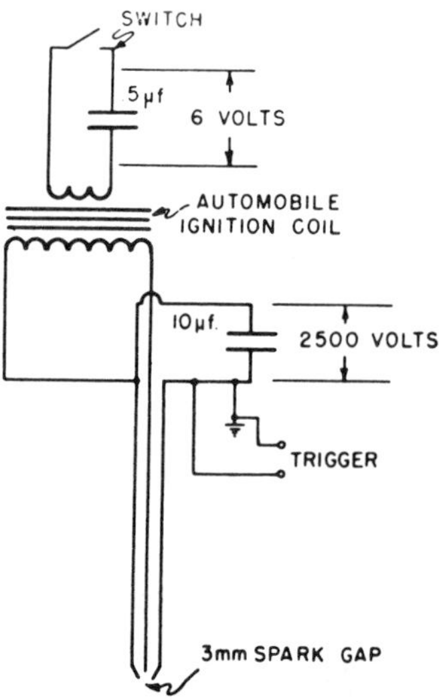


Fig. 1. Spark discharge circuit.

The recording system consists of a variable-reluctance type transducer, amplifiers, and a Tektronix 511AD cathode ray oscilloscope plus suitable trigger shaping and unblanking circuits. The transducer is

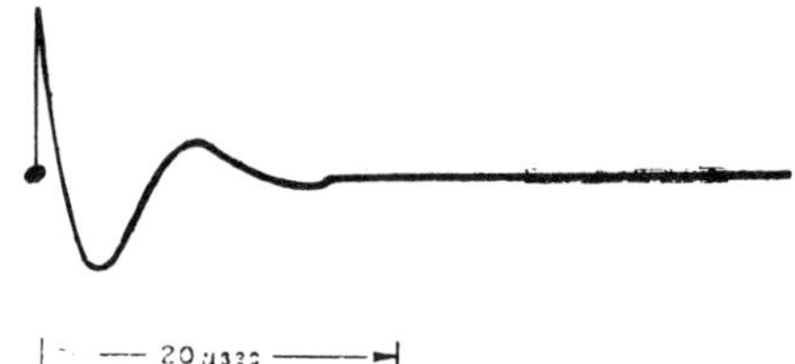

Fig. 2a. Trigger pulse from spark discharge circuit.

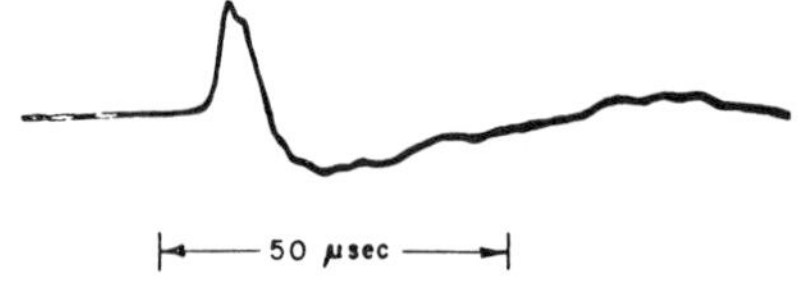


Fig. 2b. Pressure pulse in water.

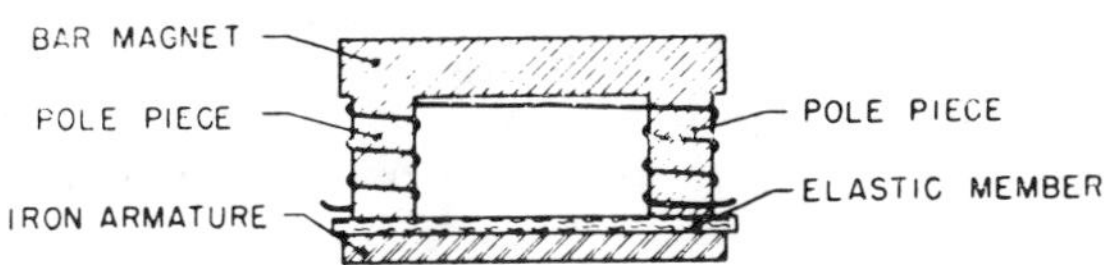


Fig. 3. Schematic drawing of transducer.

shown schematically in Figure 3. The armature is iron, 0.3 mm by 3 mm by 10 mm, and is securely imbedded at the surface of the model material. The bar magnet, pole pieces, and windings weigh 10 grams. The choice of this design, as well as the choice of the elastic member situated between the pole pieces and the armature, is made so that the natural resonance frequencies of the system are not close to the frequencies expected through the elastic medium under study. With a single layer of "Kleenex" tissue as the elastic member between the pole pieces and the armature, the mechanical resonance frequency of the transducer is at about 1500 cps; the electrical resonance of the transducer is at about 70 kcps. With an amplifier whose gain is 2500 inserted between the transducer and oscilloscope, adequate sensitivity was obtained for the experiments.

The transducer is seen to be of the velocity type, and is sensitive, essentially, to the component of velocity at right angles to the plane of the armature. Its output, after passing through the amplifier, is sent to the vertical deflection plates of the oscilloscope whose horizontal sweep and unblanking circuits are synchronized with the spark discharge by the trigger pulse obtained from the spark discharge circuit. Permanent recording of the transient output is obtained by photographing the single-sweep pattern on the face of the cathode ray tube. The time scale is obtained by photographing the single horizontal sweep with the output of a crystal-controlled oscillator substituted in place of the transducer output.

The selection of a suitable material for the elastic medium involves many considerations. It was desired to have a material that could be easily modified so that a variety of experimental procedures could be effected, and whose elastic properties were such that the frequencies and times of arrival of the propagated disturbances would lie within measurable ranges. Since any wave disturbance emanating from the initial source is subject to reflection from the boundaries of the medium, the total observation time at any point of the medium is limited to the time of arrival of the highest velocity component of the initial disturbance which arrives at the observation point by way of the shortest reflected path. It is required, therefore, that all events of interest emanating from the initial source pass the observation point within this period of time, a condition easily met for very large blocks of elastic material, but somewhat difficult to satisfy with blocks of reasonable dimensions. Furthermore, the several components, or types of waves, created by the initial disturbance travel with different velocities. Unless the elastic constants of the material are such that sufficient velocity contrast exists between the components, they will not be separable at the observation point, and the results would be too complex for clear interpretation.

Tests were made with several types of gels and waxes, and a refinery residue wax of melting point 54°C was selected for the elastic material. The choice was influenced by convenience, and by factors concerned with the design and testing of the measuring apparatus. Although the contemplated early experiments were not intended to be scale-model studies, the effect of scaling on the design parameters had to be considered. The elastic constants and the density of the residue wax are close to the values of near-surface earth material and, for a scaling factor of $10^{-3}$ for length, the resulting factors for time and frequency are $10^{-3}$ and $10^{3}$, respectively, placing these quantities within a readily measurable range. In several other respects, however, the choice was not an ideal one; the cast block of wax was not perfectly homogeneous, and the propagation losses for the frequencies involved were somewhat larger than desirable *.

The elastic-medium was prepared by casting the residue wax into a block 150 cm by 148 cm by 32 cm. Small air bubbles are scattered in portions of the block but their effect on the propagation properties is believed to be small since the wavelengths ** encountered are relatively large. It is also apparent from color differences existing along a cored column that the block is not ideally homogeneous. Evidently, gravitational separation of components of the chemically heterogeneous wax occurred during the cooling process, the denser fractions tending to migrate downward. However, cooling occurred from all sides of the block, causing the outside surfaces to harden before the interior, so that the denser material settled in a plane region nearly

* We have recently been informed by Dr. C. van der Poel of his work at the Laboratory N.V. De Bataafsche Petroleum Maatschappij, Amsterdam, on the visco-elastic properties of bitumens and paraffins, and it appears that certain tars have many properties more suitable for our purposes. A study of these materials will be undertaken.

** This term is used in its most general sense and there is no question of oscillatory waves in the present connection.

midway between the top and bottom surfaces of the block. (As will appear from the time-distance data, the block exhibits propagation characteristics which can be attributed to a slight velocity contrast within the material.) The block was contained by steel plates 0.5 cm in thickness, and supported above the ground by posts set along the outer edge.

### Experimental Observations and Discussion

The experiments to be described were undertaken not only to make observations of transient elastic waves under relatively simple conditions, but also to evaluate the performance of the apparatus. For the first series of experiments the spark source was set 3 cm above the surface of the wax block and, with the detector position fixed, the source-detector distance was varied in 2 cm steps from 6 cm to 90 cm. This method was used because of convenience since it was found that the results from successive spark discharges were almost exactly repeatable. A portion of such a series of records is shown in Figure 4, together with a photograph of the timing markers. The records shown have not been retouched. The high-frequency transient appearing at the beginning of each record is due to electrical pickup of the spark discharge by the recording equipment. It is highly damped and disappears before any of the seismic energy appears at the detector. The amplifier gain for the several records was not held constant, but was increased with distance in order to bring out the features on the records. The number at the right of each record indicates the relative gain of the amplifier for the different source-detector distances.

Figure 5 is a plot of the time-distance relation for the events appearing in the early portion of the records. For source-detector distances greater than 30 cm it was necessary to reduce the horizontal sweep speed of the indicating oscilloscope, and the double values for the points at 30 cm distance were obtained from separate readings of the same events for the two sweep speeds.

The line of points labeled A represents the first observable arrival on the records and is attributed to the compressional wave arriving through the wax block. The first vertical motion of the surface of the wax block is upward but the records were made in such a manner that the oscilloscope deflection is inverted with respect to the surface motion of the wax block. The points B, K and L represent later developments of the event, indicating the time of arrival of the maximum upward velocity, zero velocity, and maximum downward velocity features, respectively. It is evident that these features spread apart in time up to a distance of about 20 cm, after which the spacing relative to A becomes more constant.

This effect has long been noted on refraction records made in seismic prospecting, and has been investigated theoretically by L. Cagniard (1)*. It is a real effect due to the type of propagation encountered along a boundary for excitation at a point above the boundary, for the case where the velocity of the lower medium is the greater one. To some extent, however, the observed effect in this case is due to selective absorption or to dispersion in the wax block. This was checked by an auxiliary experiment in which the spark and the pressure-sensitive detector were set in a large tank of water 22 cm apart. Photographs of the pressure pulse were taken with several thicknesses of wax inserted between the spark and the detector. The wax insertions were cylindrical in shape, 22 cm in diameter, and of heights 5.1 cm, 9.9 cm, and 20.1 cm. Figure 6 shows the results, and it can be observed that the pulse shape becomes more rounded as the thickness of the wax is increased, indicating a loss of the higher frequencies of the pulse in passing through the wax.

The event C appears to be a refraction from an acoustic discontinuity within the wax block. (The points D are a later development of the event.) It has a higher velocity than that of the initial compressional wave, and its time-distance curve projected to zero distance does not pass through the origin. The mechanism for the transfer of energy is believed to be by shear waves to the refracting layer, compressional along the layer, and shear to the detector. Calculations made for this mechanism indicate an effective boundary 4.4 cm below the top surface of the wax block, a value quite conceivable when consideration is given to the cooling process involved in the formation of the wax block, and a critical minimum distance of 3.4 cm for the appearance of this event. If the mechanism of refraction were entirely by compressional waves the critical distance would be 90 cm, and if the mechanism were by reflection from a boundary within the block, with a time-distance curve fitting the refraction curve at the longer distances, we would expect more curvature in the time-distance curve at distances less than 90 cm.

The line of points labeled J represents the first upward velocity maximum of the most striking event on the records shown in Figure 4. It is fol-

*References given at end of paper.

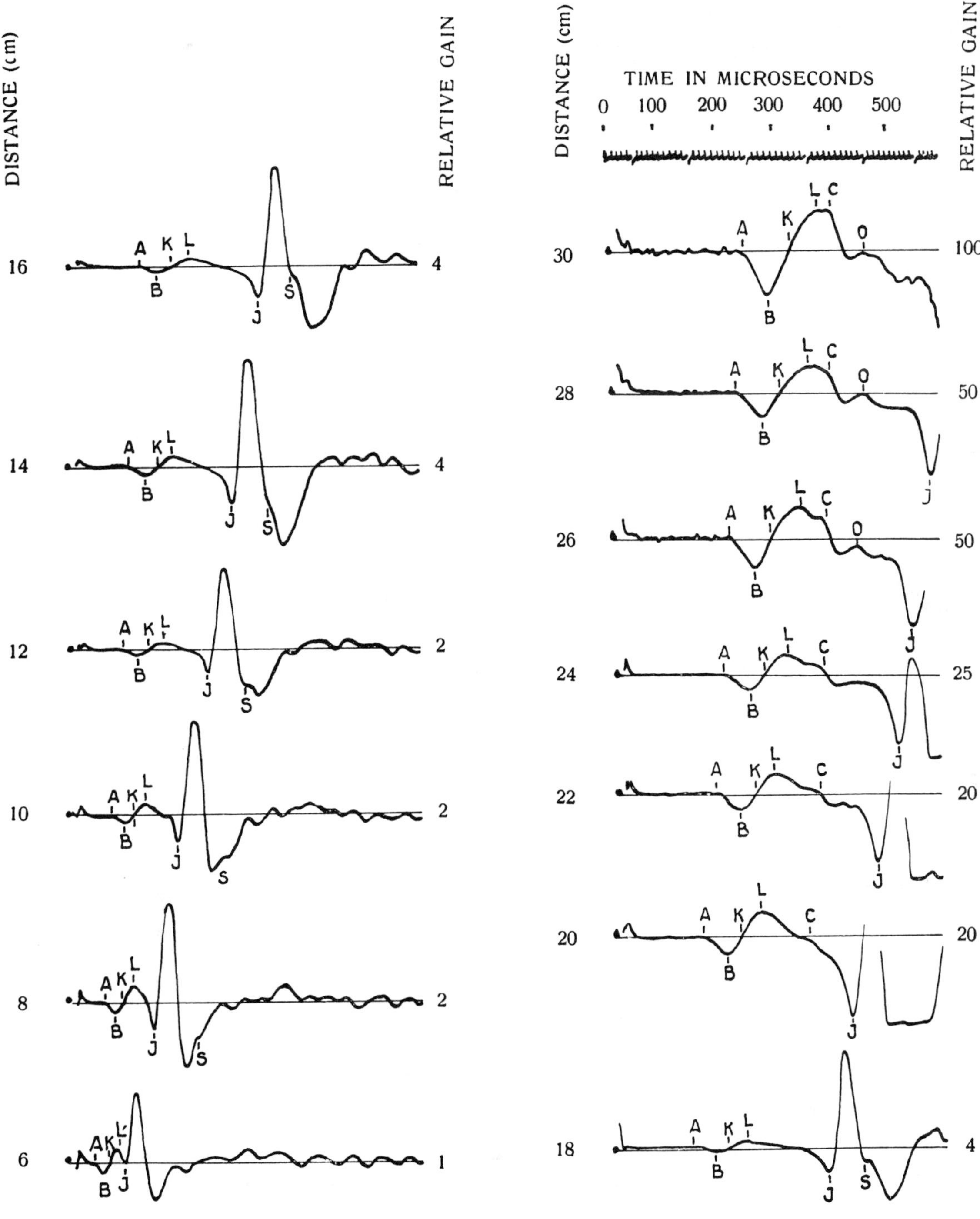


Fig. 4. Portion of a Typical Series of Records.

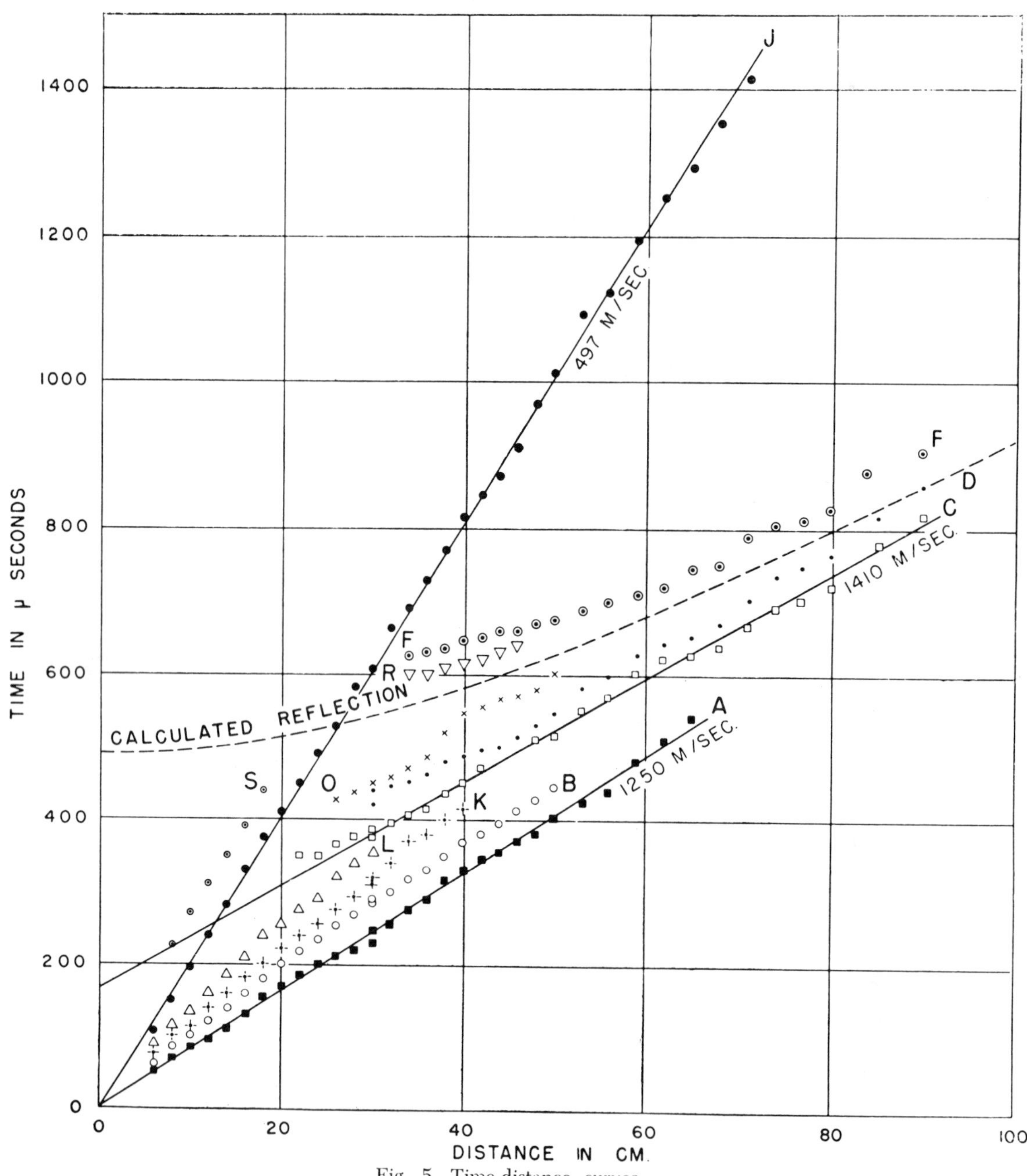


Fig. 5. Time-distance curves.

lowed by the event S whose development does not appear clearly, and whose existence is probably related to the inhomogeneities of the wax block.

The event J is attributed to the Rayleigh wave mainly because of its amplitude development. The amplitude of this event relative to the amplitude of the compressional wave increases with distance, an unlikely development for a body shear wave. (The velocity of the shear wave should be very slightly higher than that of the Rayleigh wave. Although no event corresponding to the shear wave can be observed on these records it is hoped that further work with a more homogeneous model material, which will eliminate the refracted energy, might aid

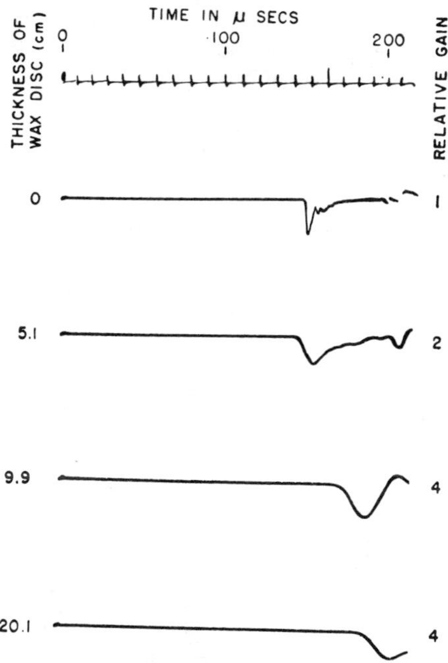


Fig. 6. Change in shape of pressure pulse travelling through wax.

the interpretation.) If a shear wave velocity 1.06 times the Rayleigh wave velocity is assumed, an assumption which appears reasonable since this factor varies between 1.09 and 1.04 for a range of Poisson's ratio between 0.25 and 0.5 (2) the calculated elastic constants of the wax are $0.7 \times 10^{10}$ dynes/sq cm for Young's modulus, $2.52 \times 10^{9}$ dynes/sq cm for the modulus of rigidity, and 0.39 for Poisson's ratio.

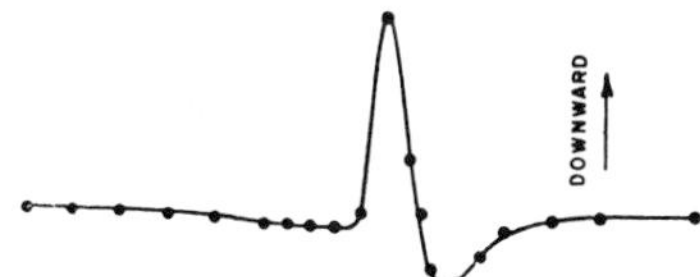


Fig. 7a. Derived curve of Rayleigh wave (from Lamb).

The results of an auxiliary experiment were very revealing. A cylinder of loosely packed glass wool was prepared with outer diameter of 6.5 cm, and a hole of 2 cm diameter was formed along the central axis. This "muffler" was placed around the spark source with its long axis colinear with the vertical line between the spark and wax, and with the muffler in contact with the wax surface. The area of impingement of the sound energy of the spark upon the wax is thus greatly reduced. A record taken under these conditions for a source-detector distance of 20 cm is shown in Figure 7b. It can be noted that the height of the Rayleigh wave relative to the compressional wave is decreased, that the trough one wavelength after J is reduced, and that the event S appears to be unaffected. Although the event S appears at about the expected time for a direct sound wave from the source, it is not attributed to this cause for the following reasons: direct sound should have been eliminated or greatly reduced by the muffler; the detector is very insensitive to any component of velocity except in the vertical direction; at greater distances, where the amplifier gain must be increased, the sound wave is apparent and its character is very different from that of S; and, finally, the time-distance curve for S indicates a velocity slightly greater than 500 meters/second.

The character of the Rayleigh wave has been predicated theoretically (3, 4) for idealized assumptions concerning the source of the disturbance and the type of elastic medium. Conclusions from the observed character of the Rayleigh wave can be made not only with regard to how closely the experimental results check the theory, but also with regard to the question of how closely the experimental conditions approximate the idealized assumptions made in the theoretical approach. — Figure 7a is a derived curve of the vertical velocity due to the Rayleigh wave, based upon Lamb's (4) theoretical curve for the vertical displacement to be expected for the Rayleigh wave at large distances from the source. Lamb's conditions are that a homogeneous elastic half-space is subjected to a short impulse of force, applied at a point of the surface.

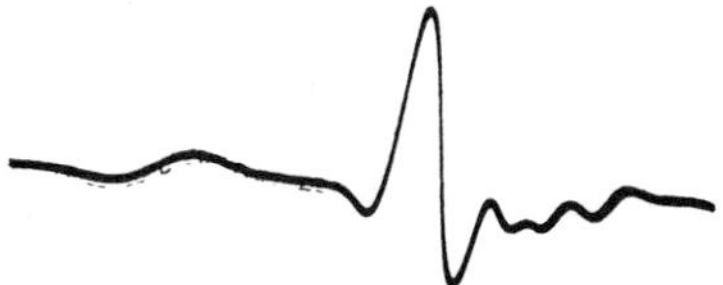

Fig. 7b. "Muffled" record at 20 cm.

It is seen that Figure 7b is in substantial agreement, especially if the event S is imagined to be absent.

These results are not in agreement with statements made by Leet (5) in his report of the observations made of the atomic bomb test on July 16, 1945. His observations of the Rayleigh wave, as do the observations of earthquake and field exploration seismologists, indicate a multicycle and almost sinusoidal character for the Rayleigh wave. Leet contends that Lamb's conditions have been met, hence observation does not check theory. During the course of these model experiments each advance made towards the idealized conditions has given results closer to the predicted ones, and it is believed that the oscillatory character of the Rayleigh wave observed in earthquake and prospecting seismology is a result of either a distributed source, rather than a point source, or of a layered nonhomogeneous medium, rather than an ideally homogeneous medium.

The event F shown on Figure 5 appears distinctly at the longer source-detector distances. It follows closely the calculated curve for a reflection from the bottom of the block. Proof that F is the bottom reflection was afforded by the following auxiliary experiment: a run of records was made at a number of locations over the block, keeping the source-detector distance constant. Then the bottom boundary conditions were changed by placing a thick steel plate in contact with a portion of the bottom boundary, and a repeat run was made. Those records where the reflection point fell on the steel plate showed an amplitude and character change for F only; elsewhere F was unaffected. The other events on the records were not affected by the steel plate.

Several shorter series of events are also shown in Figure 5. These, like event S, may result from reflections or refractions from other inhomogeneities in the wax, or may be interference phenomena between the events already discussed.

Since no shear wave was observed in the series of records of Figure 4, different experimental procedures were undertaken to separate the body shear wave from the surface wave. One type of experiment was to imbed a pressure-sensitive detector within the wax at a depth of 16.5 cm. Since the detector density is different from that of the wax, a shear wave traversing the detector will give rise to compressional energy, and the detector will respond. A series of records made with the source placed at various positions on the surface show an event arriving with a velocity slightly higher than that previously observed for the surface wave; but its amplitude, relative to the amplitude of the compressional wave, did not increase as the source-detector distance increased. In a second type of experiment the detector was set at a point on the bottom surface (the bottom is covered with a steel plate 0.5 cm thick, and is therefore not a completely free boundary), and the spark source was moved along the top surface of the wax block. A diagram of the set-up, together with the time-distance curves of the results, is shown in Figure 8.

The first event, P, arrives with the velocity of the compressional wave, and the later arrival, S, which also appears to take a direct path from the source to the detector, has a velocity near that of the Rayleigh wave. Since the thickness of the wax block is of the order of 6 Rayleigh wavelengths, and since the Rayleigh wave is of negligible amplitude at that depth below the surface, the event S is identified as the body shear wave. It was not noticeable in the previous experiments probably because its amplitude is small compared with the Rayleigh wave. The records also show event SP, identified as a bottom boundary wave excited by the body shear wave.

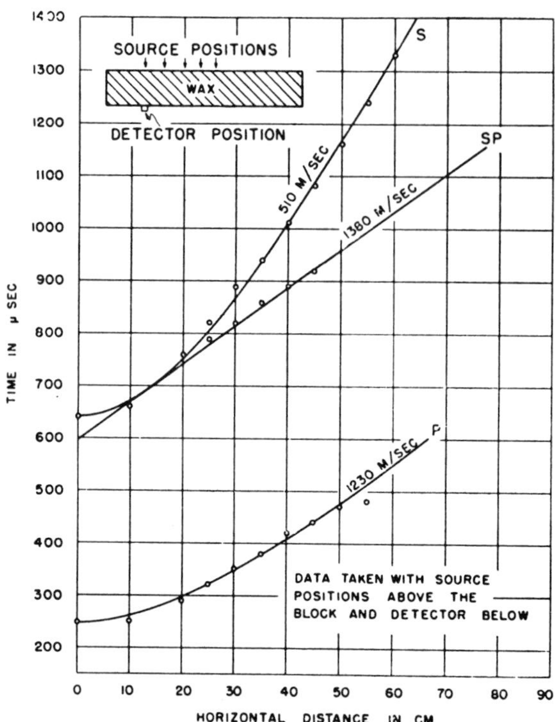


Fig. 8. Results of experiment to detect body shear wave.

## Conclusion

It is evident that the work done so far has been exploratory. It has been shown, however, that model experiments of this kind are quite feasible and that results of practical interest can be obtained. The classical theory seems to be adequate for the simple homogeneous case. It is planned to continue this work with other media, both homogeneous and layered.

Manuscript received Dec. 14, 1950.

## Bibliography

(1) Cagniard, L., Reflexion et Refraction des Ondes Seismiques Progressives, Gauthier-Villars, Paris, 1939, Chapters X, XI.

(2) Timoshenko, S., Theory of Elasticity, McGraw Hill Book Co., 1934.

(3) Rayleigh, "On Waves Propagated along the Plane Surface of an Elastic Solid", Proc. Lond. Math. Soc., vol. XVII (1885), 4—11.

(4) Lamb, H., "The Propagation of Tremors over the Surface of an Elastic Solid", Phil. Frans. Roy. Soc. Lond., vol. CCIII (1904), 1—42.

(5) Don Leet, L., "Earth Motion from the Atomic Bomb Test", American Scientist, vol. XXXIV (1946), 198—211.

## DISCUSSION

In complimenting the authors on their work Mr. H. Closs (Geological Survey, Hanover, Germany) stressed the importance of experimental studies on seismic events taking place from the shot to the detector. He then referred to the time-distance diagram of Figure 5, and asked if the observed velocity of C, 1410 m/sec., was perhaps affected by a possible inclination of the acoustic discontinuity that appeared to be present within the wax block.

In a written comment, Mr. N. Ricker (Carter Oil Company, Tulsa, Oklah, U.S.A.) referred to the authors' concern about the rather large propagation losses in their experiments. In this connection he made an estimate of the so-called transition frequency ($f_0$)* of the wax: He pointed out that for a sensibly homogeneous medium the square of the wavelet breadth ($b^2$) should scale along with the travel time by the relation:

$$f_o = \frac{6}{\pi} \cdot \frac{\triangle\, t_o}{\triangle\, (b^2)}$$

By enlarging the authors' Figure 5 he estimated the observed wavelet breadths b by taking twice the time interval between the picks B and L; the travel times $t_0$ were given by the source-to-detector distance divided by the dilatational velocity 1250 m/sec. Thus, by making a suitable graph of the relationship between $t_0$ and the observed values of b, he found a value of 21,500 cycles/second for the transition frequency of the wax block. In order that the wavelet breadth be small compared with the travel time, so that the various disturbances be free from overlapping, he recommended the use of a material with a higher transition frequency. In his opinion there should be little trouble finding a suitable medium as transition frequencies vary widely; for aluminium, for instance, it is about 2 x $10^{10}$ cycles/second.

Mr. F. Gassmann (Eidg. Techn. Hochschule, Zürich, Switzerland), in a written comment, expressed his appreciation of the experimental work undertaken by the authors. He felt that the observation of travel time curves as shown in Fig. 5 was very promising; but if in addition to these curves also wavelet shapes were studied (Fig. 6 and 7), it might be necessary to be very cautious in the interpretation of the records, because these records are influenced by the mechanical response of the transducers (Fig. 3) to the passage of elastic waves; several degrees of freedom in the transducer should be considered.

The authors were represented by Mr. N. D. Smith (Shell Oil, Houston, Texas), who in reply to Mr. Closs stated that according to his recollection the profiles were shot from both directions, and hence the effect of a possible slope in the discontinuity would have been observed.

He thanked Mr. Ricker for his comments and felt that the application of the latter's theory might aid in the selection of suitable model materials.

Referring to Mr. Gassmann's inquiries about the response characteristics of the electro-mechanical transducers, he stated that these were not known in detail. Attempts were made to investigate mechanical resonances, through various types of shaking tables and by means of sensitive strain gauges, but the results were not very conclusive. However, by varying the masses of the components of the transducers, and the type of elastic coupling, no significant changes were observed in the recorded velocities and wave shapes, which gave a feeling of confidence that no significant resonances could exist in the pertinent frequency region.

One of the authors, Mr. Kaufman, replied in writing to the question raised by Mr. Closs. He confirmed that profiles were shot in several directions across the wax block; no change in the velocity of C was noted, so that the high velocity for this event could not be attributed to a sloping layer. Aside from this direct test he considered it unlikely that sloping layers could have been formed during the cooling of the wax block from a molten state.

* Refer preceding paper.

# Seismic Model Study*

E. T. Howes, L. H. Tejada-Flores, and Lee Randolph
*United Geophysical Company, Inc., Pasadena, California*
(Received February 24, 1953)

A method has been developed for studying seismic sound pulses in reduced scale models which simulate geophysical seismic exploration methods. Use is made of a tank of water for transmission of the acoustic wave from a source to a faulted limestone stratum and return to a pressure sensitive receiver.

Results obtained demonstrating the geometric aspects of the propagation, reflection, refraction, and diffraction of sound pulses are described and illustrated on accompanying plates.

Geologic structures such as faulted rock strata or salt domes, when traversed by reflection seismograph surveys, may produce seismograms of mixed and confusing patterns. The study of type geologic structures by the use of seismic scale models with known and controlled conditions can establish some general criteria for seismic record interpretation.

A relatively common geological situation of interest in prospecting for oil "traps" is one in which a succession of approximately horizontal rock strata are broken or faulted into two blocks separated by a plane fault surface. In a typical situation the fault plane might be oriented at an angle of 45° with respect to horizontal, and the two blocks slipped up and down relative to each other on the fault plane to produce horizontal elongation. Such a slippage is often referred to as "normal faulting."

It would be very difficult to construct a model with complete similitude to its geological counterpart in the earth. An attempt was made rather to illustrate the geometrical aspects of propagation, reflection, refraction, and diffraction of a simple sound pulse radiated from a point source and generated by a miniature explosion. Somewhat similar work has been done by Frank Rieber,[1] using a spark source in air, and by photographing the instantaneous configuration of the wave front by high-speed spark photography. In the present instance, water rather than air was used to simulate shallower strata, and a faulted limestone stratum was used to simulate "normal" fault slippage at greater depth. The traveling sound impulse was received with a small pressure-sensitive hydrophone, and the corresponding pressure variations recorded photographically from the screen of a cathode ray oscilloscope. No definite scale factor was used, but the model may be considered as having about 1/2000th of the linear dimensions of a possible geological counterpart in the earth.

A corrugated iron water tank $10\frac{1}{2}$ ft in diameter by $5\frac{1}{2}$ ft in depth was used. For the sake of greater convenience, the entire model was "turned on end." Thereby a horizontal plane surface in the earth becomes a vertical plane in the model tank. The geological structure was simulated by a slab of Bedford limestone 15 in. thick, $53\frac{1}{2}$ in. high, and 86 in. long. A cut was made through the slab at an angle of 45° with its large surface, providing a smooth, tightly fitting fault plane.

A diagram of the tank, model reflector, and general arrangement of power source, seismic source, and hydrophone receiver is shown in Figs. 1(a) and 1(b). For siesmic reflection profiles the source and receiver were fixed in relation to each other, the source spaced $5\frac{5}{8}$ in. above the receiver. This assembly was movable, on a track, parallel to the face of the model reflector. The source was a spark gap 19 in. below the water surface, and both source and receiver were 12 in. from the face of the model reflector. The receiver is acoustically isolated from the source by a 2-in. sheet of plastic Styrofoam. This isolation was necessary so that the high energy of the initial explosion would not greatly disturb the front part of the reflection record.

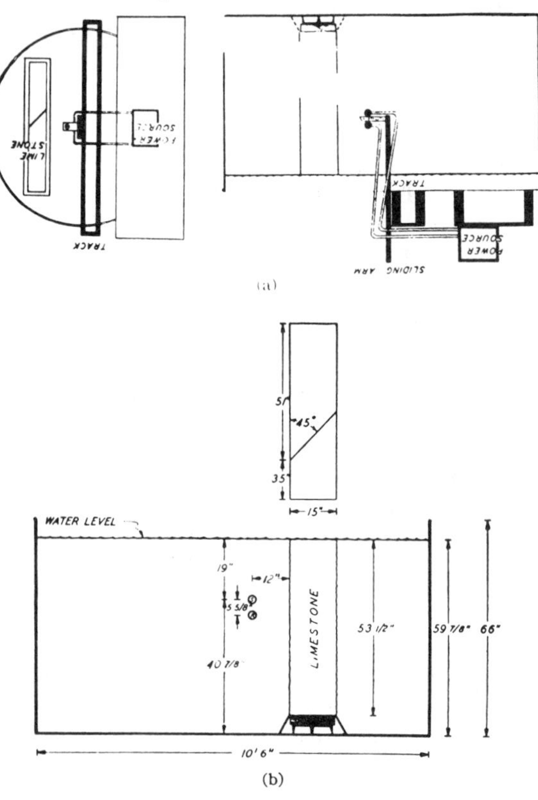

(a)
(b)
Fig. 1.

* This paper was presented before the Acoustical Society of America at the November 13, 1952 meeting in San Diego, California.

[1] Frank Rieber, Geophysics I, 196 (1936).

Reprinted from Journal Acoustical Society of America, Volume 25, 915-921 with permission from American Institute of Physics

A photograph of the tank setup is shown in Fig. 2. The model limestone blocks in this picture have been separated at the fault plane.

The explosion can easily be heard 50 ft from the tank with the gap 2 ft underwater. The intensity of the seismic wave usually varies about 10 percent on repeated trials. There is an extremely high-frequency electrical oscillation that takes place across the gap during the period of discharge which, however, does not appear in the seismic wave.

Figure 3 shows the underwater spark gap and mounting arrangement. Bakelite sheets clamp two structural steel bars which rigidly hold two $\frac{1}{4}$-in. diameter balls. The Bakelite sheets fasten to a wood strip which in turn is mounted on top of a 2-in. Styrofoam sheet. The screened tube at the right of the spark gap shields a tourmaline crystal hydrophone whose output triggers the oscilloscope sweep.

The electrical power supply for the spark gap consists of an electronic circuit continuously charging a 6.25-mf condenser through a current limiting resistor to about 2500 volts. The condenser is discharged nearly instantaneously across the spark gap through a hydrogen-filled thyratron.

The hydrophone used for receiving the sound impulse in the water is a barium titanate pickup $\frac{3}{4}$ in. in diameter by 2 in. in length, covered by molded Neoprene.

To insure having an accurately known time scale on the film records, a photo is taken periodically of a standard 20-kc or 50-kc sine wave. The recording instruments used in this model study work are shown in Fig. 4(a).

As a check after all the instrumentation was properly functioning, the velocity of sound propagation in water was measured. Measurements gave a value of 4975 ft per sec at 72°F.

Fig. 2.

Fig. 3.

(a)

0 200 400 600 800
TIME - μ SEC

0 200 400 600 800
TIME - μ SEC

(b)

Fig. 4.

The wave shape of the acoustic pulse radiated by the spark gap is shown in Fig. 4(b). The upper photograph was recorded with the hydrophone and amplifier response essentially flat from 10 cps to 100 kc. The lower photograph was made using a low pass electronic-type filter cutting off 36 db per octave above 50 kc. The principal signal at the left is the pulse radiated from the spark gap, and recorded at a distance of 6 in. It has a very steep front, the rise time being less than 10 $\mu$sec. Following shortly after this pulse is a small rarefaction pulse reflected from the Styrofoam member of the spark gap support.

The next measurement was determination of the velocity of sound propagation in Bedford limestone. The spark gap was placed against the end of the model and the receiving hydrophone placed at several stations along an adjacent side of the model (Fig. 5). The oscillograms showed pulses in addition to that trans-

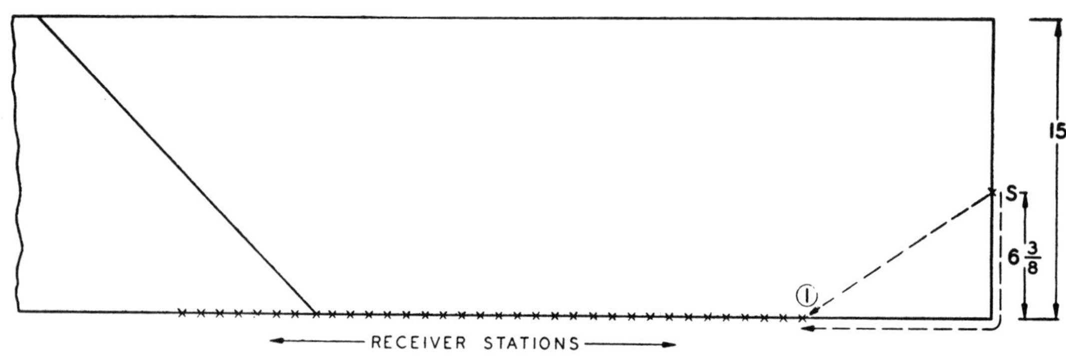


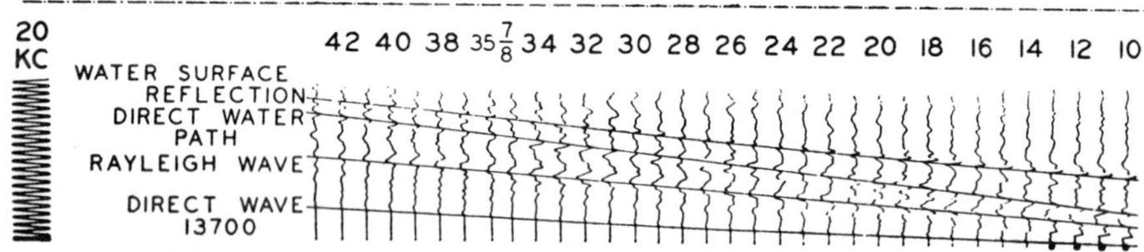


Fig. 5. Velocity measurement.

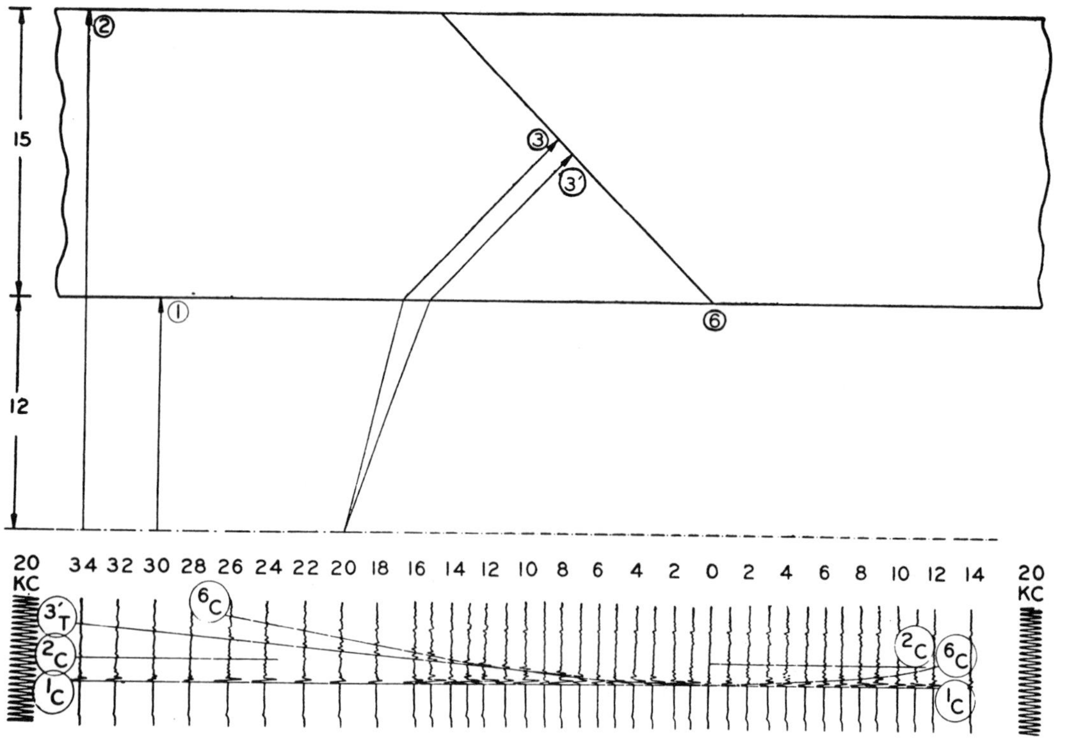


Fig. 6. Plane front fault reflection profile.

mitted directly through the limestone. These included a Rayleigh wave along the surface of the model, the water path pulse transmitted by diffraction around the corner of the model, and a pulse reflected from the water surface which also had to travel around the corner of the model.

From the measured velocity value of 13 700 ft/sec for the compressional wave, the transverse wave velocity in limestone was calculated to be 7900 ft per sec, assuming a Poisson's ratio of 0.25, and density of 2.6 g/cc. The Rayleigh wave velocity was likewise calculated as 0.92 times the transverse velocity, or 7200 ft/sec.

It is interesting to note the low amplitude of the direct wave through the limestone compared to the Rayleigh or direct water path waves. Even though the spark source was very close to the face of the model, a large part of the energy still entered the water.

Figure 6 shows a profile series of oscillograms taken with the spark source and hydrophone offset 12 in. from the face of the limestone slab. The spark source and hydrophone were kept in fixed relationship, the spark source being $5\frac{5}{8}$ in. above the hydrophone. This assembly was moved parallel to the model face with stations taken either side of the fault trace. The physical distances are noted on Fig. 6. On the several oscillograms shown at the bottom, lines join the more prominent events which are numbered and identified as follows:

*Event* $1_C$—Compressional wave reflected from the front face of the model slab. About 50 percent of the incident energy is reflected.

*Event* $2_C$—Wave reflected from the rear face of the model, showing reversal in phase.

*Event* $3'_T$—A wave traveling from the source as a compressional wave through the water, entering the limestone slab with refraction in accordance with Snell's law, traveling as a shear or "transverse" wave through the limestone, being reflected at the fault surface, and returning through the limestone as a shear wave reversed in phase, re-entering the water and returning to the receiver as a rarefactional wave, reversed in phase from the original radiated pressure impulse. It is interesting to note that this wave is slightly stronger than the wave reflected from the back surface of the limestone slab. However, both these waves are quite weak because of the high reflection coefficient of the front water-limestone interface.

*Event 3*—A possible event similar to $3'_T$, except that travel within the limestone is as a compressional wave rather than as a shear wave. This event could not be recognized on the oscillograms, probably because of very low and insufficient energy.

*Event* $6_C$—A possible event arising from wave diffraction phenomena at the trace of the fault plane with the front face of the model at 6 in Fig. 6. Because of the break in continuity of the slab at this point, stress differences are set up within the limestone as the cusp of the incident and reflected wave fronts pass over the fault trace. This fault trace then acts like a virtual line source for a diffracted wave whose travel time has been calculated and plotted on Fig. 6. However, the event is probably very weak and cannot be identified with certainty on the oscillograms. This event, however, shows clearly on Figs. 7 and 9 under different circumstances.

A second reflection profile was made next and is

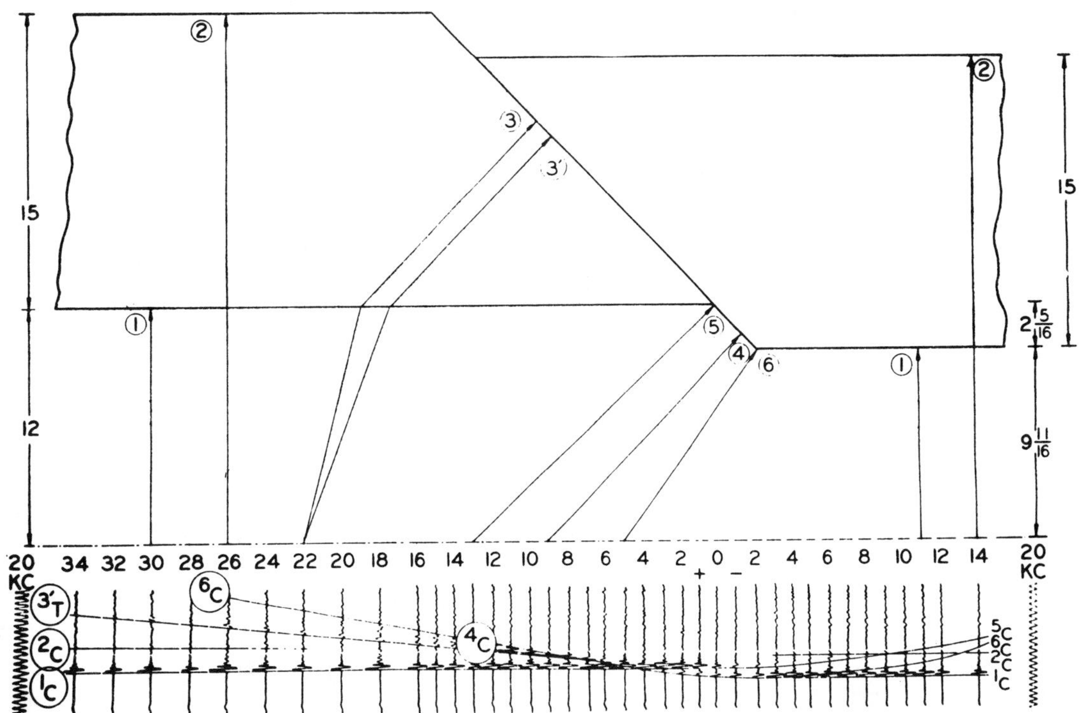


FIG. 7. Displaced front fault reflection profile.

shown in Fig. 7, where the two blocks of the limestone model have been displaced along the fault plane. The right block front face is parallel to the left block face, but closer to the traverse line by $2\frac{5}{16}$ in. Again several prominent events on the oscillograms are joined by curved lines which are identified as follows:

*Event* $1_C$—Compressional wave reflected from the front faces of the limestone blocks.

*Event* $2_C$—Wave reflected as a rarefactional pulse from the rear face of the limestone blocks. Note the low amplitude due to the high reflection coefficient of the front limestone-water interface.

*Event* $3'_T$—Wave reflected from the fault plane and traveling as a shear wave in the limestone, as described in Fig. 6.

*Event* $4_C$—Wave reflected from the portion of the fault plane exposed to water.

*Event* $5_C$—A diffracted wave having a "virtual" line source at edge 5. This event can be recognized particularly at stations $-1$ and $-2$.

*Event* $6_C$—A diffracted wave having a "virtual" line source at edge 6. This event can be clearly recognized from stations $+4$ to $-2$.

A refraction profile was made next with the two halves of the limestone model separated, but with the front faces of the model halves in the same plane, as shown in Fig. 8. This does not correspond to any geological situation, but is interesting from the standpoint of wave propagation phenomena.

This profile was made by moving the hydrophone along a traverse line parallel to the model face while keeping the source fixed at one end of the traverse line. Prominent events were again joined by curved lines which are identified as follows:

*Event* $1_C$—A wave traveling as a compressional wave from the source to the limestone surface, along the limestone slab as a compressional wave at 13 700 ft/sec, and reradiating into the water and returning as a compressional wave to the receiver. This wave is very weak, however and can hardly be recognized beyond station 24.

*Event* $1_R$—A wave having traveled partly along the surface of the limestone slab as a Rayleigh and/or shear wave. Because of the relatively small difference in velocity of these two kinds of waves, it is difficult to distinguish between the two. The amplitude of this wave is greater than that of wave $1_C$, which traveled as a compressional wave in the limestone.

*Event 2*—A compressional wave having traveled by direct path through the water.

*Event 3*—A wave having been reflected as a rarefactional wave from the surface of the water.

A spaced model reflection profile was next made as shown in Fig. 9. Again this configuration has no geological counterpart, but wave diffraction phenomena are well illustrated. The front faces of the two halves of the model are in the same plane. The spark source and hydrophone were in fixed relation, one above the other, as in Figs. 6 and 7. Of particular interest are the diffracted waves from edges 6 and $8_C$ as virtual line sources. The principal events of interest are joined by lines on the oscillograms and are identified as follows:

*Event* $1_C$—A wave reflected from the front face of the model.

*Event* $3_C$—A wave traveling as a compressional wave in the limestone and reflected from the fault surface.

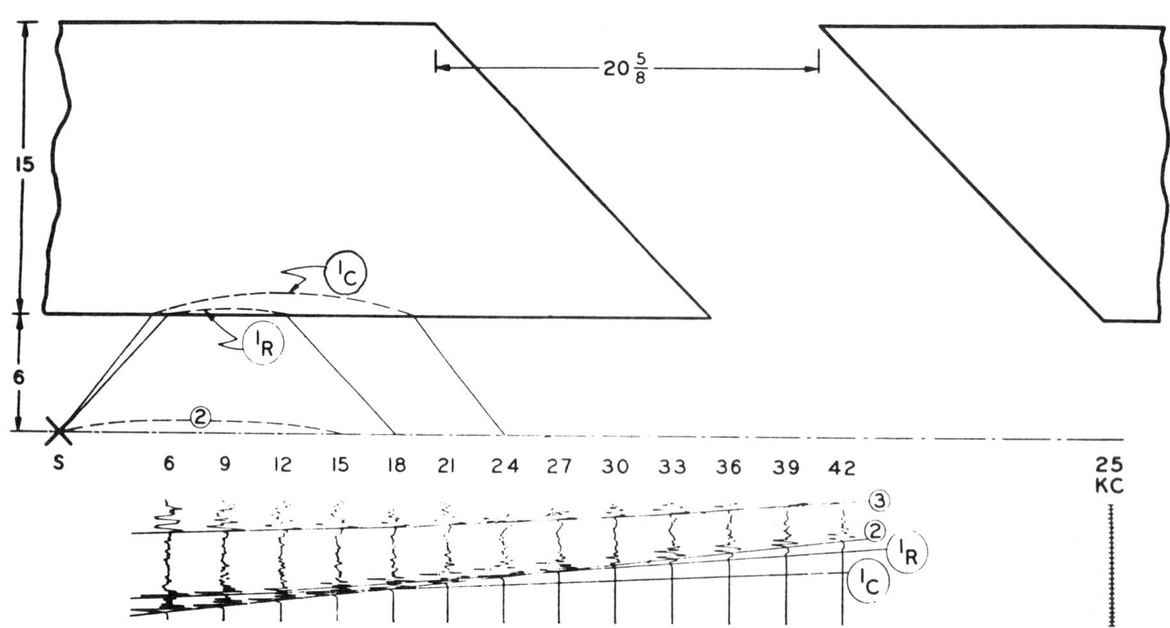


Fig. 8. Spaced model refraction profile.

This event is very weak and hardly discernible.

*Event 3'$_T$*—A wave traveling as a shear wave in the limestone and reflected from the fault surface, as described in Fig. 6.

*Event 6$_C$*—A wave resulting from diffraction phenomena at edge 6 of the limestone slab. As the cusp formed by the initial incident wave front and its reflected portion moves down the front surface of the slab and passes edge 6, a diffracted wave is radiated with edge 6 acting as a virtual line source. This event can be clearly seen from stations −1 to −8. It is hardly discernible, if at all, at stations to the left of station 0.

*Event 7$_C$*—A compressional wave reflected from fault surface 7$_C$ of the right-hand limestone block.

*Event 8$_C$*—An event similar to 6$_C$, but originating from diffraction phenomena at edge 8$_C$.

The preceding plates have illustrated the transmission, reflection, and diffraction of an acoustic wave pulse in a simple water-limestone model. It is not intended to imply that this model simulates many of the complex conditions existing in the actual earth. However, there is enough similarity to make the results of interest in analyzing seismograms recorded in seismic prospecting. Although the initial wave radiated from the spark gap appears at a distance of 6 in. to be a unidirectional compressional pulse, the wave reflected from a water-limestone interface at a distance of 12 in. appears to have the more complicated shape shown in Fig. 10. The reasons for this change in pulse form have not been fully investigated, and are not fully understood at this time. It is interesting to note that this reflected pulse is generally similar in form to seismic pulses in the earth, as illustrated in a recent published article by Norman Ricker.[2]

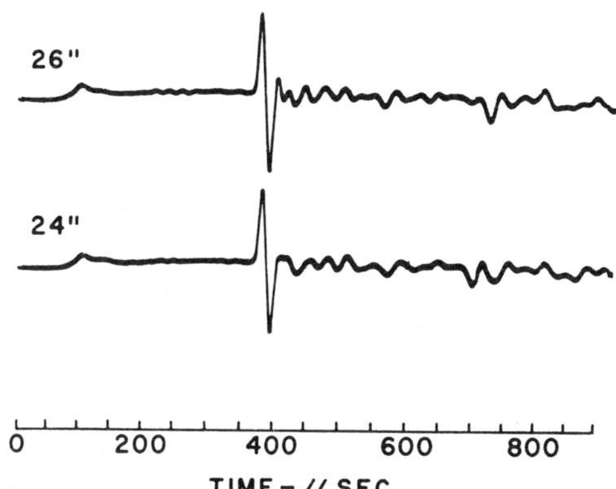


FIG. 10. These oscillograms are from Fig. 6.

The results obtained to date from use of the model study technique indicate a reliable method is available for study in detail of many seismic field problems involving wave propagation. Numerous phenomena can be clearly demonstrated, which will enlarge concepts of seismic record interpretation. Recording the diffracted waves at considerable distances either side of their virtual sources is significant and also pertinent to field interpretation. Work is continuing with this tool in the hope of bettering the understanding of the complex problems encountered in seismic record interpretation.

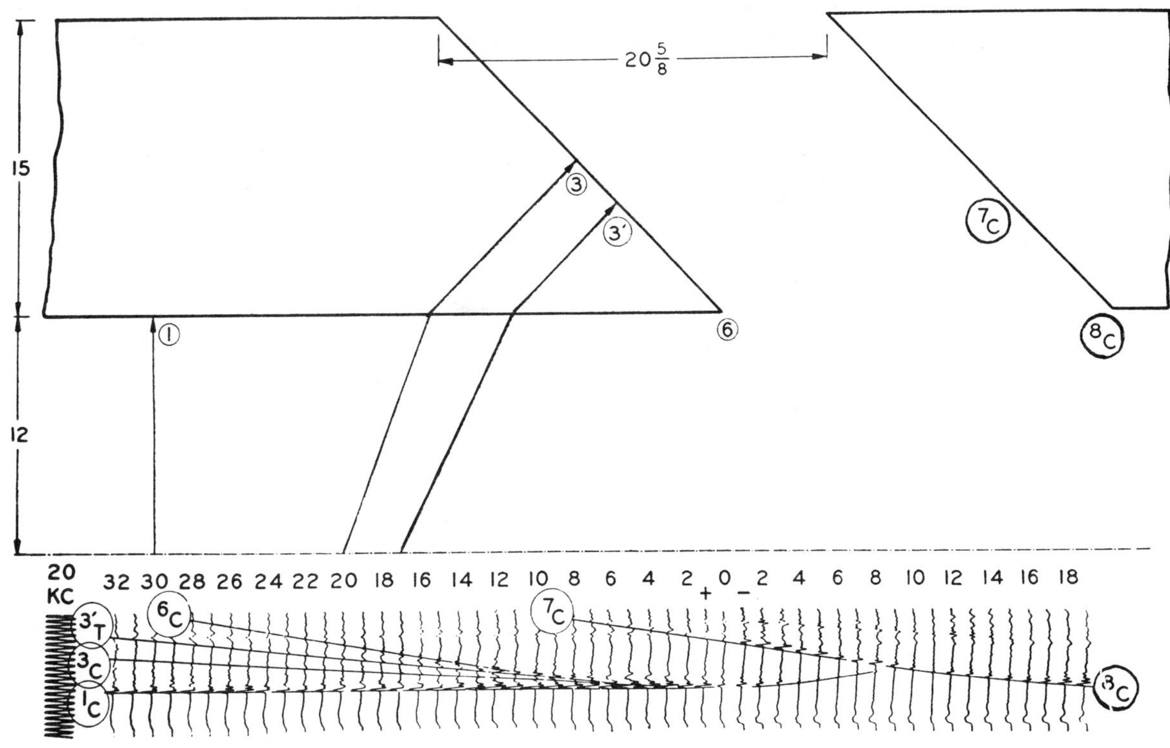


FIG. 9. Spaced model reflection profile.

[2] Norman Ricker, Geophysics **XVIII**, 10 (1953).

# TWO-DIMENSIONAL MODEL SEISMOLOGY*

JACK OLIVER,† FRANK PRESS,† AND MAURICE EWING†

ABSTRACT

The solutions of many problems in seismology may be obtained by means of ultrasonic pulses propagating in small scale models. Thin sheets, serving as two dimensional models, are particularly advantageous because of their low cost, availability, ease of fabrication into various configurations, lower energy requirements, and appropriate dilatational-to-shear-velocity ratios. Four examples are presented: flexural waves in a sheet, Rayleigh waves in a low velocity layer overlying a semi-infinite high velocity layer, Rayleigh waves in a high velocity layer overlying a semi-infinite low velocity layer, and body and surface waves in a disk.

## INTRODUCTION

In recent years investigators in both the commercial and academic fields have attacked the problems of elastic wave propagation by studying waves of ultrasonic frequencies traveling through small scale three-dimensional models, and a few papers have been published on the subject (Kaufman and Roever, 1951; Northwood and Anderson, 1953; Terada and Tsuboi, 1927). Although promising results were obtained, widespread application of the method to significant problems has been slow, chiefly because of the difficulties in procuring suitable model materials and in the fabrication of desirable configurations. This paper describes the use of two-dimensional models, which virtually eliminates these and other difficulties without detracting from the usefulness of the method in a great majority of cases.

The models take the form of thin sheets and the propagation takes place along directions lying in the plane of the sheet. In practice these sheets are generally 1/16 inch thick, and only wavelengths long compared to this thickness are employed.

## THEORY

We wish to determine the velocities of propagation of dilatational and shear waves in a thin plate and of Rayleigh waves on the edge of a thin plate. We restrict ourselves to waves in which the particle motion is symmetrical about the median plane of the plate, thus excluding any type of wave motion which involves bending of the thin plate.

Love's concept of a "generalized plane stress" (Love, 1944) is applicable and most convenient. The $x$ and $y$ directions are taken in the plane of the plate, the $z$ direction perpendicular to it. $X_y$ is the stress in the $x$ direction on the face normal to the $y$ axis, etc. Love's concept supposes the stress $Z_z$ to vanish throughout the plate and the tangential stress $Z_x$ and $Y_z$ to vanish at $z=\pm h/2$ only,

* Manuscript received by the Editor December 7, 1953.

† Lamont Geological Observatory, Columbia University, Palisades, N. Y.

Reprinted from Geophysics, 19, 202-219

where $h$ is the plate thickness. For a thin plate, Love takes average values of the components of displacement, strain and stress, e.g.

$$\bar{u} = \frac{1}{h}\int_{-h/2}^{h/2} u\,dz.$$

$u$ and $v$ are the displacements in the $x$ and $y$ directions respectively.

In the following we use only average values for these quantities and the bar is omitted.

The equations of motion are

$$\frac{\partial X_x}{\partial x} + \frac{\partial X_y}{\partial y} = \rho \frac{\partial^2 u}{\partial t^2}, \qquad \frac{\partial Y_x}{\partial x} + \frac{\partial Y_y}{\partial y} = \rho \frac{\partial^2 v}{\partial t^2} \tag{1}$$

where $\rho$ is the density.

The average stress components are given by Love as

$$X_x = \frac{2\lambda\mu}{\lambda + 2\mu}\left(\frac{\partial u}{\partial x} + \frac{\partial v}{\partial y}\right) + 2\mu \frac{\partial u}{\partial x} \tag{2a}$$

$$Y_y = \frac{2\lambda\mu}{\lambda + 2\mu}\left(\frac{\partial u}{\partial x} + \frac{\partial v}{\partial y}\right) + 2\mu \frac{\partial v}{\partial y} \tag{2b}$$

$$X_y = \mu\left(\frac{\partial v}{\partial x} + \frac{\partial u}{\partial y}\right). \tag{2c}$$

Substituting (2) into (1),

$$\rho \frac{\partial^2 u}{\partial t^2} = \left(\frac{2\mu\lambda}{\lambda + 2\mu} + \mu\right)\frac{\partial \tau}{\partial x} + \mu D^2 u \tag{3}$$

and similarly in $v$, where

$$\tau = \frac{\partial u}{\partial x} + \frac{\partial v}{\partial y} \quad \text{and} \quad D^2 = \frac{\partial^2}{\partial x^2} + \frac{\partial^2}{\partial y^2}.$$

Differentiate equation (3) with respect to $x$ and the similar equation in $v$ with respect to $y$ and add the results to obtain

$$\rho \frac{\partial^2 \tau}{\partial t^2} = \frac{4\mu(\lambda + \mu)}{\lambda + 2\mu} D^2 \tau \tag{4}$$

Equation (4) checks the velocity given by Lamb for dilatational waves in a thin plate, i.e.

$$V_p = \left[\frac{4\mu(\lambda + \mu)}{\rho(\lambda + 2\mu)}\right]^{1/2}. \tag{5}$$

Take the curl (two dimensional) of equation (3) to obtain

$$\rho \frac{\partial^2}{\partial t^2}\left(\frac{\partial u}{\partial y} - \frac{\partial v}{\partial x}\right) = \mu D^2\left(\frac{\partial u}{\partial y} - \frac{\partial v}{\partial x}\right). \tag{6}$$

Thus the shear wave velocity is the same as in an infinite solid, i.e.

$$\beta = \left(\frac{\mu}{\rho}\right)^{1/2}.$$

We now consider a thin plate which occupies the $xy$ plane from $y=0$ to $y=+\infty$. Let

$$u = \frac{\partial \phi}{\partial x} + \frac{\partial \psi}{\partial y}, \qquad v = \frac{\partial \phi}{\partial y} - \frac{\partial \psi}{\partial x}; \tag{7}$$

then

$$D^2\phi = \tau, \qquad D^2\psi = \frac{\partial u}{\partial y} - \frac{\partial v}{\partial x}$$

where $\phi$ and $\psi$ satisfy

$$\frac{\partial^2\phi}{\partial t^2} = V_p^2 D^2\phi, \qquad \frac{\partial^2\psi}{\partial t^2} = \beta^2 D^2\psi. \tag{8}$$

Choose plane wave solutions of the form

$$\begin{aligned} \phi &= Ae^{ik(-ry+x-ct)} \\ \psi &= Be^{ik(-sy+x-ct)} \end{aligned} \tag{9a}$$

where

$$r = \left(\frac{c^2}{V_p^2} - 1\right)^{1/2}, \qquad s = \left(\frac{c^2}{\beta^2} - 1\right)^{1/2}. \tag{9b}$$

$c$ is the velocity of propagation of constant phase in the $x$ direction and $k$ the wave number. Thus the wavelength in the $x$ direction, $l$, is given by $l=2\pi/k$ and the period $T$ by $T=2\pi/kc$.

At the free edge $y=0$,

$$X_y = Y_y = 0. \tag{10}$$

Using equation (2), (7), (9), and (10) we get

$$[V_p^2(1 + r^2) - 2\beta^2]A + 2\beta^2 sB = 0 \tag{11}$$

$$2rA + (1 - s^2)B = 0. \tag{12}$$

Eliminating $A$ and $B$ from (11) and (12) and using (9b)

$$\left(2 - \frac{c^2}{\beta^2}\right)^2 = 4\left(1 - \frac{c^2}{V_p^2}\right)^{1/2}\left(1 - \frac{c^2}{\beta^2}\right)^{1/2} \tag{13}$$

This is identical with the equation for Rayleigh waves on the surface of a semi-infinite solid (Bullen, 1947) with the exception that the plate dilatational velocity replaces the infinite solid dilatational velocity. It is clear that no difficulties will arise in more complicated cases. Thus most of the problems on propagation of plane waves in a stratified earth may be related to the two dimensional models simply by replacing $\alpha$ by $V_p$.

## EQUIPMENT AND TECHNIQUES

The principal features of the apparatus and procedure are described here briefly. The transmitting and receiving equipment is adapted from similar equipment in use in several laboratories, but the use of thin plates for construction of the models is apparently novel.

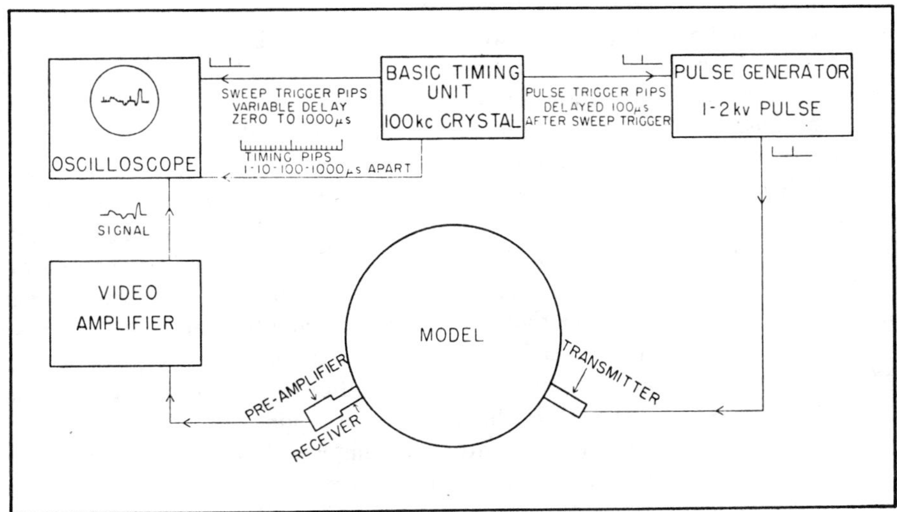


FIG. 1. Block diagram of apparatus.

A block diagram of the equipment is shown in Figure 1. Electrical pulses of approximately 15 microsecond duration are applied to the transmitter, a small piezoelectric element in contact with the model. This transducer delivers an acoustical pulse, shown as traced from the seismogram in Figure 2, to the model. The resulting acoustical energy arriving at various points on the model is detected by a similar transducer, amplified, and then displayed on the cathode ray oscilloscope. The pulse repetition rate is adjustable from 2 to 100 pulses per second. The oscilloscope sweep is triggered at a fixed time interval ahead of the pulse so that a standing wave pattern, including the time break, is observed. The basic

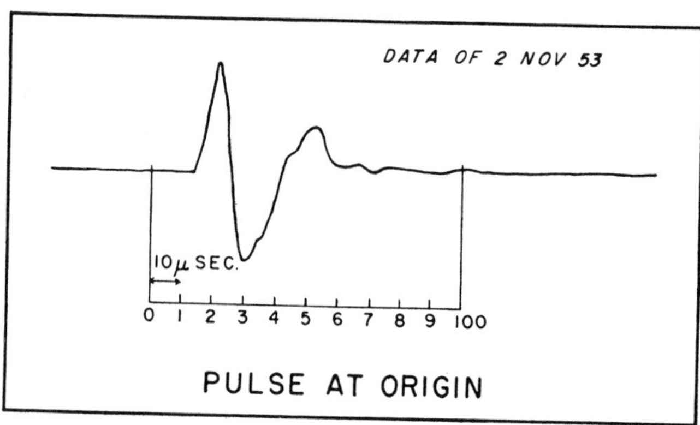


FIG. 2. Pulse at origin.

timing unit which initiates the trigger pulses for both the oscilloscope and the pulse also supplies timing markers at convenient intervals. The resulting pattern on the oscilloscope is then photographed by a Polaroid Land camera and the miniature seismogram is obtained immediately.

The transmitting transducer is a barium titanate cylinder 1/10 inch thick and $\frac{1}{4}$ inch in diameter. To prevent internal reflections in the source mount, the transducer is backed by a long aluminum rod of the same diameter. The part of the energy that enters the backing and then reflects from the end of the rod back toward the source is too late and too greatly attenuated to confuse the seismogram. Difficulties due to free oscillations or "ringing" of the source are avoided by utilizing a frequency range that is below the ringing frequency of the transducer. The receiving transducer is also of barium titanate 1/25 inch thick and 3/32 inch in diameter and is mounted in a similar but less elaborate manner. The smaller thickness further improves the reduction of ringing.

The electronic equipment is currently being improved and will be thoroughly described in a later report. The present unit delivers a pulse of about 1,000–2,000 volts to the source crystal. The voltage amplification of the receiving unit is less than 100,000.

The two-dimensional models are usually built in the form of disks. These are particularly advantageous for the study of surface waves on the edge of the disk since there are no reflections of the surface waves and since a very long path may be obtained by using the multiple trips around a relatively small disk. The curvature of the disk may be made small compared to a wavelength so that the usual surface wave equations for flat-lying strata will be applicable with the slight modification of the dilatational wave velocity, or it may be made larger for studying the effect of sphericity of the earth upon long period surface waves. The body phases of earthquake seismology may be studied in the usual manner very conveniently. Any degree of complexity in the layering may be obtained by means of concentric rings of various materials glued together in the desired rela-

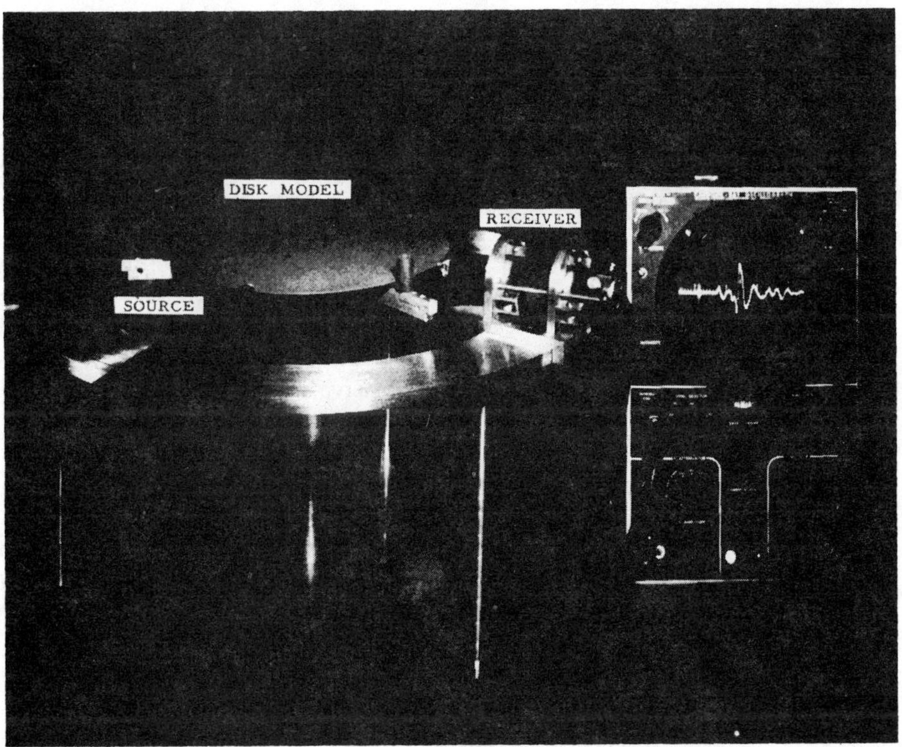


FIG. 3. Model, transducers, and oscilloscope.

tion. The diameters of the disks used in this study are about 20 inches. The thickness is at present standardized at 1/16 inch. This is small compared to all the wavelengths involved so that there is no dispersion due to finite disk thickness. For some types of problems shapes other than the circular might be desirable; e.g., refraction and reflection problems of seismic prospecting may be studied by constructing a cross section of the configuration in question. Table 1 lists the materials that have been sampled to date with the compressional, shear and Rayleigh velocities.

The disks are held in position by three rubber-covered supports which touch them at three points on the periphery (Figure 3). Experiments show that for a disk of this thickness the supports do not affect the surface wave propagation appreciably. The disk is mounted concentric with a graduated ring on which the receiver travels, allowing accurate measurement of all arc distances. The transmitting transducer is set at an arbitrary zero which may be determined accurately by running profiles from the transducer in both directions. A plot of the dilatational wave travel times vs. distance immediately gives the zero of the transmitted pulse in both time and distance.

## COMPARISON OF TWO- AND THREE-DIMENSIONAL MODELS

Two-dimensional models (2-D) have the following advantages over three-dimensional (3-D) models.

1. The materials are far less expensive. 3-D models are usually constructed of some material which can be poured, e.g., wax, concrete, tars, or various chemical setting cements, or of some material which can be purchased in large slabs, e.g., limestone, marble, etc. All of these are expensive and become increasingly difficult to handle when layered media are desired. It is particularly difficult to pour materials in large blocks and maintain homogeneity throughout. The 2-D models, on the other hand, may be constructed of any material which can be bought in 1/16 inch sheets. This includes most metals and alloys and many types of plastics, fibers, etc. They are not prohibitively expensive and are generally far more homogeneous than poured materials.

2. The fabrication is greatly simplified. Layered configurations may be built up from concentric rings, easily cut to accurate specifications on a large lathe. Almost any sort of hard glue may be used as a binding agent.

3. Storage is simplified, a drawer of a map case may be used for storage of a large number of 2-D models. 3-D models frequently require about 50 sq ft of floor space for each model.

4. The energy requirements are greatly diminished in the 2-D model. For body waves the intensity varies as $1/R$, as opposed to $1/R^2$ in the 3-D case. Surface waves undergo no geometrical attenuation, instead of the $1/R$ variation in the 3-D case. This is a tremendous advantage as it allows the use of pulses of 1,000 volts or less as opposed to 5,000–10,000 volts in the 3-D case. Similarly lower gains may be used in the amplifying system, which permits use of a simple receiving circuit.

5. Most materials commercially available do not make good models of rock because Poisson's ratio is too large, usually about 0.33 as opposed to about 0.25 for rocks. In the 2-D model the dilatational wave velocity is altered (see theory) so that a pseudo-Poisson's ratio (the same relation of the dilatational and shear velocities as for Poisson's ratio) now falls in the range appropriate to rocks (see Table 1).

6. Multiple trips around the disk by the surface waves may be used to give large path distances. The present paper includes waves that have made $2\frac{1}{2}$ trips (about 13 feet) but waves have been observed that have made $7\frac{1}{2}$ complete circuits (about 40 feet). Although a similar scheme might be followed three-dimensionally by using a sphere, the construction of spheres, particularly with concentric layers, is very difficult.

The multiple paths also provide a means of measuring attenuation that is free of instrument calibration and differences in coupling of the model to the transducer. This approach is identical to that used in the measurement of the absorption coefficient of the mantle by means of long Rayleigh waves (Ewing and Press, in press).

**Table I**

**Measured Elastic Wave Velocities and Densities**

| Material | Designation | Sample | Plate Dilatation Velocity $V_p$ in ft/sec | Shear Velocity, $\beta$ in ft/sec | Rayleigh Velocity, $V_r$ in ft/sec | Density $\rho$ in gm/cc | Pseudo-Poisson's Ratio: $\frac{\frac{1}{2}-(\beta/V_p)^2}{1-(\beta/V_p)^2}$ | Remarks |
|---|---|---|---|---|---|---|---|---|
| Aluminum Alloy | 24S-T4 | Disk $\frac{1}{16}$"×10.56" | 18,300 | 10,400 | 9,800 | 2.77 | .26 | |
| Plexiglass | Colorless, Transparent | Disk $\frac{1}{16}$"×10.75" | 7,750 | 4,500 | 4,200 | 1.22 | .25 | |
| Panelyte | Grade 550 | Disk $\frac{1}{16}$"×19.522" | 10,650 | 6,000 | 5,600 | 1.44 | .27 | Anisotropic* 2% |
| Copper | Cold rolled | $\frac{1}{16}$"×10.75" | 13,000 | 7,500 | 6,900 | 9.03 | .25 | |
| Steel | Cold rolled | $\frac{1}{16}$"×10.75" | 17,850 | 10,650 | 9,700 | 7.78 | .22 | |
| Steel | Hot rolled | $\frac{1}{16}$"×10.75" | 17,600 | 10,300 | 9,700 | 7.80 | .24 | |
| Brass | Half hard | $\frac{1}{16}$"×10.75" | 12,500 | 7,000 | 6,500 | 8.34 | .28 | |
| Zinc | | .052"×10.75" | 13,100 | 7,700 | 7,300 | 7.07 | .24 | |
| Vinyl Plastic | Phonograph record | $\frac{1}{16}$"×11.875" | 6,420 | 3,870 | 3,540 | 1.47 | .21 | |
| Fish paper | Franklin Fibre Lamitex | $\frac{1}{16}$"×10.875" | 8,970 | 5,490 | 5,110 | 1.14 | .22 | Anisotropic* 22% |
| Fibre | Franklin Hard Vulcanized | $\frac{1}{16}$"×10.875" | 8,650 | 5,270 | 4,900 | 1.06 | .21 | Anisotropic* 15% |
| Formica | Linen Laminate | $\frac{1}{16}$"×10.75" | 8,450 | 5,050 | 4,700 | 1.30 | .24 | Anisotropic* 7% |

* For anisotropic materials slowest velocities are tabulated. These are generally perpendicular to the grain. For Panelyte the tabulated velocity is an average.

At present 3-D models are necessary only when one or more layers of a stratified configuration is a liquid or when it is desirable to model three-dimensional effects such as reflections not in the vertical plane of the shot and detector, which might be encountered in seismic prospecting.

## EXAMPLES

*Flexural Waves*

The experimental model for studying the propagation of flexural waves in a plate, as worked out by Lamb (1917), was a ring cut from 24 ST Aluminum sheet, 1/16 inch thick. The outside diameter $2R_2$ of the ring was 19.112 inches and the inside diameter $2R_1$, $\frac{1}{2}$ inch less. Radial motions of the ring were investigated as the 2-D analog of ordinary flexural waves. Figure 4 is a tracing of the oscillogram obtained with the receiver at the antipodes of the source. At this point there is constructive interference between the waves arriving from opposite directions. The first arrival following the time break corresponds to a dilatational wave through the ring. The first dispersive train of flexural waves corresponds to $R_1$ and $R_2$, the second to $R_3$ and $R_4$, etc., where the $R$'s with subscripts are used in the same sense as in earthquake seismology. The values obtained from the experimental data are plotted as the small circles of Figure 5.

The theoretical equations for wave motion in an elastic plate in a vacuum were presented by Lamb (1917). The dispersion of flexural waves is determined by his period equation for asymmetrical motion. In our notation this equation takes the form

$$\frac{\tanh \frac{kh}{2}\left(1 - \frac{c^2}{\beta^2}\right)^{1/2}}{\tanh \frac{kh}{2}\left(1 - \frac{c^2}{\alpha^2}\right)^{1/2}} = \frac{\left(2 - \frac{c^2}{\beta^2}\right)^2}{4\left(1 - \frac{c^2}{\alpha^2}\right)^{1/2}\left(1 - \frac{c^2}{\beta^2}\right)^{1/2}} \tag{14}$$

where $h$ is the layer thickness.

To adapt this to our 2-$D$ models we need only replace $\alpha$ by $V_p$ and $h$ by $R_2 - R_1 \equiv H$. Group velocity is obtained from the well known formula

$$U = c + \kappa H \frac{\partial c}{\partial(kH)} \tag{15}$$

A family of theoretical curves of group velocity vs. period computed with the constants for 24 ST Aluminum alloy is shown in Figure 5. Thickness is the parameter and very close agreement with the observed points is evident. Thus we have a good check on both the theory and the experimental method.

Slight complications have arisen due to the shape of the source pulse, in particular the later origin times of the longer periods. This difficulty may be completely circumvented by measuring arrival times of each period on successive

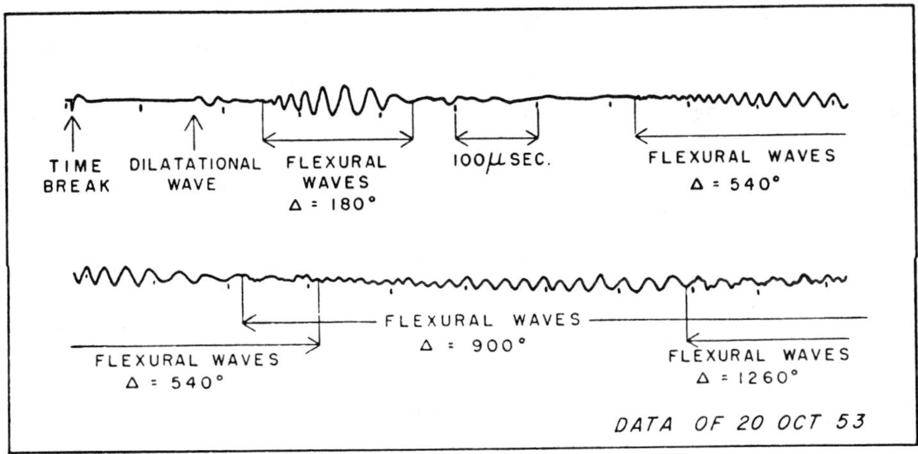


FIG. 4. Seismogram—flexural waves.

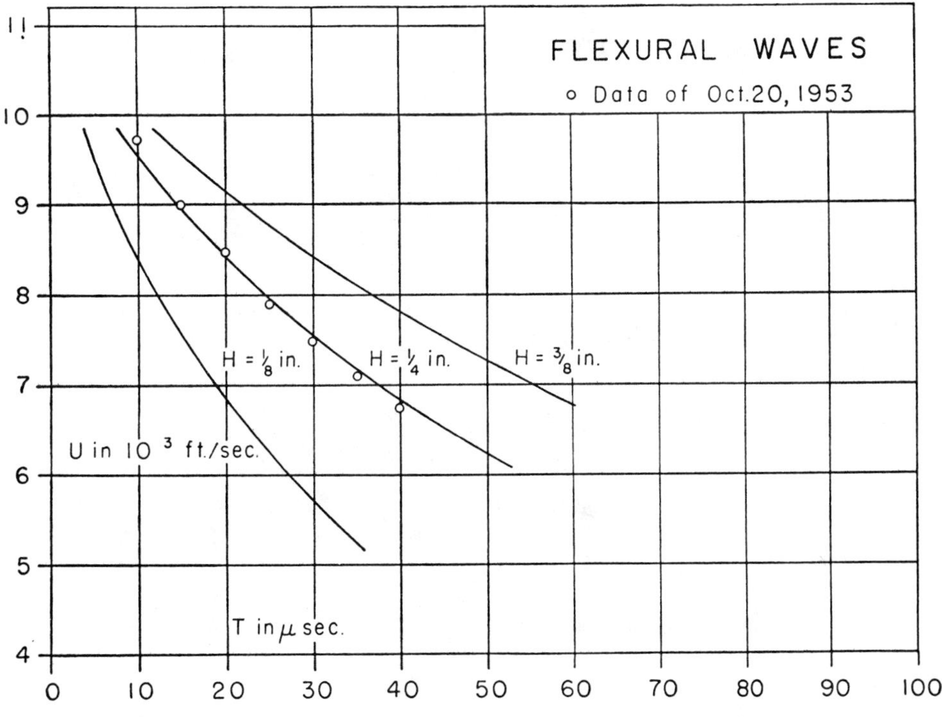


FIG. 5. Flexural wave dispersion.

trips around the ring and calculating the group velocity over the interval. The points of Figure 5 were obtained by this method. If the initial rise of the source is used as a time break, a maximum error of 2% in the velocity might result. This effect is interesting and will be the subject of future study but it does not affect the present results.

*Rayleigh Waves—Low Velocity over High Velocity*

The most common configuration in nature which produces dispersion in Rayleigh waves is comprised of a low velocity layer of thickness $h$ over a high velocity semi-infinite layer. Lee (1932) has presented the theoretical equations for this case. His period equation is

$$\xi\eta' = \xi'\eta \tag{16}$$

where

$$\xi = \left(2 - \frac{c^2}{\beta^2}\right)\left[\left(\frac{c^2}{\beta'^2}\frac{\mu'}{\mu} - \phi\right)\cos kh\eta_1 + \frac{\eta_1'}{\eta_1}\left(\frac{c^2}{\beta^2} + \phi\right)\sin kh\eta_1\right)$$

$$+ 2\eta_2\left[\eta_1'\phi \sin kh\eta_2 - \frac{1}{\eta_2}\left(\frac{c^2}{\beta'^2}\frac{\mu'}{\mu} - \phi - \frac{c^2}{\beta^2}\right)\cos kh\eta_2\right]$$

$$\xi' = \left(2 - \frac{c^2}{\beta^2}\right)\left[\eta_2'\phi \cos kh\eta_1 + \frac{1}{\eta_1}\left(\frac{c^2}{\beta'^2}\frac{\mu'}{\mu} - \phi - \frac{c^2}{\beta^2}\right)\sin kh\eta_1\right]$$

$$+ 2\eta_2\left[\left(\frac{c^2}{\beta'^2}\frac{\mu'}{\mu} - \phi\right)\sin kh\eta_2 - \frac{\eta_2'}{\eta_2}\left(\frac{c^2}{\beta^2} + \phi\right)\cos kh\eta_2\right]$$

$$\eta = \left(2 - \frac{c^2}{\beta^2}\right)\left[\eta_1'\phi \cos kh\eta_2 - \frac{1}{\eta_2}\left(\frac{c^2}{\beta'^2}\frac{\mu'}{\mu} - \phi - \frac{c^2}{\beta^2}\right)\sin kh\eta_2\right]$$

$$+ 2\eta_1\left[\left(\frac{c^2}{\beta'^2}\frac{\mu'}{\mu} - \phi\right)\sin kh\eta_1 - \frac{\eta_1'}{\eta_1}\left(\frac{c^2}{\beta^2} + \phi\right)\cos kh\eta_1\right]$$

$$\eta' = \left(2 - \frac{c^2}{\beta^2}\right)\left[\left(\frac{c^2}{\beta'^2}\frac{\mu'}{\mu} - \phi\right)\cos kh\eta_2 + \frac{\eta_2'}{\eta_2}\left(\frac{c^2}{\beta^2} + \phi\right)\sin kh\eta_2\right]$$

$$+ 2\eta_1\left[\eta_2'\phi \sin kh\eta_1 - \frac{1}{\eta_1}\left(\frac{c^2}{\beta'^2}\frac{\mu'}{\mu} - \phi - \frac{c^2}{\beta^2}\right)\cos kh\eta_1\right],$$

and where

$$\phi = 2\left(\frac{\mu'}{\mu} - 1\right)$$

and

$$\eta_1 = \left(\frac{c^2}{\alpha^2} - 1\right)^{1/2}, \qquad \eta_2 = \left(\frac{c^2}{\beta^2} - 1\right)^{1/2}$$

$$\eta_1' = \left(1 - \frac{c^2}{\alpha'^2}\right)^{1/2} \qquad \eta_2' = \left(1 - \frac{c^2}{\beta'}\right)^{1/2}$$

and the other symbols have the same meaning as before. Again we replace the $\alpha$ by $V_p$ and the theoretical curves are shown in Figure 6. The computed points are tabulated in Table II. The constants were chosen to fit the case of Plexiglass overlying Panelyte. (See Table I.)

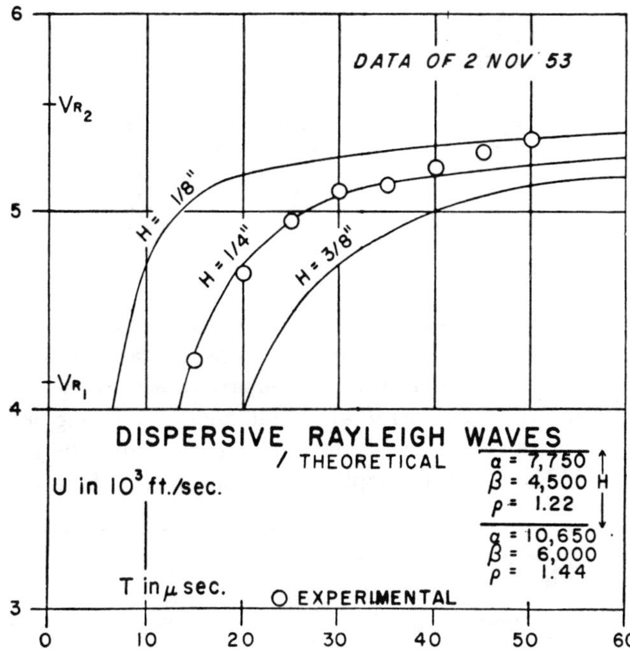


FIG. 6. Rayleigh wave dispersion—low velocity layer/high velocity substratum.

The disk was 19.522 inches in diameter with outside diameter of the outer ring $\frac{1}{2}$ inch greater. Figure 7 is a record taken at $\Delta = 180°$. Because of the small contrast between these two materials and the high attenuation in the plastic, more precision was obtained by using seismograms at several different $\Delta$'s rather than the one from the antipodes alone. As before, this eliminates the difficulty with the origin time of various periods in the pulse by making the velocity determination independent of pulse time. Figure 6 shows the experimental results in good agreement with the theory.

TABLE II

PHASE AND GROUP VELOCITY VS $kH$ FOR A LAYER WITH $\alpha=7,750$ FT/SEC, $\beta=4,500$ FT/SEC, $\rho=1.219$ GM/CC OVERLYING A SEMI-INFINITE LAYER WITH $\alpha=10,650$ FT/SEC, $\beta=6,000$ FT/SEC, $\rho=1.436$

| Computed values | | Values obtained graphically | | |
|---|---|---|---|---|
| $c$ in ft/sec | $kH$ | $kH$ | $U$ in ft/sec | $c$ in ft/sec |
| 5,535 | 0 | 0 | 5,535 | 5,535 |
| 5,500 | .075 | .05 | 5,488 | 5,510 |
| 5,450 | .203 | .1 | 5,449 | 5,490 |
| 5,400 | .358 | .2 | 5,379 | 5,453 |
| 5,530 | .550 | .3 | 5,323 | 5,418 |
| 5,300 | .765 | .4 | 5,276 | 5,388 |
| 5,200 | 1.152 | .5 | 5,233 | 5,362 |
| 5,100 | 1.450 | .75 | 5,123 | 5,302 |
| 5,000 | 1.722 | 1,00 | 4,987 | 5,241 |
| 4,900 | 1.955 | 1.25 | 4,766 | 5,168 |
| 4,800 | 2.196 | 1.50 | 4,543 | 5,083 |
| 4,700 | 2.425 | 1.75 | 4,283 | 4,977 |
| 4,600 | 2.689 | 2.00 | 4,020 | 4,838 |
| 4,400 | 3.392 | 2.25 | 3,842 | 4,973 |
| 4,300 | 3.965 | 2.50 | 3,685 | 4,672 |
| 4,200 | 5.107 | 2.90 | 3,623 | 4,528 |
| | | 3.0 | 3,626 | 4,498 |
| | | 4.0 | 3,770 | 4,296 |
| | | 5.0 | 3,886 | 4,207 |

The Panelyte is anisotropic, having a velocity about $1\frac{1}{2}\%$ higher in one direction than in the perpendicular direction. This was not taken into account in the calculations and is a source of error in the comparison with the theoretical curves. The points for the two longest periods both fall high. These long periods can only be distinguished in the dispersive train after a considerable length of path has been traversed. At that time the amplitudes are low and, in fact, the relative am-

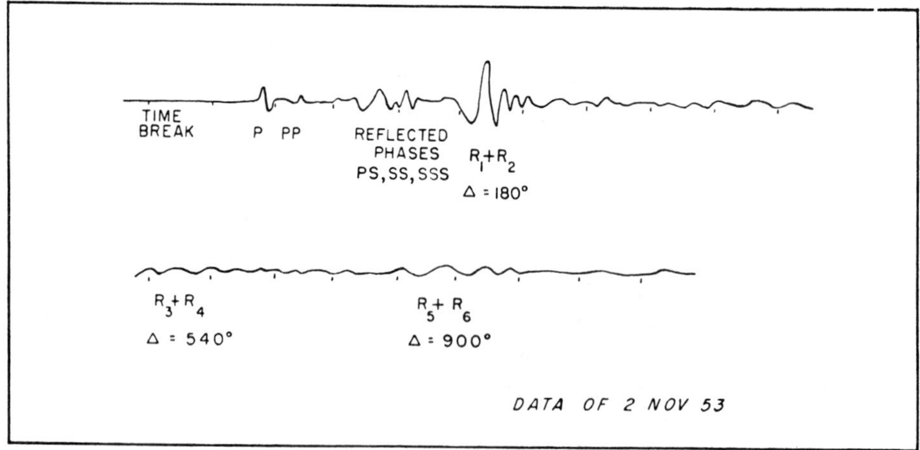


FIG. 7. Seismogram—low velocity layer/high velocity substratum.

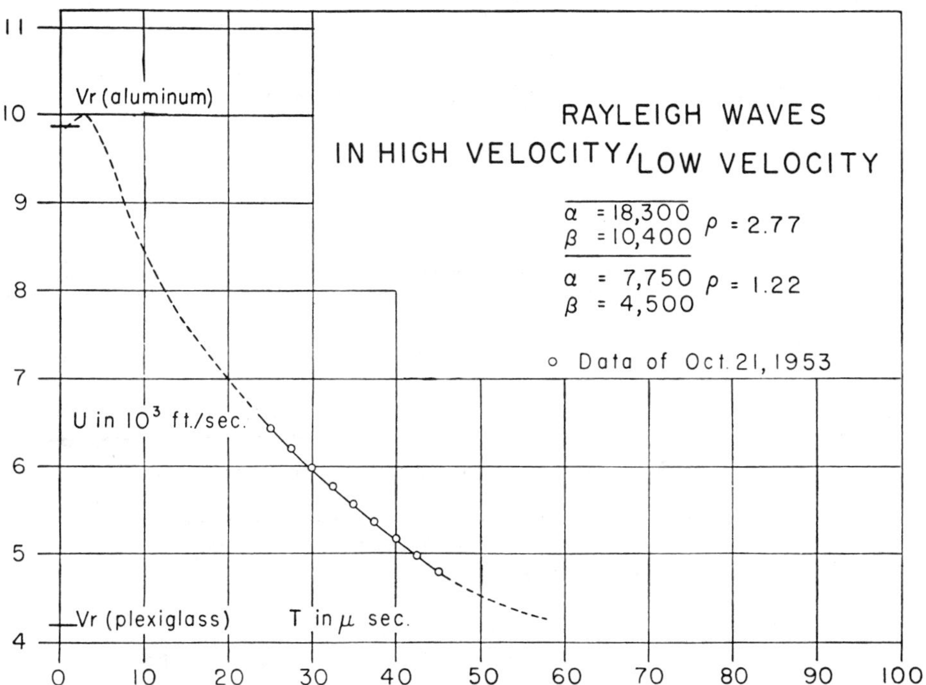


FIG. 8. Rayleigh wave dispersion—high velocity layer/low velocity substratum.

plitudes of such periods are always low due to the pulse shape. Thus the accuracy of velocity measurement is reduced for these waves. Furthermore, it is possible that these waves may be long enough to be affected by the curvature of the disk. The discrepancy is in the proper direction, but this effect has not yet been studied. At most, however, the error is 2%.

Experiments in higher mode Rayleigh waves (shear modes) and on Stoneley waves are planned.

*Rayleigh Waves—High Velocity over Low Velocity*

The previous examples have shown a comparison of experimental data with theoretical curves. In both cases the theoretical calculations were relatively simple and within the range of an ordinary desk calculator. In the case of a high velocity layer over a semi-infinite low velocity stratum the computations increase greatly in complexity. On the other hand, there is no difficulty in modelling this case. Figure 8 is a plot of the Rayleigh wave dispersion for the case of a ring of aluminum alloy overlying a disk of Plexiglass. Figure 9 is a tracing of the seismogram at the antipodes. The ring of aluminum is the same as that use for the flexural waves. This configuration corresponds to a layer of rock with elastic constants $\alpha = 18{,}300$ ft/sec and $\beta = 10{,}400$ ft/sec, overlying a semi-infinite layer with

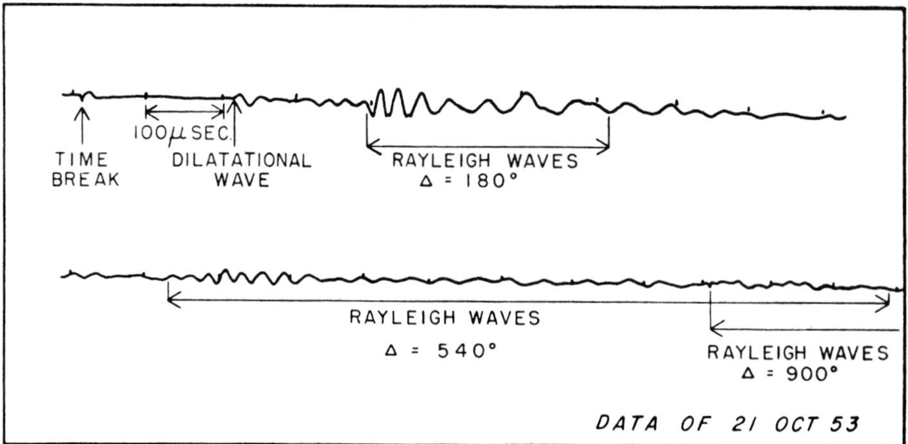


FIG. 9. Seismogram—high velocity layer/low velocity substratum.

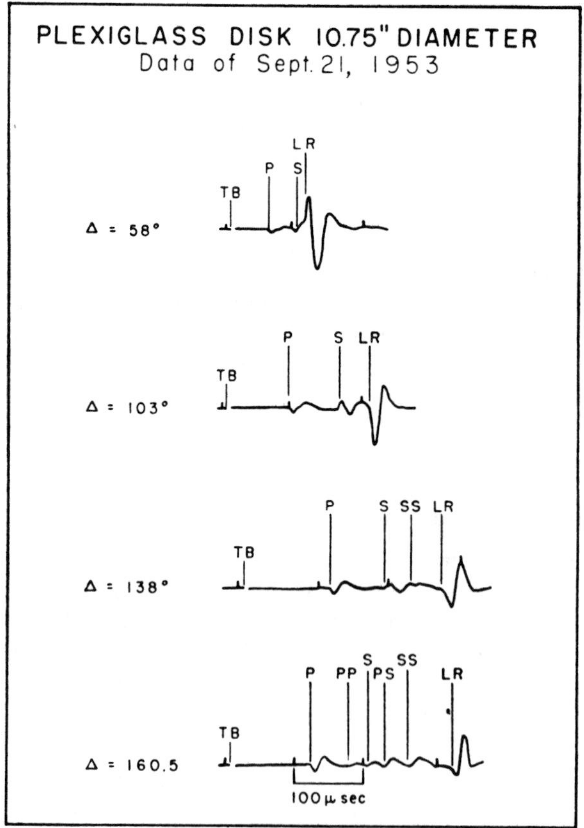


FIG. 10. Seismogram—Plexiglass disk.

$\alpha = 7{,}750$ ft/sec, $\beta = 4{,}500$ ft/sec. The ratio of the densities is $\rho/\rho' = 2.77/1.22$. This is a fair approximation to the case of Palisades diabase overlying Triassic sandstones and shales along the west bank of the Hudson River. It is also similar to structure in the permafrost areas in the Arctic.

Because of the limited spectrum of the pulse only a portion of the complete dispersion curve is obtained. Other segments may be studied by varying the thickness of the layer.

*Body and Surface Waves in Disks*

The previous examples, all concerning surface waves, used wavelengths which were small enough compared to the curvature of the disks so that the waves were effectively traveling along flat surfaces. Disks were chosen merely as a

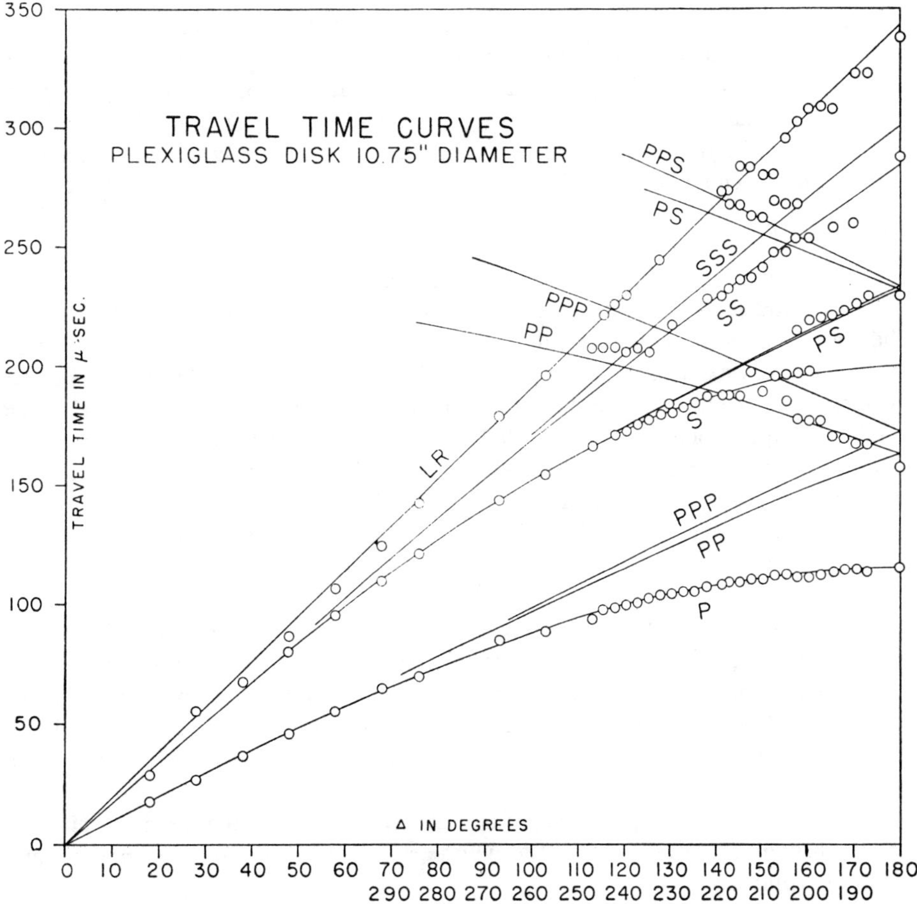


**Fig. 11.** Travel time curves—body and surface waves in a disk.

convenient shape to work with because of the multiple paths and lack of surface wave reflections. However the disk as a two-dimensional model of the earth also appears promising for the study of body waves. The seismograms of Figure 10 were taken on a uniform Plexiglass disk of diameter 10.75 inches. Figure 11 is a plot of the experimental points read from the records, compared with theoretical travel time curves based on velocities determined from the *P* arrivals. Poisson's ratio is taken as $\frac{1}{4}$. Phases definitely identifiable are *P*, *PP*, *S*, *PS* and *PPS*, *SS*, and *LR*. At the larger Δ's there is some difficulty in determining the exact arrival time and phase of the later arrivals due to overlapping of other phases. The present picks for the later arrivals are nearly always late. Clearly, this can be avoided by the use of larger disks and future studies will be made on disks of at least twice this diameter. This should improve the resolution of phases and accuracy of the arrival times measurably.

The polarity of the first break of the Rayleigh wave changes with distance. Such is not the case for Rayleigh waves propagating along the straight edge of a plate, which rules out the possibility of selective absorption of the high frequencies as the cause. The change of polarity is possibly the result of the generation of the waves by the curved *P* and *S* wave fronts at the surface. The phenomenon is not understood at present.

### *Scaling to Related Problems*

The dispersion curves of the preceding examples are presented in the dimensions as modeled. Scaling to other problems is a relatively simple matter, as can be seen from an inspection of the period equation for flexural waves or Rayleigh waves. Once a fixed relation is established between all the dilatational and shear velocities the equations may be solved for the ratio $c/\beta$ or $c/\alpha$ and similarly $U/\beta$ or $U/\alpha$. Thus velocities may be scaled to any other problem for which the elastic constants have the same ratios merely by multiplying the measured velocities by the appropriate ratio $\beta'/\beta$ or $\alpha'/\alpha$. Similarly, the period associated with a given phase and group velocity will vary directly with $H$, the layer thickness.

## ACKNOWLEDGMENTS

The authors wish to acknowledge the assistance of many co-workers at the Lamont Geological Observatory. In particular, Mr. Charles Kershaw constructed all the electronic gear and designed the new components. Mr. James Dorman and Mr. Donald Schiller carried out the computations, and Mr. Louis Collyer built the mechanical parts of the apparatus. Mr. Paul Pomeroy made many of the velocity determinations.

The Carter Oil Company generously gave assistance in the design of circuits used, through Dr. Franklyn Levin, who participated in the seismic model work for that company.

## REFERENCES

Bullen, K. E., 1947, An introduction to the theory of seismology: Cambridge University Press.

Ewing, M. and Press, Frank. An investigation of mantle Rayleigh waves: in press, Bull. Seis. Soc. Am.

Kaufman, S. and Roever, W. L., 1951, Laboratory studies of transient elastic waves: The Hague, Proceedings of Third World Petroleum Congress, Section I, p. 537–545.

Lamb, H., 1917, On waves in an elastic plate: Proc. Roy. Soc. Lond. A 93, p. 114.

Lee, A. W., 1932, The effect of geological structure on microseismic disturbances: Mon. Not. Roy. Astr. Soc., Geoph. Suppl., v. 3, p. 83–105.

Love, A. E. H., 1944, A treatise on the mathematical theory of elasticity: New York, Dover Publications.

Northwood, T. D. and Anderson, D. V., 1953, Model seismology: Bull. Seis. Soc. Am., v. 43, p. 239–246.

Terada, T. and Tsuboi, C., 1927, Experimental studies on elastic waves (Parts I and II): Tokyo Imperial University, Bull. Earthquake Res. Inst., v. 3, p. 55–65, v. 4, p. 9–20.

# SEISMIC MODEL EXPERIMENTS WITH SHEAR WAVES*

J. F. EVANS†

ABSTRACT

The experimental study of shear waves in the earth has been limited by the difficulty of producing them in sufficient strength. However, sensitive piezoelectric shear plates can now be made which enable experimentation with shear waves using small-scale seismic models. Seismic model experiments serve to demonstrate the simplicity of SH-shear wave reflections in a single homogeneous layer, the production of SH waves by an impulsive horizontal thrust, and the development of relatively high amplitude Love waves in a low-velocity surface layer. The results of these model experiments with shear waves are in general agreement with and confirm theory. They also agree with the results of field experiments in the scattered cases for which comparison is available.

## INTRODUCTION

The fact that the compressional waves in seismic prospecting are subject to partial conversion into shear waves when obliquely incident at a reflecting boundary has received frequent mention in recent years (e.g., Ricker and Lynn, 1950; Dix, 1952, p. 343 and 347–354). The shear waves resulting from such conversions are of the SV type, i.e., they have a component of particle motion perpendicular to the reflecting surface. Conversely, SV waves when obliquely incident at a reflecting boundary are partially transformed into compressional waves.

It has also been pointed out (e.g., Jolly, 1956) that SH waves, i.e., shear waves having horizontally-directed particle motion, are not subject to transformation into compressional energy when reflected or refracted at horizontal surfaces; and they might, therefore, be expected to give a simpler reflection pattern than the compressional waves ordinarily employed in reflection prospecting.

Experimental studies of shear waves in the earth have been limited in the past by the difficulty of producing them in sufficient strength for propagation over substantial distances. A similar limitation has applied to experimentation with shear waves in small seismic models composed of elastic solid materials. However, by proper pre-polarization of ceramic barium titanate (Cherry and Adler, 1947; Mason, 1950), piezoelectric shear plates can be made whose sensitivity is comparable to that of thickness-expanding plates of the same material. Such shear plates when incorporated into transducers are as convenient to manipulate as the compressional-wave sources and receivers more commonly employed in seismic model research (Evans et al., 1954; Oliver et al., 1954; Clay and McNeil, 1955; Levin and Hibbard, 1955). An applied force of ten pounds is sufficient for reproducible coupling between small shear transducers and the surface of a seismic model.

* Paper read at the 28th Annual International Meeting of the Society at San Antonio, October 16, 1958. Manuscript received by the Editor July 9, 1958.

† Pan American Petroleum Corporation, Tulsa, Oklahoma.

Reprinted from Geophysics, 24, 40-48

## RESULTS AND DISCUSSION

### *SH-Wave Reflections*

In the seismic model experiments with shear waves to be described, the theoretically-predicted simplicity of SH-wave reflections under ideally simple conditions was first demonstrated. Figure 1 shows two identical models consisting of homogeneous and isotropic blocks. (The material was rolled 24-ST aluminum.)

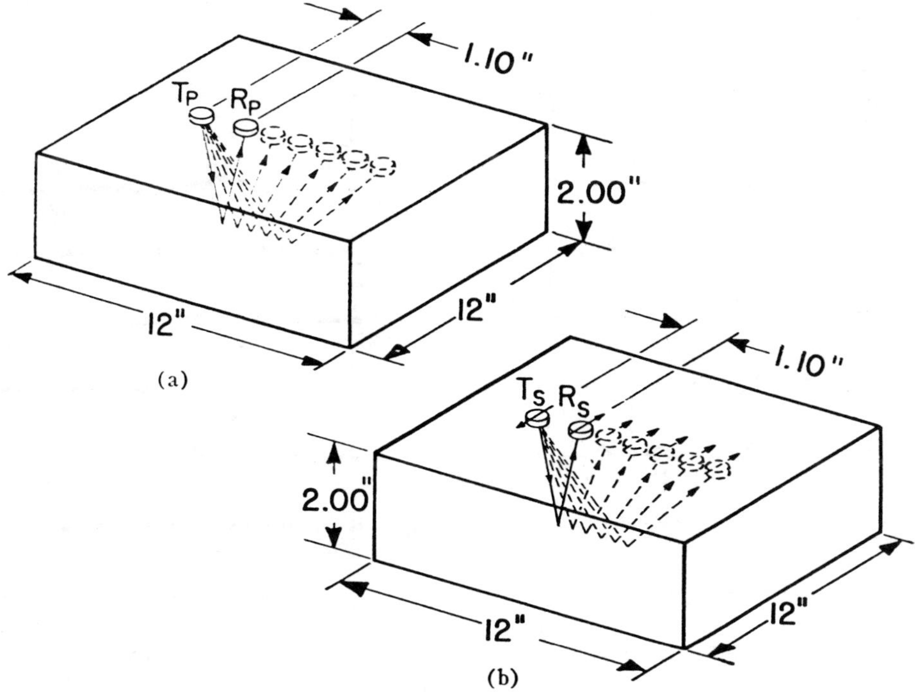


FIG. 1. Models and setups for comparing compressional and SH-wave reflections in a single homogeneous block. Shear-wave transducers are distinguished by an arrow, the arrow pointing in the direction of applied particle motion, if a wave source, or in the direction of maximum sensitivity to particle motion, if a detector.

On the block at the left, a compressional wave source $T_P$ and a compressional receiver $R_P$ were set up. The receiver was successively moved out to more distant positions, as shown by the disks drawn in dashed lines, to obtain a multi-trace record similar to the seismograms of reflection prospecting. On the block at the right, a shear-wave source $T_S$ and receiver $R_S$ were deployed with the same geometry as for the compressional transducers. The arrow placed on the shear-wave source indicates its direction of applied particle motion, and the arrow on the shear receiver its direction of sensitivity to particle motion.

Figure 2, parts (a) and (b), has records comparing the model responses to compressional and to shear excitation, Figure 2a being that obtained with compressional transducers. The top and bottom traces of each record have time markers 10 microseconds (0.000010 sec) apart which are connected vertically to give time lines. The second trace on each record shows the time break (marked T.B.) and the form of the electric voltage applied to the wave sources for the initiation of elastic waves. The driving voltage can be seen to approximate a step function. Subsequent traces show the wave motion at the first, second, third, etc., receiver positions.

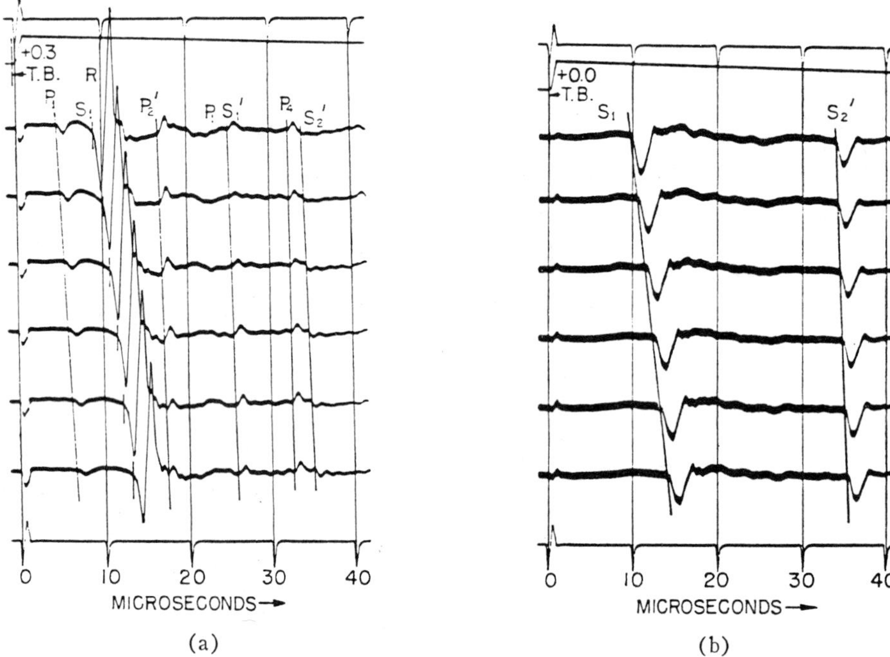


Fig. 2. (a) Reflection record obtained with compressional transducers with the model and setup shown in Figure 1. (b) Corresponding reflection record obtained with shear transducers oriented to maximize the transmission and reception of SH waves.

The compressional-wave record (Figure 2a) has numerous events: $P_1$, the direct compressional wave; $S_1$, the direct surface shear wave, immediately followed by and mixed up with the Rayleigh wave $R$; the compressional reflection, $P_2'$, from the bottom of the model; a transformed reflection $P_1S_1'$ due to conversion of the downgoing compressional wave front into shear energy by reflection; $P_4$, the first multiple compressional reflection; and $S_2'$, a shear reflection. (The presence of this shear reflection is explained by the fact that the compressional-wave source applied a localized downward thrust on the surface of an elastic solid; such a thrust develops relatively strong shear waves in oblique directions.)

Figure 2b, made with shear sender and receiver aligned to maximize the sending and reception of SH waves, has two events only; the direct SH wave marked $S_1$, and its reflection from the bottom of the model, marked $S_2'$. Records of longer duration (not shown) exhibit multiple shear reflections from the bottom of the model.

Thus the predicted theoretical advantages of SH-wave reflections under ideal conditions are experimentally confirmed, and it is concluded that SH waves under ideal conditions produce unusually simple and clear reflections.

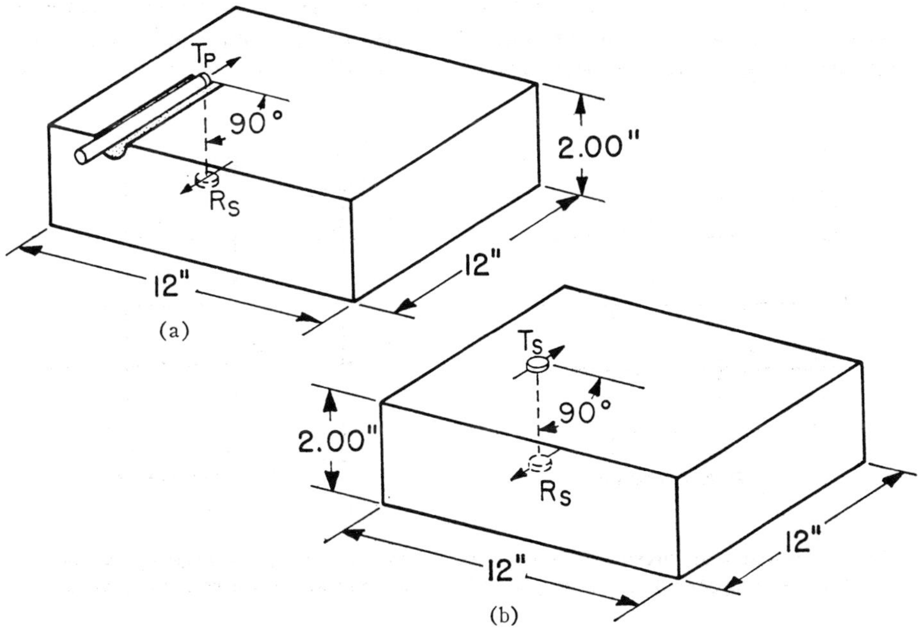


FIG. 3. (a) Model and experimental arrangement for demonstrating the production of SH waves by application of a horizontal thrust to the surface. (b) Corresponding arrangement for producing shear waves with a shearing source.

*SH Waves from Compressional Source*

How may SH waves of substantial amplitude be conveniently produced in the earth? The mechanical principle employed by several experimenters in the field (White et al., 1956; Jolly, 1956) can be demonstrated with a model. Figure 3a shows a semi-cylindrical trench, or groove, with a flat vertical end, machined into the surface of an aluminum block. A compressional-wave source, i.e., a thickness-expanding transducer, was laid in the groove and pressed against the flat end. When electrically pulsed, the compressional-wave source applied particle motion to the model in a direction parallel to the plane of the surface. If this plane is called "horizontal," then the wave source applied a horizontal thrust to the model at a point just below its surface. The waves thus produced were received, as shown in the figure, on the other side of the model at a point directly opposite that of application of the thrust.

For purposes of comparison, a shear-plate transducer was employed, as indicated in Figure 3b, to send waves across a geometrically similar path in a model similar except for the groove. The waves from both sources were received with a 3-component set of receivers. The first of the set was a shear detector aligned with its direction of maximum sensitivity parallel to the applied particle motion of the source; the second was the same detector with its direction of sensitivity rotated 90° with respect to its first position; and the third was a compressional detector. As between shear and compressional detectors, the receiver amplifier gain was adjusted to give approximately the same over-all sensitivity to wave motion.

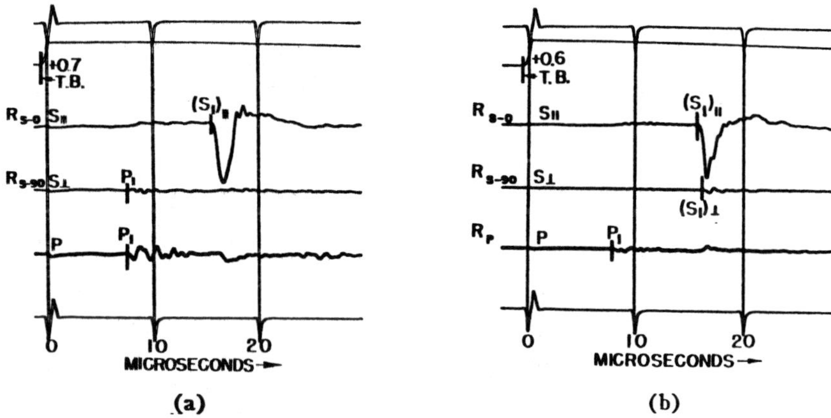


FIG. 4. Three-component records of SH waves sent from two sources: (a) A compressional transducer horizontally applied to the surface, and (b) a standard shear transducer normally applied.

This 3-component set of receivers roughly corresponded to the use of 3-component seismometers in field practice.

Records comparing the wave outputs from the two sources in the perpendicular direction are shown in Figure 4 (a and b). The record of Figure 4a was made with the horizontal thrust source, that of Figure 4b with the shear-plate source. The first detector traces on each record show the SH-wave component, with the horizontal thrust source comparing favorably with the shearing source. The second detector traces show the shear motion at right angles to the SH-wave motion; again the horizontal thrust source shows up favorably. On the last detector traces, obtained with compressional ($P$) receivers, the compressional-wave disturbance is seen to have slightly more amplitude from the thrust source than from the shear source. It is possible, however, that this indicated greater amplitude of compressional wave energy is associated with the presence of the surface groove. Thus these records show that the horizontal compressional-wave source is approximately the equivalent of the shear-wave source. It is concluded that

SH waves can be produced in an elastic solid by horizontally applying an impulsive force to the surface of the solid.

This conclusion receives support and confirmation from Figure 5, which shows an SH-wave reflection record made with the same model geometry as the SH-

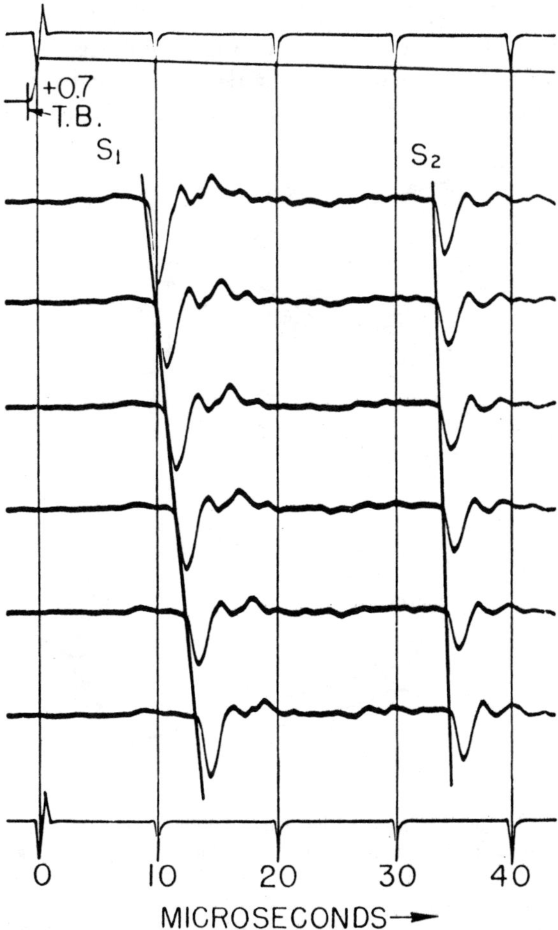


Fig. 5. SH-wave reflection record in a single homogeneous block, obtained with a horizontal compressional-wave source. (To be compared with the SH-wave record of Figure 2(b), which was made with a shear-plate transducer.)

wave record of Figure 2b, but using the horizontal compressional transducer to initiate waves. The record of Figure 5 is to be compared with that of Figure 2b. The form of the shear waves shown on Figure 5 is somewhat more oscillatory than that of Figure 2b, but again the extra oscillations may be due to the presence of the groove necessary for planting the wave source. As on Figure 2b,

Figure 5 shows only two distinct events: the direct SH wave, $S_1'$ and its reflection $S_2'$ from the bottom of the model.

*Effect of a Surface Layer*

The behavior of SH waves and also a method of producing them have been demonstrated in models having a single layer only. What would be the effect on the transmission of SH waves if the model was complicated by the addition of a thin surface layer of lower-velocity material? Such a model is shown in Figure 6a, together with a transducer layout for making an SH-reflection record. Here

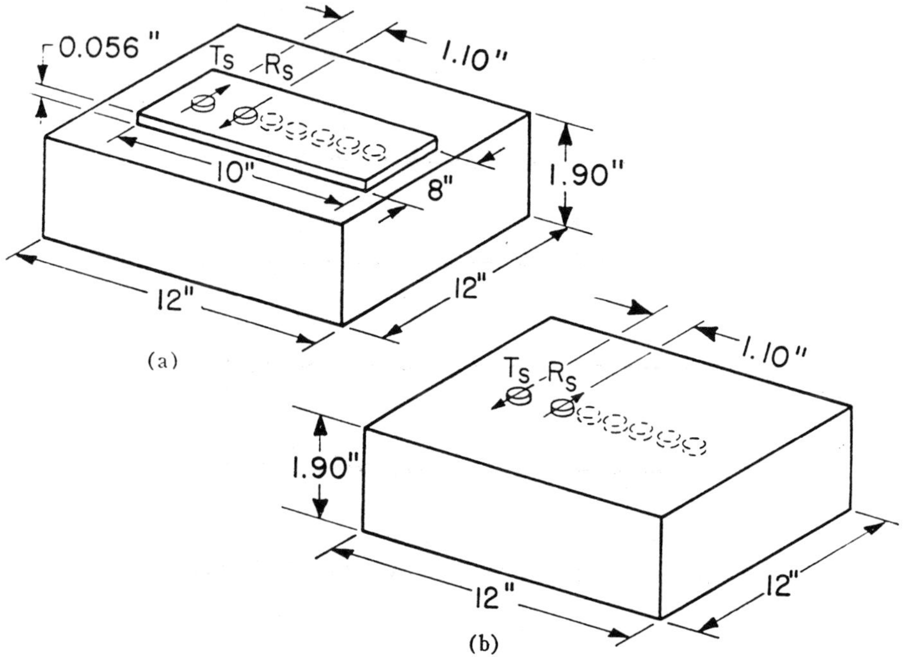


**Fig. 6. Models and setups for showing the effect of adding a low-velocity surface layer on the transmission of SH waves. (a) Two-layer model (50-50 solder on steel). (b) Single-layer model (steel only).**

the thin layer is made of solder,* and the thick underlying layer is rolled steel, to which solder can be readily bonded. Figure 6b shows the corresponding model and transducer layout without the surface layer.

Results with and without a surface layer are shown in Figures 7a and 7b. With the surface layer in place (Figure 7a), relatively high amplitude, oscillatory waves $L$ occur, which are believed to be Love waves. (Love waves are shear waves having particle motion parallel to the free surface and channeled within a surface layer.)

* The thickness of this layer was about one-third of the wave length of body waves in solder.

The comparative record of Figure 7b was made with the surface layer absent. It should be mentioned that this record was made with the same transducers and the same sensitivity as Figure 7a. The time of the expected SH-wave reflection, as based on the one-layer record, the known thickness of the surface layer, and the measured velocity of shear waves in the material of the surface layer, is indicated by the line marked $*S_2'$ on the left-side record of Figure 7a. The reflection is weakly indicated on one or two traces, but it is totally obscured on the other traces. It is concluded that the presence of a thin, lower-velocity surface layer gives rise to Love waves of such amplitude as to interfere greatly with the otherwise clear SH-wave reflections.

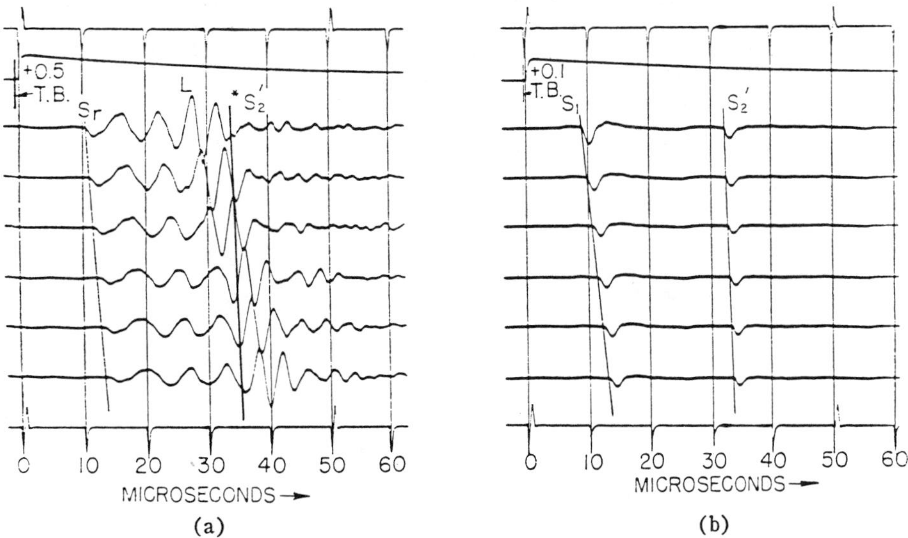


Fig. 7. (a) SH-wave pattern obtained with 2-layer model of Figure 6a. (b) Corresponding record obtained with the single-layer model of Figure 6b.

Jolly (1956, p. 914) has published field records having Love-wave patterns bearing a strong general resemblance to the Love-wave train of Figure 7a.

## CONCLUSIONS

With seismic model records it has been demonstrated: (1) that SH waves under ideal conditions produce unusually simple and clear reflections; (2) that SH waves can be produced in an elastic solid by horizontally applying an impulsive force to the surface of the solid; and (3) that the presence of a thin, lower-velocity surface layer gives rise to Love waves, of such amplitude as to interfere greatly with the otherwise clear SH-wave reflections.

All of these points demonstrated with the model are in agreement with theory as well as with field experiments in the earth in the scattered cases for which comparison is available. It is concluded, therefore, that seismic models and accompanying techniques afford an excellent means for the experimental study of shear waves.

## REFERENCES


Clay, C. S., and McNeil, Halcyon, 1955, An amplitude study on a seismic model: Geophysics, v. 20, p. 766–773. (Model technique; converted waves.)

Dix, C. H., 1952, Seismic prospecting for oil: New York, Harper and Brothers, p. 343 and 347–354.

Jolly, R. N., 1956, Investigation of shear waves: Geophysics, v. 21, p. 905–938. (SH waves; Love waves.)

Levin, F. K., and Hibbard, H. C., 1955, Three-dimensional seismic model studies: Geophysics, v. 20, p. 19–32. (Model technique; converted waves.)

Press, Frank, Oliver, Jack, and Ewing, Maurice, 1954, Seismic model study of refractions from a layer of finite thickness: Geophysics, v. 19, p. 388–401.

Ricker, Norman, and Lynn, Ralph D., 1950, Composite reflections: Geophysics, v. 15, p. 30–50.

White, J. E., Heaps, S. N., and Lawrence, P. L., 1956, Seismic waves from a horizontal force: Geophysics, v. 21, p. 715–723. (Generation of SH waves.)

# TWO-DIMENSIONAL MODELING AND ITS APPLICATION TO SEISMIC PROBLEMS*

F. A. ANGONA†

ABSTRACT

Laboratory seismograms of a fault model demonstrate the mechanism for diffraction and clearly show the difference in amplitude decay and moveout between a reflection and a diffraction. The inverted order and the deflection to the side of reflected energy from a curved reflector with a buried focus is demonstrated. A comparison is made of seismograms from a simple fault model and from one combining a fault with curved reflectors leading to it. The curved surfaces increase the overlap of the reflected events and mask the fault. Modeling techniques, involving the control of the reflection coefficient between layers by thickness variations and the control of the propagation velocity through a layer by combining two or more materials into a laminated sheet, are demonstrated.

## INTRODUCTION

In recent years the petroleum geophysicist has become more concerned with the basic fundamentals of wave propagation in solid media because of his need for more information from field seismograms. Since the turn of the century, numerous workers have made theoretical studies of wave motion in solids; however, because of the complexity of the problem, only those cases with the simplest of geometries have been solved successfully. Consequently, there is a wide gap between the problems of interest to the seismic field party and those which can be solved by the theoretical geophysicist. A laboratory seismic model can bridge this gap. The intent of this paper is to demonstrate the application of two-dimensional seismic modeling to field problems.

Seismic modeling has been the subject of a large number of papers in the past ten years. A partial list of these papers is included in the references. As can be noted from the titles, a wide range of problems can be solved by model techniques.

Seismic models can be classified into three groups depending on the number of dimensions employed. A one-dimensional model may be constructed in different forms such as the sound tube described by Woods (1956), or the plastic rod as used by Berryman et al. (1958), or the metal rod with cross-sectional area changes in scaled correspondence to velocity changes of a velocity log as described in the author's U. S. Patent No. 2,834,422. The one-dimensional model is ideal for the studying of the relationships of many reflections and the problem associated with ghosts and multiples. This type of model is concerned only with normal incidence of acoustic waves, and assumes that each layer is flat and horizontal.

Oliver et al. (1954), introduced the two-dimensional model. Thin sheets of

* Manuscript received by the Editor September 1, 1959.

† Field Research Laboratory, Socony Mobil Oil Company, Dallas, Texas.

Reprinted from Geophysics, 25, 468-482

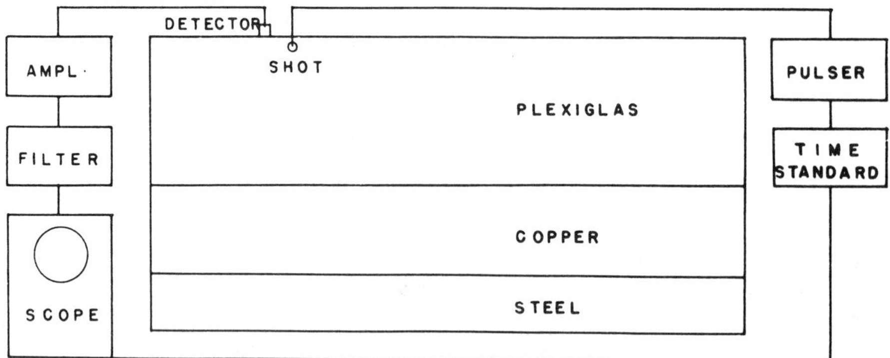


FIG. 1. A schematic diagram of a typical seismic model.

elastic materials cut into the form of a disc or cut into strips and bonded together to form a layered cross-section are examples of this type. Reflection, refraction, and surface wave problems can be solved with the two-dimensional seismic model. Both the two- and three-dimensional models are well suited for studying configurations where geometrical effects are of predominant interest. In areas where the stratigraphy does not change appreciably in the horizontal direction perpendicular to the geophone spread, the two-dimensional model can be used without any serious loss of generality.

Three-dimensional models as described by Howes et al. (1953) and by Evans et al. (1954), are representative of the third type of seismic model. Only a limited number of materials are sufficiently homogeneous to be used as components for these models. In addition, the fabrication of all but the most simple geometries is more difficult than for the other two types of models. Consequently, the three-dimensional model is best suited to those problems concerned more with the basic processes of reflection, refraction, and diffraction than with geometrical effects or where stratigraphic changes perpendicular to the spread cannot be neglected.

## TWO-DIMENSIONAL MODEL

The essential components of any seismic model are (1) one or more elastic materials arranged in a desired geometry, (2) a source of acoustic pulses, and (3) a means for detecting the resulting motions. The means by which these essentials are realized differ only slightly from one model system to another. The model system developed by the Socony Mobil Field Research Laboratory is shown in schematic form in Figure 1.

The materials used for the models are $\frac{1}{16}$ in thick sheets of plexiglas, copper, aluminum, and steel. These sheets are cut to the desired geometry and bonded together with epoxy resin. The source and receiver units are made of barium titanate. The shape and size of these units are chosen to produce an elas-

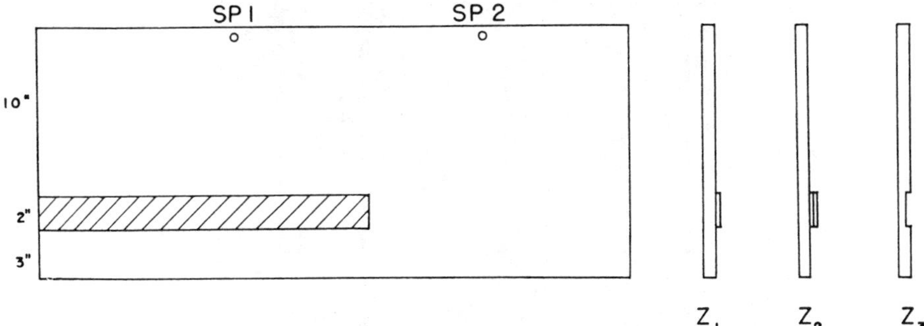


FIG. 2. A schematic drawing of model with variations in thickness.

tic pulse of reasonably short duration and with a peak frequency in the vicinity of 200 kc/sec. The pulser unit consists of a thyratron and pulse forming network which yields an electric pulse consisting of a single peak having a width of approximately five microsec and with an amplitude ranging from 100 to 1,000 volts. The received signal is amplified by a broad band (five cps to two Mcs) amplifier and filtered by an SKL electrical filter. This signal is presented to a Tektronix 535 oscilloscope and photographed with a Polaroid camera. A Tektronix Time-Mark Generator is used to Z-axis modulate the scope beam at 10 or 100 microsec intervals. The source is actuated every 20 millisec and the detector is moved from point to point to give any spread configuration desired. The picture is then enlarged and timing lines are superimposed to obtain a model seismogram resembling a field record.

## MODELING TECHNIQUES

As mentioned above, models consisting of more than one layer are usually fabricated by bonding together materials of different acoustic properties. However, there are only a dozen or so materials readily available in sheet form for models. Consequently, the reflection coefficient at the boundaries of a layer and the propagation velocity through the layer are restricted by the materials available. These restrictions, however, can be eliminated.

### *Reflection Coefficient*

The reflection coefficient at the junction of two materials is determined by the contrast in the acoustic impedance of the materials. The acoustic impedance for the plate is the product of the acoustic velocity, the density, and the thickness; consequently, a change in any one of these parameters will introduce a reflection in an otherwise uniform plate. A multi-layered model with *prescribed* reflection coefficients can thus be fabricated from a single sheet of material by either increasing or decreasing the thickness of the sheet at prescribed depths.

The influence of thickness change on the amplitude and polarity of a reflec-

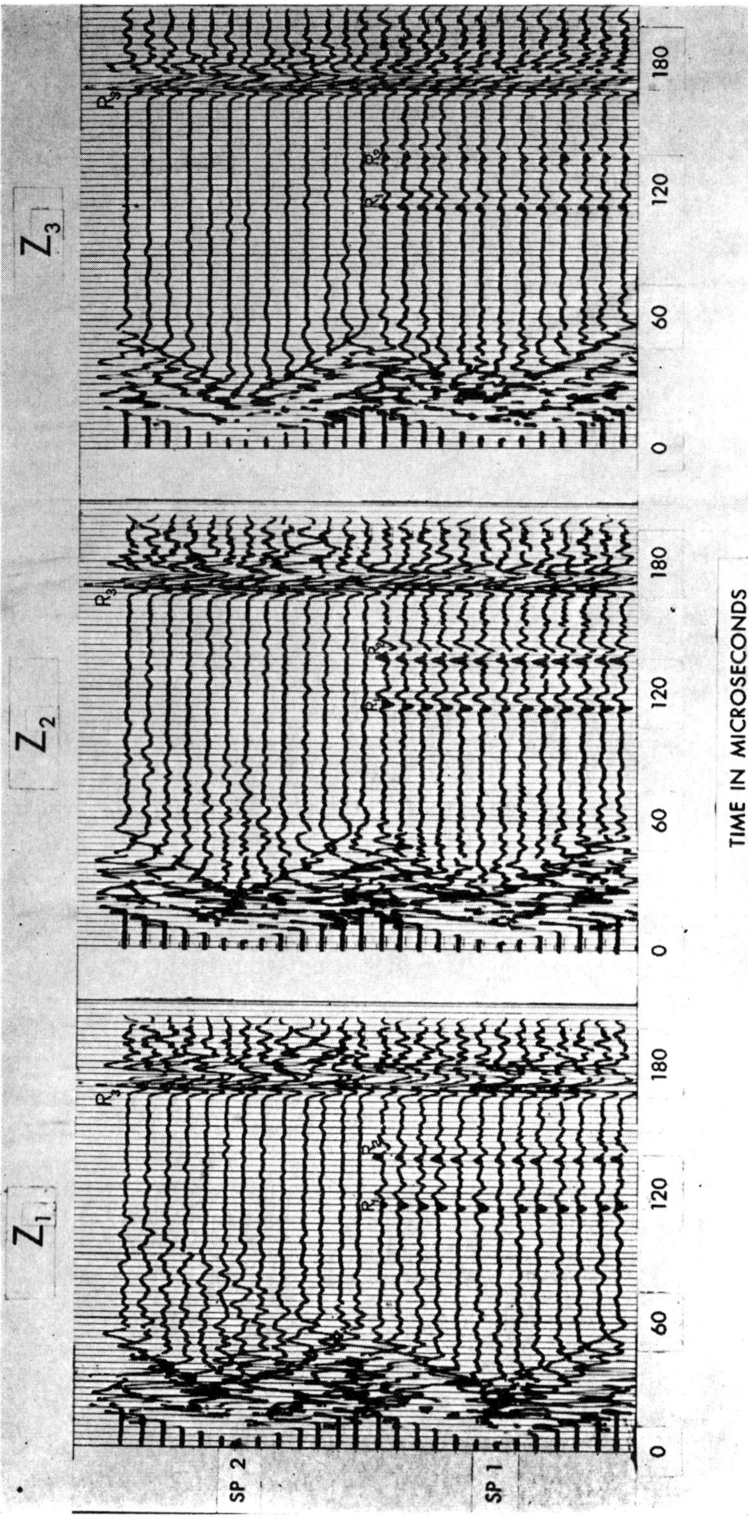


FIG. 3. Model records showing influence of thickness on the polarity and amplitude of a reflection.

tion is shown in Figures 2 and 3. The base material of this model (Figure 2) is an aluminum sheet of .063 inch thickness. Model $Z_1$ has an additional strip of .016 inch aluminum bonded to it, model $Z_2$ has two such strips, and model $Z_3$ has a recess of .013 inch depth. The records from these three models are presented in Figure 3. Each of these records was shot under identical conditions with conventional split spreads about SP 1 and SP 2. The largest reflection, $R_3$, is from the base of the model while reflections $R_1$ and $R_2$ are from the top and bottom of the layer of thickness change.

The observed polarity and amplitude of each of these reflections agree with those predicted. Reflection $R_1$ and $R_2$ are of opposite polarity for each of the models. The amplitudes of $R_1$ and $R_2$ for model $Z_2$ are approximately twice those for $Z_1$ and $Z_3$. Also the polarity of the first two reflections of model $Z_3$ are opposite to those of model $Z_1$ and $Z_2$ due to the decrease instead of the increase in thickness.

*Composite Velocity*

The restriction of only those propagation velocity values as are possessed by the few materials readily available for modeling is eliminated by combining two or more materials into a composite sheet. This composite sheet is made by bonding face to face thin sheets of different materials. The materials chosen and their respective thicknesses allow a composite velocity ranging continuously between the extremes of the materials used. For wave lengths long compared to the overall thickness of the composite sheet, the velocity can be expressed as

$$\overline{C}^2 = \frac{\alpha_1\rho_1C_1^2 + \alpha_2\rho_2C_2^2 + \cdots \alpha_n\rho_nC_n^2}{\alpha_1\rho_1 + \alpha_2\rho_2 + \cdots \alpha_n\rho_n}, \tag{1}$$

where

$\alpha_n$ is the fraction of material $n$
$\rho_n$ is the density of material $n$
$C_n$ is the acoustic velocity of material $n$.

A number of composite strips measuring 3×12 in were made of various pairs of materials. The transit time for a number of transversals across the three-inch dimension of the strips was accurately measured with the aid of a Rutherford Time Delay Generator. The composite velocity was determined by dividing this transit time into twice the width of the strip. A comparison of the measured and the calculated composite velocities in inches per microsecond is shown in

Table I

| Material 1 | Material 2 | $\alpha_1$ | $\alpha_2$ | $\overline{C}_{calc.}$ | $\overline{C}_{meas.}$ |
|---|---|---|---|---|---|
| Aluminum (0.213) | Epoxy (0.077) | 0.24 | 0.76 | 0.151 | 0.149 |
| Aluminum (0.213) | Plexiglas (0.090) | 0.20 | 0.80 | 0.147 | 0.148 |
| Steel (0.210) | Solder (0.084) | 0.41 | 0.49 | 0.146 | 0.146 |

Table I. The plate velocity for the individual component materials is given in the parenthesis in units of inches per microsecond.

A larger test section was made from a sheet of steel 16×36 in with a recessed section 4×26 in and with a depth of 0.040 in. Normal model records were taken over the model before and after a layer of solder was added. A sketch of the model is shown in Figure 4. The side-view drawings show model $Z_4$ to be steel only and model $Z_6$ to have a 0.017 in layer of solder added. Reflections from the top and bottom of the 4-in strip, denoted as $R_1$ and $R_2$, and the bottom reflection $R_3$, can be carried across the records of Figure 5. The propagation velocity of the 4-in strip can be determined from the time difference of events $R_1$ and $R_2$. This time

FIG. 4. Model composed of steel and solder to show change in velocity for composite section.

interval for model $Z_4$ yields a velocity of 0.205 in/microsec which is within three percent of the more accurately measured value of steel. The $R_1$ to $R_2$ time interval for $Z_6$ is seen to be appreciably longer and gives a velocity of 0.150 in/microsec. This value is approximately eight percent lower than that calculated for the composite layer of steel and solder. Some machining difficulties were encountered because of the flimsiness of the steel sheet. Consequently, the thickness of the steel and of the solder varied as much as 0.002 in. This uncertainty of the thickness could account for the discrepancy in the measured composite velocity.

## APPLICATIONS OF SEISMIC MODELING

Sections containing faults and folds are of paramount interest to the explorationist; however, the seismograms from these regions are complex and at times tend to defy detailed interpretation. Consequently, simple faults and geometries containing curved surfaces are fruitful subjects for seismic modeling.

### *Fault Model*

A fault model consisting of a sheet of aluminum and copper is shown in Figure 6. In addition to the reflections $R_H$, $R_L$, and $R_b$ from the interfaces labeled $H$,

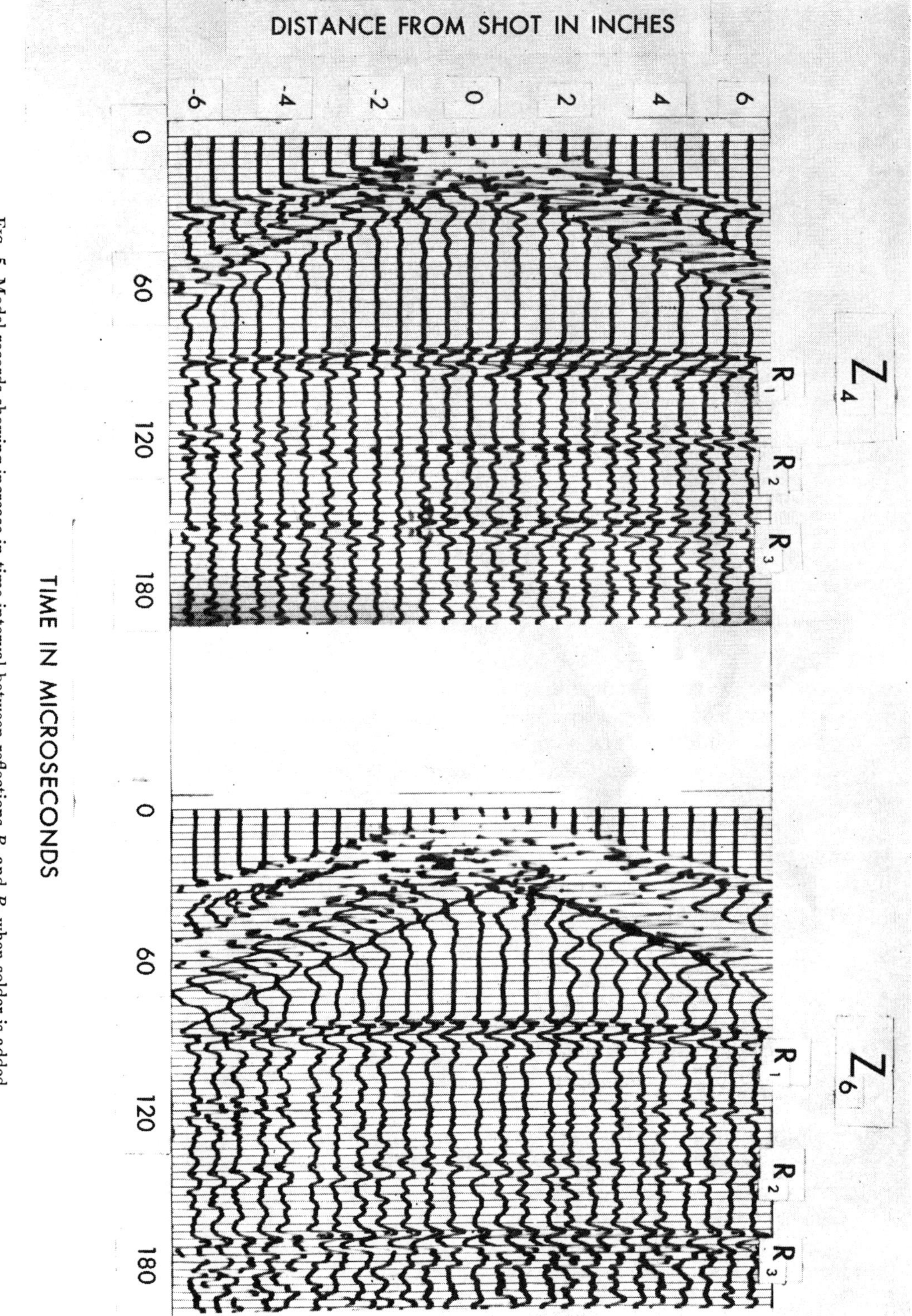


Fig. 5. Model records showing increase in time interval between reflections $R_1$ and $R_2$ when solder is added.

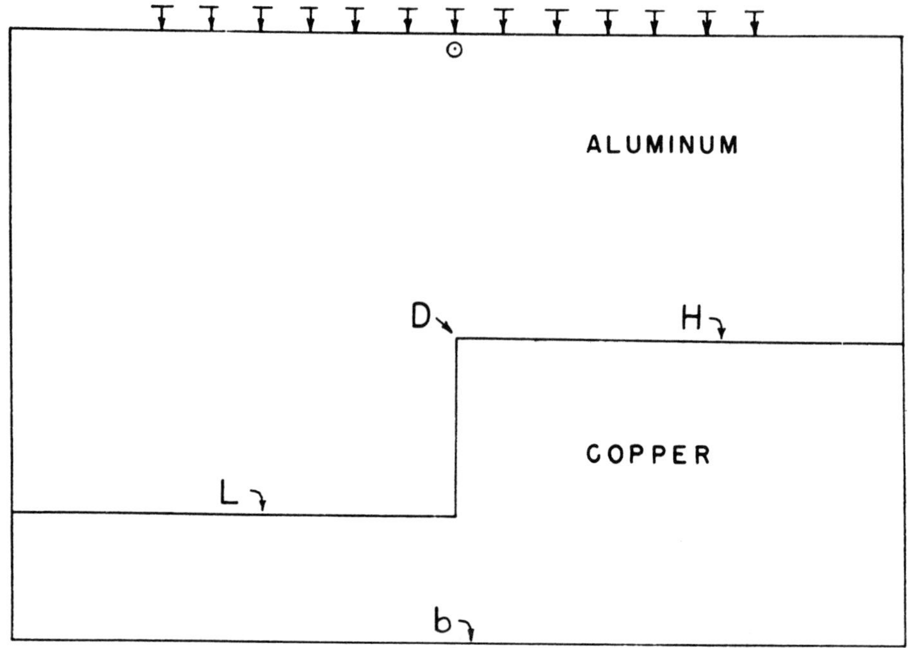


FIG. 6. A simple fault model.

$L$, and $b$, respectively, four diffraction events, two converted events, and two multiple events can be picked on the record of Figure 7.

The diffractions can be identified by both the increase in moveout and the amplitude decay with distance away from the fault. The phenomena of diffraction can be understood by employing the concepts of Huygen's principle and of limitation of a wave front. When a wave front is limited by the termination of a reflector, such as at a fault or at a wedgeout, a diffraction will occur. Hence, it is not too surprising to observe four different diffractions from the top edge, labeled $D$, of the fault plane. The diffraction $D_1$ is due to the abrupt termination of reflector $H$. The event $D_1+$ is generated by the limitation of the reflected wave front from interface $L$. The reflection from the bottom of the model separates into two events because of the difference in transit time on the two sides of the fault. Each of these wave fronts is limited by the top edge of the fault plane and thus causes the diffractions labeled $D_1^{++}$.

### *Curved Surface Model*

A reflecting interface at sufficient depth and with sufficient curvature to develop a buried focus as discussed by Dix (1952) is modeled in Figure 8. The general characteristic of a buried focus is that energy reaching the surface is deflected to the side and the order in which energy is received at the surface is in-

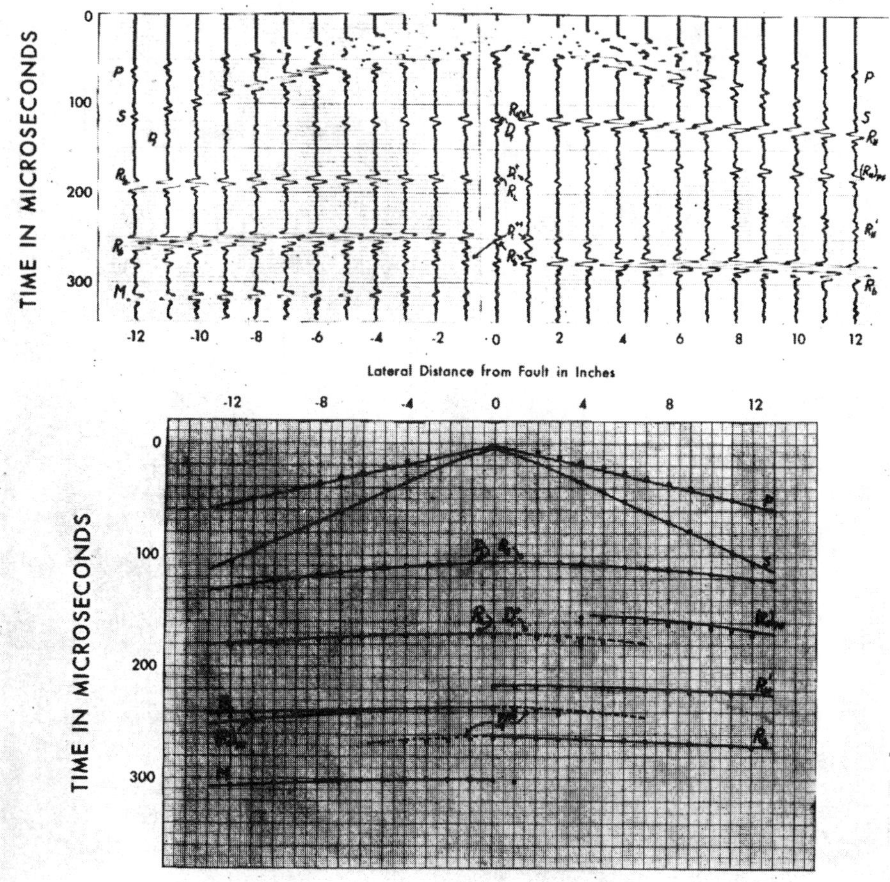


FIG. 7. Model record and time section of a simple fault.

verted from that arriving at the interface. These characteristics can be observed by following event $R_1$ across the record of SP 1 to the middle of SP3. At this point the curved surface reflection $R_{C_1}$ is deflected back toward the left. The further toward the right is the reflecting point, the more to the left is the energy received at the surface. This inversion continues to the inflection point between curves $C_1$ and $C_2$. If the reflection $R_{C_2}$ had greater amplitude, it would tie to $R_{C_1}$ at the extreme left of the record. This event continues across the SP 5 spread. The small event $D_1$ is a diffraction from the region where the reflecting interface changes from a flat surface to a concave surface.

*Complex Fault Model*

Model records from a simple fault and from curved reflectors have been presented. These records are relatively simple and from them one can reconstruct

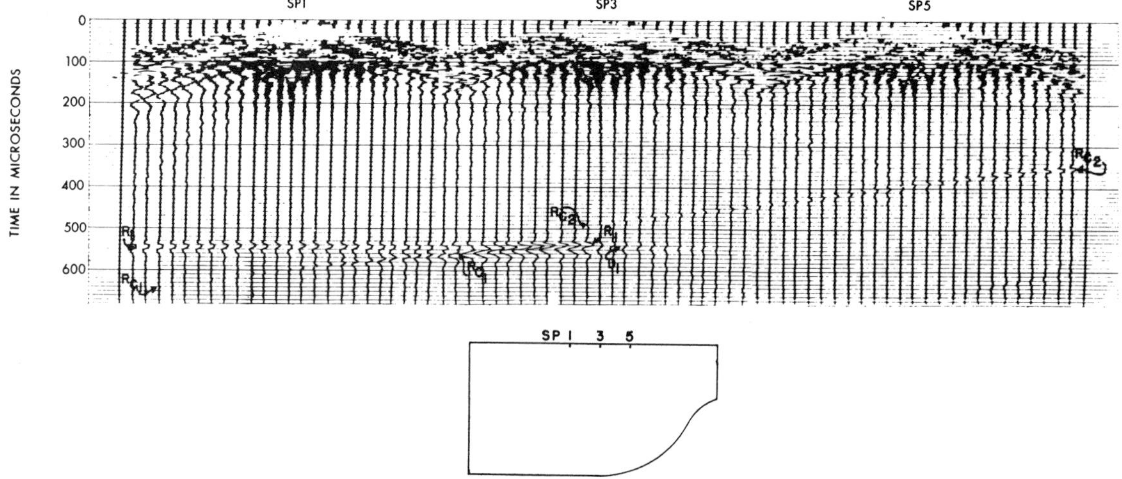


FIG. 8. A continuous profile shot over a curved surface with a buried focus.

the geometry of the model which produced them. However, when the geometries are combined, the resulting model records approach the complexity of a field seismogram. The two geometries of Figure 9 were shot under identical conditions. Shotpoints were located directly above the fault and 6 in to either side of it. A continuous profile was shot over each of the models. These records are presented in Figure 10. As before, the reflections from the flat reflectors are labeled as $R_H$, $R_L$, and $R_b$, the curved surface reflections as $R_{C_1}$ and $R_{C_2}$, and the diffractions by $D$.

The increased complexity of the model C record is quite apparent. The fault can be easily located below shotpoint 0 for model A; however, a migration or wave front technique must be employed to locate the fault of model C at shotpoint 0.

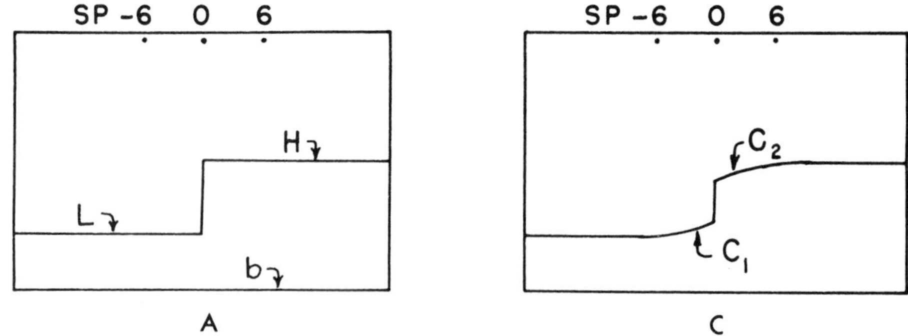


FIG. 9. Models A and C are similar except for the addition of curved interfaces leading to and from the fault for model C.

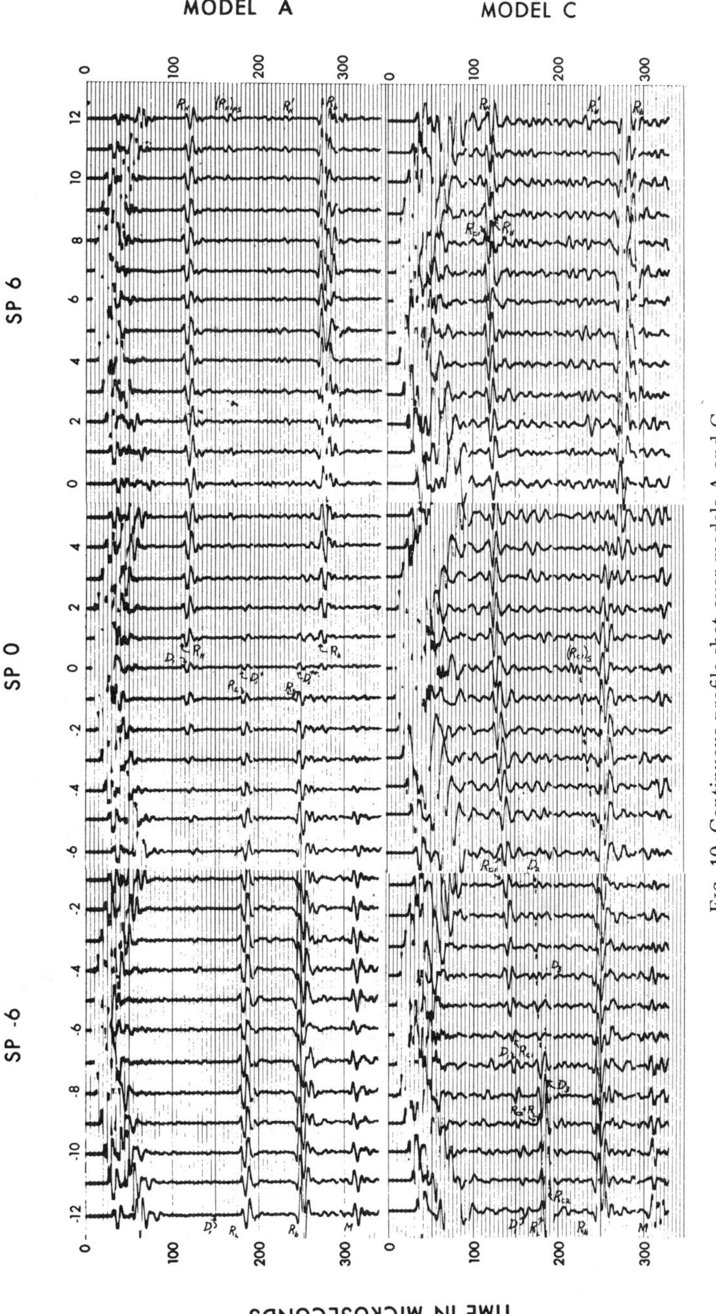


Fig. 10. Continuous profile shot over models A and C.

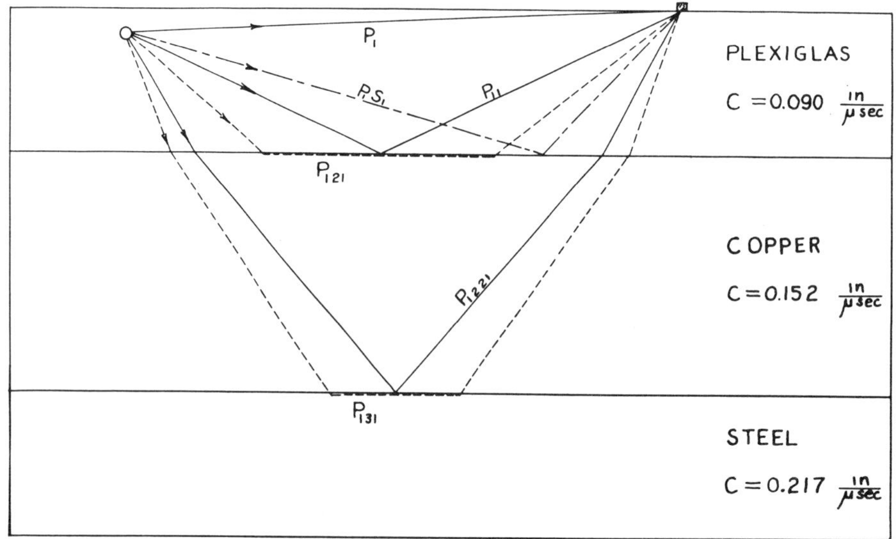


FIG. 11. A simple refraction model.

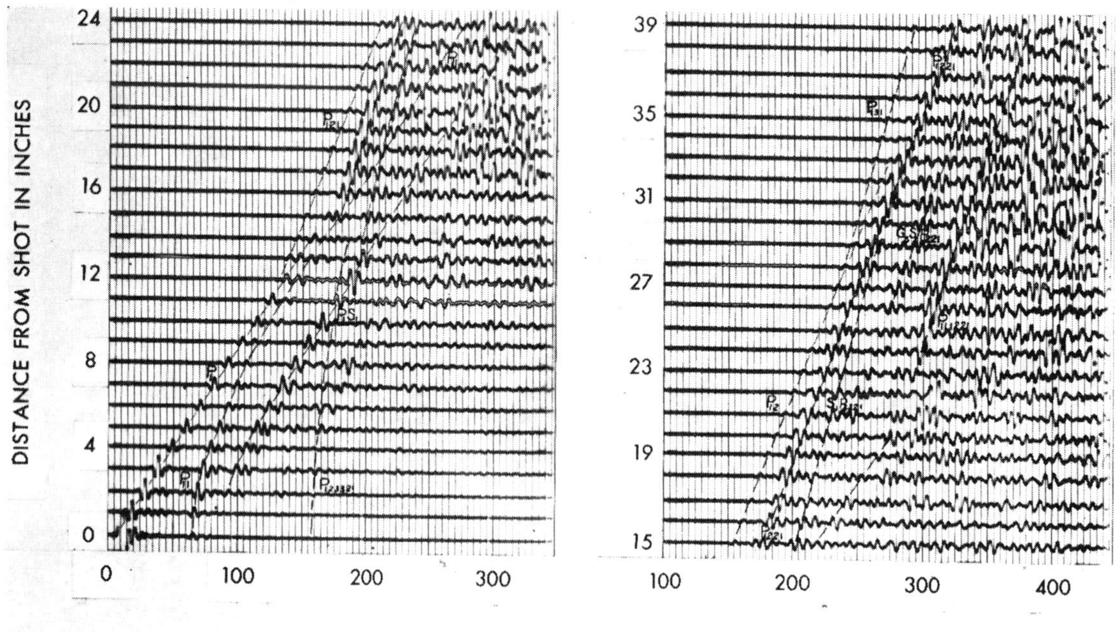


FIG. 12. Refraction records showing later arrivals.

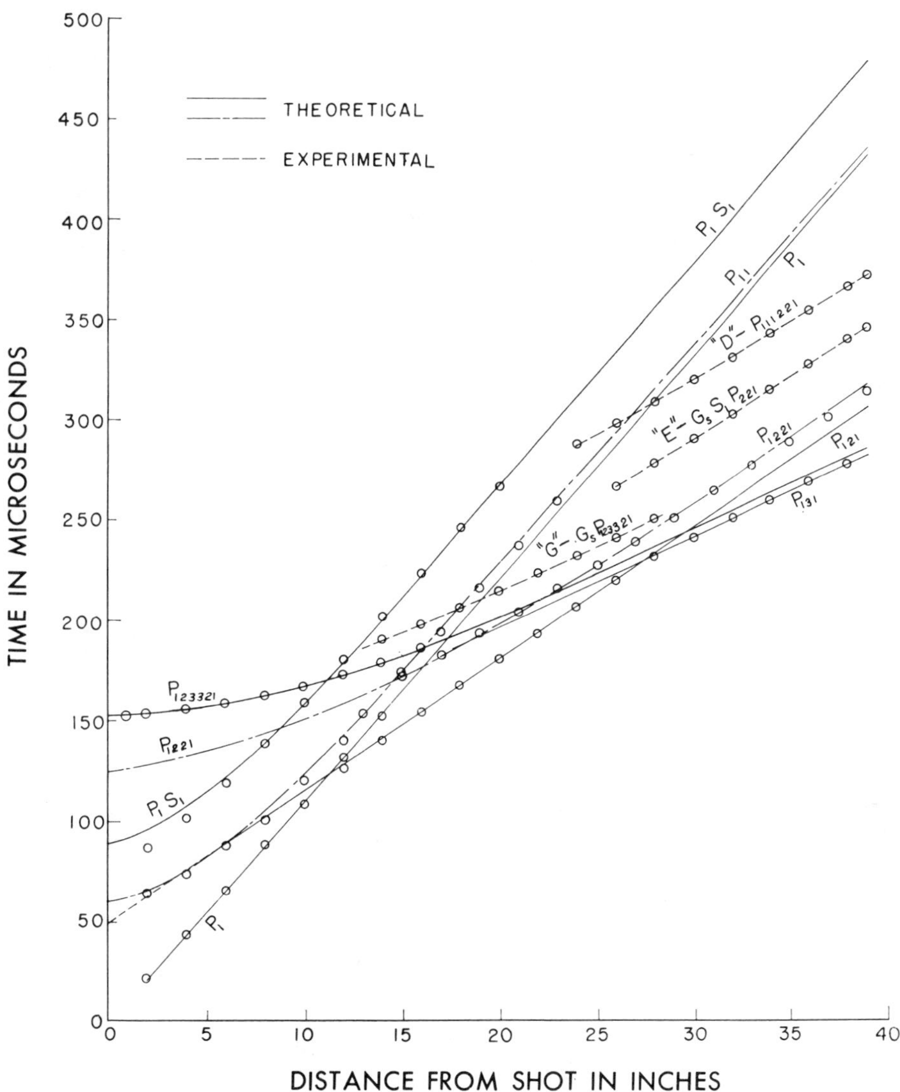


Fig. 13. Travel time curves from refraction records.

*Refraction Model*

The final model to be discussed demonstrates the application of two-dimensional modeling to the refraction process. A simple three-layer model is shown in Figure 11. The shot was fixed at SP 1 and records were taken with spreads extending from the shotpoint to a point 24 in to the right and another extending from 15 to 39 in to the right. The refraction records are presented in Figure 12.

The copper refraction comes in as the first arrival from station 12 to 28 where the refraction from the steel takes over. These refractions give velocities of 0.151 in/microsec for copper and 0.215 in/microsec for steel. These both are within one percent of the value measured directly by the transit time technique. A number of later events are picked and identified on the time-distance plot of Figure 13. The largest events on the records are the copper reflection $P_{1221}$ and the events $E$ and $D$ which are this same reflection but delayed due to a portion of the path traversed at shear velocity.

## ACKNOWLEDGMENT

The author wishes to thank the Socony Mobil Company for permission to publish this material. Appreciation is also expressed to J. Podhrasky for assistance in the fabrication of the models and to C. W. McRight in the preparation of the model seismograms.

## REFERENCES

1. Berryman, L. H., Goupilland, P. L., and Waters, K. H., 1958, Reflections from multiple transition layers: Geophysics, v. 23, p. 223–252.
2. Bremaecker, J. C., 1958, Transmission and reflection of Rayleigh waves at corners: Geophysics, v. 23, p. 253–266.
3. Carabelli, E. and Folicaldi, R., 1957, Seismic model experiments on thin layers: Geophysical Prospecting, v. 5, p. 317–327.
4. Clay, C. S., and McNeil, H., 1955, An amplitude study on a seismic model: Geophysics, v. 20, p. 766–773.
5. Dix, C. H., 1952, Seismic prospecting for oil: New York, Harper Bros., p. 138.
6. Evans, J. F., Hadley, C. F., Eisler, J. D., and Silverman, D., 1954, A three-dimensional seismic wave model with both electrical and visual observation of waves: Geophysics, v. 19, p. 220–236.
7. Evans, J. F., 1958, Seismic model experiments with shear waves: Geophysics v. 24, p. 40–48.
8. Grannemann, W. W., 1956, Diffraction of a longitudinal pulse from a wedge in a solid: JASA, v. 28, p. 494–497.
9. Howes, E. T., Tejada-Flores, L. H., Randolph, L., 1953, Seismic model study: JASA, v. 25, p. 915–921.
10. Kaufman, S. and Roever, W. L., 1951, Laboratory studies of transient elastic waves: The Hague, proceedings of third world petroleum congress, Sect. 1, p. 537–545.
11. Koepoed, O., Van Ewijh, J. G., and Bakker, W. T., 1958: Seismic model experiments concerning reflected refractions: Geophysical Prospecting, v. 6, p. 382–393.
12. Levin, F. K., and Hibbard, H. C., 1955, Three-dimensional seismic model studies: Geophysics, v. 20, p. 19–32.
13. Northwood, T. D., and Anderson, D. V., 1953, Model seismology: Bull. Seis. Soc. Am., v. 43, p. 239–246.
14. O'Brien, P. N. S., 1955, Model-seismology—the critical refraction of elastic waves: Geophysics, v. 20, p. 227–242.
15. Oliver, J., Press, F., and Ewing, M., 1954, Two-dimensional seismology: Geophysics, v. 19, p. 202–219.
16. Press, F., Oliver, J., and Ewing, M., 1954, Seismic model study of refractions from a layer of finite thickness: Geophysics, v. 19, p. 388–401.

17. Press, F., 1957, A seismic model study of the phase velocity method of exploration: Geophysics, v. 22, p. 275–285.
18. Riznichenko, Yu V. and Bugroff, V. R., 1951, The modeling of seismic waves. Izvestiya Akad Nauk SSSR, v. 5, p. 1–30.
19. Riznichenko, Yu V. and Shamina, O. G., 1957. Elastic waves in laminated solid medium investigated on two-dimensional model: Bull. of Acad. of Sci. USSR, No. 7, p. 17–38.
20. Roever, W. L., 1952, Experimental laboratory studies of elastic wave propagation: Phys. Rev., v. 85, p. 744.
21. Saraffian, G. P., 1956, Marine seismic model: Geophysics, v. 21, p. 320–336.
22. Woods, J. P., 1956, Composition of reflections: Geophysics, v. 21, p. 261–276.

# TWO-DIMENSIONAL SEISMIC MODELS WITH CONTINUOUSLY VARIABLE VELOCITY DEPTH AND DENSITY FUNCTIONS*

JOHN H. HEALY† AND FRANK PRESS†

ABSTRACT

A method for fabricating two-dimensional ultrasonic seismic models with variable velocity and density is described. The method is justified theoretically. It is tested by comparing the experimental and theoretical dispersion of Rayleigh waves in a model of a two-layered earth crust.

## INTRODUCTION

Many problems of interest to the seismologist are impossible to solve with available mathematical techniques. Where exact solutions are not available, seismic models can facilitate the development of approximate solutions that are almost as useful as the exact solution, and the seismic model provides the additional advantage of a clear illustration of the results with the model seismograms. Thus, seismic model studies can play a very important role in the study of wave propagation problems.

The assembly of the electronic part of a seismic model system has become a simple task as a result of advances in design and the ready availability of the components. However, practical difficulties involved in the fabrication of the seismic model itself seriously limit the scope of problems that can be studied.

One large class of problems that is important both in exploration seismology and in earthquake seismology is concerned with media that have a continuously variable velocity depth function. This paper describes a method of modeling a medium with variable velocity depth and density functions and illustrates the technique by a study of a problem in Rayleigh wave dispersion. Two dimensional model techniques, as described by Oliver, Press and Ewing (1954), are the basis of this method which makes use of the fact that waves propagating in thin sheets obey the two-dimensional wave equation. A variable velocity is obtained by the use of a laminated sheet in which the relative proportions of the laminae are varied so as to change the average elastic constants (hence the average velocities) of the sheet. The density can be modeled by varying the total thickness of the sheet.

This technique of modeling was suggested originally by Jack Oliver and the method of fabrication used was suggested by H. O. Walker.

## EQUIPMENT

The model set up was essentially the same as described in earlier papers (Oliver, Press and Ewing, 1954). One-eighth inch by one-quarter inch solid cylin-

* Contribution No. 969, Division of Geological Sciences, California Institute of Technology. Manuscript received by the Editor January 25, 1960.

† Seismological Laboratory, California Institute of Technology, Pasadena, California.

Reprinted from Geophysics, 25, 987-997

ders of barium titanate were used as sources and barium titanate bimorph transducers were used as receivers. The bimorph transducers were clamped between rubber sheets to provide sufficient damping to prevent ringing. This simple device proved to be very effective in eliminating the ringing problem.

## MODEL CONSTRUCTION

The models were fabricated from aluminum sheets and an epoxy resin plastic that was poured onto the aluminum and allowed to solidify. The aluminum used was 24ST in .020 inch sheets, and the plastic used was Shell Epon No. 828 with diethylenetrianamene hardener.

The aluminum sheets were cut in circles 24 inch in diameter and mounted to a 26-inch face plate with parafin wax.

A horizontal milling machine available in our shops was converted to a stub lathe so that a contour could be cut in the surface of the aluminum sheet. After the desired contour was cut in the aluminum, the face plate was removed from the lathe and a layer of plastic was poured over the aluminum and allowed to harden.

A final contour was cut in the surface of the plastic so as to provide control of two parameters, the total thickness of the model, and the ratio of plastic to aluminum. Figure 1 shows a portion of the model with source and receiver.

This process, while simple in concept, is a difficult machine shop problem. The tolerances desired exceed the capabilities of the standard machine tools and extreme care is required to obtain a satisfactory model. A great deal of credit is due our experimental machinest, Mr. Carl Holmstrom, for the successful construction of these models.

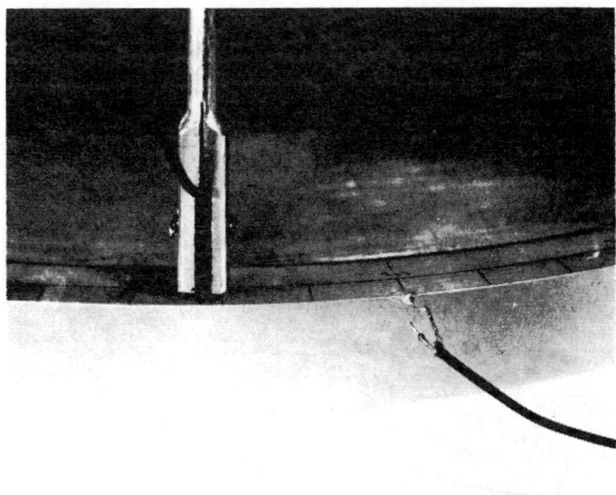

FIG. 1. Layered plate model with a damped bimorph receiver and cylindrical source.

THEORY OF TWO-DIMENSIONAL LAMINATED PLATE MODELS

If the the three-dimensional equations of motion and boundary conditions are solved to yield the period equation for a laminated plate with two sheets, it can be shown that for wave lengths, long compared to plate thickness, the phase velocity reduces to a constant value, reflecting the average elastic parameters of the plate.

Another approach is to proceed from the assumption of plane strain and derive the wave equation for a laminated plate. Consider a plane wave propagating in a layered plate in the $X$ direction so that the particle motion is perpendicular to the surface of the plate. Figure 2 shows the cross section of the plate with the differential element which we will consider. With the following definitions,

$u$ = particle motion in $X$ direction
$T_i$ = thickness of $i$th lamination
$\bar{\rho}$ = density of laminated plate (mass per unit area)
$\beta$ = shear velocity in laminated plate
$C_p$ = plate longitudinal velocity in the laminated plate
$\alpha_i$ = compressional velocity of $i$th lamination
$\beta_i$ = shear velocity of $i$th lamination
$E_i$ = Young's modulus of $i$th lamination
$\sigma_i$ = Poisson's ratio of $i$th lamination
$\rho_i$ = density (mass per unit volume) of $i$th lamination
$C_{pi}$ = plate longitudinal velocity in a homogeneous plate of the composition of the $i$th lamination

the equation of motion in the $X$ direction is,

$$\frac{\partial}{\partial X}\left[\frac{\partial U}{\partial X}\left(\frac{E_1}{1-\sigma_1^2}T_1+\frac{E_2}{1-\sigma_2^2}T_2\right)\right]=(\rho_1T_1+\rho_2T_2)\frac{\partial^2U}{\partial t^2}. \tag{1}$$

This reduces to a wave equation,

$$\frac{\partial^2U}{\partial X^2}=\frac{\rho_1T_1+\rho_2T_2}{\dfrac{E_1}{1-\sigma_1^2}T_1+\dfrac{E_2}{1-\sigma_2^2}T_2}\frac{\partial^2U}{\partial t^2}. \tag{2}$$

Thus, the plate longitudinal velocity is given by,

$$C_p^2=\frac{\dfrac{E_1}{1-\sigma_1^2}T_1+\dfrac{E_2}{1-\sigma_2^2}T_2}{\rho_1T_1+\rho_2T_2}=\frac{\rho_1C_{p1}^2T_1+\rho_2C_{p2}^2T_2}{\rho_1T_1+\rho_2T_2}. \tag{3}$$

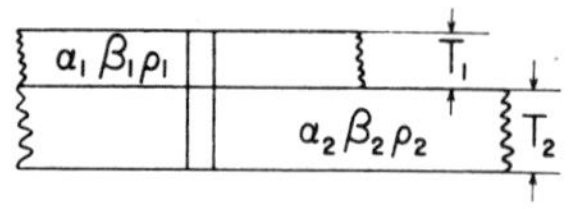

FIG. 2. Cross section layered plate.

Similarly for shear waves one finds,

$$\beta^2 = \frac{\rho_1\beta_1{}^2T_1 + \rho_2\beta_2{}^2T_2}{\rho_1T_1 + \rho_2T_2} \,. \tag{4}$$

From comparison with the homogeneous plate it is evident that $\bar{\rho}=\rho_1T_1+\rho_2T_2$ replaces the density factor and $\rho_1\rho_{p1}^2T_1+\rho_2\rho_{p2}^2T_2$ or $\rho_1\beta_1^2T_1+\rho_2\sigma_2^2T_2$ replaces the elastic factor.

We have verified equation (3) by solving the exact three-dimensional equations of motion for an infinite laminated plate subject to the exact boundary condition at the free surfaces and the interface. Equation (3) is a limiting form of the solution for phase velocity which emerges from the long wave length condition $\lambda >> T_i$. This result has also been experimentally verified as will be seen in the next section.

It is more difficult to establish the use of equations (3) and (4) and to specify exact boundary conditions when the thicknesses of the lamina are varied to model the velocity and density parameters. One might justify the use of these equations by arguing intuitively from the results of Oliver et al. (1954) for homogeneous plates. We prefer to justify these equations experimentally by study of Rayleigh wave dispersion in a model of a double layered crust overlying a mantle.

Note that the mass per unit area will be the density that must be considered when we vary the thickness of the lamina to model the velocity and density parameters. By suitable choice of $T_1$ and $T_2$ one can model density and one of the elastic constants provided it falls between the corresponding values of the model materials chosen.

## EXPERIMENTAL VERFICATION OF THE THEORY

An experimental verification of the results derived was obtained with a series of layered plates constructed, as described above, of a layer of plastic over aluminum. The total thickness of the plate was $\frac{1}{8}$ inch and the ratio of plastic to aluminum was varied in 11 steps from all aluminum to all plastic. Velocities of compressional and shear waves in these plates were measured and the results are shown in Figure 3. Good agreement was found for the shear waves, but a marked variation from the theoretical values was shown by the compressional waves.

The cause of this discrepancy was thought to be a violation of the assumption that the wave length of the waves was sufficiently long as compared to the plate thickness. A second series of plates was constructed whose total thickness was only $\frac{1}{16}$ inch and good agreement was found with the theoretical values.

As further check on the effect of lamina thickness, the exact phase velocity curve for a two-layer plate with a plastic to aluminum ratio of .040/.085 was computed. The exact period equation was formulated according to Haskell's matrix method (1953) and solved numerically on the Seismological Laboratory's electronic computer, the Bendix G15D. The results are shown in Figure 4. These

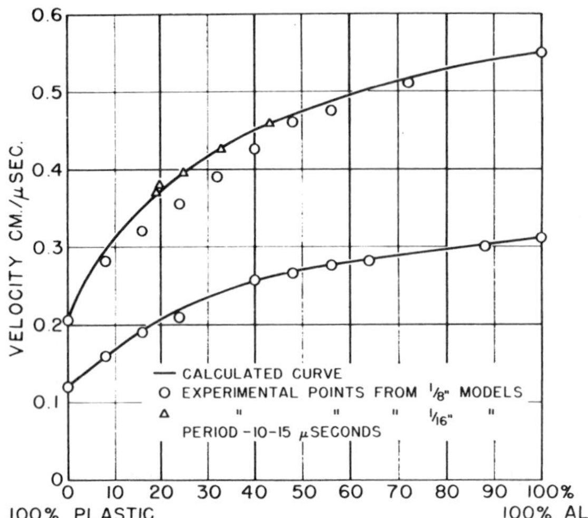


FIG. 3. Compressional and shear wave velocity in plastic and aluminum laminated plates.

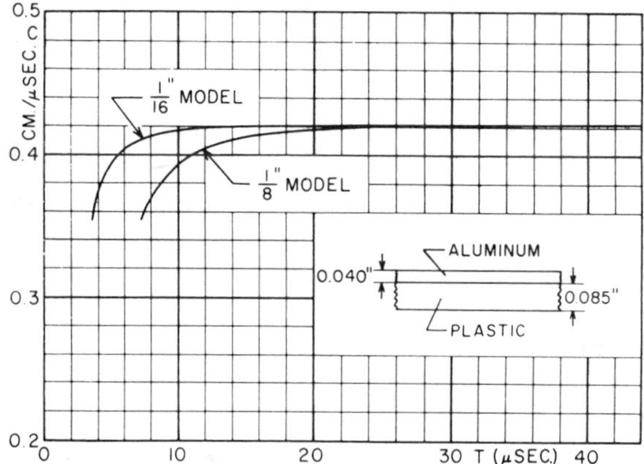


FIG. 4. Phase velocity for a two layer plate .040 inch aluminum and .085 inch plastic.

curves verify that for the periods used, 10–15 micro/sec, there is marked dependence of phase velocity on period in a $\frac{1}{8}$ inch plate. It is interesting to note that the dispersion is more severe for the laminated plate than for the homogeneous plate. We believe this is a result of coupling between the symmetric and anti-symmetric waves in the laminated plate. The dispersion in the $\frac{1}{16}$ inch plate is negligible for these periods.

It was concluded that with the materials and frequencies used in our experiment, total plate thickness should be less than $\frac{1}{16}$ inch. In the particular experiment described below, all models were less than $\frac{1}{32}$ inch in thickness.

### PHASE VELOCITY IN A TWO-LAYER CRUST

As a preliminary to use of this method in the study of wave propagation in the earth's crust, an attempt was made to reproduce the Rayleigh wave phase velocity in a two-layer crust. Since dispersed surface waves may be regarded as multiply reflected and refracted body waves, a powerful test of a model is a comparison of experimental and theoretical dispersion curves. For this comparison we have selected a two-layered crust overlying a mantle, as computed by Haskell. The first model constructed modelled only the shear wave velocity, neglecting the density and compressional wave velocity.

As the Rayleigh wave velocity is predominately controlled by the rigidity of the medium, one would expect that variations in density and compressional velocity would have only secondary effects on the Rayleigh wave velocity.

The model was a disk 24 inch in diameter with thickness as shown in the top cross section of Figure 5. Both phase and group velocity were measured on this model and the results are shown in Figure 6. The phase velocity falls below the expected values, which was anticipated since we failed to model density.

In the second attempt, we had improved our technique so that we were able to cut a contour in the aluminum sheet and in the plastic, and were able to model both the density and the shear velocity. The lower cross section of Figure 5 shows the thicknesses used in this model.

For the second model, the phase velocity was measured in three ways. One method was to excite the disk model with a continuous sine wave source so as to set up a standing wave on the edge of the disk composed primarily of the Rayleigh wave trains traveling in opposite directions around the disk. This is analo-

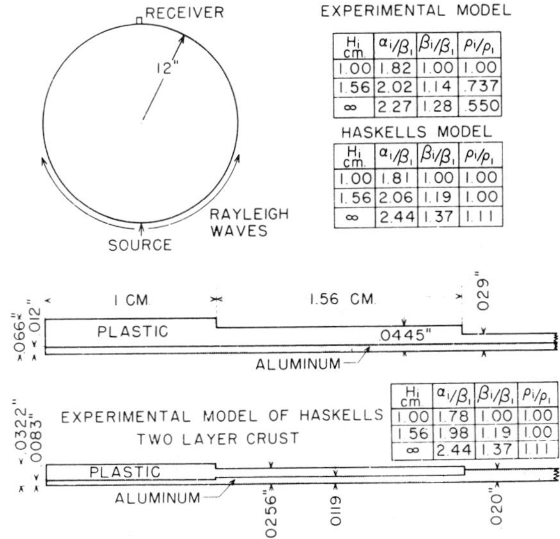


EXPERIMENTAL MODEL

| $H_i$ cm | $\alpha_i/\beta_1$ | $\beta_i/\beta_1$ | $\rho_i/\rho_1$ |
|---|---|---|---|
| 1.00 | 1.82 | 1.00 | 1.00 |
| 1.56 | 2.02 | 1.14 | .737 |
| ∞ | 2.27 | 1.28 | .550 |

HASKELLS MODEL

| $H_i$ cm | $\alpha_i/\beta_1$ | $\beta_i/\beta_1$ | $\rho_i/\rho_1$ |
|---|---|---|---|
| 1.00 | 1.81 | 1.00 | 1.00 |
| 1.56 | 2.06 | 1.19 | 1.00 |
| ∞ | 2.44 | 1.37 | 1.11 |

EXPERIMENTAL MODEL OF HASKELLS TWO LAYER CRUST

| $H_i$ cm | $\alpha_i/\beta_1$ | $\beta_i/\beta_1$ | $\rho_i/\rho_1$ |
|---|---|---|---|
| 1.00 | 1.78 | 1.00 | 1.00 |
| 1.56 | 1.98 | 1.19 | 1.00 |
| ∞ | 2.44 | 1.37 | 1.11 |

FIG. 5. Cross-section of two models with tabulated values of constants.

gous to exciting a free mode of the earth. The phase velocity could be determined by counting the nodes around the disk. This method was most effective with the longer wave lengths (Figure 7). In the second method the Rayleigh wave train traveling in one direction from the source was damped so that the wave train traveling in the other direction propagated without interference, and the phase velocity could be determined by following a particular peak or trough around the disk (Figure 8).

The results of these two methods are shown in Figure 9. Good agreement exists between the theoretical and experimental results. Since the theoretical curve was for a half space it was necessary to apply a correction for the curvature of the disk model. An approximate correction was made for each theoretical phase velocity point by multiplying the phase velocity by a factor which would be correct for a homogeneous model. The correction for cylindrical curvature was originally given by Sezawa and later corrected by Oliver (Ewing, Jardetzky and Press, p. 265).

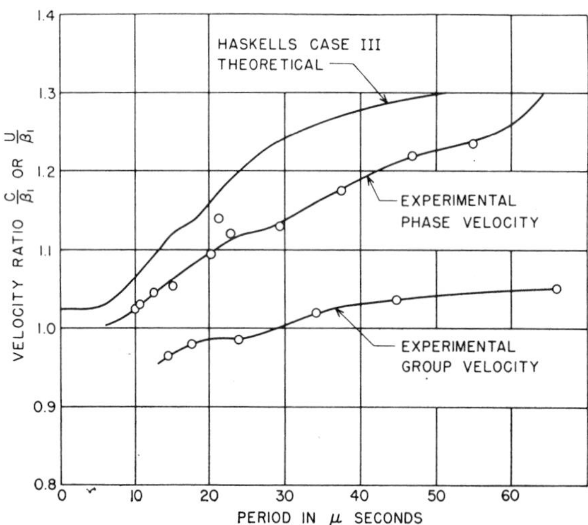


FIG. 6. Rayleigh wave phase and group velocity for a two layer crust (see upper Figure 5 for constants).

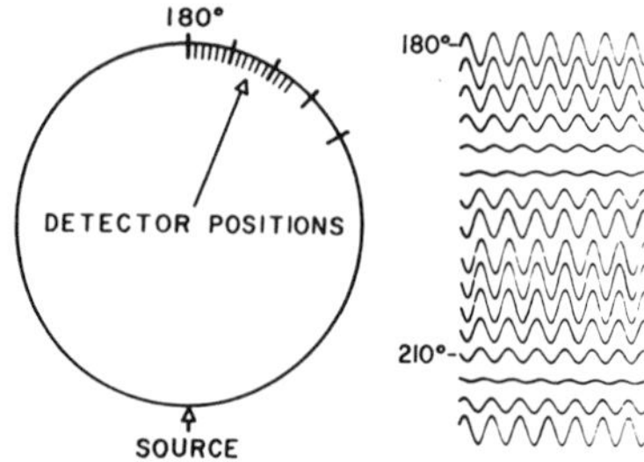


FIG. 7. Seismogram showing standing Rayleigh wave.

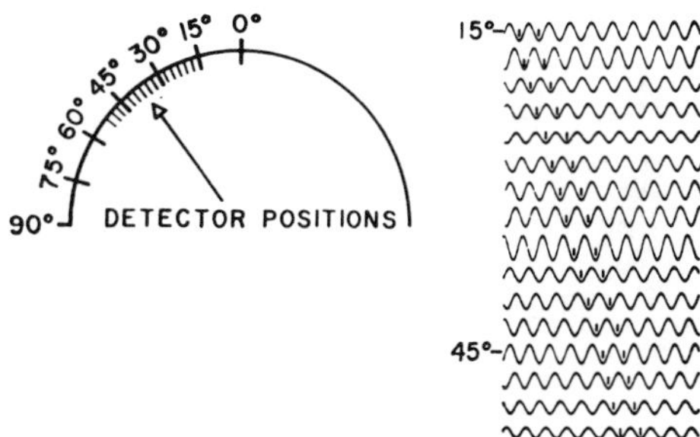


FIG. 8. Traveling Rayleigh wave from continuous source.

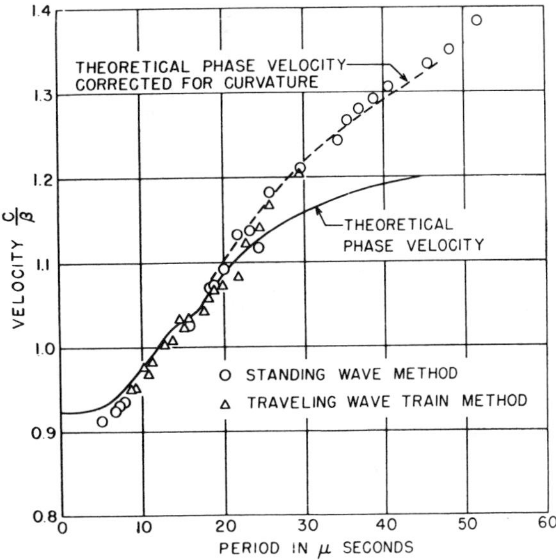


FIG. 9. Theoretical and experimental phase velocity (constants given in lower part of Figure 5).

While these two methods give fairly good agreement with the theoretical phase velocity, it is desirable to improve these results further to distinguish the finer features of the phase velocity curve. Two additional methods were tried; the method of Brune and a Fourier analysis method. The precision of the Fourier analysis method indicates that a high accuracy can be achieved by this modeling technique, and the comparison of the four methods of phase velocity determination is a good illustration of an important problem that can be studied with this type of model. Brune's method, based on the stationary phase technique, makes use of the arrival times of peaks and troughs of the dispersed wave train to determine phase velocity (Brune, Nafe and Oliver, in press).

The Fourier analysis made use of the following relationship between the phase of a Fourier component at two distances:

$$C = \frac{(X_b - X_a)\omega}{\phi_b - \phi_a + 2n\pi} = \text{Phase Velocity}$$

$X_b$, $X_a$—are distances to points $a$ and $b$.
$\phi_b$, $\phi_a$—apparent phase angles of frequency component $\omega$ measured with respect to origin time.
$\omega$—angular frequency.

$$\frac{\phi_b - \phi_a + 2n\pi}{\omega} = \text{phase delay time between points } a \text{ and } b.$$

The seismograms in Figure 10 were digitized and subjected to a Fourier analysis on our electronic digital computer. This analysis yielded amplitude and phase of the Fourier spectral components. The latter was used to derive the experimental phase velocity dispersion curve (Figure 11). Excellent precision is obtained by this method as evidenced by the agreement between the experimental data and theoretical curves. The small bend in the theoretical curve at 17 micro/sec is indicated by the experimental phase velocities. The systematic discrepancy that does exist is less than one half of one percent and is well within the error expected from the inaccuracies in the machining process.

## CONCLUSION

This study shows the feasibility of constructing two-dimensional models in which a body velocity and density can be made to vary continually with depth. A successful test was made by comparing theoretical and experimental dispersion curves of Rayleigh waves in a double layered crust.

In future papers we will report on further applications of this modeling technique: (1) the effect of the low velocity upper mantle layer on body wave amplitudes (the shadow zone problem); (2) methods of deducing the properties of the source by operating on dispersed surface waves.

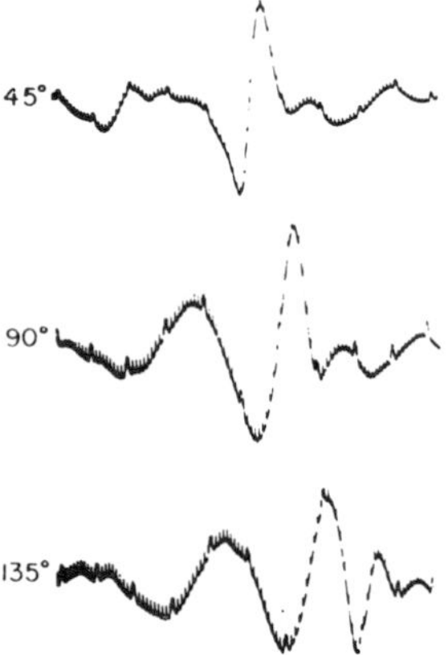


Fig. 10. Seismograms showing Rayleigh wave dispersions at three distances.

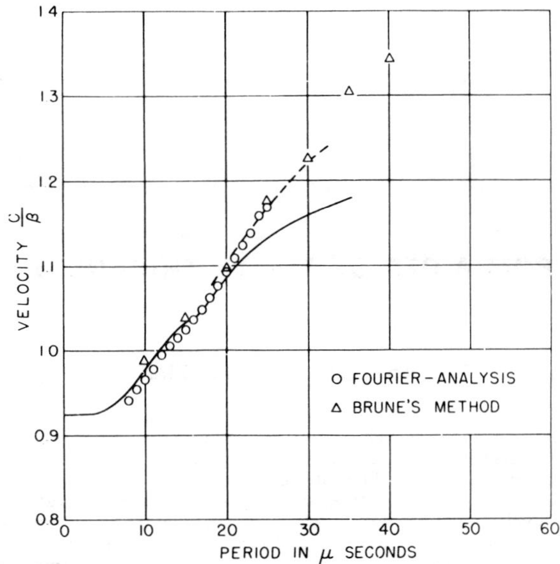


FIG. 11. Experimental dispersion curves according to Fourier analysis method and Brune's method together with theoretical curve. Dashed curve includes correction for curvature.

## ACKNOWLEDGMENTS

This research was partially supported by grants from the Office of Ordnance Research and the American Petroleum Institute.

## REFERENCES

Brune, J., Nafe, J., and Oliver, J., in press, A simplified method of analyzing and synthesizing dispersed surface waves: Jour. Geophys. Res.

Ewing, M., Jardetzky, W., and Press, F., 1957, Elastic waves in layered media: New York, McGraw-Hill.

Haskell, N. A., 1953, The dispersion of surface waves in multilayered media: Bull. Seis. Soc. Amer., v. 43, p. 17–34.

Oliver, J., Press, F., and Ewing, M., 1954, Two-dimensional model seismology: Geophysics, v. 19, p. 202.

# HEAD WAVES FROM A BED OF FINITE THICKNESS*

FRANKLYN K. LEVIN† AND JOHN D. INGRAM†

The behavior of the head wave from a high-speed layer embedded in a low-speed half-space has been investigated with two-dimensional seismic models. Twelve layer thicknesses ranging from four wavelengths to one-tenth wavelength were used. A simple theory based on interference between the head wave and the reflections from the bottom of the layer gave amplitude-distance values which agreed with the observations for layer thicknesses down to about one-third of a wavelength. For thick layers, the experimental amplitude dependence on distance was different from the theoretical law (−3/2 power of the distance). The velocity minimum for thin layers discovered by Lavergne was confirmed and a possible slight velocity maximum at intermediate layer thicknesses noted. The velocity for zero layer thickness appeared to be greater than the free bar velocity. Systematic variations of head-wave spectra with layer thickness occurred.

## INTRODUCTION

Although the behavior of the head wave (refracted wave, lateral wave) from a layer over a half-space has been understood for some time (Heelan, 1953; Ewing et al, 1958), properties of head waves from more complicated systems are still largely unknown. A system of some practical importance is a layer embedded in a half-space. This system is difficult to handle mathematically (Rosenbaum, 1961) and most of the investigators have used seismic model techniques. Over a period of years, Riznichenko, Shamina, and Davydova studied with models both solid layers in fluid half-spaces and solid layers in solid half-spaces (Riznichenko and Shamina, 1957 and 1959; Davydova, 1959; Shamina, 1960). Somewhat earlier, Press, Oliver, and Ewing (1954) investigated the related system of a thin solid layer under another layer having a lower velocity and over a higher-velocity solid half-space. O'Brien (1957) used field data to study multiply-reflected head waves in a shallow layer. Most recently, Lavergne (1961) in a beautiful series of experiments looked at the velocity and amplitude dependence of the head wave. His models were two-dimensional and consisted of butt-jointed thin sheets of plexiglass and duralumin. Of the investigations mentioned above, only that of Lavergne involved a continuous series of models spanning the full range from layers much thicker to those much thinner than a wavelength. At the time Lavergne reported his results, we had just completed an identical study with models made of lucite (plexiglass) and duralumin. Our results largely confirmed those of Lavergne but there were some differences. Also, we were able to explain quantitatively the observed amplitude dependence with distance and qualitatively, the observed frequency spectra.

## EXPERIMENTAL PROCEDURES

The models were constructed of 7- to 8-ft long, 3/16-inch thick sheets of duralumin and lucite (plexiglass); these were cemented together along

* Manuscript received by the Editor April 23, 1962.

† Jersey Production Research Company, Tulsa, Oklahoma.

Reprinted from Geophysics, 27, 753-765

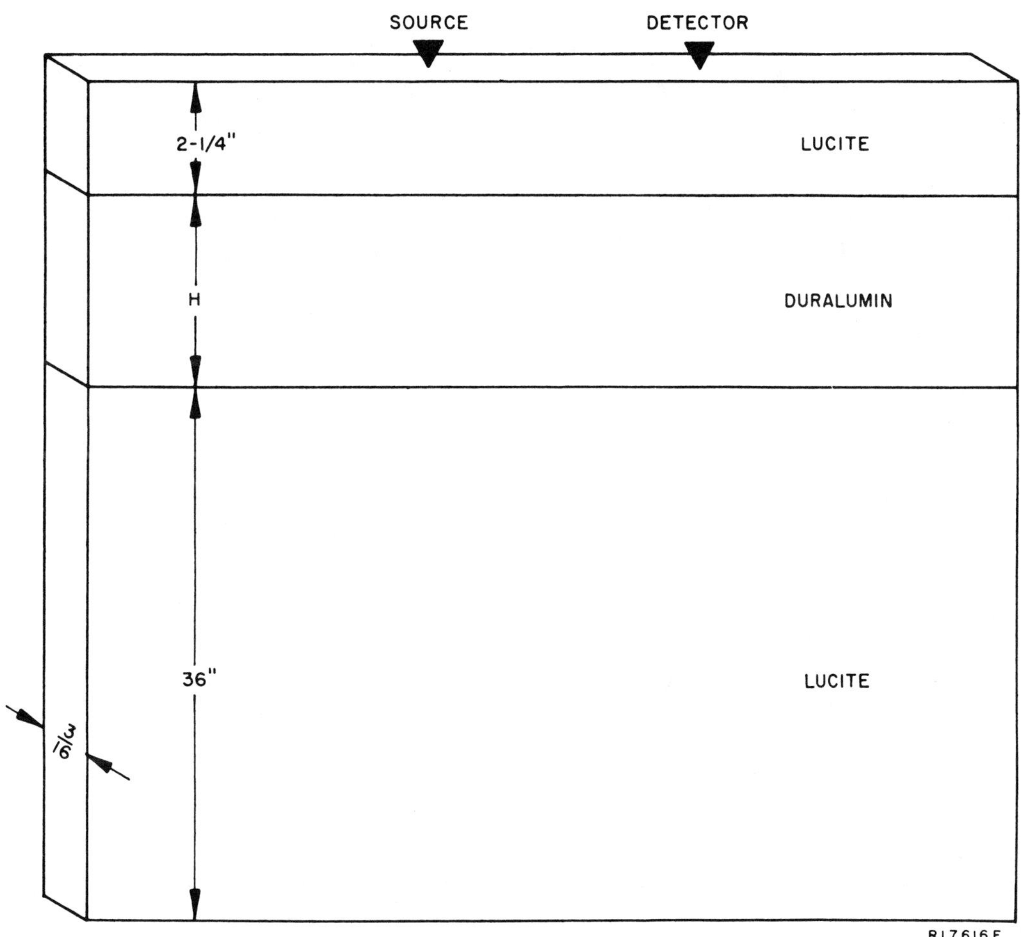


FIG. 1. A typical two-dimensional model. $H$ values from $10\frac{1}{16}$ inches to $\frac{1}{4}$ inch were used.

their edges with Armstrong C-7 epoxy cement (Figure 1). Velocities and densities for lucite and duralumin have been given by Oliver et al (1954). Because of the sheet thickness used, the compressional wave velocities we found (17,600 ft/sec and 7,600 ft/sec) were lower than the true plate velocities, but for duralumin, the medium of most importance in our study, the difference was negligibly small. Ultrasonic pulses were generated and detected with standard seismic model equipment (Levin and Hibbard, 1955). Our transducers were made of lead zirconate titanate.

The top lucite layer was $2\frac{1}{4}$ inches, the bottom lucite layer, 36 inches in depth (Figure 1). Duralumin layers from 10 1/16 inches to $\frac{1}{4}$ inch were used. A typical head wave arrival and its spectrum appears in Figure 2. If we assume the equivalent wavelength is the velocity divided by the frequency at the peak of the spectrum, we find wavelengths of about one and $2\frac{1}{4}$ inches in the lucite and duralumin respectively. Thus, our duralumin layers covered the range of four wavelengths to one-tenth wavelength while the bottom lucite layer was effectively an infinite half-space.

In analyzing our data we took as the pulse amplitude the height from the first valley to the first peak ($A_{23}$ in the notation of Riznichenko and Shamina (1959)). To reduce amplitude scatter to a reasonable value, the detector was coupled and recoupled at least four times and the values averaged. As the experiments continued, the coupling of the source changed from day to day

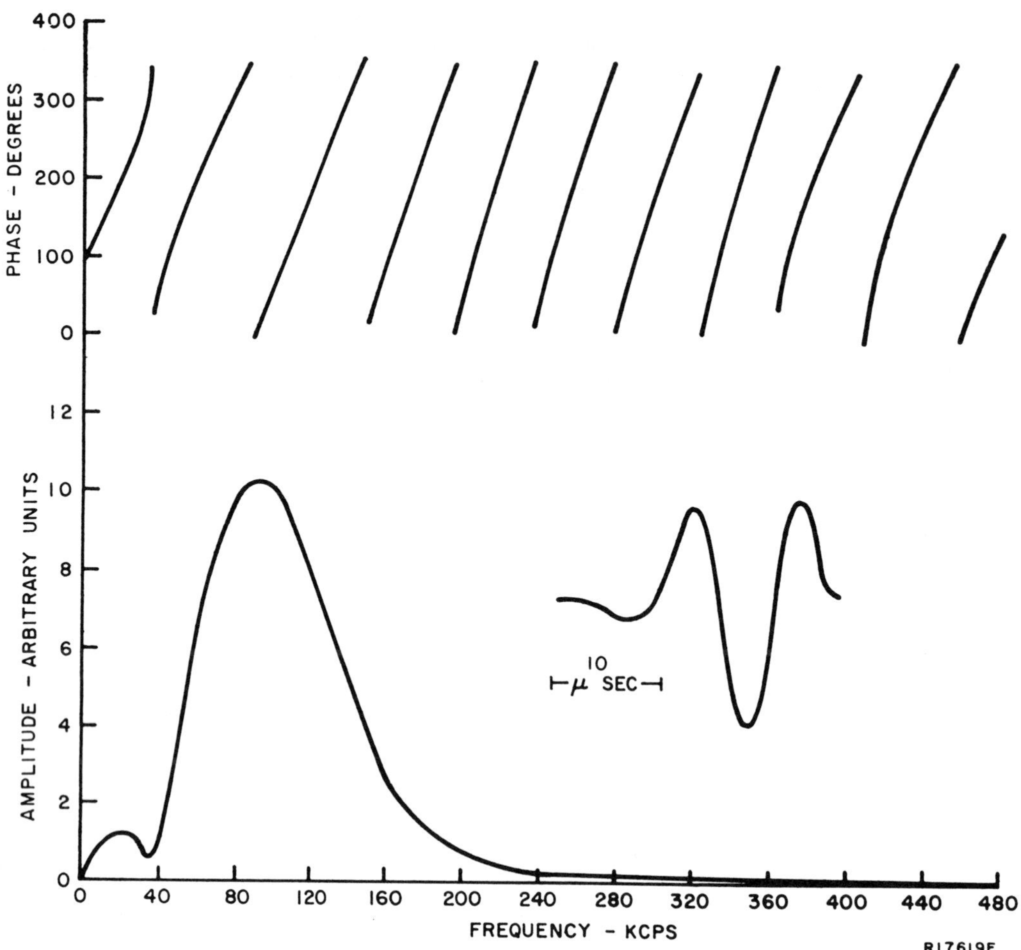


FIG. 2. A typical head-wave arrival with its spectrum.

and the electronic components of the equipment aged slowly. Consequently, while the data from any given run were consistent internally, an absolute comparison of values from layers of different thickness could not be made. We did detect changes in the amplitude-distance plots but were unable to say where one plot would have fallen relative to another if the source conditions and system sensitivity had remained constant.

For each of the duralumin layers, the refraction waveform was recorded at a source-detector separation of 15 inches. We digitized the waveforms on a $\frac{1}{2}$-microsecond spacing and computed their spectra with an 800-point computer program. For the thicker duralumin layers, $3\frac{1}{2}$ inches and above, the refraction arrival was distinct and there was no question as to where it ended. Waveforms from the thinner layers did not stop so obviously. These we terminated smoothly but arbitrarily, requiring that the pulse duration be the same as that for a thicker layer within plus or minus one microsecond. Since the amplitude was small over the region of termination, the precise curve chosen to end a given waveform was unimportant.

Arrival times of the first break and the first few valleys and peaks were measured directly on the face of the cathode ray oscilloscope. Through the time values for each cycle we passed least-square straight lines. The first reliable "pick" was that of the first valley; at great distances and in thin layers the first breaks were uncertain.

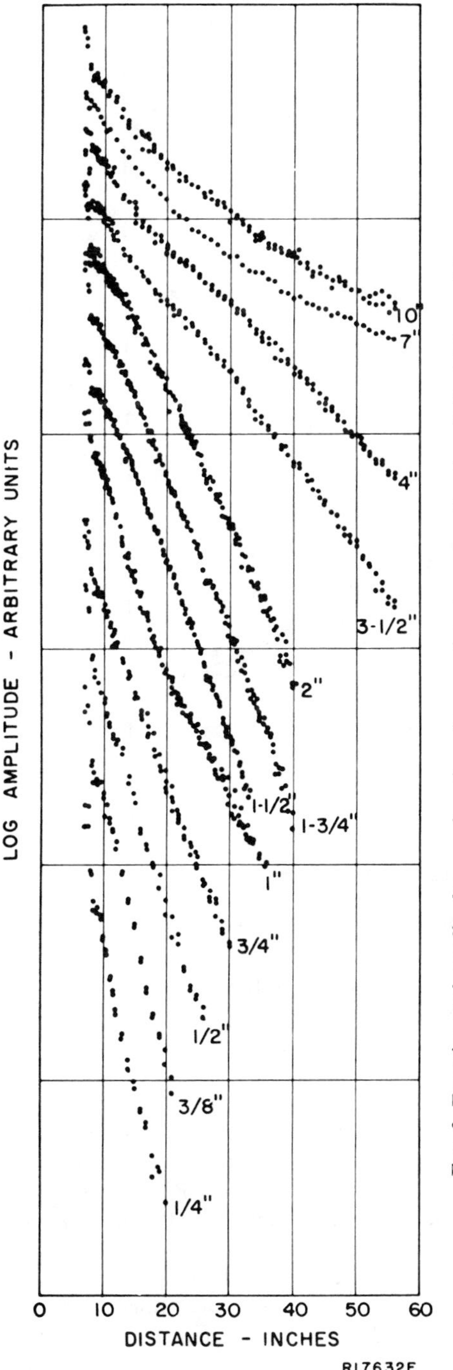


FIG. 3. Experimental amplitudes as a function of source-detector separation for all the high-speed layers.

Although the standard deviations of the slopes of first-break straight lines were not unduly large (Figure 5), it is probable that this indicates consistent, not accurate picks.

### EXPERIMENTAL RESULTS

Figure 3 shows all the amplitude-distance data. As mentioned above, each point represents the average of four or more independent measurements and the shapes, but not the relative positions, of the curves are significant. (We did note a marked decrease in the absolute amplitude at a given distance as the layer became thinner.) The 10- and 7-inch plots are similar; probably they represent the amplitude-distance relationship for a lucite layer over a duralumin half-space. Attempts to fit the 10-inch plot with a $-3/2$ power of the distance law (Heelan, 1953) failed. The data were fitted quite closely (Figure 4) by

$$A = A_0 x^{-1.042} e^{-0.0140x}. \qquad (1)$$

The cause of this discrepancy between theory and experimental results is being investigated. Tentatively, we attribute it to the fact that we were not measuring the amplitude of the very first energy to arrive.

The 4-inch curve of Figure 3 breaks at about 30 inches. Beyond 30 inches, the amplitude falls off more quickly than at shorter distances. As the duralumin layer gets thinner, the break in the curve moves toward smaller source-detector separations while the rate of amplitude decrease beyond the break increases. For layers thinner than one inch, the break is missing and the rate of amplitude decrease is large. However, the correspondence between amplitude decay and layer thickness is not a simple one. The $1\frac{1}{2}$-, $1\frac{3}{4}$-, and 2-inch curves are all steeper than the one-inch curve. The plots for layers thinner than $\frac{3}{4}$ inch again show a rapid amplitude decay. Within experimental error, the $\frac{1}{4}$- and $\frac{3}{8}$-inch curves are the same.

Velocities were computed as the reciprocal slopes of the time-distance curves. Figure 5 is a plot of the velocities of the first break and first valley against layer thickness; the deviations indicated are standard deviations of the slopes computed from the usual expression. Where no deviations are indicated, they are smaller than the diameter of the plotted point. In agreement with

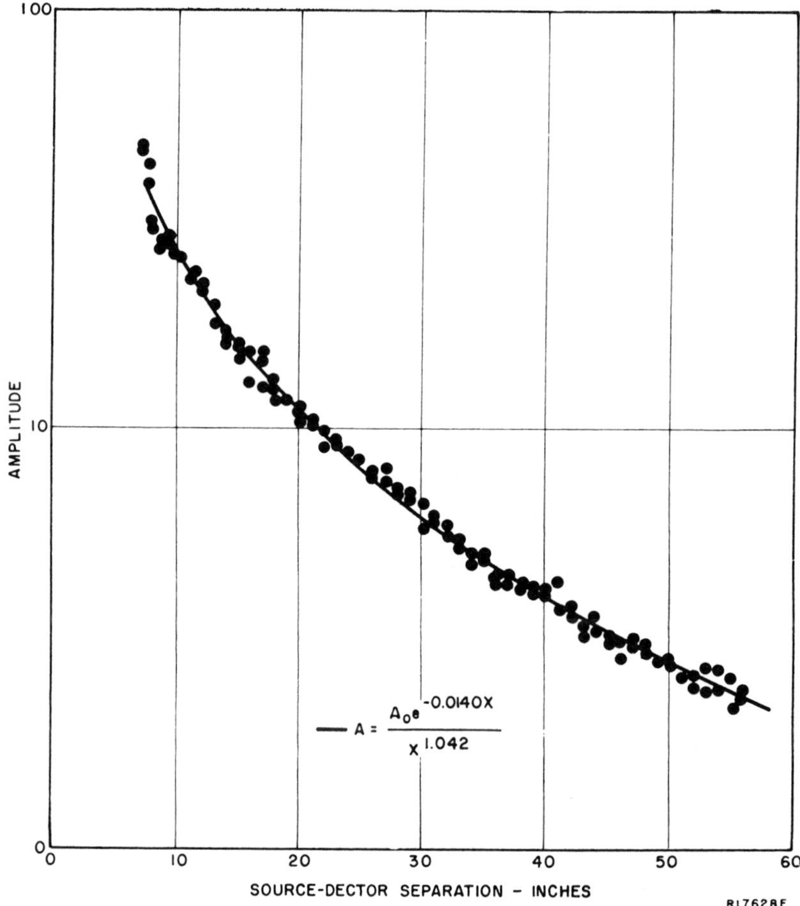


FIG. 4. A curve fitted to the experimental amplitude-distance data for the 10-inch layer.

the results of Lavergne (1961), the velocity goes through a minimum at small thicknesses. The lowest point is about $4\frac{1}{2}$ percent less than the velocity for a 10-inch layer. In addition there is an apparent velocity increase in going toward the minimum from thick layer values. The rise, about one percent, is well outside the spread of the standard deviations but is still small. We have no explanation for the rise and are inclined to discount it.

Confirmation of the overall shape of the velocity versus layer thickness curve was provided by the later peaks and valleys (Figure 6). These showed somewhat deeper minima (6 to $8\frac{1}{2}$ percent) and greater rises (1 to $2\frac{1}{2}$ percent) than did the first valley curve. With increasing cycle number, the minima deepened and the lowest point of the curves moved from a layer thickness of $\frac{3}{8}$ inch for the first valley plot to $\frac{3}{4}$ inch for the third valley data.

The end point of the curves of Figure 5, corresponding to a duralumin layer of zero thickness, could not be determined experimentally. In contrast to the conclusions of Riznichenko and Shamina (1959) and Lavergne (1961), we found that the velocity in a free duralumin bar, about 16,700 ft/sec, was too low to fit our data. From Figure 5, we would expect a zero thickness limiting velocity higher than that of a free bar but lower than the velocity of compressional waves in an infinite medium. It seems reasonable that the

(*Text continued on page 760*)

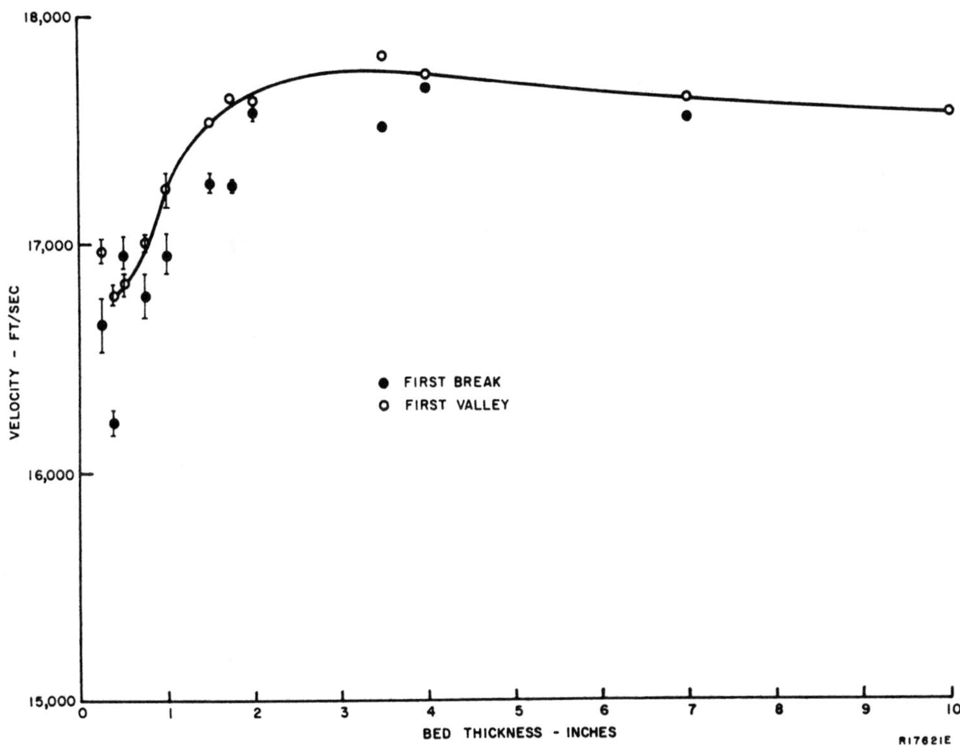


FIG. 5. Velocity from the slope of time-distance straight lines as a function of layer thickness—early cycles.

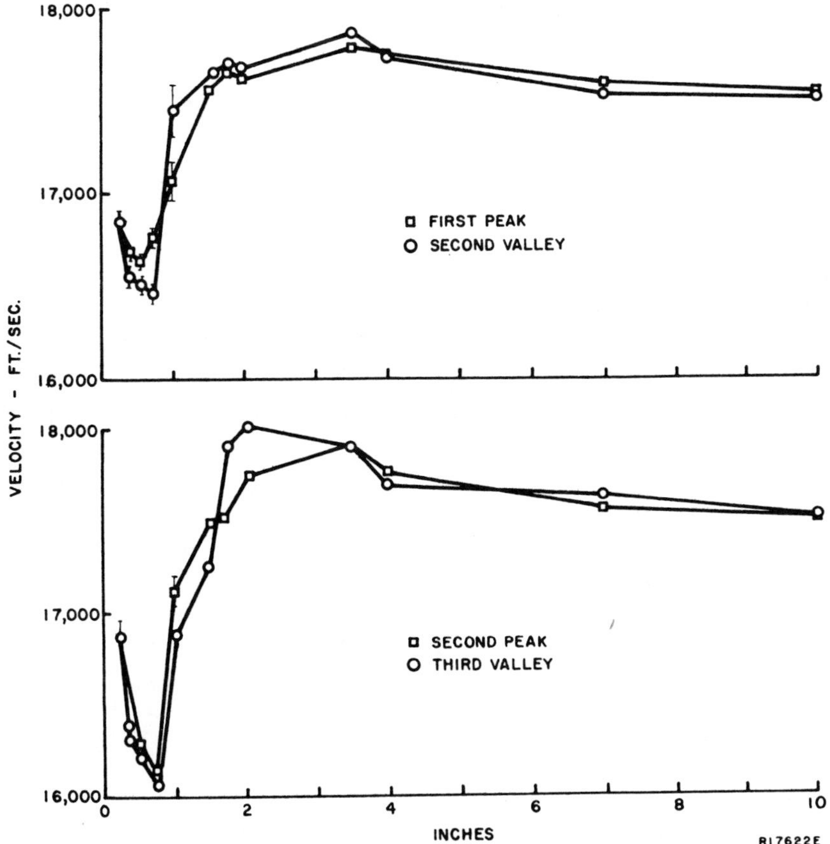


FIG. 6. Velocity from the slopes of time-distance straight lines as a function of layer thickness—later cycles.

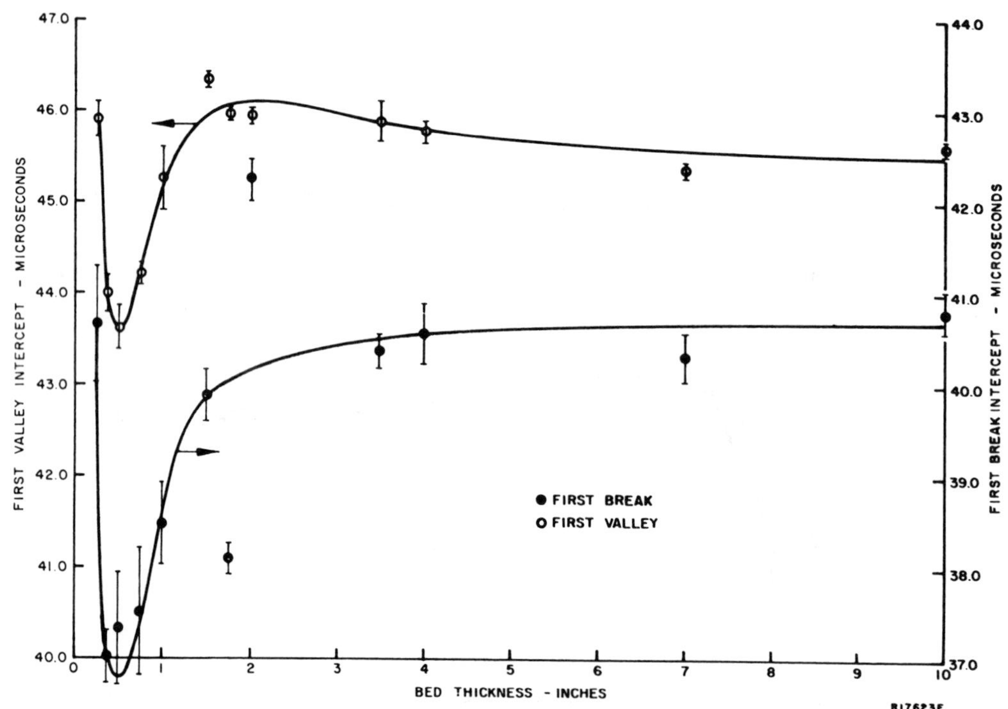


FIG. 7. Intercepts of the time-distance lines as a function of layer thickness—early cycles.

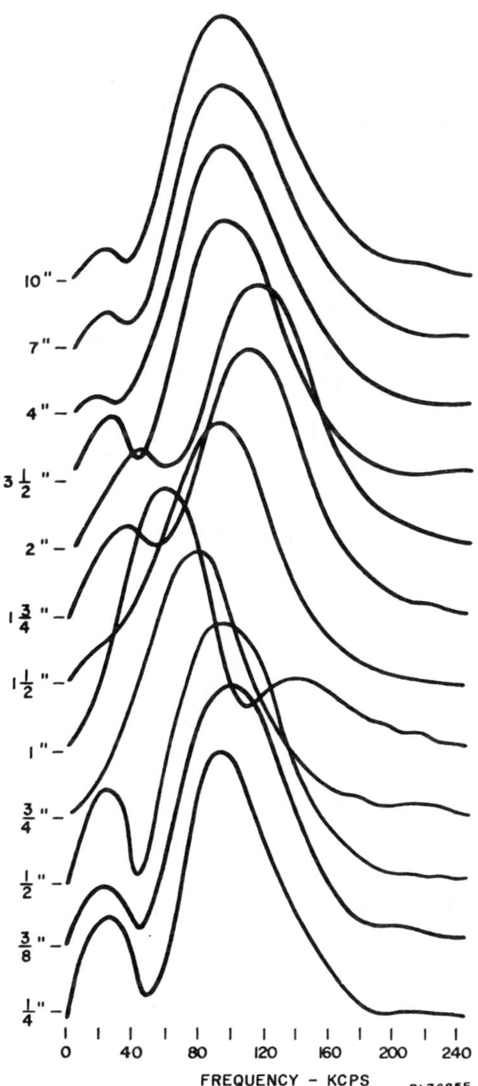


FIG. 8. Frequency spectra of the head-wave arrivals at a source-detector separation of 15 inches for different layer thicknesses.

limiting value might be different from the free bar velocity. Whether it should be higher than the free bar velocity, indicating stiffening of the duralumin by the lucite, or lower, corresponding to a mass loading of the duralumin, is not obvious. We hope to examine this point at a later time.

Confirming the results of Lavergne (1961), the intercepts of our least-squares lines also went through a minimum (Figure 7). For the first breaks and first valleys, this occurred at a layer thickness of $\frac{1}{2}$ inch. Thus, not only did the slopes of the time-distance lines increase as the duralumin layer thinned, but the lines intercepted the time axis at lower values. Unlike the slopes which demand only that a given cycle be picked in a consistent manner, the intercepts depend critically on the absolute values of the source-detector separation and travel time. Hence, even though the standard deviations indicated in Figure 7 are not excessive, the first valley data points scatter more than those of Figures 5 and 6. The scatter of the first-break intercepts is even worse, again a reflection of our inability to decide just where the deviation from the base line occurs.

Figure 8 displays the frequency spectra at a 15-inch source-detector separation for the different duralumin layer thicknesses. Basically, all the spectra are bell-shaped with one prominent maximum and, in some cases, a smaller maximum at lower frequency. The most notable variation is a smooth shift in the position of the major maximum from 93 kcps for the thicker layers, up to 110 kcps, down to 58 kcps, and finally to 94 kcps for the thinnest layer. Equally systematic but less striking are the changes in the amplitude and position of the low-frequency maximum and minimum. The phase curves corresponding to the amplitude spectra of Figure 8 are linear over the major peak.

#### INTERPRETATION OF EXPERIMENTAL RESULTS

Riznichenko and Shamina (1959) and Shamina (1960) have emphasized that for thin layers at any appreciable distance from the source, the first arrival will be a combination of the head wave and the reflections within the layer. Apparently no one previously has tried to use this concept to predict the observed amplitude-distance relationships. For our models, the interfering events are the head wave and the reflection and multiple reflections from the lower duralumin boundary. The reflections and head waves are turned over relative to each other and they have different shapes. In theory, the reflection pulse is the time derivative of the head-wave pulse. In practice, we found it necessary to determine the exact pulse shapes by experiment. The head-wave shape was taken to be that detected on the top lucite surface when the duralumin layer was 10 inches thick. To find the shape of the reflection event, we built a special model consisting of a $2\frac{1}{4}$-inch

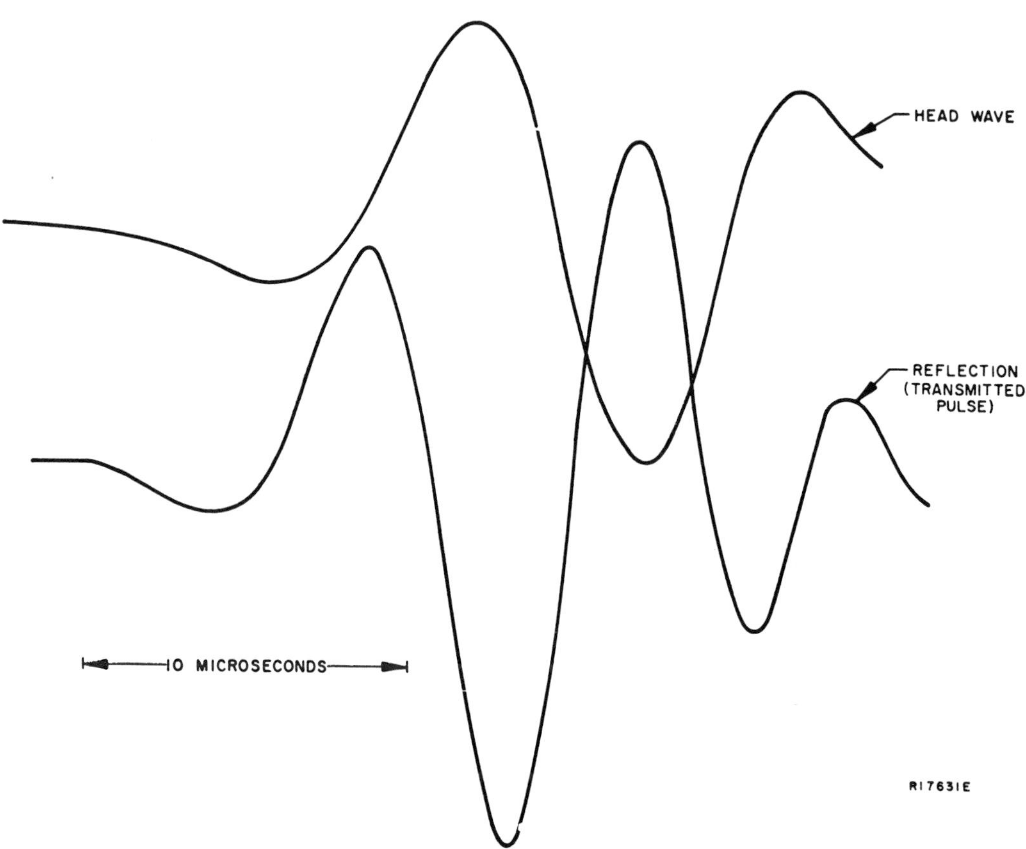


FIG. 9. Comparison of the head wave and reflection (transmitted pulse) from the bottom of the high-speed layer.

lucite layer over a 10-inch duralumin layer. With our source on top of the lucite and detector on the bottom of the duralumin, we determined not only the shape of the transmitted pulse (which for a thick layer is the same as that of the reflected pulse) but also the amplitude-distance relationship for the transmitted (reflection) pulse. After the application of appropriate corrections (see Appendix), the amplitude decay of the transmitted pulse turned out to be the same as that of the head wave within experimental error.

Figure 9 is a comparison of the head-wave and reflected (transmitted) pulses; since we did not know precisely where the first breaks were, we have arbitrarily made the first zero crossing of the reflection coincide with the first minimum of the head wave.

In Figure 10 the amplitude-distance data for the transmitted pulse are superimposed on a line drawn through the 10-inch layer data of Figure 3. Because of the widely different acoustic impedances of the lucite and duralumin, we could not use the transmitted pulse to find relative initial amplitudes of the head wave and reflection. Fortunately, over a range of source-detector separations, the head wave and the reflection from the bottom of the duralumin were separated from one another and from all other arrivals. By direct comparison, it was found that the initial amplitude of the reflection was 6.7 times that of the head wave. Plane-wave reflection and transmission coefficients were assumed and the amplitude decrease with distance was computed for duralumin layer thicknesses down to $\frac{3}{4}$ inch. Details of the calculations are given in the Appendix.

Figure 11 shows the computed points superimposed on curves drawn by eye through the data of Figure 3. The agreement is very good. The computed values begin to deviate from the experimental curves for the $\frac{3}{4}$-inch layer. For thinner

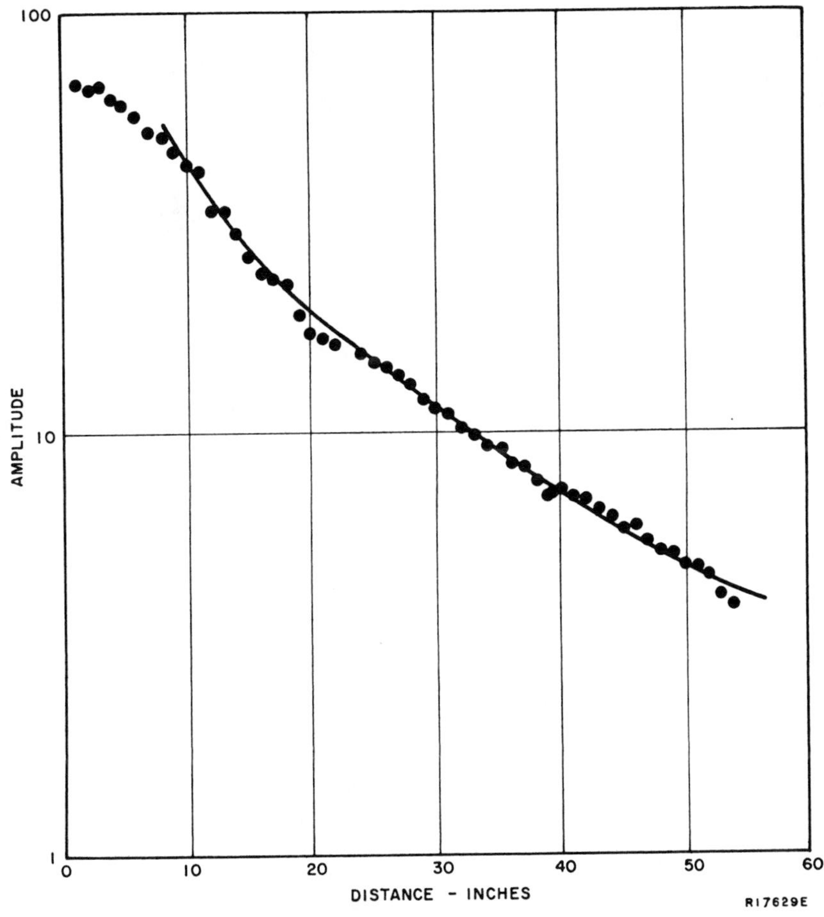


FIG. 10. Amplitude of the pulse detected along the bottom of the high-speed layer as a function of horizontal distance. The curve is a line drawn through the 10-inch layer data of Figure 3.

layers, the number of contributing multiples became so great that computation by this method became impractical. A different approach is required. We presume that formulation in terms of the leaking modes (Ewing et al, 1957) would yield an amplitude-distance expression suitable for thin layers. We have not examined this possibility.

At the time the amplitudes were computed, the velocity and frequency changes of Figures 5, 6, and 8 had not been found. Consequently, no attempt to predict them was made. It seems clear that careful summing of the complete head-wave and reflection waveforms as described above and in the Appendix would have reproduced the observed anomalies; however, to reproduce the amplitude-distance curves, the simplification of summing only at those times required to give amplitude $A_{23}$ was adopted. We have some evidence that interference between the head wave and reflections does produce frequency as well as amplitude changes. For the 10-inch layer, the distance of first interference was greater than our maximum source-detector separation and, except for a slight loss of high frequencies, the spectra of all head-wave pulses are the same (Figure 12). For the 4-inch layer, the distance of first interference was about 32 inches. Head-wave spectra for source-detector separations less than 32 inches are similar to those from the 10-inch layer but spectra at greater distances show

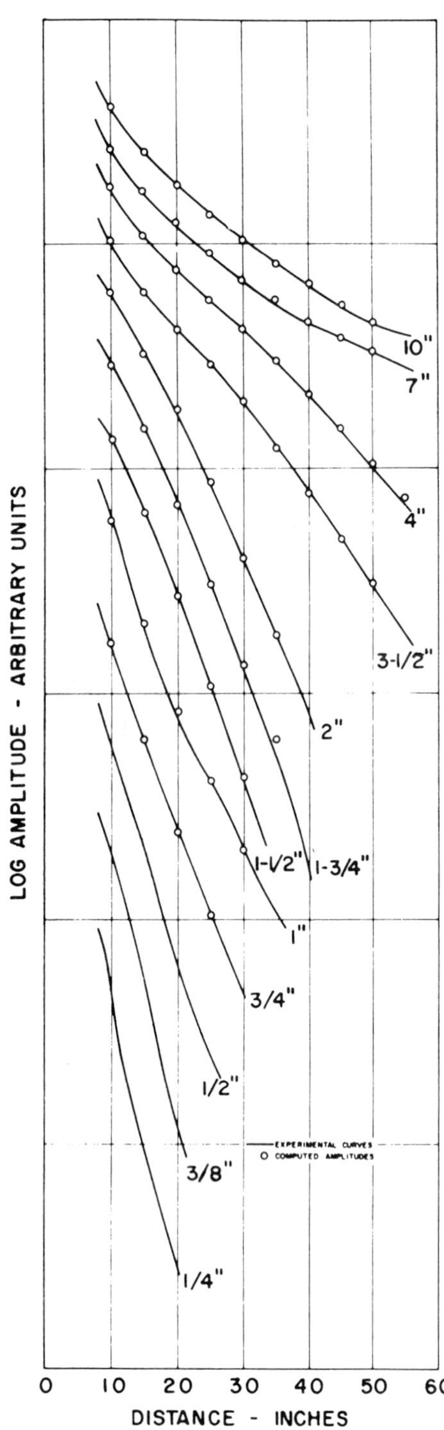


FIG. 11. Computed amplitude points superimposed on lines drawn by eye through the experimental amplitude-distance plots of Figure 3.

maxima shifted toward higher frequencies and a relatively larger proportion of high frequencies (Figure 12). Since the completion of the model studies, an exact theory of head waves from a bed of finite thickness has been developed (Dunkin, 1962). It is being used to examine a number of cases. As a result, we have not tried to fit the observed velocity and frequency data by the approximate, summing method.

## DISCUSSION

When we undertook the model program, we hoped to find some property or combination of properties of the head wave that would characterize uniquely the thickness of a high-speed layer embedded in a low-speed half-space. This hope was not realized. Where the velocity deviates from its thick-layer value, it is double-valued. The amplitude-distance curves are complex and do not drop uniformly as the layer thins. Finally, the spectra of the head wave depend on layer thickness but, unless one had a complete suite of spectra and knew the shape of the input pulse, the spectral variations would hardly be usable as a thickness indicator. Consequently, we found no good way to determine the thickness of an embedded layer.

On the other hand, we did answer some interesting questions. The velocity minimum discovered by Lavergne was confirmed and found to include later cycles of the head-wave complex. Previous investigators had hypothesized interference between the head wave and reflections as an important factor in thin-layer phenomena. We used this mechanism to explain quantitatively the observed amplitude-distance relationships and, qualitatively, the observed frequency changes. For layers not more than half a wavelength thick, experimental data can now be matched by a rather simple theory.

Some points remain to be investigated. One is the reality of the slight maximum in the velocity-versus-layer thickness curve; another, the value of velocity for a layer of zero thickness. Implicit in the latter is the development of a head-wave theory for very thin layers. As reported above, a theory good for thick layers and layers of intermediate thickness has been developed at this laboratory and will be reported later (Dunkin, 1962). With this, we expect to check apparent deviations from the $-3/2$ power of distance law.

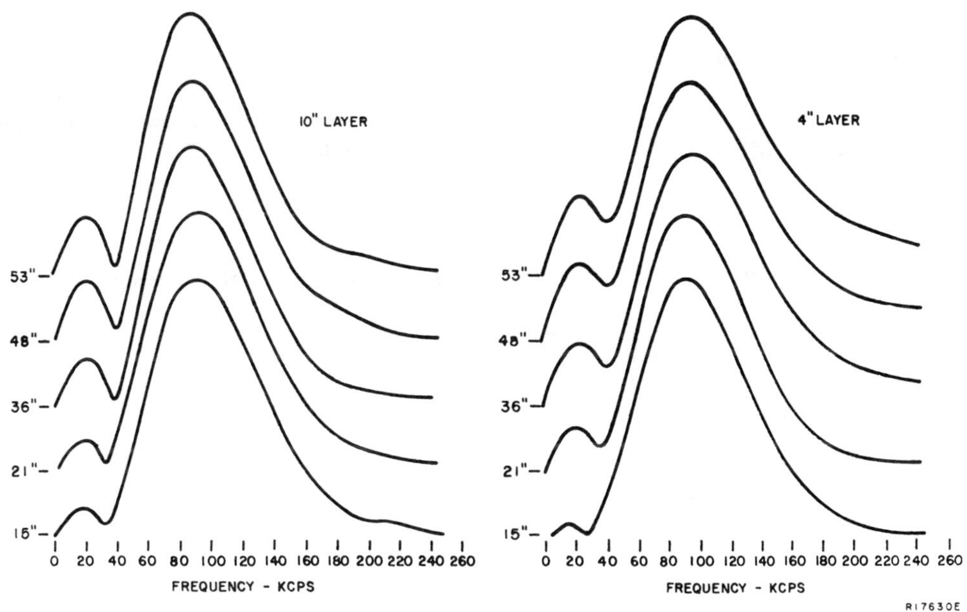


FIG. 12. Spectra for several source-detector separations for 10-inch and 4-inch high-speed layers.

ACKNOWLEDGMENTS

We wish to thank Jersey Production Research Company for permission to report this paper. We should like to note particularly the work of Donald J. Robinson who developed a technique of manufacturing two-dimensional, layered seismic models and took a large part of the data. Without his help, the project would have been impossible.

REFERENCES

Davydova, N. I., 1959, On the dependence of the amplitude of longitudinal head waves, associated with thin layers, from the velocity contrast of the media: Bull. (Izvestiya) Acad. of Sci., U.S.S.R., Geophysics Series, May, pp. 462–468.

Dunkin, J. W., 1962, A study of two-dimensional head waves in fluid and solid system: To be published.

Ewing, W. M., Jardetzky, W. A., and Press, F., 1957, Elastic waves in layered media: New York, McGraw-Hill Book Co., Inc.

Heelan, P. A., 1953, On the theory of head waves: Geophysics, v. 18, pp. 871–893.

Lavergne, M., 1961, Etude sur modele ultrasonique du probleme des couches minces en sismique refraction: Geophysical Prospecting, v. 9, pp. 60–73.

Levin, F. K., and Hibbard, H. C., 1955, Three-dimensional seismic model studies: Geophysics, v. 19, pp. 19–32.

O'Brien, P. N. S., 1957, Multiply reflected refractions in a shallow layer: Geophysical Prospecting, v. 5, pp. 371–380.

Oliver, H., Press, F., and Ewing, M., 1954, Two-dimensional model seismology: Geophysics, v. 19, pp. 202–219.

Press, F., Oliver, J., and Ewing, M., 1954, Seismic model study of refractions from a layer of finite thickness: Geophysics, v. 19, pp. 388–401.

Riznichenko, Yu. V., and Shamina, O. G., 1957, Elastic waves in a laminated solid medium, as investigated on two-dimensional models: Bull. of the Acad. of Sci. of the U.S.S.R., Geophysics Series, No. 7, pp. 17–37.

———, 1959, Elastic waves in layers of finite thickness: Bull. of the Acad. of Sci. of the U.S.S.R., Geophysics Series, March, pp. 231–243.

Rosenbaum, J. H., 1961, Refraction arrivals along a thin elastic plate surrounded by a fluid medium: J. Geophys. Research, v. 66, pp. 3899–3906.

Shamina, O. G., 1960, An investigation of the dynamic features of longitudinal waves in layers of different thickness: Bull. (Izvestiya) Acad. of Sci., U.S.S.R., Geophysics Series, August, pp. 754–760.

APPENDIX

The theoretical model chosen to correspond to the two-dimensional lucite and duralumin models was that of a line source on the surface of a layered half-space (Figure 1). A bundle of rays from the source striking the I–II boundary at an angle breaks up into four ray bundles. (Medium I is lucite; medium II, duralumin.) Compressional and shear waves are both reflected back into medium I and transmitted into medium II. The first of the transmitted waves to return to the surface is the compressional $P_1P_2P_2P_1$ return. The waves which will concern us shall be represented as

$$w_{\text{head}} = \frac{A_0}{X^{\gamma_0}} e^{-\alpha_0 x} f_0(t), \tag{2}$$

and

$$w_{\text{refl.}} = \frac{A_1}{X^{\gamma_1}} e^{-\alpha_1 x} f_n(t). \tag{3}$$

We shall assume that $f_n(t) = f_1(t - \delta_n)$, where $\delta_n$ is the difference in arrival times of the head wave and reflected waves. Experimentally, we found that $\gamma_0 = \gamma_1$, $\alpha_0 = \alpha_1$. Thus,

$$\begin{aligned} w &= w_{\text{head}} V_{10}' + \sum_{n=1}^{\infty} w_{\text{refl.}}{}^{(n)} T_{12}{}^{(n)} R_{21}{}^{(n)} T_{21}{}^{(n)} V_{10}{}^{(n)} \\ &= \frac{A_0}{X^{\gamma}} e^{-\alpha x} \left[ V_{10}' f_0(t) + \sum_{n=1}^{\infty} \frac{A_n}{A_0} f_1(t - \delta_n) T_{12}{}^{(n)} R_{21}{}^{(n)} T_{21}{}^{(n)} V_{10}{}^{(n)} \right] \\ &= \frac{A_0}{X^{\gamma}} e^{-\alpha x} \left[ V_{10}' f_0(t) + \frac{A_n}{A_0} \sum_{n=1}^{\infty} f_1(t - \delta_n) T_{12}{}^{(n)} T_{21}{}^{(n)} T_{21}{}^{(n)} V_{10}{}^{(n)} \right], \end{aligned} \tag{4}$$

where

$T_{12}{}^{(n)}$ = plane wave transmission coefficient of the I–II boundary for the $n$th reflection,

$R_{21}{}^{(n)}$ = plane wave reflection coefficient of the II–I boundary raised to the $(2n-1)$ power,

$T_{21}{}^{(n)}$ = plane wave transmission coefficient back into medium I for the $n$th reflection,

$V_{10}{}^{(n)}$ = vertical amplitude factor for the $n$th reflection.

As described earlier in the paper, we found that $A_n/A_0 = 6.7 \cdot f_0(t)$ and $f_1(t)$ were also determined using the model itself (Figure 9). Thus,

$$w = \frac{A_0}{X^{\gamma}} e^{-\alpha x} \left[ V_{10}' f_0(t) + 6.7 \sum_{n=1}^{\infty} f_1(t - \delta_n) T_{12}{}^{(n)} R_{21}{}^{(n)} T_{21}{}^{(n)} V_{10}{}^{(n)} \right] \tag{5}$$

Instead of computing $\alpha$ and $\gamma$, we used the experimental amplitude-distance curve corresponding to a 10-inch duralumin layer. All that remained was to obtain $T_{12}{}^{(n)}$, $R_{21}{}^{(n)}$, $T_{21}{}^{(n)}$, $V_{10}{}^{(n)}$, and $\delta_n$. The first four were calculated from the well-known equations for reflection and transmission at a solid-solid interface. The calculation was laborious but straightforward. We obtained $\delta_n$ as follows. Let

$T_0$ = arrival time of the head wave

and

$T_1$ = arrival time of the first reflected wave.

Then,

$$T_0 = \frac{4.50}{12 \cdot 7{,}600 \cos \xi_c} + \frac{(x - 4.50 \tan \xi_c)}{12 \times 17{,}600}$$

and

$$\delta_1 = \frac{4.50}{12 \cdot 7{,}600} \left( \frac{1}{\cos \xi} - \frac{1}{\cos \xi} \right) + \frac{1}{12 \cdot 17{,}600} \left( \frac{2H}{\cos \eta} - x + 4.50 \tan \xi_c \right).$$

Obviously, $\delta_n$ can be obtained by choosing the proper $n$ and multiplying $H$ by $n$:

$$\delta_n = \frac{4.50}{12 \cdot 7{,}600} \left( \frac{1}{\cos \xi} - \frac{1}{\cos \xi_c} \right) + \frac{1}{12 \cdot 17{,}600} \left( \frac{2nH}{\cos \eta} - x + 4.50 \tan \xi_c \right). \tag{6}$$

Here $\xi$, $\xi_c$, and $\eta$ are angles of incidence. A computer program was written to give $\delta_n$ in terms of $n$, $x$, and $H$.

# Chapter 3.
# Tests of Wave Theory

The distinguishing feature of the articles selected for this chapter is that they demonstrate the use of physical modeling as an independent check on predictions of mathematical models of wave propagation. (Based on this criterion, Healy and Press' 1960 article of the last chapter should be placed in this chapter. The tone and objectives of that paper seemed, however, more in spirit with Chapter 2.) Checking mathematical predictions is to be distinguished from experiments in which numerical results are obtained, as in Terada and Tsuboi (1927) wherein the dispersion curve is shown, but in which the results are not compared to a mathematical model.

A quick summary of the essential features of each paper follows.

+ Tatel (1954) was directly inspired by Lamb's original paper (1904), detailing "Lamb's problem" of the elastic response of an infinite half-space to a line source. Not only did Tatel produce a vertical component seismogram in remarkable agreement with Lamb's predictions but he also provided experimental verification of *P*-wave/Rayleigh-wave interconversion, and source-receiver reciprocity.

+ Clay and McNeil (1955) demonstrated the necessity of correction for sphericity when calculating reflection coefficients when the source is near the reflector.

+ O'Brien (1955) showed experimentally the theoretically predicted attenuation of headwave amplitudes as (r**-0.5)*(L**-1.5).

+ Granneman (1956) matched the diffraction of a *P*-wave from a wedge with the mathematically predicted amplitudes.

+ Sherwood (1958) tested the Haskell-Thomson matrix method for calculating full synthetic seismograms with his experiments on a block of aluminum.

+ Roever et al. (1956) tested the Cagniard-de Hoop method of calculating full synthetic seismograms with their water tank experiments.

+ Donato (1960), in a charmingly concise paper, demonstrated that the headwave wavelet is the time integral of the body-wave wavelet.

+ White (1960) not only demonstrated seismic reciprocity, but also illustrated reciprocity's practical utility in computing low-frequency radiation patterns.

+ Sorge (1965) demonstrated that the vertical and horizontal components of Rayleigh-wave particle motion decay with depth as predicted by theory. Sorge also showed that the particle motion hodograms were elliptical, again as predicted by theory.

+ Hilterman (1970) showed that diffraction modeling could be extended from individual wedges (a la Granneman) to entire irregular and discontinuous surfaces. The success of the numerical modeling was a tribute to the power of the Kirchhoff approximation for acoustic wave propagation.

+ Gangi and Mohanty (1971) performed an entirely different sort of test of the Kirchhoff approximation through their verification of the elastic analog to the optical principle known as *Babinet's principle.*

+ Melia and Carlson (1984) provided a test of layering-induced anisotropy in the framework of Postma's long-wavelength theoretical model. They also showed the dispersion that results when the wavelengths get short enough to be comparable to the layering.

+ Hsu and Schoenberg (1990) performed a model-driven set of experiments to prove out Schoenberg's theory of fracture-induced anisotropy.

+ Rathore et al. (1991) tested Hudson's theory of microcrack-induced anisotropy by creating synthetic sandstones with controlled void and spaces to simulate microcracks.

+ Ass'ad et al. (1992) showed that Hudson's first-order theory (and second-order, as well) may not be correct for rocks whose crack density exceeds 7 percent.

+ Hood and Mignogna (1993) also performed experiments on what is perhaps the only orthorhombic physical model ever built to specification.

# NOTE ON THE NATURE OF A SEISMOGRAM—II

By Howard E. Tatel

*Department of Terrestrial Magnetism, Carnegie Institution of Washington, Washington 15, D. C.*

(Received April 29, 1954)

ABSTRACT

Model experiments have experimentally verified Lamb's calculation. Also they have corroborated the conversion hypothesis of Part I, in which compressional waves produce strong Rayleigh waves at a surface discontinuity. In addition, the reciprocity theory is shown to be valid, and the generation of Rayleigh waves from deep focus sources has been studied. Apparently, any obstacle in the path of a plane (compressional) wave acts as a scattering center, and if at the surface produces Rayleigh waves.

Single unidirectional impacts are made at a point on a steel block by brief voltage pulses applied to a piezoelectric crystal of barium titanate. The seismic waves are received and converted to electric pulses by another barium titanate crystal. Pulse amplifiers ("500" type) with good response from a few kilocycles per second to 6 Mc/sec amplify the pulses, which are displayed by an oscilloscope triggered by the transmitter pulse. Pictures are taken by an exposure to the repeated traces of from 500 to 2,000 repeated pulses. The electrical pulses are of 0.4 $\mu$ sec duration for the velocity seismogram or a Heaviside-type step pulse with a rise time of about 0.4 $\mu$ sec for the displacement seismograms. The crystals have effective dimensions of 1 to 2 mm, but lengths of 5 cm to eliminate reverberation effects. The separation of transmitter and receiver is from 5 to 15 cm, usually. The compressional pulses travel 0.61 cm/$\mu$ sec (6.1 km/sec). The wavelength of the P wave appears to be about 0.7 cm or less.

## *Surface waves*

Figure 1 is a reproduction from Lamb's original publication of the surface disturbance due to a point impulse. On Figure 2 is a description of the salient parts of the seismogram. Figure 3 is a seismogram taken with transmitter and vertical component receiver spaced 5 cm apart ($\Delta$) on a large steel block. The upper left trace, *a*, is the displacement as a function of the time, and *b* is the same function with amplification 10 times that of *a*. Trace *c* is the velocity of the surface particles, and *d* is the same function amplified 30 times. Note that the essential details agree well with Lamb's theoretical descriptions. There is a single P pulse at about 8 $\mu$ sec, followed by a long smooth wave. In the experimental case, there is no easily observed S wave. Then there occurs at 15 $\mu$ sec a large Rayleigh pulse. The record after this is fairly quiescent for a few microseconds; then all manner of reverbera-

Reprinted from Journal of Geophysical Research, 59, 289-294 with permission from American Geophysical Union

tions set in, detectable for as long as 0.3 second. Similar seismograms have been made out to $\Delta$ = 15 cm, which are quite like these, though the long tail after the compressional wave takes on a more complex form, for reasons not yet understood.

The next test was to see the effect upon the seismograms of surface discon-

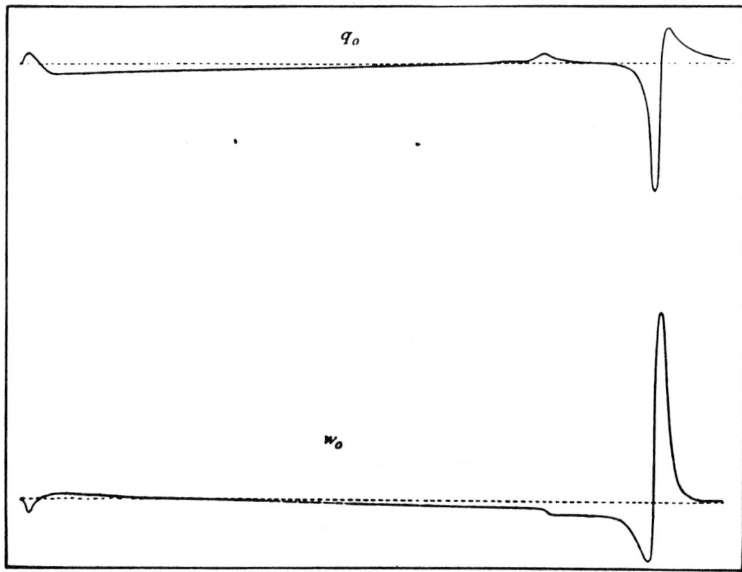


FIG. 1—Lamb's (1903) ground motion from a distant impulse; upper curve = horizontal motion, lower curve = vertical motion

tinuities. The first test, on a small steel block with many small holes of diameter 3 to 7 mm drilled into the surface to a depth of 3 to 7 mm between transmitter and receiver, resulted in a completely changed seismogram, with several oscillations between P and R—approaching the ones found in the usual field observation. Although this test showed the importance of the surface, it could not be used to determine the mechanism. A disturbing center, inserted to introduce disturbing conversion or scattering wave components, made in the form of a brass bar, 3/8″ by 1-1/4″ and 4″ long, was placed with its small face in contact with the smooth (undrilled) surface of the large steel block, midway between transmitter and

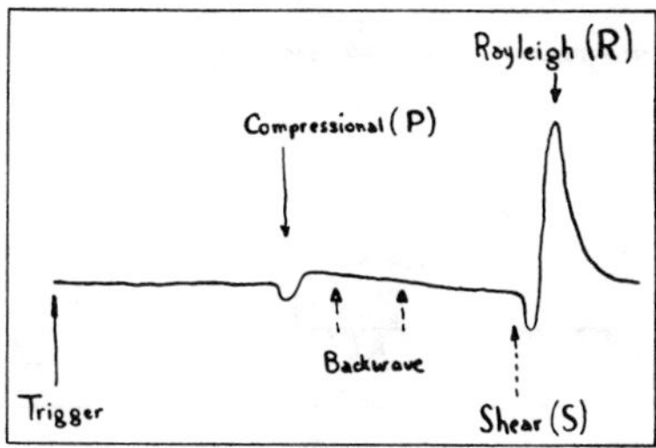


FIG. 2—Diagram of seismogram; the point "trigger" denotes the application of the voltage pulse to the "send" crystal

receiver, as shown in Figure 4. In order to facilitate surface contact, a soft grease was used between the surfaces. When this disturbing block or "artificial mountain" was in place, the seismograms were appreciably altered, as shown in Figure 4, in which are presented seismograms of the undisturbed and "brass block" disturbed

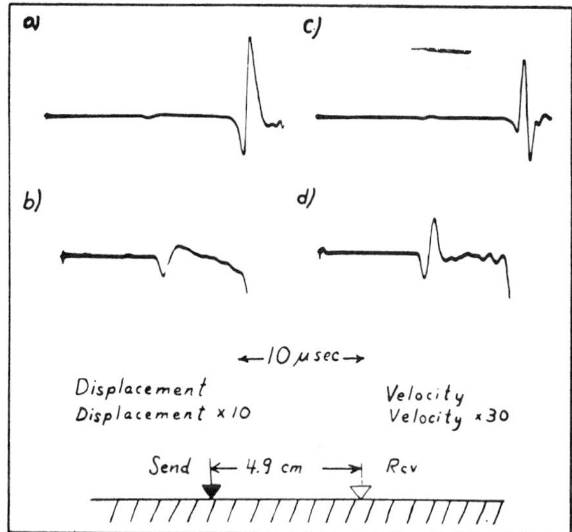


FIG. 3—Model seismograms showing the surface motion from a point impact, as in Lamb's theory; source size much less than the wavelength and reception distance much greater than the wavelength

seismograms. Since there was in this case only one scattering or conversion center, certain properties of the mechanism can be deduced. Most of the disturbance appears just before the Rayleigh wave, and, since the source was not disturbed, the P wave must have produced at the scattering center a slower wave. Assuming

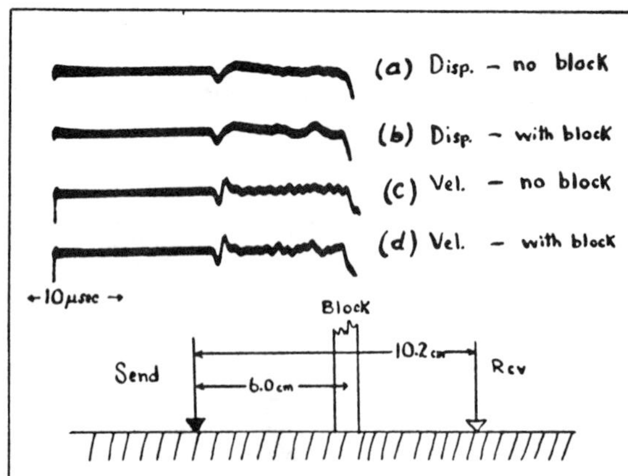


FIG. 4—Seismograms showing the effect of a surface disturbance by a surface obstacle of wavelength dimensions; the compressional wave produces a Rayleigh wave at the obstacle

this is a Rayleigh wave, the time of first disturbance is computed to arrive at 22.5 $\mu$ sec. It is observed at 21.7 $\mu$ sec. The agreement is good. Therefore, the major effect of a surface discontinuity is to produce a converted Rayleigh wave. This is in agreement with our hypothesis, which was based upon our field experience.*

*Transmitter at depth*

Experiments were conducted with the transmitter in a hole up from the bottom of a steel block. Seismograms are shown in Figure 5 for these cases. The characteris-

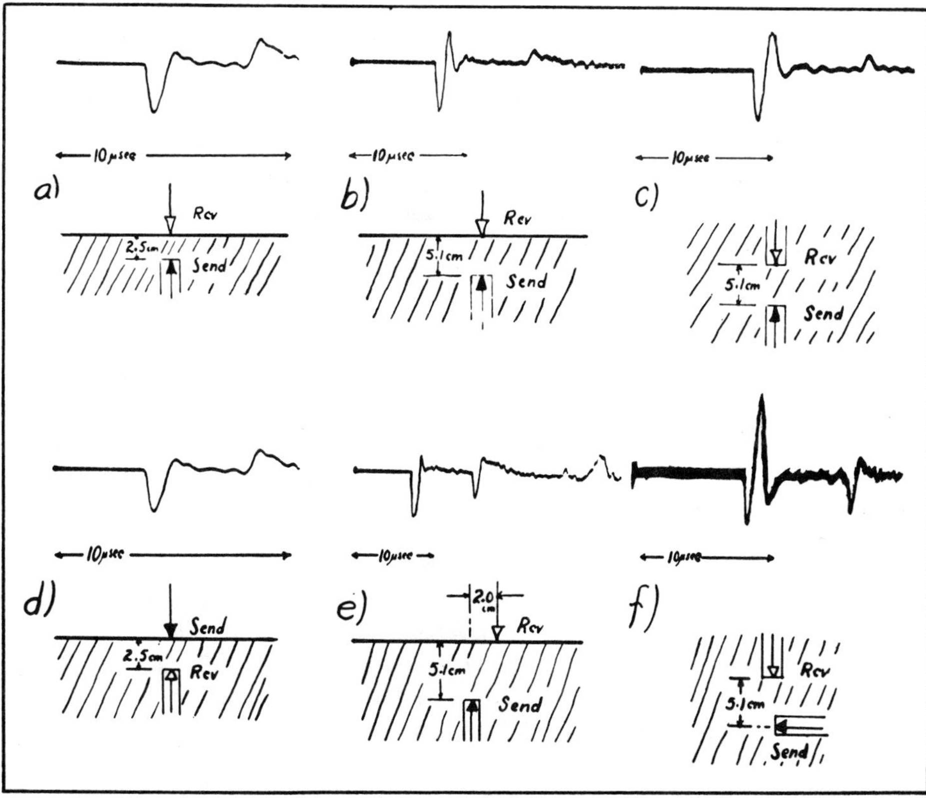


**FIG. 5—Various experiments with deep focus source (and receiver); *a* and *d* show "reciprocity"; all displacement seismograms**

**Note added in proof:* For the case of the surface transmitter and the surface receiver, in addition to $P \rightarrow R$ conversion at a surface discontinuity, the inverse $R \rightarrow P$ conversion at a surface discontinuity has been verified. If the scattering center is placed almost midway between transmitter and receiver, the amplitudes of the transformed waves at the receiver are apparently identical. Consider the amplitude of the $P$ wave at the detector as a standard. Then in this experimental arrangement the $P$ wave incident at the scatterer transforms into an $R$ wave whose amplitude at the detector is slightly less than that of the standard. The $R$ wave incident upon the scatterer (of much greater amplitude than the incident $P$ wave) transforms into a $P$ wave whose amplitude at the detector is again slightly less than that of the standard.

These conversion waves appear to have marked directional characteristics which need to be investigated.

tics of the traces with transmitter 2.5 cm (Fig. 5*a*) and 5.1 cm (Fig. 5*b*) below the surface are markedly different from that with transmitter on the surface. The Rayleigh wave is diminished by a factor of 20 to 50 relative to the P wave. The disturbance (as diagramed in Fig. 6) after the P wave is greatly diminished (a trace of it is observable at high amplification). Thus, as Nikano [1] had shown, the Rayleigh wave diminishes with focal depth. In addition, the behavior of the wave just after the P wave indicates it too is a surface wave—probably a Rayleigh wave.

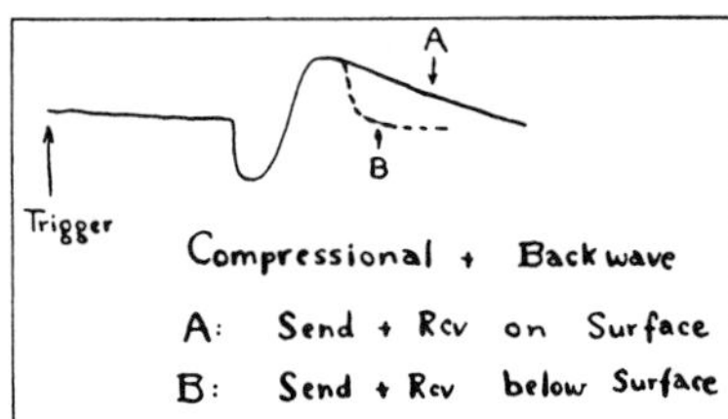


FIG. 6—Diagram indicating disappearance of the backwave as receiver is changed from surface to depth, thereby indicating surface nature of this wave

### *Reciprocity*

At this point, the electrical connection to the transmitter in the hole and the receiver on the surface were interchanged without changing any setting in the rest of the equipment. The resultant seismogram is shown in Figure 5*d* and is a close replica of the seismogram in Figure 5*a*. The result is that the Helmholtz reciprocity relationship, as for the sound in air [2], holds for the conditions in this experiment in a solid.

### *Interior transmitter and receiver*

The seismograms from transmitter and receiver, both in deep holes and separated 2″, are also shown in Figure 5. There are two cases: the crystals coaxial (Fig. 5*c*) and perpendicular (Fig. 5*f*). The received P pulse at 8 $\mu$ sec is single and shows no trace of the long wave seen in surface transmission and reception, further indication of the surface nature of this wave. Figure 5*f* shows, perhaps, a small shear wave.

### *Deep focus, surface scattering and reception*

The transmitter had a focal depth of 5.1 cm, the receiver a $\Delta$ of 9 cm. The brass block again as a scattering center, placed almost at the epicenter, made a disturbance of the seismogram in much the same way as for the surface experiment. Here again, a time calculation shows the disturbance should occur in 16.2 $\mu$ sec if a P-P effect, 22.6 $\mu$ sec if a P-R effect. It occurred at 21.8 $\mu$ sec, showing it is, indeed, a P-R effect. Thus, an interior wave at a surface scattering center makes a surface Rayleigh wave. By reciprocity, a surface Rayleigh wave scatters conversion P waves at a disturbance point. Thus, the concept of the interaction of plane P or S waves upon surfaces must be broadened to recognize the effect of scattering centers as new sources producing not only P and S non-plane waves, but also more complex waves, such as Rayleigh waves.

## *Conclusion*

This note is intended to be brief and only a small number of conclusions may be stated.

The estimate by Lamb of the surface effect of a point disturbance is examined experimentally. The results are found in all respects to be in qualitative agreement with Lamb.

Surface irregularities are found to alter the simple wave patterns, making them conform to field seismograms. This is in agreement with a hypothesis previously deduced from field experience. The P wave makes R waves at a surface scattering center.

The long tail after the first arrival in Lamb's pictures is shown to be a surface wave, not a body wave.

Reciprocity is shown to hold for the particular models used.

Scattering centers at a surface cause conversion of P to R and R to P waves, respectively.

Thus, the complex nature of a seismogram can be attributed in part to the interaction of seismic waves with surface irregularities and the resultant generation of surface waves. Just what fraction is due to surface and volume scatterings remains for future investigation to determine.

The bearing of these observations on the practical matter of interpreting various "phases" observed in seismograms after the first wave arrivals, whether from shots or earthquakes, is reasonably obvious. Observations where source and receiver are both deeply buried offer hope for observing vertical reflections and other low intensity returns.

It is a pleasure to acknowledge the stimulating discussions with Mr. Martin Greenspan, of the National Bureau of Standards, concerning the impedance matching of piezoelectric crystals.

## *References*

[1] H. Nikano, Jap. J. Astr. Geophys., **2**, 233-236 (1925).
[2] Lord Rayleigh, The theory of sound, **2**, 145, London, Macmillan and Co. (1878).

# AN AMPLITUDE STUDY ON A SEISMIC MODEL*

C. S. CLAY†‡ AND HALCYON McNEIL†

ABSTRACT

The measured amplitudes of two seismic events which have traveled through a two layer seismic model are compared with the amplitudes calculated from plane wave reflection and transmission theory.

The relative amplitudes of the events, one a dilatational-dilatational and the other a dilatational-to-shear conversion event, are found to be in agreement with calculations based on reflection theory for plane waves, after correction for $1/r^2$ divergence.

## INTRODUCTION

It has become possible with the development of seismic models to attack certain seismic problems in the laboratory. In some cases, these experiments can be regarded not only as model studies but also as actual seismic experiments on a small scale. Thus we can test the applicability of the present elastic wave theory to seismic problems. Evans et al. (1954) have made a study of the events transmitted along a thin plate and have shown good agreement with the normal-mode theory as applied to the thin plate by Tolstoy and Usdin (1953). Similarly, Press and Oliver (1955) have demonstrated in the laboratory many properties of the air-coupled flexural wave.

An attractive feature of a seismic model is that a single repeating source and a single detector may be used for making comparisons of the amplitudes of seismic events. The seismic model studies of Levin and Hibbard (1955) have shown that the dilatational to shear conversion events are very prominent. For this reason, it seemed desirable to make a quantitative amplitude study of a dilatational event and a dilatational-to-shear conversion event and to compare the experimental results with amplitudes calculated from theory. Since it was not practical to follow a more rigorous analysis of the problem such as was made by Cagniard (1939), comparison of experimental data with the transmission amplitudes calculated from plane wave theory was made. Actually the problem to be solved was the case of spherically diverging waves and, as an approximation, the plane wave amplitudes were corrected by a $1/r^2$ divergence factor.

Levin and Hibbard have shown that the measurement of reflection amplitudes would be very difficult on this model because of the interference between reflections and other events, such as ground roll. However, the first arriving transmitted events are relatively free from interferences, so the amplitude study has been made on these events. The dilatational event is the dilatational wave through

* Manuscript received by the Editor February 21, 1955.

† The Carter Oil Company, Research Department, Tulsa, Oklahoma

‡ Now at the Hudson Laboratories of Columbia University, Dobbs Ferry, New York.

Reprinted from Geophysics, 20, 766-773

cement and dilatational wave through marble, $P_1P_2$. The conversion event is the dilatational wave through cement converted at the interface to a shear wave ($SV$) through marble, $P_1S_2$.

EXPERIMENTAL AMPLITUDES

The amplitude measurements were made with the two-layer seismic model used by Levin and Hibbard.[1] A spark gap was used as the seismic source and a barium titanate ceramic was used for the detector.[2] It seemed desirable to make

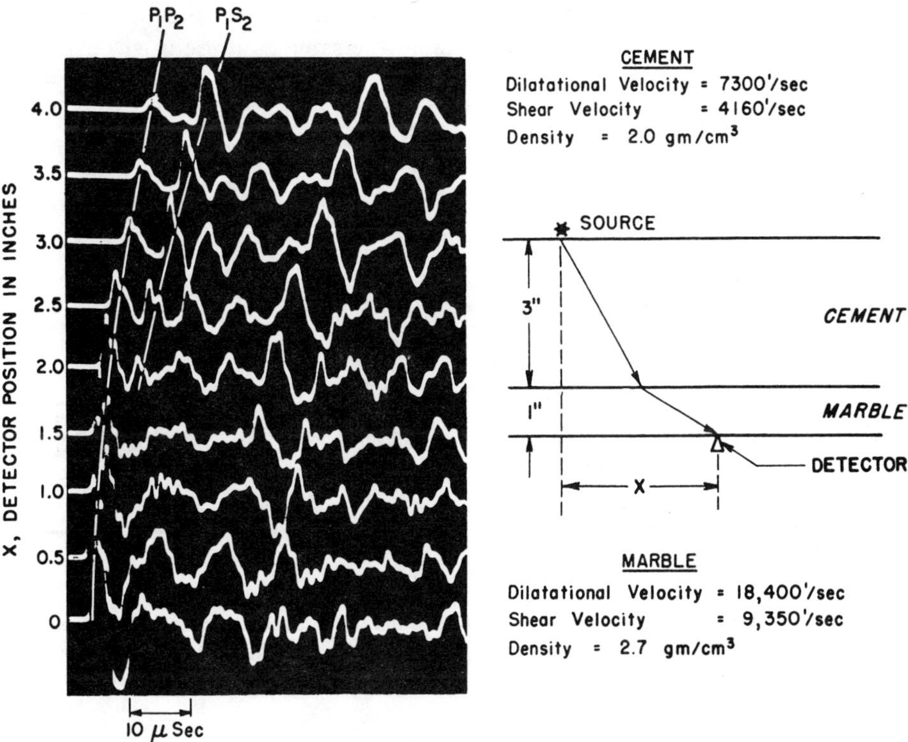


FIG. 1. Transmission through two layer seismic model (with constant gain).

the amplitude study with the transient from a spark so as to simulate the phase distortion from a shot that might be expected in seismic events for incident

[1] The model consisted of a layer of 3" Sauereisen #31 cement over a 1" layer of marble. The model had an areal extent of approximately 5'×7'. It was possible to place source and detector on either the top (cement) or bottom (marble) so as to make reflection and transmission studies.

[2] The spark gap was the open end of a coaxial line after Kaufman and Roever (1951). The outer conductor of the line was thin-wall ⅛-inch-diameter stainless steel tubing; the inner conductor, .045 inch steel wire; and the insulation, plastic radio spaghetti. A rotary switch was used to place a 2-μf capacitor charged to 800 volts across the spark gap. The spark repetition frequency was about 4 pulses per second.

angles greater than critical. Distortion of the pulse was discussed theoretically by Fisher (1948). Arons and Yennie (1950) have done experiments in shallow water in which phase distortion of oblique reflections was observed.

The detector, the same as that used by Levin and Hibbard, was a barium titanate ceramic which worked into a cathode follower with very high input impedance. The size of the detector was kept as small as practical, approximately ⅛-inch square, and the thickness of the ceramic was 0.1 inch. Since barium titanate ceramics seem to be largely sensitive to vertical motion with low sensitivity to horizontal motion, the amplitude data were not corrected for detector response. The amplifying system and oscilloscope were the same as those used by Levin and Hibbard, with flat response from 10 kc to 5 mc.

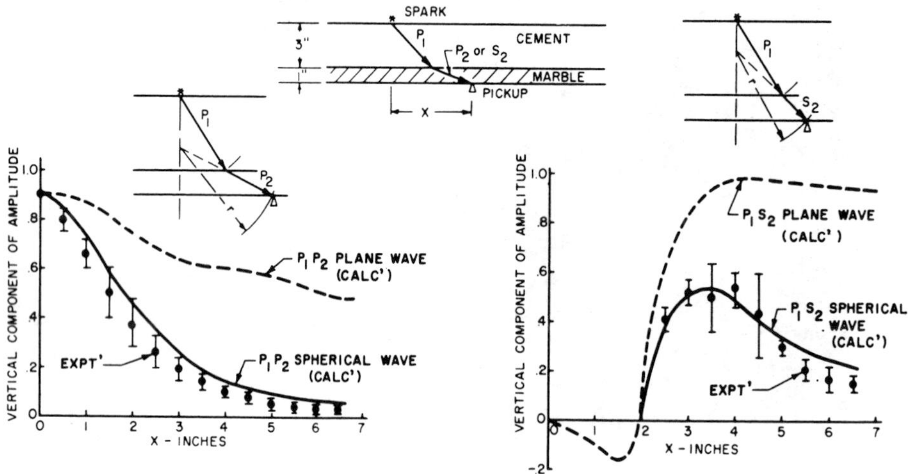


FIG. 2. Comparison of the experimental amplitudes and the vertical components of the theoretical amplitudes for the transmitted events. The theoretical plane wave amplitudes are shown uncorrected and with $r^{-2}$ correction for spherical spreading. $r$ is the apparent radius of curvature of the wave front.

A model seismogram, taken with the spark, is shown on Figure 1. The seismogram was made by taking a series of multiple exposures, one exposure for each detector position. The film in the camera was moved for each exposure, making it possible to leave the oscilloscope trace in the same position and to use the same field of view of the camera lens. Each trace of the seismogram was made with an exposure of about 5 sec with the spark repeating 4 times per second.

The transmission events $P_1P_2$ and $P_1S_2$, which were chosen for the amplitude study, have been identified on the seismogram shown in Figure 1. The first event, dilatational through cement-dilatational through marble ($P_1P_2$), is transmitted for incident angles of the dilatational wave $P_1$ in the cement which are less than the critical angle. However, at detector positions less than two inches from the source (corresponding to a less-than-critical angle of incidence for the $P_1P_2$

event), the converted shear wave $P_1S_2$ was not identifiable. For angles of incidence greater than critical or detector positions farther than 2 inches from the source, the dilatational wave that was converted to shear at the cement-marble interface has been identified on Fig. 1 as the $P_1S_2$ event.

The experimental amplitudes for each event are shown on Figure 2. The amplitudes were measured on enlarged photographs similar to the seismogram on Figure 1. In addition to the detector which was moved, another detector was held stationary during the recording of a profile as a monitor of the source amplitude. Inasmuch as the cement layer contained a large number of small air bubbles, the spark and detector were moved to different positions on the model in order to obtain a better average value for the transmission amplitudes and to reduce the effect of random local inhomogeneities in the model. Each data point on Figure 2 represents the average value of the amplitudes from three profiles and the spread that is shown is the standard deviation of the values. All amplitude data were normalized to a transmitted dilatational amplitude of 0.91 at zero detector distance so that they would agree, at that position, with the theoretical transmission amplitudes for plane waves.

## THEORETICAL AMPLITUDES

The plane-wave amplitude calculations were made by the Zoeppritz equations described by Macelwane and Sohon (1936, p. 147–178). The actual calculation was carried out in steps as follows: A unit dilatational plane wave in the cement was assumed to strike the cement-marble interface at a given angle and the components reflected from and refracted into the marble were computed. The amplitudes of the incident wave, the reflected dilatational, reflected shear conversion, refracted dilatational, and refracted shear conversion waves were fitted together mathematically at the interface so as to satisfy the boundary conditions. Since there are four unknown amplitudes, it was necessary to solve four simultaneous equations. This process was repeated for each angle of incidence to obtain the amplitudes of the reflected and refracted waves. The direction of travel or angle of incidence of each wave was obtained from Snell's law.

The calculation of the vertical component at the free surface of the cement or marble was a similar but less complex problem in two unknowns. In this case, the incident dilatational wave was reflected both as a dilatational and as a shear wave, as was also the incident shear wave. The boundary condition of zero stress on the free surface, together with the use of Snell's law, gave the amplitude of the reflected events. The vertical component of displacement, observed on the surface, was found by calculating the vertical component of the sum of the incident and reflected waves. For example, the dilatational wave at vertical incidence was reflected with the same amplitude but opposite phase. Thus, the amplitude of the displacement at the free surface was twice that of the incident wave. The assumptions made were first, that the plane waves are of single frequency; second, that absorption may be neglected; and third, that there is no body motion

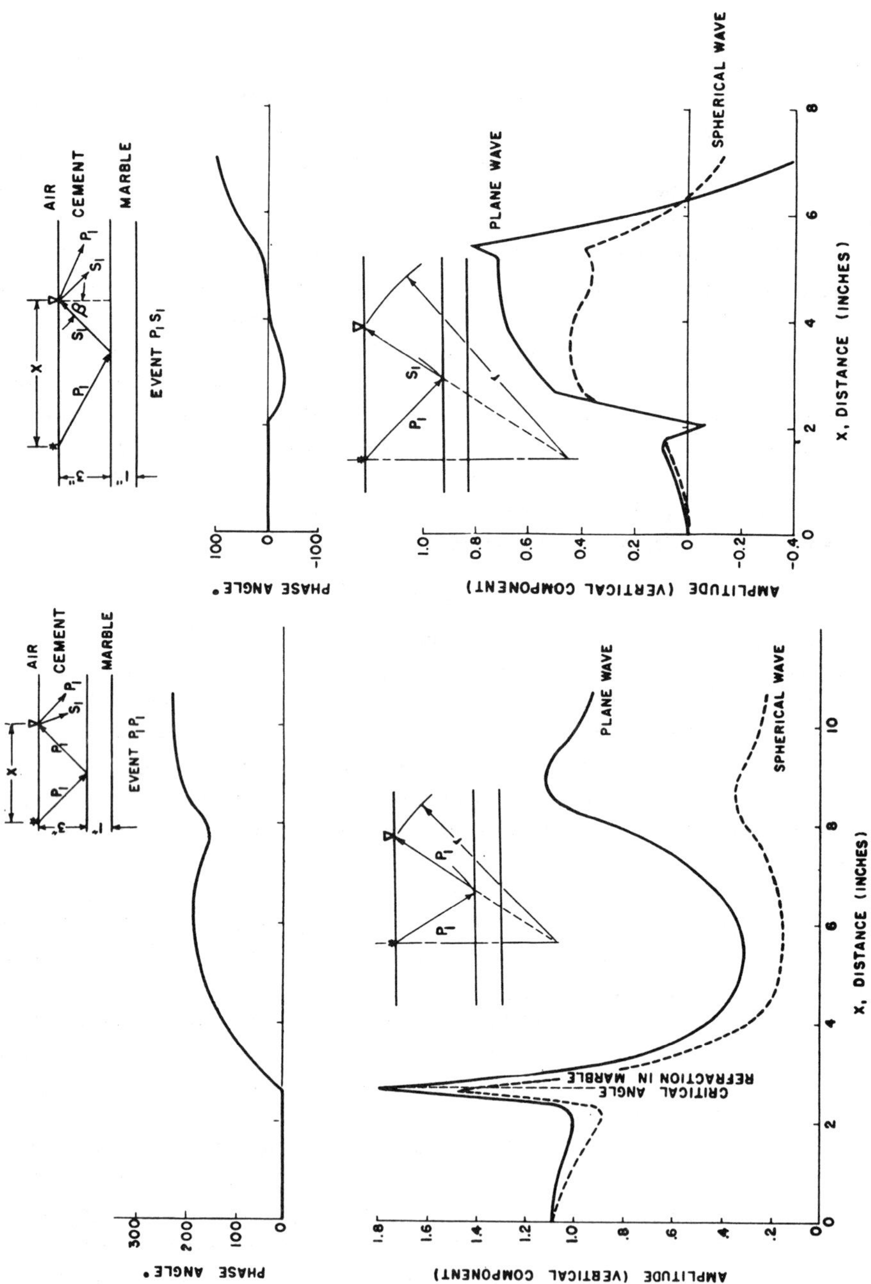


FIG. 3. Theoretical vertical components and phase of reflected events. The plane wave amplitudes and the same with $r^{-2}$ correction for spherical spreading are shown. $r$ is the apparent radius of curvature of the wave front.

of the whole layer. The relative amplitudes of the vertical displacements for the plane-wave calculations are shown in Figures 2 and 3. The null at about the 2-inch position for the dilatational-shear transmitted event corresponds to the dilatational-reflection critical angle.

The plane-wave amplitudes were approximately corrected for spherical spreading in the following manner. The amplitudes were assumed to decay as a function of the radius of curvature of the wave front. It can be seen that the curvature of the wave front changed upon passing from the cement into the marble and that the position of the apparent center of curvature was not constant except for the dilatational-reflection case. The refraction of spherical waves by a plane surface has been treated as an optical problem by Robertson (1941) in section 20 of his text. Figure 3 shows the wave-front curvature for both the $P_1P_1$ and the $P_1S_1$ reflections.

The decay function of the amplitudes was determined experimentally. The ratios of the experimental amplitudes to the corresponding plane wave amplitudes were fitted by least squares to the functions $Ae^{-ar}/r$ and $Be^{-br}/r^2$ in order to better determine the form of the expression for amplitude decrease. The sum of residuals squared was calculated for each case. The average of the residuals squared for the expression involving $r^2$ was about one third of the average for the inverse $r$ expression. For this reason, reciprocal $r^2$ was chosen to modify the plane wave amplitudes for spherical divergence.[3]

Mencher (1952, 1953) and Dix (1954, 1955) have shown that the displacement potential function due to a step function in a displacement-type point source yields, as the solution of the wave equation, an amplitude decreasing as the inverse square of the distance and a derivative which decreases inversely as the distance. Thus, near the source the displacement decreases as $r^{-2}$ and at large distance the form of the disturbance is changed and its displacement decreases as $r^{-1}$. The experimental amplitudes were best fitted by $r^{-2}$, which indicates that the model experiments were made under near-source conditions.

If the transmission amplitudes with spherical approximation are compared with the corresponding plane-wave amplitudes, as shown in Figure 2, it can be seen that the amplitude of both $P_1P_2$ and the converted shear $P_1S_2$, are considerably reduced at distances greater than 3 inches. In a similar manner the reflected events are also altered. If the reflection amplitudes with spherical wave front corrections are compared with the corresponding plane wave amplitudes, as shown in Figure 3, it can be seen that the dilatational reflection maximum at about 8 inches is very much smaller and that the reflected shear amplitude is also reduced. The effect of directivity of the source has not been included in these calculations.

The directivity of the source was similar to the directivity given by Sezawa and Kanai (1936) for a piston-type source. The theoretical amplitude of the

[3] The ratios of experimental to theoretical amplitudes were also considered as functions of travel time. The results of this indicated that the decay was $t^{-6}$ for the $P_1P_2$ event and $t^{-3.8}$ for the $P_1S_2$ event. These results were different from the $t^{-2.5}$ relation obtained by Ricker (1953) in a single medium.

$P_1P_2$ event is almost unchanged by correction for the source directivity. The departure angle of $P_1$ is less than 24° from the normal to the surface and the correction is less than 2%. The theoretical amplitude of the $P_1S_2$ event is changed at large $x$. At $x=5''$, the departure angle of $P_1$ is about 45° and the correction factor for source directivity is 0.6. The source correction would reduce the theoretical values still more at larger $x$. The agreement of the theoretical and experimental amplitudes is improved by correction for the source directivity.

## CONCLUSION

For the transmission case, the agreement between approximate theoretical calculations and the experimental measurements is quite good. The theoretical work was done under the assumption that the source of dilatational waves was uniform in all directions. The experimental source was directive vertically and this may account for the reduced experimental amplitude at large distances. The amplitudes of the dilatational event at zero detector position, and the shear wave at 3 inches, as determined experimentally were in good agreement with the theoretical amplitudes at these distances and this comparison should give more confidence in the use of such theoretical approximations. Of course, approximate calculations such as these cannot apply at the critical angles and one must use more exact treatments, such as those of Cagniard (1939) and Heelan (1953), which consider the coupled nature of the vector and scalar displacement potential functions.

The reflection amplitudes have not been studied in these experiments. It is believed, however, that the theoretical amplitudes shown on Figure 3 would be observed if it were possible to obtain the reflections free of interferences, because the incident, reflected, and transmitted events are connected by boundary conditions at the interface. Thus, the procedure of calculating the plane-wave reflected and transmitted amplitudes and then making an approximate correction for spherical divergence of the wave front may be an adequate treatment of this problem.

## ACKNOWLEDGMENT

The authors wish to thank the Carter Oil Company for permission to publish this work.

## REFERENCES

Arons, A. B. and Yennie, D. R., 1950, Phase distortion of acoustic pulses obliquely reflected from a medium of higher sound velocity: J. Acoust. Soc. Am., v. 22, p. 231–237.

Cagniard, L., 1939, Reflection et refraction des ondes seismiques progressives: Paris, Gauthier-Villars.

Dix, C. H., 1954, The method of Cagniard in seismic pulse problems: Geophysics, v. 19, p. 722–738.

———, 1955, Mechanism of generation of long waves from explosions: Geophysics, v. 20, p. 87–103.

Evans, J. F., Hadley, C. F., Eisler, J. D., and Silverman, D., 1954, A three-dimensional seismic wave model with both electrical and visual observation of waves: Geophysics, v. 19, p. 220–236.

Fisher, F. A., 1948, Über die total Reflection von ebenen Impulswellen: Annalen der Physik. 6 Folge, Band 2, S.211.

Heelan, P. A., 1953, On the theory of head waves: Geophysics, v. 18, p. 871–893.

Kaufman, S. and Roever, W., 1951, Laboratory studies of transient elastic waves: The Third World Petroleum Congress, The Hague, Proceedings, Leiden, E. J. Brill, Section 1, p. 537–545.

Levin, F. K. and Hibbard, H. C., 1955, A three-dimensional seismic model: Geophysics, v. 20, p. 19–32.

Macelwane, J. B. and Sohon, F. W., 1936, Introduction to theoretical seismology: New York, John Wiley and Sons, Inc.

Mencher, A. G., 1952, Ph.D. Thesis, University of California at Los Angeles.

———, 1953, Epicentral displacement caused by elastic waves in an infinite slab: Journal of Applied Physics, v. 24, p. 1240–1246.

Press, F. and Oliver, J., 1955, Model study of air coupled surface waves: J. Acoust. Soc. Am., v. 27, p. 43–46.

Ricker, N., 1953, The form and laws of propagation of seismic wavelets: Geophysics, v. 18, p. 10–40.

Robertson, J. K., 1941, Introduction to physical optics, 3rd Edition: New York, D. Van Nostrand Company, Inc.

Sezawa, K. and Kanai, K., 1936, Polarization of elastic waves generated from a plane source: Bull. Earth. Res. Inst., Tokyo, v. 14, p. 489–505.

Tolstoy, I. and Usdin, E., 1953, Dispersive properties of stratified elastic and liquid media: a ray theory: Geophysics, v. 18, p. 844–870.

# MODEL SEISMOLOGY—THE CRITICAL REFRACTION OF ELASTIC WAVES*

P. N. S. O'BRIEN†

ABSTRACT

Model experiments on the head wave are described. Quantitative data are obtained for three different pairs of media: light lubricating oil on top of a saturated solution of calcium chloride, water on top of wax, and water on top of concrete. Within the limits of experimental error these data agree with the theoretical prediction that the head wave should decay with distance as $r^{-1/2}L^{-3/2}$ (Heelan, 1953).

## INTRODUCTION

It has long been known that in the case when a medium of higher velocity underlies one of lower velocity, energy follows a ray path which is incident on the interface at the angle of critical reflection, travels along the interface with a velocity characteristic of the lower medium, and continually re-emerges into the upper medium, again at the critical angle. This is illustrated in Figure 1.

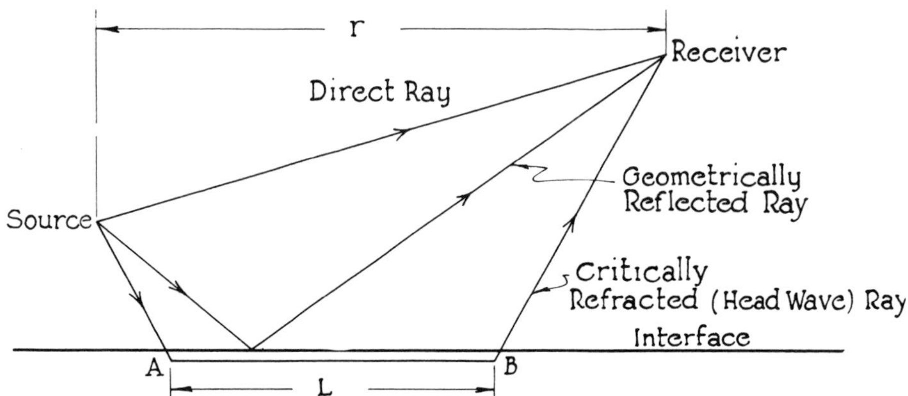


FIG. 1. The direct, geometrically reflected, and critically refracted (head wave) rays.

In exploration seismology these waves are usually called refracted or, perhaps more correctly, critically refracted waves. However, as the adjective "refracted" normally refers to the lower medium, the term *head wave* (Heelan, 1953; von Schmidt, 1938) is used in this paper.

Cagniard (1939) and others—Ott, 1941 and 1949; Gerjuoy, 1953; Heelan, 1953—have shown analytically that the decay of amplitude with distance should

* Presented before Section III of the Royal Society of Canada at its annual meeting in June, 1954. Manuscript received by the Editor September 30, 1954.

† Department of Physics, University of Toronto, Toronto, Ontario.

Reprinted from Geophysics, 20, 227-242

be proportional to $r^{-1/2}L^{-3/2}$, where $r$ is the orthogonal projection on the interface of the distance between the source and detector and $L$ is the distance the wave has travelled in the lower medium.

These workers have also shown that the form of the head wave is related to the form of the initial generating wave, as a function is related to its time derivative (see, for instance, Heelan, 1953, p. 882), and that the displacements in the case of a dilatational head wave are directed along the ray.

Excellent experiments on the head wave have been reported in the literature. Von Schmidt (1938) obtained schlieren photographs which clearly showed an acoustic head wave, and H. Maecker (1949) obtained many quantitative data on an optical head wave.

In order to check the predictions of theory for elastic head waves—in particular the law of decrease of amplitude with distance—some laboratory experiments were conducted which are reported on below.

### EXPERIMENTAL METHOD

Most of the theoretical analyses of the problem of head waves have been developed for homogeneous, isotropic, infinitely extensive slabs in contact

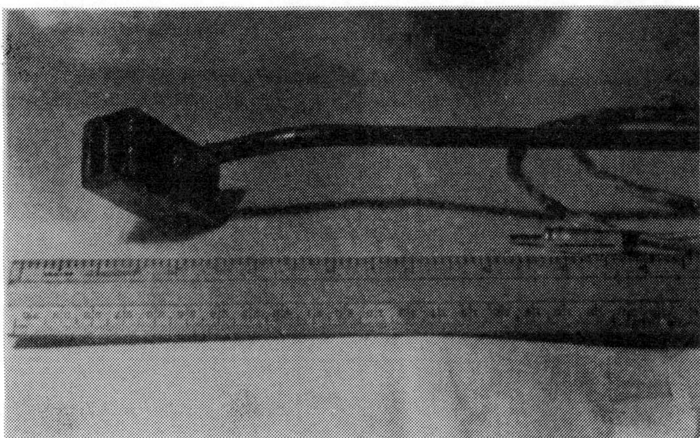

FIG. 2. The source unit.

along a plane surface. In order to produce experimental conditions approximating those called for in the theories, a wax block of dimensions 120 cm×40 cm ×35 cm was held at the bottom of a large tank of water. Most of the experiments were carried out with this block serving as the lower medium; however, some results were obtained with water overlying a concrete block about 3 m×2 m×1 m, and with a light lubricating oil overlying a saturated aqueous solution of calcium chloride contained in a glass tank 125 cm×25 cm×25 cm.

Rochelle salt transducers, of resonant frequency about 100 kcps, were used to act as the source and detector of elastic energy in the upper medium. The source unit consisted of two rochelle salt 45° $X$-cut bricks, size 5.1 cm×1.9 cm ×0.6 cm, connected in parallel. It was necessary to use such a large unit because the head wave is a second order effect. A photograph of the source unit is shown in Figure 2. The receiver occupied about one quarter the volume of the source, and had an active face of about 1.0 cm×0.3 cm. Because the upper medium was always liquid, it was easy to alter the position of the transducers, and to keep them well away from the outer boundaries of the tank. In order for the transducers to operate at maximum sensitivity, the liquid in which they were immersed was kept at 24°C.

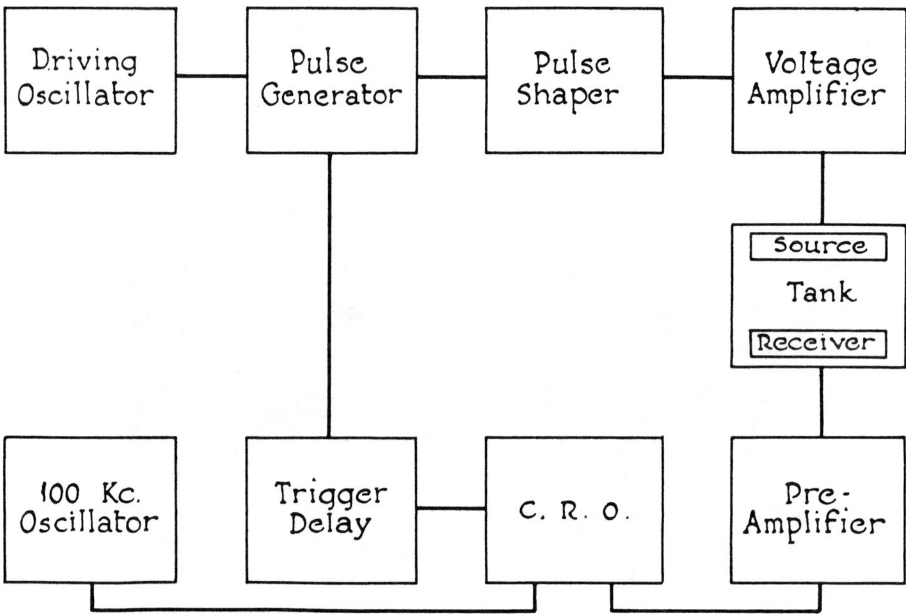


FIG. 3. Block diagram of electronic apparatus.

The experimental set-up and electronic circuitry is very similar in most seismic model studies. The technique used in these experiments was almost identical to that used by Anderson and Northwood (1953) and may be explained by reference to Figure 3. A pulse generator was driven by an audio oscillator and its output fed into the source transducer, via a shaping circuit and voltage amplifier. The shape of the transmitted pulse is a function of the shape of the voltage pulse applied to the transmitting transducer, the characteristics of the transducer, the bonding of the transducer to the medium and the properties of the medium. The shaping circuit was used to exercise some control over the shape of the transmitted pulse by controlling the shape of the applied voltage pulse. The output

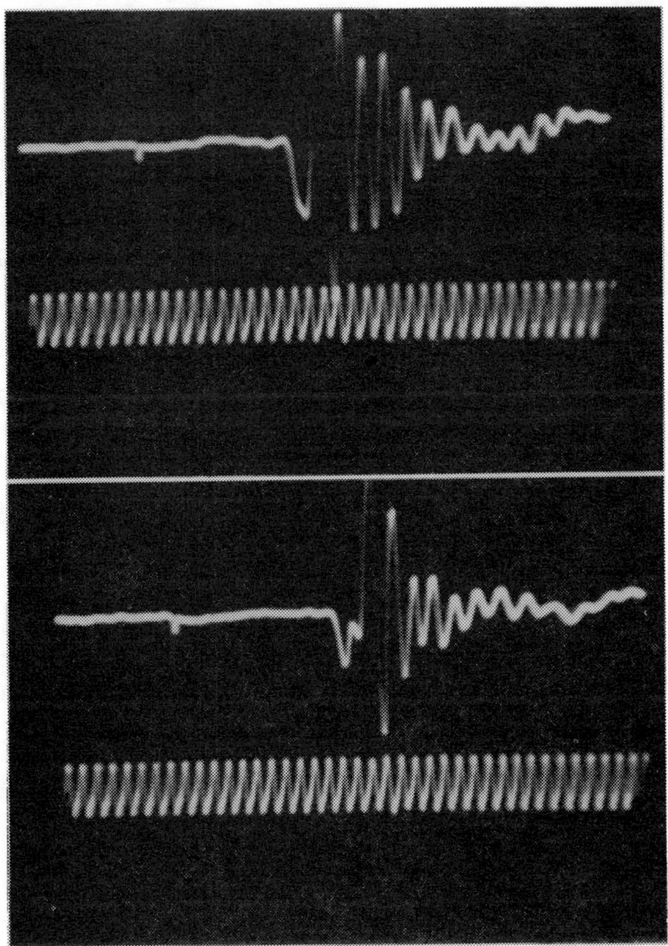

FIG. 4. Photographs of the head wave ($P_1P_2P_1$) and direct pulses for water overlying wax. The upper photograph shows the head-wave and direct pulses completely merged. The lower photograph shows the separation of the first half-cycle of the head-wave pulse. The downward pip on the upper trace is the initiation time of the source.

The lower trace shows the 100-kc timing signal.

of the receiving transducer was fed through a pre-amplifier to the vertical plates of one beam of a split-beam oscilloscope.

A triggering signal of variable delay time was taken from the pulse generator and applied to the synchronizing terminals of the oscilloscope on which one could view a stationary pattern of the received disturbance.

The amplitude of the signal generated by the receiving crystal varied from about 0.5 mv to 0.5 v, and was measured, in arbitrary units, to within about $\pm 5\%$ by means of a graticule superimposed on the screen of the oscilloscope.

To ensure that all elastic reverberations in the medium due to one pulse had ceased before the onset of the next pulse, the repetition rate had to be kept lower than about 60 times per second.

To time the interval between the initial and received pulses the output of a 100-kc crystal oscillator was displayed on the second trace of the oscilloscope, as illustrated in Figure 4. It was possible to measure times to ±one microsecond.

Both the transducers were held in clamps which could be moved along a guide bar graduated in centimeters. It was possible to measure relative distances to ±1 mm, although absolute measurements were only meaningful to ±1 cm, owing to the finite size of the transducer elements.

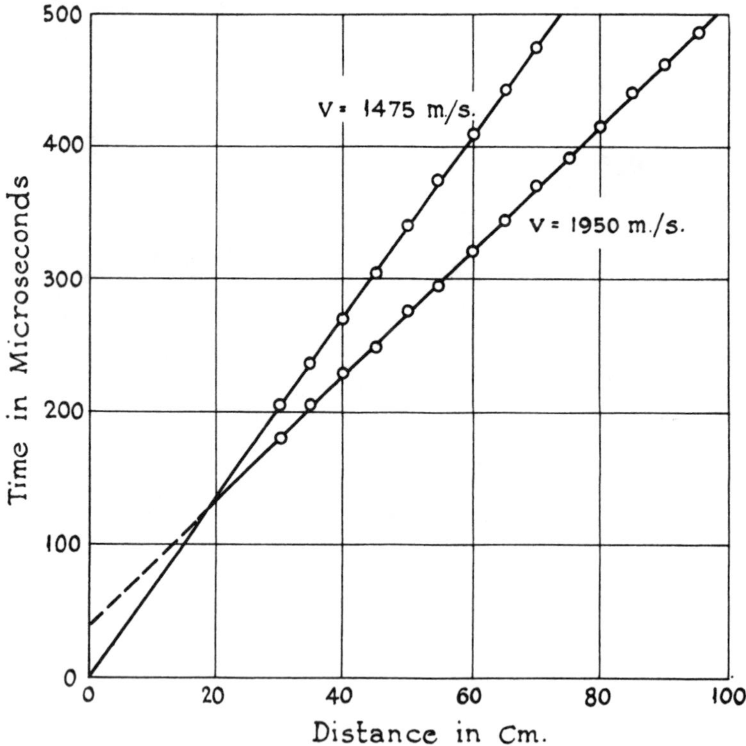


FIG. 5. Time-distance plot for water overlying wax.

## EXPERIMENTAL RESULTS

During the past few years several interesting and important papers on model seismology have been published (e.g., Anderson and Northwood, 1953; Oliver et al., 1953; Evans et al., 1954). One of them (Oliver et al.) was directly concerned with seismic refraction studies. The cases they investigated are more complex

TABLE I

| Medium | Velocity by Direct Measurement | Velocity from Time-Distance Plot |
|---|---|---|
| Calcium chloride solution | 1,850 m/s | 1,850 m/s |
| Wax (dilatational velocity) | 1,950 m/s | 1,950 m/s |
| Concrete (dilatational velocity) | 3,890 m/s | 3,860 m/s |
| Concrete (distortional velocity) | 2,250 m/s | 2,270 m/s |

than the one treated in this paper, for they were interested in head-wave pulses associated with a layer of finite thickness. Previous experimenters on model seismology have been mainly concerned with the recognition and identification of the various "events" on their seismograms. This paper differs from others in presenting quantitative data on the amplitudes of a particular "event"—the head-wave pulse.

Three different parameters were measured: the distance of the source and receiver from the interface and from each other, the time of arrival of the relevant pulse with respect to the initial pulse, and the amplitude of the first half-cycle of the observed pulse.

Two factors render it impossible to make reliable observations on the wave form of the head wave. In the first place, the wave form displayed on the oscilloscope is largely determined by the characteristics of the receiving transducer and amplifying circuits; and, secondly, it was not possible to completely separate the head wave from the subsequent pulses, with the result that it could never be observed in its entirety.

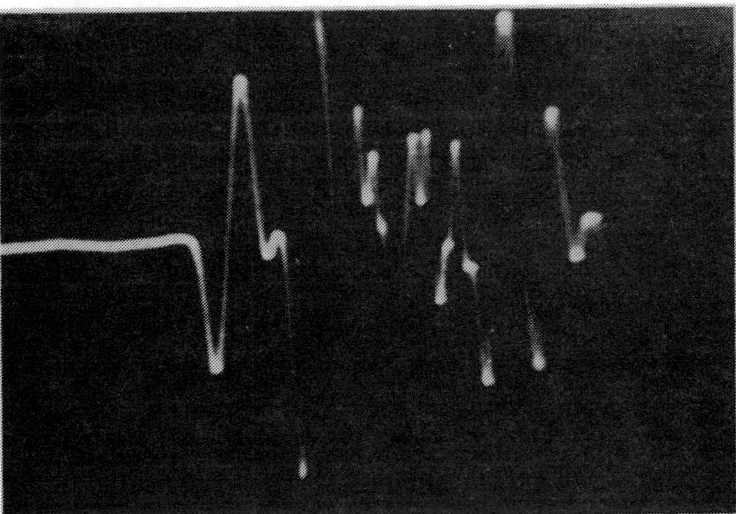

FIG. 6. The head wave ($P_1P_2P_1$) and subsequent pulses for water overlying wax. The direct pulse arrives after about one complete cycle of the head-wave pulse.

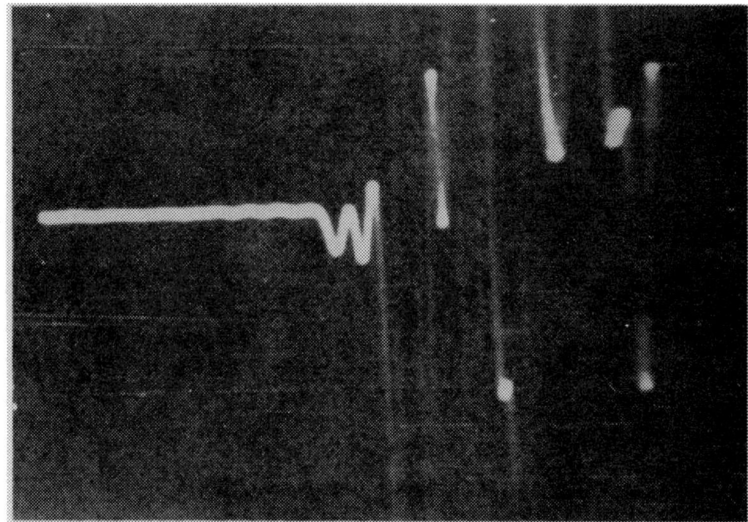

FIG. 7. The head wave and subsequent pulses for oil overlying calcium chloride solution. The onset of the direct pulse is marked by the first strong downward movement of the trace.

Time-distance graphs have been drawn for the three sets of media; a typical one for water overlying wax is shown in Figure 5. As there is nothing novel in head-wave time-distance graphs the others have not been reproduced here. They all show that, within the limits of experimental error, the velocities of the lower medium as found from the head wave agree precisely with those found by direct measurement. A comparison may be seen in Table I. In the case of water overlying wax there is no head wave travelling with the distortional velocity for wax, as this is less than the velocity through water.

It was found that maximum response was obtained when the sensitive face of the receiver was parallel to the front of the head wave. This is in agreement with the theoretical prediction that the displacements in the wave are along the ray path.

In all cases the first movement of the head wave was in the same direction

TABLE II

| Media | Ratio of Amplitudes | Distance between Source and Receiver |
|---|---|---|
| Water/wax | 1:20 | 40 cm |
| Oil/calcium chloride | 1:15 | 42 cm |
| Water/concrete | | |
| ($P_1P_2P_1$)* | 1:100 | 30 cm |
| ($P_1S_2P_1$) | 1:10 | 30 cm |

* Each letter stands for a segment of a ray path. Dilatational waves are denoted by the letter P, distortional waves by the letter S. The subscript denotes the medium in which the wave travels.

as that of the incident pulse—see for example Figures 6, and 7.

When two liquids were used, it was possible to compare the head wave at points just above the interface with the refracted pulse at points just below the interface. It was found that the times of arrival of the two events were coincident at the boundary, and that their waveforms and amplitudes respectively approached each other as the receiver was moved close to the boundary (Fig. 9).

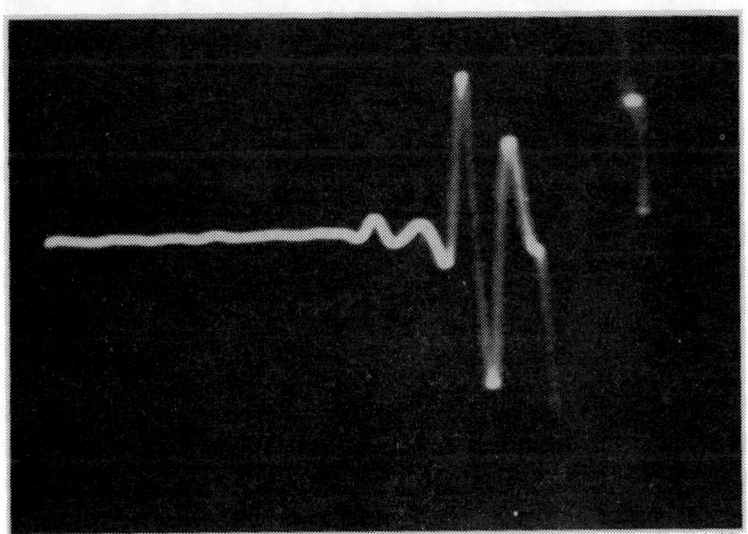

FIG. 8. The two head-wave pulses ($P_1P_2P_1$ and $P_1S_2P_1$) and the direct pulse for water overlying concrete.

The onset of the head wave travelling along the interface with the distortional velocity of concrete occurs after about two cycles of the earlier head wave pulse, which travels with the dilatational velocity. The onset of the direct pulse is given by the strong downward movement of the trace occurring after about one complete cycle of the second head-wave pulse.

## THE AMPLITUDE OF THE HEAD-WAVE PULSE WITH RESPECT TO THE DIRECT PULSE

Along the interface the head wave exists only at points that are farther from the source than the point $A$ in Figure 1. The ratios of the amplitude of the head wave to that of the direct pulse at points close to the interface are given in Table II.

It is interesting to note that in the case of water overlying concrete the wave $P_1S_2P_1$ is considerably stronger than the wave $P_1P_2P_1$. This result would seem to tie in with the fact, already known but perhaps not given sufficient attention, that the velocity for a particular layer obtained by seismic refraction shooting may not agree with those obtained by reflection shooting and by well surveys—even when the layer is thick, homogeneous, and isotropic. More important, this

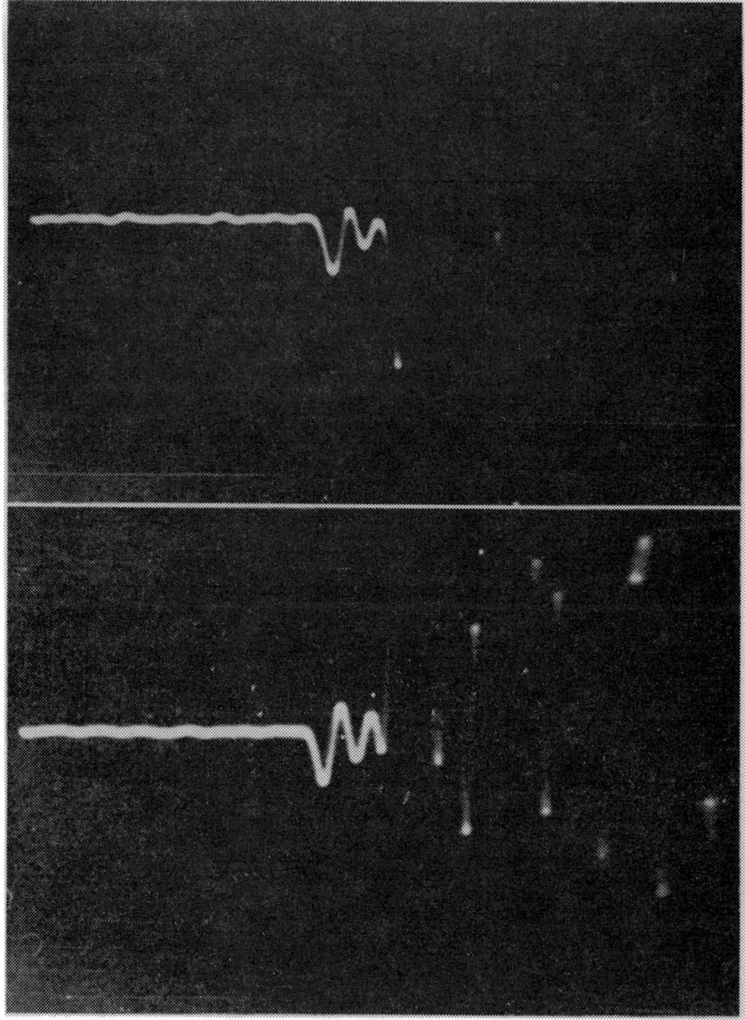

FIG. 9. Upper photograph: the receiver just above a liquid-liquid interface. Lower photograph: the receiver just below a liquid-liquid interface. *Note:* The strong upward movement in the lower photograph is due to a reflection from the bottom of the tank.

velocity may not be the one required in calculating the critical angle and delay time for a ray critically refracted at a deeper interface. An illustration of how this may affect depth determination in refraction shooting is given in the Appendix.

It should be noted that in the photographs reproduced in this paper, the relative amplitudes of the head wave and direct pulses cannot be obtained by simple scaling, because the transducers were oriented to have their direction of maxi-

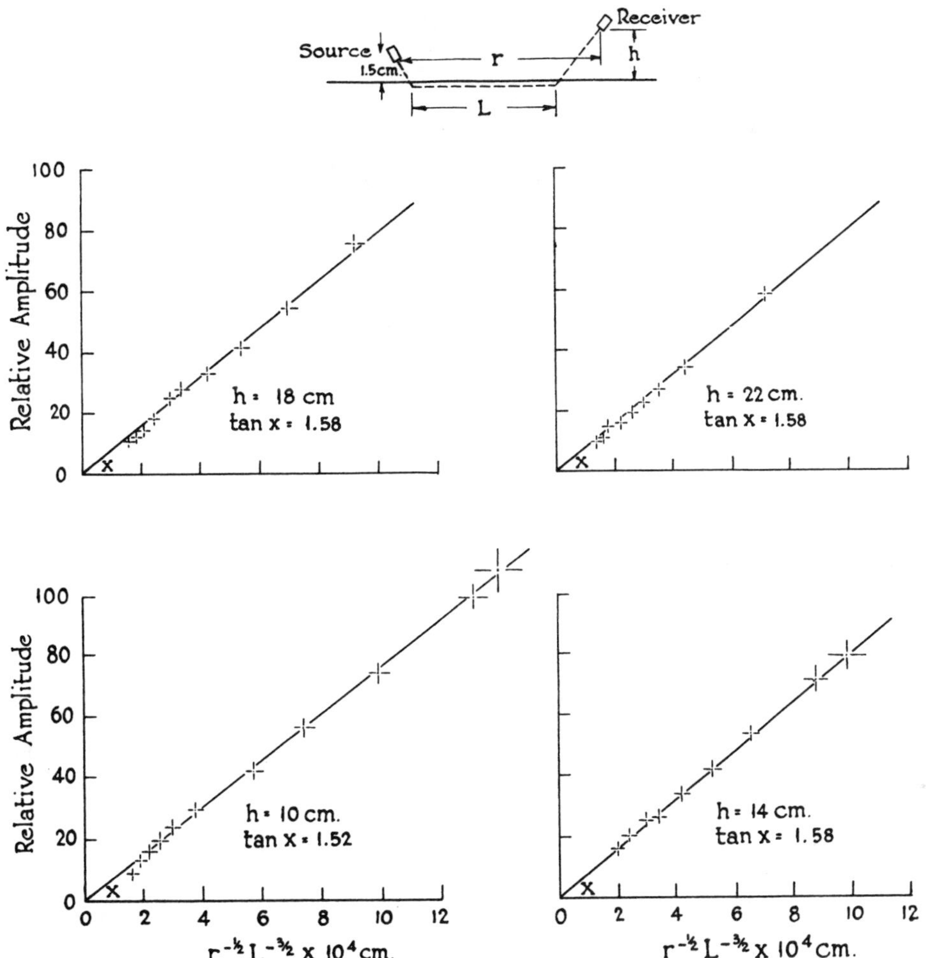


FIG. 10. Decrease of amplitude of the head-wave pulse ($P_1P_2P_1$) along lines parallel to a wax-water interface.

mum sensitivity perpendicular to the front of the head wave.

In order to measure the amplitude of the head wave, it is necessary that sufficient time should elapse for at least its first half-cycle to have been received before the arrival of the direct pulse. This is illustrated in Figure 4, where an amplitude measurement could be made on the lower photograph but not on the upper one. Because of this it was not possible to measure amplitudes at points closer to $A$ than about 10 cm, which, assuming that the predominant frequency in the pulse is equal to the resonant frequency of the transducer, corresponds to four or five wavelengths of the major Fourier component of the pulse.

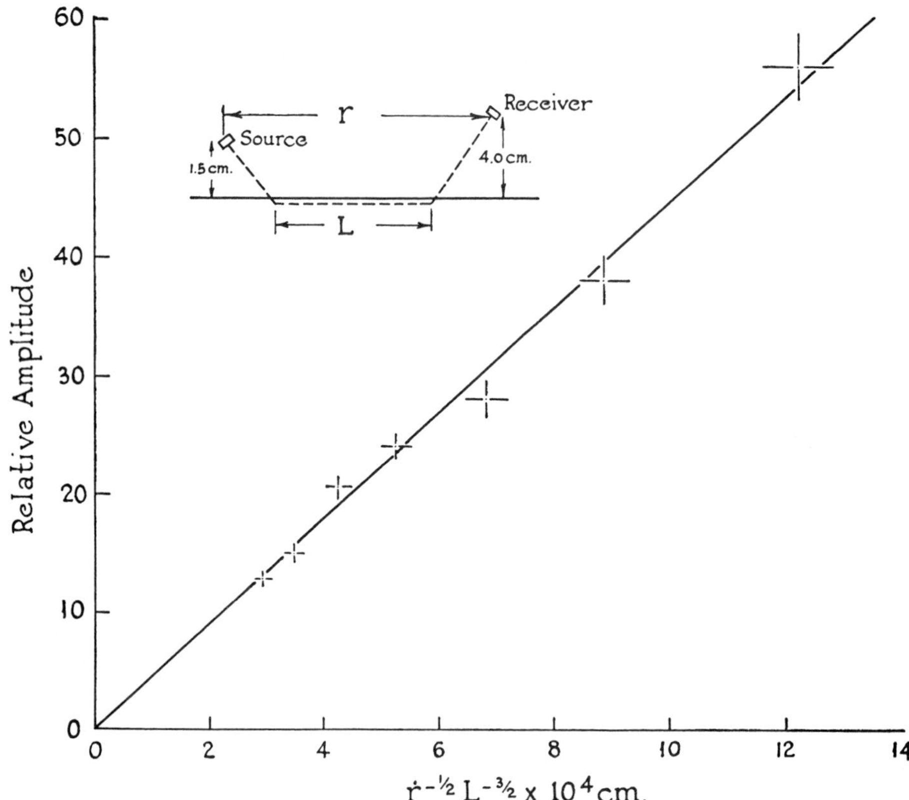


FIG. 11. Decrease of amplitude of the head-wave pulse ($P_1P_2P_1$) along a line parallel to a liquid-liquid interface.

VARIATION WITH DISTANCE OF THE AMPLITUDE OF THE HEAD WAVE

According to theory, the decrease of amplitude of the head-wave pulse should be given by the equation

$$A = Kr^{-1/2}L^{-3/2}$$

where,

$A$ = amplitude of the head wave,
$K$ = a constant depending upon the source,
$r$, $L$ = distances defined in Figure 1.

Measurements of amplitudes were made along lines parallel and along lines perpendicular to the interface. In the resulting graphs, Figures 9 to 12, the horizontal and vertical lines forming a cross at each point give an estimate of the maximum observational error in the measurements of the two variables, distance

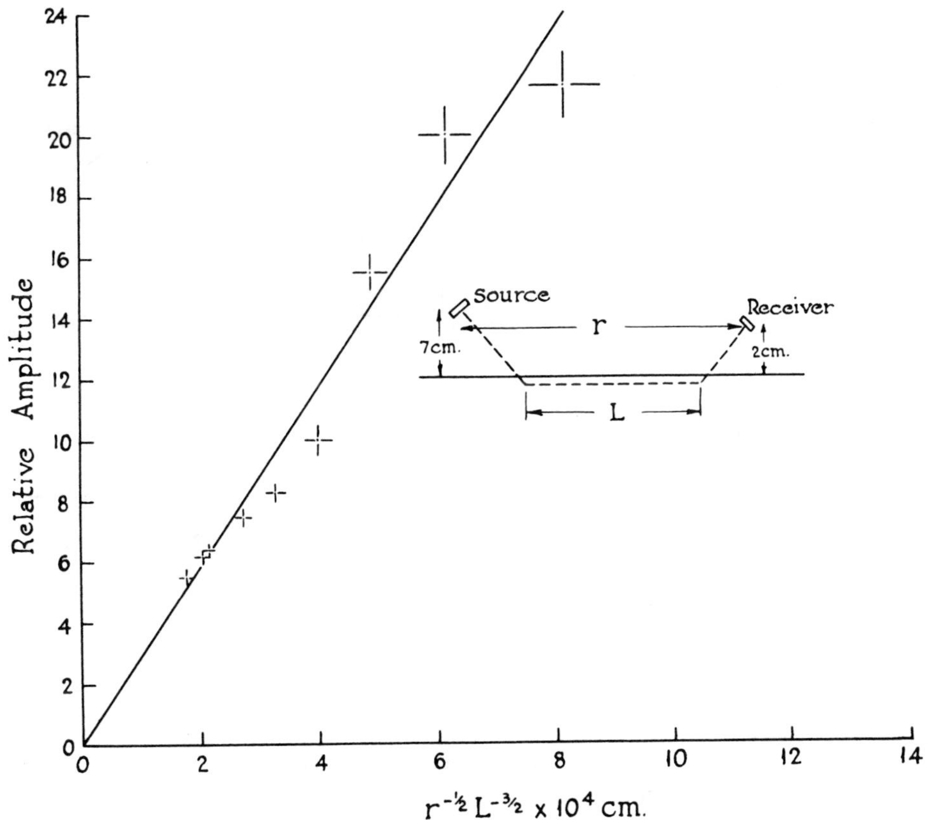


FIG. 12. Decrease of amplitude of a head-wave pulse ($P_1S_2P_1$) along a line parallel to a water-concrete interface.

and amplitude. They do not indicate the total experimental error, which includes other factors not easily estimated, such as mis-orientation of the transducers, inhomogeneities of the media, etc.

### VARIATION OF AMPLITUDE ALONG A LINE PARALLEL TO THE INTERFACE

Figures 10, 11, and 12 show the amplitude of the head wave plotted against the factor $r^{-1/2}L^{-3/2}$. As required by theory, straight lines were drawn through the points and through the origin—as obviously the amplitude should be zero at infinity. The scatter about the lines is considered to be within the total experimental error; thus these data agree with the theory.

In Figure 10, four sets of data are plotted, each for a different height $h$ of the receiver above the interface, all the other conditions being identical. It may be seen that the gradients of the lines, given by tan $X$ in the figure, agree to within 5 percent. This shows further agreement with the theory.

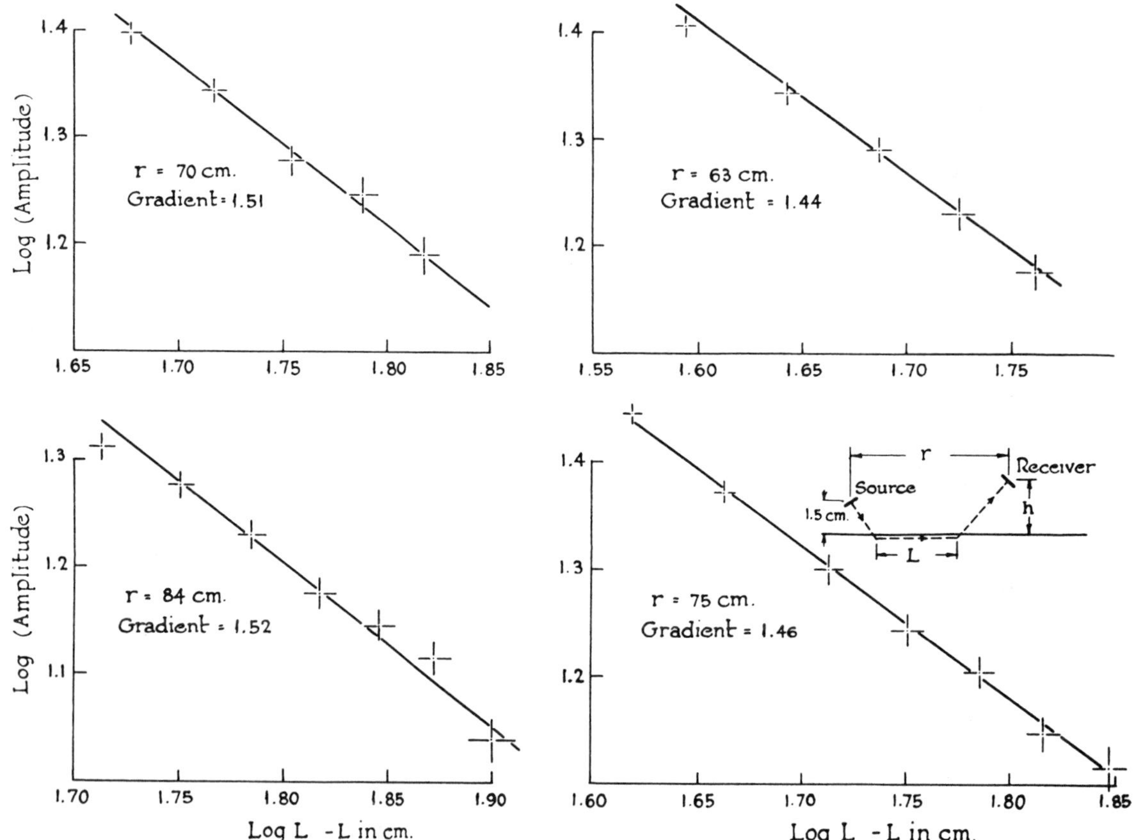


FIG. 13. Plot showing decrease of amplitude of the head-wave pulse ($P_1P_2P_1$) along lines perpendicular to wax-water interface.

In Figure 12, which is a plot for water overlying concrete, the scatter in the points is greater than in the corresponding plots for water overlying wax and for oil overlying calcium chloride solution. This is mainly because the observations were made on the pulse which travelled along the interface with the distortional wave velocity. This pulse coincides in time with the tail of the earlier pulse which travelled along the interface with the dilatational wave velocity; the reliability of the amplitude measurements is thus decreased. Inhomogeneities in the concrete may also account for some of the scatter in the points, although in mixing and pouring the concrete special care was exercised to obtain a fine, homogeneous mixture.

## VARIATION OF AMPLITUDE ALONG A LINE PERPENDICULAR TO THE INTERFACE

Four sets of data were obtained with the receiver set at different cylindrical distances $r$ (see Fig. 1) from the source, all other conditions remaining constant. By varying the height $h$ of the receiver above the interface, while keeping $r$ constant, the detected head wave must follow a ray path whose distance $L$ along the interface decreases as $h$ increases. Figure 13 shows the logarithm of the relative amplitude plotted against the logarithm of $L$; straight lines have been drawn through the points. As $r$ is constant for each set of data, these lines should have a gradient of $-1.50$; in fact, their gradients vary from $-1.44$ to $-1.52$, averaging $-1.49$. The experimental data are certainly not sufficiently reliable to attach any significance to this difference between the arithmetical average of the experimental gradients and their theoretical value.

Further, if $A$ is the intercept on the log $L$ axis and the amplitude really does vary as $r^{-1/2}L^{-3/2}$, then $3A/2+(\log r)/2$ should be constant. The data show that this is so to within 5 percent.

## SUMMARY AND CONCLUSIONS

Head waves were observed with three different pairs of media: oil overlying calcium chloride solution, water overlying wax, and water overlying concrete. In the first pair no distortional wave is possible and the observed head wave travelled along the interface with the dilational velocity of the lower medium. In the second pair the distortional velocity in wax is less than the velocity through water and no head wave was observed which travelled with the distortional velocity through wax. In the third pair, both the dilatational and distortional velocities through concrete are greater than the velocity through water and head waves travelling with both velocities were observed.

Quantitative measurements were made on the amplitude of the head wave and, within the experimental error, the data agree with the predictions of Cagniard and others that the decrease with distance should be proportional to $r^{-1/2}L^{-3/2}$, where $r$ is the orthogonal projection on the interface between the source and detector and $L$ is the distance the wave has travelled in the lower medium.

## ACKNOWLEDGMENTS

This work is part of a doctoral thesis written at the University of Toronto. Financial assistance from the National Research Council of Canada is gratefully acknowledged.

## APPENDIX

In the laboratory experiments reported above, it is shown that for water overlying concrete, the head-wave pulse travelling along the boundary with the $S$-wave velocity of the lower medium has much greater energy than the head wave pulse travelling with the $P$-wave velocity. If this should occur in exploration refraction shooting, then the head wave travelling with the $P$-wave velocity might be missed, especially if secondary arrivals are being used, and consequently the thickness of the layer might be erroneously calculated.

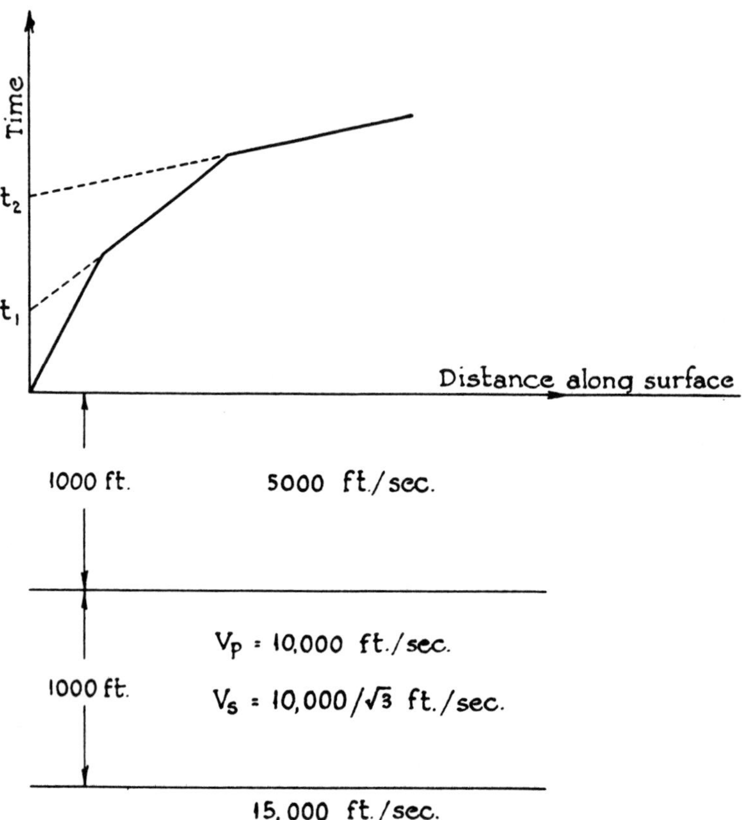


FIG. 14. A hypothetical three-layer case.

The following simple example may serve as an illustration of how such an error could result.

Consider the section drawn in Figure 14. Two successive layers, each of thickness one thousand feet, rest on top of another layer. The uppermost layer has a *P*-wave velocity of 5,000 feet per second, the second layer a *P*-wave velocity of 10,000 feet per second, and the third layer a velocity of 15,000 feet per second. A refraction time-distance graph for such a section would consist of three straight lines, as shown in the figure.

If the arrivals recorded have travelled the whole of their paths as *P* waves, then the gradients of the lines would be 0.00020 sec/ft, 0.00010 sec/ft, and 0.000067 sec/ft. The intercept times $t_1$ and $t_2$ would be 0.346 sec and 0.526 sec respectively.

Now suppose that the first critically refracted wave had travelled along the interface with the *S*-wave velocity. Assuming that Poisson's ratio for the middle layer was 0.25, the gradient of the second limb of the time-distance graph would be 0.00017 sec/ft. If the wave critically refracted at the lower boundary travels through the second layer as an $S$-wave, $t_2$ would also change and no error in thickness would result. However, if this second critically refracted wave had travelled down to the second boundary as a $P$-wave, then calculation of depth would be made using the irrelevant $S$-wave velocity $10{,}000/\sqrt{3}$ ft/sec, instead of the $P$-wave velocity 10,000 ft/sec.

In this latter case the intercept times will be 0.200 sec and 0.526 sec. The thickness of the first layer will, of course, be calculated correctly, but that of the second layer will be calculated as 466 feet, and an error of 534 feet will result.

It would be interesting if data were available on the division of head wave energy between the $P_1P_2P_1$ and $P_1S_2P_1$ phases for actual geologic boundaries.

## REFERENCES


Anderson, D. V. and Northwood, T. D., 1953, Propagation of a pulse over the surface of a semi-infinite solid: Bull. Seis. Soc. Amer., v. 43, p. 239–245.

Cagniard, L., 1939, Reflexion et refraction des ondes seismiques progressive: Paris, Gauthier-Villars.

Evans, J. F., et al., 1954, A three dimensional seismic wave model with both electrical and visual observation of waves: Geophysics, v. 19, p. 220–236.

Gerjuoy, E., 1953, Total reflection of waves from a point source: Communications on Pure and Applied Maths., v. 6, p. 73–91.

Heelan, P. A., 1953, On the theory of head waves: Geophysics, v. 18, p. 871–893.

Maecker, H., 1949, Quantitativer Nachweis von Grenzschichtwellen in der Optik: Ann. Physik., v. 4, p. 409–431.

Oliver, J., Press, F., and Ewing, M., 1954, Seismic model study of refraction from a layer of finite thickness: Geophysics, v. 19, p. 388–401.

Ott, H., 1942, Reflexion und Brechung von Kugelwellen; Effekte 2. Ordnung: Ann. Physik., v. 51, p. 443–446.

Ott, H., 1949, Zur Reflexion von Kugelwellen: Ann. Physik., v. 4, p. 432–440.

von Schmidt, O., 1938, Über Knallwellenausbreitung in Flüssigkeiten und festen Körpern; Physik. Z., v. 39, p. 868–874.

# Diffraction of a Longitudinal Pulse from a Wedge in a Solid

W. W. Grannemann
*California Research Corporation, La Habra, California*

(Received January 20, 1956)

The diffraction theory developed by Sommerfeld and extended by Pauli is used to approximate the relative amplitude of the diffracted pulses from a wedge in a solid. The geometric terms are neglected and the amplitude of the diffracted wave is obtained from the contour integral. Barium titanate transducers are used for sources and receivers on tar and concentrate models to check the predictions of the theory. Several relative amplitude *versus* distance curves are given for regions where the diffracted pulse is separated from other waves. The experimental and theoretical curves agree to ±4 decibels in the regions tested.

## INTRODUCTION

The problem of the diffraction of longitudinal waves from a rigid wedge in a gas was solved by Sommerfeld.[1] Pauli[2] made an extension of this solution in the electromagnetic case, and successful comparisons between experimental and theoretical electromagnetic diffraction patterns from a conducting wedge have been made.[3,4]

The model used in making diffraction measurements in a solid imposes problems not accounted for in the theoretical method of Sommerfeld and Pauli. The transmission property of the wedge, the attenuation of the waves, the shear waves, the refracted waves, the surface Rayleigh wave, and other free boundary conditions must be considered for diffraction in a solid. If a pulsed source is used, the longitudinal pulse may be separated in time from the other waves in certain regions, because there is a difference in velocity of the various waves. The effect of the free boundary of the surface of the model can be largely included in the source and receiver patterns. The models consist of a $10\frac{1}{2}$ degree concrete wedge embedded in tar and a 90-degree concrete step embedded in tar. Experimental and theoretical curves of the amplitude of the diffracted pulse will be given for regions where the diffracted pulse is easily separated from the other waves. Press, Oliver, and Ewing[5] have done some seismic modeling of layered models.

## THEORY

In order to simplify the theoretical considerations of the problem, a number of assumptions have been made. The relative amplitude of the diffracted pulse *versus* angle around the wedge is assumed to be the same as for a perfectly rigid wedge. The diffracted pulse is assumed not to change in shape with distance. The boundary condition for a rigid wedge in a homogeneous medium is the same as for $E$-plane polarization in the electromagnetic case; i.e., $\partial V/\partial \bar{n}=0$, where $\bar{n}$ is the unit normal to the wedge, and $V$ is the amplitude of the diffracted wave from the wedge. From Pauli's work it can be shown that the amplitude of the diffracted wave from the wedge is

$$V=(2\pi kr)^{-\frac{1}{2}}\{\exp[-i(kr+\pi/4)]\}F,$$

where $r$ is the distance from the corner of the wedge to the receiver,

$$k=2\pi/\lambda,$$

$\lambda$ is the wavelength,

$$F=\frac{n^{-1}\sin(\pi/n)}{\cos(\pi/n)-\cos[(\psi-\psi_0)/n]}+\frac{n^{-1}\sin(\pi/n)}{\cos(\pi/n)-\cos[(\psi+\psi_0)/n]},$$

and

$$n=(2\pi-\theta)/\pi.$$

Cylindrical coordinates with the origin at the edge of the wedge are used. The angles are measured from the top edge of the wedge and the distances are measured as shown in Fig. 1. The above expression for the amplitude of the diffracted wave is not valid near the boundaries of the regions, that is, along the shadow line or the line of a reflected ray from the corner of the wedge. An asymptotic series obtained from the contour integral must be used near these boundaries. It should be remembered that this is a steady-state solution and the direct wave is assumed to be a plane wave.

An expression for the relative amplitude of the diffracted pulse may be obtained from the above solution, when the attenuation factor, a source pattern

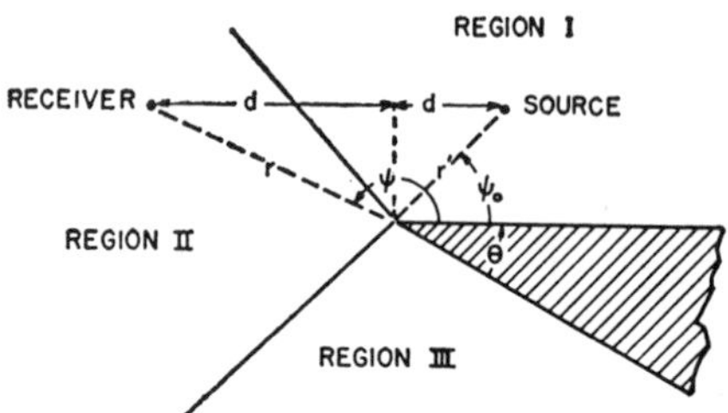


Fig. 1. Diagram of the wedge.

[1] A. Sommerfeld, Math. Ann. **45**, 263 (1894); **47**, 317 (1896).
[2] W. Pauli, Phys. Rev. **54**, 924 (1938).
[3] R. B. Watson and C. W. Horton, J. Appl. Phys. **21**, 802 (1950).
[4] W. W. Grannemann and R. B. Watson, J. Appl. Phys. **26**, 392 (1955).
[5] Press, Oliver, and Ewing, Geophysics, **19**, 388 (1954).

Reprinted from Journal Acoustic Society of America, 28, 494-497 with permission from American Institute of Physics

factor for the pulse, and a receiver pattern factor for the pulse are combined with the solution. A pulse may be thought of as consisting of a number of frequency components. The relative amplitude of the diffracted pulse is

$$V'=\frac{\sum_{p=1}^{m} A_p(2\pi k_p r_1)^{-\frac{1}{2}}\exp[-i(k_p r_1+\pi/4)]F_1 P_{r1} P_{s1}\exp[-\gamma(r_1+r_1')]}{\sum_{p=1}^{m} A_p(2\pi k_p r_2)^{-\frac{1}{2}}\exp[-i(k_p r_2+\pi/4)]F_2 P_{r1} P_{s1}\exp[-\gamma(r_2+r_2')]},$$

where $\gamma$ is the attenuation constant, $r$ is the distance from the corner of the wedge to the receiver, $r'$ is the distance from the corner of the wedge to the source, $P_s$ is the source pattern factor at the angle arc sin $(d'/r')$, $P_r$ is the receiver pattern factor at the angle arc sin $(d/r)$, $m$ is the number of components considered in the pulse, and $A_p$ is the amplitude of the component. The numerical subscripts refer to specific source or receiver positions. If $r_1$ equals $r_2$, the terms in the equation which contain frequency and are constant multipliers cancel, and the equation is still valid for pulses. In this case,

$$V'=\frac{F_1 P_{r1} P_{s1}\exp-\gamma r_1'}{F_2 P_{rs} P_{s2}\exp-\gamma r_2'}.$$

In cases where $r_1$ is not equal to $r_2$, the calculation can be made much simpler by assuming the pulse is contained in a narrow band of frequencies and can be approximated by a single component. The relative amplitude is then approximately

$$V'\approx\frac{r_2^{\frac{1}{2}} F_1 P_{r1} P_{s1}\exp[-\gamma(r_1+r_1')]}{r_1^{\frac{1}{2}} F_2 P_{r2} P_{s2}\exp[-\gamma(r_2+r_2')]}.$$

Under these conditions it is also necessary to assume that the pulse does not change shape or spread with distance.

## EXPERIMENTAL PROCEDURE

The experimental models are 92 cm in width, 184 cm in length, and 28 cm in depth. A cross section of the step model is shown in Fig. 2. The model containing the $10\frac{1}{2}$-degree concrete wedge is of similar construction. A cross section of the wedge model is shown in Fig. 3. The models are in steel tanks to allow easy pouring of the tar.

The experimental arrangement of the source and receiver is shown in Fig. 2. The source consists of five barium titanate transducers placed in line over the step and perpendicular to the strike. Five transducers are used to increase the power and directivity of the source. A step voltage is applied simultaneously to all five transducers. The source produces a pressure impulse on the tar, which approximates a $1\frac{1}{2}$ cy sin wave of about 50 kc. The waves are picked up by a single barium titanate transducer, which is used as the receiver. The signal is amplified by a low noise wide band amplifier, filtered, and displayed on a calibrated oscilloscope.

A typical received signal consists of a longitudinal pulse, a diffracted pulse, and a surface wave, as shown

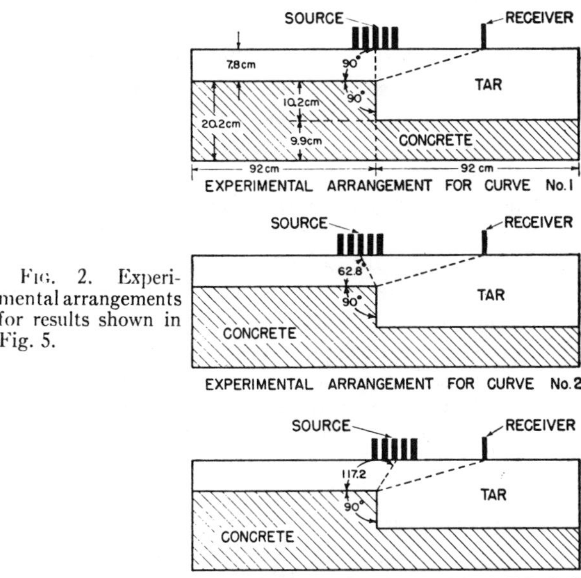


Fig. 2. Experimental arrangements for results shown in Fig. 5.

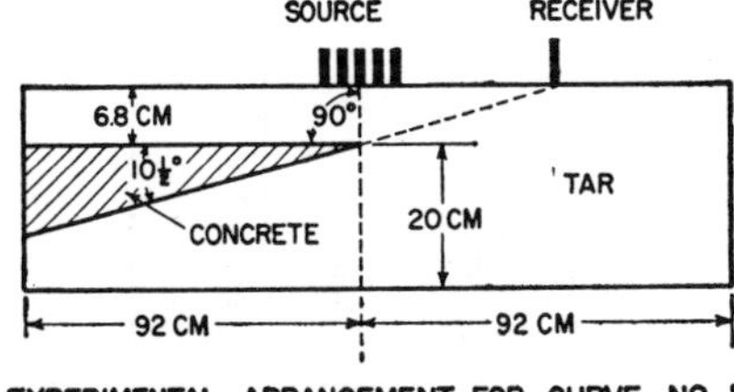


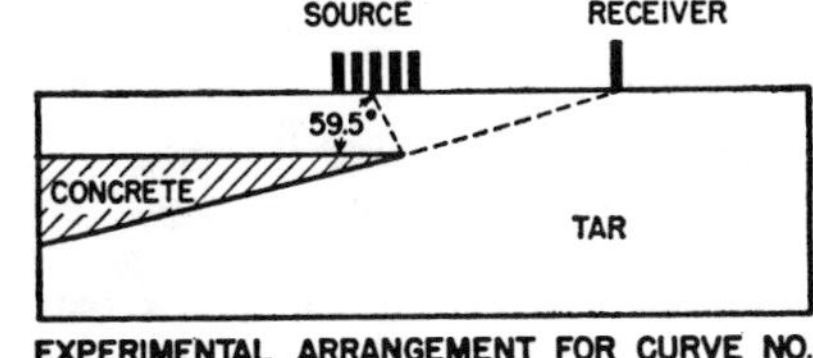


Fig. 3. Experimental arrangements for results shown in Fig. 6.

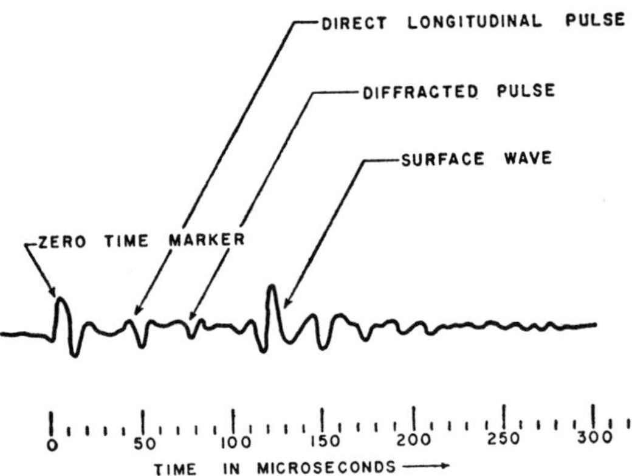


FIG. 4. Typical received trace.

in Fig. 4. The longitudinal pulse moves from the source directly to the receiver; this is the direct *P*-wave. The second event to occur at the receiver is the diffracted longitudinal pulse, which has traveled from the source to the step and up to the receiver. The velocity of the longitudinal pulse is 0.23 cm/$\mu$sec. The next large event is a combination of shear and surface Rayleigh wave, which travels at a much lower velocity than the longitudinal wave. As the receiver is moved close to the source, the surface wave interferes with the diffracted pulse and makes good amplitude measurements impossible. As the receiver is moved away from the source, the path length of the direct wave, and the path length of the diffracted wave approach the same length and interference occurs. As the amplitude of the diffracted pulse decreases, noise also becomes a problem. The experimental points shown in Fig. 5 represent the range of good amplitude measurements on the step model for the source positions shown in Fig. 2. Figure 6 shows the curves for the wedge model. Each experimental point is the average of three readings. On the high side of the step, interference occurs with the reflected wave and the refracted wave, as well as with the direct wave and the surface wave. Amplitude measurements on the diffracted pulse cannot be made without interference for any appreciable distance in this region.

In order to compute the theoretical diffraction patterns, it is necessary to know the source and receiver directivity patterns. To obtain the directivity patterns, the source or receiver was placed at the center of a hemispherical tar model, which had a diameter of 94 cm. The amplitude of the direct longitudinal pulse was measured on the outside of the model at five degree intervals. The round dots in Fig. 7 represent the experimental points obtained from the hemispherical model. The triangular points were obtained from a half-cylinder tar model, which had a diameter of 13.6 cm. The data from the small model would be expected to differ from that of the large model because of the near

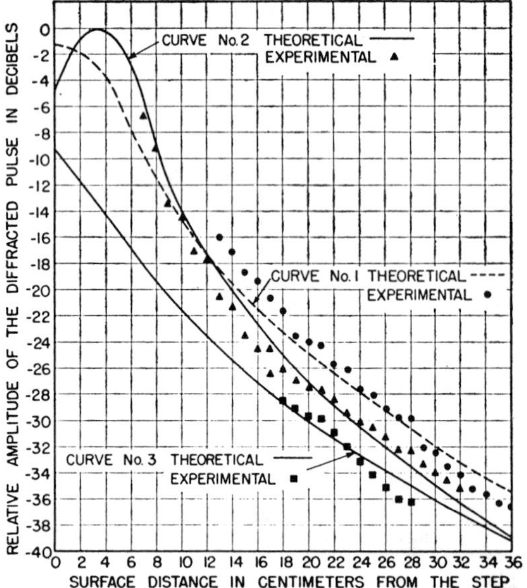


FIG. 5. Relative amplitude of the diffracted pulse *vs* surface distance from the step.

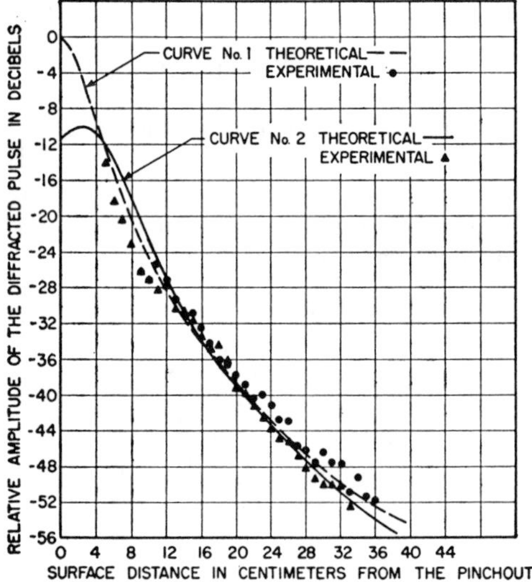


FIG. 6. Relative amplitude of the diffracted pulse *vs* surface distance from the wedge.

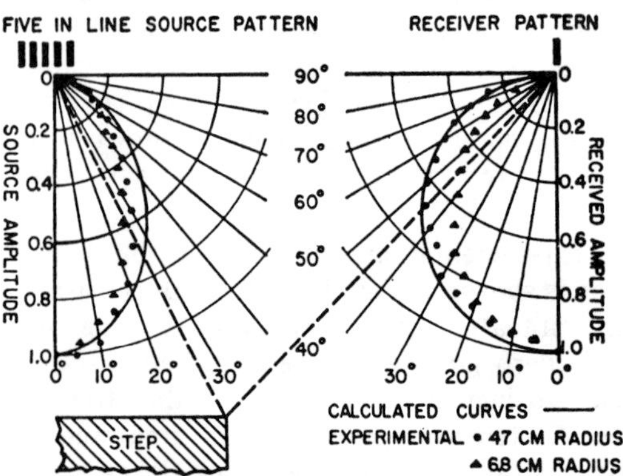


FIG. 7. Source and receiver patterns.

zone effects. If a single barium titanate transducer is assumed to be a dipole on the surface of the tar, the solid line in Fig. 7 is obtained. The five in line source pattern was calculated for a large distance from the source, using steady-state conditions, and a 50-kc frequency.

The attenuation constant for tar was obtained from the hemispherical model by comparing the amplitude of the first event to the amplitude of the second event. The first event is the direct longitudinal pulse which has traveled from the source to the receiver a distance of 47 cm. The second event is a longitudinal pulse which has traveled to the opposite side of the hemisphere and undergone total reflection and focusing. The pulse is focused at the center of the hemisphere, where total reflection occurs again. The pulse then goes to the receiver after traveling 141 cm. The spreading effect of the spherical wave is canceled and the attenuation factor may be determined. The amplitude ratio is 4 to 1, hence $\gamma = 0.015$.

## RESULTS

The theoretical curve and the experimental points for the step model agree within experimental error. The relative amplitude of the diffracted pulse is plotted against distance from the step in Fig. 5. The theoretical curve was computed using the experimental source pattern from the small half-cylinder model, since the radius of this model was about equal to the source to step distance. The attenuation factor was determined from the hemispherical model. The receiver pattern was assumed to be a cosine curve in all calculations. If the source is moved from directly over the step to 4 cm from the step, the path length from the source to the step changes by about 0.9 cm. Some spreading of the energy can be expected. However, a correction for this would be less than 1 db, so it was neglected.

The amplitude of the diffracted pulse is plotted against distance from the wedge edge in Fig. 6. The depth of tar over the concrete was 6.8 cm for the wedge model. The spreading of energy is somewhat greater in this case than with the step model. A correction factor of $1/r'$ was used. This has the effect of lowering the solid theoretical curve by 1.3 db. The experimental points tend to fall away from the theoretical curves near the edge of the wedge. The directivity pattern of the receiver differs from a cosine curve in the near zone, and this, together with experimental errors, is sufficient to cause the lower experimental values.

# Elastic Wave Propagation in a Semi-Infinite Solid Medium

By J. W. C. SHERWOOD
Physics Department, Imperial College, London
and
Applied Physics Division, National Research Council, Ottawa

*Communicated by R. W. B. Stephens; MS. received 13th September 1957, and in final form 5th November 1957*

*Abstract.* A simple physical picture is given of plane waves possessing complex angles of propagation, which play an important role in the theory of elastic wave radiation. Their utility is illustrated by studying continuous sinusoidal wave propagation parallel to the unstressed, plane boundary of a semi-infinite, homogeneous and isotropic solid medium, the Rayleigh and head waves being particular features of the investigation. A novel study of the field due to an impulsive force acting at a line in the surface of a semi-infinite medium indicates a general method of solving important transient propagation problems encountered in seismology. The equivalent problem of an impulsive force acting at the edge and in the plane of a semi-infinite thin sheet has been simulated experimentally by detonating small explosive charges at the edge of an aluminium sheet. The displacements detected by a condenser microphone technique are in excellent agreement with the theoretical determinations.

## § 1. Introduction

The problem of elastic pulse transmission through the earth has been steadily gaining in interest, from the earlier investigation of earthquake waves to the more recent use of controlled explosions in the search for oil and ore bodies. The interpretation of the elastic disturbances, which are complicated by the effects of multiple reflections, refraction, diffraction and scattering arising from the heterogeneous nature of the medium, depends upon associating the salient features with familiar propagation characteristics. Until recent years detailed knowledge of such characteristics has been restricted in two ways: firstly by the involved mathematical analyses required for all but the simplest elastic wave problems and secondly by the difficulty of performing controlled experiments. The advent of laboratory model seismology techniques (e.g. Levin and Hibbard 1955, Oliver *et al.* 1954) has eliminated the latter limitation, but theoretical difficulties still remain. In this study emphasis is placed on the theoretical attack, novel model seismology techniques being employed to substantiate the work.

## § 2. Plane Wave Theory

It is well known that elastic disturbances can propagate with slower velocities than those associated with simple dilatation, P, and rotation, S, waves. Plane waves possessing this very property are simple solutions of the equation of motion for an infinite, isotropic and homogeneous medium. They are of vital importance in the general theory of radiation, as they may be superposed to synthesize any wave field, just as in Fourier synthesis sine waves may be superposed to construct any arbitrary wave. These plane waves will be discussed here in some detail, as an adequate description does not appear to exist in the present literature.

The equation of motion, in the absence of external forces, may be expressed in the form

$$(\lambda+2\mu)\,\nabla(\nabla\cdot\mathbf{s})-\mu\,\nabla\times\nabla\times\mathbf{s}=\rho\partial^2\mathbf{s}/\partial t^2, \qquad \ldots\ldots(1)$$

where $\lambda$, $\mu$ are the Lamé constants, $\rho$ is the density, $t$ is the time and $\mathbf{s}$ is the vector displacement. For a plane dilatation or rotation wave generality is not lost by taking the $y$ axis to lie perpendicular to the displacement $\mathbf{s}$. The dilatation $\Delta$ and the rotation $W$ (strictly speaking $W$ is twice the rotation) may then be defined as

$$\Delta=\nabla\cdot\mathbf{s}=\frac{\partial u}{\partial x}+\frac{\partial w}{\partial z}, \quad W\mathbf{j}=\nabla\times\mathbf{s}=\left(\frac{\partial u}{\partial z}-\frac{\partial w}{\partial x}\right)\mathbf{j}, \qquad \ldots\ldots(2)$$

where $\mathbf{j}$ is the unit vector in the $y$ direction, and $(u, w)$ are the displacement components in the $x$, $z$ directions. Assuming sinusoidal time dependence, the equation resulting from the substitution of definitions (2) in equation (1) is

$$(\lambda+2\mu)\,\nabla\Delta-\mu\,\nabla\times\mathbf{j}W=-\rho\omega^2\mathbf{s}, \qquad \ldots\ldots(3)$$

where $\omega$ is the pulsatance. Successively eliminating $W$ and $\Delta$, by taking the divergence and curl of equation (3) and manipulating, we obtain the two independent scalar wave equations

$$\nabla^2\Delta+k_1^{\,2}\Delta=0; \qquad \nabla^2 W+k_2^{\,2}W=0, \qquad \ldots\ldots(4)$$

where

$$k_1=\frac{\omega}{\alpha}=\omega\left(\frac{\rho}{\lambda+2\mu}\right)^{1/2}, \qquad k_2=\frac{\omega}{\beta}=\omega\left(\frac{\rho}{\mu}\right)^{1/2}, \qquad \ldots\ldots(5)$$

$\alpha$ and $\beta$ being the velocities associated with dilatation and rotation waves, respectively. Equations (3) and (4) indicate that the plane wave solution for the dilatation and its associated displacement components may, omitting the time factor $\exp(i\omega t)$, be expressed by

$$\begin{bmatrix}\Delta\\ u\\ w\end{bmatrix}=A\exp\{-ik_1(x\sin\theta+z\cos\theta)\}\begin{bmatrix}-ik_1\\ \sin\theta\\ \cos\theta\end{bmatrix}, \qquad \ldots\ldots(6)$$

where $\theta$ is defined as the propagation angle. When $\theta$ is real it denotes the angle between the direction of propagation and the $z$ axis. The wave travels with a velocity $\alpha$ and possesses a uniform amplitude of motion, with the displacement in the direction of propagation. However $\theta$ may be complex and the imaginary term imparts some interesting properties to the plane wave. We may investigate these properties by assuming $\theta=i\theta'$, where $\theta'$ is real. Then equation (6) becomes

$$\begin{bmatrix}\Delta\\ u\\ w\end{bmatrix}=A\exp\{-ik_1 z\cosh\theta'+k_1 x\sinh\theta'\}\begin{bmatrix}-ik_1\\ i\sinh\theta'\\ \cosh\theta'\end{bmatrix}.$$

This describes a plane wave propagating parallel to the $z$ axis with velocity and wave number given respectively by

$$c=\alpha\operatorname{sech}\theta'<\alpha; \qquad k'=k_1\cosh\theta'. \qquad \ldots\ldots(7)$$

The amplitude of the wave varies along the wave front as $\exp(k'x\tanh\theta')$ and at any point the particle motion is elliptical with the major axis in the direction of propagation.

In general it is seen that the real part of $\theta$ gives the direction of propagation of the dilatation wave, while the imaginary part specifies the physical characteristics of the wave. Analogous results may be obtained for plane rotation waves.

### 2.1. *Propagation Parallel to a Stress Free Boundary*

The problem of plane wave propagation parallel to a stress-free boundary has received considerable attention in recent years (see Ewing *et al.* 1957, chap. 2, Roesler 1955). This particular analysis is being presented as it clearly illustrates the utility of the preceding plane wave concepts and provides a refined treatment of the problem.

A dilatation wave $\Delta_1$ incident on a plane stress free boundary $z=0$ will generate reflected dilatation, $\Delta_2$, and rotation, $W_3$, waves (Kolsky 1953, p. 24). If $\theta$ is the complex angle of incidence and reflection of the dilatation waves and $\psi$ is the angle of reflection of the rotation wave

$$\left.\begin{aligned}\Delta_1&=A_1(-ik_1)\exp\{-ik_1(x\sin\theta-z\cos\theta)\}\\ \Delta_2&=A_2(-ik_1)\exp\{-ik_1(x\sin\theta+z\cos\theta)\}\\ W_3&=A_3(-ik_2)\exp\{-ik_2(x\sin\psi+z\cos\psi)\}\end{aligned}\right\}, \qquad \ldots\ldots(8)$$

where

$$\sin\theta/\sin\psi=\alpha/\beta=k_2/k_1. \qquad \ldots\ldots(9)$$

Since both the tangential and normal surface stresses are zero, two relationships exist between $A_1$, $A_2$, $A_3$, $\theta$ and $\psi$ (Kolsky 1953, p. 27), namely

$$\left.\begin{aligned}2(A_1-A_2)\cos\theta\sin\psi-A_3\cos2\psi&=0\\ \text{and}\quad (A_1+A_2)\cos2\psi\sin\theta-A_3\sin\psi\sin2\psi&=0\end{aligned}\right\}. \qquad \ldots\ldots(10)$$

Interest here is restricted to propagation parallel to the boundary and this may be obtained by expressing $\theta$ in the form

$$\theta=\tfrac{1}{2}\pi+i\theta' \qquad \ldots\ldots(11)$$

where $\theta'$ is real. The complete wave system is equivalent to a combined dilatation disturbance $\Delta_1+\Delta_2$ travelling along the boundary with velocity $c$ and wave number $k'$ (see equation (7)), together with an associated rotation disturbance $W_3$ propagated in the direction designated by $\psi$. Equations (7) and (10) in conjunction with equation (9) indicate that when the dilatation velocity $c$ is between $\alpha$ and $\beta$, $\psi$ is real and the rotation travels away from the boundary at an angle $\arcsin(\beta/c)$ and with a velocity $\beta$. On the other hand, when the dilatation velocity $c$ is less than $\beta$, $\psi$ is complex, with the real part equal to $\frac{1}{2}\pi$ radians, and the rotation then propagates along the boundary with the velocity $c$.

Equation (8) also indicates that the dilatation $\Delta_1+\Delta_2$ and the rotation $W_3$ only have the same simple exponential form if $A_1$ is zero. For equation (10) to be consistent this requires that

$$\begin{vmatrix}-2\cos\theta\sin\psi & -\cos2\psi\\ \cos2\psi\sin\theta & -\sin\psi\sin2\psi\end{vmatrix}=0. \qquad \ldots\ldots(12)$$

Using equations (7), (9) and (11), the variables $\theta$ and $\psi$ may be eliminated from equation (12) in favour of the variable velocity $c$ and the known velocities

$\alpha$ and $\beta$. This yields the well-known Rayleigh wave velocity equation. Further investigation shows that the individual dilatation and rotation parts of the disturbance decrease or increase exponentially with increasing $z$, according to whether $\theta'$, $\psi'$ assume positive or negative values, respectively. The solution yielding a disturbance that decreases with increasing $z$ is generally termed the Rayleigh wave.

This brief analysis shows that a dilatation wave system can propagate along a stress-free boundary with a velocity $c$, between $\alpha$ and $\beta$, continuously generating a simple rotation wave (this is often called the von Schmidt or head wave for the particular case where $c=\alpha$) into the medium at an angle arc sin $(\beta/c)$. Combined dilatation and rotation disturbances can also propagate parallel to the boundary with velocities less than $\beta$. The Rayleigh wave is the particular system which does not increase in amplitude at large distances from the boundary.

## § 3. The Radiation from Localized Transient Forces

In the preceding account only plane waves possessing sinusoidal time dependence have been considered, whereas in seismology interest lies mainly in the diverging disturbances created by localized transient forces. The extensive bibliography compiled by Ewing, Press and Jardetzky (1957) lists the diverse theoretical contributions to this wide field of study up to the year 1955. Various methods of solution have been employed in these investigations and there is a consequent demand for a standard system of development consisting of a number of major steps, each step being clearly related to a basic physical process. It is considered that a suitable method can be provided by a combination of the work of Eason, Fulton and Sneddon (1955) and Sauter (1950). The proposed technique will be broadly indicated in the following section by a novel investigation of an impulsive force acting on a line in the surface of a semi-infinite solid medium, a problem which was initially considered by Lamb (1904) and again by Sauter (1950). The technique provides an alternative to the methods employed in the many relevant investigations listed by Ewing *et al.* (1957) and in several more recent works, including those of Garvin (1956), Jeffreys and Lapwood (1957) and Pekeris (1955).

## § 4. The Field Radiated by an Impulsive Force Acting on a Line in the Surface of a Semi-Infinite Medium

The general nature of the field generated by an impulsive force acting on a line in the surface of a semi-infinite, homogeneous and isotropic medium can be conceived from the simple analysis contained in § 2.1. Dilatation and rotation disturbances radiate from the force with velocities ranging from $\alpha$ to 0 and $\beta$ to 0, respectively. In addition, the dilatation disturbances travelling along the surface with velocities between $\alpha$ and $\beta$ generate rotation waves into the medium. At a time $t$ after the initiation of the impulsive force, dilatation exists in the regions a, b and c shown in figure 1, while rotation exists in regions b and c.

In the detailed analysis Cartesian $(x,y,z)$ and cylindrical $(R,\phi,y)$ coordinates are employed, the $y$ axis being coincident with the source while the positive $z$ axis and $\phi=0$ half-plane lie in the medium, perpendicular to the surface. As a first step the radiation from the impulsive line force into an infinite medium is expressed as an integral superposition of plane waves possessing complex angles of propagation. The stress free boundary is then introduced, causing the plane waves radiating into the negative $z$ half-space to be completely reflected into the positive $z$ half-space. In this manner the required field is expressed as an integral superposition of plane waves radiating only into the positive $z$ domain, and is finally evaluated using standard integration techniques.

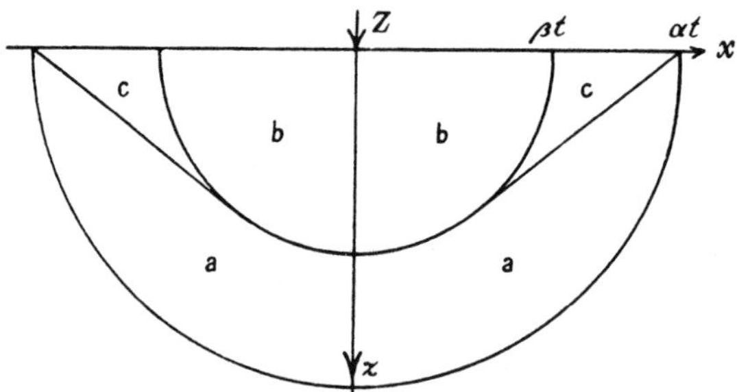


Figure 1. The field radiated by an impulsive force acting on a line in a surface of a semi-infinite solid.

### 4.1. *Radiation into an Infinite Medium*

The equation of motion for a medium upon which forces per unit volume $\mathbf{F}$ are being exerted is

$$(\lambda+2\mu)\,\nabla(\nabla\cdot\mathbf{s})-\mu\,\nabla\times\nabla\times\mathbf{s}+\mathbf{F}=\rho p^2\mathbf{s} \qquad \ldots\ldots(13)$$

where $p$, for the present moment, may be considered as the operator $\partial/\partial t$. If the dilatation and rotation potentials $\Delta'$ and $\mathbf{W}'$ are defined as

$$\nabla\cdot\mathbf{s}=\Delta=p^2\Delta'/\alpha^2=h^2\Delta'; \quad \nabla\times\mathbf{s}=\mathbf{W}=p^2\mathbf{W}'/\beta^2=k^2\mathbf{W}', \qquad \ldots\ldots(14)$$

equation (13) becomes

$$\nabla\Delta'-\nabla\times\mathbf{W}'+\mathbf{F}/\rho p^2=\mathbf{s}. \qquad \ldots\ldots(15)$$

Taking the divergence and then the curl of equation (15) we obtain

$$\nabla^2\Delta'+\nabla\cdot\mathbf{F}/\rho p^2=h^2\Delta' \qquad \ldots\ldots(16)$$

and

$$-\nabla\times\nabla\times\mathbf{W}'+\nabla\times\mathbf{F}/\rho p^2=k^2\mathbf{W}'. \qquad \ldots\ldots(17)$$

With the vector identity $-\nabla\times\nabla\times\mathbf{W}'+\nabla(\nabla\cdot\mathbf{W}')=\nabla^2\mathbf{W}'$ and the fact that, by definition, $\nabla\cdot\mathbf{W}'=\nabla\cdot(\nabla\times\mathbf{s})/k^2=0$, equation (17) becomes

$$\nabla^2\mathbf{W}'+\nabla\times\mathbf{F}/\rho p^2=k^2\mathbf{W}'. \qquad \ldots\ldots(18)$$

Equations (16) and (18) are the inhomogeneous wave equations for the dilatation and rotation wave potentials. The problem now is to solve these equations for the particular force that is being considered, and then substitute in equation (15) to obtain the displacement field. The components of the impulsive force acting in the $x$ and $z$ directions can be treated independently. Consider first the component acting in the $z$ direction, with a strength $Z$ per unit length of the $y$ axis; the force per unit volume becomes

$$\mathbf{F}=Z\delta(x)\delta(z)\delta(t)\mathbf{k}, \qquad \ldots\ldots(19)$$

where $\mathbf{k}$ is the unit vector in the $z$ direction and $\delta(x)$ is the Dirac delta function which will be loosely defined here as

$$\lim_{\varepsilon\to 0}\int_0^{\varepsilon}\delta(x)\,dx=1.$$

Since symmetry considerations show that there is no $y$ component of displacement, the rotation must be in the $y$ direction and may be written as

$$\mathbf{W}' = W'\mathbf{j}. \qquad \ldots\ldots(20)$$

Substitution of equations (19) and (20) in equations (16) and (18) yields

$$\left.\begin{aligned} &\nabla^2\Delta' + Z\frac{\partial}{\partial z}\delta(x)\delta(z)\delta(t)/\rho p^2 = h^2\Delta', \\ \text{and} \quad &\nabla^2 W' - Z\frac{\partial}{\partial x}\delta(x)\delta(z)\delta(t)/\rho p^2 = k^2 W' \end{aligned}\right\}. \qquad \ldots\ldots(21)$$

The solution of these equations can be effected by reducing them to simple algebraic form, using standard integral transformation techniques. Now $p$ has previously been considered as the operator $\partial/\partial t$, but it also corresponds to the parameter involved in the direct and inverse Laplace transformations (McLachlan 1953)

$$g\langle p\rangle = p\int_0^\infty g(t)\exp(-pt)\,dt, \qquad g(t) = \int_{\mathrm{Br}} g\langle p\rangle \exp(pt)\,dp/2\pi i p,$$

where $g\langle p\rangle$ is defined as the Laplace transformation of $g(t)$ and Br denotes the Bromwich integration contour from $c-i\infty$ to $c+i\infty$, $c\geqslant 0$. It is thus convenient to employ Laplace transformations for the $x$, $z$ and $t$ coordinates, giving three-dimensional transformations defined by

$$g\langle\xi,\eta,p\rangle = \xi\eta p\int_0^\infty dt\int_0^\infty dz\int_0^\infty dx\, g(x,z,t)\exp(-\xi x-\eta z-pt)$$

and
$$g(x,z,t) = \int_{\mathrm{Br}} dp\int_{\mathrm{Br}} d\eta\int_{\mathrm{Br}} d\xi\, g\langle\xi,\eta,p\rangle\exp(\xi x+\eta z+pt)/(2\pi i)^3\xi\eta p. \qquad \ldots\ldots(22)$$

Applying the direct transformation to equations (21) we have

$$\left.\begin{aligned} &(\xi^2+\eta^2)\Delta'\langle\xi,\eta,p\rangle + Z\xi\eta^2/\rho p = h^2\Delta'\langle\xi,\eta,p\rangle \\ \text{and} \quad &(\xi^2+\eta^2)W'\langle\xi,\eta,p\rangle - Z\xi^2\eta/\rho p = k^2 W'\langle\xi,\eta,p\rangle, \\ \text{or} \quad &\Delta'\langle\xi,\eta,p\rangle = -Z\xi\eta^2/\rho p(-h^2+\eta^2+\xi^2) \\ \text{and} \quad &W'\langle\xi,\eta,p\rangle = Z\eta\xi^2/\rho p(-k^2+\eta^2+\xi^2) \end{aligned}\right\}. \qquad \ldots\ldots(23)$$

Performing the inverse Laplace transformations for the $x, z$ coordinates,

$$\begin{bmatrix}\Delta'\langle p\rangle \\ W'\langle p\rangle\end{bmatrix} = \frac{Z}{\rho p(2\pi i)^2}\int_{\mathrm{Br}} d\xi\int_{\mathrm{Br}} d\eta\begin{bmatrix}-\eta/(-h^2+\eta^2+\xi^2) \\ \xi/(-k^2+\eta^2+\xi^2)\end{bmatrix}\exp(\xi x+\eta z). \quad \ldots(24)$$

Each integral with respect to $\eta$ possesses simple poles at $\pm\sqrt{(m^2-\xi^2)}$, as indicated in figure 2, where $m$ has the values $h$ and $k$ for the $\Delta'\langle p\rangle$ and $W'\langle p\rangle$

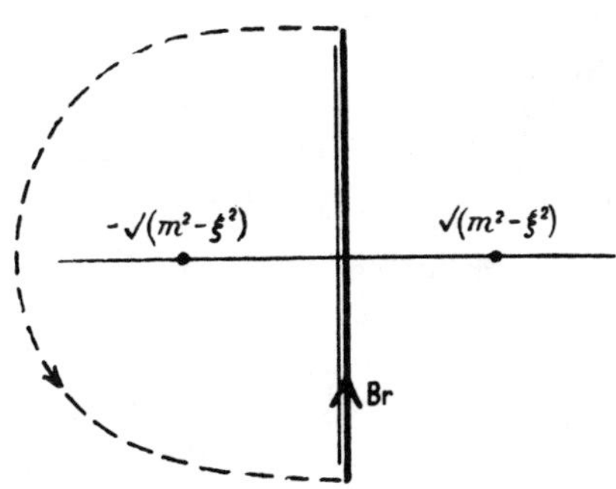


Figure 2. Complex $\eta$ integration plane.

expressions, respectively. When $z>0$, each integral around the dotted contour from $+i\infty$ to $-i\infty$ is zero. Thus, from the theorem of residues, the value of each integral along the Br contour is $2\pi i$ times the residue at the pole $-\sqrt{(m^2-\xi^2)}$. Equations (24) then become

$$\begin{bmatrix}\Delta'\langle p\rangle\\ W'\langle p\rangle\end{bmatrix}=-\frac{Z}{4\pi i p p}\int_{\mathrm{Br}}d\xi\begin{bmatrix}\exp\{\xi x-z\sqrt{(h^2-\xi^2)}\}\\ \xi\exp\{\xi x-z\sqrt{(k^2-\xi^2)}\}/\sqrt{(k^2-\xi^2)}\end{bmatrix}. \quad \ldots\ldots(25)$$

Complex angles are now introduced by performing the coordinate transformations

$$\xi=-h\sin\theta=-k\sin\psi,\qquad d\xi=-h\cos\theta\,d\theta=-k\cos\psi\,d\psi \quad \ldots\ldots(26)$$

giving

$$\begin{bmatrix}\Delta'\langle p\rangle\\ W'\langle p\rangle\end{bmatrix}=\frac{Z}{4\pi\rho p}\int_{i\infty}^{-i\infty}i\begin{bmatrix}-h\cos\theta\exp\{-h(x\sin\theta+z\cos\theta)\}\,d\theta\\ k\sin\psi\exp\{-k(x\sin\psi+z\cos\psi)\}\,d\psi\end{bmatrix}, \quad \ldots\ldots(27)$$

since $p$ (and thus $h$ and $k$) may be treated as a real and positive parameter.

The introduction of the inverse Laplace transformation for the time coordinate $t$ yields the potentials

$$\begin{bmatrix}\Delta'\\ W'\end{bmatrix}=\frac{Z}{8\pi^2\rho}\int_{\mathrm{Br}}\frac{dp}{p}\int_{i\infty}^{-i\infty}\begin{bmatrix}-\cos\theta\exp\{pt-h(x\sin\theta+z\cos\theta)\}\,d\theta/\alpha\\ \sin\psi\exp\{pt-k(x\sin\psi+z\cos\psi)\}\,d\psi/\beta\end{bmatrix}. \quad \ldots\ldots(28)$$

The potentials propagating into the half-space $z>0$ have now been expressed as double integral superpositions of plane waves possessing complex angles of propagation and angular frequencies of $-ip$. Due to the symmetry of the problem, the potentials propagating into the half-space $z<0$ may be written as

$$\begin{bmatrix}\Delta'\\ W'\end{bmatrix}=\frac{Z}{8\pi^2\rho}\int_{\mathrm{Br}}\frac{dp}{p}\int_{i\infty}^{-i\infty}\begin{bmatrix}\cos\theta\exp\{pt-h(x\sin\theta-z\cos\theta)\}\,d\theta/\alpha\\ \sin\psi\exp\{pt-k(x\sin\psi-z\cos\psi)\}\,d\psi/\beta\end{bmatrix}. \quad \ldots\ldots(29)$$

### 4.2. *Normal Impulsive Force on a Semi-Infinite Medium*

The introduction of the stress free boundary at $z=0$ results in the generation of reflected dilatation and rotation potential waves by each of the incident plane waves contributing to the dilatation potential in equation (29), the reflection coefficients being $A$, $B$ and the propagation angles $\theta$, $\psi$ respectively. Reflected dilatation and rotation potential waves are also generated by each incident rotation potential wave, the reflection coefficients in this case being given by $C$, $D$ and the propagation angles by $\theta$, $\psi$. An extension of the analysis in §2.1 yields the following expressions for the reflection coefficients

$$\left.\begin{aligned}A=D&=(4\sin^3\psi\cos\theta\cos\psi-\sin\theta\cos^2 2\psi)/E\\ B&=(4\sin^2\psi\cos\theta\cos 2\psi)/E\\ C&=(-4\sin\theta\sin\psi\cos\psi\cos 2\psi)/E\end{aligned}\right\}, \quad \ldots\ldots(30)$$

where

$$E=4\sin^3\psi\cos\theta\cos\psi+\sin\theta\cos^2 2\psi.$$

The combination of these reflected waves with the waves specified in equations (28) gives the complete disturbance propagating into the half-space $z>0$, the relevant potentials becoming (with the aid of definitions (26))

$$\begin{bmatrix}\Delta' \\ W'\end{bmatrix} = \frac{Z}{8\pi^2\rho}\int_{\mathrm{Br}}\frac{dp}{p}\int_{i\infty}^{-i\infty}\begin{bmatrix}\cos\theta\{(A-1)+C\sin\psi/\cos\psi\} \\ \times\exp\{pt-h(x\sin\theta+z\cos\theta)\}\,d\theta/\alpha \\ \{(A+1)\sin\psi+B\cos\psi\} \\ \times\exp\{pt-k(x\sin\psi+z\cos\psi)\}\,d\psi/\beta\end{bmatrix}.$$

When the reflection coefficients (30) are inserted into the expressions for the potentials it is evident that the integrands of the double integrals over $p$ and $\theta$ or $\psi$ are real functions (i.e. $i$ does not occur explicitly). The following indicated steps

$$\begin{bmatrix}\Delta' \\ W'\end{bmatrix} = \int_{\mathrm{Br}}\int_0^{-i\infty} - \int_{\mathrm{Br}}\int_0^{i\infty} = \int_{\mathrm{Br}}\int_0^{-i\infty} + \int_{(\mathrm{Br})^*}\int_0^{i\infty} = 2\mathscr{R}\left\{\int_{\mathrm{Br}}\int_0^{-i\infty}\right\},$$

plus a transformation to cylindrical coordinates $(y, R, \phi)$ then yield the potentials

$$\begin{bmatrix}\Delta' \\ W'\end{bmatrix} = \frac{Z}{2\pi^2\rho\beta}\mathscr{R}\left\{\int_{\mathrm{Br}}\frac{dp}{p}\int_0^{-i\infty}F(\theta,\psi)\begin{bmatrix}-\cos 2\psi\exp\{pt-hR\cos(\theta-\phi)\}\,d\theta \\ \sin 2\psi\exp\{pt-kR\cos(\psi-\phi)\}\,d\psi\end{bmatrix}\right\} \qquad \ldots\ldots(31)$$

where

$$F(\theta,\psi) = \cos\theta\sin\psi/E(\theta,\psi). \qquad \ldots\ldots(32)$$

The poles $r$ and branch points $p$, $s$ of the integrands of equations (32) are shown in figure 3. The poles occur at the values of $\theta$, $\psi$ that make $E(\theta,\psi)$ zero and correspond physically to the Rayleigh wave propagated along the free surface (see equation (12)). Inspection of equations (25) and (26) indicates that branch points ($p$, $s$) occur when $\cos\theta$ and $\cos\psi$ are respectively zero.

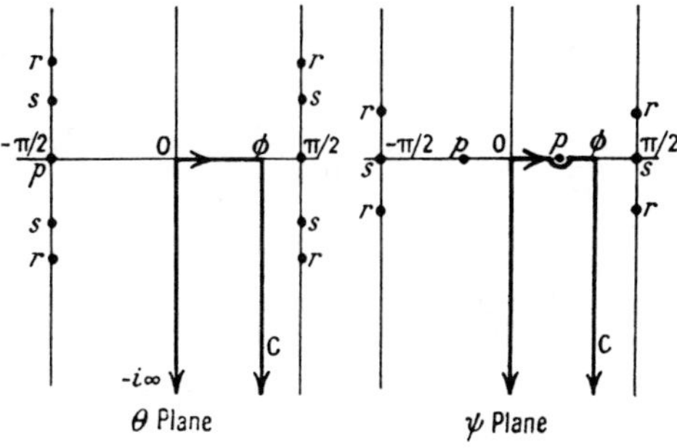


Figure 3. Complex $\theta$ and $\psi$ integration planes.

It is evident that $\cos(\theta-\phi)$ and $\cos(\psi-\phi)$ may be made real throughout the $\theta$ and $\psi$ integrations by replacing the imaginary axis integration contour with a contour C from 0 to $\phi$ and $\phi$ to $\phi-i\infty$, a step that is justified because the integration from $\phi-i\infty$ to $-i\infty$ is zero. It is now valid to interchange the order of the $p$ and $\phi$, $\psi$ integrals and the subsequent execution of the $p$ integration yields

$$\begin{bmatrix}\Delta' \\ W'\end{bmatrix} = \frac{Z}{\pi\rho\beta}\mathscr{R}\left\{\int_C iF(\theta,\psi)\begin{bmatrix}-\cos 2\psi\, H\{t-R\cos(\theta-\phi)/\alpha\}\,d\theta \\ \sin 2\psi\, H\{t-R\cos(\psi-\phi)/\beta\}\,d\psi\end{bmatrix}\right\}, \qquad \ldots\ldots(33)$$

where $H$ denotes the unit step function. The displacements corresponding to these potentials are, from equation (15),

$$\begin{bmatrix} s_R^{\Delta} \\ s_{\phi}^{\Delta} \end{bmatrix} = \frac{Z}{\pi\rho\beta}\mathscr{R}\left\{\int_C iF(\theta,\psi)\cos 2\psi\,\delta\{t - R\cos(\theta-\phi)/\alpha\}\begin{bmatrix}\cos(\theta-\phi) \\ \sin(\theta-\phi)\end{bmatrix}\frac{d\theta}{\alpha}\right\}$$

and

$$\begin{bmatrix} s_R^{W} \\ s_{\phi}^{W} \end{bmatrix} = \frac{Z}{\pi\rho\beta}\mathscr{R}\left\{\int_C iF(\theta,\psi)\sin 2\psi\,\delta\{t - R\cos(\psi-\phi)/\beta\}\begin{bmatrix}\sin(\psi-\phi) \\ -\cos(\psi-\phi)\end{bmatrix}\frac{d\psi}{\beta}\right\}, \quad \ldots\ldots(34)$$

where the subscripts $R$, $\phi$ denote radial and tangential displacement components, and the superscripts $\Delta$, $W$ indicate whether the disturbance is due to the dilatation or rotation.

Finally, the evaluation of the integrations along the contour C gives

$$\begin{bmatrix} s_R^{\Delta} \\ s_{\phi}^{\Delta} \end{bmatrix} = \frac{-Z}{\pi\rho\beta R}\mathscr{R}\left\{iF(\theta,\psi)\cos 2\psi\begin{bmatrix}\cot(\theta-\phi) \\ 1\end{bmatrix}\right\}_{\theta=\phi-i\theta'}, \quad \frac{\alpha t}{R} > \cos\phi \quad \ldots\ldots(35)$$

$$\begin{bmatrix} s_R^{W} \\ s_{\phi}^{W} \end{bmatrix} = \frac{-Z}{\pi\rho\beta R}\mathscr{R}\left\{iF(\theta,\psi)\sin 2\psi\begin{bmatrix}1 \\ -\cot(\psi-\phi)\end{bmatrix}\right\}_{\psi=\phi-i\psi'}, \quad \frac{\beta t}{R} > \cos\phi \quad \ldots\ldots(36)$$

where $\theta' = \operatorname{arc\,cosh}(\alpha t/R)$, $\psi' = \operatorname{arc\,cosh}(\beta t/R)$ and, as indicated, the expressions are to be evaluated at $\theta = \phi - i\theta'$ or $\psi = \phi - i\psi'$.

In equation (35) the displacement components when $\alpha t/R < 1$ are zero, since $\theta$ and $\psi$ are both real. When $\alpha t/R > 1$ the displacement due to the dilatation becomes

$$\begin{bmatrix} s_R^{\Delta} \\ s_{\phi}^{\Delta} \end{bmatrix} = \frac{Z}{\pi\rho\beta R}\mathscr{R}\left\{F(\theta,\psi)\cos 2\psi\begin{bmatrix}\coth\theta' \\ -i\end{bmatrix}\right\}_{\theta=\phi-i\theta'}. \quad \ldots\ldots(37)$$

When $\beta t/R > 1$, expression (36) corresponds to the direct rotation wave and yields

$$\begin{bmatrix} s_R^{W} \\ s_{\phi}^{W} \end{bmatrix} = \frac{Z}{\pi\rho\beta R}\mathscr{R}\left\{F(\theta,\psi)\sin 2\psi\begin{bmatrix}-i \\ -\coth\psi'\end{bmatrix}\right\}_{\psi=\phi-i\psi'}. \quad \ldots\ldots(38)$$

When $\beta t/R < \cos(\phi-\gamma)$, where $\gamma$ is the critical angle $\arcsin(\beta/\alpha)$, $\theta$ and $\psi$ in equation (36) are real and the displacements consequently zero. However, $\theta$ becomes complex when $1 > \beta t/R > \cos(\phi-\gamma)$ and expressions (36) then correspond to the displacements of the rotation disturbance that is generated by the dilatation propagating along the surface of the semi-infinite solid, giving

$$\begin{bmatrix} s_R^{W} \\ s_{\phi}^{W} \end{bmatrix} = \frac{Z}{\pi\rho\beta R}\mathscr{R}\left\{-iF(\theta,\psi)\sin 2\psi\begin{bmatrix}1 \\ \cot\psi''\end{bmatrix}\right\}_{\psi=\phi-\psi''}, \quad \ldots\ldots(39)$$

where $\psi'' = \arccos(\beta t/R)$.

### 4.3. *Tangential Impulsive Force on a Semi-Infinite Medium*

In §§ 4.1 and 4.2 only the impulsive force component in the $z$ direction was considered. A similar analysis involving only the $x$ component $x$ of the impulsive force per unit length of the $y$ axis yields the following field. The displacement components due to the dilatation disturbance are

$$\begin{bmatrix} s_R^{\Delta} \\ s_{\phi}^{\Delta} \end{bmatrix} = \frac{X}{\pi\rho\beta R}\mathscr{R}\left\{G(\theta,\psi)\sin 2\theta\frac{\beta^2}{\alpha^2}\begin{bmatrix}\coth\theta' \\ -i\end{bmatrix}\right\}_{\theta=\phi-i\theta'} \quad \ldots\ldots(40)$$

where $G(\theta, \psi) = \cos\psi \sin\theta / E(\theta, \psi)$. The direct rotation wave displacements are

$$\begin{bmatrix} s_R^W \\ s_\phi^W \end{bmatrix} = \frac{X}{\pi\rho\beta R} \mathscr{R} \left\{ G(\theta, \psi) \cos 2\psi \begin{bmatrix} i \\ \coth\psi' \end{bmatrix} \right\}_{\psi = \phi - i\psi'}. \qquad \ldots\ldots (41)$$

The disturbance with which the head wave is connected is given by

$$\begin{bmatrix} s_R^W \\ s_\phi^W \end{bmatrix} = \frac{X}{\pi\rho\beta R} \mathscr{R} \left\{ iG(\theta, \psi) \cos 2\psi \begin{bmatrix} 1 \\ \cot\psi'' \end{bmatrix} \right\}_{\psi = \phi - \psi''}. \qquad \ldots\ldots (42)$$

The expressions (37) to (42) provide the complete theoretical solution to the problem under consideration in this section. The displacements in the directions $\phi = 0$, 9, 18, 27, 31·5, 36, 40·5, 45, 49·5, 54, 58·5, 63, 67·5, 72, 76·5, 81, 84, 86, 88, 89, 90 degrees have been numerically evaluated as a function of normalized time

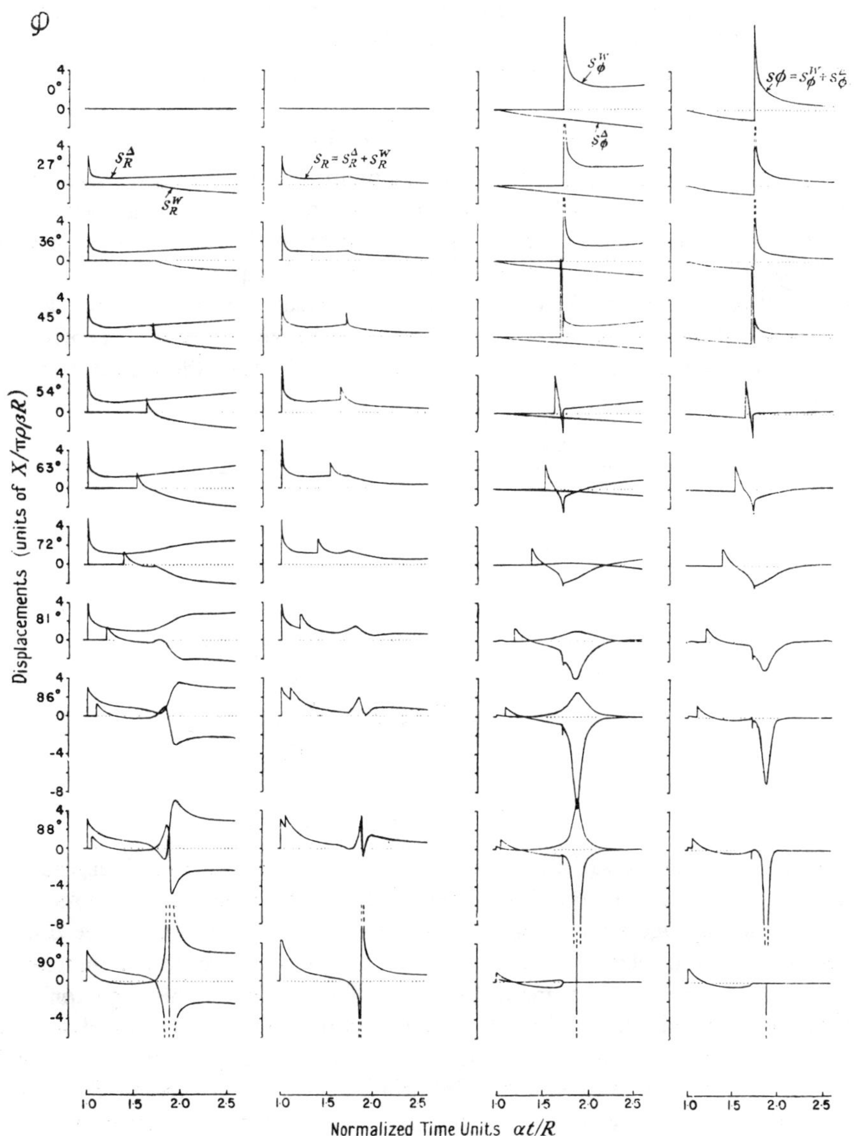


**Figure 6.** Theoretical radial and tangential displacements due to an impulsive force acting along the surface of the semi-infinite medium.

$\alpha t/R$, using the specific velocity ratio $\alpha/\beta = \sqrt{3}$. The dilatation, rotation and total radial and tangential displacement components in several of these directions are plotted in figures 4, 5 and 6. Also shown in figures 4 and 5 are the displacement recordings obtained in an experimental simulation (described in §5) of the impulsive force acting normally to the surface, the symbols P, S and H indicating the discontinuities occurring at the arrival times of the direct dilatation and rotation waves and the head wave, respectively.

It is of interest to note that, by the general theory of reciprocity (Rayleigh 1877), the solution developed here also yields the surface disturbance of a semi-infinite medium due to a directional impulsive force which acts on an internal line lying parallel to the surface. Pekeris (1955) has solved an analogous three-dimensional problem of a transient force acting at an internal point in a direction perpendicular to the surface. When his detailed computations are published it will be interesting to compare them with the results presented here.

## § 5. Experimental Investigation

An experimental investigation of the problem examined in §4.2 was assisted by means of an analogy which avoided the difficulties of simulating a line source and detecting displacements in the interior of a bulky medium. Since the problem is two-dimensional in nature it appears that it might be equivalent to an impulsive force acting at the edge and in the plane of a thin sheet. Oliver, Press and Ewing (1954) have rigorously established this equivalence, subject to the replacement of the infinite medium dilatation wave velocity $\alpha$ by the thin sheet dilatation wave velocity.

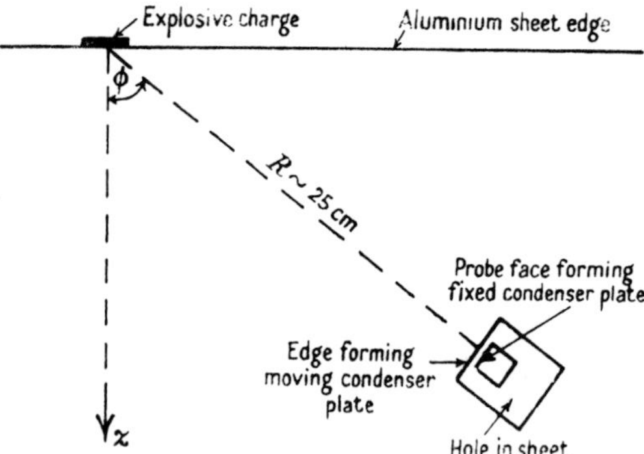


Figure 7. Experimental technique.

Two criteria must be satisfied in order clearly to resolve the rapid displacement variations in the plane of the thin sheet. Firstly, the linear dimensions of the force and the displacement detector must be small compared with the distance separating them and secondly, the time duration of the force must be much smaller than the time taken for the main disturbance to travel to the detector. The experimental system finally adopted is indicated in figure 7. The explosive cap, about 1·3 mm in diameter, of a Christmas cracker was placed symmetrically at the straight edge of a 0·5 mm thick aluminium sheet and detonated by means of a heated nichrome wire. This provided an impulsive force with a time duration of about 2 microseconds (Kolsky 1953, p. 189). The radial and tangential displacements in a specific direction were detected by a condenser along the edge of the aluminium sheet, the impulse per unit area was $5{\cdot}4 \times 10^5\,\mathrm{g\,cm^{-1}\,sec^{-1}}$. Then, since the impulse was found to be equivalent to the maximum force operating for $1{\cdot}5\,\mu$sec, the value of the maximum explosive pressure must have been about $4 \times 10^{11}$ dyn cm$^{-2}$, a value that is fairly typical of high explosives (Pack, Evans and James 1948).

As a result of this high pressure plastic flow of the aluminium occurred in the immediate vicinity of an explosive cap, causing a permanent indentation at the edge of the sheet. Since this violates the theoretical assumption of linear elastic processes, it is perhaps surprising that such excellent agreement exists between the theoretical and experimental displacement fields. This is of direct interest in connection with exploration seismology, where the elastic behaviour around the source (usually an explosive charge) is also generally non-linear.

## § 6. Conclusion

The solution of a particular transient elastic wave problem has been derived here using simple physical concepts, the results being substantiated by an experimental analogue. The technique may clearly be adapted to provide a method of solution for similar but more complex studies, including the particular case where a localized impulsive force is situated in a stratified medium. This is the basic problem encountered in seismic prospecting and it will be the subject of a subsequent publication.

## Acknowledgments

The author wishes to express his deep appreciation of the invaluable aid and encouragement of Dr. R. W. B. Stephens during the course of this research, the main part of which was performed in the Physics Department of the Imperial College of Science and Technology, with the financial assistance of the Department of Scientific and Industrial Research. The author is also indebted to the National Research Council of Canada for the award of a Postdoctorate Fellowship and the provision of facilities enabling both refinements in the theoretical analysis and the computation of expressions (40) to (42) in the text.

## References

Eason, G., Fulton, J., and Sneddon, I. N., 1955–56, *Phil. Trans. Roy. Soc.* A, **248**, 575.

Ewing, W. M., Jardetzky, W. S., and Press, F., 1957, *Elastic Waves in Layered Media* (New York: McGraw-Hill).

Garvin, W. W., 1956, *Proc. Roy. Soc.* A, **234**, 528.

Jeffreys, H., and Lapwood, E. R., 1957, *Proc. Roy. Soc.* A, **241**, 455.

Kolsky, H., 1953, *Stress Waves in Solids* (Oxford: Clarendon Press).

Lamb, H., 1904, *Phil. Trans. Roy. Soc.* A, **203**, 1.

Levin, F. K., and Hibbard, H. C., 1955, *Geophysics*, **20**, 19.

McLachlan, N. W., 1953, *Complex Variable Theory and Transform Calculus* (Cambridge: University Press).

Oliver, J., Press, F., and Ewing, W. M., 1954, *Geophysics*, **19**, 20.

Pack, D. C., Evans, W. M., and James, H. J., 1948, *Proc. Phys. Soc.*, **60**, 1.

Pekeris, C. L., 1955, *Proc. Nat. Acad. Sci. U.S.A.*, **41**, 469, 629.

Rayleigh, Lord, 1877, *The Theory of Sound*, **1**, §107 (New York: Dover).

Roesler, F. C., 1955, *Phil. Mag.*, **46**, 517.

Sauter, F., 1950, *Z. angew. Math. Mech.*, **30**, 203.

microphone technique. A square hole, of side 1·0 mm, was formed in the aluminium sheet, the edges being parallel and perpendicular to the line connecting the hole to the explosive about 25 cm away. One edge was used as the moving condenser plate, the parallel stator plate being provided by a face of a probe (0·6 mm square) inserted into the hole. A fixed 500 pF condenser was placed in parallel with the condenser detector, whose plate separation was about 0·1 mm, and a polarizing voltage of 200 volts provided from a source of 0·5 MΩ output impedance. This gave a circuit with a time constant ($CR \simeq 250\,\mu$sec) sufficiently long to maintain the condenser charge tolerably constant over a time of 40 $\mu$sec, which was the approximate duration of the significant elastic disturbance. The voltage variation of the condenser was amplified by a 1·5 Mc/s wide-band amplifier and displayed on the single stroke time base of a high speed oscilloscope, this being initiated by the pulse from a photocell irradiated with the light flash from the explosive. The oscilloscope display corresponded directly to the time variation of the relevant displacement in the aluminium sheet, the displacement amplification factor $A$ being

$$A \simeq SGEk_0a/Cd^2 \sim 1{\cdot}6 \times 10^3 \qquad \ldots\ldots(43)$$

where $S$ is the deflection sensitivity of the cathode-ray tube ($4{\cdot}0 \times 10^{-4}$ m/v), $G$ is the overall gain of the electronic amplifier ($4{\cdot}0 \times 10^4$), $E$ is the condenser microphone polarizing voltage ($2{\cdot}0 \times 10^2$ volts), $k_0$ is the permittivity of free space ($8{\cdot}85 \times 10^{-12}$ F), $a$ is the effective area of the condenser plates ($3{\cdot}0 \times 10^{-5}$ m$^2$), $C$ is the total capacity of the microphone circuit ($5{\cdot}0 \times 10^{-10}$ F), and $d$ is the gap between the condenser plates (approximately $0{\cdot}1 \times 10^{-3}$ m).

Photographs of the oscilloscope display, corresponding to the radial and tangential displacements for various values of $\phi$, are shown in figures 4 and 5 (Plates). The rapid variations corresponding to the arrival times of the various disturbances are clearly resolved, and the initial displacement features are in excellent agreement with those resulting from the theoretical analysis. Features occurring in later parts of the experimental records are due to disturbances reflected from various edges of the aluminium sheet.

### 5.1. *Quantitative Analysis*

The orders of magnitude of the physical characteristics of an explosive cap were determined by comparing the theoretical and experimental observations. The theory predicted that an impulsive force would initiate an impulsive radial displacement travelling along the edge of the sheet with Rayleigh wave velocity (see the radial displacement for $\phi = 90°$, in figure 4, also Lamb 1904). Since the explosive impulse duration (about 2 $\mu$sec) was considerably greater than the time of passage (about 0·4 $\mu$sec), of a Rayleigh wave across the face of an explosive cap this meant that the experimentally observed Rayleigh wave displacement had approximately the same time variation as the explosive force. The display of this displacement on a 20 $\mu$sec time base indicated that the impulsive force had a roughly Gaussian dependence on time and was equivalent to the maximum force operating for about 1·5 $\mu$sec.

The experimental displacement record was calibrated by means of the displacement amplification factor estimated in equation (43). Then from a direct comparison with the theoretical displacement expressed in units of $Z/\pi\rho\beta R$ (see figure 4), the impulse per centimetre sheet width $Z$ was determined to be $7 \times 10^4$ g sec$^{-1}$. As this typical explosive cap extended over a length of 1·3 mm

# PROPAGATION OF ELASTIC WAVE MOTION FROM AN IMPULSIVE SOURCE ALONG A FLUID/SOLID INTERFACE†

## I. EXPERIMENTAL PRESSURE RESPONSE

By W. L. ROEVER and T. F. VINING

## II. THEORETICAL PRESSURE RESPONSE

By E. STRICK

## III. THE PSEUDO-RAYLEIGH WAVE

By E. STRICK

(*Communicated by Sir Edward Bullard, F.R.S.—Received* 11 *August* 1958)

Parts I and II of this report compare the experimentally observed pressure response for the impulse-excited fluid/solid interface problem with that derived from a corresponding theoretical investigation. In the experiment a pressure wave is generated in the system by a spark and detected with a small barium titanate probe. The output of the probe is displayed on an oscilloscope and photographed. Two cases are investigated: one where the transverse wave velocity is lower than the longitudinal wave velocity of the fluid and the other where the transverse wave velocity is higher. Both of these observed responses are shown to agree even as to details of wave-form, with exact computations made for a delta-excited line source. This comparison is justified by making an approximate calculation for the decaying point source and showing that at these distances it does not differ appreciably from the delta-excited line source.

In the case of low transverse wave velocity one finds, besides critically refracted *P*, direct, and reflected waves, a Stoneley type of interface wave. Although the emphasis in recent years has been towards minimizing the importance of Stoneley waves, the evidence here is that a Stoneley wave can be the largest contributor to a response curve.

In the case of high transverse wave velocity the critically refracted *P* wave is smaller, and the Stoneley wave, though it tends to maintain a rather constant amplitude, becomes compressed in time and arrives very soon after the reflexion. Between the critically refracted *P* wave and the direct arrivals one finds both experimentally and theoretically a pressure build-up preceding the arrival time that might be expected for a critically refracted transverse wave.

In part III this pressure build-up is investigated and found to consist of the superposition of three arrivals. The most prominent of these is a pseudo-Rayleigh wave. The others are the critically refracted transverse wave and the build-up to the later arriving Stoneley wave. Detailed investigation of the pseudo-Rayleigh wave shows it to have the velocity of a true Rayleigh wave which is independent of the existence of the fluid. Furthermore, it has the same retrograde particle motion as the true Rayleigh wave. However, it is radiating into the fluid as it progresses and therefore has many of the properties of a critically refracted arrival when measurements are made in the fluid. Mathematically it differs from the true Rayleigh wave in that its origin is not from a pole on the real axis of the plane of the variable of integration, but rather from a pole which lies on a lower Riemann sheet in the complex plane. In the high transverse wave velocity case this pole is not too far removed from the real axis and the imaginary part of the pole location might be interpreted as a decay factor. The real part, however, yields only approximately the velocity of the pseudo-Rayleigh

† Publication No. 173, Shell Development Company, Exploration and Production Research Division, Houston, Texas.

Reprinted from Philosophical Transactions of the Royal Society, Series A, 251, 455-488 with permission from Royal Society of London

wave, for the actual velocity as pointed out above is precisely that of the true Rayleigh wave velocity. The migration of this complex pole explains why such a pseudo-Rayleigh wave was not observed in parts I and II in the low transverse velocity case.

The problem under discussion is intimately related to the classic work of Horace Lamb *On the propagation of tremors over the surface of an elastic solid.* One need make only a minor re-interpretation of the source function in order to compare directly the wave-forms (excluding of course the Stoneley wave contribution).

Finally, a method is suggested for obtaining the solid rigidity of bottom sediments in water-covered areas from *in situ* measurements of the pseudo-Rayleigh wave and/or Stoneley wave velocities and arrival times.

CONTENTS†



# I. EXPERIMENTAL PRESSURE RESPONSE

By W. L. ROEVER AND T. F. VINING

## 1. INTRODUCTION

The purpose of this investigation is to obtain complete solutions to a problem in elastic wave propagation both experimentally and theoretically with emphasis on experimentally duplicating as nearly as possible the parameters and conditions assumed in the theory. It is

† The system of numbering the sections is based on (*a*) the correspondence between part I and part II and (*b*) the dependence of part III on part II.

believed that such a dual attack may lead to a better understanding of the solution and give an idea of the relative merits of the two methods in obtaining solutions to more complicated problems. This type of comparison between model seismology and theoretical seismology has been made before for the 'Lamb problem' by Kaufman & Roever (1951), Northwood & Anderson (1953), and most recently by Tatel (1954), who has shown good wave-form agreement between theory and experiment.

We have chosen for this investigation the somewhat more complicated problem of propagation in the vicinity of a fluid/solid interface. The theory for plane waves reflected from such an interface goes back to Knott (1899), but since true plane waves are difficult to approximate in the laboratory this theory has not been experimentally verified. Weinstein (1952) has performed experiments on an aluminium/water interface using a harmonic beam; he finds a discrepancy between his results and the theory which may result from the finite width of beam used.

Arons & Yennie (1950) have made field observations of the wave-form of spherical pulses reflected from the sea bottom for various geometries. These wave-forms are compared with the wave-form predicted from the theory for reflexion of a plane pulse from a fluid/fluid interface. The predicted phase shift on reflexion results in a change in shape of the reflected pulse which is in good agreement with that observed.

Von Schmidt (1938) investigated, by the schlieren visual technique, the wave-fronts generated by a spark which arrived before the reflected wave. In his paper one finds a beautiful demonstration of the existence of the critically refracted† wave-fronts which had been utilized for many years in the interpretation of earthquake and seismic prospecting data.

Very recently, O'Brien (1955), in a seismic model study somewhat similar to ours, has investigated the dependence on distance of the peak amplitude of critically refracted waves. He varied the distance both along and normal to water/wax, water/concrete, and oil/sodium chloride interfaces and showed that the theoretical expression derived for the amplitude in the neighbourhood of the wave-front also appears to give good agreement when applied to the amplitude maximum.

Although certain discrete arrivals such as the critically refracted and reflected wave-fronts have been studied in recent years, the complete wave-form has not been investigated. The Stoneley interface wave, although it has been studied theoretically, has not, to our knowledge, been demonstrated experimentally. It is our goal in this paper to obtain complete pressure-response curves both experimentally and theoretically in order to indicate the relative importance of the various arrivals.

## 2. Statement of the problem; technique of solution

We consider here the problem of the pressure wave generated by an impulse-excited point source in a large, three-dimensional fluid medium in contact with a similarly large solid medium. The particular objective of these studies will be to observe accurately the effect of

† We shall use critically refracted wave to refer to the arrival corresponding to the minimum time path which has variously been called the 'head wave' (Von Schmidt, 1938; O'Brien 1955) and conical wave (Cagniard 1939). We shall designate the critically refracted longitudinal wave as $P_{cr}$ and the critically refracted transverse wave as $S_{cr}$ from the terms 'primary' and 'secondary' used in earthquake seismology.

the presence of solids of various elastic constants on the pressure wave-form observed at a point in the fluid. By a large medium we mean one with linear dimensions sufficiently great that reflexions from the outer boundaries will reach the detector after all events associated with the interface have been recorded. To generate a suitable pressure pulse we use a spark because this enables us to release a relatively large amount of energy from a small source. Otherwise the technique is very similar to that used in many of the earlier model experiments mentioned previously.

## 3. Apparatus (experimental)

A capacitor discharging across the spark gap under the water creates a pressure pulse and, at the same time, by inductive coupling, triggers the horizontal sweep of a Tektronix-type 511-AD oscilloscope. A pressure-sensitive, barium titanate probe measures the resulting pressure which, after suitable amplification, is presented to the vertical amplifier of the oscilloscope. The detector and pre-amplifier can be manually replaced by a time mark generator as the need arises (see figure 1).

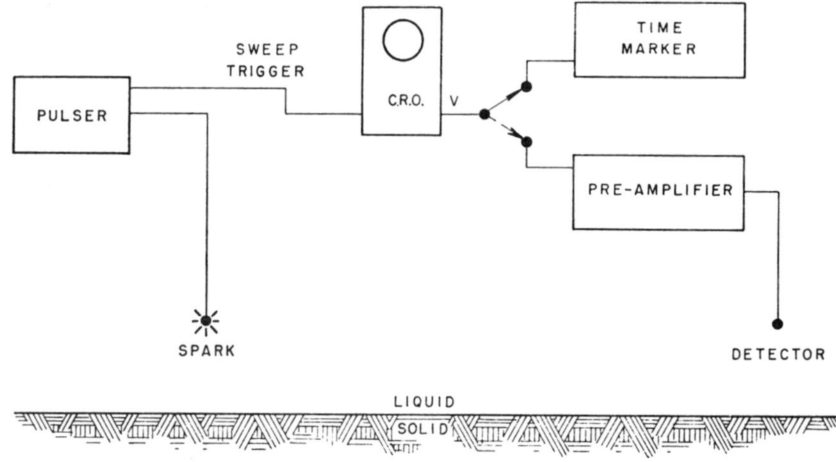


Figure 1. Block diagram of equipment.

### (a) *Nature of the source*

The power source for the spark consists simply of an 800 V supply which charges a 1 $\mu$F capacitor through a 10 k$\Omega$ resistor. A type OA 5 thyratron is used to discharge the capacitor across the spark gap (see figure 2*a*). A peak discharge current of the order of 800 A flows across the gap when the thyratron is triggered.

The spark gap is made by squeezing a $\frac{1}{8}$ in. o.d., thick-walled copper tube onto a 0·035 in. diameter copper wire having extruded nylon insulation. In this manner, the outside diameter of the tip is reduced to $\frac{1}{16}$ in. and the nylon insulation is reduced to about 0·001 in. The gap, which becomes eroded after 50 to 100 firings, can be rejuvenated by filing a small amount of metal off the tip. Observations of the current flowing in the spark circuit as a function of time indicate that most of the energy is dissipated in the first 2 $\mu$s. Figure 2*b* shows the pressure response as a function of time as observed in a large tank of water (to which sodium chloride has been added to increase the conductivity) at a distance of 10 cm from the spark. From figure 2*b* one observes that the pressure pulse can be approximated by an instantaneous rise in pressure, followed by an exponential decay to $1/e$ of the peak value in about 5 $\mu$s. The amplitude of the frequency spectrum of such a pulse is given by $\frac{1}{2}\pi\sqrt{(\alpha^2+\omega^2)}$, where $\alpha$ is the decay constant which in this case is $1/5\times10^{-6}=2\times10^5$/s. Thus the frequency where the amplitude is down to one-half its low-frequency value is given by

$$\omega=\sqrt{3}\,\alpha=2\sqrt{3}\times10^5\,\text{rad/s}\quad\text{or}\quad 2\sqrt{3}/2\pi\times10^5\sim55\,\text{kc/s},$$

which corresponds to a wavelength in the water of 2·7 cm. Since the diameter of the source is about 3 mm, the source and also the detector to be described will appear to be good point sources and point detectors when placed in a practical fluid such as water. The spark creates an oscillating bubble, the first oscillation having a period of about 600 $\mu$s, so that the complication of multiple arrivals can be avoided by making sure that all interesting events occur well within that time.

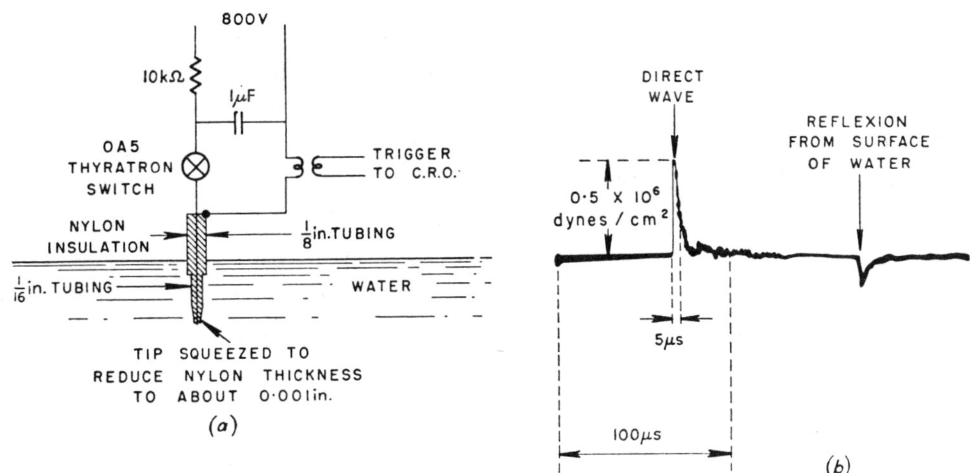


FIGURE 2*a*. Diagram of spark source. FIGURE 2*b*. Pressure pulse observed at 10 cm.

### (*b*) *Nature of the detector*

The pressure probe consists of a $\frac{1}{8}$ in. by $\frac{1}{8}$ in. barium titanate cylinder mounted on the end of a 4 in. long piece of $\frac{1}{16}$ in. hypodermic needle tubing which is in turn mounted on the end of an 18 in. piece of $\frac{1}{8}$ in. tubing. An insulated wire drawn through the hole in the tubing provides the electrical contact to the inner surface of the sensitive element. The aim in this construction is to minimize the effect of the probe on the pressure field and also to minimize acoustic energy travelling along the support. A very similar construction is described by Heuter & Bolt (1955) in more detail. By the reciprocity technique of calibration, it has been established that the probe has a response of $-130\pm4$ db with respect to 1 V dyn$^{-1}$cm$^{-2}$ within the range of 5 to 100 kc/s, with some response well above 100 kc/s.

### (*c*) *Nature of the fluid and solid media*

The fluid for our experiments is ordinary tap water having a compressional velocity of about 1500 m/s. The exact value depends upon the temperature, the air content, and the type and concentration of salt. In order to obtain a value accurate to within 10 m/s, we make the velocity measurement on the same fluid used in the experiment. This is accomplished by making a time-distance profile in the fluid with the solid sufficiently far removed. Since the plot of arrival time against source-detector distance must be a straight line with the reciprocal of the slope yielding the desired velocity of compressional wave propagation in the fluid, the velocity is obtained by carrying out a straight-line least-squares fit with readings taken in, say, 1 cm intervals. Such an analysis shows that twenty to thirty such readings will yield the desired accuracy in velocity measurement. One consequence of this straight-line least-squares fit is that the resulting straight line does not necessarily go through zero when the centre-to-centre source-detector distance is taken to be zero. This non-zero intercept of the time axis is applied as a zero-time correction to the time axis of the interface response curves. The numerical values and probable errors for the fluid velocity and zero-time correction are included below in the data for the solid media because, as we have already stated, the fluid velocity does vary from one experiment to another.

The choice of solid is dependent upon whether or not its rotational velocity $b$ is greater or less than the fluid velocity $c$, for only when $b$ is greater than $c$ will one observe $S_{cr}$ (i.e. a critically refracted wave-front which has apparently travelled along the solid with the transverse velocity of the solid) in addition to $P_{cr}$. It would, therefore, be desirable to have a solid medium for which this refracted transverse wave is present, and one for which it is not. After some consideration, the following selections for the solid media were made:

Case 1, $b < c$. The solid is a very high viscosity pitch with the trade designation Kopper's 85 °C ring and ball softening point pitch.

Case 2, $b > c$. The solid is plaster of paris with additives (trade designation, Halliburton's Cal-Seal type LT 60) mixed three parts powder to one part water by weight.

Table 1. Physical constants measured for cases 1 and 2

| | case 1 | case 2 |
|---|---|---|
| specific gravity | 1·27 | 1·89 |
| longitudinal wave velocity, $a$ (m/s) | 2463 | 3192 |
| transverse wave velocity, $b$ (m/s) | 1003 | 1814 |
| attenuation of longitudinal waves at 500 kc/s (approximate) (db/cm) | 1 | 0·6 |
| attenuation of transverse waves at 25 kc/s (approximate) (db/cm) | 0·6 | 0·2 |
| fluid compressional wave velocity, $c$ (m/s) | $1496 \pm 5$ | $1484 \pm 6$ |
| zero-time correction ($\mu$s) | $-0{\cdot}7 \pm 0{\cdot}1$ | $-1{\cdot}2 \pm 0{\cdot}2$ |

Although at low frequencies pitch behaves as a viscous fluid it can be seen from table 1 that at the frequencies involved in these experiments it behaves as an elastic solid and has, in fact, properties quite similar to those of Lucite. The value of attenuation as measured for the pitch is of the same order of magnitude as that for the plaster of paris, and indicates that losses should not be too serious at distances of the order of 10 cm.

The characteristic velocities $a$ and $b$ of cases 1 and 2 were measured by pulse techniques which are independent of the method diagrammed in figure 1, although the physical principle involved is similar. Cylindrical rods of an appropriate length, say about 10 cm, and of 2 to 3 cm diameter were cored from the solid block on which the propagation study had been made. The compressional velocity is determined by measuring the time for a longitudinal pulse to travel between a barium titanate transducer used to generate the wave at one end of the cylinder and a similar transducer used to detect the wave at the other end.

The transverse wave velocity is measured in a similar arrangement where a torsional wave instead of a longitudinal wave is transmitted along the cylinder. The transducers used to generate this torsional wave are made by arranging eight polarized, circular sectors of barium titanate to approximate a circumferentially polarized disk. When voltage is applied between the faces of the disk there is a rotation of one face relative to the other. This twisting motion is used to generate a torsional wave in the sample and, since these are reciprocal transducers, another disk of the same type is used to detect the torsional wave. A construction similar to this has been discussed by Kennel (1955). The velocities measured by this method are reproducible to within 2 to 3 %.

Attenuation is measured by plotting the logarithm of the amplitude of the received pulse against sample length for a number of different length samples. The slope of the straight line averaging the resulting points is taken as the attenuation. Although this is not a very accurate method because the coupling varies from one sample to another, it is sufficient for determining the relative desirability of various solid materials. The frequency at which the attenuation is measured is estimated from the period of the first cycle of the observed pulse.

## 4. Pressure response determination

Referring to the experimental arrangement of figure 1, and the co-ordinate designations as shown in figure 3, we must select the source-detector distance $r$, source-interface distance $H$, and detector-interface distance $H-z$ in such as manner that the experimental and

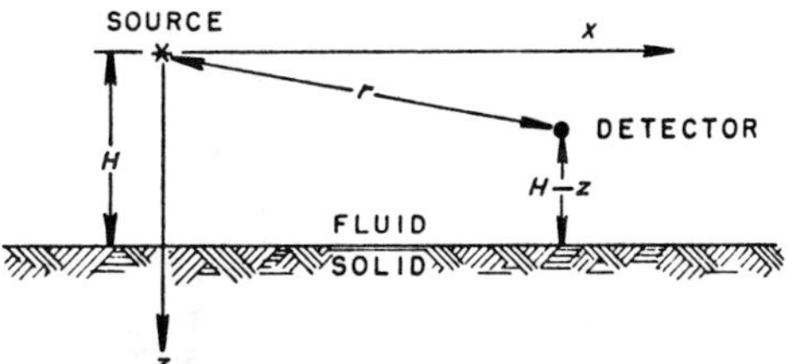


Figure 3. Co-ordinate designations.

theoretical investigations of this and the following part will describe similar situations. Since we should like to be able to attach characteristic wave-forms to those parts of the pressure response curve which follow the corresponding arriving wave-fronts, it is apparent that we must select a source-detector distance which is sufficiently great that the arriving fronts are well separated in time. On the other hand, the attenuation measurements on rods of the solid media as listed in § 3 are of the order of magnitude of $\frac{1}{2}$ db/cm and probably increase with frequency. This means that we cannot use too large a source-detector distance if we desire to keep losses and the consequent high-frequency attenuation from severely limiting our correlation of theory with experiment. Since the source-detector combination has a pass band from 5 to 100 kc/s and the materials used have velocities ranging from 900 to 4000 m/s, we are able to observe wavelengths in the range of 80 to 0·9 cm. Choosing the source-detector spacing ($x$) as 10 cm makes $x$ fall in the cross-over range where

$$10\lambda > x > 0{\cdot}1\lambda.$$

This cross-over region is of great interest because the most widely used theoretical techniques, i.e. the asymptotic and normal mode methods, are not accurate within this region. At 10 cm, the attenuation in plaster of paris and pitch will not be too severe, and the initial requirement, i.e. that $x$ be sufficiently large that the events of physical interest be well separated in time, will be shown to be satisfied both by this experiment and by the theoretical analysis to follow.

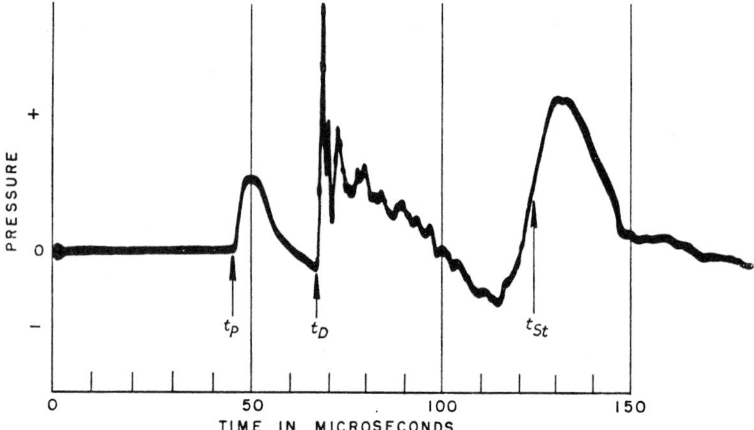


FIGURE 4. Pressure response due to water/pitch interface, $b < c$.

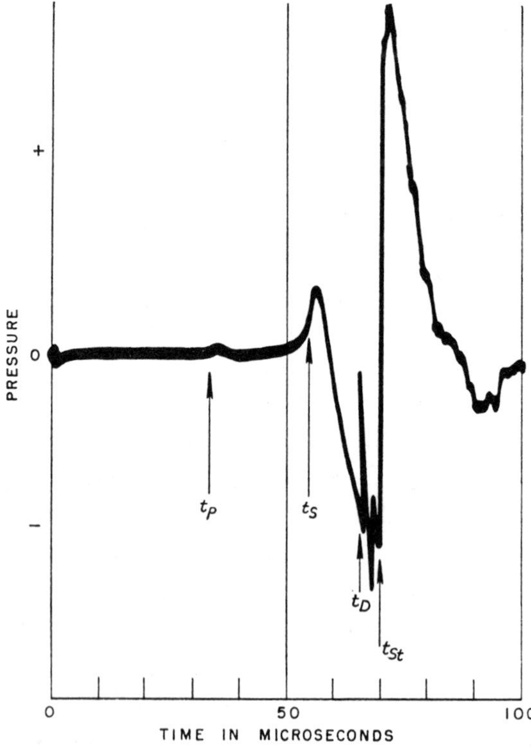


FIGURE 5. Pressure response for a spark source near a plaster of paris interface, $b > c$.

The limitations on the choice of source-interface distance $H$ and detector-interface distance $H-z$ are not as severe. Preliminary experiments indicated that the amplitude of interface waves decreased rapidly as either the source or detector or both were removed from the interface. At a distance of about 0·5 cm from the interface for both source and detector, the amplitude of the interface wave was such that all parts of the wave-form could be observed with a single attenuation setting.

With the source and detector located at the distances stated above, the pressure response curves of figures 4 and 5 have been obtained. Figure 4 was made with pitch as the solid medium and figure 5 with plaster of paris. Shown below each curve is a time scale which has been zero-corrected in accordance with the discussion in § 3.

These response curves are reproduced in figures 8 and 9 of the theoretical treatment in part II, where one notes that a close wave-form correlation exists between the experimental and theoretical solutions of the problem. In the following § (5), figures 4 and 5 of this paper will be interpreted in the light of the existing wave-front arrival time technique that has become rather standardized in recent years, and in part II, a complete wave-form analysis will be shown to answer many of the questions that cannot be treated in the arrival time analysis.

## 5. Discussion of results

### (a) *Wave-form interpretation*

In accordance with a previous statement, the interpretation of the wave-forms of figures 4 and 5 will be limited here to a discussion of the arrival times of the various wave-fronts and the surface waves. Since the water velocity is determined to $\pm 10$ m/s and the source-detector distance is measured to $\pm 1$ mm, the computed arrival time $t_D$ for the direct wave-front is immediately located to approximately $\pm 1$ μs on the response curves. The direct wave is, in both cases, a distinct arrival at the computed time characterized by high frequency and some oscillation which probably represents resonances in the detector system above 100 kc/s. The reflected wave cannot be recognized as a separate arrival in this geometry since for the 10 cm spacing and 1500 m/s velocity in the water, the time difference between the direct and reflected arrivals is only of the order of $\frac{1}{8}$ μs, which is less than the resolution of our apparatus. Since the pressure wave suffers a phase reversal upon reflexion, the net effect is to reduce the amplitude observed in the direct wave region.

The first arrival, $t_P$, on both response curves can be identified as a critically refracted wave-front from the well-known minimum travel-time formula

$$t_P = \frac{x}{a} + \frac{2H - z}{c} \sqrt{\left\{1 - \left(\frac{c}{a}\right)^2\right\}}.$$

These critically refracted rays apparently travel from the source to the interface with the fluid velocity $c$ and at the critical angle $\theta = \sin^{-1}(c/a)$ for total reflexion and then along the interface, with the compressional velocity of the solid, generating a front in the fluid at the same angle of total reflexion. Application of the above minimum travel-time formula to the plaster of paris and pitch data of cases (1) and (2) in § 3 yields $t_P = 35{\cdot}71$ μs for plaster of paris and $t_P = 46{\cdot}5$ μs for pitch, these values being within 1 μs of the observed arrival times of the responses in figures 4 and 5. The velocity of this arrival was determined by the usual time-distance plot for comparison with the velocity obtained from a small sample.

It is also well known that if the rotational velocity $b$, as well as the compressional velocity $a$, of the solid is greater than the fluid velocity $c$, then a second minimum time arrival can exist travelling along the solid part of the interface with the rotational velocity instead of the compressional velocity of the solid. The same minimum travel-time formula holds, except that $a$ is replaced by $b$. Since the data of case (2) for plaster of paris are consistent with $b > c$, they can be applied to the modified travel-time formula and yield an arrival time for the front of the so-called critically refracted transverse wave of 58·3 μs. This arrival time is designated as $t_S$ on figure 5 and falls just ahead of the peak of large positive pressure response occurring before the direct wave-front arrival. The identification of this peak pressure response with that of the $S_{cr}$ wave raises a question concerning the nature of the pressure build-up occurring just before this arrival, which in a sense seems to anticipate this minimum time arrival. This and experiments on other materials, where the arrival falls still farther ahead of the peak, indicate that there is no visible 'break' on the curve that can be associated with the $S_{cr}$ wave. Therefore, in general, true shear velocities cannot be obtained from the usual time-distance plot, although in all cases examined the velocity of the pressure maximum was a good approximation to $b$.

The extremely large pressure variation which occurs after the arrival of the direct wave-front has the general appearance of a boundary wave. If we identify this part of the response as a Stoneley interface wave and insert the velocity data for plaster of paris into the Stoneley wave equation (Stoneley 1924) we arrive at a velocity of 1345 m/s for the Stoneley wave, which for an interface distance of 10 cm yields an arrival time of 74·3 μs. This is very close to the observed arrival time of a point of zero pressure amplitude in figure 5.

Table 2. Characteristic velocities in metres/second

| | rod data (§3) | profile data | average data |
|---|---|---|---|
| *case* 1 | | | |
| $b < c$ (pitch) compressional velocity $a$ | 2463 | 2470 | 2467 |
| $V_{St}$ | — | 824 | — |
| rotational velocity, $b$ | 1003 | 1008† | 1006 |
| fluid velocity, $c$ | — | 1496 | 1496 |
| *case* 2 | | | |
| $b > c$ (plaster of paris) compressional velocity, $a$ | 3192 | 3370 | 3281 |
| $V_{St}$ | — | 1365 | — |
| rotational velocity, $b$ | 1814 | 1889† | 1852 |
| fluid velocity, $c$ | — | 1484 | 1484 |

† Computed from the Stoneley wave equation using measured value of $V_{St}$.

When $c > b$, as in the pitch of case 1, the minimum travel-time formula for the $S_{cr}$ wave becomes complex and such a wave-front does not exist. However, we can make use of the measured characteristic velocities for pitch as listed in § 3 and of the Stoneley wave equation to arrive at a Stoneley wave velocity of 818 m/s and a corresponding travel time of 122·2 μs for the arrival time of the zero-pressure amplitude of the Stoneley wave. This is within about 2 μs of the zero as observed on the response for pitch in figure 4. Therefore, in order to obtain a value for the rotational velocity from the profile data which can be compared with the rod data listed in § 3, we can resort to the profile for the velocity $V_{St}$ of the point of symmetry of the Stoneley wave and, making use of the compressional velocities from such profile data and the previously mentioned Stoneley wave equation, we can determine the rotational velocity of the solid. Profile data obtained in this way are listed in table 2, together with the average of these values which will be used in computation of the results of the theoretical analysis of part II.

Extensive charts of the solutions of the Stoneley wave equation, which enable one to determine quickly one of the parameters, $a$, $b$, $c$, $V_{St}$, or $\rho^f/\rho^s$ when all others are known, have been prepared in this laboratory for a wide range of values of the parameters and have been published elsewhere (Strick & Ginzbarg 1956).

### 6. Conclusions

Measurements of the pressure-time history resulting from a transient point source in the vicinity of a fluid/solid interface show the arrival of a Stoneley interface wave in addition to the arrivals anticipated from simple ray path theory. This to our knowledge is the first experiment to confirm the existence of the Stoneley wave.

The measured response curves indicate that, although there is no 'sharp break' associated with the arrival of the critically refracted transverse wave, the pressure maximum occurring near this time travels with approximately the transverse wave velocity.

The velocities of the Stoneley wave (where it can be identified), the direct wave, and the critically refracted longitudinal wave may be substituted into the Stoneley wave equation to obtain a supplementary check on the transverse wave velocity. This procedure can be used regardless of whether the transverse wave velocity is greater than or less than the fluid velocity and in fact, in the latter case, is the only method of obtaining a value of transverse wave velocity from profile data.

## II. THEORETICAL PRESSURE RESPONSE

By E. STRICK

### 1. Introduction

As in the preceding, experimental part, the main purpose of this investigation is to obtain accurate pressure response curves from an impulsive point source in the vicinity of the fluid/solid interface of homogeneous isotropic elastic media. Although the procedure for obtaining the exact response to a non-periodic point source exciting a solid/solid interface was given by Cagniard (1939) as early as 1939, the complexity of his solution has up to the present time limited the applications of his method to simpler boundary-value problems such as the source located on the free boundary of a semi-infinite elastic solid. In this paper we shall call this latter problem the Lamb problem, since it was first discussed with a rather complete mathematical development by Horace Lamb (1904). Lamb considered both line and point sources on the surface having an impulse excitation. Although his results were qualitatively very good, his method was very intricate. Recently Pekeris (1955*a*) gave a solution in closed form to this Lamb problem for the case of a point pressure source with a Heaviside step excitation in the vertical stress. The Lamb problem becomes somewhat more involved if the location of the source is modified to be within the semi-infinite solid. Using the mathematically simpler line rather than point source, Nakano (1925) and Lapwood (1949) have given qualitative asymptotic solutions of this problem. Garvin (1956) and Gilbert (1956) have obtained the exact closed form solution for the same problem utilizing the method of Cagniard, whereas Takeuchi & Kobayashi (1955) have obtained exactly the same result by means of the Fourier integral. For a point source, numerically incomplete solutions have since been given by Pinney (1954) and Pekeris (1955*b*).

## 2. Statement of the problem; technique of solution

The procedure for constructing integrals of the solutions of elastic wave boundary-value problems is closely related to the technique originally developed by Sommerfeld (1949) for solving problems of radio wave propagation. An integral representation for a periodic source alone having the form of a Fourier integral is first established and the integrand is then interpreted as a periodic elementary plane or cylindrical wave of complex propagation constant. With boundaries present, other convergent Fourier integrals are similarly constructed to account for the wave amplitudes in the various media due to the reflexion and refraction of these periodic inhomogeneous plane waves at the boundaries. A sufficient number of amplitude coefficients are introduced into these integrands to allow the boundary conditions to be satisfied. If the wave equations are also satisfied and the integrals are convergent, then the uniqueness theorem is employed to prove that this is the desired solution. In order to obtain a non-periodic from this periodic solution one can invoke the unit function integral operator (Muskat 1933; Lapwood 1949) to determine the response to a Heaviside step excitation of the source, or one can introduce the Laplace transform variable $s$ by $s = \pm i\omega$ and utilize this powerful tool as Cagniard does to avoid the double integration as required in the former method. The details of Cagniard's procedure are rigorously and elaborately discussed in his book (Cagniard 1939). Dix (1954) has recently summarized the method of Cagniard and remarks that the method appears to be particularly attractive in that it permits one to formulate an elastic wave-boundary value problem in the case of non-periodic source excitations that is free of the *a priori* assumption that steady-state solutions exist. As Dix further points out, Cagniard shows that these periodic solutions form a class from which solutions of pulse problems can be synthesized. However, even though one can in this manner formulate the problems without reference to the existence of the periodic waves, in this paper we shall adhere to the procedure of first establishing the integral for the periodic solution that satisfies both wave and boundary equations and then obtain the non-periodic solution from it. That is, by considering the coefficient of $e^{\pm i\omega t}$ with $\omega$ replaced by $\pm is$ to be the $s$-multiplied (McLachlan 1953) Laplace transform of the response to a step function excited source we can immediately go over into the Cagniard formalism. In this way we can facilitate comparison of the integral with those already established in the literature.

The desired solution of the problem of the response to a step function point source exciting a fluid/solid interface would be to evaluate Cagniard's integrals in the limit as the rotational velocity of the elastic medium in which the point source is located is made to vanish. However, the resulting evaluation would still be quite involved, and in order to simplify calculations as well as to obtain information on the effect that the geometry of the source has upon the resulting wave-form a slightly different approach was indicated. It was decided to carry out the Cagniard calculation first for the case of a delta-function-excited line source, for in this case, as in all problems involving line sources with plane parallel boundaries, the solution can be expressed in a closed algebraic form, i.e. not requiring further integration. Then a similar calculation will be made using instead a source that closely approximates at large distances to a delta-function-excited point source, but still has the form of the previous line source representation. In this way it can be shown that at large distances the geometric difference of the two sources will not appreciably affect the wave-form of the pressure response. The close agreement of these calculations with the experimental response curves of part I will further strengthen this conclusion. As a result, it was not considered necessary to carry out the tedious Cagniard problem for the exact exponentially decaying point source.

Because our problem is in some respects almost a special case of the solid/solid interface problem treated by Cagniard, many of his arguments for solving the problem on a rigorous basis are also valid here. For this reason, our early mathematical development will be rather brief. Since the essential difference is in the nature of the source, special emphasis will be placed on its mathematical representation.

## 3. Apparatus; theoretical

### (a) *Nature of the source*

(i) *The line source*

Consider a cylindrical pressure source of radius $r_0$ immersed in an infinite fluid of velocity $c$. For a step-function excitation of the source the radially symmetric pressure response is shown in appendix 1 to have the form†

$$p_0^f(r,t;\text{step}) = \frac{P_0}{2\pi i}\int_\Omega \frac{d\omega}{\omega}\frac{H_0^{(2)}(\omega r/c)}{H_0^{(2)}(\omega r_0/c)}e^{i\omega t}. \tag{3·1}$$

Considering $(1/2\pi i)\int_\Omega (d\omega/\omega)$ as a unit function operator (Lapwood 1949) it is apparent that the periodic response to a line source is correctly given by

$$p_0^f(r,t;\, e^{i\omega t}) = P_0 \lim_{r_0\to 0}\frac{H_0^{(2)}(\omega r/c)}{H_0^{(2)}(\omega r_0/c)}e^{i\omega t} \to \tfrac{1}{2}\pi i\, P_0 \lim_{r_0\to 0}\frac{H_0^{(2)}(\omega r/c)}{\ln(\omega r_0/c)}e^{i\omega t}. \tag{3·2}$$

The zero-frequency singularity in (3·1) is avoided by the path of the contour $\Omega$. If the upper limit of the angular frequency $\omega$ contained in the source is such that $\omega \ll \lim_{r_0\to 0} c/r_0$, the logarithm term has the form of a slowly varying large negative number whose magnitude can be absorbed into the constant $P_0$, and we have for the simple line source

$$p_0^f(r,t;\, e^{i\omega t}) = -\tfrac{1}{2}\pi i P_0 H_0^{(2)}(\omega r/c)e^{i\omega t}. \tag{3·3}$$

In his text, Sommerfeld (1949) shows that our reduced form (3·3) of the line source does indeed represent physically an outgoing cylindrical wave at large radial distances. If we make use of the Sommerfeld representation for the Hankel function as given by

$$H_0^{(2)}\left(\frac{\omega r}{c}\right) = \frac{1}{\pi}\int_{a-i\infty}^{b+i\infty}\exp\{-i(\omega r/c)\cos w\}\,dw, \quad \text{with} \quad \begin{cases} -\pi < a < 0, \\ 0 < b < \pi, \end{cases} \tag{3·4}$$

† In this paper we have adopted the convention that the symbol which follows the semicolon in the functional parenthesis of the response variable shall describe the source excitation for which that particular response is being determined.

then upon applying the above unit function operator it is easy to show (Lapwood 1949) by elementary integration that the response to the step function excited simple line source is

$$p_0^f(r,t;\,\text{step}) = P_0\,\text{arc cosh}\,(ct/r)\,\mathbf{H}\{t-(r/c)\}, \tag{3·5}$$

where $\mathbf{H}(t-r/c)$ is the step function which is zero for $t<r/c$ and unity for $t>r/c$. By time differentiation we obtain the response to a delta-function excitation of the simple line source

$$p_0^f(r,t;\,\text{delta}) = \frac{c}{r}P_0\frac{\mathbf{H}\{t-(r/c)\}}{\sqrt{\{(ct/r)^2-1\}}}. \tag{3·6}$$

Now, our choice of harmonic line-source representation, as given by inserting (3·4) with $a=-\frac{1}{2}\pi$, $b=+\frac{1}{2}\pi$ into (3·3), yields

$$p_0^f(r,t;\,\mathrm{e}^{\mathrm{i}\omega t}) = \frac{1}{2\mathrm{i}}\int_{-\frac{1}{2}\pi-\mathrm{i}\infty}^{\frac{1}{2}\pi+\mathrm{i}\infty}\exp\{-\mathrm{i}(\omega r/c)\cos w\}\,\mathrm{d}w\,\mathrm{e}^{\mathrm{i}\omega t}, \tag{3·7}$$

which is not suitable as a basis for constructing integrals for the potentials which can satisfy boundary conditions on plane boundaries. A procedure which has been successfully carried

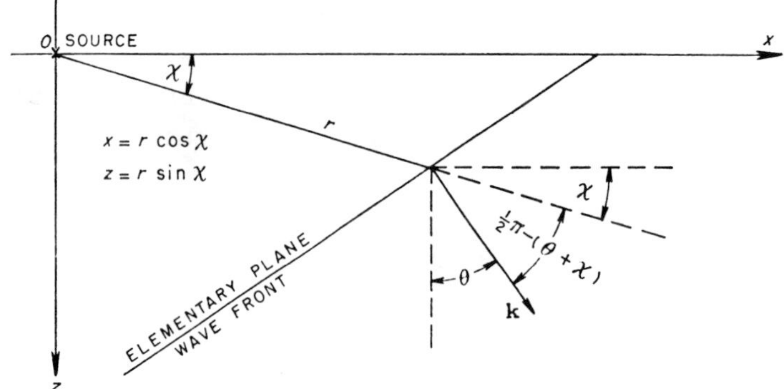


FIGURE 6. Propagation of an elementary plane wave-front in rectangular co-ordinates with $z$ axis positive downwards.

out by Lamb (1904) and Lapwood (1949) is to introduce the new integration variable [say $q=(\omega/c)\sin w$] and express the exponent in rectangular co-ordinates. An equivalent, but better physical approach is to follow to some extent the procedure of Sommerfeld (1949) in deriving the desired expression by considering the line source response to be due to the superposition in the polar co-ordinates $(r,\chi)$ of figure 6, of plane waves of propagation constant $k=\omega/c$ travelling at an angle $\theta$ with respect to the $z$ axis. The $z$ axis taken to be positive downwards is taken as the reference direction, because later in the development a plane boundary will be placed at $z=H$, and it is customary in plane wave-propagation problems to define angles of incidence with respect to the normal to the boundary. Then at the point $D$ in the fluid we have the elementary plane wave

$$p_0^f = \rho^f\omega^2\phi_0^f = P_0\exp\{-\mathrm{i}[(\omega r/c)\cos(\tfrac{1}{2}\pi-\theta-\chi)-\omega t]\} = P_0\exp\{-\mathrm{i}[(\omega r/c)\sin(\chi+\theta)-\omega t]\}, \tag{3·8}$$

which is a periodic solution to

$$\nabla^2\phi_0^f = \ddot{\phi}_0^f/c^2 \tag{3·9}$$

and corresponds to outgoing plane waves as in figure 6.

If we form the integral superposition of plane waves of this type over a range of angles of incidence $\theta$ (the use of the word incidence here will become clear when we apply this source integral to the boundary value problem of §4) as follows

$$p_0^f(r,\chi;\, e^{i\omega t}) = \frac{P_0}{2i}\int_a^b \exp\{-i(\omega/c)\,r\sin(\chi+\theta)\}\,d\theta\, e^{i\omega t}, \quad \text{where} \quad \begin{cases} -\pi < a < 0 \\ 0 < b < \pi \end{cases} \tag{3·10}$$

and permit $\theta$ to enter the complex plane, and if we also make the transformation of variable to $w = \frac{1}{2}\pi - \theta - \chi$, we arrive at a Sommerfeld type integral. Expanding the exponent $r\sin(\chi+\theta)$ and identifying as in figure 6

$$x = r\cos\chi, \quad z = r\sin\chi \tag{3·11}$$

we obtain (since the $z$ axis is positive downwards, the range $0 \leqslant \chi \leqslant \pi$ corresponds to the bottom half of figure 6 and therefore increasing $\chi$ corresponds to a clockwise rotation about the orgin of figure 6):

$$p_0^f(x,z;\, e^{i\omega t}) = \frac{P_0}{2i}\int_{-\frac{1}{2}\pi - i\infty}^{\frac{1}{2}\pi + i\infty} \exp\{-i(\omega/c)\,(z\cos\theta + x\sin\theta)\}\,d\theta\, e^{i\omega t}. \tag{3·12}$$

Although we shall confine our calculations in this paper to $0 \leqslant z \leqslant H$, i.e. to positive $z$, it is desirable that we indicate how (3·12) need be interpreted should we decide to extend our calculations to negative $z$. By definition our pressure source must be invariant to $z \to -z$. Since our line source consists of plane waves travelling away from the origin, we must simultaneously require that $\theta \to \pi - \theta$. Thus, $z\cos\theta$ and therefore $p_0^f$ is invariant to $z \to -z$. With this convention we find that the vertical component of displacement changes sign on $z \to -z$ as it should.

The form (3·12) of the integral contains the desired separation of variables in rectangular co-ordinates that will permit us to satisfy the boundary conditions at $z = H$ for all $x$. Since this is equivalent to matching the propagation constant in the $x$ direction along the interface, an algebraically simpler procedure obtained by introducing the new variable of integration $q$ by

$$q = (1/c)\sin\theta \tag{3·13}$$

is indicated. Then (3·12) becomes

$$p_0^f(x,z;\, e^{i\omega t}) = \frac{P_0}{2i}\int_{-\infty}^{+\infty} \frac{\exp\{-i\omega(qx + \Gamma z)\}}{\Gamma}\,dq\, e^{i\omega t}, \tag{3·14}$$

where

$$\Gamma = \frac{1}{c}\cos\theta = \sqrt{\left(\frac{1}{c^2} - q^2\right)}. \tag{3·15a}$$

We have implicitly assumed here that our radical (3·15$a$) has a phase zero for $|q| > 1/c$. The radical $\sqrt{\{(1/c^2) - q^2\}}$ is subject to the convergence condition

$$\mathscr{I}\sqrt{\{(1/c^2) - q^2\}} \lesseqgtr 0 \quad \text{for} \quad \omega \lesseqgtr 0. \tag{3·15b}$$

Under the transformation (3·13) the saddle points at $\theta = \pm\frac{1}{2}\pi - \chi$ are replaced by branch points at $q = \pm 1/c$. For convergence the contour passes below the branch point at $q = -1/c$ and above that at $q = +1/c$. Note that (3·14) has the desired exponential form of the Fourier integral. As was previously indicated, the transform of the response to a step function excitation of the line source is obtained from (3·14) by introducing $s = i\omega$ into the coefficient of $e^{i\omega t}$ and then distorting the contour 90° to follow the imaginary axis. We then obtain for $s$ positive real:

$$\bar{p}_0^f(x, z; \text{step}) = \frac{P_0}{2i} \int_{-i\infty}^{+i\infty} \frac{\exp\{-s(qx+\Gamma z)\}}{\Gamma} dq, \tag{3·16}$$

where we now require, instead of (3·15*b*), that

$$\mathscr{R}\,\Gamma \geqslant 0.$$

In order to avoid confusion due to the fact that we are going to treat simultaneously the response to both line and point sources, and because it is customary in seismic wave-propagation problems to consider the positive $z$ direction as pointing downward into the earth, we consider the choice of co-ordinate systems for the two sources in some detail. In the line source construction which has the form given by equation (3·3), we imagine an expanding physically significant cylindrical wave of radius $r$ whose axis is normal to the plane of figure 6 and in particular to the downward $z$ direction, that is, the axis of this source lies along the $y$ axis of figure 6. In equation (3·14) we obtained an integral representation of this radially symmetric line source which could be interpreted as a superposition of non-physical (i.e. elementary) plane waves travelling along the horizontal $x$ axis and having for $q > 1/c$ an exponential decay along the $z$ axis.

(ii) *The point source*

In a discussion of the response from a harmonic point source the radially symmetric and physically significant point source function in three dimensions is decomposed into an integral sum of non-physical cylindrical waves of radius $\rho$ travelling in the horizontal $\rho$ direction and again having an exponential decay in the $z$ direction. Note that the axis of these cylindrical waves is taken to be along the $z$ axis and is therefore perpendicular to the axis of the previously discussed two-dimensional cylindrical wave from the line source. Owing to the angular symmetry of these elementary cylindrical waves, it has been shown by Sommerfeld that the integral decomposition for the point source can be expressed in the form

$$p_0^f(\rho, z; e^{i\omega t}) = \frac{P_0 \exp\{-i\omega[(R/c)-t]\}}{R} = P_0 \int_0^\infty \frac{J_0(\lambda\rho)\exp[-\surd\{\lambda^2-(\omega/c)^2\}|z|]}{\surd\{\lambda^2-(\omega/c)^2\}} \lambda\, d\lambda\, e^{i\omega t}, \tag{3·17}$$

where $\mathscr{R}\surd\{\lambda^2-(\omega/c)^2\} > 0$.

Dix (1954) has shown by direct application of the Cagniard formalism to (3·17) that the transform of (3·17) does indeed lead to the response to a step-function-excited point source in an infinite medium as given by

$$p_0^f(R; \text{step}) = \frac{P_0 \mathrm{H}(t-R/c)}{R}, \tag{3·18}$$

where

$$R^2 = \rho^2 + z^2. \tag{3·19}$$

The form (3·17), however, corresponds to standing cylindrical waves in the radial direction and for this reason is physically not desirable for use in our simple interface problem.

In order to arrive at an alternate representation we introduce the new variable of integration $q$ by

$$\lambda = -isq, \tag{3·20}$$

together with $s = \mathrm{i}\omega$ so that

$$\bar{p}_0^f(\rho, z;\, \text{step}) = -sP_0 \int_0^{\mathrm{i}(\infty/s)} \frac{I_0(sq\rho)\exp(-s\Gamma|z|)}{\Gamma}\, q\,\mathrm{d}q, \tag{3·21}$$

where

$$\Gamma = \surd\{(1/c^2) - q^2\}$$

as in (3·15*a*). The branch cut is taken to be $(1/c, \infty)$ with phase of $\Gamma = 0$ for $q < 1/c$.

Upon transformation to the $\Gamma$ plane we obtain the branch radical

$$q = \surd\{(1/c^2) - \Gamma^2\}, \tag{3·22}$$

and (3·21) becomes

$$\bar{p}_0^f(\rho, z;\, \text{step}) = +sP_0 \int_{1/c}^{\infty/s} I_0(sq\rho)\exp(-s\Gamma|z|)\,\mathrm{d}\Gamma. \tag{3·23}$$

Using the well-known relation (Watson 1952)

$$I_0(sq\rho) = \pm(\mathrm{i}/\pi)\,[K_0(s\rho a\,\mathrm{e}^{\pm \mathrm{i}\pi}) - K_0(sq\rho)], \tag{3·24}$$

we find

$$\bar{p}_0^f(\rho, z;\, \text{step}) = \mp\frac{\mathrm{i}sP_0}{\pi}\left[\int_{\infty/s}^{1/c} K_0(s\rho q\,\mathrm{e}^{\pm\mathrm{i}\pi})\exp(-s\Gamma|z|)\,\mathrm{d}\Gamma + \int_{1/c}^{\infty/s} K_0(s\rho q)\exp(-s\Gamma|z|)\,\mathrm{d}\Gamma\right]. \tag{3·25}$$

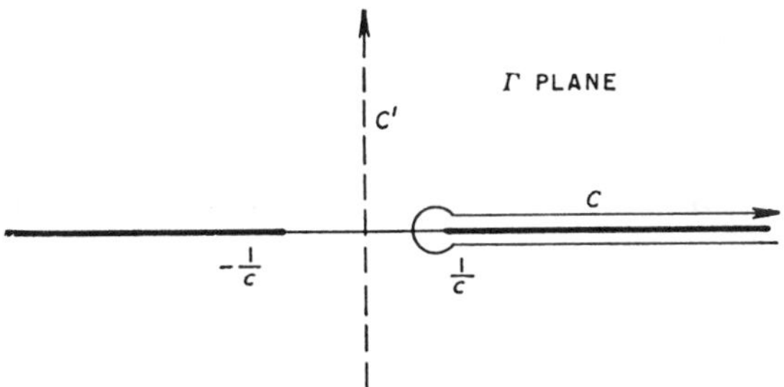


FIGURE 7. Location of the branch points and branch cut in the $\Gamma$ plane.

Just as in the $q$ plane, the radical (3·22) in the $\Gamma$ plane has phase zero to the left of the branch point at $\Gamma = 1/c$ on the real $\Gamma$ axis. The contours of both integrals in (3·25) are now taken above the branch cut from $\Gamma = 1/c$ to $\Gamma = \infty$. But since $q = \surd\{(1/c^2) - \Gamma^2\}$ has phase $-\mathrm{i}$ above and $+\mathrm{i}$ below the cut in the $\Gamma$ plane, it is apparent that the first integral may be set equal to one whose path is taken below the cut provided $q\,\mathrm{e}^{\mathrm{i}\pi}$ is replaced by $q$. We then have the single integral

$$\bar{p}_0^f(\rho, z;\, \text{step}) = \mp\frac{\mathrm{i}sP_0}{\pi}\int_C K_0(sq\rho)\,\mathrm{e}^{-s\Gamma z}\,\mathrm{d}\Gamma, \tag{3·26}$$

taken over the contour $C$ of figure 7. This contour may then be distorted into the imaginary axis $(-\mathrm{i}\omega, 0, +\mathrm{i}\omega)$ as is designated by $C'$ in Figure 7. Note that in accord with our established procedure, letting $s = \mathrm{i}\omega$ and distorting the contour 90° shows (3·26) to be a Fourier integral in the $\Gamma$ plane.

Over the contour $C'$ the argument of $K_0$ is positive real. The form (3·26) for the integral representation of the step-function transform is, of course, singular for $\rho = 0$. However, this is also true of the step transform of the line source as given by (3·16).

We can now proceed to apply the method of Cagniard to show that the integral (3·26) also reduces to (3·18). This proof is given in appendix 3, where it is apparent that the major portion of the proof consists of elementary algebraic transformation of real variables of real integrals, and it is not necessary, as in the case of the point source, to make the second transformation of variable back into the $q$ plane in order to obtain the solution (3·18) for $\rho \neq 0$. For $\rho \to 0$ it can easily be shown that the integral (3·26) converges if we do not perform the limiting process until after the integration is carried out. When (3·26) is used in boundary value problems involving layered media, the integral representations within the layered media will involve the modified Bessel function $I_0$ instead of $K_0$, which is well behaved for $\rho \to 0$. This follows since with $s = \mathrm{i}\omega$, the $K_0$ in (3·26) goes into $H_0^2$ with positive argument corresponding to only outgoing cylindrical waves from the source, whereas $I_0$ goes into $J_0$ with its corresponding standing wave interpretation.

Although we have stated that it is not necessary to make a transformation of variable back into the $q$ plane in order to reduce (3·26) to (3·18), it is desirable to work in the $q$ plane if we are to obtain a relation between the point source and line source responses. The transition to the $q$ plane is most simply carried out by applying (3·22) to (3·26) obtaining

$$\bar{p}_0^f(\rho, z; \text{step}) = \mp \frac{\mathrm{i}sP_0}{\pi} \int_{\infty,\, 1/c,\, \infty} K_0(sq\rho) \exp(-s\Gamma|z|) \frac{q}{\Gamma} \mathrm{d}q, \tag{3·27}$$

where the contour of integration which surrounds the branch cut from $q = 1/c$ to $q = \infty$ can easily be distorted to $[-\mathrm{i}(\infty/s), \mathrm{i}(\infty/s)]$. For large values of the argument, the well-known asymptotic expansion is valid,

$$K_0(sq\rho) = \mp \sqrt{\left(\frac{\pi}{2s\rho q}\right)} \mathrm{e}^{-sq\rho} \left[1 - \frac{1}{8sq\rho} + \ldots\right]. \tag{3·28}$$

Considering only the first term which is good for large $\rho$ such that

$$\rho \gg \left|\frac{1}{8sq}\right| \quad \text{or} \quad \rho \gg \left|\frac{1}{8\omega q}\right|, \tag{3·29}$$

the following approximation is obtained

$$\bar{p}_0^f(\rho, z; \text{step}) = \mp \mathrm{i}P_0 \sqrt{\left(\frac{s}{2\pi\rho}\right)} \int_{-\mathrm{i}\infty/s}^{+\mathrm{i}\infty/s} \frac{q^{\frac{1}{2}}}{\Gamma} \exp(-sq\rho - s\Gamma|z|)\, \mathrm{d}q, \tag{3·30}$$

which for $s = \mathrm{i}\omega$ becomes a Fourier integral in the $q$ plane.

Comparing (3·30) for the transform of the response to a step-excited point source with the corresponding relation (3·16) for the line source it is apparent that one can obtain (3·30) from (3·16) by merely replacing $x$ by $\rho$ and multiplying by the factor

$$-\left[\frac{2}{\pi} \frac{sq}{\rho}\right]^{\frac{1}{2}}. \tag{3·31}$$

In a comparison of point and line source responses, the existence of the factor $\rho^{-\frac{1}{2}}$ in (3·31) is to be expected. The factor $q^{\frac{1}{2}}$ is a very slowly varying quantity; for our simple interface problem where the major contributions come from $q$ essentially between $1/a$ and $3/c$, we can easily neglect it. However, retaining this term will not complicate the calculations appreciably. The expression (3·30) contains an absolute value sign on the co-ordinate $z$ which is not present in (3·16). Sommerfeld introduced this modification in order that his integral would be invariant to $z \to -z$. Sommerfeld did not, however, attempt to retain the identification of his variable of integration as an angle of incidence with respect to a later appearing boundary and consequently did not require that $\Gamma \rightarrow -\Gamma$ as $z \rightarrow -z$. Since we are making our calculations only for positive $z$, we need not concern ourselves with any further modification due to the existence of this absolute value symbol. The remaining factor $s^{\frac{1}{2}}$ will require some discussion.

There are two ways we can deal with this factor $s^{\frac{1}{2}}$ which corresponds operationally to half-order differentiation. The first way is to make use of the product theorem for $s$-multiplied Laplace transforms to arrive at an exact interpretation of our first-order response as given by (3·30). This approach is discussed in appendix 4 and will not be used here. The second and much simpler approach is to reason along the following lines. If the factor $s^{\frac{1}{2}}$ were not present, then just as in appendix 2 the method of Cagniard would naturally lead to the response to a delta-function excitation of the source. On the other hand, if a factor $s$ instead of $s^{\frac{1}{2}}$ were present in (3·30), then in order to avoid a final time integration we would be naturally led to interpret the response as being due to a step function excitation of the source. Therefore, since $s^{\frac{1}{2}}$ lies between these two extremes, we should expect (3·30) to correspond to the response to an excitation which lies somewhere between that of a step function and a delta function. Thus the response to a point source with such an intermediate excitation should resemble the response due to a delta-function-excited line source.

In many seismological problems many investigators go to great labour to set up the response to a point source and using asymptotic methods are not able to obtain very accurate results. Since the excitation of the wave-form of an earthquake is not known, it is just as well to assume the above intermediate excitation instead of a delta or step. Then use can be made of the rigorous line-source calculations to obtain a response which at large radial distances although only an approximation in a quantitative sense shows all essential features of the response in the proper time scale. In this paper, for a comparison with experimental results for point sources, we shall make use of the approximation obtained by dropping the factor $s^{\frac{1}{2}}$, realizing at the same time that our resulting wave-form will be incorrect by an amount corresponding to a half-order time integration of the response to a delta-function-excited source. In §4 we shall see that this approximation leads to a surprisingly good agreement between the theoretical and experimental results.

### (*b*) *Nature of the detector*

For a step- or delta-function excitation of the pressure source, closed form solutions of the response (i.e. not requiring further time integration or differentiation) exist for only certain physically significant response variables. In particular, for a step-function excitation of the line source such a closed-form solution would appear for a particle acceleration detector. On the other hand, a delta-function-excitation of the same source would lead to closed-form solutions for both pressure and particle velocity detectors. Although we shall initially construct the transform of the response to a step-function excitation of the source, owing to our utilization of the $s$-multiplied Laplace transform the inversion of this transform will lead directly to closed-form expressions for pressure and particle velocity responses to a delta-function excitation of the line source. This interpretation as the response to a delta-function excitation is desirable, because after some high-frequency filtering such a response as (3·6) would roughly resemble an exponential decay or the excitation utilized in the experiment in part I.

### (c) *Nature of the fluid and solid media*

In our study of elastic wave propagation, both fluid and solid media are assumed to be homogeneous and isotropic. The customary equations of motion and boundary conditions are valid; that is, in two dimensions we have:

(1) For the fluid with the displacement potential $\phi^f$ defined by displacement

$$(U^f, W^f) = \text{grad}\,\phi^f, \tag{3·32}$$

the wave equation

$$\nabla^2\phi^f - \frac{1}{c^2}\frac{\partial^2\phi^f}{\partial t^2} = 0, \tag{3·33}$$

where $c$ is the sound velocity of the fluid, i.e.

$$c = \surd(\lambda^f/\rho^f). \tag{3·34}$$

The fluid pressure is obtained from the displacement potential by

$$p^f = -\sigma^f = -\rho^f(\partial^2\phi^f/\partial t^2). \tag{3·35}$$

(2) Similarly, for the solid we have in two dimensions in terms of the displacement potentials $\phi^s$ and $\psi^s$ defined by

$$U^s = \frac{\partial\phi^s}{\partial x} + \frac{\partial\psi^s}{\partial z}, \quad W^s = \frac{\partial\phi^s}{\partial z} - \frac{\partial\psi^s}{\partial x}, \tag{3·36}$$

the wave equations

$$\nabla^2\phi^s = \frac{1}{a^2}\frac{\partial^2\phi^s}{\partial t^2} \quad \text{and} \quad \nabla^2\psi^s = \frac{1}{b^2}\frac{\partial^2\psi^s}{\partial t^2}, \tag{3·37}$$

where $a$ and $b$ are the compressional and shear (rotational) velocities, respectively, of the solid and are given by

$$a = \left[\frac{\lambda^s + 2\mu^s}{\rho^s}\right]^{\frac{1}{2}} \quad \text{and} \quad b = \left[\frac{\mu^s}{\rho^s}\right]^{\frac{1}{2}}. \tag{3·38}$$

The solid stresses are obtained from

$$\sigma^s_{zz} = \rho^s\left[\left(1 - 2\frac{b^2}{a^2}\right)\frac{\partial^2\phi^2}{\partial t^2} + 2b^2\left(\frac{\partial^2\phi^s}{\partial z^2} - \frac{\partial^2\psi^s}{\partial x\,\partial z}\right)\right], \tag{3·39}$$

$$\sigma^s_{xz} = \rho^s\left[\frac{\partial^2\psi^s}{\partial t^2} + 2b^2\left(\frac{\partial^2\phi^s}{\partial x\,\partial z} - \frac{\partial^2\psi^s}{\partial x^2}\right)\right]. \tag{3·40}$$

The expressions for the wave equations and boundary conditions for use with the point source in three dimensions can be found in Cagniard (1939). For their application to the fluid/solid interface problem we can refer to the thesis by Spencer (1956). The numerical evaluation of the exact solution of the transient excitation of the fluid/solid interface by a point source for points off the normal to the interface passing through the source would be too laborious to attempt at the present time; therefore, in accord with our approximate development in §3 (*a*) we are going to proceed under the assumption that since the response in an infinite fluid is alike for both point and line sources the responses will remain alike (under the assumptions of §3 (*a*)) when a solid boundary is present. Under this assumption it will not be necessary to make use of the three-dimensional equations and so the following pressure-response determination will be devoted solely to the exact two-dimensional problem.

4. Pressure response determination

The displacement potential due to a line source in an infinite fluid is by (3·14) and (3·35)

$$\phi_0^f(x, z; e^{i\omega t}) = Q_0 \int_{-\infty}^{+\infty} \frac{\exp\{-i\omega(\Gamma z + qx)\}}{\Gamma} dq\, e^{i\omega t}, \tag{4·1}$$

where $\Gamma$ is defined by (3·15) and
$$Q_0 = \frac{P_0}{2i\rho^f \omega^2}. \tag{4·2}$$

When a fluid/solid interface is present at $z = H$, one has the modification of figure 6 as shown in figure 8. The additive potential $\phi_1^f$ due to a superposition of elementary plane waves reflected from the interface and solid potentials $\phi^s$, $\psi^s$ due to refraction into the solid are constructed with (4·1) as a model as follows. For $\phi_1^f$, in order to construct a source function at the image location one need simply replace $z$ in (4·1) by $2H - z$ and insert a reflexion

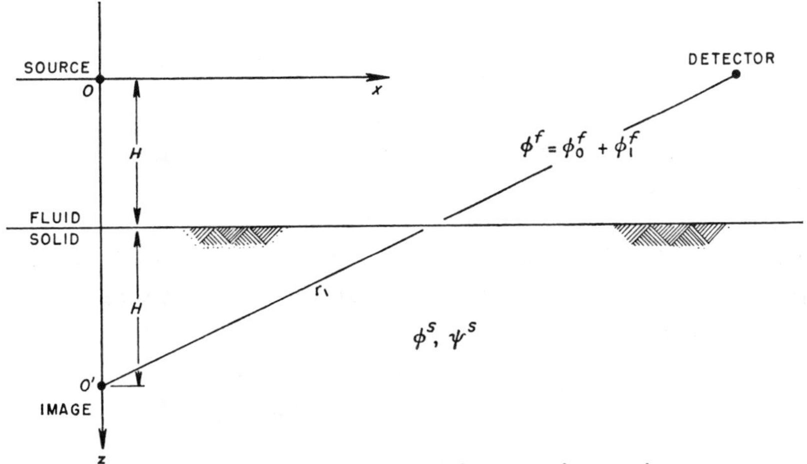


Figure 8. Location of the image point at O′.

coefficient $A(q)$ into the integrand. For $\phi^s$ the term $\Gamma\, z$ in the exponent is replaced by $\surd\{(1/a^2) - q^2\}z$ and the transmission coefficient $B(q)$ incorporated into the integrand. Just as in (3·15b) we must have

$$\mathscr{I}\surd\{(1/a^2) - q^2\} \lesseqgtr 0 \quad \text{for} \quad \omega \gtrless 0. \tag{4·3a}$$

Similarly, $\psi^s$ is constructed by replacing $\Gamma\, z$ by $\surd\{(1/b^2) - q^2\}z$ with $C(q)$ as the undetermined coefficient. Also

$$\mathscr{I}\surd\{(1/b^2) - q^2\} \lesseqgtr 0 \quad \text{for} \quad \omega \gtrless 0. \tag{4·3b}$$

In this way we have constructed the set of potentials

$$\phi_1^f(x, z; e^{i\omega t}) = Q_0 \int_{-\infty}^{+\infty} \frac{A(q) \exp\{-i\omega[\Gamma(2H - z) + qx]\}}{\Gamma} dq\, e^{i\omega t}, \tag{4·4a}$$

$$\phi^s(x, z; e^{i\omega t}) = Q_0 \int_{-\infty}^{+\infty} \frac{B(q) \exp(-i\omega[\surd\{(1/a^2) - q^2\}z + qx])}{\Gamma} dq\, e^{i\omega t}, \tag{4·4b}$$

$$\psi^s(x, z; e^{i\omega t}) = Q_0 \int_{-\infty}^{+\infty} \frac{C(q) \exp(-i\omega[\surd\{(1/b^2) - q^2\}z + qx])}{\Gamma} dq\, e^{i\omega t}, \tag{4·4c}$$

where the coefficients $A$, $B$ and $C$ are determined in the usual manner from the three conditions that the normal stress $\sigma_{zz}$ and normal displacement $w$ be continuous and the tangential stress $\sigma_{zz}$ vanish at the interface at $z = H$ for all $x$.

For our purposes, only the determination of the reflexion coefficient $A(q)$ is necessary, and since the algebraic manipulation is identical to that in the derivation of the well-known reflexion coefficient for plane waves impinging upon a fluid/solid interface, only the final result will be given, i.e.

$$A(q) = \Gamma g(q) - \frac{\rho^f}{\rho^s}\sqrt{\left(\frac{1}{a^2}-q^2\right)} \Big/ \Gamma g(q) + \frac{\rho^f}{\rho^s}\sqrt{\left(\frac{1}{a^2}-q^2\right)}, \tag{4·5}$$

where
$$g(q) = (1-2b^2q^2)^2 + 4b^4q^2\sqrt{\left\{\left(\frac{1}{a^2}-q^2\right)\left(\frac{1}{b^2}-q^2\right)\right\}}, \tag{4·6}$$

whose vanishing yields $q = 1/V_R$, i.e. the reciprocal of the Rayleigh wave velocity $V_R$ for a free boundary semi-infinite elastic solid.

Hence, at any point $(x, z)$ in the fluid the harmonic pressure response is given by

$$p^f = p_0^f + p_1^f, \tag{4·7}$$

with
$$p_1^f(x, z; e^{i\omega t}) = \frac{P_0}{2i}\int_{-\infty}^{+\infty}\frac{A(q)}{\Gamma}\exp\{-i\omega[\Gamma(2H-z)+qx]+i\omega t\}\,dq, \tag{4·8}$$

and $p_0^f$ is given by (3·14).

In order to obtain a non-dimensional form for further calculation let us introduce the new integration variable $u$ by

$$u = cq, \tag{4·9}$$

so that according to (3·13) the variable $u$ is simply the sine of the angle that the propagation vector of the elementary plane waves makes with the normal to the fluid/solid interface. Then $A(q)$ can be re-expressed as

$$A(u) = \frac{N(u)}{D(u)} = \frac{\gamma f(u) - (\rho^f/\rho^s)\,K_2^4\alpha}{\gamma f(u) + (\rho^f/\rho^s)\,K_2^4\alpha}, \tag{4·10}$$

where
$$f(u) = (K_2^2 - 2u^2)^2 + 4u^2\alpha\beta, \tag{4·11}$$

and we have introduced

$$K_1 = \frac{c}{a}, \quad K_2 = \frac{c}{b}, \quad \alpha = \sqrt{(K_1^2-u^2)}, \quad \beta = \sqrt{(K_2^2-u^2)}, \quad \chi = c\Gamma = \sqrt{(1-u^2)}. \tag{4·12}$$

The reflected part of $p_1^f$ of the pressure response becomes in the $u$ plane

$$p_1^f(x, z; e^{i\omega t}) = \frac{P_0}{2i}\int_{-\infty}^{+\infty}\frac{A(u)}{\gamma}\exp\{-i(\omega/c)(ux+\gamma z)\}\,du\,e^{i\omega t}. \tag{4·13}$$

At this point it is convenient to introduce the $s$-multiplied Laplace transform by considering the coefficient of $e^{i\omega t}$ in (4·13) with $s = i\omega$ to be the transform of the reflected part of the response to a step-function excitation of the source. At the same time it is convenient to rotate the path of integration to the imaginary axis as designated by $C$ in figure 9.

$$\bar{p}_1^f(x, z; \text{step}) = \frac{P_0}{2i}\int_{-\infty}^{+\infty}\frac{A(u)}{\gamma}e^{-s\tau}\,du, \tag{4·14}$$

with
$$\tau = u\frac{x}{c} + \gamma\frac{2H-z}{c}. \tag{4·15}$$

Another distortion of the contour of interest is that designated $C'$ which surrounds the cuts on the positive real $u$ axis.

In order to co-ordinate this theoretical development with the experimental discussion of part I, we shall carry along separate numerical investigations as in figure 9 for the pitch and plaster of paris solids. Theoretically, the division into cases 1 and 2 is made according to the possible existence of a critically refracted $S$ wave. Thus, case 2 is defined by the requirement that the critically refracted $S$ wave arrive before the reflected spherical wave-front, i.e. that $b > (r_1/x)\,c$. This condition is not exactly the same as the $b > c$ condition of part I, but for $x = 10\,(2H-z)$ as assumed in the experimental work it follows that $r_1/x = 1{\cdot}005$ and the difference between the two conditions is not important here.

Since we shall carry along exact theoretical curves for pitch and retarded plaster of paris, the necessary parameters† reproduced from part I are listed in table 3 ($x = 10\,(2H-z)$ in both cases).

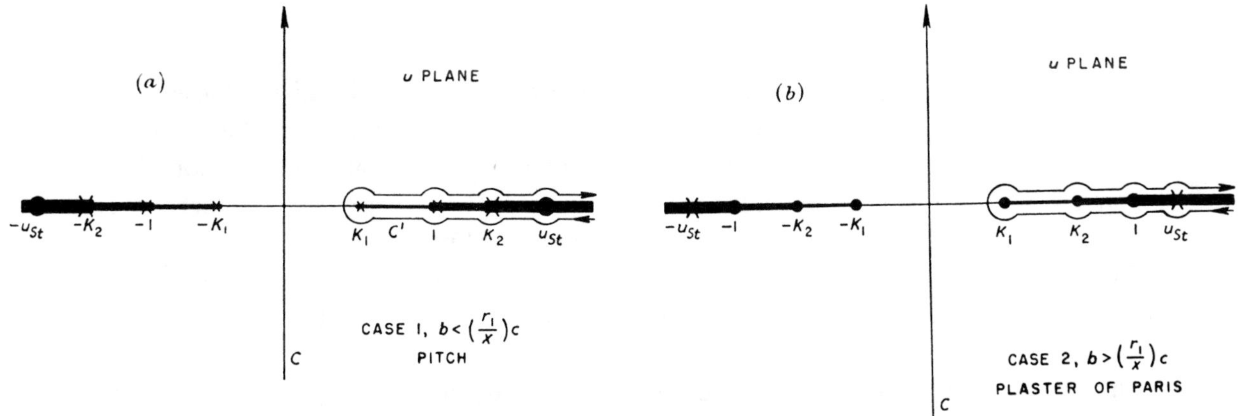


FIGURE 9. Contour for integral (4·14).

TABLE 3. VELOCITY AND SPECIFIC GRAVITY DATA FOR PITCH AND PLASTER OF PARIS

| case 1, pitch, $\sigma = 0{\cdot}401$ | case 2, plaster of paris, $\sigma = 0{\cdot}275$ |
|---|---|
| $K_1 = 0{\cdot}6144$ | $K_1 = 0{\cdot}4401$ |
| $K_2 = 1{\cdot}497$ | $K_2 = 0{\cdot}8052$ |
| $u_R = c/V_R = 1{\cdot}588$ | $u_R = c/V_R = 0{\cdot}8714$ |
| $\rho^f/\rho^s = 0{\cdot}787$ | $\rho^f/\rho^s = 0{\cdot}524$ |
| $c = 1501$ m/s | $c = 1484$ m/s |

The integrand of (4·14) contains branch points at $\pm K_1$, $\pm K_2$, and $\pm 1$, and simple poles $\pm u_{St}$ on the axis as determined by the vanishing of the denominator of the integrand. These latter poles give rise to Stoneley interface waves for the fluid/solid system.

Just as in the development of the source problems in § 3, figure 7, we here find it desirable to choose our branch cuts along the real $u$ axis from the branch points to infinity. The phases of the radicals $\alpha$ and $\beta$ are like $\gamma$ chosen to be zero in between the pairs of the respective

† The parameters of table 3 do not agree exactly with those of table 2 of part I, but except for the compressional velocity of plaster of paris they are in better than 1% agreement. The exception corresponds to profile rather than average data. Owing to various factors such as ageing of the solid material it was frequently necessary to modify the experimentally measured velocities and it was not convenient to modify so frequently the theoretical computations.

branch points at $\pm K_1$, $\pm K_2$ and $\pm 1$. This implies a phase $-$i above and $+$i below the respective cuts on the positive real $\tau$ axis. This range of phase will define the upper Riemann sheets for all radicals. In this convention the phase $-$i will be on the upper sheet.

The existence of a pole in $A(U)$ for the fluid/solid interface yielding a Stoneley-type interface wave has been known (Scholte 1948) for some time. Sato (1954) has discussed some of the harmonic properties of this interface wave, and extensive tabulated data have been prepared by Strick & Ginzbarg (1956). Referring to these latter tables we find that $u_{St} = c/V_{St} = 1\cdot830$ for pitch and $u_{St} = c/V_{St} = 1\cdot092$ for plaster of paris.

It is also important for our theoretical development to know that no other poles in $A(U)$ other than the above Stoneley pole and its negative exist on the top Riemann sheet as defined above. Cagniard (1939) has shown that only a single Stoneley pole and its negative exist for a solid/solid interface, and Spencer (1956) has carried out a similar analysis with similar results for our fluid/solid interface. However, these investigations were carried out under the assumption of the branch cuts directly joining the respective positive with negative branch points whereas our cuts proceed from the branch points to infinity. Consequently, our choice of the top Riemann sheet differs from that selected by Cagniard and Spencer, and as a result it is necessary to carry out an analogous proof for our investigation. This is shown in appendix 5 with numerical curves for pitch and plaster of paris. The generalization to an arbitrary solid follows just as in the afore-mentioned papers.

Following Cagniard we next consider (4·15) as a transformation of variable from the $u$ plane to the $\tau$ plane after which the positive real value of the complex variable $\tau$ can be interpreted as the time variable $t$.

The branch points at $K_1$ and $K_2$ go over by (4·15) directly into $t_P$ and $t_S$, respectively, where

$$t_P = K_1\frac{x}{c}+\frac{2H-z}{c}\surd(1-K_1^2), \quad t_S = K_2\frac{x}{c}+\frac{2H-z}{c}\surd(1-K_2^2). \qquad (4\cdot16a, b)$$

We can easily invert (4·15) to obtain $u$ as a function of $\tau$, i.e.

$$u = u(\tau) = \sin\theta = \frac{c\tau}{r_1}\frac{x}{r_1}-\frac{2H-z}{r_1}\delta, \qquad (4\cdot17)$$

where
$$\delta = \surd\{1-(c\tau/r_1)^2\}, \qquad (4\cdot18a)$$

and where
$$r_1 = \surd\{x^2+(2H-z)^2\} \qquad (4\cdot18b)$$

is the image-detector distance. Also

$$\alpha_\tau = \surd(K_1^2-u^2)|_{u=u(\tau)} \quad \text{and} \quad \beta_\tau = \surd(K_2^2-u^2)|_{u=u(\tau)}, \qquad (4\cdot18c, d)$$

where $u = u(\tau)$ is an abbreviation for (4·17).

The branch point at $u = 1$ associated with the radical $\gamma$ disappears in the transformation (4·15), being replaced by $\tau = r_1/c$ in the radical $\delta$. For $\gamma$ it follows that

$$\gamma = \surd(1-u^2) = \cos\theta = \frac{c\tau}{r_1}\frac{2H-z}{r_1}+\frac{x}{r_1}\delta, \qquad (4\cdot19)$$

and
$$\frac{du}{\gamma} = \frac{c}{r_1}\frac{d\tau}{\delta}. \qquad (4\cdot20)$$

A sign ambiguity in $\delta$ appearing in equation (4·18 $a$) due to the algebraic inversion of (4·15) has been resolved by the convergence requirements $\mathscr{R}\,\gamma \geqslant 0$ applied to (4·19). Thus, $\delta$ has phase zero for $\tau < r_1/c$.

Finally, in the $\tau$ plane the integral (4·14) takes on the form

$$p_1^{r}(x, z; \text{step}) = \frac{P_0 c}{2\mathrm{i} r_1} \int_{C_t} \frac{A(u)|_{u=u(\tau)}}{\delta} \mathrm{e}^{-s\tau} \,\mathrm{d}\tau, \tag{4·21}$$

where $u(\tau)$ is given by (4·17).

Numerical values of the branch points and poles as they appear in the right half of the $\tau$ plane are listed below in table 4 for cases 1 and 2.

TABLE 4. NUMERICAL VALUES OF BRANCH POINTS AND POLES IN THE $\tau$ PLANE IN MICROSECONDS

| | case 1, pitch | case 2, plaster of paris |
|---|---|---|
| $t_P$ | 44·64 | 35·71 |
| $t_S$ | 99·73 − i7·42 | 58·26 |
| $r_{1/c}$ | 66·96 | 67·72 |
| $\tau_{St}$ | 121·93 − i10·21 | 73·58 − i2·96 |

The contours $C$ and $C'$ obtained respectively from $C$ and $C'$ of figure 9 are shown in figure 10. Considering the transformation $C$ to $C_\tau$ first we note that $C_\tau$ is the same curve for both cases 1 and 2 if one assumes the same fluid velocity for both cases, since the transformation (4·15) is independent of the parameters of the solid. $C_\tau$ crosses the $t$ axis at $t = 3H - z/c$. The contour curves away from the imaginary axis for large $\tau$.

The transition from $C'$ to $C'_\tau$ is somewhat more complicated than that from $C$ to $C_\tau$. For $u < x/r_1$ the contour $C'$ which hugs the cuts in the $u$ plane transforms into a portion of $C'_\tau$ which similarly hugs the cuts in the $\tau$ plane up to $\tau = r_1/c$. Then, as $u$ proceeds in a semicircle about $u = x/r_1$ in a direction of increasing positive real $u$, the contours in the $\tau$ plane loop about branch point at $t = r_1/c$ crossing the cut for the radical $\delta$, thereby entering the lower sheet of this radical.† As $u$ proceeds further from $x/r_1$ to 1, the contour in the $\tau$ plane still hugging the cuts travels along the $t$ axis towards the origin from $t = r_1/c$ to $t = x/c$ still on the lower $\delta$ sheet. Then, as $u$ goes through a semicircle about $u = 1$, $\tau$ travels through an infinitesimal arc away from the real $t$ axis. Increasing $u$ above 1 takes $\tau$ farther away from the real $t$ axis on the bottom $\delta$ sheet in the manner of the dotted lines of figure 10.

A detail of contour $C'_\tau$ for plaster of paris of case 2 with a choice of $x = 2(2H - z)$ instead of $x = 10(2H - z)$ is shown in figure 11. The point $u = x/r_1$ now occurs at 0·895 instead of 0·995 and is now sufficiently removed from $u = 1$ to allow a corresponding separation of $r_1/c$ from $x/c$ in the $\tau$ plane. For $x$ equal to 10 cm we find, using data from table 3, that the path above the positive real axis in $(K_2, \infty)$ (shown in the upper part of figure 11) is mapped into the path shown in the bottom of figure 11. The corresponding part of the contour $C'$ in the $u$ plane that travels below the cuts transforms into the mirror image about the $t$ axis of that shown above for the above cut transformation. The locations of the points on the contours in the complex $\tau$ plane for $C'$ above the cut are complex conjugate to those below the cut.

† The complete specification of a Riemann sheet requires that we state the phase of all three radicals. However, in order to simplify notation here and elsewhere in this paper we shall adopt the understanding that unless there is a statement to the contrary, the phase of the unmentioned radicals will refer to their upper sheets.

Referring back to our discussion of the transformation of the unprimed contour $C$ to $C_\tau$ which showed a tendency to enclose the $t$ axis of the $\tau$ plane, we observe that the transformation of the primed contour $C'$ into the $\tau$ plane not only retains enclosure of the area on the top $\alpha$ and $\beta$ sheets exterior to the positive real $t$ axis, but also gathers in a portion of the

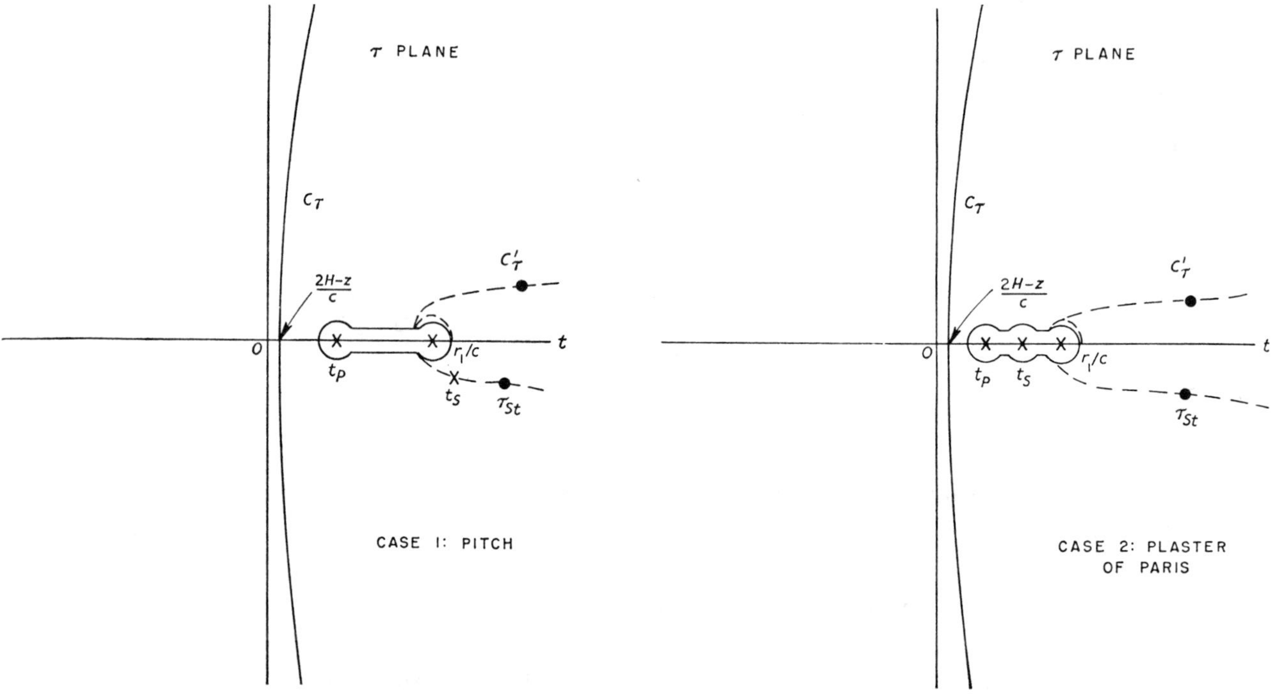


FIGURE 10. Contours and branch cuts in the $\tau$ plane (not to scale).

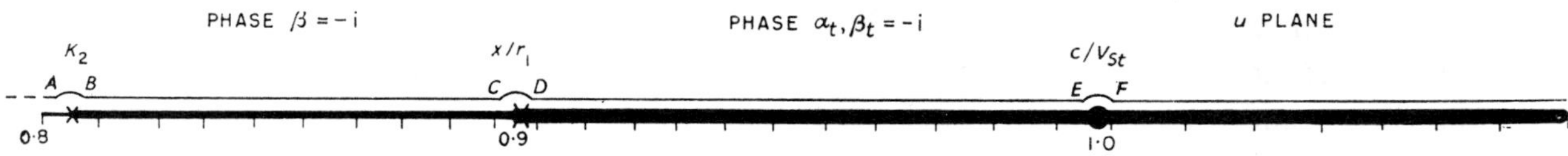


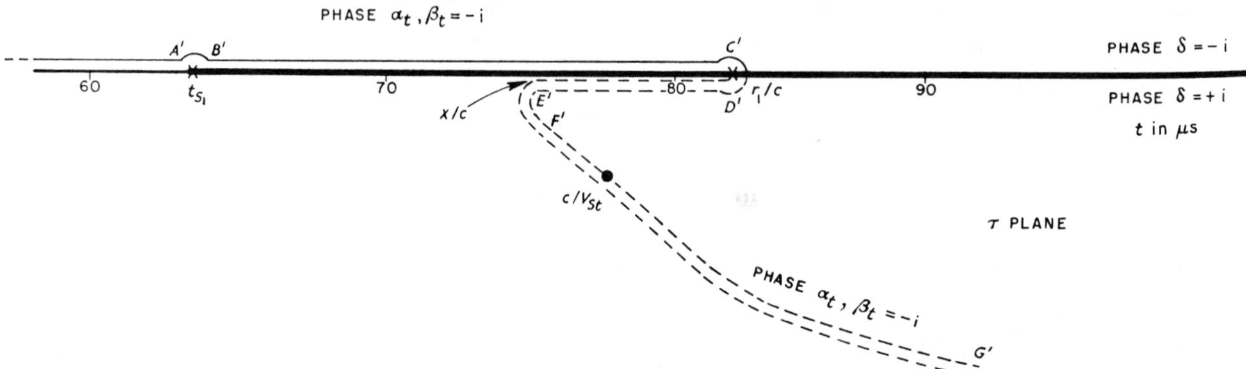


FIGURE 11. Detail of figure 10*b* with $x = 2(2H-z) = 10$ cm (not to scale).

area on the lower $\delta$ sheet on both sides of the axis. In order to see this double overlapping effect a little better, let us consider the phase behaviour of the radical $\alpha_\tau$ as we follow a complete circle of finite radius centred at $\tau = \tau_P$ on the top sheet defined by that radical having a phase in the range $-\mathrm{i}$ to $+\mathrm{i}$. Starting at the point in the $\tau$ plane designated $G'$, where the phase of $\alpha_\tau$ is $-\mathrm{i}$ and proceeding in a counterclockwise fashion to the right of the branch point of $\delta$ at $t = r_1/c$ we note that the area between the curve $D'E'F'G'$ and the cut for the radical $\delta$ is on the lower $\delta$ sheet. As our loop crosses the $\delta$ cut, where $\alpha_\tau$ has a phase in its fourth quadrant, it comes to the top $\delta$ sheet. We now follow a complete rotation of $2\pi$ about the $\alpha_\tau$ branch point in the $\tau$ plane during which, because of (4·15), the phase change of $\alpha_\tau$ is less than $\pi$. During this rotation the loop does not pass through the point $G'$ (which is on the lower $\gamma$ sheet) from which it started, but instead passes just above it on the top $\gamma$ sheet. In a similar vein it does not cross the cuts for $\alpha_\tau$ and $\beta_\tau$ since they are also here on the lower $\gamma$ sheet. Continuing on across the positive real $t$ axis the loop goes back down into the lower $\gamma$ sheet where the phase of $\alpha_\tau$ is now in its first quadrant. Finally, when the point in the $\tau$ plane which is complex conjugate to $G'$ is reached, the phase loop of $\alpha_\tau$ from $-\mathrm{i}$ to $+\mathrm{i}$ is completed.

In appendix 5 it is shown the existence of only one pair of poles, i.e. $\pm c/v_{St}$, on the top Riemann sheet of the $u$ plane, and since $\tau$ is a single-valued function of $u$ on these overlap regions on the lower sheet, there can be no other poles on these overlap regions other than the Stoneley pole. Therefore, the half of the $C'_\tau$ contour as illustrated in figure 11 *b* can be distorted to follow just above the positive real $t$ axis now being wholly on the top $\alpha_\tau$, $\beta_\tau$ and $\delta$ sheets. Along this path to the right of $t = r_1/c$ the phases $\alpha_\tau$ and $\beta_\tau$ are in their fourth quadrants, whereas $\delta$ is of course always $-\mathrm{i}$. Similarly, the other half of the contour $C'_\tau$, which is complex conjugate to the half which we have just distorted, can also be distorted downward where it will come to the top sheets on the lower side of the $\delta$ cut on the $t$ axis. The phases of $\alpha_\tau$ and $\beta_\tau$ to the right of $r_1/c$ are in their first quadrants, and the phase of $\delta$ is always $+\mathrm{i}$. Note that in making such a distortion of the contour $C'_\tau$, the Stoneley pole has been left on the lower $\delta$ sheet of the $\tau$ plane. Also, in case 1 we see in figure 10 *a* that the branch point $t_S$ as well as its cut falls completely on the lower $\delta$ sheet, and is consequently left behind when the contour is distorted to the real $t$ axis.

Along our new contour which hugs the positive real $t$ axis from $t_P$ to infinity in a clockwise direction the radicals below the axis are complex conjugate to those above the axis. It follows that $A(u)$ is similarly related about the positive real axis in the $\tau$ plane. We now find it advantageous to extend the contour from $t_P$ to the origin of the $\tau$ plane so as to hug completely the whole positive real $t$ axis. Now, making use of the fact that the integrand of (4·21) beneath the positive real $t$ axis is complex conjugate to that above, the integral (4·21) can be re-expressed in the form where the contour is above the $t$ axis:

$$\bar{p}_1^f(x, z;\, \text{step}) = \frac{P_0 c}{r_1} \mathscr{I} \int_0^\infty \frac{A(u)|_{u=u(\tau)}}{\delta} e^{-s\tau}\, \mathrm{d}\tau. \tag{4·22}$$

In this integral, since $\tau$ is always positive real, we may replace $\tau$ by the time $t$. Similarly, $t$ will be used in place of $\tau$ in all expressions to follow.

In order to get the integral into the form of an $s$-multiplied Laplace transform integral we multiply (4·22) through by the factor $s$. Since the elastic wave system is initially quiescent, we can interpret $s\bar{p}_1^f(\text{step})$ as the transform $\bar{p}_1^f(\text{delta})$ of the response to a delta-function excitation of the source. Then, making use of Lerch's uniqueness theorem, we can write

$$p_1^f(x, z; \text{delta}) = \frac{P_0 c}{r_1} \mathscr{I}\left[\frac{A(u)|_{u=u(t)}}{\delta}\right] \mathbf{H}(t - t_P), \tag{4·23}$$

where use has been made of the fact that the bracketed term is real for $t \leqslant t_P$. On this positive real $t$ axis, the phases of the radicals $\alpha_t$, $\beta_t$ and $\delta$ where

$$\alpha_t = \surd(K_1^2 - u^2)|_{u=u(t)}, \quad \beta_t = \surd(K_2^2 - u_2)|_{u=u(t)},$$

and

$$\delta = \surd\{1 - (ct/r_1)^2\},$$

are shown in figure 12.

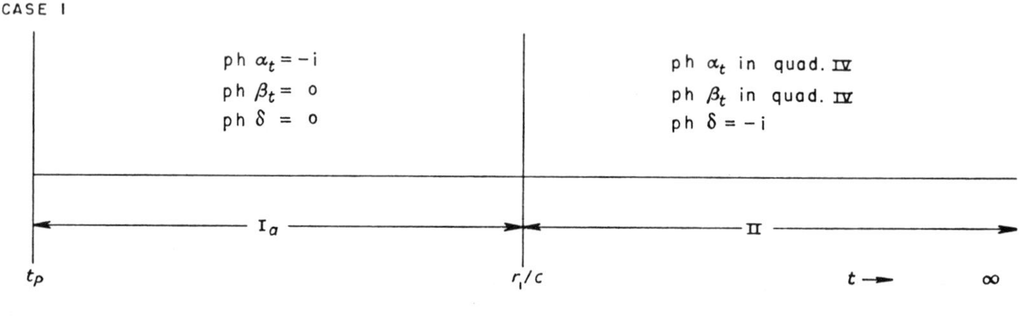


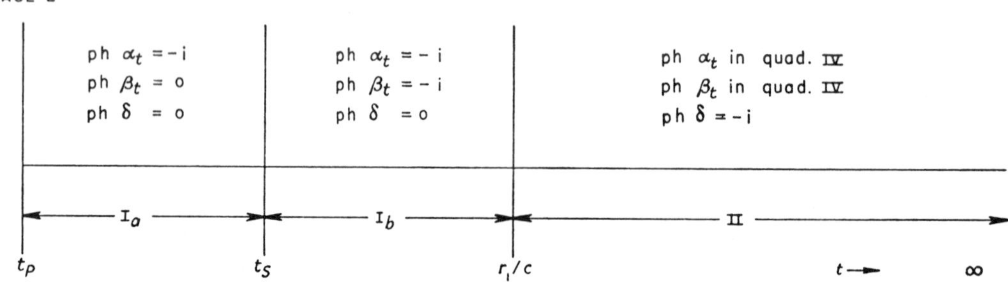


FIGURE 12. Phases of the radicals $\alpha_t$, $\beta_t$ and $\delta$ for ranges of the time $t$.

To response (4·23) we must according to (4·7) add the direct wave $p_0^f$ as given by (3·6) in order to obtain the total response $p^f$. In this way we find

$$p^f(x, z; \text{delta}) = \frac{P_0 c}{r} \frac{\mathbf{H}(t - r/c)}{\surd\{(ct/r)^2 - 1\}} + \frac{P_0 c}{r_1} \mathscr{I}\left[\frac{A(u)|_{u=u(t)}}{\delta}\right] \mathbf{H}(t - t_P), \tag{4·24}$$

where $r^2 = x^2 + z^2$ and $r_1^2 = x^2 + (2H - z)^2$. Also, $A(u)$ is defined by (4·10), $u(t)$ by (4·17) and $t_P$, $t_S$, by (4·16).

It is interesting to note that $A(u)$ has the algebraic form of the complex reflexion coefficient for reflexion from the fluid/solid interface if the intermediate variable $u$ is interpreted by (3·13) and (4·9) as the sine of the angle of incidence $(\theta)$ between the direction of the propagation vector of the elementary plane waves and the normal to the interface.

In figure 12 there has been a decomposition of the response into regions I and II, depending upon whether or not the response time is less than or greater than $r_1/c$. The time $t_P$ as given by (4·16$a$) is, of course, the classic minimum time expression for the arrival of the critically refracted $P$ wave as utilized in part I. Likewise, $t_S$ corresponds to the arrival time for a critically refracted $S$ wave. Finally, $r_1/c$ being the time for the arrival of a wave-front from the image point is the classic expression for the true reflexion arrival.

Thus, the response before the arrival of the direct and reflected wave-fronts is proportional to the imaginary part of the reflexion coefficient divided by a function which is related to the image source function. The angle of incidence $\theta$ is here positive real and a monotonically increasing function of the time $t$. Therefore, in this region, there exists a one-to-one correspondence between the angle of incidence $\theta$ and observed events on a wave-form.

After the time $t > r_1/c$ the response becomes proportional to the real part of the reflexion coefficient divided by the image source function. In this region where we expect to find a reflected cylindrical wave as well as the Stoneley interface wave we note that the angle of incidence becomes complex and is no longer easily interpretable.

Making use of the phase description in figure 12 we can further express the response $p_1^f$ in the following real form:

$$\textit{Region } I_a \begin{cases} \text{case 1:} \quad t_P \leqslant t \leqslant \dfrac{r_1}{c} \quad \text{or} \quad K_1 \leqslant u \leqslant \dfrac{x}{r_1}, \\ \text{case 2:} \quad t_P \leqslant t < t_S \quad \text{or} \quad K_1 \leqslant u \leqslant K_2, \end{cases}$$

so that

$$p_1^f(x, z; \text{delta}) = \frac{2(\rho^f/\rho^s)\, K_2^4 P_0 c}{r_1 \delta} \left[ \frac{\gamma (K_2^2 - 2u^2)^2 \surd(u^2 - K_1^2)}{\gamma^2 (K_2^2 - 2u^2)^4 - \alpha_i^2 \{4u^2 \beta_i \gamma + (\rho^f/\rho^s)\, K_2^4\}^2} \right]_{u = \frac{x}{r_1}\frac{ct}{r_1} - \frac{2H-z}{r_1}\delta} . \tag{4·25}$$

$$\textit{Region } I_b \begin{cases} \text{case 1:} \quad p_1^f \equiv 0 \text{ in region } I_b, \\ \text{case 2:} \quad t_S < t < \dfrac{r_1}{c} \quad \text{or} \quad K_2 < u < \dfrac{x}{r_1}, \end{cases}$$

so that
$$p_1^f(x, z; \text{delta}) = \frac{2P_0 (\rho^f/\rho^s)\, K_2^4 c}{\delta r_1} \left[ \frac{\gamma \surd(u^2 - K_1^2) f(u)}{\gamma^2 \{f(u)\}^2 - \alpha_i^2 (\rho^f/\rho^s)^2 K_2^8} \right]_{u = \frac{x}{r_1}\frac{ct}{r_1} - \frac{2H-z}{r_1}\delta}, \tag{4·26}$$

where
$$f(u) = (K_2^2 - 2u^2)^2 - 4u^2 \surd(u^2 - K_1^2) \surd(u^2 - K_2^2). \tag{4·27}$$

*Region II*, both cases: $t \geqslant \dfrac{r_1}{c}$ or $u \geqslant \dfrac{x}{r_1}$,

$$p_1^f(x, z; \text{delta}) = \frac{P_0 c}{r_1 \delta} \left[ \frac{|\gamma|^2 |f(u)|^2 - K_2^8 (\rho^f/\rho^s)^2 |\alpha|^2}{|D|^2} \right]_{u = \frac{x}{r_1}\frac{ct}{r_1} + i \frac{2H-z}{r_1} \sqrt{\left\{\left(\frac{ct}{r_1}\right)^2 - 1\right\}}}, \tag{4·28}$$

where $f(u)$ is defined by (4·11) and

$$D(u) = \gamma f(u) + (\rho^f/\rho^s)\, K_2^4 \alpha, \tag{4·29}$$

so that
$$|D|^2 = DD^* = |\gamma|^2 |f(u)|^2 + K_2^8 (\rho^f/\rho^s)^2 |\alpha|^2 + 2K_2^4 (\rho^f/\rho^s)\, \mathscr{R}\, [\gamma f(u)\, \alpha^*]. \tag{4·30}$$

By means of the parameters of table 3, the theoretical pressure response curves for pitch (case 1) and plaster of paris (case 2) have been calculated and are illustrated in figures 13 and 14, respectively. In each case the corresponding experimental curve from part I is reproduced for comparison. It should be noted here that what is actually shown is a comparison of the theoretical response from a delta-function excitation of a line source with that due experimentally to a 6 to 12 $\mu$s exponentially decaying excitation of a point source. However, the discussion of the point source representation in § 3 (*a*) (ii) has led us to expect similar response wave-forms as is borne out by figures 13 and 14.

The computation time of a complete pressure response curve of the type shown in figures 13 and 14 is about 2 h for 100 time points using a medium speed computer (Elecom 120). It was originally intended to carry out a similar theoretical calculation using a point source. However, the numerical work would have been much greater, being roughly 100 h

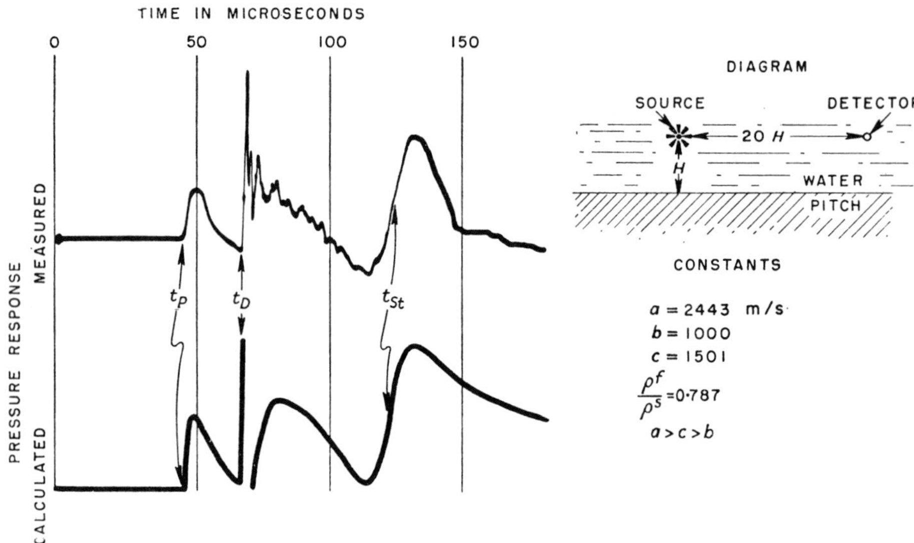


FIGURE 13. Pressure response at a water/pitch interface.

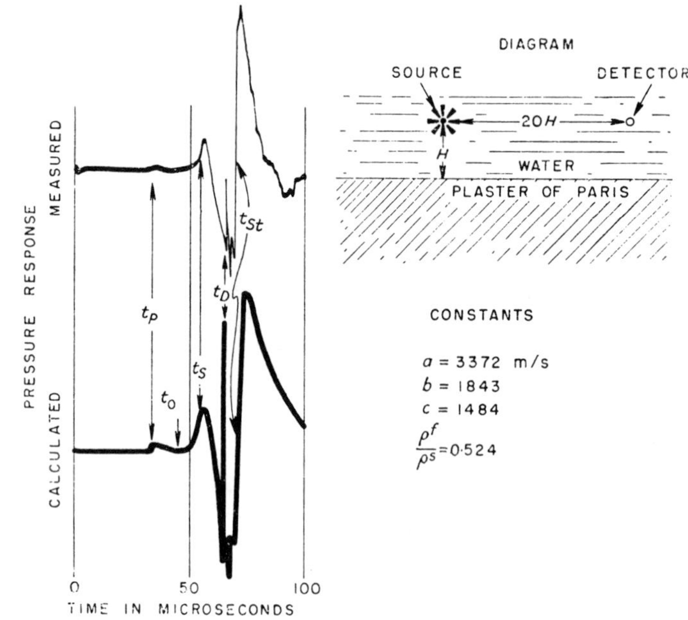


FIGURE 14. Pressure response at a water/plaster of paris interface.

for the same curve; and considering the close agreement obtained using our approximate viewpoint, it was decided not to make the more exacting point source calculations at this time.

One major disadvantage attendant with the use of the mathematically simple line source (3·6) is the inability, due to the singularity, of assigning an amplitude to the excitation. This difficulty could have been avoided by the use of equation (A 4·10) in appendix 4. However, this requires an extra time integration which, although elementary, would not materially aid in the interpretation of the results of this paper. For our discussion we have found it sufficient to equate the theoretical and experimental amplitudes at an arbitrary point, which is usually the peak-to-peak amplitude of the Stoneley wave.

## 5. Discussion of results

Now let us analyze in detail the wave-forms of figures 13 and 14. The very first arrival at time $t_P$ is of course the critically refracted $P$ wave, being much more pronounced for pitch in figure 13 than for plaster of paris in figure 14. The amplitude variation of the critically refracted $P$ wave in figure 13 is described mathematically by (4·25). For $t \sim t_P$, the pressure builds up as $\surd(u-K_1)$ with $u(t)$ given by (4·17). The positive maximum can be obtained from the vanishing of the time derivative of (4·25), but such a lengthy expression is not very useful. The wave-form of this critically refracted $P$ wave with its single positive maximum is in good accord with other theoretical investigations beginning with Jeffreys's (1949) early work which first indicated its possible existence. The critically refracted $P$ wave is terminated at $t = t_D = r/c$ at 66·62 $\mu$s by the direct arrival with its positive singularity as given by (3·6). Very soon thereafter at $t = r_1/c = 66{\cdot}96$ $\mu$s region II starts with the simple reflexion of the spherical wave-front accompanied by a similar but negative singularity due to the vanishing of $\delta$. In order to simplify comparison with experiment, this portion of the wave-form containing the two singularities has been omitted. The curve starts again with the decay of the simple reflexion. It appears as though the whole of region II of the complete wave-form has been shifted upwards which we might expect from the tail effect of the two-dimensional delta-function-excited line source. Continuing on, we note that at $t = t_{St} = 122{\cdot}2$ $\mu$s corresponding to the arrival of the point of symmetry between the negative and positive extremums in the pressure response, occurs what we identify as the Stoneley arrival for the water/pitch interface. As mentioned in the discussion following table 3, the Stoneley equation for such an interface yields for a corresponding harmonic excitation the Stoneley velocity $V_{St} = c/1{\cdot}830$. Assuming in the usual fashion the instantaneous beginning of the generation of the interface wave at the epicentre, the arrival time $t_{St}$ is simply $x/V_{St}$ which yields 121·9 $\mu$s. This agreement, although close, is not exact. Referring to (4·28) we note that the vanishing of the numerator yields an equation which, although similar, is not the Stoneley equation for harmonic wave propagation, the latter being given by the vanishing of the denominator of (4·10). For our pulsed wave propagation the intermediate variable $u$ is in general a complex function of $x$, $2H-z$, $c$ and $t$ instead of being simply $c/V_{St}$, and one has to calculate $t_{St}$ for varying $x$ subject to the vanishing of the numerator in order to determine the incremental Stonely velocity. However, if we assume that $x \gg 2H-z$ so that we can neglect the imaginary part of $u(t)$, then $u = ct/x = c/(x/t)$ is real and we obtain the same relation as for harmonic excitation. It should be noted that the arrival time $t_{St}$ of the Stoneley wave for pulsed excitation is not the same as the absolute value nor the real part of the complex value of $\tau_{St}$ given in table 4. The latter is the location in the $\tau$ plane of the Stoneley pole in the $u$ plane as transformed according to (4·15). In our method of solving the problem, the solution in region II appears as a superposition of the reflected and interface waves, the latter not appearing as an isolated pole contribution. The effect of the pole which is on the lower $\delta$ sheet in the complex $\tau$ plane is to distort the integrand of (4·22) into the wave-form of the interface wave. For $x \gg 2H-z$ this complex pole approaches the real $t$ axis.

The most apparent discrepancy between the theoretical and experimental wave-forms of figure 13 lies in the rapid oscillations that occur in the experimental wave-form between $t_D$ and $t_{St}$. These rapid oscillations do not appear in the theoretical calculations which were taken to an accuracy of seven significant figures and seem to lie in some experimental aspect as discussed in part I.

Replacing the pitch by plaster of paris whose shear wave velocity $b$ is greater than the water velocity $c$ results in (see figure 14) a diminishing of the amplitude of the critically refracted $P$ wave and an increase in the amplitude of the interface wave. Furthermore, there is a definite compression of the wave-form in time of both of these arrivals. The critically refracted $P$ wave for plaster of paris starts at 35·71 $\mu$s and lasts apparently until a zero in the response curve is reached at the point designated $t_0$ at 44·02 $\mu$s. The width of the critically refracted $P$ wave here is 8·31 $\mu$s as compared with 21·98 $\mu$s for pitch. If we define the width of the Stoneley interface wave as the time difference in the arrivals of the negative and positive peaks, then the width of the interface wave for plaster of paris in figure 14 is 6·23 $\mu$s as compared with 16·8 for pitch. Owing to a slightly different water velocity in table 3 the direct arrival is observed as in table 4 at 67·72 $\mu$s. The zero of the Stoneley interface wave is calculated to be at 73·75 $\mu$s. Just as in the case for pitch, use of the Stoneley wave charts (Strick & Ginzbarg 1956) gives $c/V_{St} = 1{\cdot}092$ for plaster of paris which yields an arrival time for harmonic wave propagation of 73·59, which is in rather good agreement. The reason for this close agreement here is that, as given in table 4, the Stoneley pole in the time plane has a rather small imaginary component.

The remaining feature of the pressure wave-form of figure 14 is the rather large amplitude variation between the critically refracted $P$ wave and the direct arrival. We noted earlier that for case 2 region I is divided into two parts with the separation time given by $t_S$ which is defined in (4·16$b$). This latter expression is, however, precisely that for a critically refracted $S$ wave which travels along the interface with the shear wave velocity $b$ of the plaster of paris, With the use of (4·16$b$), table 4 shows such an arrival should occur at 58·26 $\mu$s. However, such a value of $t_S$ occurs well up and near, but not on, the peak of the positive pressure swing in region $I_b$. It is difficult to see from this that a critically refracted $S$ head wave exists in this wave-form. The beginning of this large positive pressure build-up apparently occurs at the afore-mentioned time $t_0 = 44{\cdot}02$ $\mu$s, which is well within region $I_a$, which is much too early to explain by any minimum time path of the type associated with $t_S$.

Then there is the large negative (but finite) swing in the pressure response occurring after the positive peak is reached. Since this is not characteristic of the critically refracted $P$ wave, we might also not expect it to be characteristic of the critically refracted $S$ wave. Comparing the Stoneley wave of figure 14 with that of figure 13 suggests the possibility that this negative swing might be attributed to the very beginning of the Stoneley interface wave, which occurs before the direct arrival even though the arrival of the zero of the Stoneley wave occurs later.

Another complication in the interpretation occurs when we consider the behaviour of the point of zero amplitude on the wave-form which occurs between $t_S$ and $t_D$. Referring back to equation (4·26) we note that the only point of zero amplitude in region $I_b$ occurs when $f(u)$

vanishes. But the vanishing of $f(u)$ as defined by (4·27) is, except for the presence of $c$ instead of $b$ in the independent variable, precisely the known Rayleigh wave equation for surface waves travelling with a velocity $V_R$ along a semi-infinite elastic solid. If we let $u = c/V_R$ in (4·15) and take the time derivative holding $z$ = constant, then we find that

$$\partial x/\partial t = V_R, \tag{5·1}$$

i.e. this point of zero amplitude travels along the interface with exactly the true Rayleigh wave velocity of the solid as though the fluid were not present. Since one would certainly expect a true Rayleigh wave to exist if we were to go to the limit of vanishing fluid density, it is further suggestive that the build-up before the minimum time $t_S$ is related in some way to the Rayleigh wave. Since we have already shown in appendix 5 that only the pole associated with the Stoneley interface wave (and its negative) can exist on the top Riemann sheet, the pseudo-Rayleigh† pole must occur on a lower Riemann sheet of the $u$ plane.

Thus, we reach the rather attractive hypothesis that the large pressure variation in region $I_b$ is due to a linear superposition of pseudo-Rayleigh, Stoneley, and critically refracted shear waves. This hypothesis will be verified in part III of this paper.

Another point of interest is associated with the horizontal velocity of the zero amplitude point $t_0$ of region $I_a$ along the interface. Referring to equation (4·25) we note that the pressure vanishes for $u = K_2^2/\surd 2$. Again applying the time derivative to (4·15) holding $z$ = constant we find that

$$\partial x/\partial t = \surd 2\, b, \tag{5·2}$$

i.e. that the point of zero amplitude occurring at time

$$t_0 = \frac{x}{\surd 2\, b} + \frac{2H-z}{c}\sqrt{\left(1-\frac{c^2}{2b^2}\right)}, \tag{5·3}$$

travels along the interface with a horizontal phase velocity equal to $\surd 2$ times the shear wave velocity of the solid being independent not only of the existence of the fluid but also of the density and compressional velocity of the solid.

## 6. Conclusions

In figures 13 and 14 are given examples of the very close wave-form agreement between theory and experiment that can be obtained when a fluid/solid interface is excited by a transient point or line source located in the neighbourhood of the interface. For a delta-function excitation of a line source the complete pressure response solution is given in an exact closed form. It is also shown that for large distance as measured along the interface (i.e. $x \gg 2H-z$) the above exact solution for the line source can be reinterpreted as an approximate first-term asymptotic expansion for the pressure response due to the excitation of a point source that lies intermediate between that of a step and a delta function.

For a pitch/water interface (figure 13) there are four distinct arrivals: (1) critically refracted $P$ wave, (2) direct, (3) reflected, and (4) Stoneley interface waves. The critically

† Scholte (1949) has designated as a pseudo-Rayleigh wave the behaviour of the true Rayleigh wave when one takes into account the existence of the atmosphere over the semi-infinite solid. In order to avoid confusion we are extending his definition to that for a fluid/solid interface. Although we find this use of the adjective pseudo to be somewhat unsatisfactory (we prefer the adjective 'radiating'), we believe the whole nomenclature of interface waves to be in need of revision and are therefore not introducing new terminology at this time.

refracted $P$ wave and the Stoneley interface wave are both large in amplitude and extended in time, being the most significant events on the response curve. The direct and reflected arrivals seem to have a much higher frequency content.

For a plaster of paris/water interface there is a compression of the wave-form in time of the critically refracted $P$ wave and the Stoneley interface wave, but they are still major events on the response curve. Since the shear wave velocity of the plaster of paris is greater than the water velocity, we are led to expect a critically refracted $S$ wave to exist between the critically refracted $P$ wave and the direct arrival. However, the wave-form variation in this region does not show any indications of the existence of such a critically refracted $S$ wave. If it exists at all, it must be rather small. There is, however, a large pressure wave-form in this region which will be investigated in part III of this paper. For the present, we suggest that it is a combination of critically refracted $S$, pseudo-Rayleigh, and Stoneley interface waves.

It has been shown (Cagniard 1939) that when a step-excited point source is located in the vicinity of an interface between two media, there will be a logarithmic singularity in the response curve just before the arrival of the reflexion. In our exact solution for the response due to a delta-function excitation of a line source the only singularity that we observe that occurs before the reflexion is of the same nature as that of the source itself; furthermore, if we apply a time integration to our solution in order to recover the response to a step-function excitation, we do not find singularities of any kind except at $t =$ infinity.

## Research Note

# Seismic Model Experiments on the Shape and Amplitude of the Refracted Wave

**R. J. Donato**

It has been known for many years from theoretical consideration that the first arrival part of a refracted seismic wave should have a waveform given by the time integral of the original wave (Jeffreys 1926). Quantitatively this is difficult to verify experimentally from field refraction records, although it is one of the causes of the "rounding" of the first part of the refracted waveform. However, by using seismic model techniques the experimental verification both of the wave shape and of the amplitude predicted by equation (1) becomes possible, and this note describes such experiments.

Liquid media were used throughout the experiments. An electric spark immersed in the medium overlying the refractor simulated the seismic explosive source, and a lead zirconate transducer horizontally separated from the source by distance $R$ (see figure) received the refracted wave. The receiver was designed to have a flat frequency response throughout the spectrum covered by the pressure pulse. The relationship between the received pressure of the refracted wave $p_r(T)$ and the incident pressure $p_0(t)$ referred to unit distance from the source is given in terms of the elastic properties of the media by

$$p_r(T) = \frac{2c_2}{\left[\left(\frac{c_2}{c_1}\right)^2 - 1\right]\frac{\rho_2}{\rho_1}(RL^3)^{\frac{1}{2}}} \int_{\tau}^{T} p_0(t-\tau)dt \qquad (1)$$

where

$$\tau = \frac{1}{c_2}\left[R + 2D\sqrt{\left(\frac{c_2}{c_1}\right)^2 - 1}\right]$$

$D$ = the height above refractor of the source and the receiver.
$L$ = the distance travelled in the refractor.
$c_1$, $c_2$ = the velocities of the overburden and the refractor.
$\rho_1$, $\rho_2$ = the densities of the overburden and refractor.

The figure shows photographic records of (*a*) the original pulse, (*b*) the pulse after passing through an electronic integrator, (*c*) the refracted pulse. The last waveform of the series gives the result of integrating (*a*) numerically. Curves (*b*), (*c*) and (*d*) compare favourably.

All the terms in equation (1) can be measured and substituted in the equation. The calculated and measured peak amplitudes of the refracted wave for two

Reprinted from Geophysical Journal, 1960, 3, 270-271 with permission from Royal Astronomical Society, Blackwell Scientific Publications, Ltd.

liquid/liquid combinations agree to better than 5 per cent which is within the experimental accuracy.

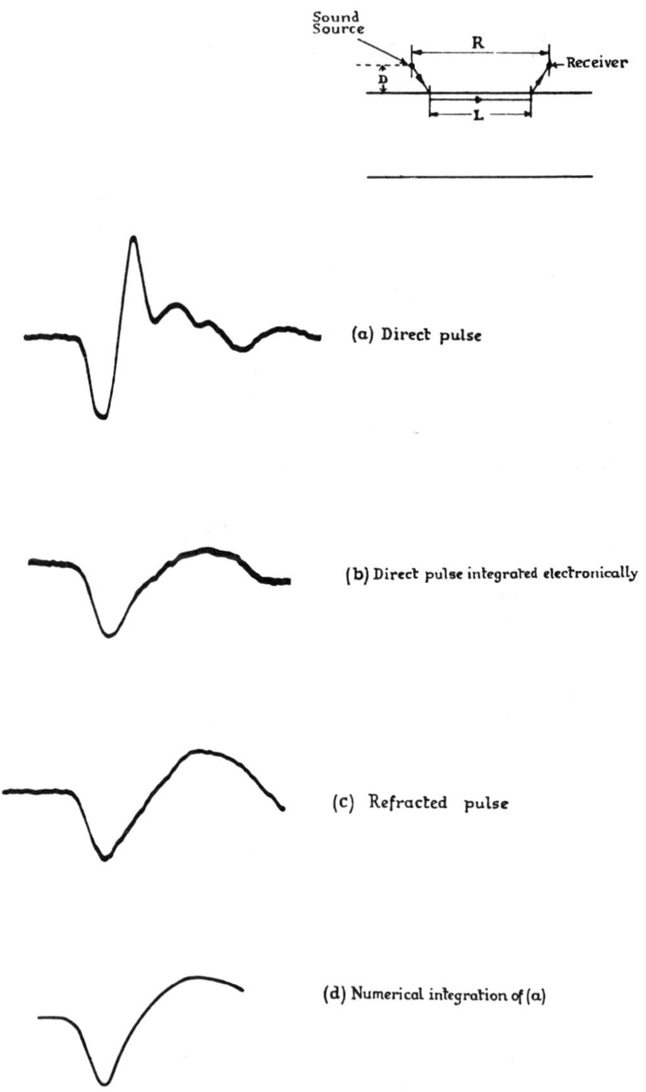


## Acknowledgments

The author is indebted to the Chairman and Directors of The British Petroleum Co. Ltd for permission to publish this note.

*British Petroleum Research Centre,*
*Sunbury-on-Thames:*
1960 *January* 5

## Reference

Jeffreys, H., 1926. *Proc. Camb. Phil. Soc.*, **22**, 472.

# USE OF RECIPROCITY THEOREM FOR COMPUTATION OF LOW-FREQUENCY RADIATION PATTERNS*

J. E. WHITE†

ABSTRACT

Starting with a simple word statement of the reciprocity which exists between forces and displacements in a general elastic solid, it is shown that low-frequency radiation of shear and compressional waves from relatively complex sources can be obtained by solving relatively simple problems in static elasticity. Illustrative examples include radiation from radial and tangential pairs of forces acting on the wall of a cylinder, pressure in a finite cylinder, and a pair of radial forces in a hole in a plate. For the last case, measurements in a plexiglas plate compare favorably with computations.

## INTRODUCTION

Various forms of symmetry or reciprocity between sources and resulting disturbances have long been recognized in electromagnetic fields, elasticity, and acoustics, and the intimate relationship between source strength and receiver sensitivity for a reversible transducer has been described. Hence, it is to be expected that any attempt to compute the disturbance due to a particular source configuration should automatically raise the question, "Would it be easier to compute the behavior as a receiver and apply a reciprocity relation?"

In connection with a field program aimed at systematically measuring the speeds of shear and compressional waves in near-surface earth layers, the writer became interested in the radiation of elastic waves due to localized forces on the interior of a cylindrical borehole in an elastic solid. Since direct solution of the radiation problem seemed unattainable, attention was turned to the possibility of solving the problem in reverse, which would require then a reciprocity relation for elastic solids, explicit and quantitative.

Primakoff and Foldy (1947) proved that for two transducers of the same type coupled to an inhomogeneous, bounded, acoustical (fluid) medium, a statement like the following holds: If a current into one transducer causes an open-circuit voltage at the terminals of a second transducer, then the same current into the second transducer will cause an identical open-circuit voltage at the first transducer. To support the strong implication that this would apply also when the medium included elastic solids as well as gases or liquids, the rather extreme "medium" shown in Figure 1 was assembled. The first geophone is mounted on a pipe which rests on the bottom of a glass desiccator. The second geophone is attached to the glass with a chunk of modelling clay, below the water level. The top pair of traces shows the current into the first geophone and the voltage at the second; the bottom traces show the current in the second geophone and the

* Manuscript received by the Editor January 7, 1960.

† Denver Research Center, The Ohio Oil Company, Littleton, Colorado.

Reprinted from Geophysics, 25, 613-624

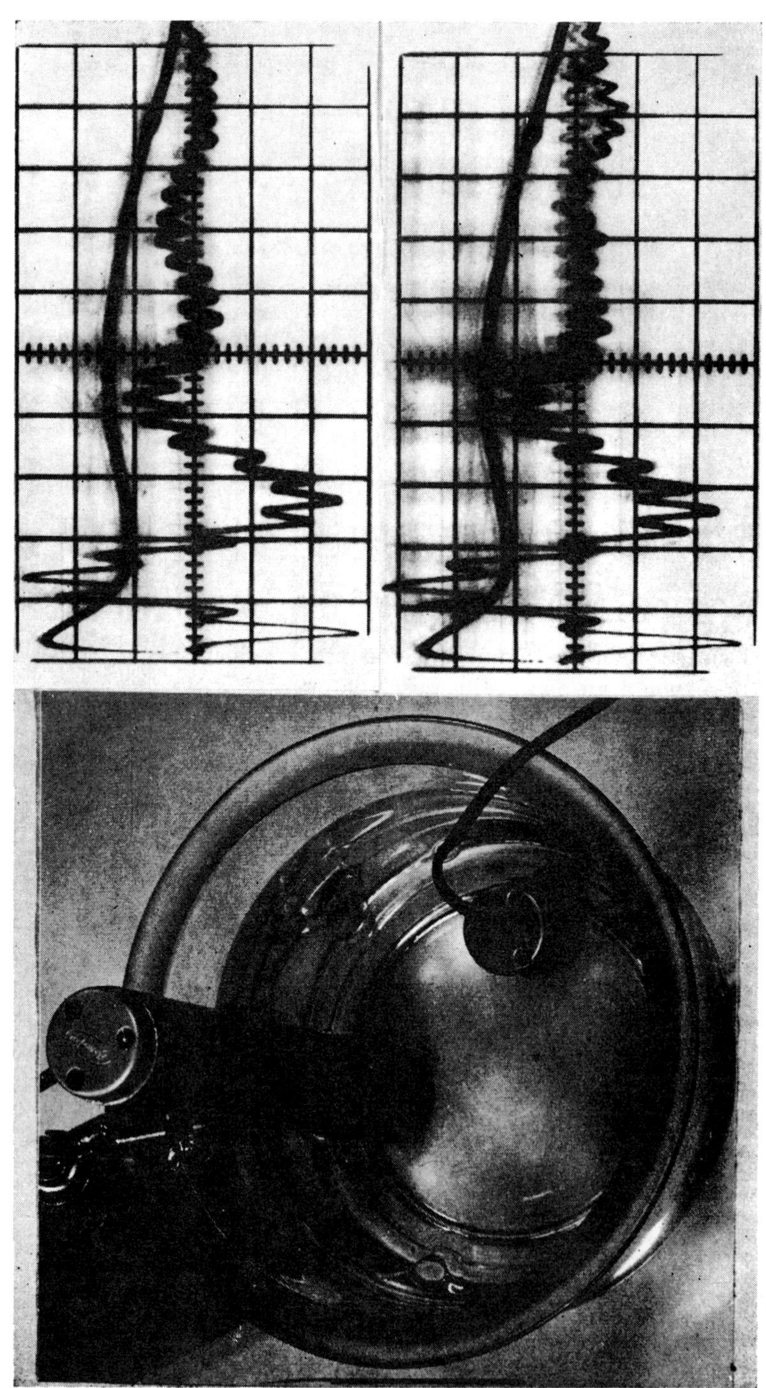

FIG. 1. Demonstration of reciprocity for "inhomogeneous elastic medium."

voltage at the first. This is convincing evidence that the reciprocity relation makes no stringent demands on the nature of the elastic medium. If the two transducers referred to above are small, light geophones imbedded in an elastic solid, an interesting conclusion can be drawn. When a current is fed into such a geophone, it will exert a localized force in the solid. The second geophone will move with the solid surrounding it, and an open circuit voltage will appear at its terminals which is proportional to the local particle velocity. If the roles are reversed, the force will be exerted at the second location and the velocity be sensed at the first. Hence, the above statement concerning transducers is entirely equivalent to a statement of reciprocity between force and particle velocity.

Morse and Feshbach (1953) discuss reciprocity conditions for elastic waves in an isotropic medium, but their notation is "compact," to say the least, and the application of their expressions to a particular case would require some further manipulation. Knopoff and Gangi (1959) show that a reciprocal relation between a point force and particle displacement holds for a bounded, inhomogeneous, anisotropic, elastic solid.

## STATEMENT OF RECIPROCITY

Although none of the above writers makes the statement in so many words, their proofs certainly justify the following:

If, in a bounded, inhomogeneous, anisotropic, elastic medium, a transient force $f(t)$ applied in some particular direction $\alpha$ at some point $P$ creates at a second point $Q$ a transient displacement whose component in some direction $\beta$ is $u(t)$, then the application of the same force $f(t)$ at point $Q$ in the direction $\beta$ will cause a displacement at point $P$ whose component in the direction $\alpha$ is $u(t)$.

Note that the statement applies to the whole disturbance, whether it be body waves, surface waves, or whatever. It is not restricted to "far-field" geometry. The usefulness of this reciprocity condition will be demonstrated in the sections which follow.

## METHOD OF APPLICATION

In each example treated here, the elastic solid is taken to be homogeneous and isotropic. The forces making up the source are in a small region: that is, the transit time across this region is short compared with the time required for an appreciable change in the transient disturbance; or, in steady-state terms, the dimensions of this region are small compared with the shortest wavelength of interest. Only the "far-field" radiation will be considered, that is, the particle displacement of the waves radiated into the interior of the solid.

With the source at the origin of spherical coordinates, the displacement at a distant point can be specified by the three perpendicular components $u_r$, $u_\theta$, and $u_\phi$. More explicitly, the collection of point forces making up the source, all of time dependence $f(t)$, will cause a radial displacement $u_r(t)$ at the observation point. The reciprocity relation states that if a radially-oriented point force $f(t)$

now acts at the observation point, the summation of displacements at the source points in the directions of the original forces will exactly equal $u_r$. The known solution for a point force, Love (1927), shows that the displacement at a point in line with the force is itself entirely in this direction also, the tangential components being zero, and it is recognized as a compressional wave.

$$u_c = \frac{1}{4\pi r\rho V_c^2} f(t - r/V_c). \tag{1}$$

This compressional wave can be described in terms of normal stresses; the displacements at the source due to these stresses computed; and the sum of the appropriate displacement set equal to $u_r$. In the far-field, plane wave relations apply approximately, and the normal stress in the direction of travel is,

$$N = \frac{1}{4\pi r V_c} f'(t - r/V_c). \tag{2}$$

A similar line of reasoning applies to either tangential displacement, $u_\theta$ or $u_\phi$. The displacement at a point at a distance $r$ perpendicular to the direction of the force is entirely tangential, consisting of a shear wave with displacement in the direction of the force,

$$u_s = \frac{1}{4\pi r\rho V_s^2} f(t - r/V_s). \tag{3}$$

This shear wave is accompanied by tangential stresses, from which displacements at the source can be computed, the sum of which equals $u_\theta$ (or $u_\phi$). The tangential stress is given by

$$T = \frac{1}{4\pi r V_s} f'(t - r/V_s).$$

### LOCAL FORCES ON CYLINDER WALL

The radiation patterns of many specific configurations can be obtained in very simple fashion by this procedure. For example, consider the angular dependence in a plane perpendicular to the cylinder of waves radiated by two oppositely-directed forces acting normal to the cylinder wall. The geometry is shown in Figure 2. By symmetry, there will be no shear wave with motion perpendicular to this plane. Hence, compressional waves and shear waves with motion in the plane will be considered.

To develop the radial motion of the cylinder wall due to a plane compressional wave, use is made of the relations, White (1953), that for a normal stress $R$ along the cylinder axis, $l_r/a = -\sigma R/Y$, whereas for a normal stress $S$ perpendicular to cylinder axis, $l_r/a = (1+2\cos 2\theta)S/Y$. In either case, a function representing translation or rotation of the whole body could be added to the expressions for radial displacement. A plane wave of normal stress $N$ in the direction of travel is

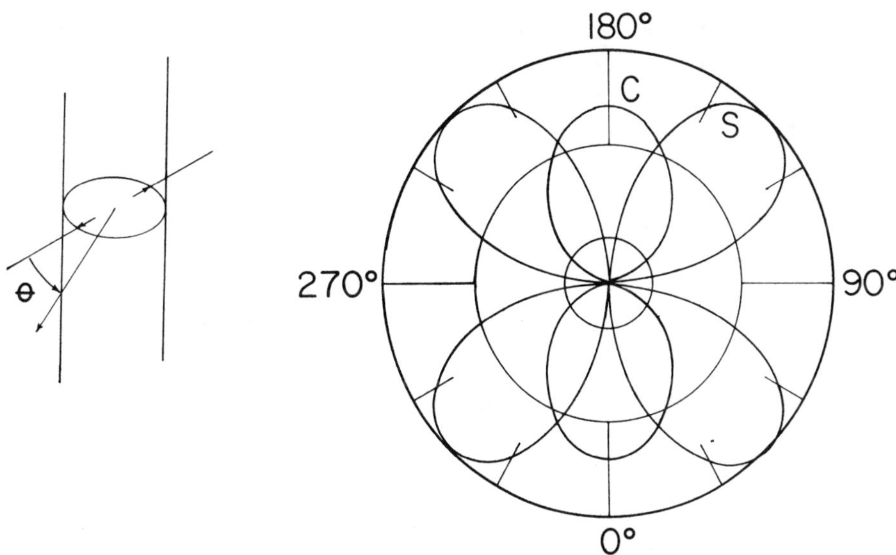


FIG. 2. Shear and compressional displacements due to radial forces in cylinder.

accompanied by a stress $N\sigma/(1-\sigma)$ in each of the two perpendicular directions. The total radial displacement due to these stresses is,

$$\frac{l_r}{a} = \frac{N}{Y}(1 + 2\cos 2\theta) + \frac{N\sigma}{Y(1-\sigma)}(1 - 2\cos 2\theta) - \frac{N\sigma^2}{Y(1-\sigma)},$$

or

$$\frac{l_r}{a} = \frac{N}{2\rho V_s^2}\left[1 + \frac{8(1 - V_s^2/V_c^2)V_s^2/V_c^2}{(3 - 4V_s^2/V_c^2)}\cos 2\theta\right]. \tag{5}$$

To this should be added the particle displacement accompanying the compressional wave. For two points located as shown in Figure 2 this bodily displacement exactly cancels, whereas the radial displacements at $\theta$ and $(\theta+180°)$ are seen to be equal. Hence, the desired sum of displacements equalling $u_r$ is twice $l_r$. Putting in the value of $N$ from equation 2,

$$u_r = \frac{a}{4\pi r\rho V_s^2 V_c}\left[1 + \frac{8(1 - V_s^2/V_c^2)V_s^2/V_c^2}{(3 - 4V_s^2/V_c^2)}\cos 2\theta\right]f'(t - r/V_c). \tag{6}$$

This is the radial displacement at a distance $r$ due to a pair of radial forces at the ends of a diameter of a cylinder.

In computing the shear wave radiated by this pair of forces, use is made of the fact that a shear stress of magnitude $T$ such as would accompany a shear wave from direction $\theta$ can be thought of as a normal stress of magnitude $-T$ at $(\theta+45°)$ plus one of magnitude $+T$ at $(\theta-45°)$. Hence the radial displacement due to the shear stress $T$ is,

$$\frac{l_r}{a} = \frac{T}{Y}[1 + 2\cos 2(\theta + 45°) - 1 - 2\cos 2(\theta - 45°)$$
$$\frac{l_r}{} = \frac{-4T}{Y}\sin 2\theta. \tag{7}$$

Without repetition of the intermediate steps, the transient displacement in the shear wave due to two opposite forces is,

$$u_\theta = \frac{a(1 - V_s^2/V_c^2)\sin 2\theta}{\pi r \rho V_s^3(3 - 4V_s^2/V_c^2)} f'(t - r/V_s). \tag{8}$$

This is the tangential displacement at a distance $r$ from a pair of radial forces on the wall of a cylindrical cavity.

The compressional and shear displacements are plotted in Figure 2, with $V_s^2/V_c^2$ set equal to $\frac{1}{3}$. Note that the compressional wave exhibits a "reverse break" phenomenon, that is, for a small angular range near 90°, the sign of the motion is opposite to the sign in the main lobes.

Consider the application of two local forces applied at the ends of a diameter and acting tangentially to the wall, as shown in Figure 3. It will be sufficient to

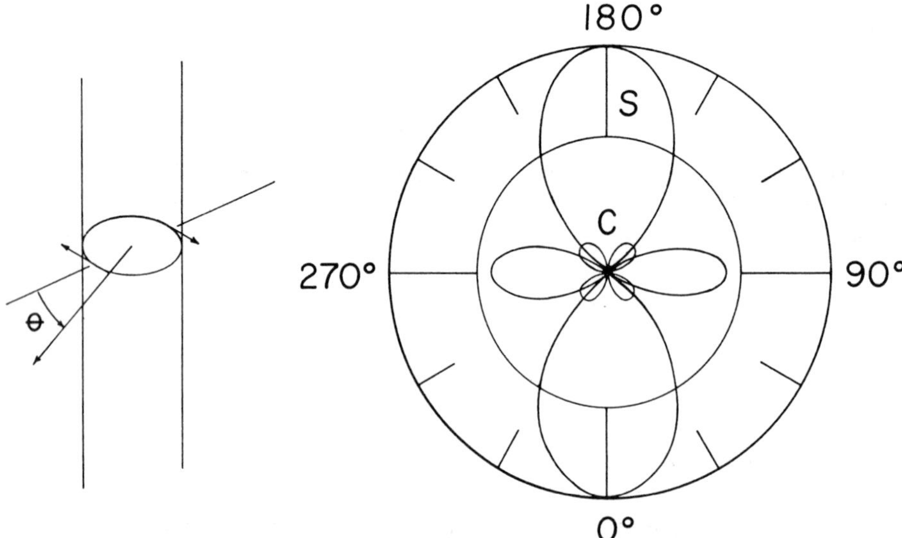


FIG. 3. Shear and compressional displacements due to tangential forces in cylinder.

compute the tangential displacement $l_\theta$ due to the application of a normal stress $S$ perpendicular to the cylinder axis. At the cylinder wall, $N_\theta = S[1-2\cos 2\theta]$ and the other stresses are zero. Then

$$\frac{N_\theta}{Y} = \frac{S}{Y}[1 - 2\cos 2\theta] = \frac{1}{a}\frac{\partial l_\theta}{\partial \theta} + \frac{l_r}{a} \cdot$$

From previous results $l_r/a = (1+2\cos 2\theta)S/Y$.

$$l_\theta = \int \frac{Sa}{Y}[1 - 2\cos 2\theta - 1 - 2\cos 2\theta]d\theta,$$

$$l_\theta = -\ 2Sa\sin 2\theta/Y$$

is the tangential displacement due to a single normal stress $S$, to which should be added any translation or rotation of the body as a whole.

The disturbance from a radially-oriented point force carries a normal stress $N$ oriented in the direction $\theta$ and a stress $\sigma N/(1-\sigma)$ perpendicular to $\theta$ and to the cylinder axis, plus a similar stress parallel to the cylinder which contributes nothing to the tangential displacements. Hence, for the arriving wave,

$$l_\theta = \frac{-2aN}{Y}\left[\sin 2\theta + \frac{\sigma}{(1-\sigma)}\sin 2\left(\theta + \frac{\pi}{2}\right)\right]$$

$$l_\theta = \frac{-2aN}{Y}\frac{(1-2\sigma)}{(1-\sigma)}\sin 2\theta = \frac{-2aN\sin 2\theta}{\rho V_c^2(1+\sigma)} \cdot \tag{9}$$

For two forces oriented as shown, corresponding tangential displacements add. Overall translation cancels, and there is no rotation. Hence the desired sum of displacements is twice $l_\theta$. Again taking $N$ from Equation 2,

$$u_r = \frac{2a(1 - V_s^2/V_c^2)\sin 2\theta}{\pi r\rho V_c^3(3 - 4V_s^2/V_c^2)} f'(t - r/V_c). \tag{10}$$

This is the radial displacement at a distance $r$ from a pair of forces acting tangentially at the ends of a diameter of a cylinder.

The radiation of shear waves can be obtained in a similar fashion. A shear wave from direction $\theta$ can be characterized by a tangential stress $T$ and a bodily rotation of magnitude $T/2\mu$. The tangential displacement at the cylinder wall due to the shear stress $T$ alone is obtained as the sum of displacements due to a normal stress of magnitude $-T$ at $(\theta+45°)$ and a normal stress of $+T$ at $(\theta-45°)$.

$$l_\theta = \frac{2Ta}{Y}[\sin 2(\theta + 45°) - \sin 2(\theta - 45°)] = \frac{4Ta}{Y}\cos 2\theta = \frac{2Ta}{\rho V_s^2(1+\sigma)}\cos 2\theta.$$

With addition of rotation term,

$$l_\theta = \frac{aT}{2\rho V_s^2}\left[1 + \frac{4}{(1+\sigma)}\cos 2\theta\right]. \tag{11}$$

Translation at the two points cancels, so the sum is twice $l_\theta$ above. Introducing $T$ from equation 4,

$$u_\theta = \frac{a}{4\pi r\rho V_s^3}\left[1 + \frac{8(1 - V_s^2/V_c^2)}{(3 - 4V_s^2/V_c^2)}\cos 2\theta\right] f'(t - r/V_s). \tag{12}$$

This is the tangential displacement at a distance $r$ from a pair of tangential forces on the wall of a cylindrical cavity.

Shear and compressional displacements for a pair of tangential forces are plotted in Figure 3, again taking $V_s^2/V_c^2$ to be $\frac{1}{3}$. Here the "reverse break" is quite pronounced, that is, for angles near 90°, the displacement is in the opposite direction from that which would be expected from the torque supplied by the pair of forces.

## RADIATION FROM FINITE CYLINDER

The radiation of elastic waves due to the application of a transient pressure on the interior of a cylinder of finite length, which is of direct interest in seismic prospecting, was treated by Heelan (1953). Through a very complex derivation, he achieved results from which the expressions for long wavelengths and great distances from the source were extracted and discussed in detail. It is the purpose of this section to show that these expressions can be derived very simply by using the reciprocity relations.

There is symmetry about the cylinder axis, so the angular dependence on $\phi$ is sought. Assume that the interior of the cylindrical length, $L$, is divided into $M$ small equal areas with a point force acting in each, radially on the cylinder wall. The average radial displacement around such a cylindrical length due to a compressional wave from a direction $\phi$ has been derived, White (1953).

$$\bar{l}_r = \frac{aN}{Y}\left[\frac{\sigma(1+\sigma) + (1 - \sigma - 2\sigma^2)\sin^2\phi}{(1-\sigma)}\right].$$

The desired sum is simply $M\bar{l}_r$. Again using equation 2,

$$u_r = \frac{aM}{4\pi r V_c Y}\left[\frac{\sigma(1+\sigma) + (1 - \sigma - 2\sigma^2)\sin^2\phi}{(1-\sigma)}\right] f'(t - r/V_c).$$

For direct comparison with Heelan's result, $f/(\pi a^2 L/M)$ is defined as a pressure $p$, and identities among elastic constants are used, leading to exactly his equation,

$$u_r = \frac{\pi a^2 L}{4\pi r\mu V_c}\left(1 - \frac{2V_s^2\cos^2\phi}{V_c^2}\right) p'(t - r/V_c).$$

Arrival of an $SV$ wave from a direction $\phi$ produces an average radial displacement,

$$\bar{l}_r = -T \sin 2\phi/\mu.$$

Following the corresponding steps as for the compressional wave, Heelan's result is again duplicated,

$$u_s = \frac{\pi a^2 L}{4\pi r \mu V_s} \sin 2\phi \, p'(t - r/V_s).$$

This agreement with the low-frequency limit of Heelan's more refined derivation gives confidence in the reasoning being followed here, and reference to Heelan's article will emphasize the relative simplicity which has been achieved.

### HOLE IN PLATE

It was felt that a direct comparison between calculation and experiment could be made most easily by means of two-dimensional seismic modelling. For this purpose, a crystal plug was mounted in a hole in a plastic plate so as to exert oppositely directed forces, as shown in Figure 4.

In order to compute the radiated displacements by reciprocity, one needs the expression for a single point force acting in the plane of the plate. Toward this end, equations (1) and (3) for sinusoidal time dependence can be integrated to give the shear and compressional waves in an infinite solid radiated by forces distributed along a line and acting perpendicular to the line (for instance, Rayleigh, 1894). As demonstrated by Oliver et al (1954), solutions in an infinite solid which have no component of motion along one axis are entirely equivalent to solutions for a plate, providing the speed of compressional waves in the plate, $V_p$, is substituted for the speed of compressional waves in the solid, $V_c$.

Hence, for a point force $Fe^{j\omega t}$ in a plate, the displacement at a distant point in line with the force is entirely radial, expressed by

$$U_R e^{j\omega t} = \frac{F\sqrt{\pi}\sqrt{V_p}}{2\pi\rho V_p^2\sqrt{2\omega r}} \, e^{-i\left(\frac{\omega r}{V_p} + \frac{\pi}{4}\right)} e^{j\omega t}.$$

The normal stress in the direction of propagation is,

$$N e^{j\omega t} = \frac{-j\omega F\sqrt{\pi}}{2\pi\sqrt{V_p}\sqrt{2\omega r}} \, e^{-i\left(\frac{\omega r}{V_p} + \frac{\pi}{4}\right)} e^{j\omega t}.$$

The displacement at a point on a line perpendicular to the force is entirely tangential, given by

$$U_\theta e^{j\omega t} = \frac{F\sqrt{\pi}\sqrt{V_s}}{2\pi\rho V_s^2\sqrt{2\omega r}} \, e^{-i\left(\frac{\omega r}{V_s} + \frac{\pi}{4}\right)} e^{j\omega t}.$$

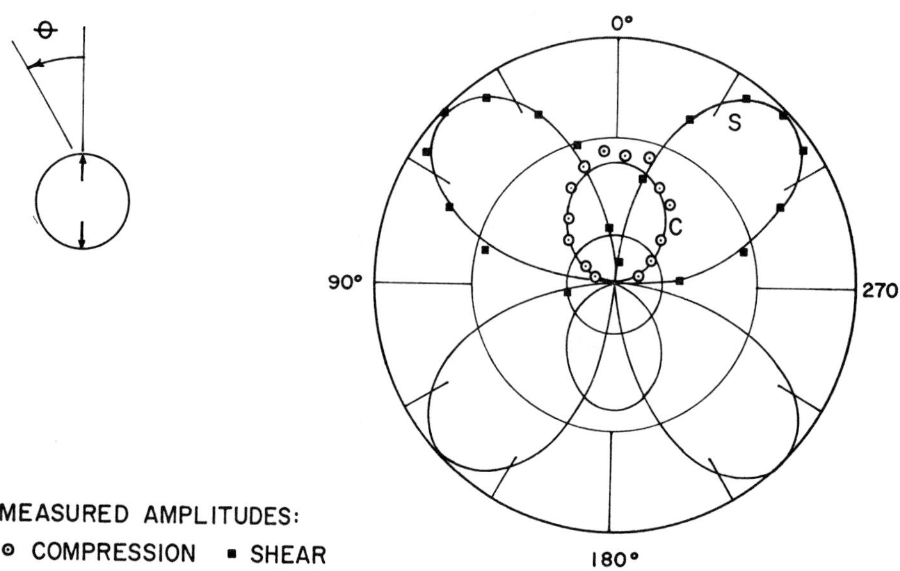


FIG. 4. Shear and compressional displacements due to radial forces on hole in plate.

The tangential stress accompanying this displacement is,

$$Te^{j\omega t} = \frac{-j\omega F\sqrt{\pi}}{2\pi\sqrt{V_s}\sqrt{2\omega r}}\, e^{-i\left(\frac{\omega r}{V_s} + \frac{\pi}{4}\right)} e^{j\omega t}.$$

Deriving the expression for compressional waves radiated by a pair of forces in a hole in a plate parallels the treatment of such a pair of forces in an infinite cylinder. The compressional wave arriving at the hole carries a normal stress $N$ in the direction of travel and $\sigma N$ at right angles. The resulting displacement of a point on the hole relative to its center is,

$$l_r = \frac{aN}{Y}\,[1 + 2\cos 2\theta + \sigma - 2\sigma\cos 2\theta],$$

$$l_r = \frac{aN}{2\rho V_s^2}\left[1 + \frac{2(V_s/V_p)^2}{1 - (V_s/V_p)^2}\cos 2\theta\right],$$

where

$$\sigma = [1 - 2(V_s/V_p)^2].$$

Again, the desired sum of displacements, corresponding to two points on opposite ends of a hole diameter, is twice $l_r$. Hence the radial displacement due to a pair of oppositely directed forces in a hole in a plate is

$$U_r e^{j\omega t} = -\frac{aF}{2\rho V_s^2}\left(\frac{j\omega}{2\pi r V_p}\right)^{1/2}\left(1 + \frac{2(V_s/V_p)^2}{1-(V_s/V_p)^2}\cos 2\theta\right) e^{j\omega(t-r/V_p)}.$$

In similar fashion, the tangential displacement is found to be

$$U_\theta e^{j\omega t} = -\frac{aF}{\rho V_s^2}\left(\frac{j\omega}{2\pi r V_s}\right)^{1/2}\frac{\sin 2\theta}{[1-(V_s/V_p)^2]} e^{j\omega(t-r/V_s)}.$$

It is to be noted that $U_r$ is a travelling compressional wave and $U_\theta$ a shear wave, each decreasing in amplitude as the square root of the distance. Each contains the term $(j\omega)^{1/2}$ which indicates that if the force is actually a transient pulse of a certain waveform, then the waveform of the displacement is different. However, the shear and compressional displacements have the same waveform, and hence their magnitudes can be compared.

With a plug of Rochelle salt contacting a half-inch diameter hole at two points, waves radiated into a quarter-inch plate of plexiglas were detected by means of a phonograph pick-up. Observed waveforms, with a dominant frequency of about 9 kc, are shown in Figure 5. Measured peak-to-peak amplitudes divided by a maximum shear amplitude are compared with calculations in Figure 4. Normalized to its maximum amplitude, the tangential function is simply $\sin 2\theta$. Dividing the radial displacement by the maximum shear displacement gives the expression

$$(V_s/V_p)^{1/2}[1-(V_s/V_p)^2+2(V_s/V_p)^2\cos 2\theta]/2,$$

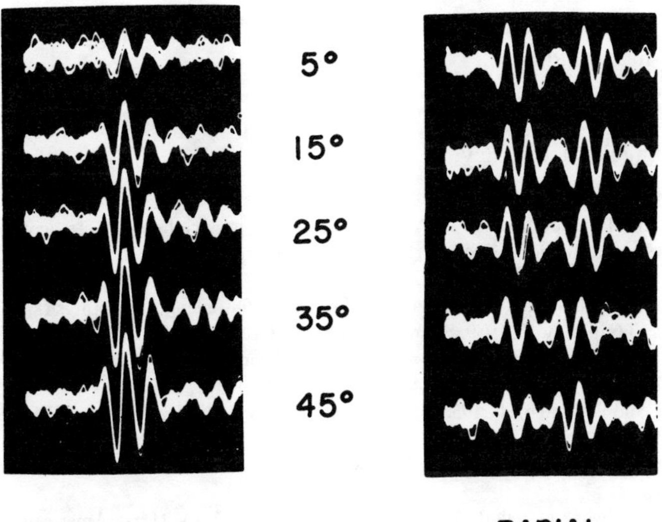


FIG. 5. Waveforms due to radial forces on hole in plate.

which is plotted as the compressional curve, using the measured value $V_s/V_p=0.57$ for the plate. Agreement is considered satisfactory, as indicated by dependence on angle and ratio of compressional to shear amplitudes.

## CONCLUSION

The very general reciprocity between forces and displacements can be used to simplify drastically the computation of the radiation of shear and compressional waves from localized sources.

## LIST OF SYMBOLS

$a$ = radius of cylinder (or hole)
$f$ = transient force
$F$ = force amplitude at frequency $\omega$
$l$ = displacement at hole or cylinder
$L$ = length of finite cylinder
$N$ = normal stress
$p$ = transient pressure
$r, \theta, \phi$ = spherical coordinates about source
$R, S$ = normal stress
$T$ = shear stress
$u$ = transient particle displacement
$U$ = displacement amplitude at $\omega$
$V_c$ = speed of compressional waves
$V_p$ = speed of compressional waves in plate
$V_s$ = speed of shear waves
$Y$ = Young's modulus
$\mu$ = shear rigidity
$\rho$ = density
$\sigma$ = Poisson's ratio
$\omega$ = angular frequency

## REFERENCES

Foldy, L. L., and Primakoff, H., 1945, A general theory of passive linear electroacoustic transducers and the electroacoustic reciprocity theorem. I–II: Journal of the Acoustical Society of America, v. 17, p. 109–20; v. 19, p. 50–58.

Heelan, P. A., 1953, Radiation from a cylindrical source of finite length: Geophysics, v. 28, p. 685–96.

Knopoff, L., and Gangi, A. F., 1959, Seismic reciprocity: Geophysics, v. 24, p. 681–91.

Love, A. E. H., 1892, A treatise on the mathematical theory of elasticity: Dover Publications, New York (First American Printing, 1944), Paragraph 212, p. 304.

Morse, P. M., and Feshbach, H., 1953, Methods of theoretical physics: McGraw-Hill Book Company, p. 882 and 1783.

Oliver, J., Press, F., and Ewing, M., 1954, Two-dimensional model seismology: Geophysics, v. 19, p. 202–219.

Rayleigh, 1894, Theory of sound: Dover Publications, New York (First American Edition, 1945), Paragraph 376, p. 421.

White, J. E., 1953, Signals in a borehole due to plane waves in the solid: Journal of the Acoustical Society of America, v. 25, p. 906–15.

# RAYLEIGH-WAVE MOTION IN AN ELASTIC HALF-SPACE†

W. A. SORGE*

Measurements made on Rayleigh waves below the surface of a simulated elastic half-space confirm in detail the behavior predicted by theory. These measurements, made by means of a two-dimensional seismic model, show that the amplitude of the Rayleigh wave falls off rapidly with increasing depth.

The theory of the motion of Rayleigh waves in a homogeneous elastic half-space is well described in the literature (Ewing, et al, 1957; Lamb, 1904; Rayleigh, 1885), and the main features of the predictions for pulse inputs have been confirmed with seismic models (Tatel, 1954). We have now confirmed the theory in detail by using two-dimensional seismic model techniques to predict the dependence on depth of Rayleigh-wave amplitudes. This dependence had not been tested prior to our investigation.

In his development of the theory of sinusoidal surface waves on a semi-infinite elastic solid, Lord Rayleigh (1885) showed that the surface-wave motion becomes small at a few wavelengths below the surface. The horizontal and vertical components of motion at the surface form a retrograde elliptical system; at 0.192 wavelength below the surface (for a Poisson ratio of 0.25) the horizontal component of motion is zero; and below this depth the motion is direct (Ewing, et al, 1957). The vertical motion, on the other hand, is maximum slightly below the surface and falls off steadily with depth.

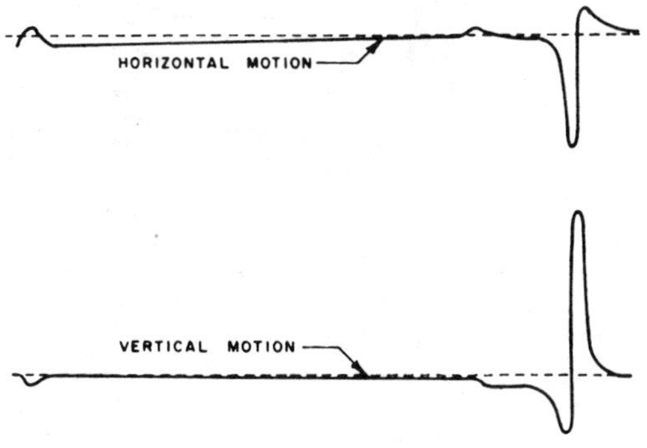


Fig. 1. Lamb's prediction of Rayleigh-wave motion from an impulse.

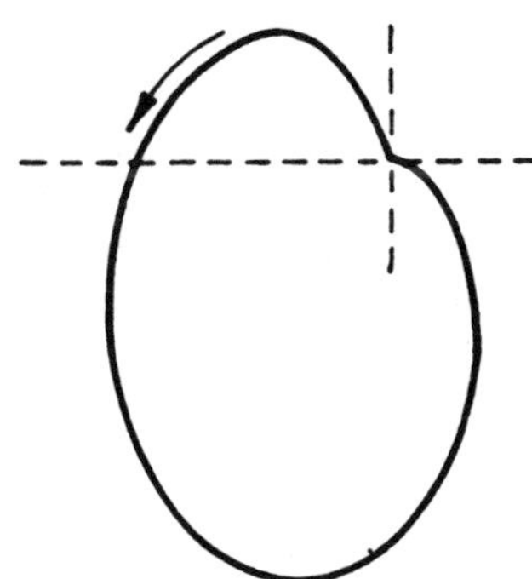

Fig. 2. Hodograph of Rayleigh-wave motion as predicted by Lamb.

Lamb (1904) considered the surface waves from an impulse and predicted the displacement at some distant point to be as shown in Figure 1. The upper curve is horizontal motion and the lower curve is vertical motion. Figure 2 is a hodograph of the motions shown in Figure 1.

The model we used to simulate the homogeneous half-space is a 1/16 inch thick sheet of Plexiglas 4 ft wide and 7 ft long. The top edge of the sheet simulates the surface of the half-space. A source crystal mounted on the top edge of the sheet is driven by a high-voltage pulse and the received signals are picked up on the side of the sheet. The source crystal is a lead zirconate titanate disc about 1/16 inch in diameter and 1/10 inch thick. (The electronic equipment is described by Levin and Hibbard, 1955.) Matched pairs of Clevite bimorph crystals (Healy and Press, 1960), which can be rotated to pick up both

† Manuscript received the Editor October 24, 1963.

* Jersey Production Research Company, Tulsa, Oklahoma.

Reprinted from Geophysics, 30, 97-101

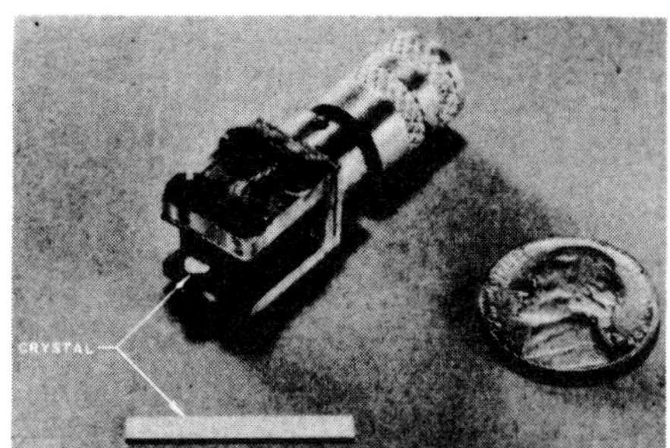


Fig. 3. Bimorph crystal and mounting.

horizontal and vertical components of motion, as well as combinations of these two components, are mounted in rubber holders on opposite sides of the Plexiglas sheet (Figure 3). They can be placed any desired distance below the top edge of the sheet and any desired horizontal distance from the source crystal.

Figures 4 and 5 show the horizontal and vertical motions of the Rayleigh wave at the surface and at depth in the model at a relatively large distance from the source, and Figures 6A and 6B are hodographs of these motions. The motion at the surface is somewhat more complex than the impulse considered by Lamb. Careful examination of the horizontal component, however, reveals that the motion does indeed reverse with depth. The horizontal motion does not become zero at any particular depth because each frequency component goes to zero at a different depth.

Frequency analyses were made of the waveforms of the vertical component of the Rayleigh wave for several distances from the source. All these spectra peaked at about 30 kc. Figures 7 and 8 show how the horizontal and vertical components of motion vary with depth at 20 inches from the source. The broken curves represent the experimental values while the solid curves represent the theoretical values. Here again we believe that the experimental curves do not exactly follow the theoretical curves because our experimental signal is a pulse rather than a continuous wave.

By using two-dimensional models we have been able to show that the Rayleigh-wave motion does behave at depth according to theory. Moreover, the results of this study add further evidence to the works of Gupta and Kisslinger (1964) and Gilbert, et al (1962) that two-dimensional seismic

Fig. 4. Horizontal components of Rayleigh-wave motion in Plexiglas model.

Fig. 5. Vertical components of Rayleigh-wave motion in Plexiglas model.

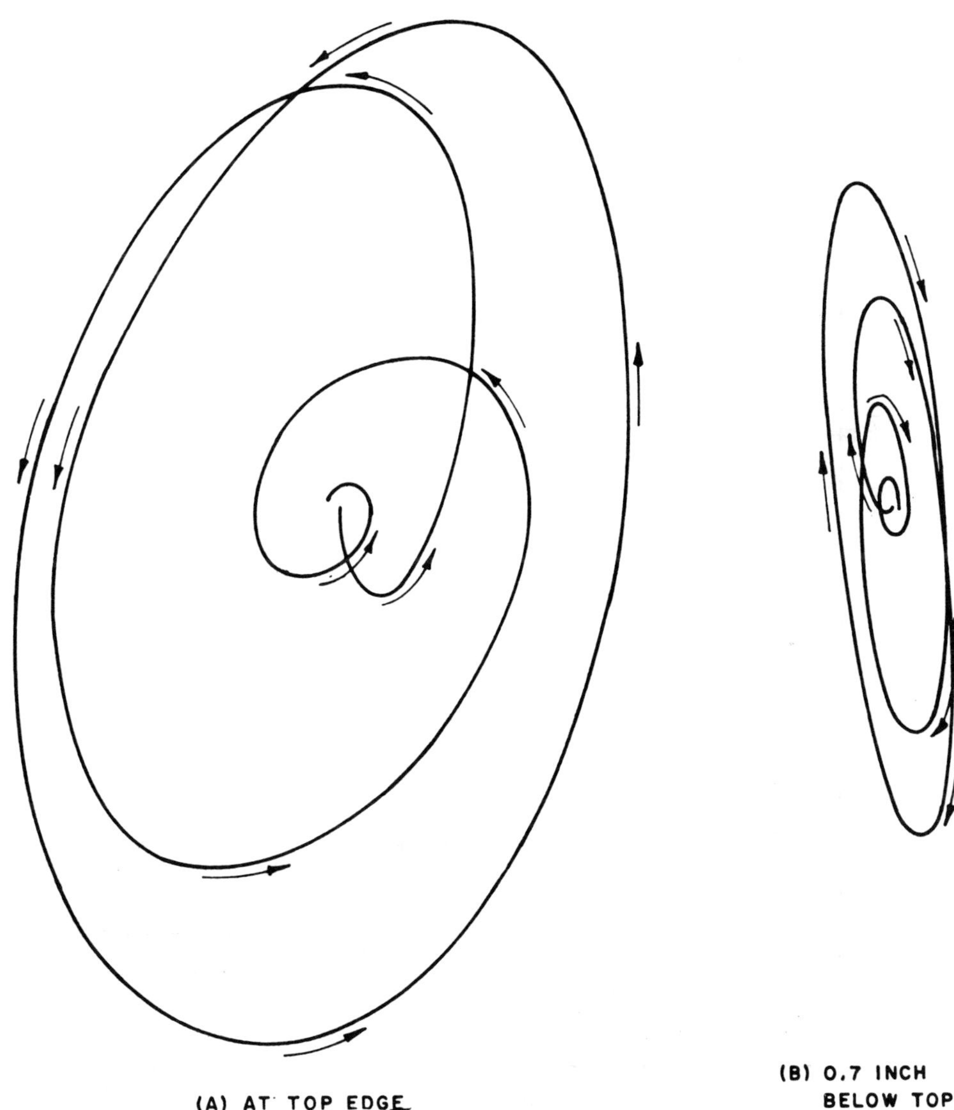


FIG. 6. Particle motion of Rayleigh wave in Plexiglas model.

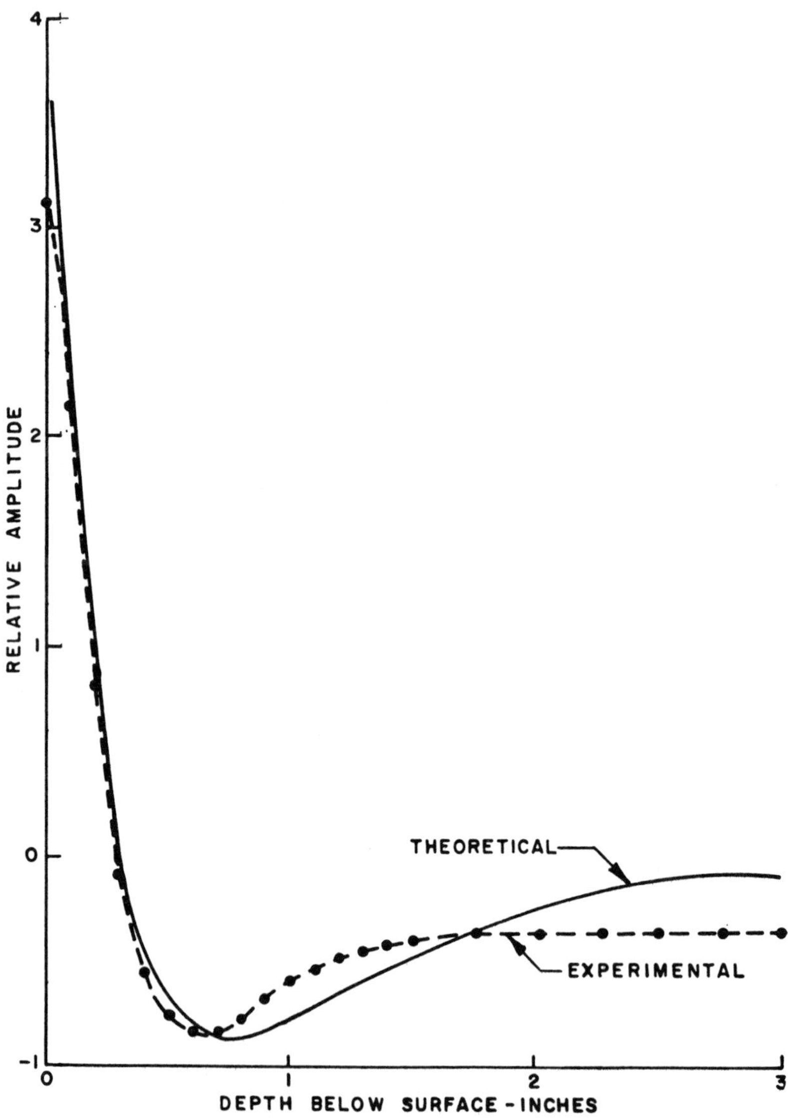


Fig. 7. Model and theoretical variations in horizontal component of Rayleigh wave.

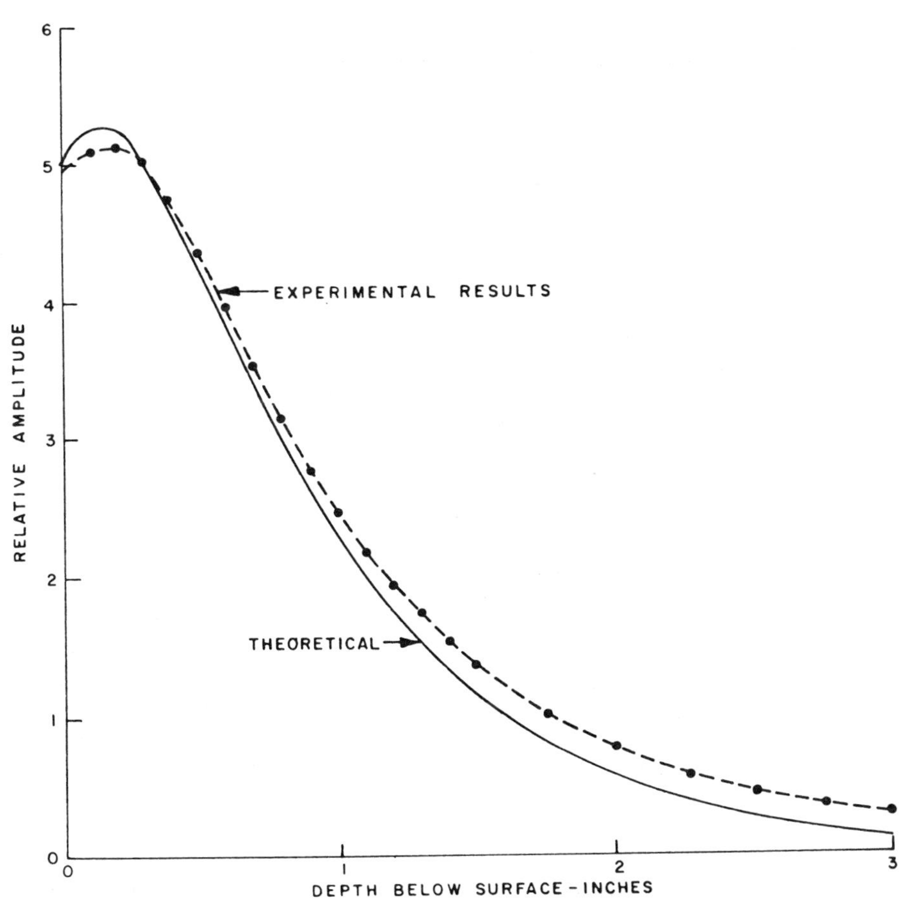


FIG. 8. Model and theoretical variations in vertical component of Rayleigh wave.

models can be used as analog computers for solving elastic-wave problems.


#### ACKNOWLEDGMENT

I wish to thank the Jersey Production Research Company, Tulsa, Oklahoma, for permission to publish the results of these studies. I also am indebted to Dr. Franklyn K. Levin for his valuable suggestions during the seismic model study and preparation of this material for publication.


#### REFERENCES


Ewing, W. Maurice, Jardetsky, W. S., and Press, Frank, 1957, Elastic waves in layered media: New York, McGraw-Hill.

Gilbert, Freeman, Laster, Stanley J., et al, 1962, Observation of pulses on an interface: Bull. Seism. Soc. Amer., v. 52, p. 847–868.

Gupta, R. N., and Kisslinger, C., 1964, Model study of explosion-generated Rayleigh waves in a half-space: Bull. Seism. Soc. Amer., v. 54, p. 475–484.

Healy, John H., and Press, F., 1960, Two-dimensional seismic models with continuously variable velocity depth and density functions: Geophysics, v. 25, p. 987–997.

Lamb, Horace, 1904, On the propagation of tremors over the surface of an elastic solid: Trans. Phil. Royal Society (London), v. 203, p. 1–42.

Levin, F. K., and Hibbard, H. C., 1955, Three-dimensional seismic model studies: Geophysics, v. 20, p. 19–32.

Oliver, Jack, Press, Frank, and Ewing, Maurice, 1954, Two-dimensional model seismology: Geophysics, v. 19, p. 202–219.

Rayleigh, Lord, 1885, On waves propagated along the plane surface of an elastic solid: Proc. London Math. Soc., v. 17, p. 4–10.

Tatel, H. E., 1954, Note on the nature of a seismogram II: Jour. Geoph. Res., v. 59, p. 289–294.

# THREE-DIMENSIONAL SEISMIC MODELING†

FRED J. HILTERMAN*

Record sections from three-dimensional acoustic models often contain diffracted events not predictable by classical raypath theory. Several observed and calculated record sections from models of typical geologic structures such as synclines, anticlines, and faults verify this diffraction phenomenon. A careful interpretation of the character and moveout of these diffracted events is required to delineate certain portions of the geologic structures.

A far-field approximation of the retarded potential equation is suitable for direct time-domain evaluation and is used to synthesize the calculated sections. The excellent comparisons between the calculated and observed record sections suggest that the mathematical modeling technique can be a useful tool for enhancing field interpretations.

## INTRODUCTION

Seismic record sections indicating faults, synclines, and other geologic structures have been an integral part of the geophysicist's interpretation for many years. The analysis of the seismic events on these record sections can, however, lead to confusing and indecisive interpretations. In this event, the geophysicist has often resorted to mathematical and experimental models.

As early as 1937, Rieber used small-scale seismic models to investigate the diffraction patterns occurring near irregular boundaries. He photographed the acoustic wavefronts diffracted from a fault model to demonstrate the geometric spreading effects of secondary propagation. To some degree, the experimental portion of the present work is an extension of Rieber's three-dimensional model studies.

The purpose of the present study was to develop a mathematical means for predicting the response to energy from a point source impinging on a three-dimensional acoustic model of arbitrary shape. Synthetic record sections displaying not only the geometric moveout but also the character of the pulse were desired. The synthetic record sections were then to be compared with experimental sections in order to verify the mathematical modeling technique.

For the experimental modeling, an eighth-inch condenser microphone and an electric spark simulated the detector array and the energy source. Paper and wood were used as model reflectors. In essence, the modeling was purely geometric, since the boundaries were rigid and shear waves were absent.

Our mathematical model utilizes a far-field approximation of a numerical technique suggested by Mitzner (1967). The development of the mathematical model from the retarded potential equations is given in the Appendix.

## EXPERIMENTAL MODEL

Initial investigations of the acoustic pulse shape indicated that a transient with a peak frequency between 30 and 50 khz could readily be generated and propagated in air. The investigations also indicated that attenuation would not significantly alter the assumption of spherical propagation from the point source if measurements were made in the far field. Model and earth prototype units are summarized in Table 1.

Figure 1 illustrates a model and the modeling

† Part of thesis T 1297 submitted for Ph.D. degree at the Colorado School of Mines, Golden, Colorado, 1970. Manuscript received by the Editor May 25, 1970; revised manuscript received August 19, 1970.

* Mobil Oil Corporation, Dallas, Texas 75221.



Reprinted from Geophysics, 35, 1020-1037

equipment that were constructed using Table 1 as a design criterion. The physical dimensions permitted the recording of a time window in which the reflections and diffractions from the geologic model boundary were uncontaminated with signals from irrelevant boundaries such as the sides of the enclosure. Electronic components are shown in the block diagram of Figure 2.

As illustrated in Figure 1, there were two viewing windows in the front of the enclosure. The right hand window contained a glass panel which reduced air turbulence when data were being collected and also acted as a safety precaution. The left-hand window was open and through this window the geologic models were placed on the movable platform. These geologic models were constructed out of wood or heavy paper; no significant change was observed between the reflected energy from the wood and from the paper.

After the geologic models were placed on the platform, the microphones and source were adjusted vertically so as to simulate the desired depth to the geologic structure. With the positions of the microphones and source fixed, successive shotpoint locations were obtained by translation of the movable platform.

**Table 1. Summary of experimental scale factors and units**

Time scale factor = $\text{time}_p/\text{time}_m$[1] = 1000
Length scale factor = $\text{length}_p/\text{length}_m$ = 12,000

| Quantity | Model | Prototype |
|---|---|---|
| Length | 1 inch | 1000 ft |
| | 1 cm | 394 ft |
| Time | 1 $\mu$sec | 1 msec |
| Velocity | 1120 ft/sec | 13440 ft/sec |
| Peak frequency | 40 khz | 40 hz |

[1] Subscript $p$ represents prototype and subscript $m$ represents model.

As the geologic models were incrementally passed beneath the source and detector, a single trace for each shotpoint was recorded by an oscilloscope camera. The traces were then manually digitized at a two microsecond sampling interval, normalized with respect to the reference microphone, and plotted as a time record section.

## EXPERIMENTAL AND MATHEMATICAL RESULTS

A difficulty which arose when discussing the record sections was the ambiguity of the words "diffraction" and "reflection" when applied to the models. It was decided, therefore, to use the definitions set forth by Dix (1952, p. 237). The

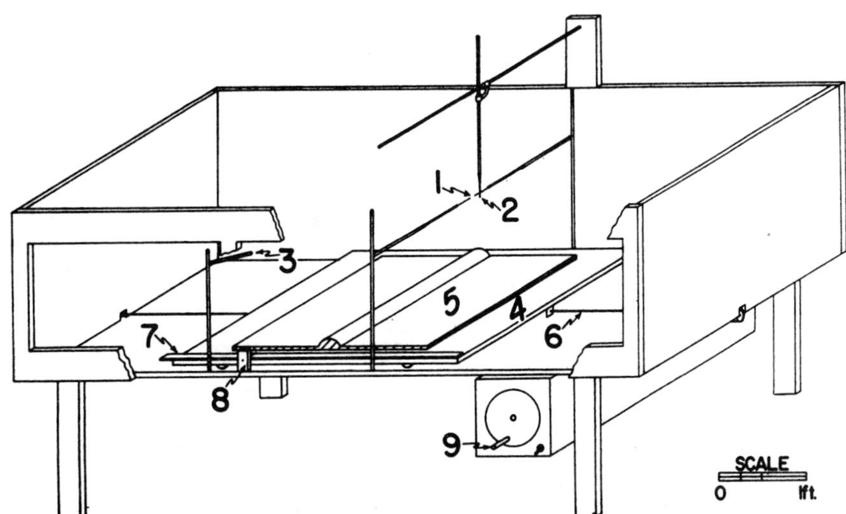


FIG. 1. Drawing of a model and associated equipment. The numbers refer to

1. Spark gap.
2. Eighth-inch condenser microphone (detector).
3. Half-inch condenser microphone (reference).
4. Horizontally movable platform.
5. Model of geologic structure.
6. Platform cable.
7. Meter stick.
8. Horizontal reference point.
9. Platform crank.

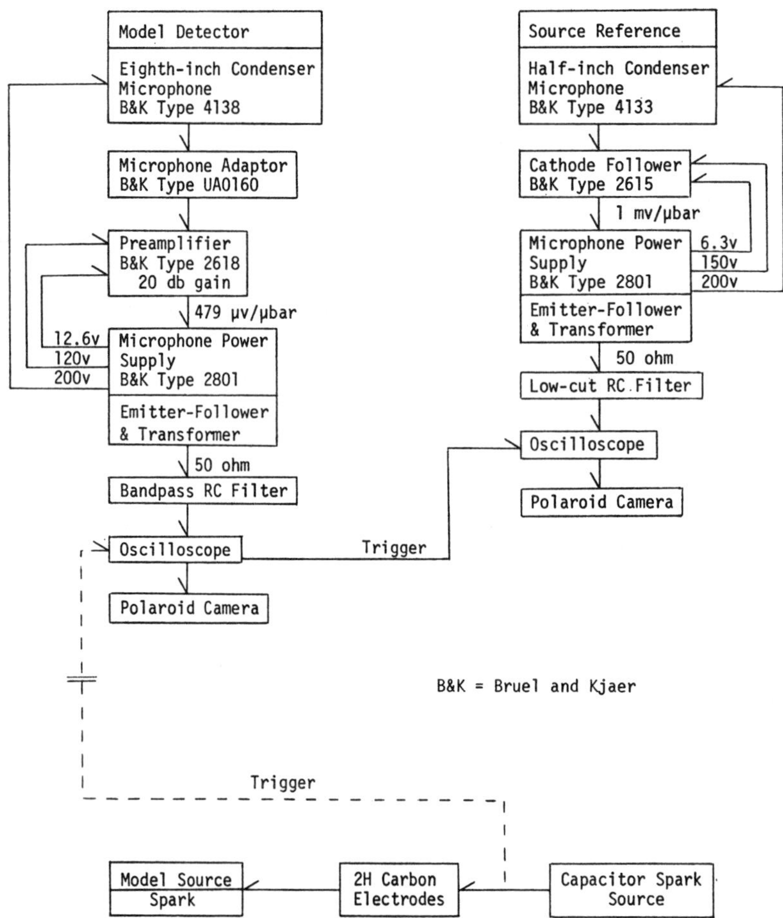


FIG. 2. Block diagram of modeling apparatus.

term reflection denotes an event which yields information concerning the curvature or dip of the boundary. Diffractions include all other events, some of which may yield information regarding the spatial location of an irregularity on the boundary.

An isometric sketch of each model accompanies the record sections. The location midway between the source and detector is shown by a small circle in the sketches and is referred to as the shotpoint location. Coordinate dimensions in the sketches are directly proportional to each other, so that distance can be represented by the shotpoint locations which are one cm apart. Two-way traveltime in microseconds is obtained by adding the appropriate time shift $T_0$ to the record section time scale.

A small time variation between the calculated and observed sections exists in several places. This small time discrepancy, introduced by experimental error, is not detrimental to the resulting interpretation.

### *Model A–Syncline*

Model A in Figure 3 consists of a concave surface with a center of curvature located below the shotpoints. This type surface configuration with its buried focal line has often plagued the geophysicist. Even after the patterns in Figure 3 are recognized as those from a concave surface, the trace analysis is still not simple. The major problem is picking the exact two-way traveltime for the point where raypath 3 in Figure 4 is normal to the boundary.

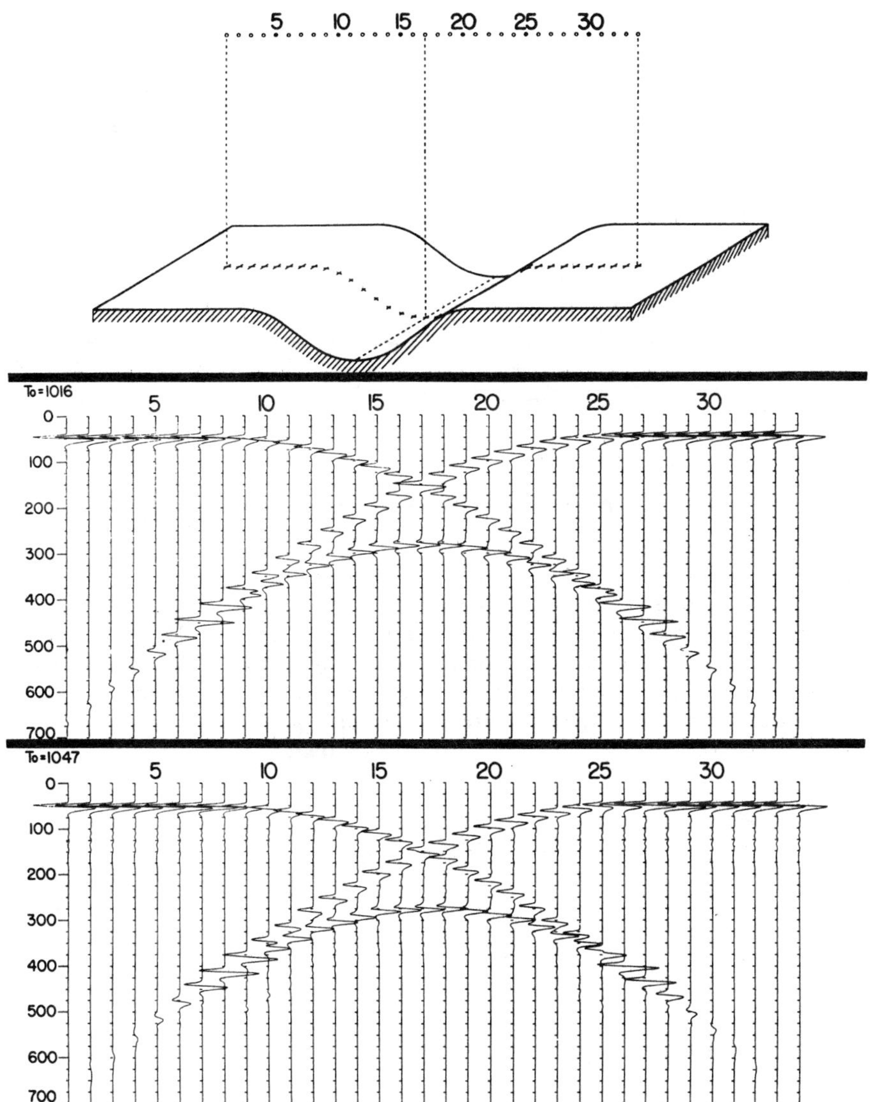


FIG. 3. Calculated (top) and observed (bottom) record sections of Model A.

Two-way traveltimes associated with raypaths 1, 2, and 3 are depicted in the trace analysis of Figure 4. However, this trace analysis was made with the knowledge of the model dimensions and shotpoint location. Also shown by two sets of arrows are the arrival times required for trough or peak reference picks. Using pulse shape alone, we found it difficult to correlate the picks of raypath 3 with those of the other two raypaths.

This dilemma is resolved if the change in pulse shape is considered. According to Dix (1952, p. 365), a reflected harmonic wave exhibits a phase change of 90 degrees every time it passes through a focal line. Fourier analyses of the incident pulse and of the returned energy for raypath 3 verified this phase phenomenon for high-frequency components. The amplitude density spectra were similar except for a constant multiplier.

Whenever a function undergoes a constant phase shift of 90 degrees, it is essentially being convolved in the time domain with the distribution function $(-\pi t)^{-1}$ (White, 1965, p. 11). The convolution of a pulse of finite duration with $(-\pi t)^{-1}$ yields a noncausal function which ex-

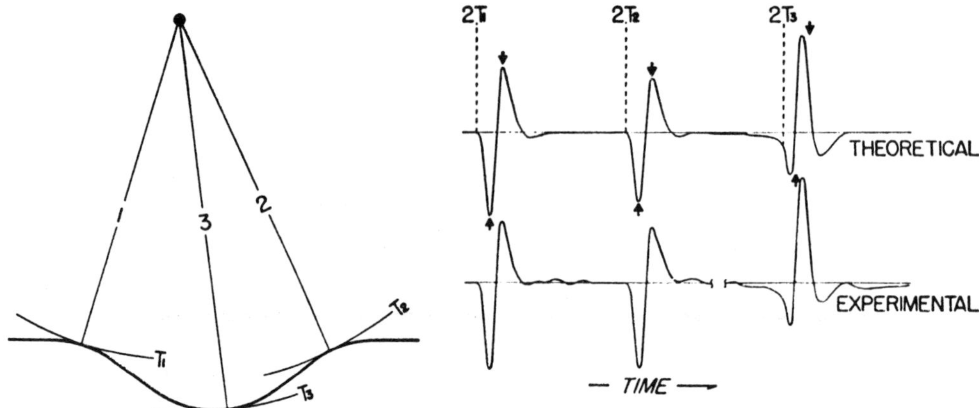


FIG. 4. Raypath geometry and trace analysis for shotpoint 15 of Model A.

tends from minus infinity to plus infinity. This noncausal effect is illustrated in Figure 4 by the nonzero onset energy at time $2T_3$. However, the phase shift for a transient incident wave cannot be exactly 90 degrees because diffracted energy would have to arrive before $2T_1$, the shortest traveltime to the surface.

Applying the reciprocity theorem to the general law that waves in an even number of dimensions leave wakes (Morse, 1948, p. 312), one concludes that the reflection at $2T_1$ has a sharp onset but has a tail that is infinite, though small in magnitude. The same principle holds for the reflection at $2T_2$, except that, due to the wake of the previous

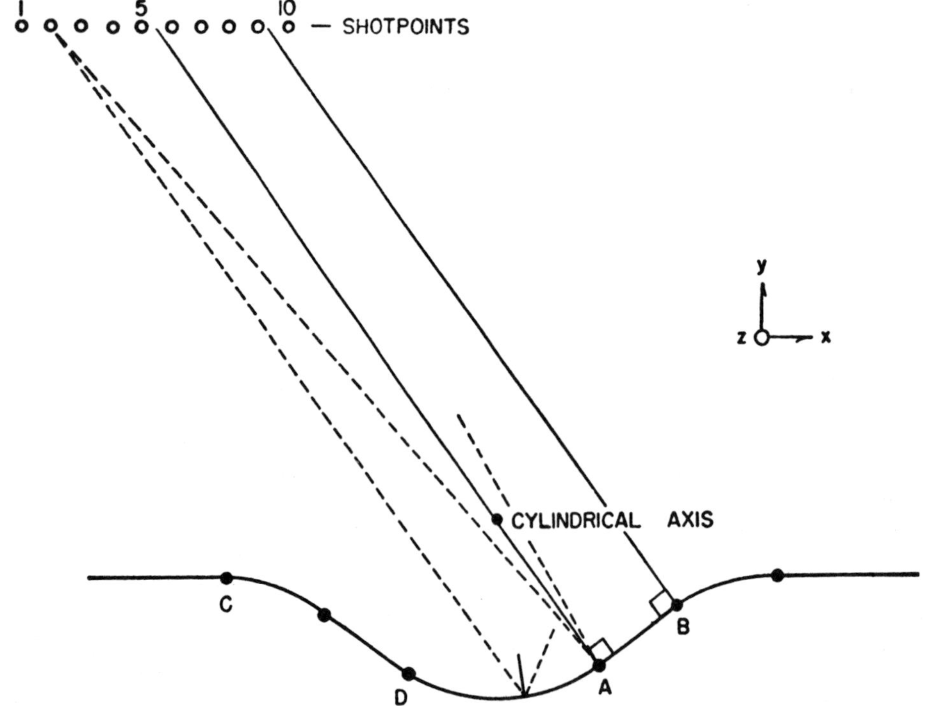


FIG. 5. Raypath limits normal to Model A.

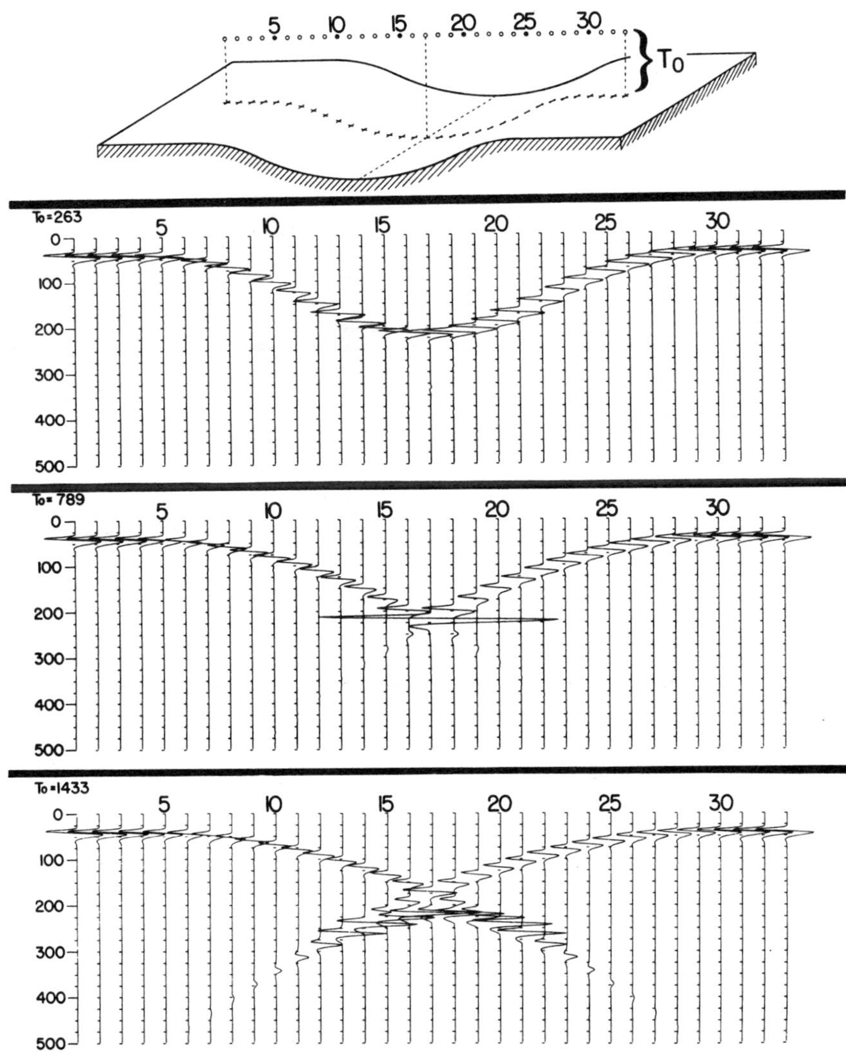


FIG. 6. Calculated record sections of Model B. Radius of curvature of concave-upward surface is 18 cm. Top section has shotpoint surface 10 cm below curvature axis; middle section, 1 cm below; and, bottom section, 10 cm above.

reflection, the reflection does not have a zero-amplitude onset. It is difficult to decide if the onset of the diffraction should be placed at $2T_1$ or at a later time associated with the surface between raypath 1 and 3. The problem is more academic than practical, since in the high-frequency approximation, the diffraction approaches a reflection with an onset time at $2T_3$. In short, this discussion suggests that classical raypath interpretation can be misleading if care is not exercised when concave boundaries are involved.

An additional feature Model A exhibits is an event that apparently transforms from a reflection to a diffraction without an appreciable increase in its time moveout. Figure 3, aided by Figure 5, illustrates this phenomenon. For shotpoints 1 to 5 (and 29 to 34), no raypaths are normal to the far flank of the syncline and hence only diffracted energy can be returned to shotpoints 1 to 5 from the far flank. The term diffraction was chosen, since shotpoints 1 to 5 do not contribute any information regarding the curvature of the anomalous surface. A trace analysis of this event is difficult because the pulse has broadened and does not have a distinct onset time.

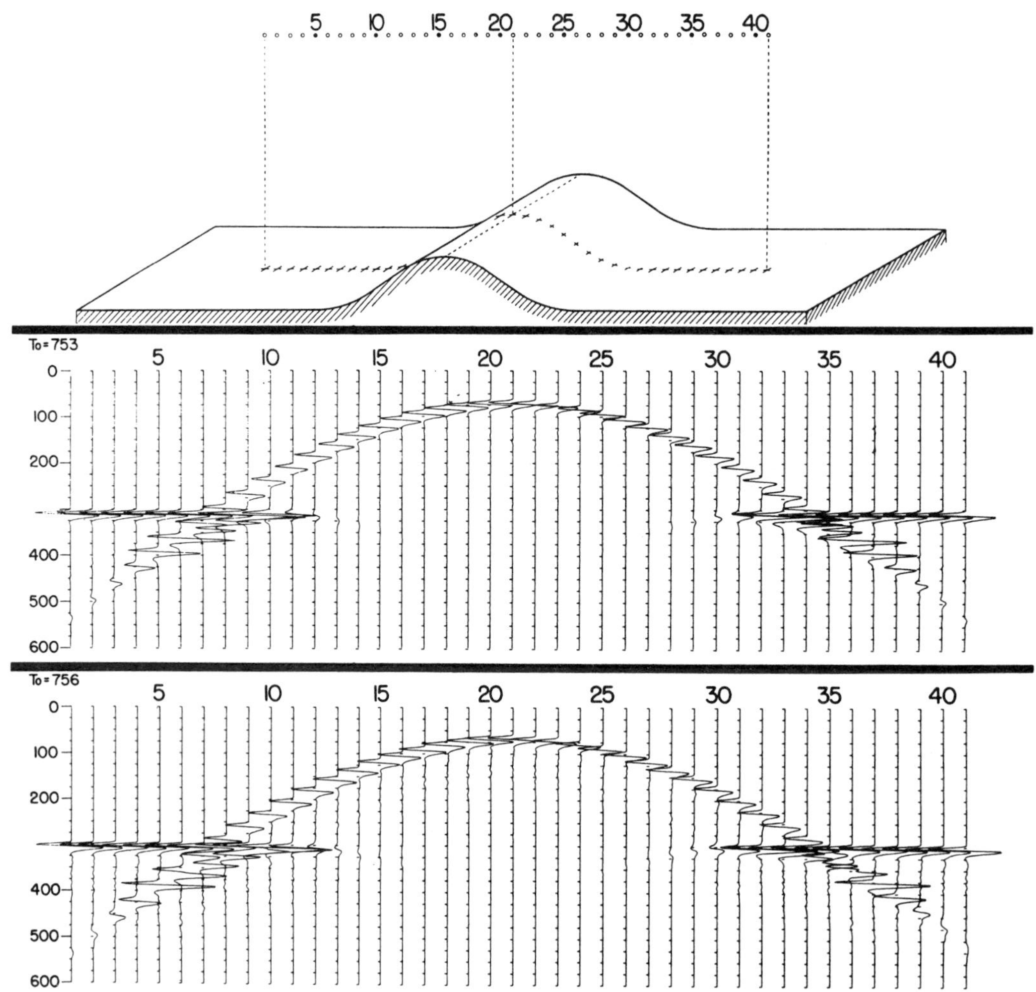


FIG. 7. Calculated (top) and observed (bottom) record sections of Model C.

*Model B–Syncline with various depths of burial*

Figure 6 illustrates the different calculated wave patterns obtained by varying the depth to the boundary. The same boundary configuration was used in all three of the calculated sections.

The amplitude of the diffraction on the 17th trace in the middle section of Figure 6 is 5 to 6 times that of the reflection amplitude on trace 1. This exact model was not tested experimentally; however, a similar model was tested and a diffraction amplitude $5\frac{1}{2}$ times that of the reflection was observed.

*Model C–Large anticline*

In Figure 7, the anticline may be considered as the boundary inverse of the syncline shown in Figure 3; that is, the size of the anomaly is the same in both models except that the anomalies are reversed in direction.

The syncline in Model A (Figure 3) was relatively easy to interpret, since the anomaly was several wavelengths high and there were zones in the time section where the diffraction was not masked by the flank reflections. On the remaining models, however, the structures were approximately two wavelengths in relief and the resolution of the events was more difficult.

*Model D–Small anticline*

The anticline illustrated in Figure 8 has two

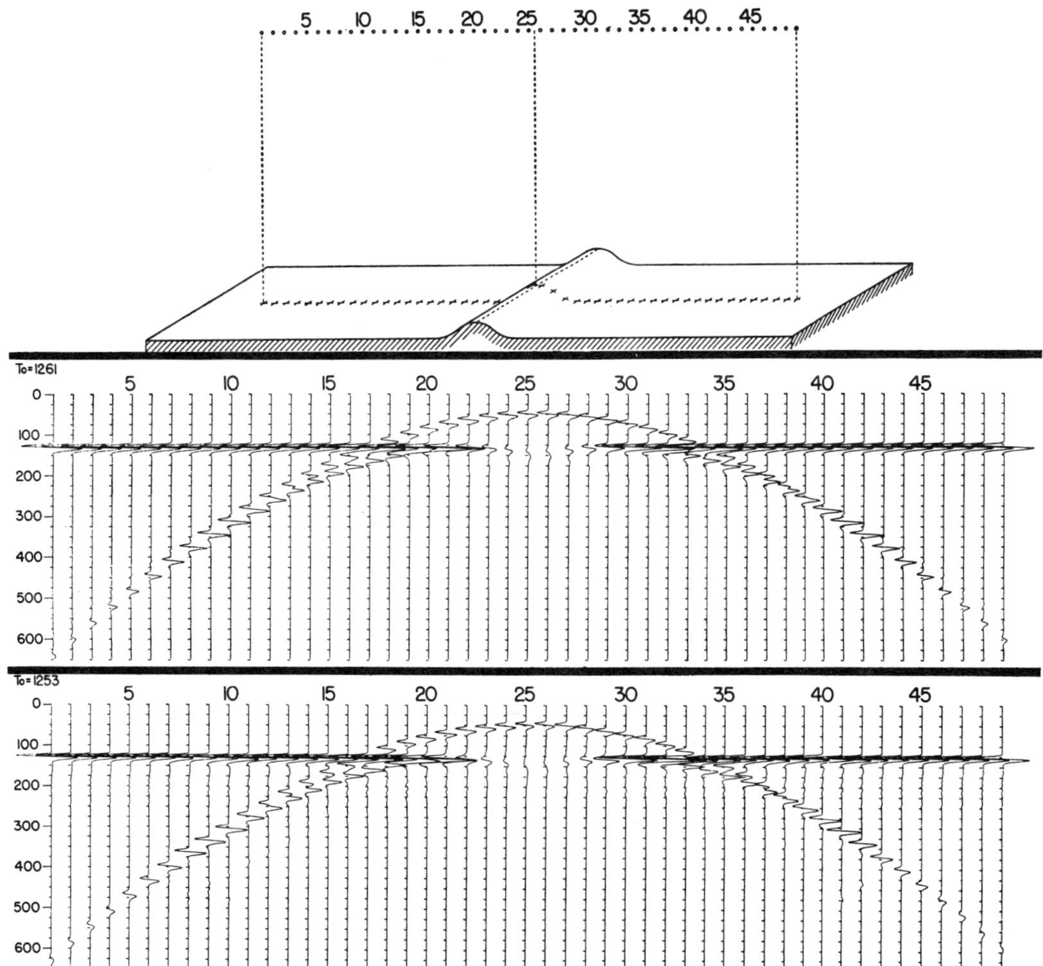


FIG. 8. Calculated (top) and observed (bottom) record sections of Model D.

surfaces that create buried foci; these are the bottom portions of the flanks. This buried foci effect is evident on the record sections if the event in the lower left-hand corner is followed upward toward the right. As the pulse reaches traces 12 and 13, a phasing effect starts. The change in pulse shape is actually a separation of a reflection from the top side of the flank and a later arriving diffraction from the lower portion of the flank. However, the small time difference between the reflection and the diffraction arrival times makes it difficult to interpret the flank of the anticline.

On traces 23 through 28, it appears as if a weak reflection at 125 μsec continues right through the anticline. This event is not a reflection but is a diffraction generated by the curvature of the flanks. In a similar manner, we can show that the energy arriving in the latter portion of traces 1 through 7 does not have a raypath which is normal to the surface. Once again, there is no sharp discontinuity to suggest a diffraction, which the event must be.

Although Model D is the same type of structure as Model C, there are several geometric differences worthy of noting in their patterns. In Model C, the buried focal lines are nearer the shotpoint surface than the buried focal lines in Model D. This geometric difference causes the diffractions from the concave surfaces of Model C to be evident on a fewer number of traces than the diffractions of Model D.

Most of the energy associated with the anomaly

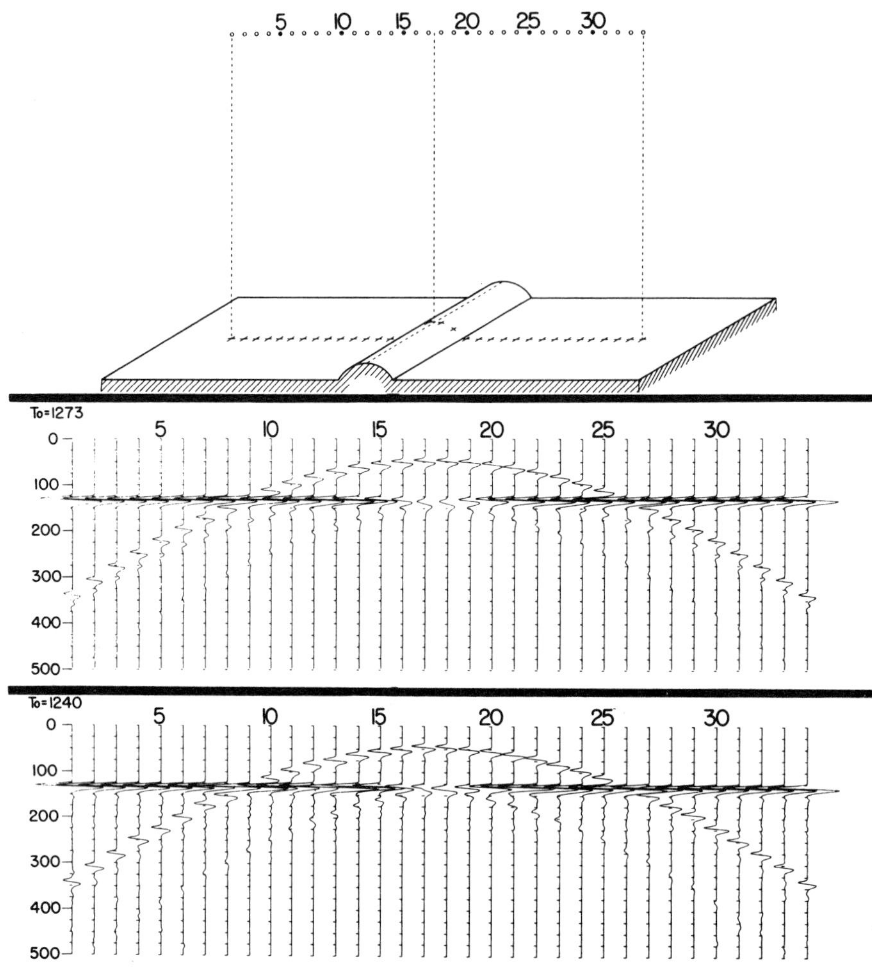


FIG. 9. Calculated (top) and observed (bottom) record sections of Model E.

arrives before the plane reflection in Model C, whereas the converse is true for Model D. However, the moveout of the convex-upward event in both model sections is approximately the same. These last two observations illustrate the importance of migration and the care that must be exercised when interpreting multilayered structures.

*Model E–Reef*

In some geologic areas it is desirable to outline accurately from seismic data the flanks of small positive anomalies. For example, what event (or events) would distinguish the two different types of flanks shown in Figures 8 and 9?

The boundary of Model E contains two discontinuities in its curvature. The magnitude of the diffracted energy from these discontinuities differs on the calculated and observed record sections. This difference is quite obvious on traces 16 to 19 at 125 $\mu$sec and also on traces 1 to 11, where the calculated diffraction from the near edge is larger than the observed diffraction. Both of these discrepancies are mainly due to the mathematical omission of the interacting surface effects. In the mathematical development, only that surface energy which is direct-incident energy is considered to have an effect on the shotpoint. Thus the energy which reflects off the side of the cylinder, strikes the flat surface, and returns to the detector is ignored. Since the boundary is rigid, this reflected-diffracted energy is sufficient

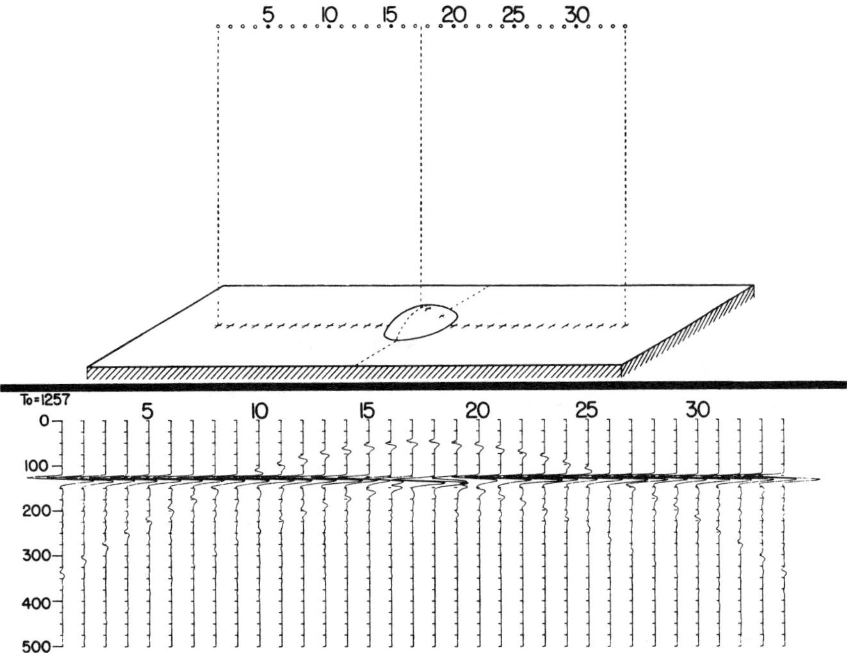


FIG. 10 Observed record section of Model F.

to cause a noticeable pressure variation at the shotpoint. This type surface energy can be incorporated in the theory but the subsequent numerical evaluation is more complicated (see Mitzner, 1967).

### *Model F–Dome*

A complete mathematical record section of Model F in Figure 10 was not computed because a satisfactory algorithm for finding the solid angle had not been developed. A pressure wave reflected from a rigid sphere was calculated and found to compare favorably with the experimental data.

In Model F, the spherical segment has a radius and altitude which are within 0.1 cm of those of the cylindrical segment in Model E. It is not surprising, therefore, that the arrival times of Model F and Model E are approximately equal.

On traces 16 to 19 of Model F, the diffractions have a magnitude approximately equal to that of the reflection on trace 1. This abnormal amplitude is associated with the in-phase reflected-diffracted energy around the base of the spherical segment.

### *Model G–Vertical fault*

Model G in Figure 11 presented unusual problems when a comparison between the observed and calculated pressure behavior was made. One of the problems was resolved when it was discovered that the shotpoint locations for the calculated section were 1.34 cm lower than the experimental shotpoint locations. This accounts for some of the moveout discrepancy between the calculated and observed diffractions in Figure 11.

Reflected-diffracted energy accounts for the amplitude difference between the calculated and the observed diffractions for shotpoints on the downthrown side. However, it is questionable if this observed energy is strongly evident on seismic field data, where the reflection coefficient is usually less than 0.3.

### *Model H–30-degree fault*

For the 30-degree fault in Figure 12, the effect of the reflected-diffracted energy is very minor as shown by the comparison of the calculated and observed responses. A crossover of the diffracted events from the fault edges occurs between shot-

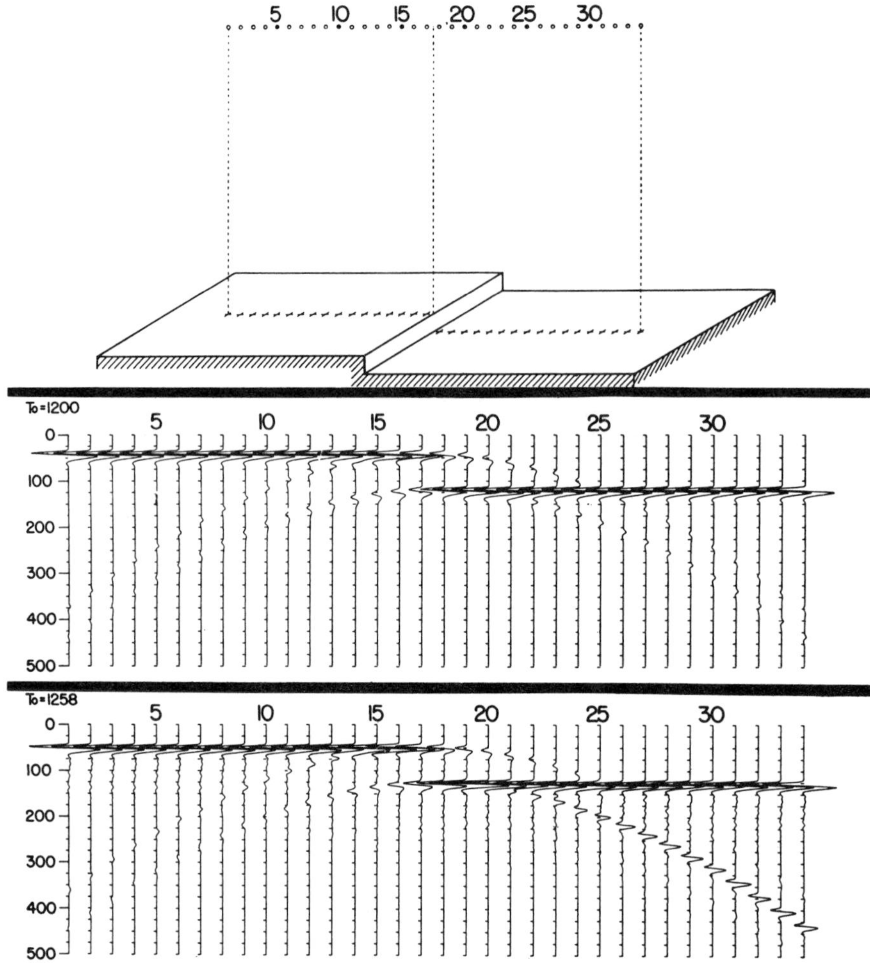


FIG. 11. Calculated (top) and observed (bottom) record sections of Model G.

points 31 and 32. The corresponding energy buildup on these traces is significant and may be used as a criterion for locating the fault plane.

*Model I–Vertical fault with drag*

The effect of introducing drag on the vertical fault is illustrated in Figure 13 by the two calculated record sections, one with drag and the other without. Angona (1960) has observed similar effects for drag on a two-dimensional model.

## CONCLUSIONS

The excellent agreement between the mathematical and experimental model sections demonstrates that the mathematical modeling technique can accurately profile a three-dimensional acoustic model. The mathematical technique is especially suitable for subsequent interpretation, since the boundary segments can be analyzed individually.

Two types of diffractions were encountered. The first had a sharp onset and yielded information regarding spatial points on the boundary, such as the top edge of a fault. The second type associated with concave-upward surfaces had no sharp onset and was not predictable in many cases by raypath theory. Diffractions of this second type are often treated in practice as unwanted noise.

It is realized that neither the mathematical nor the experimental model is an exact representation

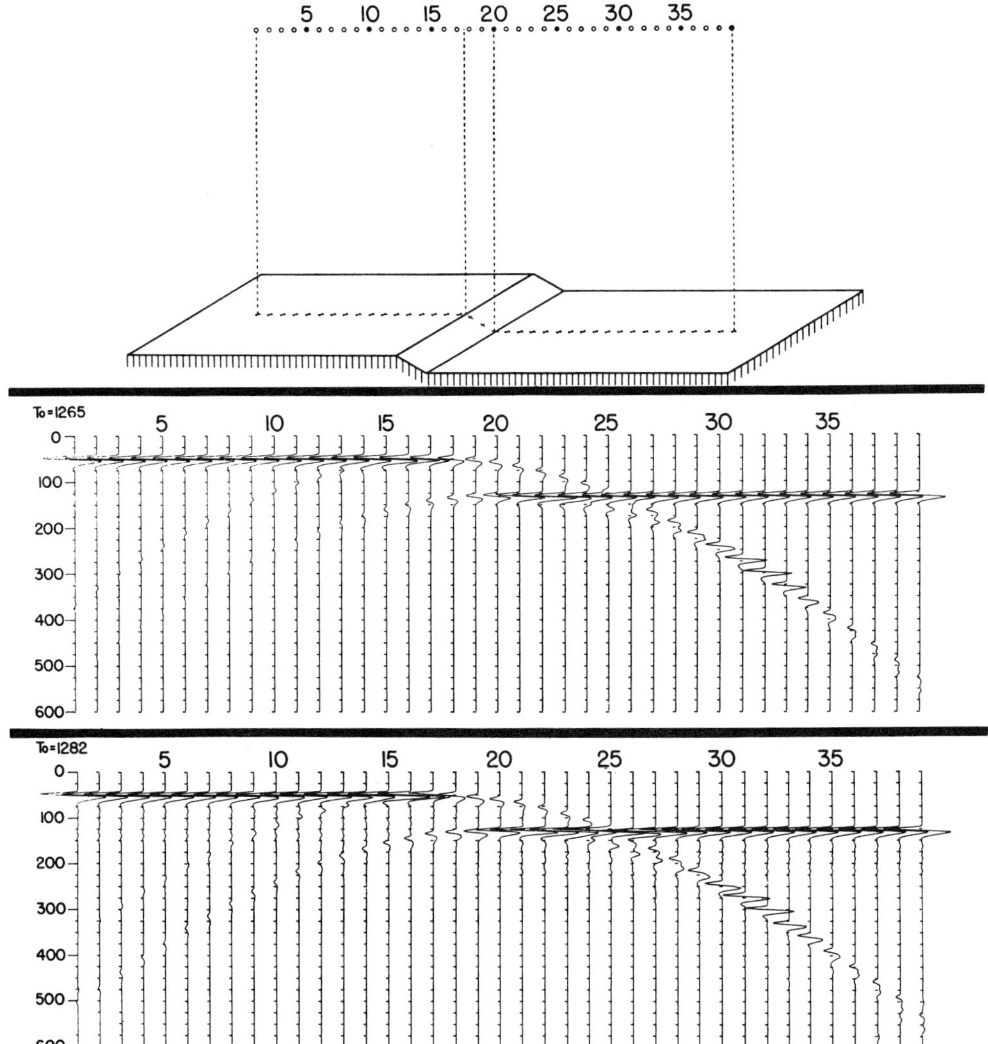


FIG. 12. Calculated (top) and observed (bottom) record sections of Model H.

of the earth prototype, but the insight gained from the reflection-diffraction complexes of the model sections is extremely useful in identifying field events which would otherwise be considered unwanted noise.


#### ACKNOWLEDGMENT

The author wishes to express his sincere appreciation to Professor John C. Hollister, who acted as thesis advisor at the Colorado School of Mines and who provided valuable suggestions during the model study and preparation of the paper.

Financial support of the work was given by NASA and Mobil Oil Corporation.

The spark-source system and circuitry were furnished by Atlantic Richfield Company and John P. Woods.


## APPENDIX

### MATHEMATICAL DEVELOPMENT

The pressure field in the fluid medium that results from the interaction of a spherical acoustic wave with a rigid boundary can be formulated several ways. We chose an integral-equation formulation because it is more amenable to sub-

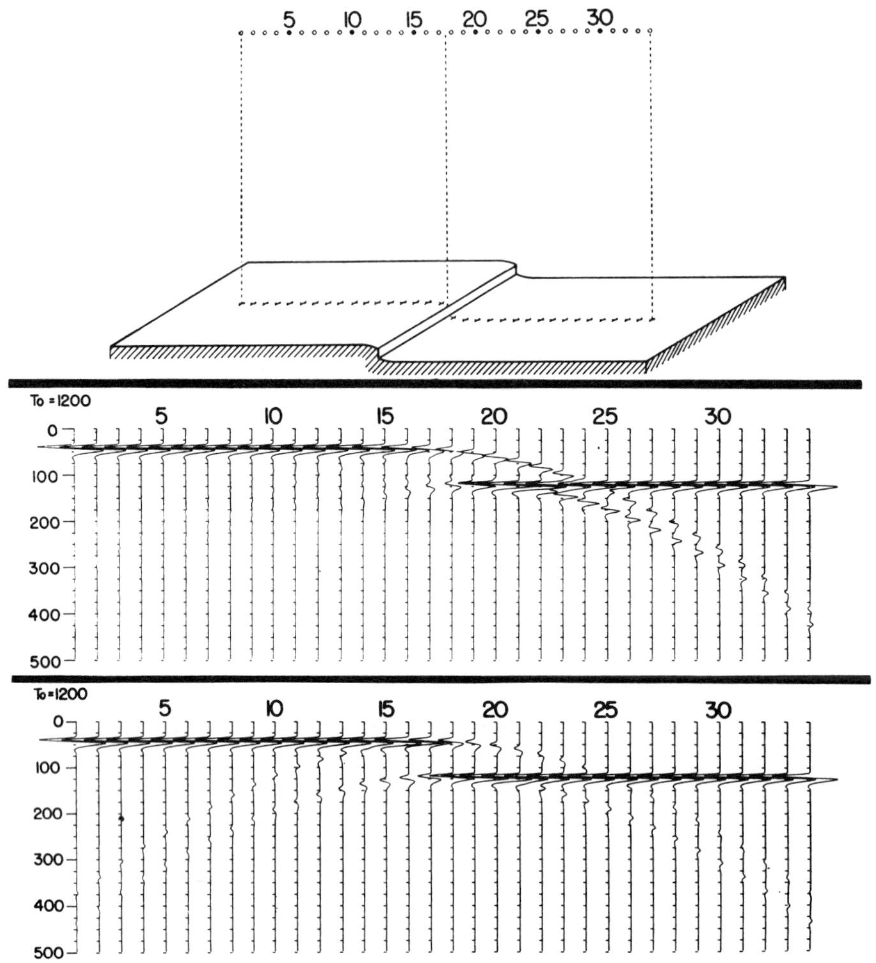


FIG. 13. Calculated (top) record section of Model I and calculated (bottom) record section of Model G. Model I is pictured above the top section; Model G appears at the top of Figure 11.

sequent numerical evaluation than the initial boundary-value method.

It has been found convenient to first express the field variation in terms of a continuous velocity potential $\phi(x, y, z; t)$, which also satisfies the wave equation. The pressure is related to the velocity potential through

$$p(x, y, z; t) = \rho \frac{\partial \phi}{\partial t}(x, y, z; t), \qquad (1)$$

where $\rho$ is the equilibrium density of the fluid.

The integral representation of the velocity potential at the field point $\phi_{fp}$ has been fully developed in terms of an incident potential $\phi_{\text{inc}}$ and a retarded surface potential $[\phi_s]$ by several authors, including Officer (1958, p. 260–265), Båth (1968, p. 192–198), and Drude (1922, p. 159–184). Imposing the rigid-boundary condition of $\partial\phi/\partial n = 0$, we get the integral equation for the velocity potential at the field point as

$$\phi_{fp}(t) = \phi_{\text{inc}}(g; t) + \frac{1}{4\pi}\int_S \cdot \left\{ \frac{1}{r^2} \frac{\partial r}{\partial n} \left( [\phi_s]_r + \frac{r}{c}\left[\frac{\partial \phi_s}{\partial t}\right]_r \right) \right\} dS, \qquad (2)$$

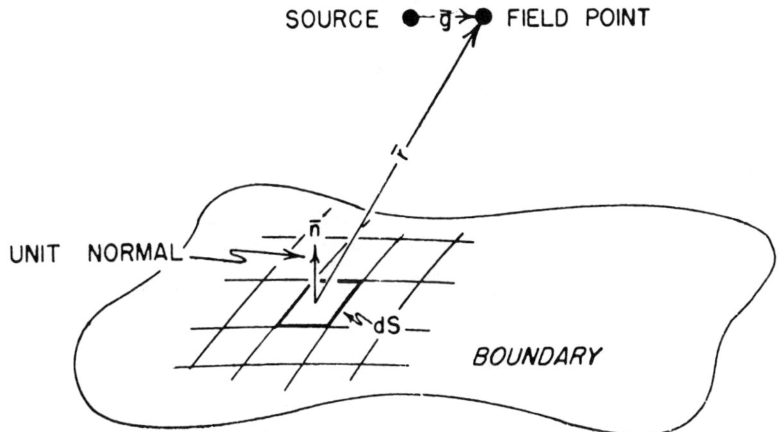


FIG. 14. General configuration of notation used in equation (2).

where Figure 14 along with the following definitions describe the notation used. $S$ = fluid-solid interface (boundary); $c$ = speed of sound in fluid; and $[\phi]_a = \phi(t - a/c)$, a time-retarded function.

Equation 2 is valid only for a point not on the boundary. For a point on $S$ where the curvature is continuous, the potential satisfies the relation

$$\phi_s(\bar{h};\, t) = 2\phi_{\text{inc}}(g';\, t) + \frac{1}{2\pi}\int_{S'} \left\{ \frac{1}{r'^2} \frac{\partial r'}{\partial n} \left( [\phi_{s'}]_{r'} + \frac{r'}{c}\left[\frac{\partial \phi_{s'}}{\partial t}\right]_{r'} \right) \right\} dS' \tag{3}$$

(see Figure 15).

If the incident pressure field is continuous, equation (1) can be applied to equations (2) and (3) to yield respectively

$$p_{fp}(t) = p_{\text{inc}}(g;\, t) + \frac{1}{4\pi}\int_{S} \left\{ \frac{1}{r^2} \frac{\partial r}{\partial n} \left( [p_s]_r + \frac{r}{c}\left[\frac{\partial p_s}{\partial t}\right]_r \right) \right\} dS, \tag{4}$$

$$p_s(\bar{h};\, t) = 2p_{\text{inc}}(g';\, t) + \frac{1}{2\pi}\int_{S'} \left\{ \frac{1}{r'^2} \frac{\partial r'}{\partial n} \left( [p_{s'}]_{r'} + \frac{r'}{c}\left[\frac{\partial p_{s'}}{\partial t}\right]_{r'} \right) \right\} dS', \tag{5}$$

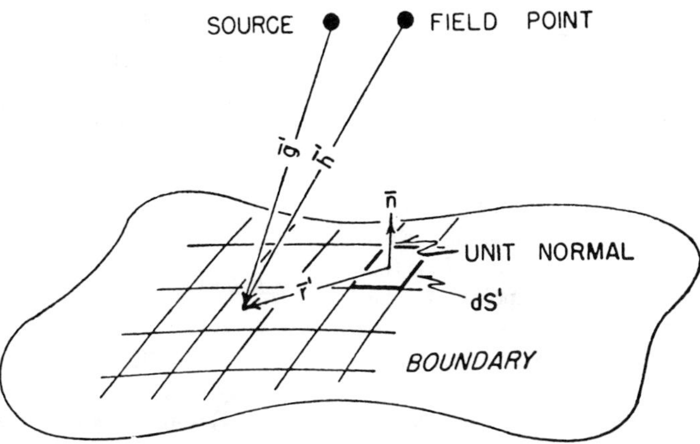


FIG. 15. General configuration of notation used in equation (3).

where the subscript identification has been retained.

At an arbitrary distance from the point source, the incident field $p_{\text{inc}}(a;t)$ can be expressed in terms of a known incident field $p_D(D;t)$ as

$$p_{\text{inc}}(a;t) = \frac{D}{a} p_D(t - a/c + D/c). \quad (6)$$

Mitzner (1967) presented a numerical technique for evaluating equations (4) and (5) in scattering problems. His technique produced excellent results but required a significant amount of computer time. In order to simplify equations (4) and (5), two approximations were introduced. The first approximation was that equation (5) can be written with the incorporation of equation (6) as

$$p_s(\bar{h};t) \simeq 2\frac{D}{g'} p_D(t - g'/c + D/c) \quad (7)$$

without unduly affecting the subsequent evaluation of equation (4).

In the second approximation, the source and field point were relocated at the same spatial point halfway between their original coordinate positions. This new spatial point is the shotpoint location. With this new coordinate definition, the incident pressure in equation (4) was dropped.

The advantage of redefining the source and field point locations is that now $g' = h = r$, so that equation (4) can be expressed in terms of equation (7) as

$$p_{fp}(t) \simeq \frac{D}{2\pi} \int_{\Omega(S)} \left[ \frac{1}{r} p_D(t - 2r/c + D/c) + \frac{1}{c} \frac{\partial p_D}{\partial t}(t - 2r/c + D/c) \right] d\Omega, \quad (8)$$

where the variable of integration was transformed to $d\Omega$, the solid-angle differential subtended at the shotpoint.

The integrand in equation (8) is a function of the variable $r$. The domain of the integrand with respect to $r$ is from $r = R_0$ to $r = \infty$, where $R_0$ is the shortest distance from the shotpoint to the boundary.

In order to arrange equation (8) in a form suitable for discrete-time analysis, the following substitutions were made: $\tau = r/c$; $g(t) = p_D(t + D/c)$; $h(t) = \partial p_D / \partial t(t + D/c)$.

Inserting the above into equation (8) yields

$$p_{fp}(t) \simeq \frac{D}{2\pi c} \int_{\Omega(S)} \left\{ \frac{1}{\tau} g(t - 2\tau) + h(t - 2\tau) \right\} d\Omega. \quad (9)$$

The continuous functions $g(t)$ and $h(t)$ were of duration $T$ and were specified in discrete time as

$$G(i\Delta) = g(t) \quad (10)$$

$$H(i\Delta) = h(t), \quad (11)$$

where $i = 0, 1, 2, \ldots, N = T/\Delta$, and $\Delta$ = sampling rate. The discrete function $H$ was actually expressed in terms of $G$ through a numerical differentiation formula (Hildebrand, 1956, p. 82), since $g(t)$ was a known time-shifted function.

In the numerical evaluation of equation (9), it was desirable first to time-contour the boundary with respect to the traveltime from the shotpoint (see Figure 16). The traveltime to the $k$th zone was averaged to give

$$T_k = t_0 + (k + 1/2)\Delta, \quad (12)$$

where $t_0 = R_0/c$.

By time-contouring the boundary in the above manner, we made both discrete functions, $G$ and $H$, constant on a given contour line. We could then replace the integral in equation (9) with a summation summed over the contour intervals. Also the observation time was specified discretely as

$$t = 2t_0 + (m + 1)\Delta. \quad (13)$$

Incorporating equations (10) through (13) into equation (9) yields

$$P(2t_0 + (m+1)\Delta) \simeq \frac{D}{2\pi c} \sum_{k=L}^{M} \Omega(k) \left[ \frac{G\{(m - 2k)\Delta\}}{T_k} + H\{(m - 2k)\Delta\} \right], \quad (14)$$

where $P(i\Delta)$ = sampled approximation of the continuous function $p_{fp}(t)$,

$m = 0, 1, 2, \ldots,$ desired number of samples,

$\Omega(k)$ = solid angle subtended at the shotpoint by the $k$th zone,

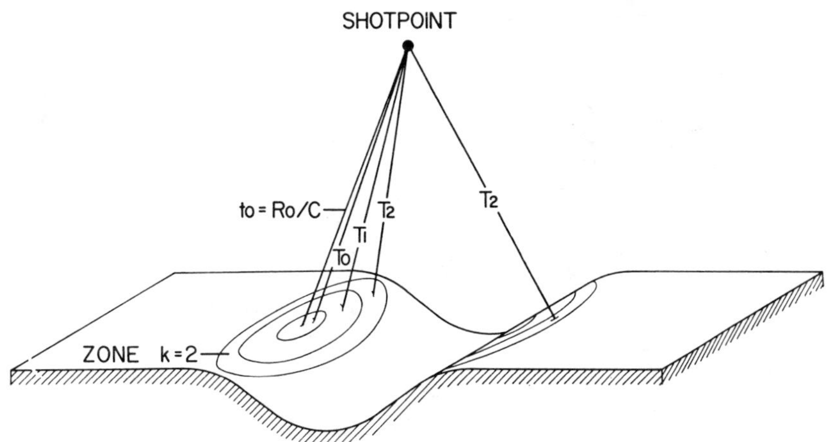


FIG. 16. Time-contouring of boundary with respect to shotpoint.

$$M = \begin{cases} m/2 & \text{for } m \text{ even} \\ (m-1)/2 & \text{for } m \text{ odd, and} \end{cases}$$

$$L = \begin{cases} 0 & \text{for } m \leq N \\ (m-N)/2 & \text{for } m > N \text{ and } (m-N) \text{ even} \\ (m-N+1)/2 & \text{for } m > N \text{ and } (m-N) \text{ odd.} \end{cases}$$

*Solid-angle calculations*

The solid angle $\Omega$ subtended at a field point by a nonsingular element is

$$\Omega = \int_S \frac{\partial r}{\partial n} \frac{1}{r^2} dS. \tag{15}$$

In the rectangular Cartesian system equation, (15) becomes

$$\Omega = \int_S \frac{[(u-x)n_x + (v-y)n_y + (w-z)n_z]}{[(u-x)^2 + (v-y)^2 + (w-z)^2]^{3/2}} dS, \tag{16}$$

(see Figure 17).

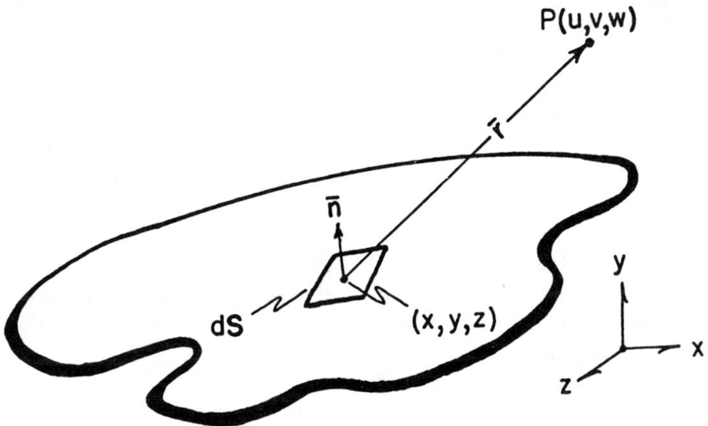


FIG. 17. Solid angle sign convention.

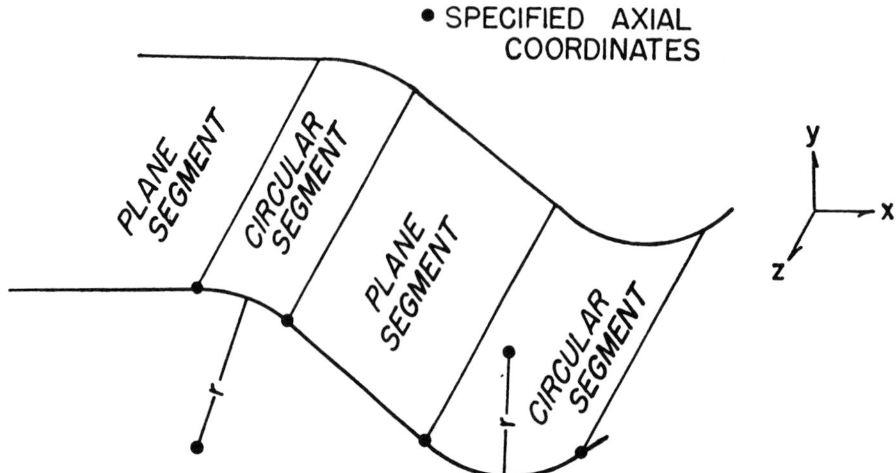


FIG. 18. Method of defining boundary.

Expressing the boundary (surface) as a function of the two independent variables $z$ and $x$ by

$$y = \beta f(x, z) + c, \tag{17}$$

we write the components of the unit normal as as discussed previously.

$$(n_x, n_y, n_z) = \frac{\left\{-\beta\left[\frac{\partial f(x, z)}{\partial x}\right], 1, -\beta\left[\frac{\partial f(x, z)}{\partial z}\right]\right\}}{\left[1 + \beta^2\left[\frac{\partial f(x, z)}{\partial x}\right]^2 + \beta^2\left[\frac{\partial f(x, z)}{\partial z}\right]^2\right]^{1/2}}. \tag{18}$$

The boundaries in this study were infinite cylindrical surfaces which we approximated with segments of planes and circular cylinders. As illustrated in Figure 18, the boundary is completely described by specifying the $x$ and $y$ coordinates of the end points of each segment and the radius and axis location of circular segments.

In equation (14), the solid angle $\Omega(k)$ is subtended at the shotpoint by the surface element inscribed by two cones whose vertices are at the shotpoint and whose elements terminate on the boundary. The length of the elements are $R_k$ for the outer cone and $R_{k-1}$ for the inner cone. The subscripted $R$ is the distance from the shotpoint to the middle of the $k$th contour zone. For $k=0$, the inner cone collapses to a straight line. This process of defining the surface of integration by two cones is similar to time-contouring the surface as discussed previously.

In order to evaluate $\Omega(k)$ associated with the $k$th zone, the solid angle corresponding to the outer cone was calculated and from this solid angle, the previously-calculated solid angle associated with the inner cone was subtracted. The process was repeated until the upper limit of the summation sign in equation (14) had been satisfied.

Since the boundary was divided into cylindrical segments, the total solid angle for a particular $k$ was found by summing the solid-angle contributions from each segment. Figure 19 illustrates the geometry involved for finding the solid-angle contribution from an arbitrary cylindrical segment. The length of the cone element is $R$.

If $P$, the shotpoint location, is in the $xy$ plane, only one-half of the surface of integration needs to be considered because of the symmetry about the $xy$ plane. Applying this principle and the relationships $n_z=0$, $w=0$, and $ds=(dx^2+dy^2)^{1/2} = [1+(dy/dx)^2]^{1/2}dx$ to equation (16) yields

$$\Omega = 2\int_{x_1}^{x_2}\int_{z_1}^{z_2} \frac{((u-x)n_x + (v-y)n_y)}{((u-x)^2 + (v-y)^2 + z^2)^{3/2}}\, dzds. \tag{19}$$

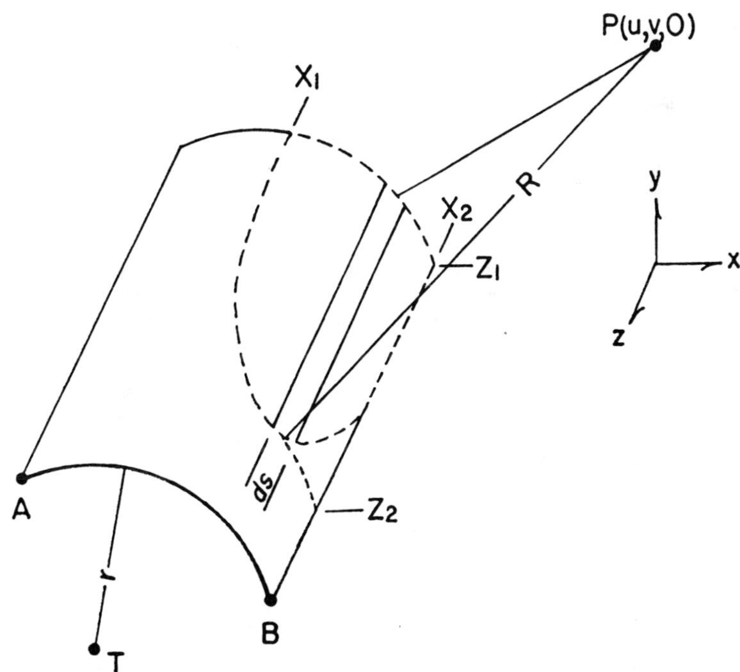


FIG. 19. Geometry and notation used in solid angle contribution from segmented surface.

Integrating equation (19) with respect to $z$ and inserting the limits $z_1=0$ and $z_2=[R^2-(u-x)^2-(v-y)^2]^{1/2}$ gives

$$\Omega = \frac{2}{R}\int_{x_1}^{x_2} \frac{((u-x)n_x+(v-y)n_y)(R^2-(u-x)^2-(v-y)^2)^{1/2}}{(u-x)^2+(v-y)^2}\, ds. \tag{20}$$

Substituting the appropriate relations for $y$, $n_y$, $n_x$, $ds$, $x_1$, and $x_2$ into equation (20) yields a single integral with respect to $x$.

## REFERENCES

Angona, F. A., 1960, Two-dimensional modeling and its application to seismic problems: Geophysics, v. 25, p. 468–482.

Båth, M., 1968, Mathematical aspects of seismology: Amsterdam, Elsevier.

Dix, C. H., 1952, Seismic prospecting for oil: New York, Harper.

Drude, P., 1922, Theory of optics [trans. by C. R. Mann and R. A. Millikan]: New York, Longmans, Green and Co.

Hildebrand, F. B., 1956, Introduction to numerical analysis: New York, McGraw-Hill.

Mitzner, K. M., 1967, Numerical solution for transient scattering from a hard surface of arbitrary shape—Retarded potential techniques: J. Acoust. Soc. Am., v. 42, p. 391–397.

Morse, P. M., 1948, Vibration and sound: New York, McGraw-Hill.

Officer, C. B., 1958, Introduction to the theory of sound transmission: New York, McGraw-Hill.

Rieber, F., 1937, Complex reflection patterns and their geologic sources: Geophysics, v. 2, p. 132–160.

White, J. E., 1965, Seismic waves: Radiation, transmissions, and attenuation: New York, McGraw-Hill.

# Babinet's principle for elastic waves*

ANTHONY F. GANGI AND BIBHUTI B. MOHANTY

*Department of Geophysics, Texas A&M University, College Station, Texas 77843*

(Received 26 August 1971)

Babinet's principle for elastic waves gives a relationship between the displacement fields due to complementary diffracting screens. The relationship has been derived using the seismic representation theorem as the starting point. The relationship for rigid complementary screens, for stress-free complementary screens, and for mixed rigid–stress-free complementary screens in each case has the same form. In deriving the relationship it is necessary to assume that the "jumps" across the complementary screens are the same as those that would exist when both complementary screens are present. This assumption has been tested in a seismic modeling experiment that utilized both stress-free and rigid screens. The experiments show that the assumption is valid within the accuracy of the experiment and, consequently, that Babinet's principle for elastic waves is valid to the same degree of accuracy.

SUBJECT CLASSIFICATION: 12.10, 12.3; 11.7.

## INTRODUCTION

Babinet's principle is very useful in optical and electromagnetic wave diffraction problems. It relates the diffraction fields set up by complementary plane screens (see, e.g., Born and Wolf[1]). In optics, a complementary screen is a plane screen with opaque areas where the original plane screen had transparent areas and with transparent areas where the original screen had opaque areas. Babinet's principle states that the sum of the fields behind the original screen and the complementary screen is just the field that would exist if there were no screens present. Alternatively, it states that the scattered field from the original screen is just the negative of the scattered field from the complementary screen.

Proofs of Babinet's principle for opaque screens and light waves are given using either the Huygens–Fresnel diffraction theory or Kirchhoff's diffraction theory (see Born and Wolf,[1] p. 381). These are approximate diffraction theories for scalar wave diffraction.

Babinet's principle takes a somewhat different form for electromagnetic waves and perfectly conducting plane screens. Starting from the known transmission and reflection properties of arrays of perfectly conducting dipoles and of arrays of slots cut out of a perfectly conducting plane, Booker[2] shows that there must be an interchange of magnetic fields and electric fields for Babinet's principle to be valid for complementary screens. Alternatively, he shows that a 90° rotation of the complementary screen must be included if there is no interchange of electric and magnetic fields. Booker[2] (p. 621) attributes this to the fact that "if one screen is a perfect conductor of electricity, its complement must be what we may describe theoretically as a perfect conductor of magnetism." Thus, we see that in the vector electromagnetic wave case, there are additional requirements for the complementary screen as compared to the scalar case. A rigorous proof of Babinet's principle for the electromagnetic case (Huxley,[3] pp. 283 ff; see also Born and Wolf,[1] pp. 559 ff) shows that an interchange of magnetic and electric fields is necessary when the original screen and its complement are both perfect conductors of electricity.

To the best of our knowledge, no attempt has been made to derive the equivalent of Babinet's principle for elastic waves. The derivation for Babinet's principle for acoustic waves would follow the derivation for light waves, i.e., the scalar case. In view of the complications introduced in the derivation for vector electromagnetic waves, it is not surprising that the elastic case has not been treated previously. In the following we will use *the seismic representation theorem* as the starting point for the derivation. This is equivalent to using *Kirchhoff's Theorem* as the starting point in the derivation for the scalar case (see Born and Wolf,[1] p. 378).

Babinet's principle for elastic waves would be of value since (1) it would allow us to obtain the solution of the complementary problem from the solution of the original problem without any additional effort, (2) it provides a check of the solutions for problems that are self-complementary (e.g., the problem of a plane wave normally incident on a half-plane), and (3) it adds to our knowledge of the complex phenomena of elastic wave diffraction.

Reprinted from Journal Acoustical Society of America, 53, 525-534 with permission from American Institute of Physics

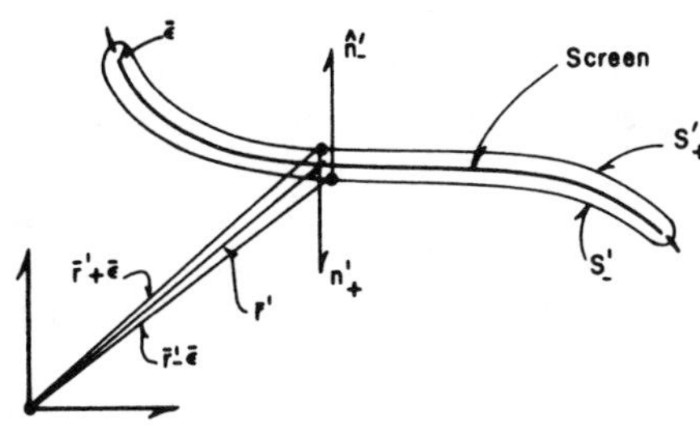


FIG. 1. Diffracting screen.

## I. THEORY

The representation theorem for homogeneous, isotropic elastic media has been given by Knopoff,[4] for homogeneous, anisotropic elastic media, by deHoop,[5] and for anisotropic, inhomogeneous media by Gangi.[6] Using the notation of the Gangi, the seismic representation theorem can be expressed as (Gangi,[6] Eq. 5, p. 2090):

$$\bar{u}(\bar{r},t)=\int_{-\infty}^{\infty}\int_{V'}\bar{\bar{G}}(\bar{r}',t'\,|\,\bar{r},0)\cdot\bar{f}(\bar{r}',\,t-t')dV'dt'$$
$$+\int_{-\infty}^{\infty}\int_{S'}[\bar{\bar{G}}(\bar{r}',t'\,|\,\bar{r},0)\cdot{}^{n}\bar{T}(\bar{r}',\,t-t')$$
$$-\bar{u}(\bar{r}',\,t-t')\cdot{}^{n}\bar{\bar{S}}(\bar{r}',t)]dS'dt,\quad(1)$$

where $\bar{u}(\bar{r},t)=$ the displacement at the point $\bar{r}$ and the time $t$ due to the force distribution $\bar{f}(\bar{r},t)$ in the volume $V'$ and due to the effects of the boundary $S'$,

$$\bar{\bar{G}}(\bar{r}',t'\,|\,\bar{r},0)=\sum_{i=1}^{3}\hat{\imath}\bar{U}(\bar{r}',t'\,|\,\bar{r},0;\,\hat{\imath})$$

is the Green's dyadic, $\bar{U}(\bar{r}',t'\,|\,\bar{r},0;\hat{\imath})$ is the displacement at the point $\bar{r}'$ and the time $t'$ due to a unit delta function point body force at $\bar{r}$ and excited at $t'=0$, i.e., due to $\bar{F}(\bar{r}',t')=\hat{\imath}\delta(\bar{r}'-\bar{r})\delta(t')$ when there is no bounding surface $S'$, ${}^{n}\bar{T}(\bar{r}',\,t-t')$ is the surface stress vector due to the body force distribution $\bar{f}(\bar{r},t)$ acting on the surface $S'$, $\hat{n}=$ the unit normal to the surface $S'$,

$${}^{n}\bar{\bar{S}}(\bar{r}',t')=\sum_{i=1}^{3}{}^{n}\bar{\tau}^{(i)}\hat{\imath},$$

the surface stress dyadic, and

$${}^{n}\bar{\tau}(i)={}^{n}\bar{\tau}(\bar{r}',t'\,|\,\bar{r},0;\,\hat{\imath})$$

is the surface stress vector at the point $\bar{r}'$ and the time $t'$ which would act on a surface whose unit normal is $\hat{n}$ and which is set up by the body force $\bar{F}(\bar{r}',t')=\hat{\imath}\delta(\bar{r}'-\bar{r})\delta(t')$.

The Green's dyadic (or second-order tensor) has the property that it is equal to its transpose if the source position and detection position are interchanged, i.e.,

$$G_{ij}(\bar{r}',t'\,|\,\bar{r},0)\equiv G_{ji}(\bar{r},t'\,|\,\bar{r}',0).\quad(2)$$

This is just a statement of seismic reciprocity (see, e.g., Knopoff and Gangi[7]).

When the force source distribution is $\bar{f}(\bar{r}',\,t-t')=\hat{a}\delta(\bar{r}'-\bar{r}_0)\delta(t'-t)$ and when the surface $S'$ includes infinitesimally thin screens (as well as a spherical surface at infinity), the displacement field is given by the representation theorem as

$$\bar{u}(\bar{r},t)=\bar{U}(\bar{r},t\,|\,\bar{r}_0,0;\,\hat{a})$$
$$+\int_{-\infty}^{\infty}\int_{S_+}[{}^{n}\bar{T}(\bar{r}',\,t-t')]_{-}^{+}\cdot\bar{\bar{G}}(\bar{r},t'\,|\,\bar{r}',0)dS'dt'$$
$$-\int_{-\infty}^{\infty}\int_{S_+'}[\bar{u}(\bar{r}',\,t-t')]_{-}^{+}\cdot{}^{n}\bar{\bar{S}}(\bar{r}',\,t-t')dS'dt',\quad(3)$$

where we have used Eq. 2 in Eq. 1 and where

$$[{}^{n}\bar{T}(\bar{r}',\,t-t')]_{-}^{+}=\lim_{\bar{\epsilon}\to0}[{}^{n}\bar{T}(\bar{r}'+\bar{\epsilon},\,t-t')-{}^{n}\bar{T}(\bar{r}'-\bar{\epsilon},\,t-t')],$$

$$[\bar{u}(\bar{r}',\,t-t')]_{-}^{+}=\lim_{\bar{\epsilon}\to0}[\bar{u}(\bar{r}'+\bar{\epsilon},\,t-t')-\bar{u}(\bar{r}'-\bar{\epsilon},\,t-t')]$$

for $\bar{r}'$ on the thin screen (see Fig. 1). These latter are the "jumps" (or differences) in the stress vector and the displacement across the diffracting screen $S'$.

To derive Babinet's principle, let us assume we have a plane rigid screen defined by $S_1$ in an elastic medium. Then the vector displacement field $\bar{u}_1(\bar{r},t)$ for a delta function source $\bar{f}_a(\bar{r},t)=\hat{a}\delta(\bar{r}-\bar{r}_0)\delta(t)$ located at $\bar{r}_0$, oriented in the $\hat{a}$ direction, and excited at $t=0$ is given by the representation theorem as

$$\bar{u}_1(\bar{r},t)=\bar{u}_0(\bar{r},t)+\int_{-\infty}^{\infty}\int_{S_1{}^+}\bar{\bar{G}}_0(\bar{r}',t'\,|\,\bar{r},0)$$
$$\cdot[{}^{n}\bar{T}_1(\bar{r}',\,t-t')]_{-}^{+}dS'dt',\quad(4)$$

where $\bar{u}_0(\bar{r},t)$ is the displacement field that would be set up by the force $\bar{f}_a(\bar{r},t)$ if the screen $S_1$ were absent.

$$\bar{\bar{G}}_0(\bar{r}',t'\,|\,\bar{r},0)$$

is the Green's dyadic for the elastic medium when $S_1$ is absent and

$$[{}^{n}\bar{T}_1(\bar{r}',\,t-t')]_{-}^{+}$$

is the "jump" or difference in the surface stress vector across $S_1$.

In the same way the displacement field $\bar{u}_2(\bar{r},t)$, in the same elastic medium for the same source $\bar{f}_a(\bar{r},t)$ but for a different rigid screen defined by $S_2$, is given by

$$\bar{u}_2(\bar{r},t)=\bar{u}_0(\bar{r},t)+\int_{-\infty}^{\infty}\int_{S_2^+}\bar{\bar{G}}_0(\bar{r}',t'|\bar{r},0)\cdot[{}^n\bar{T}_2(\bar{r}',t-t')]_-^+ dS'dt', \quad (5)$$

where all the terms are defined as above with the exception that

$$[{}^n\bar{T}_2(\bar{r}',t-t')]_-^+$$

is the "jump" in the surface stress vector across the screen $S_2$.

If we let $\bar{u}_3(\bar{r},t)$ be still another displacement field [in the same elastic medium with the same source $\bar{f}_a(\bar{r},t)$] when a rigid screen defined by $S_3$ is present and if we take the screen $S_3$ to be composed of the two screens defined by $S_1$ and $S_2$, then we find that

$$\bar{u}_1(\bar{r},t)+\bar{u}_2(\bar{r},t)\cong\bar{u}_0(\bar{r},t)+\bar{u}_3(\bar{r},t), \quad (6)$$

*provided*

$$\int_{-\infty}^{\infty}\int_{S_3^+}\bar{\bar{G}}_0(\bar{r}',t'|\bar{r},0)\cdot[{}^n\bar{T}_3(\bar{r}',t-t')]_-^+ dS'dt' \cong\int_{-\infty}^{\infty}\int_{S_3^+}\bar{\bar{G}}_0(\bar{r}',t'|\bar{r},0)\cdot\{[{}^n\bar{T}_1(\bar{r}',t-t')]_-^+ +[{}^n\bar{T}_2(\bar{r}',t-t')]_-^+\}dS'dt'. \quad (7)$$

The assumption made in Eq. 7, which is necessary to make Eq. 6 valid, can also be expressed as

$$[{}^n\bar{T}_3(\bar{r},t)]_-^+\cong[{}^n\bar{T}_1(\bar{r},t)]_-^+ \quad \text{on } S_1, \quad (8a)$$

$$[{}^n\bar{T}_3(\bar{r},t)]_-^+=[{}^n\bar{T}_2(\bar{r},t)]_-^+ \quad \text{on } S_2, \quad (8b)$$

since

$$[{}^n\bar{T}_1(\bar{r},t)]_-^+=0 \quad \text{on } S_2, \quad (9a)$$

$$[{}^n\bar{T}_2(\bar{r},t)]_-^+=0 \quad \text{on } S_1. \quad (9b)$$

The conditions expressed by Eq. 8 are more restrictive than the condition given in Eq. 7.

A close look at the assumptions given by Eqs. 7 or 8 shows that the assumptions are valid for the stress vectors set up by the incident waves and the specularly reflected (and/or converted) waves. However, it has not been possible to show that the assumptions would be valid for the stress vectors set up by the diffracted waves. But the amplitudes of diffracted waves are generally very small and, consequently, their "jumps" across the screens would be expected to be small. The contribution of these "jumps" to the displacement field away from the screen (including the diffracted field) should also be small. Consequently, we can expect that Eq. 6 is valid to a high degree of approximation.

If the screen $S_3$ represents an infinite plane surface, then $S_1$ and $S_2$ are complementary to each other and Eq. 6 becomes a statement of Babinet's principle. We note that if the screens are complementary and we are concerned with the displacement fields on the opposite side of the screens from the source, then Eq. 6 simplifies since $\bar{u}_3(\bar{r},t)$ must be zero in that region. For example, if $S_1$ represents a rigid screen defined by $z=0$, $x\geq0$, and $S_2$ is the complementary screen defined by $z=0$, $x\leq0$; then the displacements in $z>0$ with $\bar{f}_a(\bar{r},t)$ located in $z<0$ will satisfy the condition

$$\bar{u}_1(\bar{r},t)+\bar{u}_2(\bar{r},t)\cong\bar{u}_0(\bar{r},t). \quad (10)$$

Now, since the displacement field $\bar{u}_0(\bar{r},t)$ for the case of no screen will be zero everywhere except on the direct wavefront, Eq. 10 states that the diffraction fields for the complementary screens will be the negative of each other. This is equivalent to the result given in optics for scalar waves and complementary opaque diffracting screens.

The derivation of Babinet's principle for stress-free screens would follow an analogous procedure as for the rigid screens. The results would again be expressed by Eq. 6, but now the displacement fields would be those associated with the stress-free screens. The assumption, equivalent to that of Eq. 7, that would have to be made would be expressed as (analogously to Eqs. 8)

$$[\bar{u}_3(\bar{r},t)]_-^+\cong[\bar{u}_1(\bar{r},t)]_-^+ \quad \text{on } S_1, \quad (11a)$$

$$[\bar{u}_3(\bar{r},t)]_-^+\cong[\bar{u}_2(\bar{r},t)]_-^+ \quad \text{on } S_2. \quad (11b)$$

Since it has been shown in the electromagnetic wave case that the complementary screen should also have different properties (i.e., if one is a perfect conductor of electricity, the other should be a perfect conductor of magnetism), it could be expected that the complementary elastic screens would also have different properties. That is, if the original screen is rigid, the complementary screen may be stress-free (or weak) where the original screen is "transparent." Babinet's principle would once again be expressed by Eq. 6, where now, say, $\bar{u}_1(\bar{r},t)$ would be the displacement field in the presence of the rigid screen, $\bar{u}_2(\bar{r},t)$ would be the displacement field in the presence of the stress-free complementary screen, and $\bar{u}_3(\bar{r},t)$ would be the displacement field when both screens are present. Now the assumptions that must be made or, alternatively, the conditions that must be satisfied are

$$[{}^n\bar{T}_3(\bar{r},t)]_-^+\cong[{}^n\bar{T}_1(\bar{r},t)]_-^+ \quad \text{on } S_1, \quad (12a)$$

$$[\bar{u}_3(\bar{r},t])_-^+\cong[\bar{u}_2(\bar{r},t)]_-^+ \quad \text{on } S_2. \quad (12b)$$

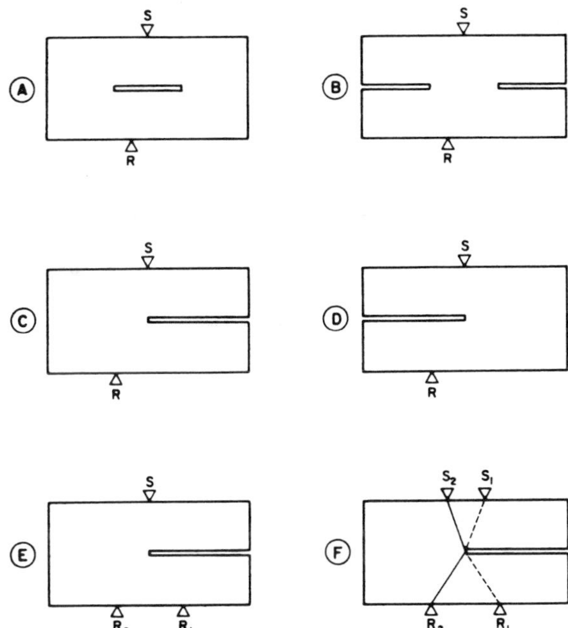


FIG. 2. Some typical complementary screens.

The above conditions can be seen to hold for the incident waves in conjunction with the reflected and converted waves. However, these conditions may not hold for the diffracted fields set up near the screens. But, the diffracted fields away from the screens depend upon the "jumps" in the incident, reflected and converted waves across the diffracting screens. Since the diffracted fields tend to be small relative to the incident, reflected, and refracted fields, it would be expected that the conditions in Eqs. 12 would be satisfied to a high degree of approximation.

The result given by Eq. 6 is more general than Babinet's principle, which requires that $S_3$ be an infinite plane screen. Therefore, Eq. 6 can be used to obtain approximate solutions for more general problems involving screens if the conditions given in Eqs. 8, 11, or 12 hold.

## II. EXPERIMENTAL RESULTS

Elastic wave experiments were performed to test the validity of the assumptions that are expressed by Eqs. 8, 11, and 12 and that were made in deriving Eq. 6. These experiments were performed using polystyrene sheets. It has been demonstrated (Love,[8] pp. 137 ff, 207 ff) that the propagation of elastic waves in thin plates or sheets is equivalent to two-dimensional wave propagation. This is valid for waves with wavelengths greater than the thickness of the plate and for waves that are symmetric in the plate (Ewing *et al.*,[9] pp. 283 ff).

Some configurations of complementary screens in plates are shown in Fig. 2. Each rectangle in this figure represents a polystyrene sheet (80 cm wide by 160 cm long and 0.15 cm thick) in which slots (weak screens) 1.0 mm wide have been made. The source and receiving transducers (poled electrostrictive ceramics) are placed along the edges of the sheet. Propagation in these sheets is equivalent to two-dimensional elastic wave propagation with the displacement and stress fields constant in the direction perpendicular to the plane of the figure. That is, the source $S$ is equivalent to a line source. The source emits relatively short duration pulses (10–20 $\mu$sec) with rise times of the order of 2–5 $\mu$sec.

The finite screen (strip) shown in Fig. 2(a) has as its complementary screen that shown in Fig. 2(b); that is, Fig. 2(a) represents diffraction of two-dimensional elastic pulses by a strip, while its complementary case [in Fig. 2(b)] corresponds to the diffraction of two-dimensional elastic pulses by a slit. The screen shown in Fig. 2(a) (the "strip") may be either weak or rigid; the same is true for the screen shown in Fig. 2(b).

Figures 2(c) and 2(d) show another set of complementary screens, this time for the geometrically simple case of a "half-plane." If the complementary screens in Figs. 2(c) and 2(d) are the same type (i.e., both rigid or both weak), then we see that the complementary case can be obtained by moving the receiving transducer $R$. This case is illustrated in Fig. 2(e), where the scattered fields detected by the receiver $R_1$ should be the negative of the scattered fields detected by the receiver $R_2$ (provided a complementary case is obtained by using the same type screen).

Figure 2(f) shows another instance in which the same elastic medium can be used to obtain the complementary case if the screen and its complement are of the same type, either both rigid or both weak. In this case both the source and receiving transducer must be moved. That is, the scattered field received at $R_1$ with the source at $S_1$ should be the negative of the scattered field received at $R_2$ with the source at $S_2$ if Babinet's principle is to hold for elastic waves.

The configurations shown in Figs. 2(e) and 2(f) have been used to demonstrate Babinet's principle. Polystyrene sheets (density$\cong$1.03 g/cm$^3$, $p$-wave velocity$\cong$1750 m/sec, and elastic impedance$\cong$1.8$\times 10^6$ kg/m$^2$ sec) were used so that it would be possible to obtain reasonable approximations to rigid screens by using screens made of stainless steel (density$\cong$7.8 g/cm$^3$, $p$-wave velocity$\cong$5200 m/sec, and elastic impedance $\cong$40$\times 10^6$ kg/m$^2$ sec). All velocity values were experimentally determined by seismic modelling techniques.[10] Rigid screens were fabricated by bonding a stainless steel strip (0.8 mm thick) in the 1.0-mm slots by means of thin layers (each approximately 0.1 mm thick) of aluminum-filled epoxy.

Figure 3 shows the elastic pulse used in the experiments. The pulses were generated using an electrostrictive ceramic (PZT-4) source transducer backed by a steel rod. The receiver transducer was constructed of PZT-5 ceramic in the same way. The steel backing to

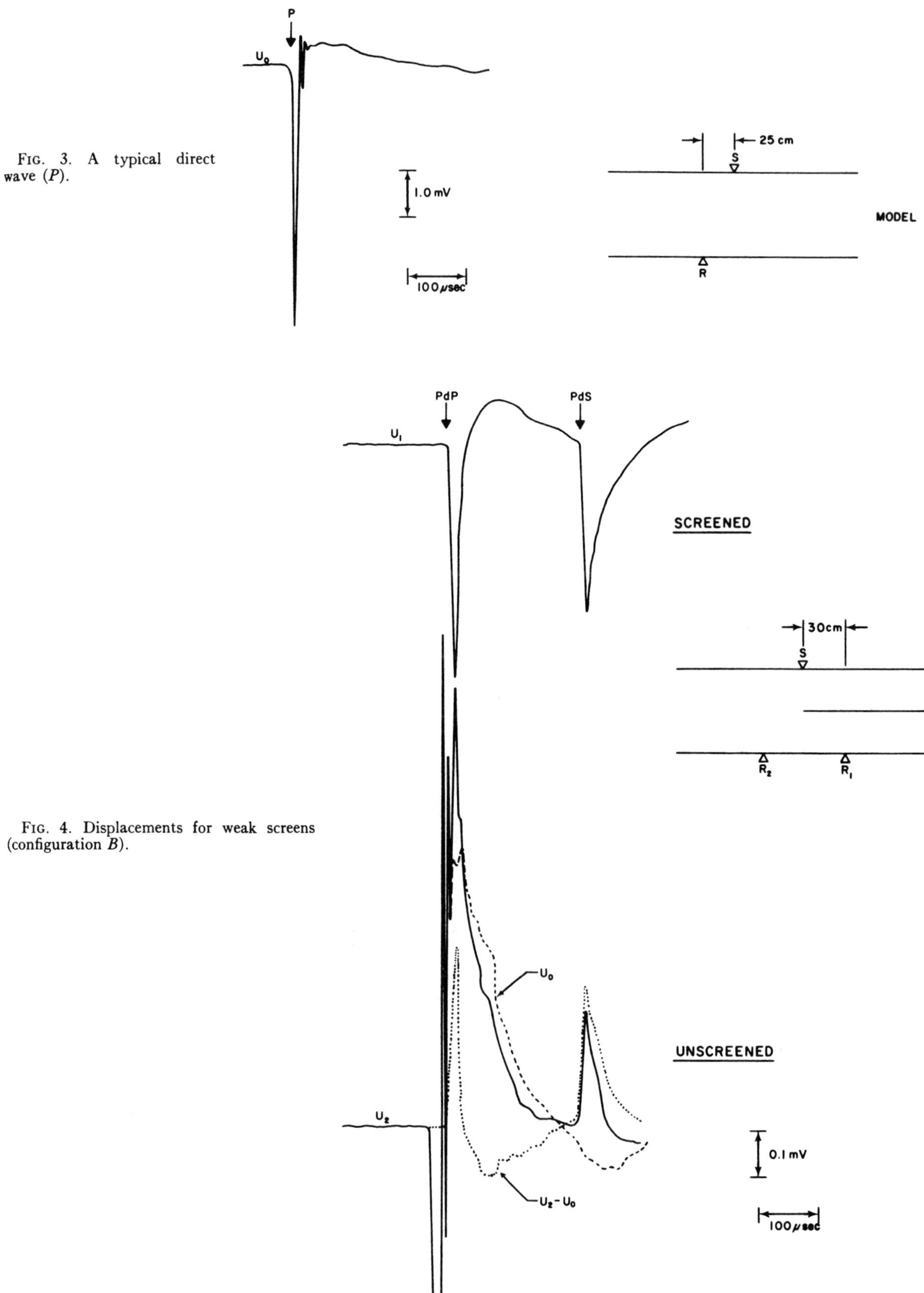


FIG. 3. A typical direct wave ($P$).

FIG. 4. Displacements for weak screens (configuration $B$).

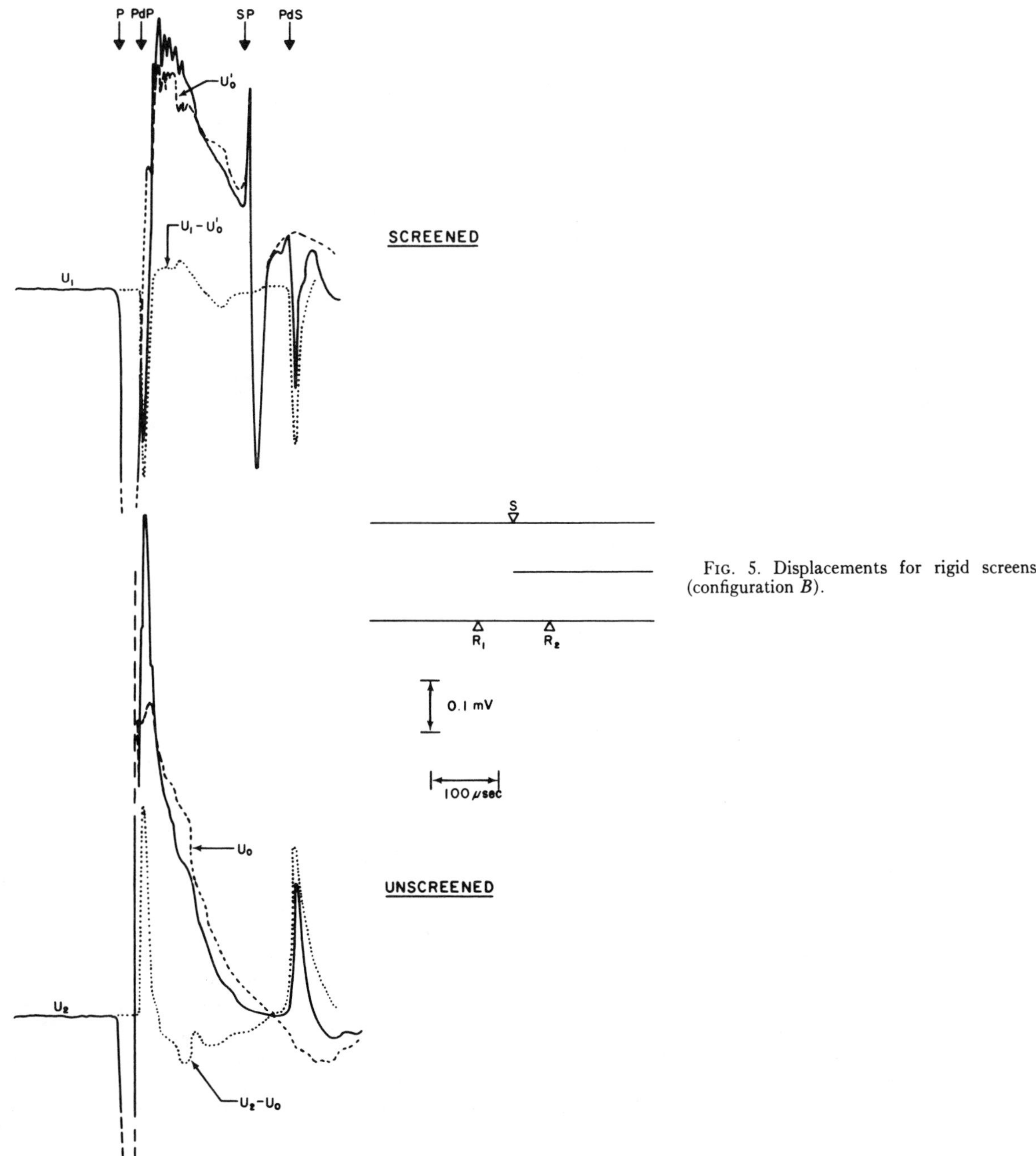


FIG. 5. Displacements for rigid screens (configuration $B$).

the transducers enables us to obtain the broad-band pulse shown in Fig. 3. (For details of the transducer assembly and associated electronics, see Mohanty[10] and Dampney *et al.*[11]) The frequency response of the elastic pulse shown in the figure is relatively flat from about 1 to 350 kHz. This sets the lower limit of the significant $p$-wave wavelengths at approximately 5 mm or approximately five times the thickness of the diffracting screen. The low frequency cutoff was introduced to decrease the amount of audio noise picked up by the transducer. The direct pulse is also used to represent the $\bar{u}_0(\bar{r},t)$ field of Eq. 6; that is, this signal was subtracted from the signals received when the complementary screens were present. Also illustrated in Fig. 3 is the seismic model (i.e., the polystyrene sheet) and the relative positions of the source $S$ and the receiver $R$.

The signals received for the case of weak complementary screens are illustrated in Fig. 4. These results were obtained for the case of configuration $B$, corresponding to the case illustrated in Fig. 2(e).

The top trace shows the signal received when the source transducer is screened by the half-plane; i.e.,

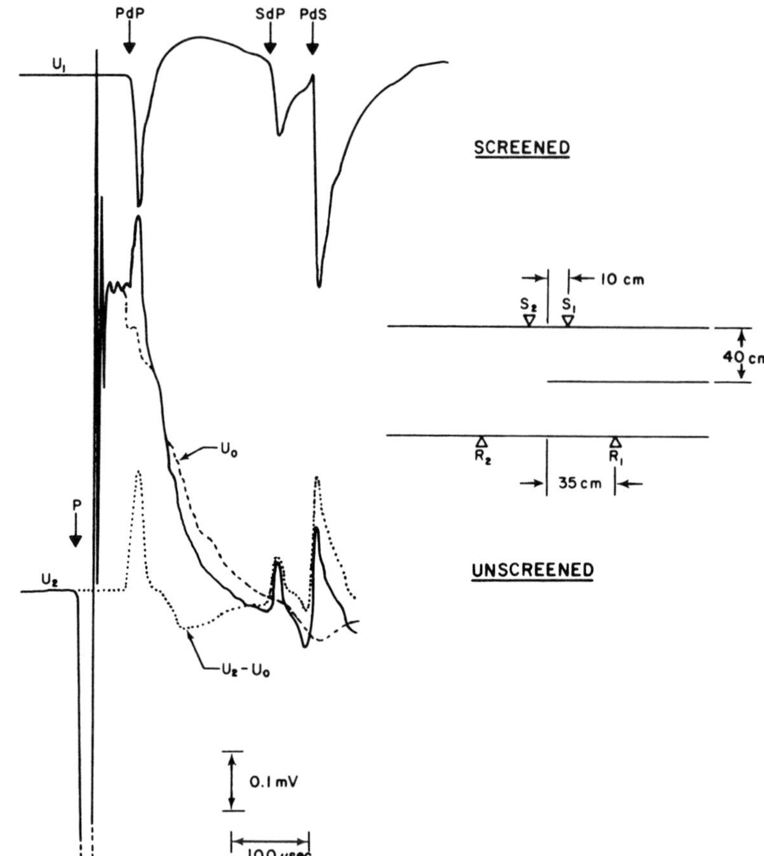


FIG. 6. Displacements for weak screens (configuration $A$).

when the receiving transducer is at the position $R_1$ shown in the inset of Fig. 4. This has been denoted the $\bar{u}_1(\bar{r},t)$ displacement field and consists of the $p$- and $s$-wave diffractions set up by the incident $p$ wave. These have been called the *PdP* and *PdS* waves, respectively.

The solid line in the bottom trace shows the displacement field $\bar{u}_2(\bar{r},t)$ obtained with the source at the same point, but with the receiver at position $R_2$. The bottom trace includes the experimentally determined displacement field $\bar{u}_0(\bar{r},t)$ that is received when there is no screen present (similar to the signal shown in Fig. 3) and it is shown as a dashed line. The dotted line gives the scattered field for this case and it is obtained by subtracting the no-screen displacement field $\bar{u}_0(\bar{r},t)$ from the unscreened displacement field $\bar{u}_2(\bar{r},t)$. Comparing the top trace with the scattered displacement in the bottom trace (dotted line), we see that the two *PdP* (and *PdS*) signals are approximately equal, have equivalent shapes (note *PdS* has more low frequency content than *PdP* in both traces) and are of *opposite sign*. These are the conditions that must be satisfied if Babinet's principle as expressed by Eq. 6 is to hold.

Figure 5 shows the results for the same configuration, this time for a "rigid" diffracting screen. From the top trace of Fig. 5 we can see that the steel strip used to simulate the rigid screen was not an accurate simulation because the direct wave is detected when the receiver is "screened" from the source by the "rigid" half-plane; i.e., when the receiver is in position $R_2$ (see the inset in Fig. 5 showing the seismic model setup). The top trace also contains the converted wave *SP* (and *PS*) which travels as an $s$ wave (a $p$ wave) from the source to the screen and then as a $p$ wave (an $s$ wave) from the screen to the receiver. This signal would not appear either if the screen were a perfectly rigid screen. Consequently, we must subtract these signals from the displacement field $\bar{u}_1(\bar{r},t)$ to obtain the scattered field. Alternatively, the displacement field denoted by $U_0'$ and shown as a dashed line in Fig. 5 can be considered to be the displacement field $\bar{u}_3(\bar{r},t)$ of Eq. 6; that is, the displacement field that would exist if both the semirigid half-plane and its complement were present. The signal due to the scattered field, $U_1-U_0'$, is shown dotted in the top trace of Fig. 5. It is composed of just the diffracted $p$ wave (*PdP*) and the diffracted $s$ wave (*PdS*) set up by the incident $p$ wave at the screen.

The solid line of the bottom trace in Fig. 5 is the signal obtained from the displacement field $\bar{u}_2(\bar{r},t)$ which exists at the receiver location $R_1$; that is, in the unscreened region. The direct signal $U_0$ that would be received if no screen were present is shown as a dashed line on this trace and the scattered field $U_2-U_0$ (obtained by subtracting these two signals) is shown as a dotted line on this trace. Again we see that the

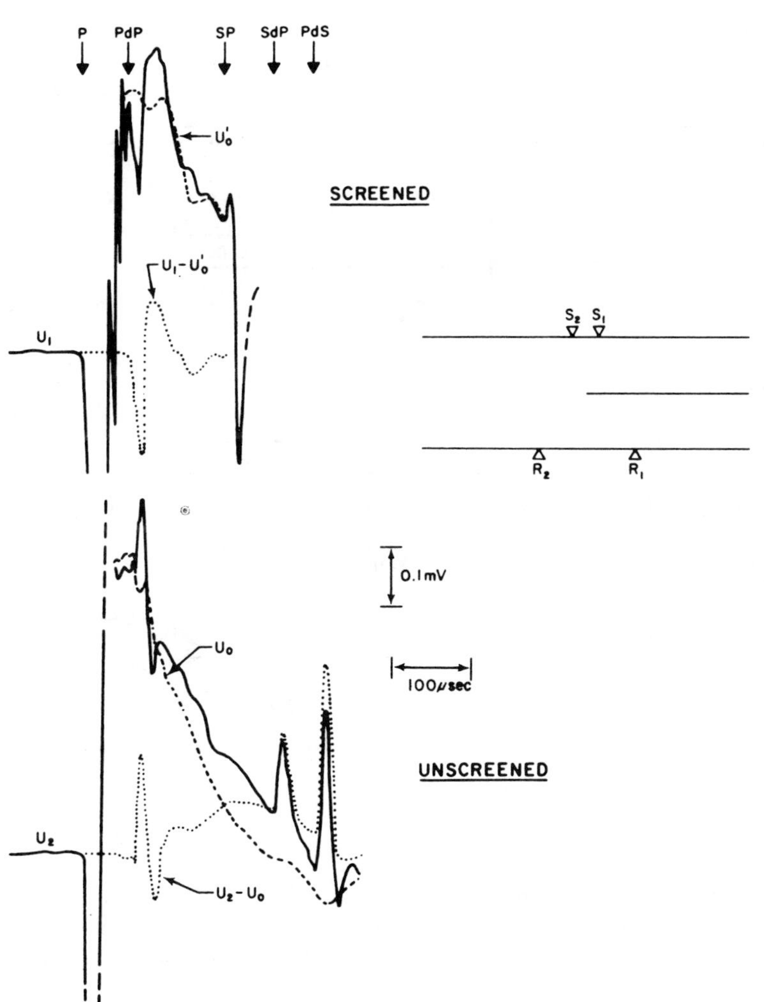


FIG. 7. Displacements for rigid screens (configuration $A$).

scattered fields ($U_1-U_0'$ and $U_2-U_0$) are approximately equal in amplitude, have the same time dependence, and are of opposite polarity. Again, this is the result predicted by Babinet's principle and expressed by Eq. 6.

Figure 6 shows the results obtained when the configuration shown in Fig. 2(f) is used. In this case, both the source and receiving transducers must be moved to obtain the complementary case on the same model (if the complementary screen is of the same type). Figure 6 shows the results for a weak screen. The top trace is the signal received when the source is screened by the receiver (i.e., the source is at $S_1$ and the receiver at $R_1$ in the inset shown in Fig. 6). In this case we can detect the diffracted $p$ wave ($SdP$) due to the $s$ wave incident on the screen. The solid line in the lower trace, is the signal received with the source at $S_2$ and the receiver at $R_2$; i.e., the complementary case for weak screens. Upon subtracting the direct signal $U_0$ (shown dashed in the bottom trace) we obtain the scattered field $U_2-U_0$ (shown dotted in this same trace). Comparing the bottom dotted trace with the upper trace we see that the signals are again approximately equal, have the same time dependence, and are of opposite polarity.

Figure 7 shows the results for the same configuration when the "rigid" screen is used. The slightly different waveshapes shown for the scattered fields (dotted curves in the figure) may be due to errors introduced in subtracting relatively equal signals.

The experimental results shown in Figs. 4–7 are summarized in Table I. The amplitudes of the two diffracted (or scattered) waves $PdP$ and $PdS$ have been tabulated for the different type screens and for the two experimental setups or configurations. The ratios of the amplitudes of these events in the screened to the unscreened cases have been calculated and tabulated in the table. If Babinet's principle holds, the ratios should equal $-1.0$. The ratios for the same type of complementary screen are given on the same line as the amplitudes and to the right of the amplitude values. That is, if a weak screen is the complement to an original weak screen, the ratio of the amplitudes for configura-

TABLE I. Summary of Experimental Results.

| EVENT | NATURE OF SCREEN | CONFIGURATION - A | | | CONFIGURATION - B | | |
|---|---|---|---|---|---|---|---|
| | | AMPLITUDE | | RATIO (S/US) | AMPLITUDE | | RATIO (S/US) |
| | | SCREENED | UNSCREENED | | SCREENED | UNSCREENED | |
| PdP | WEAK | 24.5 ±1.0 | -22.0 ±2.0 | -1.11 ±0.11 | 54.5 ±1.0 | -42.0 ±3.0 | -1.30 ±0.10 |
| | | UNSCREENED | SCREENED | | UNSCREENED | SCREENED | |
| | RIGID | -19.0 ±2.0 | 19.0 ±2.5 | -1.00 ±0.17 | -43.5 ±3.5 | -39.0 ±3.5 | -0.90 ±0.11 |
| | RATIO (S/US) | -1.29 ±0.15 | -0.86 ±0.14 | | -1.25 ±0.10 | -0.93 ±0.11 | |
| PdS | | SCREENED | UNSCREENED | | SCREENED | UNSCREENED | |
| | WEAK | 39.5 ±1.0 | -25.0 ±2.0 | -1.58 ±0.13 | 40.0 ±1.0 | -33.0 ±2.00 | -1.21 ±0.08 |
| | | UNSCREENED | SCREENED | | UNSCREENED | SCREENED | |
| | RIGID | -31.0 ±2.5 | — | — | -33.0 ±2.5 | 32.0 ±3.5 | -0.98 ±0.13 |
| | RATIO (S/US) | -1.27 ±0.11 | — | | -1.21 ±0.10 | -0.97 ±0.12 | |

tion $A$ is $-1.11\pm0.11$ for the event $PdP$. If a rigid screen is the complement of a rigid screen for the same configuration and event, the ratio of the amplitudes is $-1.00\pm0.17$. On the other hand, if we must change the type of screen in going to the complementary case, the ratios given beneath the amplitudes are to be used; for configuration $A$ and event $PdP$ we have the ratios $-1.29\pm0.15$ and $-0.86\pm0.14$.

It can be seen from the table that many of the ratios are quite close to $-1.0$; many are equal to $-1.0$ within approximately 12%, the experimental error. However, there are values that differ from $-1.0$ by more than the experimental error. This is especially true for the case of a weak complementary screen of a weak screen. The ratios for the "rigid" complementary screens of "rigid" screens seem to be more consistently close to $-1.0$.

The ratios for the unscreened rigid complementary screen amplitudes to the screened weak screen amplitudes tend to be consistently high, $-1.21\pm0.10$ to $-1.29\pm0.15$. The ratios for the unscreened weak screen amplitudes to the screened rigid complementary screen amplitudes tend to be consistently low, $-0.86\pm0.14$ to $-0.97\pm0.12$. However, these consistent differences may be due to the fact that the "rigid" screen was not sufficiently rigid. It would be expected that the diffracted waves would increase in amplitude as the screens become more rigid, that is, as the screen approaches an ideal rigid screen more closely. If the diffracted amplitudes in the rigid cases are increased, the consistently large ratios would decrease while the consistently small ratios would increase.

From the results tabulated in Table I we conclude that Babinet's principle is valid to within about 15%, which is of the order of the experimental error. Babinet's principle appears to hold, to within this accuracy, for complementary screens which are of the same type as the original screen, i.e., for both weak or both rigid complementary screens. While the results for the different kind of complementary screen (i.e., a rigid complementary screen of an original weak screen) appear to give consistently high ($\sim1.25$) or consistently low ($\sim0.93$) negative ratios, it is felt that this could be due to the fact that the "rigid" screen was not perfectly rigid. On the basis of the result obtained for the electromagnetic (vector) wave case, we would expect that the different kind of complementary screen would give the best results for Babinet's principle. We are inclined to believe that amplitude ratios closer to $-1.0$ would be obtained (consistent with Babinet's principle for elastic waves) if a more rigid screen were achievable.

In any event, we see that the amplitude ratios do come close to the $-1.0$ value predicted by Babinet's principle (we could not, *a priori*, limit the ratio to a smaller range than $\pm\infty$ without using Babinet's principle). Also, we note that the waveforms of the events for the complementary cases are similar to those predicted by Babinet's principle. Consequently, we conclude that Babinet's principle can be used to relate the diffracted elastic waves from complementary screens to a reasonably high degree of accuracy.

## ACKNOWLEDGMENTS

The research reported here was supported in part by the Air Force Cambridge Research Laboratories, Office of Aerospace Research, under Contract F19628-69-C-0083, and some of the results were given in its final report (No. AFCRL-70-0064).

* Presented at the 51st Annual Meeting of the American Geophysical Union at Washington, D. C., 23 April 1970.

[1] M. Born and E. Wolf, *Principles of Optics* (Pergammon, New York, 1965), 3rd (revised) ed., pp. 378–381, 559–560.

[2] H. G. Booker, "Slot Aerials and . . . (Babinet's Principle)," J. Inst. Elect. Engrs. (London) **95**, 620–626 (1946).

[3] L. G. H. Huxley, *A Survey of the Principles and Practice of Wave Guides* (MacMillan, New York, 1947), pp. 283 ff.

[4] L. Knopoff, "Diffraction of Elastic Waves," J. Acoust. Soc. Amer. **28**, 217 (1956).

[5] A. T. deHoop, "Representation Theorems for the Displacement in an Elastic Solid," Ph.D. Thesis, Technische Hogeschool, Delft, Netherlands, 1958.

[6] A. F. Gangi, "A Derivation of the Seismic Representation Theorem using Seismic Reciprocity," J. Geophys. Res. **75**, 2088–2095 (1970).

[7] L. Knopoff and A. F. Gangi, "Seismic Reciprocity," Geophys. **24**, 681–691 (1959).

[8] A. E. H. Love, *A Treatsie on the Mathematical Theory of Elasticity* (Dover, New York, 1944), 4th ed., pp. 137, 207 ff.

[9] W. M. Ewing, W. S. Jardetsky, and F. Press, *Elastic Waves in Layered Media* (McGraw-Hill, New York, 1957), pp. 283 ff.

[10] B. B. Mohanty, "Model Seismology," M. A. Thesis, University of Toronto, 1965 (unpublished).

[11] C. N. G. Dampney, B. B. Mohanty, and G. F. West, "A Calibrated Model Seismic System," Geophys. **37**, 445–455 (1972).

# An experimental test of *P*-wave anisotropy in stratified media

Patrick J. Melia* and Richard L. Carlson‡

ABSTRACT

In theory, stratified media in which the layers are elastically homogeneous and isotropic approximate transversely isotropic media with an axis of symmetry perpendicular to layering when the seismic wavelength is sufficiently longer than the layer spacing. The phenomenon has apparently been observed in field measurements, and acoustic anisotropy in deep-sea sediments, measured in the laboratory, has been attributed to fine-scale bedding laminations. However, to the best of our knowledge, no rigorous test of the theory has been made. We have made a partial test by making laboratory measurements of compressional-wave velocities parallel and perpendicular to layering in fabricated samples consisting of glass and epoxy. We found no statistically significant difference between observation and theory in this limited test. Further, having used several frequencies, we found that the velocities progressively change from the long-wave values toward those predicted by the time-average relation, as expected. Finally, it has been proposed that the long-wave approximation holds when the ratio of the seismic wavelength to layer thickness ($\lambda/d$) is 10–100. We found that the minimum ratio was highest in the midrange of composition (half glass, half epoxy), even though the samples in that range have the smallest combined layer thickness. This result suggests that the frequency dependence of anisotropy in layered media is a function of the proportions of the materials as well as the thickness of the layers.

## INTRODUCTION

Velocity anisotropy is a matter of some interest to seismologists because of its potential effect on the interpretation of seismic data. Seismic anisotropy has been observed in field measurements (White and Sengbush, 1953; Cholet and Richard, 1954; Hagedoorn, 1954; Urhig and Van Melle, 1955) and may reflect the presence of homogeneous anisotropic rock units or, where the seismic wavelength is long compared to the thickness of beds, stratification of rocks which have different but isotropic elastic properties. Elastic wave anisotropy is also observed in measurements of seismic properties of sedimentary rocks on a laboratory scale. Carlson et al (in press) recently concluded that acoustic anisotropy in carbonate-bearing deep-sea sediments is related to compositional layering (bedding) on a scale up to a millimeter or so.

In theory, a stratified medium consisting of laminae which are themselves homogeneous and isotropic behaves as a homogeneous, transversely isotropic solid with an axis of symmetry normal to the laminations when seismic wavelengths are sufficiently long (Postma, 1955; Krey and Helbig, 1956; White, 1965). The most general treatment is perhaps that of Backus (1962), who considered the case of nonperiodic sequences of anisotropic layers. More recent analyses (Berryman, 1979; Levin, 1979, 1980) have been stimulated in part by the ongoing development of shear-wave reflection methods.

In principle, the properties of a stratified medium can be calculated from the elastic properties of the constituents, but for reasons outlined by Levin (1979), no rigorous field or laboratory test of the theory has been conducted, at least to our knowledge. We have made a preliminary test by measuring compressional-wave velocities parallel and perpendicular to layering in fabricated samples consisting of varying proportions of glass and epoxy.

## SUMMARY OF THEORY

Referred to a coordinate system in which $x_3$ is parallel to the symmetry axis and $x_1$ and $x_2$ lie in the symmetry plane, the elastic properties of a true homogeneous, transversely isotropic medium are described by a stiffness matrix of the form

$$\begin{bmatrix} a & b & f & 0 & 0 & 0 \\ b & a & f & 0 & 0 & 0 \\ f & f & c & 0 & 0 & 0 \\ 0 & 0 & 0 & \ell & 0 & 0 \\ 0 & 0 & 0 & 0 & \ell & 0 \\ 0 & 0 & 0 & 0 & 0 & m \end{bmatrix}. \qquad (1)$$

Backus (1962) showed that in the long-wave approximation for

Texas A&M Geodynamics Research Program contribution no. 37. Manuscript received by the Editor June 13, 1983; revised manuscript received September 21, 1983.
*Formerly Texas A&M University, College Station, TX; presently ARCO International Oil and Gas Co., Los Angeles, CA 90017.
‡Department of Geophysics and Geodynamics Research Program, Texas A&M University, College Station, TX 77843.


Reprinted from Geophysics, 49, 374-378

a stratified medium in which the layers are anisotropic, the corresponding effective elastic coefficients, indicated by capital letters, are

$$\begin{aligned} A &= \langle a - f^2c^{-1}\rangle + \langle c^{-1}\rangle^{-1}\langle fc^{-1}\rangle^2, \\ B &= \langle b - f^2c^{-1}\rangle + \langle c^{-1}\rangle^{-1}\langle fc^{-1}\rangle^2, \\ C &= \langle c^{-1}\rangle^{-1}, \\ F &= \langle c^{-1}\rangle^{-1}\langle fc^{-1}\rangle, \\ L &= \langle \ell^{-1}\rangle^{-1}, \end{aligned} \tag{2}$$

and

$$M = \langle m\rangle.$$

The brackets $\langle\ \rangle$ indicate averages of the properties $(a, b, c \ldots)$ of the anisotropic layers, weighted according to their proportions in the medium.

Backus (1962) further noted that equations (2) also apply to cases in which the layers are isotropic, but the number of coefficients required to describe the properties of the layers is reduced from six to two because

$$a = c = \lambda + 2\mu, \quad b = f = \lambda, \quad \text{and} \quad \ell = m = \mu, \tag{3}$$

where $\lambda$ and $\mu$ are the Lamé parameters. If the stratified medium consists of two materials in relative proportions $P_1$ and $P_2$ $(P_1 + P_2 = 1)$, the weighted average (effective) elastic coefficients are given by, for example,

$$M = P_1 m_1 + P_2 m_2 \quad \text{and} \quad L = \left\{\frac{P_1}{\ell_1} + \frac{P_2}{\ell_2}\right\}^{-1}. \tag{4}$$

Noting that $a = \rho\alpha^2$, $\ell = \rho\beta^2$, and $f = \rho(\alpha^2 - \beta^2)$, where $\rho$, $\alpha$, and $\beta$ are the density, compressional-wave, and shear-wave velocity of the isotropic layers, Levin (1979) gave the effective elastic coefficients $(A, B, C, \ldots)$ in terms of those parameters explicitly as:

$$\begin{aligned} A &= \langle 4\rho\beta^2[1 - (\beta^2/\alpha^2)]\rangle + \langle 1 - (2\beta^2/\alpha^2)\rangle^2/\langle(\rho\alpha^2)^{-1}\rangle, \\ C &= \langle(\rho\alpha^2)^{-1}\rangle^{-1}, \\ F &= \langle 1 - (2\beta^2/\alpha^2)\rangle/\langle(\rho\alpha^2)^{-1}\rangle, \\ L &= \langle(\rho\beta^2)^{-1}\rangle^{-1}, \end{aligned} \tag{5}$$

and

$$M = \langle\rho\beta^2\rangle.$$

The density of the layered medium is given by

$$\rho = P_1\rho_1 + P_2\rho_2, \tag{6}$$

and the velocities of shear and compressional waves propagating parallel and perpendicular to the layers can be calculated from the density and effective elastic moduli $A$, $C$, $L$ and $M$:

$$\begin{aligned} V_{\mathrm{SH},h} &= (M/\rho)^{1/2}, \\ V_{\mathrm{SH},v} &= V_{\mathrm{SV},h} = V_{\mathrm{SV},v} = (L/\rho)^{1/2}, \\ V_{\mathrm{P},h} &= (A/\rho)^{1/2}, \end{aligned} \tag{7}$$

and

$$V_{\mathrm{P},v} = (C/\rho)^{1/2}.$$

Lower case letters $v$, $h$ refer to propagation perpendicular to layering (vertical) and parallel to layering (horizontal), respectively. Similarly, SH and SV refer to shear waves polarized parallel and perpendicular to the layers.

In summary, the velocities of long-wavelength shear and compressional waves propagating parallel and perpendicular to layering in a stratified medium consisting of two isotropic materials can, in theory, be calculated from the densities and elastic-wave velocities of the isotropic layers and their proportions in the composite solid. In our preliminary experimental test of the theory, we have considered only compressional-wave propagation in the vertical and horizontal directions, $V_{\mathrm{P},v}$ and $V_{\mathrm{P},h}$.

## PROCEDURES

The first problem was that of finding appropriate materials and methods of fabrication for measurements on a laboratory scale. We sought a simple and reliable means of fabricating samples from isotropic media in verifiable proportions. We also wanted the layers to have markedly different properties so the variations of velocity and density would be well above the levels of uncertainty in our measurements. We chose to use glass and epoxy.

### Sample preparation

Glass sheets for our experiment are readily available in the form of microscope slides having a nominal thickness of 1.0 mm and cover slips, with a nominal thickness of 0.15 mm. We would have preferred using only cover slips for reasons which will be discussed later, but found it necessary to use the slides as well in order to cover the full range of proportions of glass and epoxy in our samples.

The samples were prepared by interleaving the rectangular glass plates with spacers. Spacer-widths were used to vary the proportions of glass and epoxy. The edges of the glass plates were then glued in place, and the spacers were removed, leaving an open framework of glass. For casting, the glass stacks were placed in cups with their open edges down (plates vertical). The mixed epoxy was degassed in a vacuum, and poured into the molds in such a way that the glass frameworks filled from the bottom up, thereby preventing the entrapment of air between the glass plates. When the eopxy had set, the glued edges were trimmed, and the surfaces were ground and lapped smooth and parallel. The finished samples are 1 to 2 cm thick and 2 to 3 cm on a side. The arrangement of the glass and epoxy layers is periodic, and because the thickness of the glass is fixed, the thickness of one cycle in the composition ($d = 1$ epoxy layer plus 1 glass layer) varies with the proportion of glass $P_g$. In the samples fabricated from cover slips, $d$ ranges from 0.27 to 1.76 mm. The samples made from microscope slides have a combined layer thickness ranging from 1.03 to 2.74 mm.

### Velocity and density measurements

The bulk densities of the samples were determined from their weights and volumes calculated from their dimensions, and are accurate to 0.01 g/cm$^3$.

The use of spacers to fix the proportions of glass and epoxy in the samples is not very precise, and the sample densities provide a ready means of making more accurate estimates. Given the densities of the sample $\rho_s$, the glass $\rho_g$, and the epoxy $\rho_e$, the volume fraction of glass in the sample $P_g$ can be calculated

$$P_g = (\rho_s - \rho_e)/(\rho_g - \rho_e). \tag{8}$$

The accuracy of glass proportions is 0.01 fractional units.

Table 1. Theoretical and 0.2 MHz observed data.

| | | | | Theoretical | | | Observed | | |
|---|---|---|---|---|---|---|---|---|---|
| No. | $P_g$ | $d$ (mm) | Density (g/cm$^3$) | $V_{P,v}$ (km/sec) | $V_{P,h}$ (km/sec) | $A$ (%) | $V_{P,v}$ (km/sec) | $V_{P,h}$ (km/sec) | $A$ (%) |
| 1 | 0.00 | — | 1.12 | 2.534 | 2.534 | 0.00 | 2.534 | 2.534 | 0.00 |
| 2 | .085 | 1.76 | 1.24 | 2.510 | 3.194 | 23.98 | 2.509 | 3.136 | 22.21 |
| 3 | .172 | 0.87 | 1.36 | 2.506 | 3.660 | 37.44 | 2.481 | 3.774 | 41.11 |
| 4 | .205 | 0.73 | 1.40 | 2.510 | 3.804 | 40.99 | 2.438 | 3.856 | 43.65 |
| 5 | .211 | 0.71 | 1.41 | 2.511 | 3.828 | 41.57 | 2.493 | 3.718 | 39.46 |
| 6 | .258 | 0.58 | 1.48 | 2.522 | 4.007 | 45.47 | 2.524 | 3.989 | 45.01 |
| 7 | .331 | 0.45 | 1.58 | 2.553 | 4.243 | 49.74 | 2.545 | 4.314 | 51.58 |
| 8 | .365 | 2.74 | 1.62 | 2.573 | 4.340 | 51.11 | 2.636 | 4.528 | 52.81 |
| 9 | .480 | 0.31 | 1.79 | 2.673 | 4.620 | 53.38 | 2.678 | 4.678 | 54.37 |
| 10 | .495 | 0.30 | 1.80 | 2.690 | 4.652 | 53.43 | 2.702 | 4.789 | 55.72 |
| 11 | .548 | 0.27 | 1.87 | 2.760 | 4.758 | 53.15 | 2.758 | 4.881 | 55.58 |
| 12 | .593 | 1.69 | 1.93 | 2.833 | 4.842 | 52.36 | 2.896 | 5.102 | 55.17 |
| 13 | .710 | 1.41 | 2.09 | 3.101 | 5.038 | 47.61 | 3.198 | 5.128 | 46.36 |
| 14 | .882 | 1.13 | 2.33 | 3.918 | 5.302 | 30.01 | 3.721 | 5.311 | 35.20 |
| 15 | .973 | 1.03 | 2.46 | 4.976 | 5.479 | 9.63 | 5.178 | 5.602 | 7.88 |
| 16 | 1.00 | — | 2.50 | 5.560 | 5.560 | 0.00 | 5.560 | 5.560 | 0.00 |

$A$ is anisotropy, $P_g$ is fraction of glass, and $d$ is combined layer thickness (1 layer of glass + 1 layer of epoxy).

For the velocity measurements, we constructed a velocimeter with one transducer fixed and the other mounted on a gas-actuated piston. A pump from a blood-pressure cuff was used to control the axial stress on the samples during the velocity measurements. Although we noticed no dependence of velocity on the stress in the range 0–40 psi, all measurements were made under an axial stress of 24 psi. We used 0.5, 1, and 2 MHz, PZT-5A tranducers, and velocities were measured with a mercury delay line by the pulse-transmission method as described by Birch (1960).

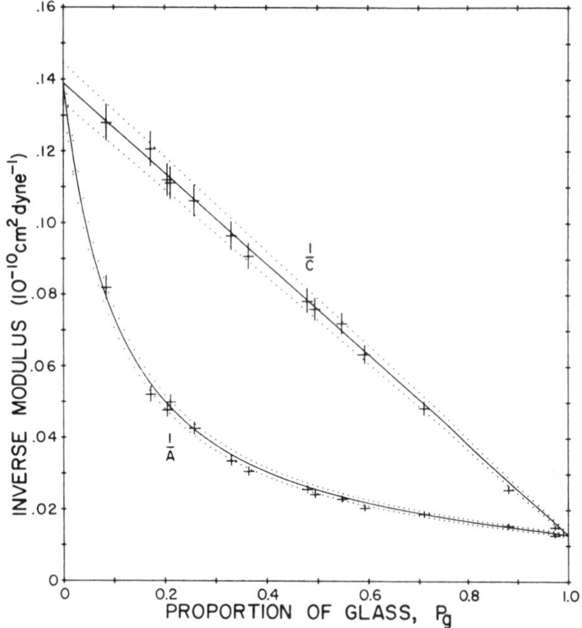


FIG. 1. Inverse moduli versus proportion of glass: $1/C$ (top) and $1/A$ (bottom). Solid lines are theoretical values, dotted lines are uncertainties in theoretical values. The uncertainty in $1/A$ is truncated at $P_g = 0.5$ to avoid obscuring observed values.

A particularly difficult problem is that the long-wave approximation is valid only when the thickness of the laminations is small compared to the acoustic wavelength. As Berryman (1979) pointed out, just how small it must be is not clear. We measured vertical velocities at several frequencies, and found that by using 0.5 MHz transducers, and passing the output signal through a Krohn-Hite filter with low- and high-pass settings of 0.1 and 0.3 MHz (a nominal center frequency of 0.2 MHz), we obtained reasonable agreement between the theoretical and observed values. We make this point in our description of the method because the low-frequency output signal has a relatively long rise time, and the accuracy of the measured velocities is thus affected. Whereas the pulse transmission technique is usually accurate to 1 percent or less, we found that the accuracy of the compressional-wave velocities is about 2 percent, and that of the shear-wave velocities is nearer 3 percent.

## RESULTS

### Glass and epoxy properties

Because the theoretical velocities are calculated from the properties of the isotropic end members, the properties of the glass and epoxy are quite important. The properties of the epoxy were measured using a machined cube of the material about 4 cm on a side. We found the epoxy to be homogeneous and isotropic with $\rho_e = 1.12 \pm .01$ g/cm$^3$, $\alpha_e = 2.53 \pm .05$ km/sec, and $\beta_e = 1.20 \pm .036$ km/sec. The properties of the glass were more difficult to determine because we were unable to obtain a solid block of either the cover slip or slide material, and measurements made on the cover slips are unreliable owing to their shape and size. We estimated the velocities in the slides by making measurements through stacks of slides with a film of water between the plates, but must point out that the water content, though minimal, has a significant effect on the measured velocity. Thus, we believe that the measured glass velocities may be slightly low. The density of the cover slips is 2.53 g/cm$^3$, while that of the slides is 2.49 g/cm$^3$; we used the average as a theoretical value. The properties of the glass are

$\rho_g = 2.51 \pm .01$ g/cm³, $\alpha_g = 5.56 \pm .11$ km/sec, and $\beta_g = 3.20 \pm .10$ km/sec. The properties of the laminated samples and their theoretical velocities calculated from equations (5) and (7) are summarized in Table 1.

### Theoretical versus observed properties

There are a number of ways of comparing the theoretical and observed properties of the samples. We note that the theoretical velocities are actually calculated from averages of the isotropic elastic constants [equations (2) and (7)]. The "observed" values of the moduli can be calculated from the measured densities and velocities of the samples:

$$C = \rho_s V_{P,v} \quad \text{and} \quad A = \rho_s V_{P,h}. \tag{10}$$

While the theoretical expression for $A$ is not simple, the inverse of $C$ is linear in $P_g$:

$$\frac{1}{C} = P_g\left\{\frac{1}{C_g} - \frac{1}{C_e}\right\} + \frac{1}{C_e}, \tag{11}$$

where $C_g = \rho_g \alpha_g^2$ and $C_e = \rho_e \alpha_e^2$. Because this linear equation lends itself to rigorous statistical analysis, we first consider the observed and predicted inverse moduli $1/C$ and $1/A$ in relation to the proportion of glass.

The theoretical and observed inverse moduli and their respective uncertainties are illustrated in Figure 1. The error in $1/C$ is about 4 percent for all values of $P_g$. That there is excellent agreement between the theoretical and observed values is evident. The theoretical linear relation between $1/C$ and $P_g$ is

$$1/C = (-.126 \pm .006)P_g + (.139 \pm .006). \tag{12}$$

The best fitting linear relation found by a least-squares fit to the 14 data points is

$$1/C = (-.128 \pm .002)P_g + (.139 \pm .001) \tag{13}$$

with $r^2 = .998$ and an rms error of .0014. Although we have not attempted to make a nonlinear least-squares fit of $1/A$ on $P_g$, it is clear that the measured and calculated properties of the samples (Figure 1) are also statistically indistinguishable. Finally, the best-fitting value of $1/C$ at $P_g = 1.0$ ($0.011 \pm .001$) yields a compressional-wave velocity in the glass of $5.97 \pm 0.7$ km/sec, suggesting that the measured value is, in fact, too low. Fortunately, the inverse modulus is relatively insensitive to the error in the velocity, so the theoretical values are little affected.

The variations of theoretical and observed velocities are shown in Figure 2. Figure 3 illustrates the variation of compressional-wave anisotropy with glass content. The anisotropy is given by

$$A(\%) = 200(V_{P,h} - V_{P,v})/(V_{P,h} + V_{P,v}). \tag{14}$$

Again we see that, given the uncertainties in the observed and predicted values, there is good agreement. We find no statistical basis for challenging the theory as developed by Postma (1955) and Backus (1962), among others. There is, however, some systematic error in the horizontal velocities and the anisotropy, particularly in the range $0.3 < P_g < 0.6$. There are several possible causes of this misfit. As noted previously, we believe that the elastic-wave velocities we have used for the glass component are too low. Clearly, a higher glass velocity would produce a closer correspondence between the calculated horizontal velocities and those observed (Figure 2). We also found that using $\alpha_g = 5.97$ km/sec yields a near-perfect fit to the anisotropy data (Figure 3). We have not used that value because it derives from

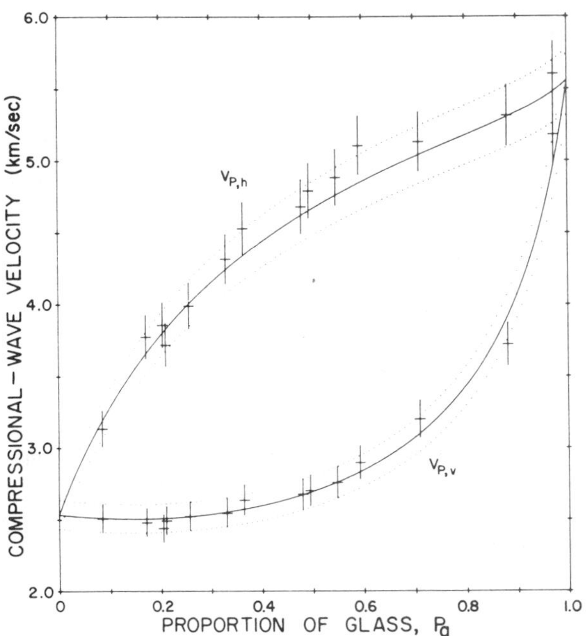


FIG. 2. Theoretical and observed horizontal (upper curve) and vertical (lower curve) velocities versus proportion of glass. Solid and dotted lines as in Figure 2.

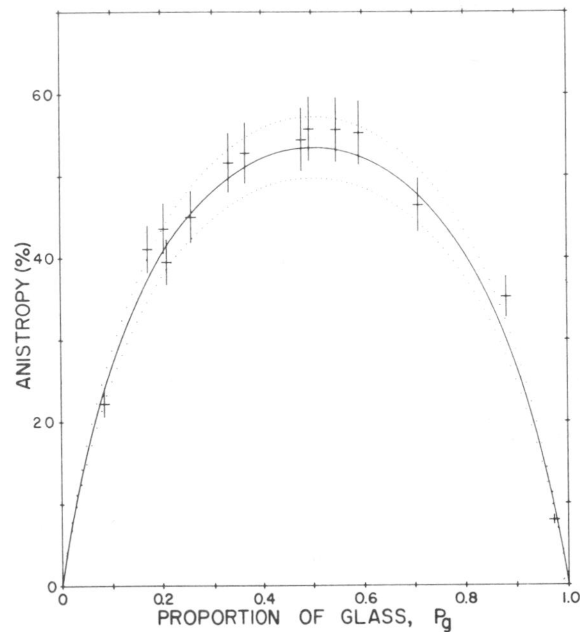


FIG. 3. Theoretical and observed compressional-wave anisotropy versus fraction of glass. Solid and dotted lines as in Figure 2.

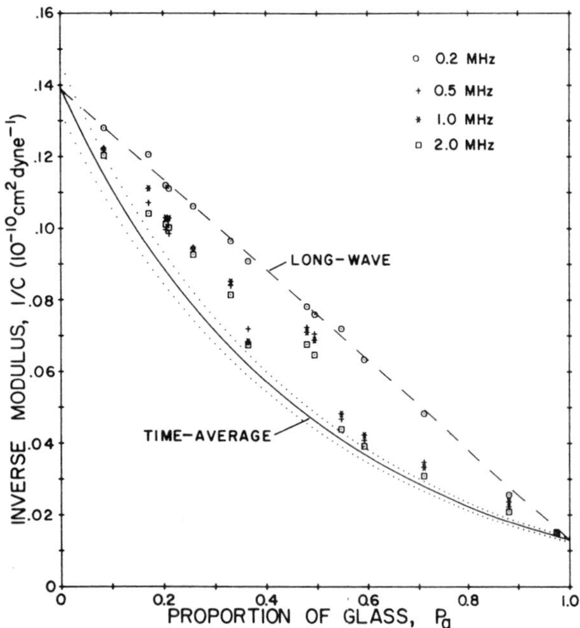


FIG. 4. $1/C$ calculated according to the long-wave approximation (upper) and time-average relation (lower), showing that measured values lie between the extremes.

the theory we are seeking to test. Further, we have assumed that the properties of the glass slides and cover slips are the same; their properties may, in fact, be slightly different. Finally, we were unable to establish that the glass layers are isotropic. A slight degree of anisotropy of the glass layers would also explain the slight misfit between the theoretical and observed velocities.

## Frequency dependence

An interesting problem related to the propagation of seismic waves in the vertical direction is the frequency dependence of velocity. Here we have been concerned primarily with the case in which the seismic wavelength is long in comparison to the bed thickness. When the wavelength is much smaller than the thickness of the laminations, ray theory applies, and the velocity of compressional-wave propagation perpendicular to the layers is given by the time-average relation

$$\frac{1}{V_{P,v}} = P_g\left\{\frac{1}{\alpha_g} - \frac{1}{\alpha_e}\right\} + \frac{1}{\alpha_e}. \tag{15}$$

Time-average velocities are higher than the long-wave velocities. As mentioned previously, we measured velocities normal to the laminations in our fabricated samples at frequencies of 0.2, 0.5, 1.0, and 2.0 MHz; the seismic wavelengths thus range over an order of magnitude. We were not able to examine this effect in detail because of limitations of our equipment, but we did find a progressive change from one limiting condition toward the other as illustrated in terms of the theoretical and observed inverse moduli ($1/C$) in Figure 4. All of the data points lie in the envelope between the time-average values and long-wave values, and the transition appears to take place over a change in wavelength of 1 or 2 orders of magnitude. This result suggests that stratified sequences may have a dispersive effect on seismic signals, which would decrease their resolving power, but might be useful for assessing in-situ long-wave anistotropy, as noted previously by Backus (1962).

A final point worth noting relates to the minimum ratio of seismic wavelength to layer spacing ($\lambda/d$) for which the long-wave approximation holds. As Berryman (1979) and others have pointed out, this question is unresolved, but it would appear that $\lambda$ must be 1 or 2 orders of magnitude greater than $d$. Although our experiment was not designed to examine the problem in detail, our results suggest that the relationship is not simple. In our sample having $P_g = .085$, $d = 1.76$ mm, and $\lambda/d = 7.3$, we found good agreement between the theoretical and observed velocities. On the other hand, the long-wave approximation does not apply to the case of $P_g = .495$, $d = .15$ mm, and $\lambda/d = 18$. In fact, our admittedly incomplete data set suggests that the minimum $\lambda/d$ for which the long-wave approximation applies reaches a maximum for $P_g = .5$, even though $d$ for our samples is minimum for this composition. This tentative result implies that the frequency dependence of compressional-wave velocities in stratified media depends upon the proportions of the layers as well as the layer spacing. The problem merits further theoretical and experimental investigation.

## ACKNOWLEDGMENT

This research was supported by Office of Naval Research Contract N-00014-80-C-0013.

## REFERENCES

Backus, G. E., 1962, Long-wave elastic anisotropy produced by horizontal layering: J. Geophys. Res., v. 67, p. 4427–4440.

Berryman, J. G., 1979, Long-wave anisotropy in transversely isotropic media: Geophysics, v. 44, p. 898–919.

Birch, F., 1960, The velocity of compressional waves in rocks to 10 kilobars, 1: J. Geophys. Res., v. 65, p. 1083–1102.

Carlson, R. L., Schaftenaar, C. H., and Moore, R. P., 1983, Causes of compressional-wave anisotropy in calcareous sediments from the Rio Grande Rise: Initial reports of the Deep Sea Drilling Project, v. 72.

Cholet, J., and Richard, H., 1954, A test on elastic anisotropy measurements at Berraine (North Sahara): Geophys. Prosp., v. 2, p. 232–246.

Hagedoorn, J. G., 1954, A practical example of an anisotropic velocity layer: Geophys. Prosp., v. 2, p. 52–60.

Krey, Th., and Helbig, K., 1956, A theorem concerning anisotropy of stratified media and its significance for reflection seismics: Geophys. Prosp., v. 4, p. 294–302.

Levin, F. K., 1979, Seismic velocities in transversely isotropic media: Geophysics, v. 44, p. 918–936.

——— 1980, Seismic velocities in transversely isotropic media, II: Geophysics, v. 45, p. 3–17.

Postma, G. W., 1955, Wave propagation in a stratified medium: Geophysics, v. 20, p. 780–806.

Urhig, L. F., and Van Melle, F. A., 1955, Velocity anisotropy in stratified media: Geophysics, v. 20, p. 744–779.

White, J. E., 1965, Seismic waves: radiation, transmission, and attenuation: New York, McGraw-Hill Book Co., Inc.

White, J. E., and Sengbush, R. L., 1953, Velocity measurements in near-surface formations: Geophysics, v. 18, p. 54–69.

# Characterization of Anisotropic Elastic Wave Behavior in Media with Parallel Fractures

SN2.2

*Chaur-Jian Hsu* and Michael Schoenberg, Schlumberger-Doll Research*

**Summary**

Ultrasonic experiments were performed to investigate the anisotropic properties of an isotropic background medium with a system of closely spaced parallel fractures. Lucite plates with roughened surfaces were pressed together to simulate a fractured medium. Wavelengths used were about ten or more times the thickness of an individual plate. Velocities were measured for compressional and shear waves both normal to and parallel to the 'fractures', and for quasi-compressional and quasi-shear waves at a variety of oblique angles. Results agree very well with a long wavelength linear slip model which postulates additional normal and tangential fracture compliances in addition to the Lame constants of the background material. Experiments were also conducted with partial or all of the fractures saturated with honey; these measured normal and tangential fracture compliances of various saturation are consistent with a simple mixing law derived based on the linear slip model. The fracture compliances of honey-filled fractures are compared with predictions based on relevant material properties and fracture thickness.

**Introduction**

Most rocks are elastically anisotropic. Aligned fractures are recognized as likely a factor contributing to the anisotropy. Aligned fractures may also relate to the producibility of a reservoir. To aid in the understanding of fracture- induced anisotropy, laboratory experiments have been conducted on a simulated fractured medium (with axisymmetric equally spaced 'fractures' in an otherwise homogeneous isotropic medium). The results are compared with a long wavelength linear slip model.

**Experiment setup**

The experiment setup is shown in Figure 1. The anisotropic medium is made of a stack of 1/32-inch-thick lucite plates. The surfaces of the plates are roughened by sand blasting in order to establish asperity coupling between the neighboring plates. Good and consistent coupling between transducers and the sample is achieved by having the sides of the stack machined smooth, and by having the transducers spring loaded against the sample. Unless otherwise stated, the fractures are air filled (dry coupled). With a hydraulic press, selected normal stress is applied to the stack to control the coupling between plates. The measured elastic constants are therefore functions of the applied stress.

**Measurements**

When the wavelengths are much greater than the spacing between fractures, this medium can be described as transversely isotropic (TI), with the axis of symmetry (defined as axis 3) perpendicular to the fracture planes. A TI medium has five independent elastic moduli: $C_{11}$, $C_{33}$, $C_{44}$, $C_{66}$, and $C_{13}$.

With a pair of compressional ultrasonic transducers in contact with the sample, waveforms were recorded and phase velocities were measured as pulses traversed the sample in two directions: 1) normal to the fractures, this gives an estimate of $C_{33}/\rho$; 2) parallel to the fractures, this gives an estimate of $C_{11}/\rho$. In Figure 2, the data are shown versus the applied normal stress up to 1,400 psi.

With a pair of shear transducers in contact with the sample, phase velocities were measured in three modes: 1) shear waves propagate normal to the fractures, giving an estimate of $C_{44}/\rho$; 2) shear waves propagate parallel to the fractures and polarized normal to the fractures (SV), giving an estimate of $C_{44}/\rho$; 3) shear waves propagate parallel to the fractures and polarized parallel to the fractures (SH), giving an estimate of $C_{66}/\rho$. Figure 3 shows the velocity data.

With two specially made point transducers and wave paths oblique to the fractures, group velocities of quasi-P waves were measured for selected angles between 50 and 90 degrees with respect to the axis 3. The parameter $C_{13}/\rho$ is then obtained from least square fitting of the quasi-P group velocity data using the already measured values of $C_{11}/\rho$, $C_{33}/\rho$, and $C_{44}/\rho$.

In order to minimize the biases due to diffraction, couplant and instruments, all of the measurements were conducted in a differential manner with reference to the waveforms through a solid block of known speed and known path length. The center frequency of the wavelets in the measurements is around 200 kHz, at which there are about 20 plates per compressional wavelength and 10 plates per shear wavelength.

**Discussion**

Several observations can be made from the data:

- Except for the SH wave propagating parallel to the fractures, there is a trend that the higher the stress the higher the velocity;
- For a P-wave traveling parallel to the fractures, the velocity approaches the plate extensional velocity at low normal stress and it approaches the intrinsic lucite P-wave velocity at high stress;
- The P-wave traveling normal to the fractures is slower than that parallel to the fractures. The velocity difference between the two directions of propagation is about 20 percent at 300 psi and 7 percent at 1,400 psi;
- The SH wave traveling parallel to the fractures has the same velocity as the intrinsic lucite shear, independent of the applied stress;

Reprinted from the 60th Annual International Meeting, Society of Exploration Geophysicists, 1990 Expanded Abstracts, 1410-1412

- The S-wave traveling perpendicular to the fractures has the same velocity as the SV wave traveling parallel to the fractures;
- The SV and SH waves traveling parallel to the fractures have different velocities: about 10 percent at 300 psi and 3 percent at 1,400 psi.

Schoenberg and Douma (1988) modeled the fractures as linear slip interfaces; at low frequencies, the five independent elastic moduli may then be expressed in terms of four basic parameters of the fractured medium: the two isotropic background parameters, $\lambda$ and $\mu$ and the dimensionless normal and transverse fracture compliances, $E_N$ and $E_T$. The relationship between these four parameters and the five TI elastic moduli can be expressed as:

$$C_{66} = \mu, \quad C_{44} = \frac{\mu}{1+E_T},$$

$$C_{33} = \frac{\lambda+2\mu}{1+E_N}, \quad C_{13} = \frac{\lambda}{1+E_N},$$

$$C_{11} = (\lambda+2\mu)\left[1 - \left(\frac{\lambda}{\lambda+2\mu}\right)^2 \frac{E_N}{1+E_N}\right].$$

Therefore, from these equations and from the measured TI elastic moduli, the values $\lambda/\rho$, $\mu/\rho$, $E_N$, and $E_T$ can be deduced. The values of $\lambda/\rho$ and $\mu/\rho$ so obtained are in excellent agreement with those obtained from independent P and S velocity measurements of a solid (not fractured) lucite block; we assume the density of the fractured sample is the same as that of the background material. This agreement indicates that the theoretical model of the fracture anisotropy describes the medium with parallel fractures very well.

To investigate the fracture compliances , $E_N$ and $E_T$, experiments with fractures filled with honey have been conducted. We have chosen honey mainly because of its high viscosity that enables the sample to support shear waves. Four sets of measurements were done for comparison: measured in sequence, one stack of sample had none, quarter, half and all of the fractures filled with honey. When the number of honey-filled fractures is quarter or half of the total fractures, honey-filled and air-filled fractures are placed periodically. Since we know the P and S velocities of the background medium from the previous measurements, we only need to measure $C_{33}/\rho$, and $C_{44}/\rho$ to obtain $E_N$ and $E_T$. The resulting values of $E_N$ and $E_T$ for all four experiments are shown in Figures 4 and 5. The data indicate that the honey-filled fractures are stiffer than the air-filled fractures for both compressional and shear waves. A simple mixing law can be derived based on the linear slip fracture model. The mixing law gives the $E_N$ and $E_T$ of the quarter and half honey-filled samples as the proportional average between the all air-filled and all honey-filled values. The data shown in Figures 4 and 5 follow the mixing law very well.

The values of $E_N$ and $E_T$ for liquid filled fractures can be predicted from elastic theory based on the properties of the liquid and the background material, and the fractional thickness of the fractures. The measured $E_N$ for honey-filled fractures agree with the prediction well. However, the agreement for $E_T$ is not conclusive because of the uncertainty of the viscosity of honey at room temperature, which is very sensitive to temperature variation, and because of the uncertainty in defining the fracture thickness in shear sense for the microscopically complicated interfaces.

In conclusion, the experiment has demonstrated the validity of the long wavelength linear slip model as applied to the special case of a periodically fractured solid. In addition, reasonable agreement has been obtained between theoretical and inferred values of the normal fracture compliance.

## References


Backus, G.E., 1962, Long-wave anisotropy produced by horizontal layering: J. Geophysical Research 66, 4427-4440.

Crampin, S., 1985, Evidence for aligned cracks in the Earth's crust: First Break 3, (3), 16-20.

Schoenberg, M., 1980, Elastic wave behavior across linear slip interfaces: JASA 68, 1516-1521.

Schoenberg, M. and Douma, J., 1988, Elastic wave propagation in media with parallel fractures and aligned cracks: Geophysical Prospecting, 36, 571-590.


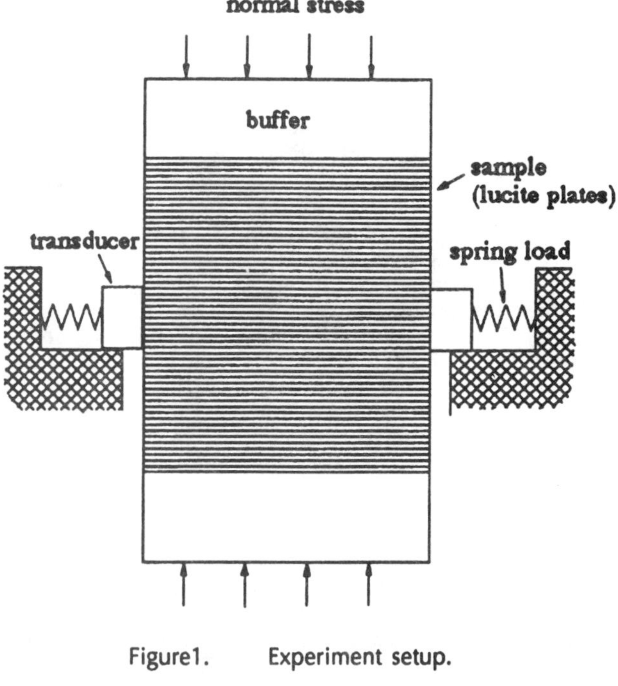


Figure1. Experiment setup.

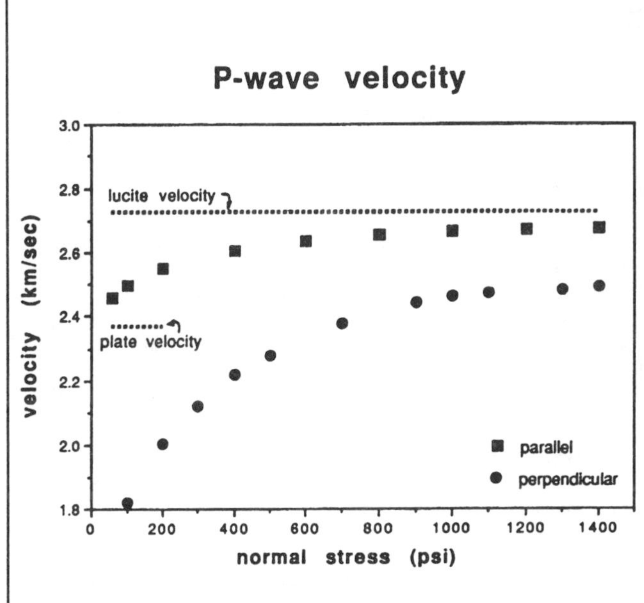


Figure 2. Data of P-wave velocity propagating normal and parallel to fractures.

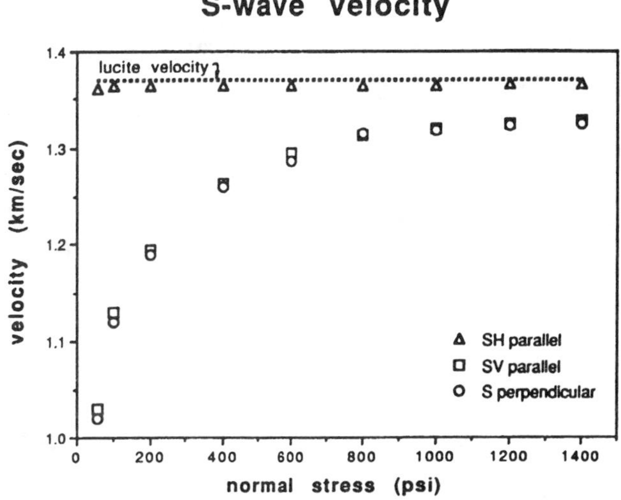


Figure 3. Data of S-wave velocity propagating normal and parallel to fractures.

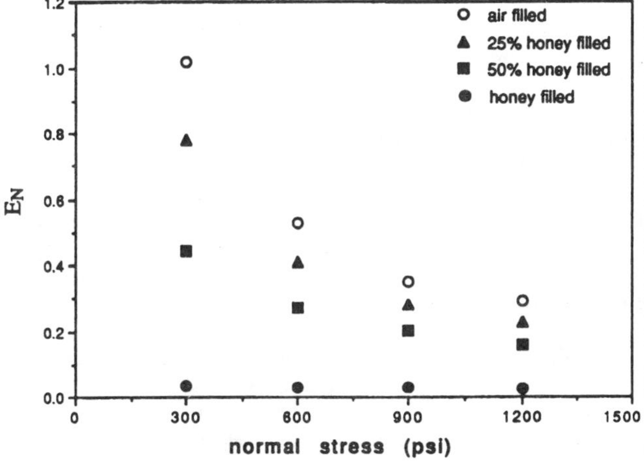


Figure 4. Data of $E_N$ for air-filled, honey-filled, and partially honey-filled fractures.

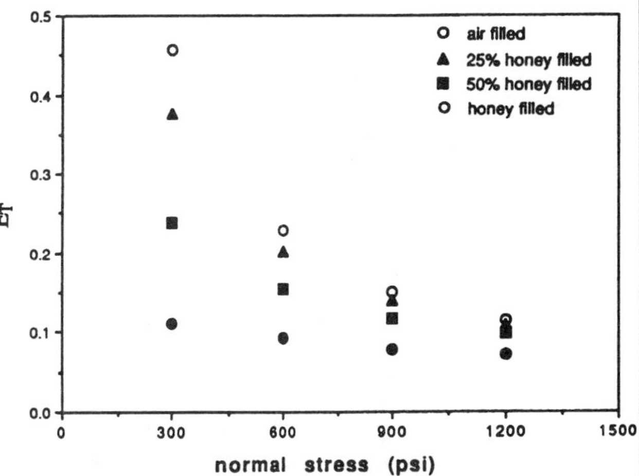


Figure 5. Data of $E_T$ for air-filled, honey-filled, and partially honey-filled fractures.

# Physical Modeling 2: Anisotropy Using Physical Models Mini-Workshop

## Wednesday Afternoon, November 13th

### Experimental versus Theoretical Acoustic Anisotropy in Controlled Cracked Synthetics

**PM2.1**

*Jaswant Singh Rathore*, Erling Fjaer, Rune Martin Holt, and Lasse Renlie,*
*Continental Shelf and Petroleum Technology Research Institute (IKU), Norway*

SUMMARY

A method has been developed for the production of synthetic sandstones containing cracks of known dimensions and geometry within a porous rock matrix. A synthetic sandstone was manufactured from sand cemented with epoxy glue. The cracks were produced by placing thin metallic discs into one half of the synthetic during manufacture. The discs were chemically leached out of the consolidated sandstone. Acoustic anisotropy and shear-wave splitting were observed in the dry and saturated synthetic half which had contained the discs. The observed data compares very well with Hudson's (1981) theoretical model for a dry cracked medium. Although the measured velocities and those predicted by Hudson's and Thomsen's (1986) models for the saturated cracked medium compare reasonably well, there is a slight discrepancy at acoustic propagation directions normal to the crack planes. The dummy half of the sandstone, which contained no discs, showed no acoustic anisotropy.

#### 1. INTRODUCTION

The knowledge of crack distributions is of particular importance for the withdrawal of fluids from reservoirs in which cracks exist as penetrating fractures, serving as primary channels for fluid flow. The presence of cracks affects the acoustic wave velocities. If the cracks are small compared to the acoustic wavelength, and if they are ordered with a preferred orientation, the acoustic waves will see an effectively homogeneous medium, but with a certain anisotropy in the acoustic wave propagation parameters. It is important to evaluate the causes of these anisotropies, for instance, to try and tell directly from theory whether the anisotropy is caused by a large number of small, isolated microcracks, or by a small number of similarly oriented, larger and possibly connected fractures. This question is of vital importance to the petroleum industry.

To improve the potential of application of acoustic anisotropy as a tool for detecting and characterising fractured systems, it is necessary to obtain an improved understanding of how fractures of different sizes affect sound propagation at different wavelengths. Even though several theoretical models for acoustic behaviour in cracked and jointed media exist (Hudson, 1981, Thomsen, 1986), none of them have yet been confirmed experimentally under controlled conditions. A controlled experiment in this context would mean an experiment in which sizes, shapes, amounts and orientational distributions of cracks are known. In geo-materials these are generally unknown parameters, although supplementary information can be obtained through microscopic investigations. A fully three-dimensional picture is not feasible. We have therefore undertaken the task of producing a synthetic material in which the above crack parameters are known.

#### 2. MANUFACTURE OF SYNTHETICS

Since a primary application area of our work will be in petroleum reservoirs, and since theory indicates (Thomsen, 1986) that porosity is important for understanding anisotropy due to cracks, we have embedded the cracks in a porous and permeable sandstone matrix.

Past experience has shown that the sand and epoxy mixture method gives well-behaved homogeneous sandstones (Rathore et al. 1989). This method was used for producing the synthetics. The sand chosen for the samples was a washed Danish beach sand, which has a narrow grain-size spectrum, with a mean value around 200μm. A sandstone mixture consisting of 25 parts sand to 1 part of (2 component) epoxy was used.

Thin metallic discs, representing the cracks were placed in this porous and permeable matrix during the manufacturing process. These discs were chemically leached out of the synthetic rocks leaving disc-voids representing the desired network of cracks.

The model was designed to produce a synthetic sample with controlled microcrack parameters and geometries embedded in a homogeneous, isotropic matrix. Thus the number of cracks, their positions and orientations, shapes, and aspect ratios are known. For practical purposes, and requirements of homogeneity, we chose a cubic sample design with 10 cm side length. 100 kHz is then an acceptable operating frequency for the acoustic measurements, with respect to sample size, wavelength and attenuation.

In a preliminary investigation, a sample was manufactured containing crack sizes which varied between circular discs of 5 mm diameter, to squares of 10 and 20 mm lengths. The metallic discs were

Reprinted from the 61st Annual International Meeting, Society of Exploration Geophysicists, 1991 Expanded Abstracts, 687-690

chemically leached out of the sample. The leaching process liberates gas, which is observed as bubbles within the percolation chamber. This acts as a good control for monitoring the process progress. When bubbling ceases, the process is complete and no more metal remains in the sample. The gas liberation stopped after about 2 hours; however the percolation process is maintained overnight. The percolation system was drained and water was percolated through the sample for 24 hours. The sample was then dried in a furnace at 110°C for over 24 hours before it was cleaved in the three disc-planes. The sample proved to be completely metal-free with clean and highly reflective smooth imprints of the metallic discs. All sized discs had been completely leached out leaving the desired disc-voids.

We have followed the recipe outlined above to manufacture a synthetic pair: samples 901C (containing cracks) and 901D (crack-free dummy) for observations of acoustic anisotropy and shear-wave splitting.

In order to have a better control of the anisotropy due to the material inhomogeneity, we manufactured a single long sample (25 cm length, 10 cm high and 10 cm wide), in which the metallic discs were concentrated in a 10 cm length at one end of the sample. The synthetic was manufactured by compressing successive layers of sand and epoxy (with the interlayered metallic discs at one end). Thus any inhomogeneity due to this manufacturing process would be observed in the disc-free end of the sample. The sample was milled into a long 16 sided regular prism and the crack containing 10 cm was parted off to give the sample 901C. A further 10 cm sample was parted off the remaining block to give the dummy sample 901D.

The disc data for sample 901C is as follows:

| | |
|---|---|
| Disc diameter | 5.50 mm. |
| Disc thickness | 0.02 mm. |
| Fixed crack density | 0.10 |
| Number of cracks in 0.001 $m^3$ | 4808 |
| Number of layers | 48 |

The crack density $\zeta = N\langle a^3\rangle$, where N is the number of cracks per unit volume and a is the crack radius.

## 3. ACOUSTIC MEASUREMENTS

The prism geometry of the samples allows for good transducer face contact for measurements of acoustic response across the faces of the prism as a function of the incidence angle, $\theta$, at 22.5° increments, starting at $\theta$=0° ,normal to the disc-planes, and finishing at $\theta$=90°, parallel with the disc-planes (Fig.1).

Acoustic velocities for P-, $S_a$- (along core axis; equivalent to the $S_h$-wave with respect to the disc-plane), and $S_t$- (tangential to the core; equivalent to the $S_v$-wave with respect to the disc-plane at $\theta$ = 90°) were measured through:

901D dry and saturated to determine matrix anisotropy
901C dry and saturated to determine total anisotropy.

The anisotropy percentages are calculated in accordance with Crampin's (1989) definition:

$$\text{Anisotropy \%} = 100 \times (V_{max} - V_{min})/V_{max}$$

where $V_{max}$ and $V_{min}$ are the maximum and minimum velocities observed. The anisotropies observed in 901D due to the manufacturing process were:

| DRY | | SAT | |
|---|---|---|---|
| P-wave : | 1 % | P-wave : | 2 % |
| $S_a$-wave : | 1 % | $S_a$-wave : | 1 % |
| $S_t$-wave : | 2 % | $S_t$-wave : | 2 % |

These anisotropies are, however, small, indicating that the manufacturing process imparts very little anisotropy to the sample.

The total anisotropies observed in 901C due to both the manufacturing process and the cracks are:

| DRY | | SAT | |
|---|---|---|---|
| P-wave : | 21 % | P-wave : | 9 % |
| $S_a$-wave : | 11 % | $S_a$-wave : | 12 % |
| $S_t$-wave : | 2 % | $S_t$-wave : | 8 % |

In general the data quality is good and is also in good agreement with the model theories of Hudson (1981) and Thomsen (1986). In both the dry and the saturated experiments the shear $S_a$- and $S_t$-waves have distinctly different velocities at impingement angles other than normal to the microcrack planes, which shows shear-wave splitting (Figs. 2d & 3d).

The dry sample data (points) have been compared with Hudson's (1981) model (curves) for acoustic anisotropy through a medium containing microcracks (Figs. 2a-2c). The input data used for the Hudson model was:

| | |
|---|---|
| Crack density, | $\zeta$ = 0.10 |
| P-wave vel. in the dry dummy | $V_{po}$ = 2485 m/s. |
| S-wave vel. in the dry dummy | $V_{so}$ = 1556 m/s. |
| Dummy bulk density | $\rho_s$ = 1696 kg/$m^3$. |

The measured data points lie a touch under the theoretical curves, indicating slightly lower measured velocities than predicted. The curvatures are in very good agreement. The data from the saturated samples (points) have been compared with Hudson's (1981) model (dotted curves) and Thomsen's (1986) model (full curves) for a system of fluid-filled microcracks in a cracked porous medium (Fig. 3a-3c). The input data used for Thomsen's model was:

Crack density, $\zeta = 0.1$
P-wave vel. in the sat. dummy $V_{po} = 2684$ m/s.
S-wave vel. in the sat. dummy $V_{po} = 1380$ m/s.
Dummy bulk density $\rho_s = 1696$ kg/m$^3$.
Bulk modulus of solid $K_s = 16.1$ GPa.
Shear modulus of solid $G_s = 15$ GPa.
Bulk modulus of fluid $K_f = 2.16$ GPa.
Equant porosity of dummy matrix $\Phi_p = 0.36$
Crack porosity $\Phi_c = 0.0023$

There is a reasonably good agreement seen for all P- and S-wave data types to both theoretical models. Within the region $\theta = 22.5°$ to $155°$ the measured P-wave acoustic data matched well with Thomsen's model (1986) for the case of equant porosity connected to a low crack porosity. The Hudson model data agrees equally well with the observed data. At normal incidence, however, both models deviate from the observed data, to approximately the same degree. For the case of S-waves, the observed data compares better with the Hudson model than with Thomsen's. The measured shear-wave splitting is seen in Fig. 3d.

## 4. DISCUSSIONS AND CONCLUSIONS

The preliminary conclusions are that the manufacture of synthetic sandstones with known crack geometries is possible with the method described above. This manufacturing process does not appear to impart any anisotropy on the synthetic rock matrix. The disc cavities produced by the percolation method give rise to measurable, and almost predictable, velocity anisotropy and shear-wave splitting.

The comparisons between experimental data with theoretical models are generally very good, in particular for the dry sample. For the saturated sample, however, there are discrepancies observed for the P-wave propagating at normal incidence to the crack plane.

We would like to point out that the measured velocities shown in Figs. 2 and 3 have not been corrected for diffraction and dispersion effects, which may be significant, in particular at incidence angles normal to the discs.

## Acknowledgments

The authors would like to thank the Amoco Norway Oil Co. for their financial support and permission to publish these results. We also thank Leon Thomsen Amoco Production Co. Tulsa, for his scientific help.

## References

Crampin, S., 1989, Suggestions for a consistent terminology for seismic anisotropy: Geophysical Prospecting, 37, 753-770.

Hudson, J.A., 1981, Wave speeds and attenuation of elastic waves in materials containing cracks: Geophys. J.R.Astr.Soc., 64, 133-150.

Rathore, J.S., Holt, R.M. and Fjær, E., 1990, Effects of stress history on petrophysical properties of granular rocks. *in* Khair, A. W., Ed., Rock mechanics as a guide for efficient utilization of natural resources: Balkema, Rotterdam, 765-772.

Thomsen, L., 1986, Elastic Anisotropy due to aligned cracks; Theoretical Models; EOS Trans. Am. Geophys. Union, **67(44)**, 1207.

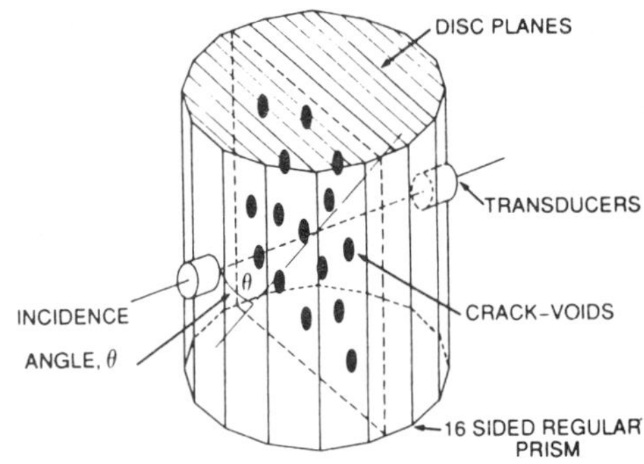


**Figure 1.** The geometry of the synthetic samples used.

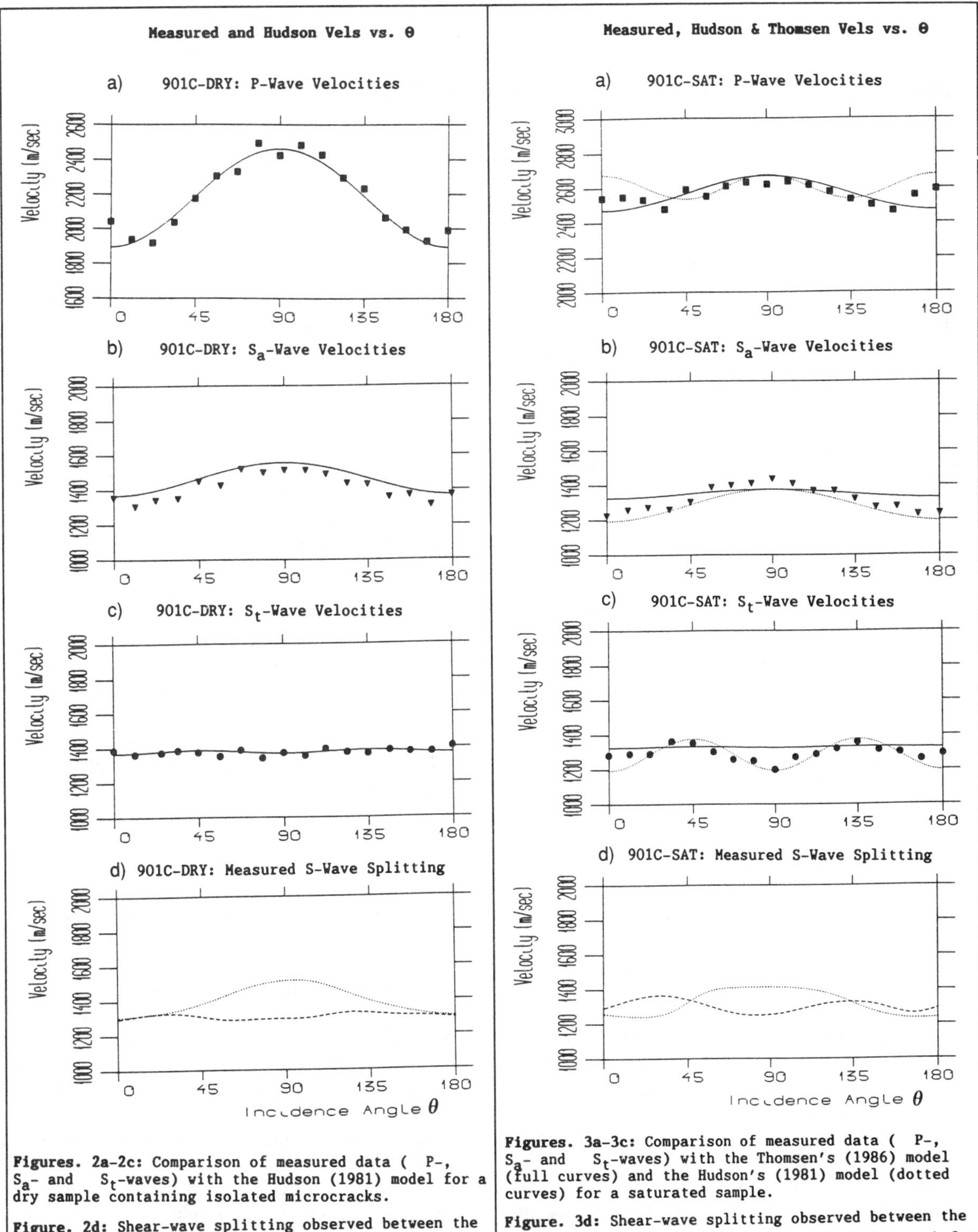


**Figures. 2a-2c:** Comparison of measured data ( P-, $S_a$- and $S_t$-waves) with the Hudson (1981) model for a dry sample containing isolated microcracks.

**Figure. 2d:** Shear-wave splitting observed between the $S_a$-wave (dotted curve) and $S_t$-waves (dashed curve) for the dry synthetic sandstone.

**Figures. 3a-3c:** Comparison of measured data ( P-, $S_a$- and $S_t$-waves) with the Thomsen's (1986) model (full curves) and the Hudson's (1981) model (dotted curves) for a saturated sample.

**Figure. 3d:** Shear-wave splitting observed between the $S_a$-wave (dotted curve) and $S_t$-waves (dashed curve) for the saturated synthetic sandstone.

# A physical model study of microcrack-induced anisotropy

Jamal M. Ass'ad*, Robert H. Tatham‡, and John A. McDonald*

ABSTRACT

A laboratory study of the effects of oriented penny-shaped inclusions embedded in a solid matrix on the propagation of seismic shear waves shows good agreement with theoretical predictions for some polarizations and poor agreement for polarizations at large crack densities. The models are constructed of a solid matrix of epoxy resin with inclusions of thin rubber discs of approximately equal cross-sectional areas. The Theoretical basis for these experiments is the theory of Hudson, in which the wavelength is greater than the dimensions of the individual cracks and their separation distance, and the cracks are in dilute concentration. By a pulse transmission method, seismograms were gathered in models free of inclusions and models with inclusions. Seismic measurements of velocity anisotropy, for variations in both a polarization and propagation direction, were performed on physical models with inclusions (cracks) representing five different crack densities (1, 3, 5, 7, and 10 percent). Variations in velocity anisotropy at different crack densities have been evaluated by using Thomsen's parameter ($\gamma$) which relates velocities to their elastic constants,

$$\gamma = \frac{1}{2}\left(\frac{vs_1^2}{vs_2^2} - 1\right) = \left(\frac{C_{66} - C_{44}}{2C_{44}}\right).$$

Comparisons between experimental and theoretical results indicate that with wave propagation parallel to the aligned inclusions, shear waves polarized parallel to the plane containing the inclusions ($S_1$) agree well with the theoretical model. However, shear waves for the same propagation direction but polarized perpendicular to the plane containing the inclusions ($S_2$) produced results that agree well with the theory for crack densities up to 7 percent, but disagree for higher crack densities. The deviation of $\gamma$ at 10 percent crack density suggests that crack-crack interaction and their coalescence may be observable and could lead to seismic techniques to differentiate between microcracks and larger macrocracks.

## INTRODUCTION

Detection of cracks and fractures in the subsurface has become an important subject in exploration and development geophysics. For example, identification and development of oil and/or gas fields, definition of hydrological parameters, and storage of toxic wastes rely on accurate detection and characterization of fractures and cracks. Anisotropy in seismic wave propagation has evolved as a tool in the characterization of systems of cracks and fractures.

Many researchers have documented the presence of seismic anisotropy in the upper crust (e.g., Crampin et al., 1984, Peacock et al., 1988, and many others). Most of these studies have focused on analyzing three-component seismograms from crustal earthquakes with various azimuths and incident angles, and determining parallel or subparallel alignments of the particle motion upon the leading shear-wave arrivals. Crampin (1987) proposed that seismic anisotropy is controlled by parallel cracks oriented parallel to the maximum contraction direction and perpendicular to the minimum contraction direction (Atkinson, 1984).

Seismic waves propagating through a fractured rock mass have generally lower velocities and greater attenuation than those propagating in an uncracked medium, depending on polarization and propagation direction. Aligned fractures are one cause of azimuthal anisotropy in rocks. Another cause is strong crystallographic-preferred orientation in deformed rocks (Nicholas and Christensen, 1987). Shear waves traveling through fractured media commonly undergo splitting: one mode is polarized parallel to the fractures, and a second

Presented at the 60th Annual International Meeting, Society of Exploration Geophysicists. Manuscript received by the Editor June 7, 1991; revised manuscript received April 21, 1992.
*Allied Geophysical Laboratories, University of Houston, Houston, TX 77204-4231.
‡Texaco, Inc., Exploration and Technology Dept., P.O. Box 770070, Houston, TX 77215-0070.


Reprinted from Geophysics, 57, 1562-1570

mode is polarized perpendicular to the fractures. Such phenomena have been reported previously in single crystals (e.g., Waterman and Teutonico, 1957; Simmons and Birch, 1963). Shear-wave splitting or birefringence occurs for all directions of propagation except along a single axis of rotational symmetry. This effect is seen in the experiments of Nur and Simmons (1969), Johnson (1989), and Ebrom et al. (1990). Crampin (1985) and Crampin et al. (1984) documented field evidence of shear-wave splitting in earthquake seismic records, verifying earlier insights by Gupta (1973). Shear-wave splitting enables us to make a reliable quantitative analysis of differential traveltimes between split shear waves, which yield information about the anisotropy and are translatable into crack parameters such as density and aspect ratio.

Several different crack theories have been published describing the effect of cracks on seismic wave propagation. Such theories are classified into two types: the first are known as crack theories (e.g., Hudson, 1980, 1981; Nishizawa, 1982) which consider distribution of small (on the scale of seismic wavelengths) unconnected, parallel, and possibly subparallel cracks, whereas the second type are known as fracture theories (e.g., Schoenberg, 1983, Schoenberg and Douma, 1988) which treat systems of infinite plane parallel cracks.

In this paper we: (1) experimentally test the results of Hudson's (1981) theory for weak material-filled microcracks; (2) experimentally evaluate the limits of crack density for which Hudson's theory is valid; and (3) analyze effects of microcrack orientations, and crack density, on shear-wave propagation. All our experimental work was carried out with the angle of incidence, perpendicular to the direction of the axis of rotational symmetry (i.e., propagation is always parallel to the crack orientation).

## EXPERIMENTAL PROCEDURE

### Model construction

Seven models were constructed: (1) a homogeneous isotropic solid block; (2) a layered solid block, both as controlled models; (3) and a series of five inclusion-filled models with different crack densities and with well defined properties (e.g., dimensions, aspect ratio, and geometry). The material used as the control and for a material containing inclusions an epoxy resin that has a density 1.20 gm/cm$^3$ was used. Thin rubber discs were used to simulate weak solid filling with a density 1.04 g/cm$^3$; these were embedded within the epoxy resin block. Each block is a cube approximately 7.6 cm (3 in.) on a side. The rubber discs are all of radius 0.318 cm (0.125 in.) and of thickness 0.0381 cm (0.015 in.). The number of discs, or inclusions ($n$), the radius of the discs ($r$) and the volume of the cube ($\nu$) determined the crack density ($\varepsilon$) (Hudson, 1981) where,

$$\varepsilon = \frac{nr^3}{\nu}.$$

Each model was constructed from 18 layers of equal thickness with thin rubber discs embedded within the model. The layer thickness was chosen to be ten times greater than the rubber disc thickness, consequently the thickness of each layer was 0.381 cm (0.15 in.). A consistency of layer thickness throughout the models was maintained by pouring the same volume of epoxy resin over the same mold area for each layer. Since the model volume 442.0 cm$^3$ (27 in.$^3$) and the number of layers (18) are known and constant, the number of rubber discs per volume can be obtained using the desired crack density ($\varepsilon$). For example, if we are interested in a 3 percent crack density and the radius of the disc is 0.318 cm (0.125 in.), the number of discs per unit volume is 15.36. The number of discs is multiplied by volume of the mold 442.0 cm$^3$ (27 in.$^3$) and divided by the number of layers (18) obtaining the total number of discs per layer (23). This number of rubber discs is distributed randomly in position per layer and without allowing any interaction among them. Next, the same method is applied for the subsequent layers until the desired number of layers are obtained (18). Table 1 shows a series of five physical models with different crack densities ($\varepsilon$) and the number of inclusions (cracks) per layer, and inclusions (crack) per unit volume.

### Data acquisition and velocity measurement method

Figure 1 shows the experimental design, a pulse transmission method, which was used to acquire shear-wave data. The receiver transducer was placed against the top surface of the model, and the source transducer was placed directly beneath the receiver against the bottom surface of the model. The transducer polarizations were aligned initially parallel to the cracks and both transducers were rotated in the same direction at 9 degree increments. In this manner, the transducers rotated 360 degrees and sampled 40 different polarizations for each block. The dimensions of the cracks and their separation distance are known and controlled. We processed our recorded seismic data using a band-pass filter (30 kHz to 60 kHz) corresponding to a wavelength from 20 mm to 40 mm. This wavelength ensures that a group of inclusions will be sampled.

Direct measurement of the first arrival of the shear-wave pulse propagating through a solid model, together with a determination of path length, are sufficient to evaluate the shear-wave velocity. Variations of the measured shear-wave velocities of both particle motions polarized parallel and perpendicular to the cracks as a function of crack density were used to test the validity of the theory of Hudson (1981).

For the traces polarized parallel to the cracks (at 0, 180, and 360 degrees), the $S_1$ (fast) arrival, and for the traces polarized perpendicular to the cracks (at 90 and 270 degrees), the $S_2$ (slow) arrival are evident and were unambiguously picked (see, for example, Figure 4). The $S_1$ and $S_2$

**Table 1. Relationship between crack density (ε), number of cracks per layer, and number of cracks per unit volume.**

| Crack density (ε) | Number of cracks per layer | Number of cracks per unit volume |
|---|---|---|
| 1% | 8 | 144 |
| 3% | 23 | 414 |
| 5% | 38 | 684 |
| 7% | 54 | 972 |
| 10% | 76 | 1368 |

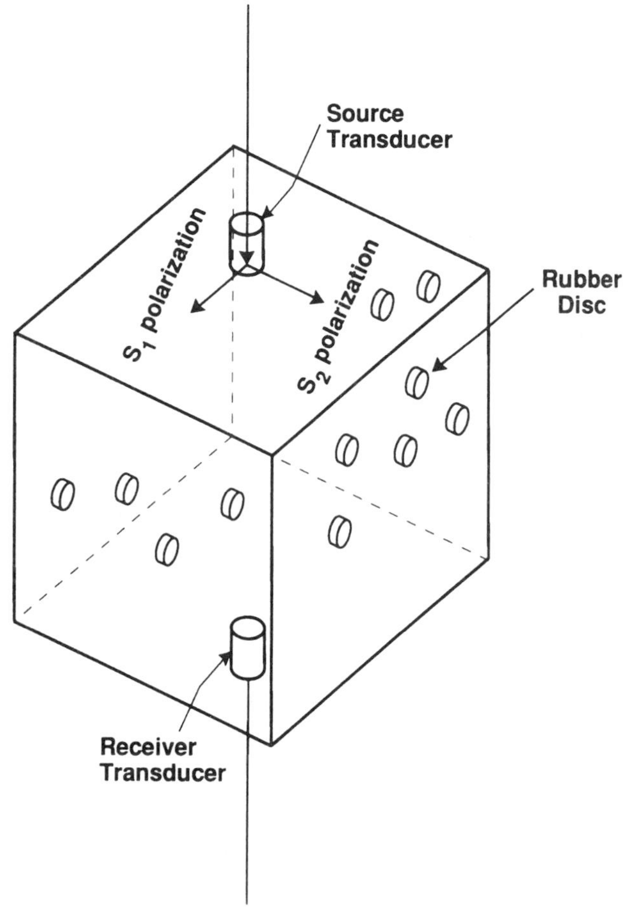


FIG. 1. An example of the cubic model showing the orientation of the discs, the inclusion of which induces anisotropy. The source was rotated through 360 degrees.

arrivals were consistently picked for a given model, and the amount of anisotropy was determined as a ratio of the two traveltimes. In this experiment the first peak of both arrivals was used to measure traveltimes.

## EXPERIMENTAL RESULTS

### Control models

Two potential problems had to be addressed before the effect of microcrack-induced anisotropy could be analyzed for models with inclusions: first, measurements were carried out on a solid block of epoxy resin to obtain control data for the subsequent experiments and to establish that the matrix material is isotropic. The data for the solid block are shown in Figure 2. The shear-wave traveltime in scaled units for all traces was 625 ms. The arrival times are consistent for all 40 traces indicating that the coupling was uniform and consistent and that the material is isotropic to $S$-wave polarization. Second, a thinly layered block of the epoxy resin without inclusions was used to study the potential effect of the anisotropy caused soley by a thinly layered medium since such a medium could behave in a transversely isotropic manner whenever long wavelength elastic waves are used (Postma, 1955). The results show only a very slight anisotropy (0.0008) due to layering, which will be neglected (Figure 3). These results are important, since they allow us to quantify the velocity anisotropy induced by cracks (rubber discs). The $S_1$ arrival time for traces at 0, 180, and 360 degrees was 628 ms, where the $S_2$ arrival time for traces at 90 and 270 degrees was 629 ms, a delay of 1 ms that is indicative of nearly negligible anisotropy.

### "Cracked" data models

Having established the isotropy of the matrix material, the effect of inclusions within the isotropic medium was studied.

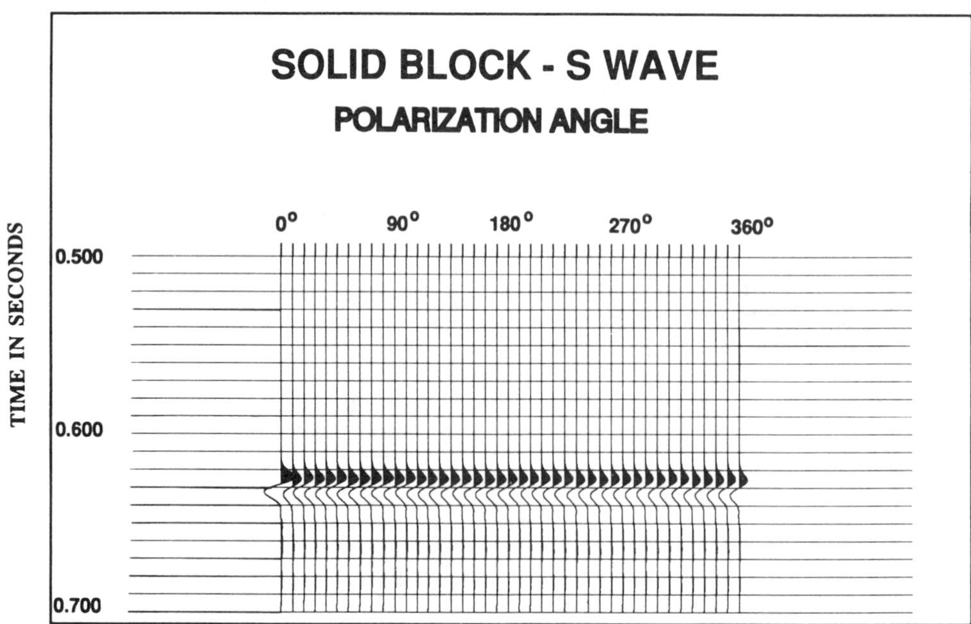


FIG. 2. Shear-wave transmission data, as a function of polarization angle, through a solid homogeneous, isotropic block of resin. Note consistency of the arrival times.

Five inclusion-filled physical models, at five different crack densities of 1, 3, 5, 7, and 10 percent were studied. The results are shown on Figures 4 through 8. For a crack density of 1 percent, the $S_1$ arrival time in scaled units for traces 0, 180, and 360 degrees was 638 ms. The $S_2$ arrival time for traces at 90 and 270 degrees was 645 ms with a delay of 7 ms indicating the presence of anisotropy, and the amount of anisotropy is 1.1 percent.

For a crack density of 3 percent, the $S_1$ arrival time for traces at 0, 180, and 360 degrees was 623 ms. The $S_2$ arrival time for traces at 90 and 270 degrees was 645 ms with a delay of 22 ms indicating an increase of anisotropy with increasing crack density and very obvious shear-wave splitting, giving an amount of anisotropy of 3.5 percent.

For the model with a crack density of 5 percent, the $S_1$ arrival time for traces at 0, 180, and 360 degrees was 636 ms. The $S_2$ arrival time for traces at 90 and 270 degrees, was 672 ms with a delay of 36 ms. This delay again clearly indicated anisotropy with increasing crack density, and the amount of anisotropy is 5.6 percent.

For the model with rubber inclusions at a crack density of 7 percent, the $S_1$ arrival time for traces at 0, 180, and 360 degrees was 628 ms. The $S_2$ arrival time for traces at 90 and 270 degrees was 670 ms, a delay of 42 ms, which is also clear evidence of anisotropy with an increase of 6 ms in traveltime from the 5 percent model with an amount of anisotropy of 6.7 percent.

For a crack model with rubber inclusions at a crack density of 10 percent, the $S_1$ arrival time for traces at 0, 180,

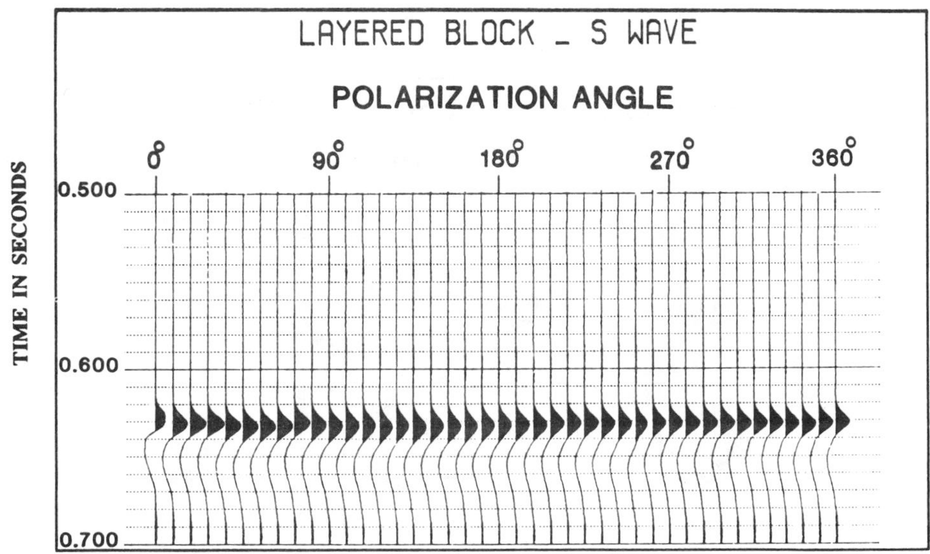


FIG. 3. Shear-wave transmission data, as a function of polarization angle, through a layered solid block of resin.

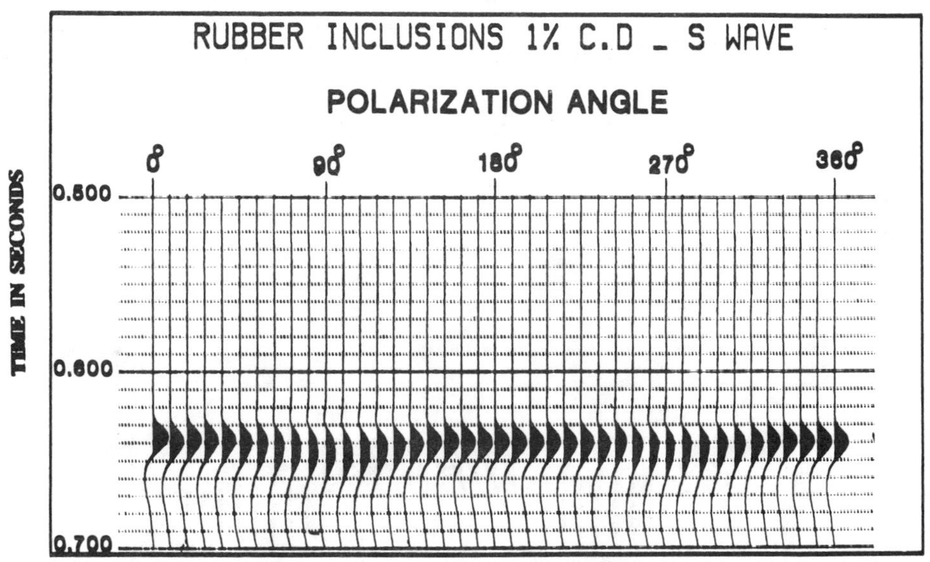


FIG. 4. Shear-wave transmission data, as a function of polarization angle, through a block with rubber inclusions that simulate a weak solid filling. The crack density is 1 percent.

and 360 degrees was 638 ms. The $S_2$ arrival time for traces at 90 and 270 degrees, was 670 with a delay of 32 ms which characterize the presence of anisotropy. This time the delay is 6 ms *less* than the previous 7 percent crack density model, and the shear wave splitting is still very prominent, and the amount of anisotropy is 5.0 percent. This effect is discussed in the next section.

## NUMERICAL RESULTS AND DISCUSSION

From these observed traveltimes and distances, along with densities, we have computed the elastic constants of the isotropic model, the layered model, and five models containing oriented penny-shaped inclusions simulating cracks. That is, $\mu = V_s^2 . \rho_b$ where $V_s$ is the shear-wave velocity in an isotropic medium, $\rho_b$ is the density of the background medium, and $\mu$ is the shear modulus of the same medium. In the case of the cracked models, we use similar equations to calculate the elastic constants ($C_{66}$, $C_{44}$), from $C_{66} = V_{s_1}^2 \cdot \rho_b$ and $C_{44} = V_{s_2}^2 \cdot \rho_b$, where $V_{s_1}$ is the fast shear velocity polarized parallel to the fractures and $V_{s_2}$ is the slow shear velocity polarized normal to the fractures.

From these numerical calculations, we evaluated the anisotropy due to aligned penny-shaped inclusions by calculating Thomsen's (1986) parameter ($\gamma$), which represents a combination of the elastic moduli ($C_{44}$, $C_{66}$) where,

$$\gamma = \frac{1}{2}\left(\frac{vs_1^2}{vs_2^2} - 1\right) = \left(\frac{C_{66} - C_{44}}{2C_{44}}\right).$$

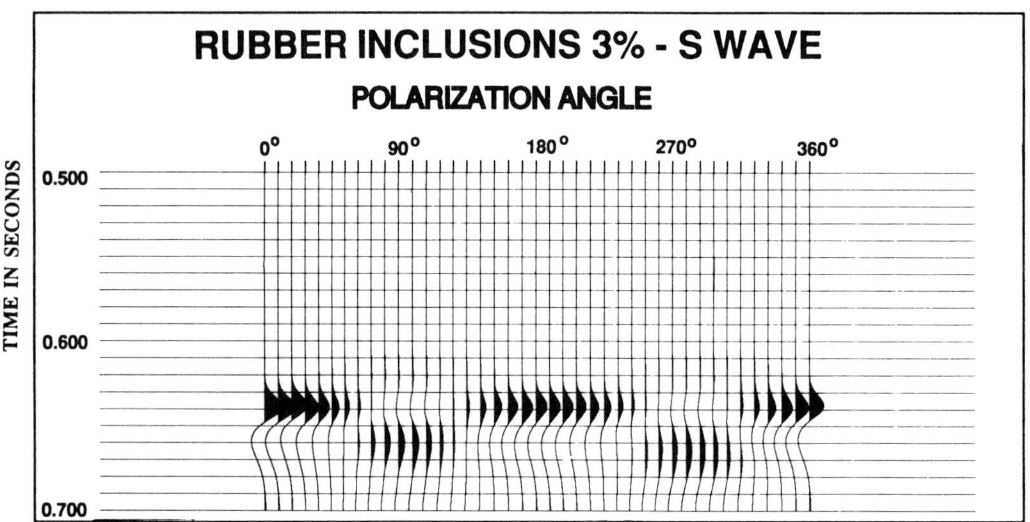


FIG. 5. Shear-wave transmission data, as a function of polarization angle, through a block with rubber inclusions that simulate a weak solid filling. The crack density is 3 percent.

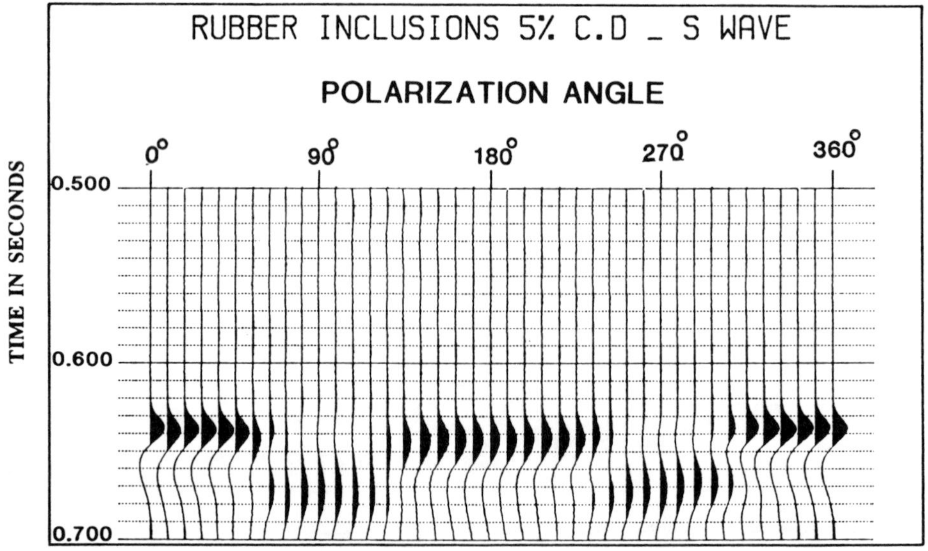


FIG. 6. Shear-wave transmission data, as a function of polarization angle, through a block with rubber inclusions that simulate a weak solid filling. The crack density is 5 percent.

Further developments have been formulated showing the relationship between the anisotropy parameter $\gamma$ and the crack density $\varepsilon$ for first and second order theories of Hudson (1980, 1981) (see Appendix). Table 2 shows the principal results of the study represented as velocities, elastic stiffness coefficients, as well as the anisotropic parameter, after three experiments were carried out on each model.

The experimental results summarized in Table 2 show several interesting features:

1) $C_{66} = C_{44}$ for an isotropic medium.
2) $C_{66} > C_{44}$ for an azimuthally anisotropic medium, as long as the propagation direction of shear-wave energy is perpendicular to the axis of rotational symmetry (i.e., parallel to the fractures).
3) $C_{66}$ is almost constant within the experimental error (0.5 percent) for all models, indicating that $V_{S_1}$ is not affected by the percentage of crack density (intensity of the fractures) as long as the polarization is parallel to the cracks. The relations between $V_{S_1}$ and the crack density are shown in Figure 9a. The mean values and standard deviations are shown by circles and vertical bars, respectively, together with the theoretical calculations for the Hudson first-order theory. Comparison between theoretical calculations and our experimental results showed that the trend in $V_{S_1}$ is a function of

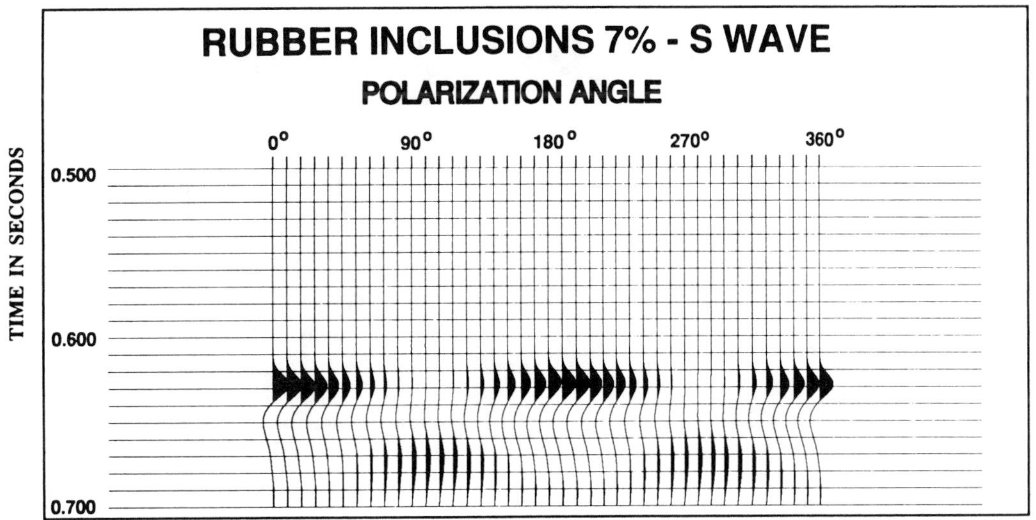


FIG. 7. Shear-wave transmission data, as a function of polarization angle, through a block with rubber inclusions that simulate a weak solid filling. The crack density is 7 percent.

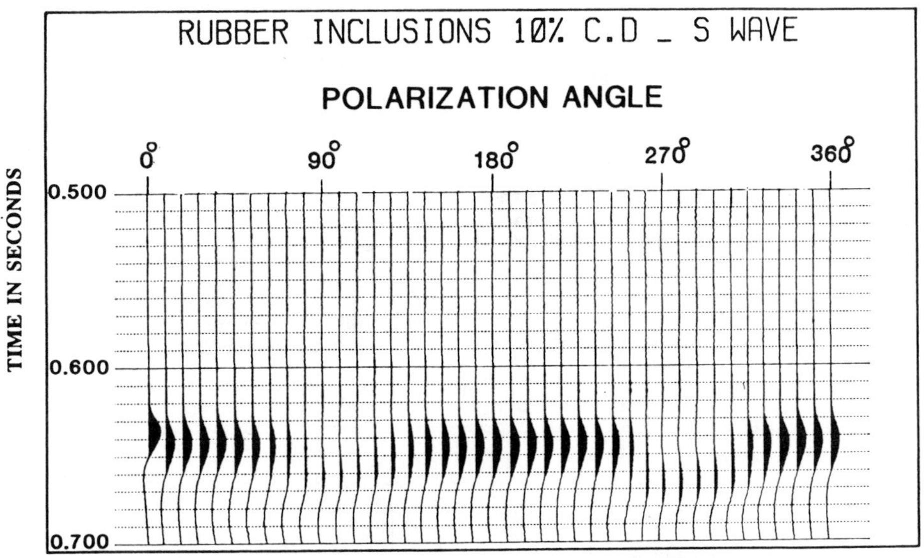


FIG. 8. Shear-wave transmission data, as a function of polarization angle, through a block with rubber inclusions that simulate a weak solid filling. The crack density is 10 percent.

crack density, agreeing well with the theoretical calculations.

4) $C_{44}$ is decreasing with increasing crack density.
5) At a 10 percent crack density an increase of $S_2$ velocity is attributed to the domination of multiple-scattering caused by crack-crack interaction (Gubernatis and Domany, 1979). This observation was also observed by Dubendroff and Menke (1990). The relations between $V_{S_2}$ and the crack density are shown in Figure 9b. The mean values and standard deviations are shown by circles and vertical bars, respectively, together with the theoretical calculations of the Hudson first-order theory. Comparing the theoretical and our experimental results, we found that $V_{S_2}$ agreed with the theory for crack densities up to around 7 percent and disagreed at higher crack densities.
6) The element $\gamma$ increases with increasing crack density and reaches a maximum at a 7 percent crack density and then decreases for the 10 percent crack density model. This suggests that crack-crack interaction becomes important for crack densities higher than about 7 percent. This observation can be explained physically as follows: when the cracks (inclusions) are getting close to each other and until their coalescence, the dimensions of the cracks are no longer smaller than the wavelength of the propagating seismic wave, thus the concept of long wavelength anisotropy is no longer satisfied, in this case the wave propagation can be viewed as a multiscattering and not a single scattering causing a large change in the velocity of shear-wave propagating perpendicular to the plane containing the cracks. Figure 10 shows a comparison between the theoretical calculations of Hudson's first and second order theories with the experimental results. There is a roll over of the experimental curve where $\gamma$ starts decreasing and the experimental results start to diverge from the theoretically predicted values at about a 5 to 7 percent crack density. Also, from the same experimental data (Figure 10), we speculate that a decrease in the percentage of velocity anisotropy at a higher crack density (10 percent) suggests an increase of crack-crack interaction which is evidence of an increase in the size of the fractures. Recent work by Ebrom et al. (1990), suggests a complimentary way of differentiating between microcrack-induced anisotropy and fractured-induced anisotropy using the dispersive effect at different frequencies.

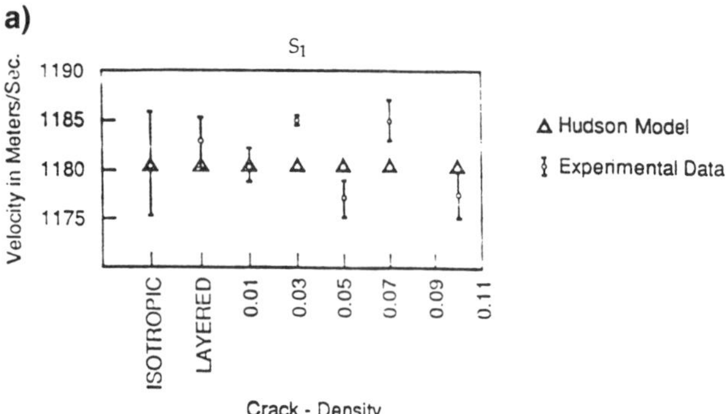


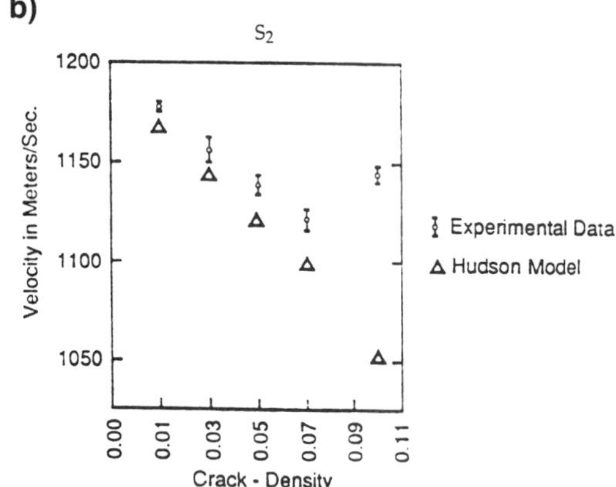


FIG. 9. Crack density versus shear-wave velocity (a) $S_1$ and (b) $S_2$. The mean values and the standard deviations are shown by circles and vertical bars, respectively, together with theoretical calculations (triangles).

## CONCLUSIONS

Physical modeling is a useful method for testing and developing processing techniques, and experimentally verifying how well theory actually predicts observed seismic processes. The feasibility of implementing these techniques and processes allow us to better understand seismic processes in the earth and may lead to field applications. In our present work, we use an epoxy resin as a matrix of a physical model and a rubber discs (inclusions) as cracks. As a result of this experiment, we inferred the following:

1) Hudson's theory agrees well with the data from models with lower crack densities ($\leq$7 percent at least approx-

**Table 2. Observed velocities, elastic constants and $\gamma$ for the models studied.**

| Model | $V_{S_1}$ (m/s) | $V_{S_2}$ (m/s) | $C_{66}$ (kPa) | $C_{44}$ (kPa) | $\gamma$ |
|---|---|---|---|---|---|
| Isotropic | 1181 ± 5.5 | 1181 ± 5.5 | $1.676 \times 10^6 \pm 13000$ | $1.676 \times 10^6 \pm 13000$ | 0.0 |
| Layered | 1183 ± 2.9 | 1182 ± 2.8 | $1.683 \times 10^6 \pm 9000$ | $1.683 \times 10^6 \pm 9000$ | 0.0008 |
| 1% C.D. | 1181 ± 1.4 | 1179 ± .64 | $1.678 \times 10^6 \pm 3265$ | $1.673 \times 10^6 \pm 1469$ | 0.0017 |
| 3% C.D. | 1185 ± .49 | 1157 ± 5.5 | $1.687 \times 10^6 \pm 1143$ | $1.609 \times 10^6 \pm 12820$ | 0.025 |
| 5% C.D. | 1177 ± 1.94 | 1138 ± 4.9 | $1.666 \times 10^6 \pm 4481$ | $1.557 \times 10^6 \pm 10945$ | 0.035 |
| 7% C.D. | 1185 ± 2.1 | 1123 ± 3.2 | $1.701 \times 10^6 \pm 4884$ | $1.518 \times 10^6 \pm 7053$ | 0.056 |
| 10% C.D. | 1178 ± 2.53 | 1146 ± 2.31 | $1.670 \times 10^6 \pm 5849$ | $1.58 \times 10^6 \pm 5196$ | 0.028 |

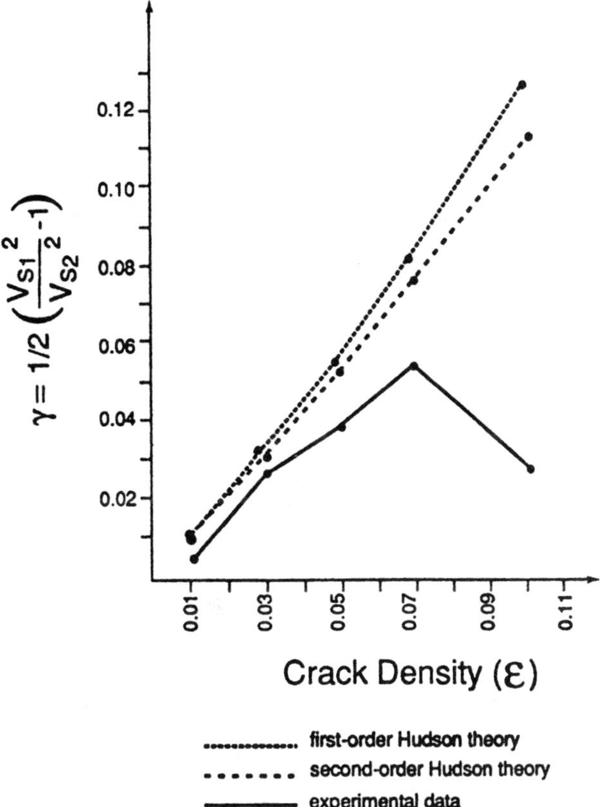


FIG. 10. The anisotropy parameter (γ) as a function of crack density (ε) showing the comparison among first and second order Hudson's calculations and experimental results.

imately) for both $S_1$ and $S_2$, but disagree with $S_2$ at higher crack densities.

2) For waves traveling parallel to the cracks, the velocity ($S_1$) is constant within the allowable experimental error regardless of the increase in the crack density, while for waves traveling normal to the cracks the velocity ($S_2$) reduces with increasing crack density. The only exception that occurs is at a very high crack density (10 percent) where the $S_2$ velocity increase is attributed to a multiple-scattering due to crack-crack interaction. This sensitivity of $S_2$ to the crack density can be taken as an indication of the lateral variability of the fracture intensity in a sedimentary basin, which is vital to the enhancement of oil recovery using horizontal drilling.

3) This study opens up a new avenue towards further study in recognizing microfractures from macrofractures. In fact, this is a research topic currently under investigation in the Allied Geophysical Laboratories at University of Houston.

## ACKNOWLEDGMENTS

The authors would like to thank Dr. Kenneth D. Mahrer for his detailed constructive comments on the manuscript. This research was supported by the industrial sponsors of Seismic Acoustics Laboratory and by an Advanced Technology Program grant No. 003652145-ATP from the State of Texas.

## REFERENCES


Atkinson, B. K., 1984, Subcritical crack growth in geological materials: J. Geophys. Res., **89**, 4077–4114.

Crampin, S., 1985., Evidence for aligned cracks in the earth's crust: First Break, **3**, 12–15.

——— 1987, Geological and industrial implications of extensive-dilatancy anisotropy: Nature, **328**, 491–496.

Crampin, S., Chesnokov, E. M., and Hipkin, R. G., 1984, Seismic anisotropy: The state of art: Geophys. J. Roy. Astr. Soc., **76**, 1–16.

Dubendorff, B., and Menke, W., 1990, Physical modelling observations of wave velocity and apparent attenuation in isotropic and anisotropic two phase media: Pure. Appl. Geoph., **132**, 363–400.

Ebrom, D. A., Tatham, R. H., Sekharan, K. K., McDonald, J. A., and Gardner, G. H. F., 1990, Dispersion and anisotropy in laminated versus fractured: An experimental comparison: 60th Ann. (Internat. Mtg., Soc. Expl. Geophys., Expanded Abstracts, 1416–1419.

Gubernatis, J. E., and Domany, E., 1979, Rayleigh scattering of elastic waves from cracks: J. Appl. Phys., **50**, 818–824.

Gupta. I. N., 1973, Preliminary variations in *S*-wave velocity anisotropy before earthquakes in Nevada: Science, **182**, (4117), 1129.

Hudson, J. A., 1980, Overall properties of a cracked solid: Math Proc. Camb. Phil. Soc., **88**, 371–384.

Hudson, J. A., 1981, Wave speeds and attenuation of elastic waves in material containing cracks: Geophys. J. Roy. Astr. Soc., **64**, 133–150.

Johnson, J. V., 1989, Obbservations of shear-wave anisotropy associated with oriented fractures: A physical modeling study: M.Sc. thesis, Univ. of Houston.

Nicolas, A., and N. I. Christensen, 1987, Formation of anisotropy in upper mantle peridotite: Rev. Geophys, **25**, 1168–1176.

Nishizawa, O., 1982, Seismic velocity anisotropy in a medium containing oriented cracks—Transversely isotropic case: J. Phys. Earth, **30**, 331–347.

Nur, A., and Simmons, G., 1969, Stress-induced velocity anisotropy in rock: An experimental study: J. Geophys. Res., **74**, 6667.

Peacock, S., Crampin, S., Booth, D. C., and Fletcher, J. B., 1988, Shear-wave splitting in the Anza seismic gap, southern California: Temporal variations as possible precursors: J. Geophys. Res., **93**, 3339–3356.

Postma, G. W., 1955, Wave propagation in a stratified medium: Geophysics, **20**, 780–806.

Schoenberg, M., 1983, Reflection of elastic waves from periodically stratified media with interfacial slip: Geophys. Prosp., **31**, 265–292.

Schoenberg, M., and Douma, J., 1988, Elastic wave propagation in media with parallel fractures and aligned cracks: Geophys. Prosp., **36**, 571–590.

Simmons, G., and Birch, F., 1963, Elastic constants of Pyrite: J. Appl. Phys., **34**(9), 2736.

Thomsen, L., 1986, Weak elastic anisotropy: Geophysics, **51**, 1954–1966.

Waterman, P. C., and Teutonico, 1957, Ultrasonic double refraction in single crystals: J. Appl. Phys., **28**, 266.

## APPENDIX A

### THEORETICAL DEVELOPMENT

Hudson (1980, 1981) showed an explicit theoretical development of the calculation of the variations in velocity of seismic waves propagating through solids containing aligned thin circular cracks (penny-shaped) filled with a weak material. For simplicity, and since our experiments were carried out when the angle of the incident wave is normal to the rotational symmetry axis ($\phi = 90$ degrees), the fast shear-wave velocity ($S_1$) is constant and is equal to the shear velocity of the isotropic material ($\beta$). The slow shear-wave velocity ($S_2$) varies depending on the crack density percentage ($\varepsilon$). Thus,

$$S_2^2 = S_1^2\left[1 - \frac{16}{3}\varepsilon\left(\frac{\lambda + 2\mu}{3\lambda + 4\mu}\right)\cdot\left(\frac{1}{1+M}\right)\right], \quad \text{for first order,} \tag{A-1}$$

and,

$$S_2^2 = S_1^2\left[1 - \frac{16}{3}\varepsilon\left(\frac{\lambda + 2\mu}{3\lambda + 4\mu}\right)\cdot\left(\frac{1}{1+M}\right) + \frac{512}{135}\varepsilon^2\frac{(3\lambda + 8\mu)}{(3\lambda + 4\mu)^2}\cdot\frac{1}{(1+M)^2}\right] \tag{A-2}$$

for second order. The parameter $M$ is given by,

$$M = \frac{4}{\pi}\left(\frac{a\bar{\mu}}{c\mu}\right)\left(\frac{\lambda + 2\mu}{3\mu + 4\mu}\right),$$

where $a$ is the radius of the crack, $c$ is the thickness of the crack, and $\bar{\mu}$ is the Lamé constant of the crack. The elements $\lambda$ and $\mu$ are the Lamé constants of the isotropic material and are known.

$$Put\ \frac{16}{3}\left(\frac{\lambda + 2\mu}{3\lambda + 4\mu}\right)\cdot\left(\frac{1}{1+M}\right) = A,$$

$$Put\ \frac{512}{135}\left(\frac{\lambda + 2\mu}{3\lambda + 4\mu}\right)\cdot\left(\frac{1}{1+M}\right)^2 = B.$$

Equations (A-1) and (A-2) become

$$S_2^2 = S_1^2[1 - A\varepsilon], \quad \text{for first order} \tag{A-3}$$

and

$$S_2^2 = S_1^2[1 - A\varepsilon + B\varepsilon^2], \quad \text{for second order.} \tag{A-4}$$

By manipulating equation (A-4) we obtain,

$$\frac{S_1^2}{S_2^2} = \frac{1}{[1 - A\varepsilon + B\varepsilon^2]}. \tag{A-5}$$

By subtracting 1 from both sides of equation (A-5) we obtain

$$\frac{S_1^2}{S_2^2} - 1 = \frac{1}{[1 - A\varepsilon + B\varepsilon^2]} - 1 \tag{A-6}$$

with further rearrangement, we get

$$\left(\frac{S_1^2}{S_2^2} - 1\right) = \frac{[A\varepsilon - B\varepsilon^2]}{[1 - A\varepsilon + B\varepsilon^2]}. \tag{A-7}$$

By multiplying both sides of equation (A-7) by 1/2, we obtain

$$1/2\left(\frac{S_1^2}{S_2^2} - 1\right) = 1/2\,\frac{[A\varepsilon - B\varepsilon^2]}{[1 - A\varepsilon + B\varepsilon^2]}. \tag{A-8}$$

The left-hand side of equation (A-8) is the same as the formula given by Thomsen (1986) for the anisotropy parameter $\gamma$.

$$\text{Hence,}\quad \gamma = 1/2\,\frac{[A\varepsilon - B\varepsilon^2]}{[1 - A\varepsilon + B\varepsilon^2]} \tag{A-9}$$

assuming we are dealing with a first-order perturbation, "$\gamma$" becomes

$$\gamma = 1/2\left(\frac{A\varepsilon}{1 - A\varepsilon}\right). \tag{A-10}$$

Thus, the relationship between anisotropy parameter ($\gamma$) and crack density ($\varepsilon$) has been established.

# Experimental separation of constituent anisotropies using contact measurements

Julie A. Hood* and Richard B. Mignogna‡

**ABSTRACT**

A material may appear homogeneous and anisotropic when the scale of its fabric is smaller than the wavelengths that measure it. These fabrics can result from a variety of causes such as thin layering or oriented stress-fractures. In the case of a stressed, layered material, the resultant anisotropy is a complex combination of the two component anisotropies. This work attempts to experimentally separate the effects of vertical fractures from a background that is already anisotropic. Vertical fracturing is superposed on a transversely isotropic material. The final material is presumed to have orthorhombic symmetry. Laboratory measurements are made at ultrasonic frequencies to determine the elastic moduli of this material. From the elastic moduli, the elastic properties of the original background material, as if it had no vertical fractures, are determined as well as the additional compliance that the fractures add to the system.

## INTRODUCTION

Fine-scale layering predominates in the Earth's crust. Horizontal layering in sedimentary basins and the upper oceanic crust generates transversely isotropic (TI) symmetry; hexagonal symmetry with a vertical symmetry axis. The elastic properties of a TI medium do not vary with azimuth. However, azimuthal variations in sedimentary basins (session on Azimuthal Anisotropy: Shear Wave Birefringence presented at the 60th Annual International Meeting of the Society of Exploration Geophysicists, 1986) and the oceanic crust (Stephen, 1981, 1985; White and Whitmarsh, 1984; Shearer and Orcutt, 1985) have been widely detected. These azimuthal variations have been presumed to be the result of stress-induced vertical fracturing (Crampin et al., 1986). Realistic Earth models must include all the significant constituent anisotropies of the fracture systems and the backgrounds in which these systems are embedded.

Hsu and Schoenberg (1990) showed that a system of stacked plates is equivalent to a homogeneous TI system in the long wavelength limit. Vertical fractures can be generated in a similar fashion. A stack of bonded plates (TI) can be cut into thin vertical wafers and bonded back together to simulate a vertically fractured, transversely isotropic medium (VFTI). In the long wavelength limit a VFTI material appears homogeneous and anisotropic with orthorhombic symmetry. This paper presents an experimental verification of the separation method presented in Hood and Schoenberg (1989) which estimates the amount of compliance that fracturing adds to a system that is already anisotropic.

The verification, using a laboratory-generated VFTI sample, was accomplished in two stages. The first stage was to build a sample with TI background. This sample was made by bonding glass plates together using an ultraviolet curing epoxy (see Figure 1 for a schematic diagram of the TI sample and Figure 2 for a photograph of the actual TI specimen). The bonding epoxy between each plate adds additional compliance to the system. As Hsu and Schoenberg (1990) showed, a material with these geometric characteristics resembles a horizontally fractured, isotropic medium.

For long wavelengths, say $\geq 10h$, where $h = h_g + h_e$, the inhomogeneous conglomerate of isotropic components appears homogeneous and anisotropic with TI symmetry (Helbig, 1984; Melia and Carlson, 1984; Carcione et al., 1991). The $h$ is the period of repeating horizontal layers, $h_g$ is the thickness of each glass plate, and $h_e$ is the thickness of the epoxy bond between

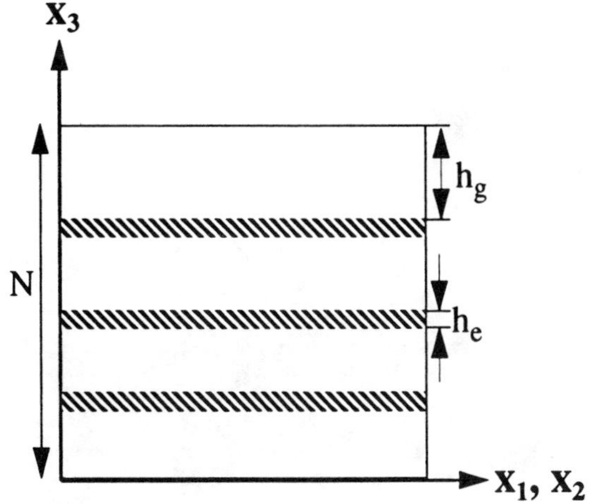


FIG. 1. Periodically layered system with two components (TI sample).

*Rosenstiel School of Marine and Atmospheric Science, University of Miami, Miami, FL 33149-1098
‡Mechanics of Materials Branch - Code 6380, Naval Research Laboratory, Washington, DC 20375-5000

plates. The extra tangential and normal compliance that the epoxy adds in the form of horizontal fractures (subscripted with $hf$) was modeled with the parameters $E_{T_{hf}}$ and $E_{N_{hf}}$, respectively. The $E_{T_{hf}}$ and $E_{N_{hf}}$ are functions of the material properties of the glass ($\lambda_g$ and $\mu_g$), epoxy ($\lambda_e$ and $\mu_e$), and the horizontal fracture thickness $h_e$.

Stage two was the addition of vertical fractures to the TI sample. The vertical fractures are simulated by cutting the stack of bonded plates into thin wafers and then rebonding the wafers (see Figure 3 for a schematic diagram of the VFTI sample and Figure 4 for a photograph of the actual VFTI specimen). The extra tangential and normal compliance that the vertical fractures (subscripted with $vf$) add is modeled by $E_{T_{vf}}$ and $E_{N_{vf}}$, respectively. Like $E_{T_{hf}}$ and $E_{N_{hf}}$, $E_{T_{vf}}$ and $E_{N_{vf}}$ are functions of the material properties as well as the amount of epoxy between the vertical wafers. The system of cut, stacked, and bonded plates appears homogeneous and anisotropic with orthorhombic symmetry when measured with sufficiently long wavelengths.

The elastic moduli of each constituent component of the VFTI system are measured during the separate stages of sample preparation. These estimates provide a means of determining how well our method of separating constituent anisotropies works. These intermediate constituent moduli calculations help us determine how advantageous it is to apply this method of separation to complex geophysical structures.

The moduli of the VFTI medium are complicated functions of all the constituent anisotropies: the Lame parameters $\lambda_g$ and $\mu_g$ of the isotropic background glass and the additional compliances $E_{T_{hf}}$, $E_{N_{hf}}$, $E_{T_{vf}}$, and $E_{N_{vf}}$ that each fracture set adds to the overall material. Our separation method, however, reveals each constituent parameter of this system as a function of the measured moduli. Therefore, the fracture properties and a description of the background medium, as if the background had no vertical fractures, are provided.

## TI FROM HORIZONTAL FRACTURES

The TI background is generated from embedding a set of parallel planar fractures horizontally into an isotropic background. Hooke's law for a general linear elastic medium can be written

$$\boldsymbol{\sigma} = \underset{\sim}{C}\,\boldsymbol{\varepsilon} \quad \text{or} \quad \boldsymbol{\varepsilon} = \underset{\sim}{S}\,\boldsymbol{\sigma} \tag{1}$$

(Auld, 1973). The $\underset{\sim}{C}$ and $\underset{\sim}{S}$ are the symmetric $6 \times 6$ matrices of stiffnesses and compliances, respectively, and are inverses of each other. Using condensed notation, the six-vectors $\boldsymbol{\sigma}$ and $\boldsymbol{\varepsilon}$ which contain the stress and strain tensor components become

$$\boldsymbol{\sigma} = [\sigma_1, \sigma_2, \sigma_3, \sigma_4, \sigma_5, \sigma_6]^t \equiv [\sigma_{11}, \sigma_{22}, \sigma_{33}, \sigma_{23}, \sigma_{13}, \sigma_{12}]^t,$$

$$\boldsymbol{\varepsilon} = [\epsilon_1, \epsilon_2, \epsilon_3, \epsilon_4, \epsilon_5, \epsilon_6]^t \equiv [\epsilon_{11}, \epsilon_{22}, \epsilon_{33}, 2\epsilon_{23}, 2\epsilon_{13}, 2\epsilon_{12}]^t \tag{2}$$

FIG. 2. Photograph of TI specimen, note U.S. penny for scale: height in $x_3$ direction is 3.062 cm.

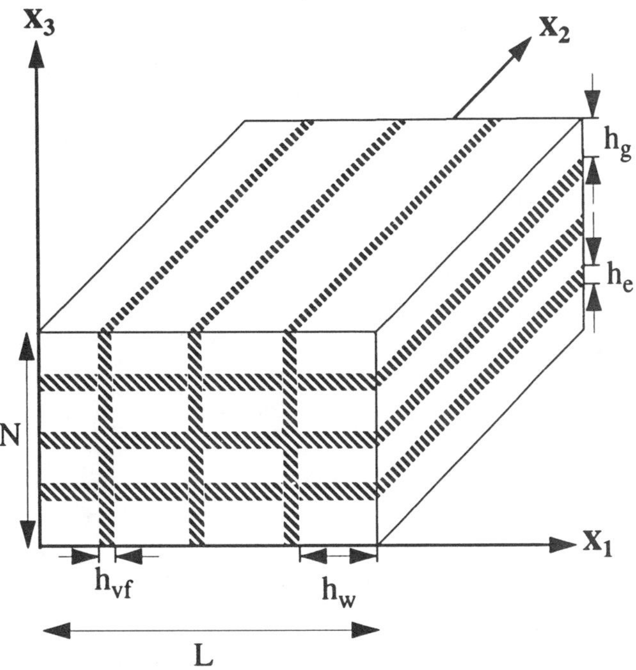


FIG. 3. Vertically fractured, transversely isotropic medium (VFTI sample).

FIG. 4. Photograph of VFTI specimen: height in $x_3$ direction is 3.062 cm; length in $x_1$ direction is 1.880 cm; length in $x_2$ direction is 3.823 cm.

Superscript $t$ denotes the transpose.

The matrix of elastic stiffness moduli for the background isotropic material is

$$\underset{\sim}{C}_I = \begin{bmatrix} c_{11_I} & c_{12_I} & c_{12_I} & 0 & 0 & 0 \\ c_{12_I} & c_{11_I} & c_{12_I} & 0 & 0 & 0 \\ c_{12_I} & c_{12_I} & c_{11_I} & 0 & 0 & 0 \\ 0 & 0 & 0 & c_{44_I} & 0 & 0 \\ 0 & 0 & 0 & 0 & c_{44_I} & 0 \\ 0 & 0 & 0 & 0 & 0 & c_{44_I} \end{bmatrix}, \tag{3}$$

where $c_{11_I} = \lambda_g + 2\mu_g$, $c_{44_I} = \mu_g$, and $c_{12_I} = c_{11_I} - 2c_{44_I} = \lambda_g$. The inverse of the stiffness matrix is the compliance matrix

$$\underset{\sim}{S}_I = \begin{bmatrix} s_{11_I} & s_{12_I} & s_{12_I} & 0 & 0 & 0 \\ s_{12_I} & s_{11_I} & s_{12_I} & 0 & 0 & 0 \\ s_{12_I} & s_{12_I} & s_{11_I} & 0 & 0 & 0 \\ 0 & 0 & 0 & s_{44_I} & 0 & 0 \\ 0 & 0 & 0 & 0 & s_{44_I} & 0 \\ 0 & 0 & 0 & 0 & 0 & s_{44_I} \end{bmatrix}, \tag{4}$$

where

$$s_{11_I} = \frac{\lambda_g + \mu_g}{\mu_g(3\lambda_g + 2\mu_g)}, \quad s_{44_I} = \frac{1}{\mu_g},$$

$$s_{12_I} = s_{11_I} - \frac{1}{2}s_{44_I} = \frac{-\lambda_g}{2\mu_g(3\lambda_g + 2\mu_g)}.$$

Schoenberg's (1980) linear slip interface LSD model is used to represent sets of fractures (see Appendix A). Fracturing $\underset{\sim}{F}$ is incorporated into a background $\underset{\sim}{S}_b$ following a method introduced in Nichols et al. (1989)

$$\underset{\sim}{S} = \underset{\sim}{S}_b + \underset{\sim}{E}^t\,\underset{\sim}{Z}\,\underset{\sim}{E} = \underset{\sim}{S}_b + \underset{\sim}{F}, \tag{5}$$

where

$$\underset{\sim}{E} = \begin{bmatrix} 0 & 0 & 1 & 0 & 0 & 0 \\ 0 & 0 & 0 & 1 & 0 & 0 \\ 0 & 0 & 0 & 0 & 1 & 0 \end{bmatrix}.$$

Using the matrix $\underset{\sim}{Z}$ from equation (A-3), a set of axisymmetric horizontal fractures can be represented as

$$\underset{\sim}{F}_{hf} = \frac{1}{\mu_g}\begin{bmatrix} 0 & 0 & 0 & 0 & 0 & 0 \\ 0 & 0 & 0 & 0 & 0 & 0 \\ 0 & 0 & E_{N_{hf}} & 0 & 0 & 0 \\ 0 & 0 & 0 & E_{T_{hf}} & 0 & 0 \\ 0 & 0 & 0 & 0 & E_{T_{hf}} & 0 \\ 0 & 0 & 0 & 0 & 0 & 0 \end{bmatrix} \tag{6}$$

$$= \begin{bmatrix} 0 & 0 & 0 & 0 & 0 & 0 \\ 0 & 0 & 0 & 0 & 0 & 0 \\ 0 & 0 & Z_N & 0 & 0 & 0 \\ 0 & 0 & 0 & Z_T & 0 & 0 \\ 0 & 0 & 0 & 0 & Z_T & 0 \\ 0 & 0 & 0 & 0 & 0 & 0 \end{bmatrix}$$

where $E_{N_{hf}}$ is the additional normal compliance that the horizontal fracturing gives the material, $E_{T_{hf}}$ is the extra tangential compliance, and the subscript $hf$ represents horizontal fracturing. Equation (6) represents the additional compliance that horizontal fracturing adds to a system. Therefore, in the long wavelength limit, a TI medium $\underset{\sim}{S}_{TI}$ is created by horizontally fracturing an isotropic background $\underset{\sim}{S}_I$ according to equation (5) as

$$\underset{\sim}{S}_{TI} = \underset{\sim}{S}_I + \underset{\sim}{F}_{hf}. \tag{7}$$

The compliance matrix representing an isotropic material with horizontal fracturing is

$$\underset{\sim}{S}_{TI} = \begin{bmatrix} s_{11_{TI}} & s_{12_{TI}} & s_{12_{TI}} & 0 & 0 & 0 \\ s_{12_{TI}} & s_{11_{TI}} & s_{12_{TI}} & 0 & 0 & 0 \\ s_{12_{TI}} & s_{12_{TI}} & s_{33_{TI}} & 0 & 0 & 0 \\ 0 & 0 & 0 & s_{44_{TI}} & 0 & 0 \\ 0 & 0 & 0 & 0 & s_{44_{TI}} & 0 \\ 0 & 0 & 0 & 0 & 0 & s_{66_{TI}} \end{bmatrix}, \tag{8}$$

where

$$s_{11_{TI}} = \frac{\lambda_g + \mu_g}{\mu_g(3\lambda_g + 2\mu_g)}, \quad s_{33_{TI}} = \frac{\lambda_g + \mu_g + E_{N_{hf}}(3\lambda_g + 2\mu_g)}{\mu_g(3\lambda_g + 2\mu_g)},$$

$$s_{44_{TI}} = \frac{1 + E_{T_{hf}}}{\mu_g}, \quad s_{66_{TI}} = \frac{1}{\mu_g},$$

$$s_{12_{TI}} = s_{11_{TI}} - \frac{1}{2}s_{66_{TI}} = \frac{-\lambda_g}{2\mu_g(3\lambda_g + 2\mu_g)}.$$

The inverse of $\underset{\sim}{S}_{TI}$ gives the stiffness matrix $\underset{\sim}{C}_{TI}$ as

$$\underset{\sim}{C}_{TI} = \begin{bmatrix} c_{11_{TI}} & c_{12_{TI}} & c_{13_{TI}} & 0 & 0 & 0 \\ c_{12_{TI}} & c_{11_{TI}} & c_{13_{TI}} & 0 & 0 & 0 \\ c_{13_{TI}} & c_{13_{TI}} & c_{33_{TI}} & 0 & 0 & 0 \\ 0 & 0 & 0 & c_{44_{TI}} & 0 & 0 \\ 0 & 0 & 0 & 0 & c_{44_{TI}} & 0 \\ 0 & 0 & 0 & 0 & 0 & c_{66_{TI}} \end{bmatrix}, \tag{9}$$

where

$$c_{11_{TI}} = \frac{\mu_g\left[\lambda_g + 2\mu_g + 4E_{N_{hf}}(\lambda_g + \mu_g)\right]}{\mu_g + E_{N_{hf}}(\lambda_g + 2\mu_g)},$$

$$c_{13_{TI}} = \frac{\lambda_g\mu_g}{\mu_g + E_{N_{hf}}(\lambda_g + 2\mu_g)}, \quad c_{33_{TI}} = \frac{\mu_g(\lambda_g + 2\mu_g)}{\mu_g + E_{N_{hf}}(\lambda_g + 2\mu_g)},$$

$$c_{44_{TI}} = \frac{\mu_g}{1 + E_{T_{hf}}}, \quad c_{66_{TI}} = \mu_g,$$

$$c_{12_{TI}} = c_{11_{TI}} - 2c_{66_{TI}} = \frac{\lambda_g\mu_g\left(1 + 2E_{N_{hf}}\right)}{\mu_g + E_{N_{hf}}(\lambda_g + 2\mu_g)}.$$

The form of equations (8) and (9) shows that equation (7) generates a system that has TI symmetry with $x_3$ as the axis of symmetry.

The entries of matrix $\underset{\sim}{C}_{TI}$ (equation 9) can be shown to be the same as those presented in Schoenberg (1983) for a horizontally fractured medium:

$$c_{44_{TI}} = \frac{\mu_g}{1 + E_{T_{hf}}}, \quad c_{66_{TI}} = \mu_g,$$

$$c_{33_{TI}} = \frac{\mu_g}{\gamma_g + E_{N_{hf}}}, \quad c_{13_{TI}} = (1 - 2\gamma_g)\,c_{33_{TI}},$$

$$\begin{aligned} c_{11_{TI}} &= 4(1 - \gamma_g)\,\mu_g + (1 - 2\gamma_g)^2\,c_{33_{TI}} \\ &= \left[1 + 4(1 - \gamma_g)\,E_{N_{hf}}\right]c_{33_{TI}} \end{aligned} \tag{10}$$

with $\gamma_g$ as defined in equation (A-1). There are only four independent parameters in this TI system: the $\lambda_g$ and $\mu_g$ of the isotropic glass background and the additional compliance that the epoxy adds to the system in the form of $E_{N_{hf}}$ and $E_{T_{hf}}$.

## ADDITION OF VERTICAL FRACTURES TO GENERATE ORTHORHOMBIC SYMMETRY

The change in compliance that horizontal fractures add to a system $\underset{\sim}{E}_{hf}$ is represented by equation (6). Matrix $\underset{\sim}{E}_{hf}$ is simply added to the compliance matrix representative of the background following equation (7). This simple relation holds only when the fracture planes coincide with the horizontal plane. A set of axisymmetric vertical fractures, however, must be transformed into the appropriate coordinate system before equation (5) can be used. Following Hood (1991), a set of axisymmetric vertical fractures has the form

$$\underset{\sim}{E}_{vf} = \frac{1}{\mu_g}\begin{bmatrix} E_{N_{vf}} & 0 & 0 & 0 & 0 & 0 \\ 0 & 0 & 0 & 0 & 0 & 0 \\ 0 & 0 & 0 & 0 & 0 & 0 \\ 0 & 0 & 0 & 0 & 0 & 0 \\ 0 & 0 & 0 & 0 & E_{T_{vf}} & 0 \\ 0 & 0 & 0 & 0 & 0 & E_{T_{vf}} \end{bmatrix} = \begin{bmatrix} Z_{N_{vf}} & 0 & 0 & 0 & 0 & 0 \\ 0 & 0 & 0 & 0 & 0 & 0 \\ 0 & 0 & 0 & 0 & 0 & 0 \\ 0 & 0 & 0 & 0 & 0 & 0 \\ 0 & 0 & 0 & 0 & Z_{T_{vf}} & 0 \\ 0 & 0 & 0 & 0 & 0 & Z_{T_{vf}} \end{bmatrix} \quad (11)$$

with respect to the coordinates of the background material where $vf$ indicates vertical fracturing. The $E_{N_{vf}}$ and $E_{T_{vf}}$ are the dimensionless compliances normal and tangential to the vertical fracture planes, respectively.

Note that although the same bonding material is used to create both the horizontal and vertical fractures, the compliances representative of the separate fracture systems are generally different. This is because the thickness of vertical fractures is not necessarily equal to the thickness of the horizontal fractures, i.e., $h_{vf} \neq h_e$ (see Figure 1) and the fracture infilling materials may have different material properties as well. Therefore, in general, $E_{N_{vf}} \neq E_{N_{hf}}$ and $E_{T_{vf}} \neq E_{T_{hf}}$.

The cut stacked plates are represented by

$$\underset{\sim}{S}_{VFTI} = \underset{\sim}{S}_{TI} + \underset{\sim}{E}_{vf}\,. \quad (12)$$

Using equation (11) to represent vertical fractures and $\underset{\sim}{S}_{TI}$ in equation (8) for the background

$$\underset{\sim}{S}_{VFTI} = \begin{bmatrix} s_{11VFTI} & s_{12VFTI} & s_{12VFTI} & 0 & 0 & 0 \\ s_{12VFTI} & s_{22VFTI} & s_{12VFTI} & 0 & 0 & 0 \\ s_{12VFTI} & s_{12VFTI} & s_{33VFTI} & 0 & 0 & 0 \\ 0 & 0 & 0 & s_{44VFTI} & 0 & 0 \\ 0 & 0 & 0 & 0 & s_{55VFTI} & 0 \\ 0 & 0 & 0 & 0 & 0 & s_{66VFTI} \end{bmatrix}, \quad (13)$$

where

$$s_{11VFTI} = \frac{\lambda_g + \mu_g + E_{N_{vf}}(3\lambda_g + 2\mu_g)}{\mu_g(3\lambda_g + 2\mu_g)},$$

$$s_{33VFTI} = \frac{\lambda_g + \mu_g + E_{N_{hf}}(3\lambda_g + 2\mu_g)}{\mu_g(3\lambda_g + 2\mu_g)},$$

$$s_{12VFTI} = \frac{-\lambda_g}{2\mu_g(3\lambda_g + 2\mu_g)}, \quad s_{22VFTI} = \frac{\lambda_g + \mu_g}{\mu_g(3\lambda_g + 2\mu_g)},$$

$$s_{44VFTI} = \frac{1 + E_{T_{hf}}}{\mu_g}, \quad s_{55VFTI} = \frac{1 + E_{T_{vf}} + E_{T_{hf}}}{\mu_g},$$

$$s_{66VFTI} = \frac{1 + E_{T_{vf}}}{\mu_g}.$$

The inverse of $\underset{\sim}{S}_{VFTI}$ gives the stiffness matrix $\underset{\sim}{C}_{VFTI}$ as

$$\underset{\sim}{C}_{VFTI} = \begin{bmatrix} c_{11VFTI} & c_{12VFTI} & c_{13VFTI} & 0 & 0 & 0 \\ c_{12VFTI} & c_{22VFTI} & c_{23VFTI} & 0 & 0 & 0 \\ c_{13VFTI} & c_{23VFTI} & c_{33VFTI} & 0 & 0 & 0 \\ 0 & 0 & 0 & c_{44VFTI} & 0 & 0 \\ 0 & 0 & 0 & 0 & c_{55VFTI} & 0 \\ 0 & 0 & 0 & 0 & 0 & c_{66VFTI} \end{bmatrix}, \quad (14)$$

where

$$c_{11VFTI} = \frac{\mu_g\left[\lambda_g + 2\mu_g + 4E_{N_{hf}}(\lambda_g + \mu_g)\right]}{D},$$

$$c_{22VFTI} = \frac{\mu_g\left[\lambda_g + 2\mu_g + 4\left(E_{N_{hf}} + E_{N_{vf}}\right)(\lambda_g + \mu_g) + 4E_{N_{hf}}E_{N_{vf}}(3\lambda_g + 2\mu_g)\right]}{D},$$

$$c_{33VFTI} = \frac{\mu_g\left[\lambda_g + 2\mu_g + 4E_{N_{vf}}(\lambda_g + \mu_g)\right]}{D},$$

$$c_{12VFTI} = \frac{\left(1 + 2E_{N_{hf}}\right)\lambda_g\mu_g}{D}, \quad c_{13VFTI} = \frac{\lambda_g\mu_g}{D},$$

$$c_{23VFTI} = \frac{\left(1 + 2E_{N_{vf}}\right)\lambda_g\mu_g}{D}, \quad c_{44VFTI} = \frac{\mu_g}{1 + E_{T_{hf}}},$$

$$c_{55VFTI} = \frac{\mu_g}{1 + E_{T_{hf}} + E_{T_{vf}}}, \quad c_{66VFTI} = \frac{\mu_g}{1 + E_{T_{vf}}},$$

where

$$D = \mu_g + \left(E_{N_{hf}} + E_{N_{vf}}\right)(\lambda_g + 2\mu_g) + 4E_{N_{hf}}E_{N_{vf}}(\lambda_g + \mu_g)\,.$$

The form of equation (14) indicates that in the long wavelength limit the system has orthorhombic symmetry. Note that the following relationships hold

$$s_{12VFTI} = s_{13VFTI} = s_{23VFTI}\,, \quad (15)$$

$$s_{44VFTI} + s_{66VFTI} - s_{55VFTI} = 2\left(s_{22VFTI} - s_{12VFTI}\right), \quad (16)$$

which reduce the number of independent parameters in this system to six; general orthorhombic media have nine. These relationships give us bases for determining how closely our constructed VFTI medium conforms to the specified symmetries imposed by the fracture model and by the background.

## SEPARATING THE VERTICAL FRACTURING FROM THE TI BACKGROUND

Our goal is to solve for the six unknown material properties $\mu_g$, $\lambda_g$, $E_{T_{hf}}$, $E_{T_{vf}}$, $E_{N_{hf}}$, $E_{N_{vf}}$ as functions of the elastic moduli $c_{ij}$ of the composite medium. One must begin with data determined by measurements in the form

$$\underset{\sim}{C} = \begin{bmatrix} c_{11} & c_{12} & c_{13} & 0 & 0 & 0 \\ c_{12} & c_{22} & c_{23} & 0 & 0 & 0 \\ c_{13} & c_{23} & c_{33} & 0 & 0 & 0 \\ 0 & 0 & 0 & c_{44} & 0 & 0 \\ 0 & 0 & 0 & 0 & c_{55} & 0 \\ 0 & 0 & 0 & 0 & 0 & c_{66} \end{bmatrix}$$

which corresponds to (17)

$$\underset{\sim}{S} = \begin{bmatrix} s_{11} & s_{12} & s_{13} & 0 & 0 & 0 \\ s_{12} & s_{22} & s_{23} & 0 & 0 & 0 \\ s_{13} & s_{23} & s_{33} & 0 & 0 & 0 \\ 0 & 0 & 0 & s_{44} & 0 & 0 \\ 0 & 0 & 0 & 0 & s_{55} & 0 \\ 0 & 0 & 0 & 0 & 0 & s_{66} \end{bmatrix}$$

If the data indicate that the medium is orthorhombic with $c_{13}(c_{22}+c_{12}) = c_{23}(c_{11}+c_{12})$ (Hood and Schoenberg, 1989) or $s_{13} = s_{23}$ (Hood, 1991), then assuming VFTI as a model is appropriate and the following procedures extract the effects of fracturing. The six relationships for these diagonal elements, the $c_{ijVFTI}$ where $i = j$ in equation (14), provide six equations in the six unknowns $\mu_g$, $\lambda_g$, $E_{T_{hf}}$, $E_{T_{vf}}$, $E_{N_{hf}}$, and $E_{N_{vf}}$. These unknowns can therefore be unambiguously determined as functions of the measured elastic moduli of equation (17).

By using the three equations $c_{44_{VFTI}}$, $c_{55_{VFTI}}$, and $c_{66_{VFTI}}$ of equation (14), the parameters $E_{T_{hf}}$, $E_{T_{vf}}$, and $\mu_g$ can be determined from measured shear-wave velocities:

$$\mu_g = \frac{1}{s_{44}+s_{66}-s_{55}} = \frac{c_{44}c_{55}c_{66}}{c_{55}c_{66}+c_{44}c_{55}-c_{44}c_{66}}, \tag{18}$$

$$E_{T_{hf}} = \frac{s_{55}-s_{66}}{s_{44}+s_{66}-s_{55}} = \frac{\mu_g}{c_{44}} - 1, \tag{19}$$

$$E_{T_{vf}} = \frac{s_{55}-s_{44}}{s_{44}+s_{66}-s_{55}} = \frac{\mu_g}{c_{66}} - 1. \tag{20}$$

If values for all the elastic moduli were known, parameters $E_{N_{hf}}$, $E_{N_{vf}}$, and $\lambda_g$ would be obtained from the longitudinal velocity measurements by using the three equations $c_{11_{VFTI}}$, $c_{22_{VFTI}}$, and $c_{33_{VFTI}}$ of equation (14):

$$\lambda_g = \frac{2\mu_g^2 s_{22} - \mu_g}{1-3\mu_g s_{22}} = \frac{2a_{13}\mu_g^2 - \mu_g}{1-3a_{13}\mu_g}, \tag{21}$$

$$E_{N_{hf}} = \frac{s_{33}\mu_g(3\lambda_g+2\mu_g) - \lambda_g - \mu_g}{3\lambda_g+2\mu_g} = \frac{a_{12}\mu_g(3\lambda_g+2\mu_g) - \lambda_g - \mu_g}{3\lambda_g+2\mu_g}, \tag{22}$$

$$E_{N_{vf}} = \frac{s_{11}\mu_g(3\lambda_g+2\mu_g) - \lambda_g - \mu_g}{3\lambda_g+2\mu_g} = \frac{a_{23}\mu_g(3\lambda_g+2\mu_g) - \lambda_g - \mu_g}{3\lambda_g+2\mu_g}, \tag{23}$$

where

$$a_{13} = \frac{c_{11}c_{33}-c_{13}^2}{X}, \quad a_{23} = \frac{c_{22}c_{33}-c_{23}^2}{X}, \quad a_{12} = \frac{c_{11}c_{22}-c_{12}^2}{X},$$

$$X = c_{11}c_{22}c_{33} - c_{11}c_{23}^2 - c_{22}c_{13}^2 - c_{33}c_{12}^2 + 2c_{12}c_{13}c_{23}.$$

Our contact measurements do not provide the off diagonal elements $c_{12}$, $c_{13}$, $c_{23}$ of $\underset{\sim}{C}$. However, since there are only six independent parameters in this VFTI system, and because the faces of the sample coincide with the symmetry planes, contact measurements provide the values for the six diagonal entries of $\underset{\sim}{C}$ in equation (17) (following the methods outlined in Appendix B) and are, therefore, sufficient to uniquely describe the elastic properties of our material.

Parameters $\mu_g$, $E_{T_{hf}}$, $E_{T_{vf}}$ were determined as functions of the measured $c_{ij}$ by using equations (18), (19), and (20), respectively. The unknowns $\lambda_g$, $E_{N_{hf}}$, and $E_{N_{vf}}$, however, needed to be determined from the longitudinal velocity measurements by first using $c_{11_{VFTI}}$ and $c_{33_{VFTI}}$ of equation (14) to provide $E_{N_{hf}}$ and $E_{N_{vf}}$ as functions containing $\lambda_g$:

$$E_{N_{hf}} = \frac{2\mu_g(\lambda_g+\mu_g) - c_{33}(\lambda_g+2\mu_g) + \sqrt{c_{11}c_{33}\lambda_g^2 + 4\mu_g^2(\lambda_g+\mu_g)^2}}{4c_{33}(\lambda_g+\mu_g)}, \tag{24}$$

and

$$E_{N_{vf}} = \frac{2\mu_g(\lambda_g+\mu_g) - c_{11}(\lambda_g+2\mu_g) + \sqrt{c_{11}c_{33}\lambda_g^2 + 4\mu_g^2(\lambda_g+\mu_g)^2}}{4c_{11}(\lambda_g+\mu_g)}. \tag{25}$$

These expressions for $E_{N_{hf}}$ and $E_{N_{vf}}$ were then substituted into $c_{22_{VFTI}}$ of equation (14) and $\lambda_g$ was solved for directly as a function of the measured $c_{ij}$:

$$\lambda_g = \frac{\frac{-\mu_1 A3}{A1} + \sqrt{\frac{A4}{A1} + \frac{\mu_1^2 A3^2}{A1^2}}}{2}, \tag{26}$$

where

$$A1 = (c_{11}-4c_{22})^2 - 2c_{11}c_{33} - 8c_{22}c_{33} + c_{33}^2 + 16c_{11}\mu_1 - 64c_{22}\mu_1 + 16c_{33}\mu_1 + 48\mu_1^2,$$

$$A2 = \sqrt{c_{11}c_{22} - 2c_{11}\mu_1 + \mu_1^2},$$

$$A3 = -8c_{11}c_{22} + 32c_{22}^2 - 8c_{22}c_{33} + 16c_{11}\mu_1 - 128c_{22}\mu_1 + 16c_{33}\mu_1 + 112\mu_1^2 - 4A2(c_{11}+4c_{22}+c_{33}-8\mu_1),$$

and

$$A4 = -16\mu_1^2\left(c_{11}c_{22} + 4c_{22}^2 - c_{22}c_{33} - 2c_{11}\mu_1 - 16c_{22}\mu_1 + 2c_{33}\mu_1 + 16\mu_1^2 + 4A2[c_{22}-2\mu_1]\right).$$

This value of $\lambda_g$ was then used in equations (24) and (25) to calculate numerical values for $E_{N_{hf}}$ and $E_{N_{vf}}$, respectively.

Note that there is a special case when an orthorhombic system has only four independent parameters. This situation could occur in a VFTI material, for example, if the vertical wafers were bonded with the same agent as that used to bond the stacked plates and the thickness between vertical wafers was the same as the thickness of the bonding agent between the stacked plates. In this state $E_{N_{vf}} = E_{N_{hf}} \equiv E_N$ and $E_{T_{vf}} = E_{T_{hf}} \equiv E_T$ so that both the vertical fracturing and horizontal fracturing have an equivalent fracture density and have the same infilling material. The resulting compliance matrix will still be orthorhombic but now there are more relationships between the $s_{ij}$ so that $S_{VFTI}$ has fewer independent parameters. Two more relationships between the measured moduli occur:

$$s_{11_{VFTI}} = s_{33_{VFTI}} \iff c_{11_{VFTI}} = c_{33_{VFTI}}, \tag{27}$$

and

$$s_{44_{VFTI}} = s_{66_{VFTI}} \iff c_{44_{VFTI}} = c_{66_{VFTI}} \, . \tag{28}$$

The four independent parameters are the $\lambda_g$ and $\mu_g$ of the background and the excess compliance in the form of $E_N$ and $E_T$ that the fractures add to the system.

## EXPERIMENTAL PROCEDURE AND RESULTS

### Transducers, physical arrangement, and electronics

Experimental longitudinal and shear-wave traveltime measurements were made using contact, through-transmission methods at 0.5 and 1.0 MHz. These frequencies were chosen to ensure that a long wavelength approximation would be a valid assumption. Longitudinal waves at both 0.5 and 1.0 MHz have no problem meeting the long wavelength requirement. However, for the case of shear at 1.0 MHz, the corresponding wavelength is only slightly more than seven times the period of the repeating layers $h_g + h_e$, glass-epoxy in the horizontal dimension or $h_w + h_{vf}$, wafer-epoxy in the vertical dimension. This may be pushing the functional limits of the long wavelength approximation according to Carcione et al. (1991), but appears to work, nonetheless.

The 0.5 MHz longitudinal and shear-wave measurements were made with corresponding pairs (transmitter and receiver) of commercial transducers[1], 2.54 cm (1.0 in) in diameter. The 1.0 MHz longitudinal measurements were made using a pair of 18.8 mm (0.75 in) diameter commercial transducers and the shear-wave measurements were made with 2.54 cm (1.0 in) diameter commercial transducers.

The transducers were mounted in a rotatable, spring-loaded fixture to ensure a constant transducer pressure against the specimen and to decrease couplant-induced variations in the measurements. Tap water was used as the longitudinal couplant. A commercial water soluble gel mixed with molybdenum disulfide was used for the shear-wave couplant.

The pulsed transducers were driven using a gated amplifier having a frequency range of 0.5 to 25.0 MHz. The source for the gated amplifier was a high-stability frequency synthesizer. The received signals were amplified with a broad band receiver also having a frequency range of 0.5 to 25.0 MHz.

### Time measurements

The traveltime was measured using an oscilloscope delta time plug-in. A reference time was established using a "bleed-off" resistor to tap off the initial RF pulse to transmitter. An initial delay was determined for each transducer pair. This delay was as a result of the transducer impedance and wearplates that are used on commercial transducers. The initial delay was measured by placing the transducer pairs face to face and measuring the time difference between the received signal and the reference signal. The specimen was then mounted in the fixture between the transducers. For shear-wave measurements, proper orientation of the transducer polarization direction was also required. The difference in time between the received signals and the reference minus the delay for the transducer wear plates was used as the measured traveltime. Using the specimen thickness measurements and the traveltime measurements, the various velocities were determined.

The estimated error in the velocity measurements was below 2 percent for $P$-waves and 1 percent for $S$-waves. The scatter in the data collected at both 0.5 Mhz and 1.0 Mhz was within the error for longitudinal and shear measurements. Therefore, the TI and VFTI materials were assumed nondispersive and the average values for the two sets of measurements were used in the data inversion.

### Sample preparation of the TI stack

The TI sample was prepared by interleaving 2 × 2 cm glass plates each with a thickness of $h_g = .04 \pm .002$ cm with an epoxy adhesive. The final sample had the shape of a parallelepiped consisting of 75 bonded glass plates. The final height of the stack was 3.062 cm. The relation between the number of layers and the total stack height was $nh_g + (n - 1)h_e = N$, where $n$ was the number of glass plates, $h_g$ the thickness of a glass plate, $n - 1$ the number of epoxy layers, $h_e$ the average epoxy layer thickness, and $N$ the total stack height. The average thickness of the bonding agent between each plate was

$$h_e = \frac{N - nh_g}{n - 1} = \frac{.062 \text{ cm}}{74} = 8.38 \times 10^{-4} \text{cm} \, . \tag{29}$$

Therefore, the epoxy constituted only 2.0 percent of the composite. Therefore, we assumed $h_e << h_g$, i.e., for $h_g + h_e = 1$, $h_e \to 0$ while $h_g \to 1$. This assumption satisfied one of the two Schoenberg criteria for the mixture of these two components to simulate a fractured medium.

### Bulk component properties

Prior to stacking and bonding, the bulk elastic moduli of the individual isotropic components were determined. From their bulk values and knowing the relative amounts of each in the composite sample, the long wavelength anisotropic moduli was calculated according to the Schoenberg LSD model (1980) using equation (9). From traveltime measurements through the bulk glass[2], the glass velocities were

$$V_{p_g} = 5.37 \times 10^5 \frac{\text{cm}}{\text{s}} \; , \; V_{s_g} = 3.37 \times 10^5 \frac{\text{cm}}{\text{s}} \; . \tag{30}$$

The density of the glass was $\rho_g = 2.51 \frac{\text{gm}}{\text{cm}^3}$. The $\mu_g$ and $\lambda_g$ are

$$\mu_g = \rho_g V_{s_g}^2 = 2.85 \times 10^{11} \frac{\text{dyne}}{\text{cm}^2} \, , \tag{31}$$

$$\lambda_g = \rho_g V_{p_g}^2 - 2\mu_g = 1.54 \times 10^{11} \frac{\text{dyne}}{\text{cm}^2} \, . \tag{32}$$

The measured density of the epoxy[3] was $\rho_e = 1.08 \frac{\text{gm}}{\text{cm}^3}$. The $\mu_e$ and $\lambda_e$ of the adhesive were calculated from traveltime measurements through bulk epoxy. The bulk epoxy wave speeds were

$$V_{p_e} = 2.24 \times 10^5 \frac{\text{cm}}{\text{s}} = \sqrt{\frac{\lambda_e + 2\mu_e}{\rho_e}} \, , \tag{33}$$

[1]Manufactured by Panametrics, Inc.
[2]Manufactured by Schott Glass Technologies, Inc., stock #D263.
[3]Manufactured by Loctite Corporation.

$$V_{s_e} = 9.56 \times 10^4 \frac{\text{cm}}{\text{s}} = \sqrt{\frac{\mu_e}{\rho_e}}, \quad (34)$$

which gives

$$\mu_e = \rho_e V_{s_e}^2 = 9.87 \times 10^9 \frac{\text{dyne}}{\text{cm}^2}, \quad (35)$$

$$\lambda_e = \rho_e V_{p_e}^2 - 2\mu_e = 3.44 \times 10^{10} \frac{\text{dyne}}{\text{cm}^2}. \quad (36)$$

Note that $\mu_e << \mu_g$, i.e., $\mu_e \rightarrow 0$ while $\mu_g \rightarrow <\mu>$ which satisfies the other Schoenberg fracture criteria. Therefore, thin layers of epoxy between glass simulates fractures according to Schoenberg's LSD model.

From the elastic moduli of the two isotropic components and their relative proportions, the fracture parameters $E_i$ were calculated as

$$E_{T_{hf}} = \frac{\eta_T \mu_g}{h} = h_e \frac{\mu_g}{\mu_e} = 2.42 \times 10^{-2}, \quad (37)$$

$$E_{N_{hf}} = \frac{\eta_N \mu_g}{h} = h_e \frac{\mu_g}{\lambda_e + 2\mu_e} = 4.41 \times 10^{-3}, \quad (38)$$

where the physical meaning of the $\eta_i$ is discussed in Appendix A. The values $\lambda_g$, $\mu_g$ of the bulk glass [equation (32), (31)] and the calculated $E_{Thf}$, $E_{Nhf}$ [equation (37), (38)] were then used to calculate the forward, expected long wavelength equivalent values for the $c_{ijTI}$ of the stack according to the Schoenberg fracture model [equation (9)]:

$$c_{44_{TI}} = \frac{\mu_g}{1 + E_{T_{hf}}} = 2.78 \times 10^{11} \frac{\text{dyne}}{\text{cm}^2}, \quad (39)$$

$$c_{66_{TI}} = \mu_g = 2.85 \times 10^{11} \frac{\text{dyne}}{\text{cm}^2}, \quad (40)$$

$$c_{33_{TI}} = \frac{\mu_g (\lambda_g + 2\mu_g)}{\mu_g + E_{N_{hf}} (\lambda_g + 2\mu_g)} = 7.16 \times 10^{11} \frac{\text{dyne}}{\text{cm}^2}, \quad (41)$$

$$c_{11_{TI}} = \frac{\mu_g \left[\lambda_g + 2\mu_g + 4E_{N_{hf}} (\lambda_g + \mu_g)\right]}{\mu_g + E_{N_{hf}} (\lambda_g + 2\mu_g)} = 7.23 \times 10^{11} \frac{\text{dyne}}{\text{cm}^2}. \quad (42)$$

**Velocity measurements of the TI stack**

The traveltimes were determined from pulse-echo contact measurements using 1.0 Mhz shear and longitudinal transducers in different propagation and polarization directions. The subscripts $i$ and $j$ in values of $V_{ij}$ and $c_{ij}$ correspond to the propagation and polarization directions, respectively. The associated material velocities are presented in Table 1 with $\rho_{TI}$, the thickness weighted average density

$$\rho_{TI} = \frac{h_g \rho_g + h_e \rho_e}{h_g + h_e} = 2.48 \frac{\text{gm}}{\text{cm}^3}. \quad (43)$$

From equations (39)-(42), using the bulk moduli of the glass and epoxy, the expected $c_{ijTI}$ were computed. Comparison of the $c_{ij}$ determined from wave speeds of the layered sample and those computed from bulk values are presented in Table 2.

Comparing the values determined from experimental data to the values computed from theory and bulk data indicates very good agreement for the longitudinal value perpendicular to the plate thickness ($c_{33}$) and the shear value parallel and in the plane of the plates ($c_{66}$). These values agree to better than 5 percent. Longitudinal values related parallel to the plates ($c_{11} = c_{22}$) vary about 10 percent and the shear values $c_{44} = c_{55}$ differ by approximately 30 percent from the values predicted by bulk measurements. There are a number of potential reasons for these differences. The theory requires very weak shear properties for one of the two component layers, which may not be completely

**Table 1. Velocities and elastic moduli of the TI stack.**

| Units of $10^5 \frac{\text{cm}}{\text{s}}$ | Data from TI stack measurements | Forward calculations where $V_{ijTI} = \sqrt{\frac{c_{ijTI}}{\rho_{TI}}}$ for $c_{ij}$ using (39)-(42) |
|---|---|---|
| $V_{11_{TI}} = V_{22_{TI}}$ | 5.64 | 5.40 |
| $V_{33_{TI}}$ | 5.28 | 5.37 |
| $V_{44_{TI}} = V_{55_{TI}}$ | 2.93 | 3.35 |
| $V_{66_{TI}}$ | 3.38 | 3.39 |

**Table 2. Elastic moduli computed from the bulk components compared with those measured from the TI stack.**

| $c_{ijTI} = \rho_{TI} V_{ijTI}^2$ units of $10^{11} \frac{\text{dyne}}{\text{cm}^2}$ | From TI stack measurements | Forward calculations from bulk measurements |
|---|---|---|
| $c_{11_{TI}} = c_{22_{TI}}$ | 7.89 | 7.23 |
| $c_{33_{TI}}$ | 6.92 | 7.16 |
| $c_{44_{TI}} = c_{55_{TI}}$ | 2.13 | 2.78 |
| $c_{66_{TI}}$ | 2.83 | 2.85 |

valid for the epoxy in our physical specimen. The epoxy may respond differently in the bulk as compared to its response as a very thin, constrained layer. Some type of layer interaction may occur that is not accounted for in the model. The assumption that we have made sufficiently long wavelength measurements may not hold as well parallel to the plates as through the plates because of property variations and averaging over the physical dimensions along the plates using shorter wavelength shear waves for $c_{44}$ and $c_{55}$. In general, the assumptions of the theory may not be well satisfied experimentally. The values measured, however, were corroborated with measurements made on a similar TI specimen. This agreement gave us a high degree of confidence in the TI measurements. Overall, the values determined from the TI sample compare well with values computed from bulk measurements.

The elastic properties $\lambda_g$, $\mu_g$, $E_{T_{hf}}$, $E_{N_{hf}}$ can be calculated from the TI stack data. The value of $\mu_g$ was determined from the calculation of $c_{66_{TI}}$:

$$\mu_g = c_{66_{TI}} = 2.83 \times 10^{11} \frac{\text{dyne}}{\text{cm}^2} . \tag{44}$$

The dimensionless tangential compliance $E_{T_{hf}}$ was determined from $c_{44_{TI}}$:

$$E_{T_{hf}} = \frac{\mu_g}{c_{44_{TI}}} - 1 = .331 , \tag{45}$$

which gives

$$\mu_e = \frac{h_e \mu_g}{E_{T_{hf}}} = 7.18 \times 10^8 \frac{\text{dyne}}{\text{cm}^2} . \tag{46}$$

By simultaneously solving the $c_{11_{TI}}$ and $c_{33_{TI}}$ of equations (9) for $\lambda_g$ and $E_{N_{hf}}$, the parameter $\lambda_g$ and the normal compliance $E_{N_{hf}}$ are determined as

$$\lambda_g = \frac{\mu_g \left(1 + 2E_{N_{hf}}\right)\left(c_{11_{TI}} - 2\mu_g\right)}{\mu_g + 4E_{N_{hf}}\mu_g - E_{N_{hf}} c_{11_{TI}}} = 2.31 \times 10^{11} \frac{\text{dyne}}{\text{cm}^2} , \tag{47}$$

and

$$E_{N_{hf}} = \frac{\mu_g - c_{33_{TI}} + \sqrt{c_{11_{TI}} c_{33_{TI}} - 2c_{33_{TI}}\mu_g + \mu_g^2}}{2c_{33_{TI}}} = .055 , \tag{48}$$

which gives

$$\lambda_e = h_e \frac{\mu_g}{E_{N_{hf}}} - 2\mu_e = 2.91 \times 10^9 \frac{\text{dyne}}{\text{cm}^2} . \tag{49}$$

The bulk values $\lambda_g$, $\mu_g$, $\lambda_e$, $\mu_e$ [equations (32), (31), (36), (35), respectively] and their corresponding values $E_{T_{hf}}$, $E_{N_{hf}}$ [equations (37), (38), respectively] are compared with the $\lambda_g$, $\mu_g$, $\lambda_e$, $\mu_e$, $E_{T_{hf}}$, $E_{N_{hf}}$ determined from the measurements of the TI stack [equations (47), (44), (49), (46), (45), (48), respectively] in Table 3.

Notice that the value calculated for $\mu_g$ is completely dependent on $c_{66}$. Therefore, any error in the measurement of $c_{66}$ causes error in $\mu_g$. This error is then carried into the subsequent calculations of $E_{T_{hf}}$, $\mu_e$, $\lambda_g$, $E_{N_{hf}}$, and $\lambda_e$. Also, since $\mu_g$ and $c_{44}$ are of the same order of magnitude, the value of $E_{T_{hf}}$ was computed from the difference of two nearly equal numbers. Therefore, $E_{T_{hf}}$ is very sensitive to $c_{44}$ and error magnification can occur. However, assuming the values of $\mu_g$ and $c_{44}$ are correct, comparing the value obtained in equation (45) with the value obtained in equation (37) indicates that the epoxy is much more compliant in the layered structure than in the bulk. One possible reason for this is differences of curing of the epoxy between the stack and the bulk.

### Sample preparation of the VFTI stack

The VFTI sample was prepared by cutting the TI stack (perpendicular to the glass plates) into vertical wafers and then rebonding the wafers with the same epoxy as used for the TI sample (see Figure 5). The average wafer thickness was $h_w = 8.80 \times 10^{-2} \pm .01$ cm. The VFTI medium consisted of 21 bonded wafers. The final length of the stack was $L = 1.880 \pm .01$ cm. A stack of $m$ vertical wafers each with thickness $h_w$ bonded and stacked side-by-side with a total length $L$ can be written as $L = mh_w + (m - 1)h_{vf}$. Therefore, the average thickness of the bonding agent between each wafer was

$$h_{vf} = \frac{L - mh_w}{m - 1} = 1.60 \times 10^{-3} \text{cm} \tag{50}$$

which constituted only 1.7 percent of the stack in the $x_1$ direction. Therefore, we assumed $h_{vf} << h_w$ so that for $h_w + h_{vf} = 1$, $h_{vf} \rightarrow 0$ while $h_w \rightarrow 1$. This assumption again satisfied one of the two criteria for the mixture of these two components to simulate a fractured medium according to Schoenberg's fracture model (1983).

**Table 3. Measured elastic constants of the bulk components compared with those determined from the TI stack.**

| Elastic moduli units of $10^{11} \frac{\text{dyne}}{\text{cm}^2}$ | Forward from bulk measurements | Determined from TI stack |
|---|---|---|
| $\lambda_g$ | 1.54 | 2.31 |
| $\mu_g$ | 2.85 | 2.83 |
| $\lambda_e$ | .344 | .0291 |
| $\mu_e$ | .0987 | .00718 |
| $E_{T_{hf}}$ | .0242 | .331 |
| $E_{N_{hf}}$ | .00441 | .055 |

**Velocity measurements of the VFTI stack**

Using the values of $\mu_g$, $\lambda_g$, $\mu_e$, $\lambda_e$ as determined from bulk measurements [equations (31), (32), (35), (36), respectively] the expected values $E_{T_{vf}}$ and $E_{N_{vf}}$ were calculated as

$$E_{T_{vf}} = \frac{\eta_T \mu_g}{h} = h_{vf}\frac{\mu_g}{\mu_e} = 4.62 \times 10^{-2}\,, \tag{51}$$

$$E_{N_{vf}} = \frac{\eta_N \mu_g}{h} = h_{vf}\frac{\mu_g}{\lambda_e + 2\mu_e} = 8.42 \times 10^{-3}\,. \tag{52}$$

Using the values of $\mu_g$, $\lambda_g$, $\mu_e$, $\lambda_e$ as determined from the TI stack measurements [equations (44), (47), (46) (49), respectively] the expected values $E_{T_{vf}}$ and $E_{N_{vf}}$ were calculated as

$$E_{T_{vf}} = \frac{\eta_T \mu_g}{h} = h_{vf}\frac{\mu_g}{\mu_e} = .631\,, \tag{53}$$

$$E_{N_{vf}} = \frac{\eta_N \mu_g}{h} = h_{vf}\frac{\mu_g}{\lambda_e + 2\mu_e} = .104\,. \tag{54}$$

From the $\mu_g$ and $\lambda_g$ of the glass [equations (31) and (32)] and the calculated $E_{T_{hf}}$, $E_{N_{hf}}$, $E_{T_{vf}}$, $E_{N_{vf}}$ [equations (37), (38), (51), (52), respectively] the long wavelength equivalent expected values for the $c_{ijVFTI}$ according to the Schoenberg fracture model were calculated from equations (14). Likewise, the forward expected $c_{ijVFTI}$ were calculated from equation (14) using the values of $\mu_g$, $\lambda_g$, $E_{T_{hf}}$, $E_{N_{hf}}$ as determined from the TI stack [equations (44), (47), (45), (48)] and the $E_{T_{vf}}$, $E_{N_{vf}}$ calculated in equations (53) and (54). Velocities calculated from the traveltimes through the VFTI stack (following Appendix B) provided the elastic stiffness moduli $c_{ij}$. The measured VFTI velocities and their associated $c_{ij}$ are presented in Table 4 with $\rho_{VFTI}$, the thickness weighted average density

$$\rho_{VFTI} = \frac{h_w \rho_{TI} + h_{vf}\rho_e}{h_w + h_{vf}} = 2.48\frac{\text{gm}}{\text{cm}^3}\,.$$

These $c_{ij}$ and the two sets of calculated $c_{ijVFTI}$ are shown in Table 5 for comparison. Notice that the $c_{ij}$ values determined from the measurements of the TI and VFTI samples are in better agreement than the values predicted from bulk values. This agreement indicates a consistency in the analysis and adds further suspicion that the epoxy behaves differently in the thin layers of the TI and VFTI structures than in bulk form.

Parameters $\mu_g$, $E_{T_{vf}}$, $E_{T_{hf}}$ are functions of $c_{ij}$ and were determined from the shear-wave measurements using solutions (18)-(20):

**Table 4. Velocity data from the VFTI stack and its associated elastic moduli.**

| Data from VFTI stack units of $10^5\ \frac{\text{cm}}{\text{s}}$ | Calculated from VFTI data $c_{ij} = \rho_{VFTI} V_{ij}^2$ units of $10^{11}\ \frac{\text{dyne}}{\text{cm}^2}$ |
|---|---|
| $V_{11} = 5.12$ | $c_{11} = 6.50$ |
| $V_{22} = 5.57$ | $c_{22} = 7.69$ |
| $V_{33} = 5.20$ | $c_{33} = 6.70$ |
| $V_{44} = 2.90$ | $c_{44} = 2.09$ |
| $V_{55} = 2.59$ | $c_{55} = 1.66$ |
| $V_{66} = 2.81$ | $c_{66} = 1.95$ |

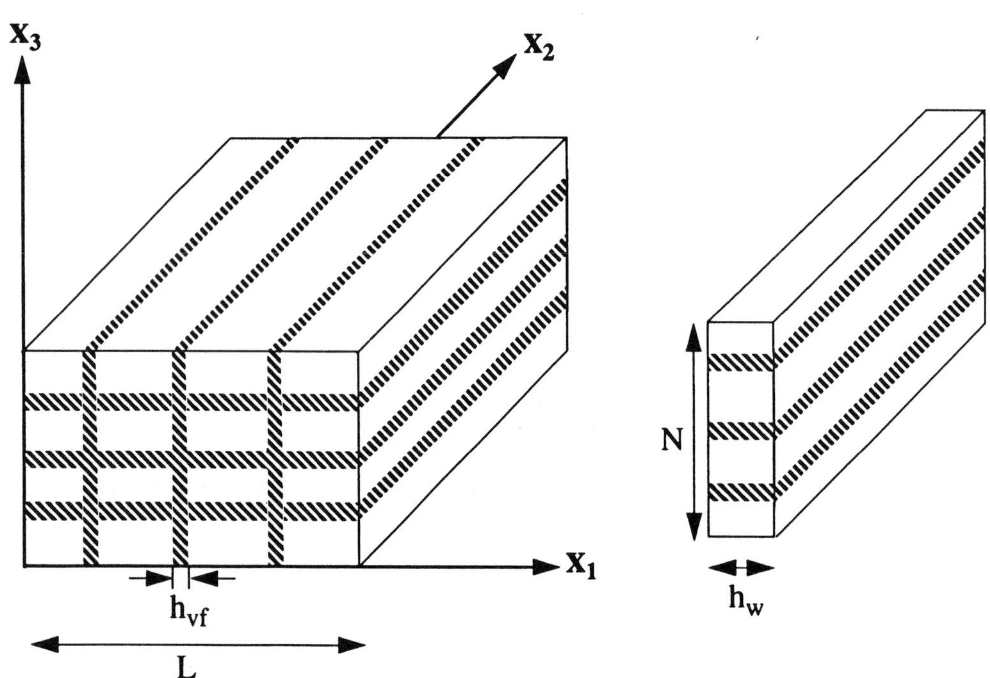


FIG. 5. Diagram of the wafers that make up the VFTI media.

$$\mu_g = 2.57 \times 10^{11} \frac{\text{dyne}}{\text{cm}^2}, \tag{55}$$

$$E_{T_{hf}} = .230, \tag{56}$$

$$E_{T_{vf}} = .315. \tag{57}$$

From the longitudinal traveltimes through the VFTI material, the unknowns $E_{N_{hf}}$, $E_{N_{vf}}$, and $\lambda_g$ were determined using the solutions in equations (24)-(26)

$$\lambda_g = 2.87 \times 10^{11} \frac{\text{dyne}}{\text{cm}^2}. \tag{58}$$

This value of $\lambda_g$ was used in equations (24) and (25), respectively, to calculate numerical values for $E_{N_{hf}}$ and $E_{N_{vf}}$:

$$E_{N_{hf}} = .055, \tag{59}$$

$$E_{N_{vf}} = .068. \tag{60}$$

Ideally we wanted to solve the four equations for $E_{T_{hf}}$, $E_{T_{vf}}$, $E_{N_{hf}}$, and $E_{N_{vf}}$ for the unknowns $\mu_e$, $\lambda_e$, $h_e$, and $h_{vf}$ using the values for $E_T$ and $E_N$ in equation (A-4) that incorporate the appropriate $h_e$ and $h_{vf}$ values. However, there is a fundamental ambiguity with Schoenberg's fracture model. The excess compliances $E_i$ are defined such that fracture density $h$ is coupled with the elastic moduli $\mu$ and $\lambda$ of the fracture infilling material. If values for $h$ are desired, then the $\mu$ and $\lambda$ must be assumed or vice versa. Therefore, the values for $\mu_e$, $\lambda_e$, $h_e$, and $h_{vf}$ can not be solved for unambiguously.

Bulk measurements for the epoxy and glass provided $\mu_e$ and $\lambda_e$ and $\mu_g$ and $\lambda_g$, respectively; the thickness of the epoxy bonds between the glass plates $h_e$ and between the vertical wafers $h_{vf}$ have been measured directly. Values for $\mu_g$, $\mu_e$, $\lambda_g$, and $\lambda_e$ were determined from the $V_{ijTI}$ measurements of the TI stack before it was wafered and rebonded to simulate the VFTI medium. Note the forward calculations using the bulk values compared with the determined TI values are provided in Table 3.

The three sets of elastic moduli, the bulk $\lambda_g$, $\mu_g$, $\lambda_e$, $\mu_e$, $E_{T_{hf}}$, $E_{N_{hf}}$, $E_{T_{vf}}$, $E_{N_{vf}}$ [equations (32), (31), (36), (35), (37), (38), (51), (52), respectively], the $\lambda_g$, $\mu_g$, $\lambda_e$, $\mu_e$, $E_{T_{hf}}$, $E_{N_{hf}}$ determined from the measurements of the TI stack [equations (47), (44), (49), (46), (45), (48), respectively], and the $\lambda_g$, $\mu_g$, $E_{T_{hf}}$, $E_{N_{hf}}$, $E_{T_{vf}}$, $E_{N_{vf}}$ determined from the measurements of the VFTI stack [equations (58), (55), (56), (59), (57), (60), respectively] are tabulated in Table 6 for comparison.

## DISCUSSION

Comparison of the VFTI elastic moduli determined from the values computed from the TI and VFTI sample data are quite good considering that these are handmade structures and the construction is tedious and difficult. The values agree to well within 10 percent. The comparison to the values computed from the bulk measurements is significantly worse and may be an indication that the epoxy does not behave the same in the bulk as it does in the thin layers of the TI and VFTI structures. To be fully conclusive, the difference in the epoxy properties will have to be confirmed.

**Table 5. Comparison of elastic moduli from measurements of the VFTI stack, computed from the bulk components, and computed with the TI stack measurements.**

| Expected $c_{ijVFTI}$ using bulk values units of $10^{11}\frac{\text{dyne}}{\text{cm}^2}$ | Expected $c_{ijVFTI}$ using values determined from TI units of $10^{11}\frac{\text{dyne}}{\text{cm}^2}$ | Calculated from VFTI data $c_{ij} = \rho_{VFTI}V_{ij}^2$ units of $10^{11}\frac{\text{dyne}}{\text{cm}^2}$ |
|---|---|---|
| $c_{11VFTI} = 7.08$ | $c_{11VFTI} = 6.11$ | $c_{11} = 6.50$ |
| $c_{22VFTI} = 7.23$ | $c_{22VFTI} = 7.75$ | $c_{22} = 7.69$ |
| $c_{33VFTI} = 7.15$ | $c_{33VFTI} = 6.80$ | $c_{33} = 6.70$ |
| $c_{44VFTI} = 2.78$ | $c_{44VFTI} = 2.13$ | $c_{44} = 2.09$ |
| $c_{55VFTI} = 2.66$ | $c_{55VFTI} = 1.44$ | $c_{55} = 1.66$ |
| $c_{66VFTI} = 2.72$ | $c_{66VFTI} = 1.74$ | $c_{66} = 1.95$ |

**Table 6. Elastic moduli computed from the bulk components and from TI stack measurements compared with those measured from the VFTI stack.**

| Elastic moduli units of $10^{11}\frac{\text{dyne}}{\text{cm}^2}$ | Bulk values from velocities | Calculations using bulk values | Determined from TI stack | Forward using TI values | Determined from VFTI stack |
|---|---|---|---|---|---|
| $\mu_g$ | 2.85 | — | 2.83 | — | 2.57 |
| $\lambda_g$ | 1.54 | — | 2.31 | — | 2.87 |
| $\mu_e$ | .0987 | — | .00718 | — | — |
| $\lambda_e$ | .344 | — | .0291 | — | — |
| Dimensionless compliances | | | | | |
| $E_{T_{hf}}$ | — | .0242 | .331 | .331 | .230 |
| $E_{N_{hf}}$ | — | .00441 | .055 | .055 | .055 |
| $E_{T_{vf}}$ | — | .0462 | — | .631 | .315 |
| $E_{N_{vf}}$ | — | .00842 | — | .104 | .068 |

Once the elastic moduli for a material are obtained, the assumption that the medium consists of a TI background embedded with parallel vertical fractures can be tested by examining whether constraint $s_{13} = s_{23}$ holds. The relationships of equations (15)-(16), (27)-(28) constrain the complexity of the background symmetry and the properties of the fracture systems. For example, the TI background is the result of axisymmetric horizontal fractures in an isotropic matrix only if $s_{12} = s_{13}$ holds; the horizontal and vertical fractures have identical properties only if $s_{11} = s_{33}$ and $s_{44} = s_{66}$.

The foregoing separation method derives explicit formulae for the fracture compliances and the background elastic moduli for media such as sedimentary basins with stress-induced fractures. The ability to separate constituent anisotropies allows us to use more complex models to represent the Earth. For example, the fracture compliances furnish information on the fracture density or on the properties of the infilling material in the fractures while the background elastic moduli provide insight into the lithologic makeup of the Earth.

## ACKNOWLEDGMENTS

Thanks to Drs. Michael Schoenberg, Dan Ebrom, and Mr. Henry Chaskelis for many useful discussions. Special thanks to JoAnn O. Sinton of the University of Hawaii Petrography Lab and Sandy L. Adams of Kai Manu Enterprises for patiently preparing laboratory samples. This research was performed while one of the authors (JAH) held a Research Associateship with the National Research Council.

## REFERENCES

Auld, B. A., 1973, Acoustic fields and waves in solids, Volume 1: John Wiley & Sons.

Backus, G. E., 1962, Long-wave anisotropy produced by horizontal layering: J. Geophys. Res., **66**, 4427–4440.

Carcione, J. M., Kosloff, D., and Behle, A., 1991, Long-wave anisotropy in stratified media: A numerical test: Geophysics, **56**, 245–254.

Crampin, S., Bush I., Naville, C., and Taylor D., 1986, Estimating the internal structure of reservoirs with shear-wave VSPs: The Leading Edge, **5**, no. 11, 35–39.

Helbig, K., 1984, Anisotropy and dispersion in layered media: Geophysics, **49**, 364–373.

Hood, J. A., 1991, A simple method for decomposing fracture-induced anisotropy: Geophysics, **56**, 1275–1279.

Hood, J. A., and Schoenberg, M., 1989, Estimation of vertical fracturing from measured elastic moduli: J. Geophys. Res., **94**, 15611–15618.

Hsu, J. C., and Schoenberg, M., 1990, Characterization of anisotropic elastic wave behavior in media with parallel fractures: 60th Ann. Internat. Mtg., Soc. Expl. Geophys., Expanded Abstracts, 1410–1415.

Melia, P. J., and Carlson, R. L., 1984, An experimental test of *P*-wave anisotropy in stratified media: Geophysics, **49**, 374–378.

Nichols, D., Muir, F., and Schoenberg, M., 1989, Elastic properties of rocks with multiple sets of fractures: 59th Ann. Internat. Mtg., Soc. Expl. Geophys., Expanded Abstracts, 471–474.

Schoenberg, M., 1980, Elastic wave behavior across linear slip interfaces: J. Acoust. Soc. Am., **68**, 1516–1521.

Schoenberg, M., 1983, Reflection of elastic waves from periodically stratified media with interfacial slip: Geophys. Prosp., **31**, 265–292.

Shearer, P., and Orcutt, J., 1985, Anisotropy in the oceanic lithosphere-theory and observations from the Ngendei seismic refraction experiment in the south-west Pacific: Geophys. J. R. Astr. Soc., **80**, 493–526.

Stephen, R. A., 1981, Seismic anisotropy observed in upper oceanic crust: Geophys. Res. Lett., **8**, 865–868.

Stephen, R. A., 1985, Seismic anisotropy in upper oceanic crust: J. Geophys. Res., **90**, 11383–11396.

White, R. S., and Whitmarsh, R. B., 1984, An investigation of seismic anisotropy due to cracks in the upper oceanic crust at 45°N, Mid-Atlantic Ridge: Geophys. J. R. Astr. Soc., **79**, 439–467.

## APPENDIX A: SCHOENBERG'S FRACTURE MODEL

Consider an infinite linear elastic medium made up of plane homogeneous isotropic layers such that the layering is periodic with period $h$ (see Figure A-1). In the long wavelength limit, the periodically stratified elastic medium appears equivalent to a homogeneous material with anisotropic properties (Backus, 1962). The symmetry class is hexagonal or TI with the symmetry axis coincident with the direction which is normal to the plane of the layers (the $x_3$ axis).

To model a fracture set, Schoenberg (1983) considered a periodically layered system composed of only two components (as in the TI sample, see Figure 1). Let layer 1 be an isotropic glass and have elastic properties $\lambda_g$, $\mu_g$ while isotropic layer 2 is an epoxy with properties $\lambda_e$, $\mu_e$. Let layer 1 have thickness $h_g$ and layer 2 have a thickness of $h_e$ such that $h_g + h_e = h$. As the epoxy layers between the glass plates become very thin ($h_e \rightarrow 0$) and very compliant ($\mu_e \rightarrow 0$) they simulate planes of weakness analogous to Schoenberg's linear slip interface LSD model (1980). The collection of all these thin soft layers over the entire system simulates a set of equally spaced planar parallel fractures. The spacing between the fracture planes (fracture density) is denoted $h$. Schoenberg then defined compliances $\eta_T$ and $\eta_N$ as

$$\lim_{\substack{h_g \to 1 \\ h_e \to 0 \\ \mu_e \to 0}} \frac{h_e h}{\mu_e} \rightarrow \eta_T\,, \quad \lim_{\substack{h_g \to 1 \\ h_e \to 0 \\ \mu_e \to 0}} \frac{\gamma_e h_e h}{\mu_e} \rightarrow \eta_N\,, \quad \gamma_i = \frac{\beta_i^2}{\alpha_i^2}, \tag{A-1}$$

where $\beta_i$ is the shear-wave velocity and $\alpha_i$ is the longitudinal wave velocity of the $i^{th}$ layer. The $\eta_T$ is the tangential slip of all fractures in distance $h$ as a result of unit shear stress; $\eta_N$ is the total

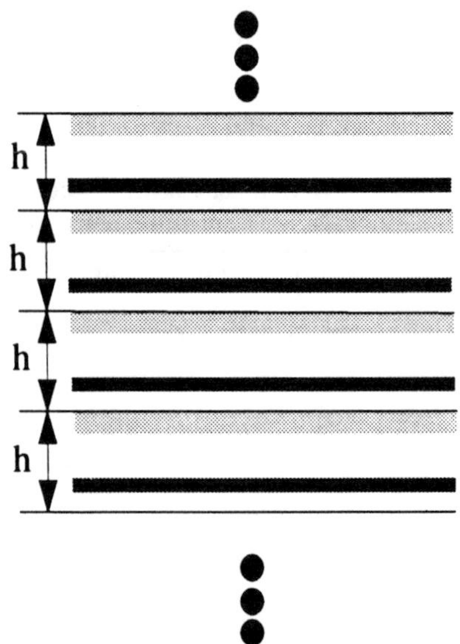


FIG. A-1. Periodically layered infinite linear elastic medium.

normal slip because of unit normal stress. So to simulate a material embedded with a set of Schoenberg's LSD planar parallel fractures the composite media must be a two-component system and the nature of the material comprising layer 2 must satisfy the requirements that $h_e \to 0$, $\mu_e \to 0$.

Across each epoxy interface the stresses ($\sigma_i$) are continuous but displacement ($u_i$) discontinuities are allowed so that

$$\begin{matrix}\Delta u_1 = \eta_T \sigma_5 \\ \Delta u_2 = \eta_T \sigma_4 \\ \Delta u_3 = \eta_N \sigma_3\end{matrix} \quad \to \quad \begin{bmatrix}\Delta u_3 \\ \Delta u_2 \\ \Delta u_1\end{bmatrix} = h\underset{\sim}{Z}\begin{bmatrix}\sigma_3 \\ \sigma_4 \\ \sigma_5\end{bmatrix}, \tag{A-2}$$

where

$$\underset{\sim}{Z} = \frac{1}{h}\begin{bmatrix}\eta_N & 0 & 0 \\ 0 & \eta_T & 0 \\ 0 & 0 & \eta_T\end{bmatrix} = \frac{1}{\mu_g}\begin{bmatrix}E_N & 0 & 0 \\ 0 & E_T & 0 \\ 0 & 0 & E_T\end{bmatrix}. \tag{A-3}$$

The $E_N$ and $E_T$ are dimensionless compliances normal and tangential to the fracture planes, respectively

$$E_T = \frac{\eta_T \mu_g}{h} = h_e \frac{\mu_g}{\mu_e}, \quad E_N = \frac{\eta_N \mu_g}{h} = h_e \frac{\mu_g}{\lambda_e + 2\mu_e} \tag{A-4}$$

(Schoenberg, 1983). The $\eta_T$ is the average tangential compliance of a fracture and $\eta_N$ is the average normal compliance of a fracture. The $E_i$ are dimensionless ratios of strain in the fractures to corresponding strain in the background material and are positive numbers. These parameters give information as to the nature of the infilling material in the fractures or the fracture density, i.e., $\mu_e$ or $h_e$, respectively.

The above calculations were used to model the excess compliance in the TI sample as $E_{T_{hf}}$ and $E_{N_{hf}}$. The vertical fracturing was modeled in a similar fashion. Identical epoxy filled the gaps between the wafers (vertical fractures) and between the glass plates (horizontal fractures). Therefore, $\mu_e$ and $\lambda_e$ were also used in the calculations for the excess compliance because of vertical fracturing, $E_{T_{vf}}$ and $E_{N_{vf}}$. The $\mu$ and $\lambda$ of the wafer replaced the $\mu_g$ and $\lambda_g$ of the TI sample. The contribution that the epoxy made to the effective moduli of the wafer was, however, so small that we used $\mu_g$ and $\lambda_g$ to represent the $\mu$ and $\lambda$ of the wafer. The $h$ representative of the periodic layering differed between horizontal and vertical, as $h_g + h_e$ and $h_w + h_{vf}$, respectively. Since the spacing between the wafers $h_{vf}$ was much smaller than the thickness of the wafers ($h_{vf} \to 0$) and the epoxy was much more compliant than the wafer itself ($\mu_e \to 0$), the Schoenberg criteria for modeling the bonded wafers as a set of vertical fractures was satisfied.

## APPENDIX B: DETERMINING THE ORTHORHOMBIC ELASTIC MODULI

Measurements sufficient to determine orthorhombic stiffness moduli were made following Hood and Schoenberg (1989). Substituting a plane wave of frequency $\omega$ and slowness components $s_1, s_2, s_3$, i.e., $u_i = U_i \exp(s_j x_j - \omega t)$, into the displacement equations of motion gives the Christoffel equation

$$(c_{ijkl} s_k s_i - \rho \delta_{jl}) U_l = 0 . \tag{B-1}$$

The set of all real eigensolutions $s_j$ of equation (B-1) is the slowness surface of an anisotropic elastic medium. A material having three perpendicular mirror or two-fold symmetry planes whose normals are coincident with the coordinate axes is usually called orthorhombic and has the characteristic elastic modulus matrix, equation (17). Using equation (17) in equation (B-1) gives the Christoffel equations for orthorhombic media:

$$\begin{aligned}&(c_{11}s_1^2 + c_{66}s_2^2 + c_{55}s_3^2 - \rho)U_1 + (c_{12} + c_{66})s_1 s_2 U_2 \\ &\quad + (c_{13} + c_{55})s_1 s_3 U_3 = 0 , \\ &(c_{12} + c_{66})s_1 s_2 U_1 + (c_{66}s_1^2 + c_{22}s_2^2 + c_{44}s_3^2 - \rho)U_2 \\ &\quad + (c_{23} + c_{44})s_2 s_3 U_3 = 0 , \\ &(c_{13} + c_{55})s_1 s_3 U_1 + (c_{23} + c_{44})s_2 s_3 U_2 \\ &\quad + (c_{55}s_1^2 + c_{44}s_2^2 + c_{33}s_3^2 - \rho)U_3 = 0 .\end{aligned} \tag{B-2}$$

Contact velocity measurements on the [0, 0, 1] face of the specimen correspond to setting $s_1 = s_2 \equiv 0$ in (B-2) which gives

$$\begin{aligned}&(c_{55}s_3^2 - \rho)U_1 = 0 , \\ &(c_{44}s_3^2 - \rho)U_2 = 0 , \\ &(c_{33}s_3^2 - \rho)U_3 = 0 .\end{aligned} \tag{B-3}$$

The $c_{33}$ was determined from longitudinal wave measurements parallel to the $x_3$ direction; $c_{44}$ was determined from shear waves polarized in the $x_2$ direction; and $c_{55}$ was determined from shear waves polarized in the $x_1$ direction.

Contact velocity measurements on the [1, 0, 0] face of the specimen correspond to setting $s_2 = s_3 \equiv 0$ in (B-2) which gives

$$\begin{aligned}&(c_{11}s_1^2 - \rho)U_1 = 0 , \\ &(c_{66}s_1^2 - \rho)U_2 = 0 , \\ &(c_{55}s_1^2 - \rho)U_3 = 0 .\end{aligned} \tag{B-4}$$

The $c_{11}$ was determined from longitudinal wave measurements parallel to the $x_1$ direction; another $c_{55}$ was determined from shear waves polarized in the $x_3$ direction; and $c_{66}$ was determined from shear waves polarized in the $x_2$ direction.

Contact velocity measurements on the [0, 1, 0] face of the specimen correspond to setting $s_1 = s_3 \equiv 0$ in (B-2) which gives

$$\begin{aligned}&(c_{66}s_2^2 - \rho)U_1 = 0 , \\ &(c_{22}s_2^2 - \rho)U_2 = 0 , \\ &(c_{44}s_2^2 - \rho)U_3 = 0 .\end{aligned} \tag{B-5}$$

The $c_{22}$ was determined from longitudinal wave measurements parallel to the $x_2$ direction; another $c_{44}$ was determined from shear waves polarized in the $x_3$ direction; and another $c_{66}$ was determined from shear waves polarized in the $x_1$ direction.

# Chapter 4. Physical Model Data Tests of Processing Algorithms

After the introduction of digital computers into exploration geophysics in the late 1960s, the need arose for methods of testing seismic data processing algorithms. Field seismic data are often complicated by unexpected geological heterogeneities and acquisition irregularities. Synthetic seismic data avoid these difficulties.

For simple systems, synthetic data *can* be obtained by numerical modeling, rather than by physical modeling. There is, however, something vaguely unsettling about using computer-generated data to test a computer algorithm. Physical model data are useful here precisely because the events recorded are not subject to prior censorship, whether such censorship be intentional or unintentional.

+ French (1974) used Gulf's physical model tank to collect data over a model designed to produce sideswipe (out-of-plane reflections). As such, the data could test whether 3-D migration could correctly remove sideswipe and produce migrated sections that were accurate indicators of the interfaces directly below a given surface line. Two follow-on papers are Gardner et al. (1974) and French (1975).

+ Macdonald et al. (1987) performed a test of Bayesian inversion on physical model data to obtain both the reflector geometries and the elastic moduli.

+ East et al. (1988) showed that a Radon-transform based tomographic inversion performs better than SIRT (Simultaneous Iterative Reconstruction Technique) for a simulated crosshole geometry investigating a pinch out.

+ Forel and Gardner (1988) demonstrated the ability of 3-D dip moveout (DMO) (followed by post-stack migration) to correctly image dipping event data collected in common-midpoint (CMP) mode.

+ Lo (1988) compared the Born and Rytov approximations in seismic tomography. He concluded that diffraction tomography (Rytov approximation) is better suited for limited aperture experiments and those situations where the wavelength is large compared to the investigated body.

+ Uren et al. (1991) used an anisotropic physical model to show that migration correctly positions events in space *if* the correct (in this case anisotropic) velocity field is employed.

# TWO-DIMENSIONAL AND THREE-DIMENSIONAL MIGRATION OF MODEL-EXPERIMENT REFLECTION PROFILES

WILLIAM S. FRENCH*

A reflection profile represents an unfocused picture of the subsurface. In areas of rapid structural change, this unfocused picture may not reveal directly the true geometry of subsurface structures. Computer processing techniques, collectively called migration, have been used by many companies to focus 2-D reflection data. A description of the migration process can be given which allows immediate generalization to three-dimensions with arbitrary source and receiver positions.

Reflection profiles digitally recorded in the laboratory over known acoustically semitransparent structural models establish the effectiveness of migration. Processed reflection data over 3-D models demonstrate that 3-D migration eliminates many of the lateral correlation ambiguities caused by "sideswipes" and "blind structures."

Structure maps developed from the results of 3-D migration of reflection data give a true and precise picture of 3-D models. When the same data are processed using 2-D migration, the mapped structures are distorted.

In structurally complex areas it is desirable to collect 3-D reflection data. Single profiles cannot, and conventional grids may not, reveal adequate cross-dip information.

## INTRODUCTION

In areas of rapid structural change, unprocessed seismic reflection profiles may not reveal directly the true geometry of subsurface structures. In some cases, it is possible to identify an isolated structure on the basis of its characteristic reflection pattern (e.g., a fault or syncline). In general, however, it may be necessary to employ some technique to "focus" the data from complex areas.

Two calculational procedures to focus seismic reflection data have been discussed in the literature: downward continuation of moveout corrected seismograms (Claerbout and Doherty, 1972) and digital migration (Schneider, 1971). In essence, the methods utilize a surface-recorded reflection profile to calculate the hypothetical profile that would be obtained if the geophones could be placed in a plane deep in the earth. Their usefulness results from the fact that the closer the profile is to a structure, the more the reflection patterns take on the geometry of the structure. Both techniques are founded on the mathematics of wave propagation and should provide identical results if properly executed. The first technique is a numerical solution to the differential equations of wave motion. The second technique is based upon numerical solution of integral equations describing the wave motion (Maginness, 1972). The integral equations admit a construct which proves to be of great value to the intuition,

Presented at the 43rd Annual International SEG Meeting October 25, 1973, Mexico City. Manuscript received by the editor December 6, 1973.

* Gulf Research & Development Co., Pittsburgh, Penn. 15230.



Reprinted from Geophysics, 39, 265-277

and such an approach will be adopted in this paper; the term migration will be used as a synonym for the construct.

In this paper the assumptions inherent to migration will be stated, the technique will be extended to handle the case of 3-D objects with arbitrary source-receiver arrays, and the results of both 2-D and 3-D migration of seismic model data will be presented.

## ASSUMPTIONS OF DIGITAL MIGRATION

The typical seismic record is the result of recording at several locations the reflected waves generated by an impulsive source. A profile consists of a collection of such records for several source locations. For convenience of discussion we will deal with the case of marine profiles. Thus the detectors are hydrophones and the profile consists of the pressure-time records $P(s_j, r_j^i, t)$, where

$s_j = (x_j, y_j, z_j)$ is the location of the $j$th shotpoint,

$r_j^i = (x_j^i, y_j^i, z_j^i)$ is the location of the $i$th hydrophone recording the $j$th shot, and

$t$ is a time variable which is reset to zero at the instant of each successive shot.

At this point, no particular shooting geometry has been assumed (i.e., the shotpoints and receiver locations do not necessarily lie in the same line or even in the same plane). The task is to interpret $P(s_j, r_j^i, t)$ in terms of subsurface structural geometry. Migration will be used to accomplish this end.

Computer migration schemes represent a search for scattering centers (i.e., diffraction or reflection points). The process involves assigning to each subsurface point a number which is a measure of the probability that scattered energy emanated from that point. The number is determined by summing the recorded data for all shotpoints and receiver locations at times where energy from that subsurface point could arrive. A number with large absolute value (positive or negative depending upon the wavelet shape) indicates that scattered (reflected or diffracted) energy probably did come from that particular subsurface point. In other words, the coherence of the recorded data along a traveltime versus distance surface appropriate to the source-receiver geometry and subsurface point determines the assigned probability that a reflector or diffractor exists at that subsurface point.

The assumptions of the migration method used in this paper are as follows:

(a) Shear waves can be ignored.

(b) Each subsurface point represents a possible scattering center.

(c) Reflecting surfaces can be considered a continuum of scattering centers.

(d) The pulse shape for the scattered waves is the same for all directions. The pulse is short enough so that the delay in arrival time of later portions can be neglected.

(e) Traveltime versus distance surfaces can be calculated to sufficient accuracy by averaging horizontal velocity variations. That is, a vertical path is considered the least-time travel path from surface to subsurface point, and moveout times for other surface points can be calculated on the basis of surface distance and a root-mean-square velocity-depth function (see Taner and Koehler, 1969).

(f) A coherent signal summed along the appropriate time-distance surface leads to an average with large absolute value. Any noise summed along the time-distance sursurface will lead to a small average value due to the equal probability of positive and negative numbers in the noise field.

It is known that some of the above assumptions are not strictly true [e.g., the shape of a pulse scattered by a small sphere depends upon the scattering angle (Morse and Ingard, 1968)]. However, they are consistent with the usual assumptions of seismic data processing.

## GENERAL EQUATIONS FOR MIGRATION

The above description of migration can be stated mathematically. We assume that conditions (a) through (f) hold and that we have available the pressure-time data $P(s_j, r_j^i, t)$. Let the time required for a pulse to travel the dashed line path of Figure 1 [from the source at $s_j$ to the subsurface point $r = (x, y, z)$ and subsequently to a receiver at $r_j^i$] be given by $t_j^i(r)$. As before, we establish a measure of the probability that a scattering center exists at $r$ by forming the sum

$$M(r) = \sum_j \sum_i P[s_j, r_j^i, t_j^i(r)]. \qquad (1)$$

Equation (1) represents 3-D migration for arbitrary source and receiver arrays. If the source and receiver lie in the $xy$-plane, then [according to condition (e)]:

$$t_j^i(r) = \left\{ T_{1/2}^2(r) + \frac{(x_j - x)^2 + (y_j - y)^2}{V^2(z)} \right\}^{1/2} + \left\{ T_{1/2}^2(r) + \frac{(x_j^i - x)^2 + (y_j^i - y)^2}{V^2(z)} \right\}^{1/2}, \tag{2}$$

where $T_{1/2}(r)$ is the one-way vertical traveltime to $r$ and $V(z)$ is the rms velocity function as described by Taner and Koehler (1969).

If we consider the special case where there is but one receiver for each source and both source and receiver are at the same location, then equation (2) becomes

$$t_j^1(r) = t_j(r) \tag{2a}$$

$$= \left\{ T_0^2(r) + \frac{4[(x_j - x)^2 + (y_j - y)^2]}{V^2(z)} \right\}^{1/2},$$

where $T_0$ is the two-way vertical traveltime to $r$.

Equation (1) involves an implicit time-to-depth conversion since the rms velocity is given as a function of depth. In practice, we know the rms velocity more accurately as a function of time. Therefore, in the following applications of equation (1) we replace $z$ by $T_0$ and plot migrated time sections rather than depth section. Equations (1) and (2a) have been used to focus data collected with both 2-D and 3-D models.

Two-D models have thickness variations only in the direction of the profile—cross-sections perpendicular to the profile line are of constant thickness. For such models, only a single profile of data is required. Three-D models have thickness variations both in-line with and perpendicular to the profile line; these models require an areal coverage of reflection data.

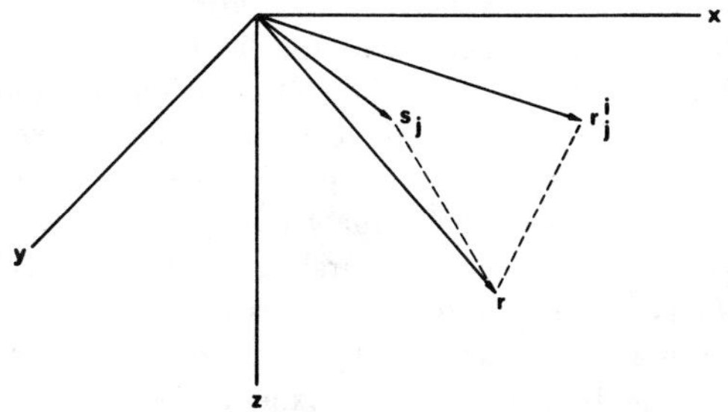


FIG. 1. Coordinate system for migration equation.

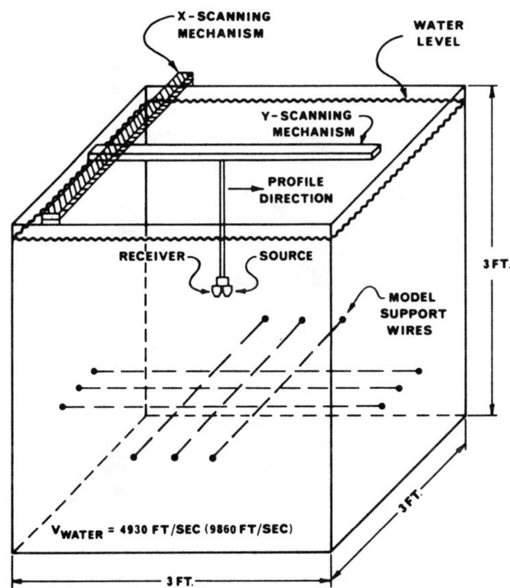


FIG. 2. Water tank arrangement.

A 3-D model was studied, and in order to set a standard for the quality of results desired from the 3-D processing, an experiment was run on a 2-D model whose cross-section displayed the same structures and dips found in certain cross-sections of the 3-D model. Results of these experiments are given in the following sections along with others for an additional 2-D experiment which displays a pertinent phenomenon.

## EXPERIMENTAL ARRANGEMENT

Ultrasonic pulse-echo electronics with digital recording was used to obtain the reflection data from a model supported in a water tank by fine wires (Figure 2). Two-D or 3-D single coverage reflection data were recorded at some predetermined height above the model. The 2-D data consisted of the usual single coverage profile while the 3-D data consisted of numerous parallel 2-D profiles resulting in a square grid of source-receiver locations. The model dimensions are such that the data represent simulated field data. The scaling is indicated by the following equivalent dimensions and velocities:

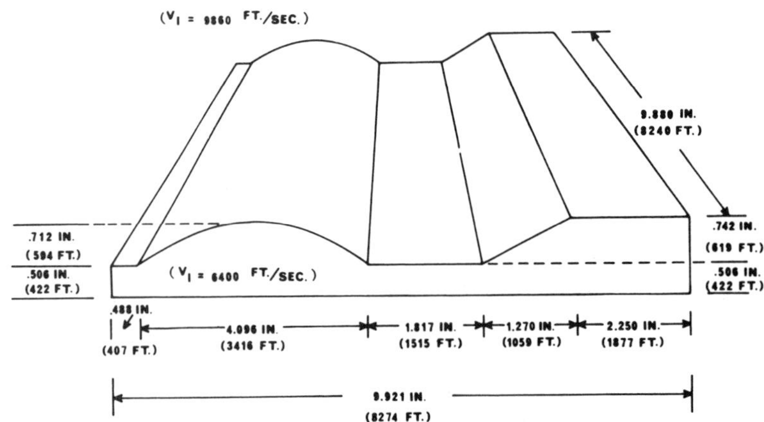


FIG. 3. Ridge and fault model with equivalent field dimensions. The equivalent field velocities are shown.

| | Model | Field |
|---|---|---|
| Time | .2 $\mu$sec | 1 msec |
| Length | 6.336 inches | 1 mile |
| Velocity | $v$ | $2v$ |

Hereafter, equivalent field values will be given in parentheses. In all cases the source-receiver separation was .827 inches (675 ft). The data were processed, however, as if the source and receiver were coincident. The actual sound speed in the material of which the models were constructed was less than that of water. Thus, the models represent structures with a velocity less than that of the overburden.

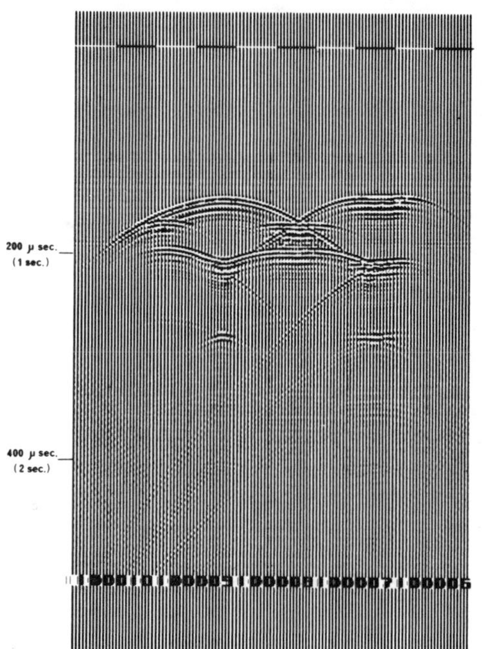


FIG. 4. Profile over ridge-and-fault model. Shotpoint separation = .133 inch (111 ft).

#### TWO-DIMENSIONAL RESULTS

The 2-D model consisted of a rounded ridge and an angular step (or normal fault) whose dimensions and corresponding field values are given in Figure 3. A single coverage reflection profile taken 4.11 inches (3430 ft) above the top of the step is shown in Figure 4. This profile was focused by 2-D migration according to equations (1) and (2a). The migration results are shown in Figure 5. The simple process described by equations (1) and (2a) generates some background noise seen in the figure. This migration was carried out using a constant field-equivalent rms velocity for all depths [i.e., $V(z) = 9860$ ft/sec in equation (2a)]. As a result, the top surface of the model is properly in focus but the velocity-anomaly generated structures of the bottom surface are slightly out of focus. Furthermore, due to the fact that the same rms velocity-depth function was used for all traces (i.e., horizontal velocity variations were ignored), the migrated profiles generated are what we have called "true vertical time sections" rather than "true structural geometry sections." Thus, velocity-anomaly generated pseudostructures are not eliminated by the process. They are, in fact, brought into approximate focus.

A clear example of this pseudostructural focus-

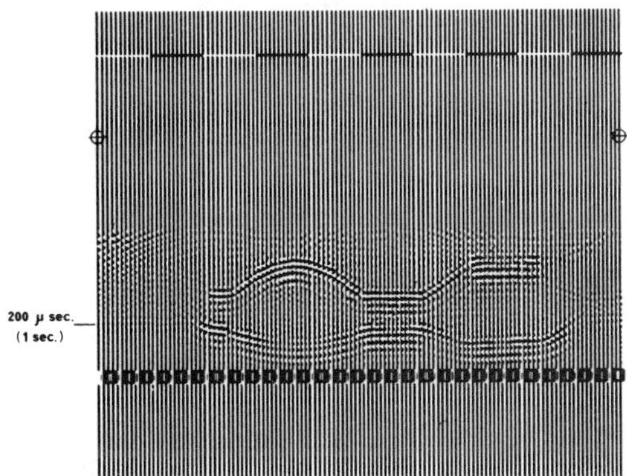


FIG. 5. 2-D migrated time section of data from ridge-and-fault model (Figure 4).

ing is presented in Figures 6, 7, and 8. Figure 6 gives dimensions for the 2-D model of a low-velocity layer with a high curvature ridge. Note that the bottom of the model is flat. A reflection profile taken at a height of 8.9 inches (7440 ft) above the center of this model is shown in Figure 7, and the migrated result using the same $V(z)$ for all traces is shown in Figure 8. A velocity-anomaly generated pseudosyncline in the reflection from the base of the model is brought into focus by the migration just as is the true ridge on the surface of the model. There is a three-to-one exaggeration of the vertical scale in these profiles which explains the stubby appearance of the migrated result when compared to the model shown in Figure 6. An incidental result (apparent in Figure 7) is the fact that the velocity-anomaly generated syncline produces the same reflection cross-over pattern obtained from a real buried focus.

Further processing is necessary in order to eliminate velocity-anomaly generated pseudostructures (i.e., in order to flatten the reflection from the bottom of the model in this case). The pseudostructures are caused by interval velocity variations and can be corrected through time-depth conversion. The methodology is known and straightforward, but accurate velocities are required for the conversion. It is not the purpose of this paper to discuss time-depth conversion problems. We are concerned here with the proper 3-D focusing of time structures.

Returning to Figure 5, we see that the migrated results give a true picture of the upper surface of the model. This was verified by comparison with measurements taken directly from the model. (From here on we will be concerned only with the upper surface of the models as the pseudostructures from the lower surfaces have been improperly focused.)

Figure 5 represents a 48 trace-scan migration (24 to each side), i.e., data from 48 traces were summed to give each migrated trace. A field equivalent 30–60 hz, 24 db/octave band-pass filter was applied to the results (the equivalent bandwidth of the somewhat "ringing" source used was 30–60 hz). The figure indicates the quality of focusing obtainable when equations (1) and (2a) are applied to simple 2-D model data. This 2-D experiment was carried out as a control for the 3-D experiment (to be described in the next section), and the processing parameters used for the 2-D migration (trace-scan distance, velocity,

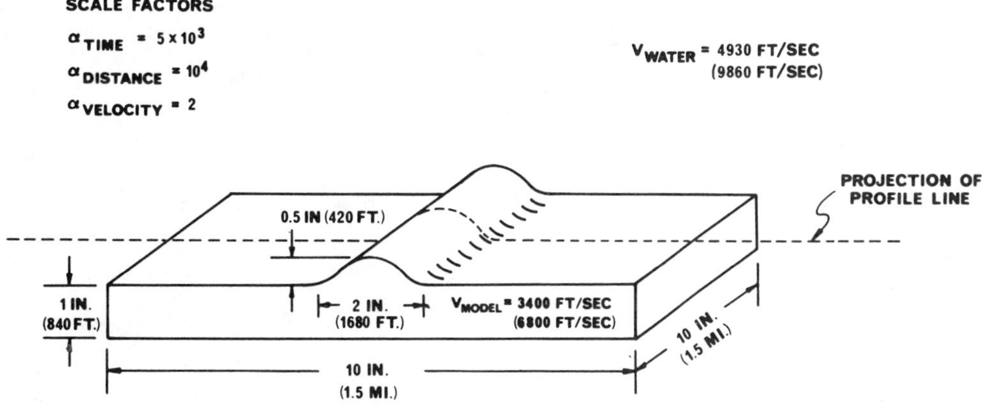


FIG. 6. 2-D high curvature ridge model.

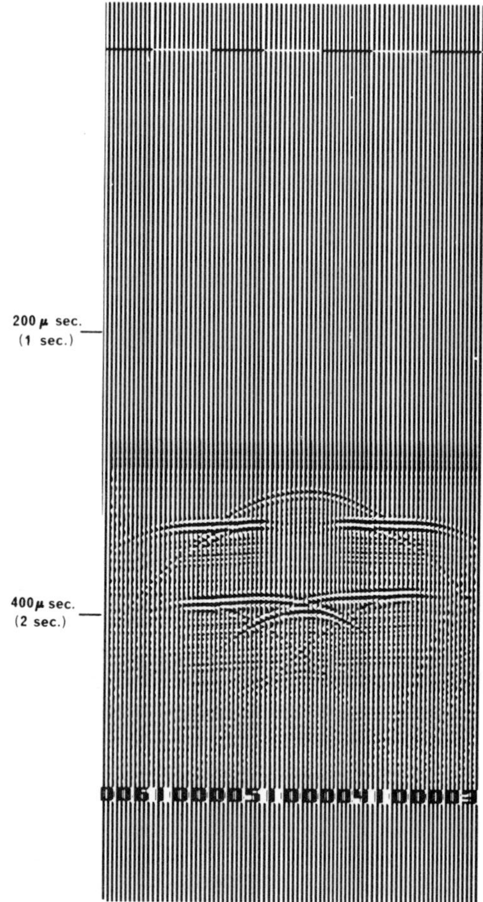


FIG. 7. Profile over model shown in Figure 6. Shotpoint separation = .2 inch (167 ft).

shotpoint distances, final filter parameters) were also used for the 3-D migration.

### THREE-DIMENSIONAL RESULTS

The model used for the 3-D experiment is shown in Figure 9 (see Figure 11 for a perspective view). The purposes of the experiment were to study 3-D complications in reflection profiles and to provide data from a known structure for testing 3-D migration as defined by equations (1) and (2a). Reflection data were recorded over the model and processed to produce 13 different 3-D migrated cross-sections of the model.

Figure 10 represents a plan view of the data collection scheme with the model shown schematically in solid lines. The numbers on the left of Figure 10 represent the successive source-receiver locations for the first of 96 profiles shot over the model. Numbers along the top of Figure 10 represent the locations of the 96 profiles. The dashed lines which are numbered at the bottom of Figure 10 show the locations of the cross-sections which were calculated for the purpose of constructing a structure map. Our choice of these thirteen 3-D migrated cross-sections was quite arbitrary. The total number of 3-D migrations performed on field data will depend only upon the number of true cross-sections required to delineate the structures. Figure 11 is a photograph of the model showing relief along the cross-sections of interest.

Results for cross-sections 3, 7, and 11 are shown in Figures 12 to 14. Each figure represents a different cross-section, and the display labeled A is the 2-D raw data profile over that cross-section. The display labeled B is a conventional 2-D migration of the raw data. Display C of each figure is the 3-D migration result for that cross-section. The single raw data profile shown is used in the 2-D migration. The 3-D migration, on the other hand, uses data from a number of profiles on either side of the reconstruction plane of interest. We chose to collapse the data from 48 consecutive profiles (24 on each side) into each of the 3-D migrated cross-sections shown in Figures 12–14.

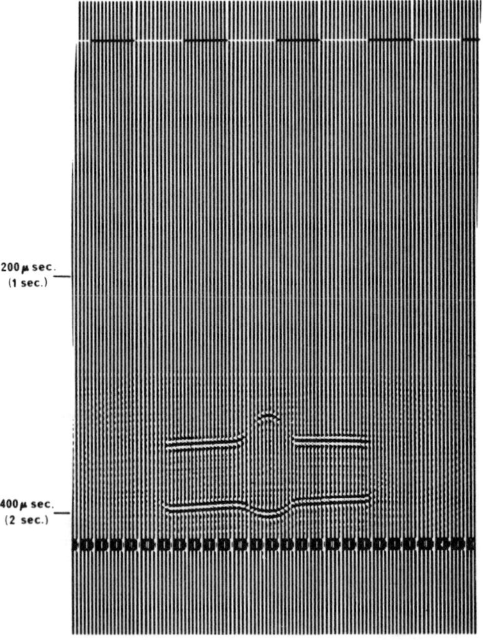


FIG. 8. Migration of profile shown in Figure 7.

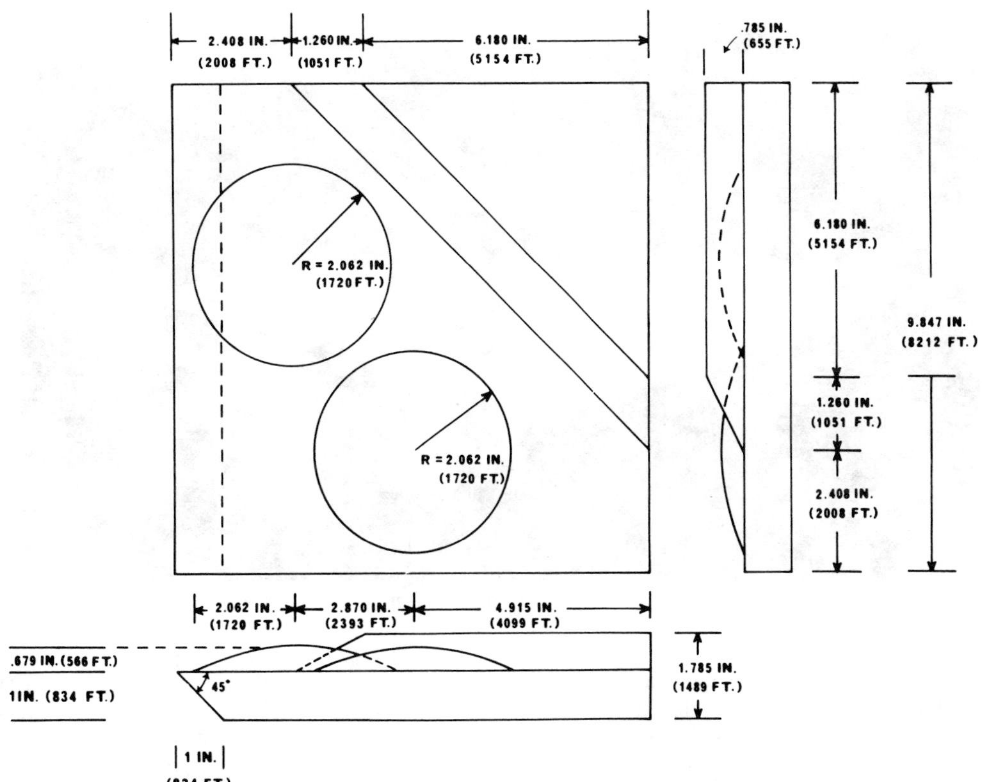


FIG. 9. 3-D model with equivalent field dimensions. See Figure 11 for perspective view.

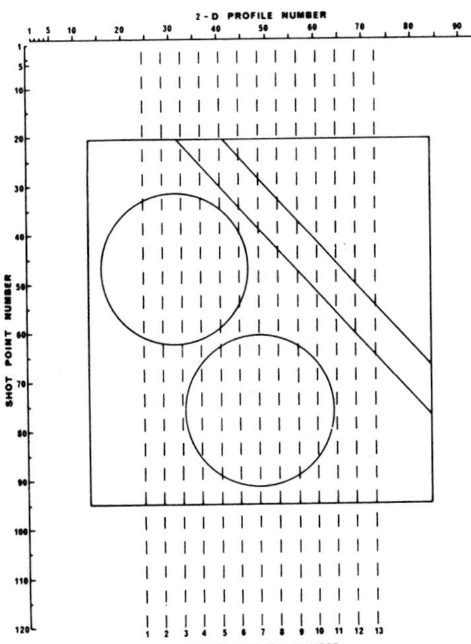


FIG. 10. 3-D data grid.

The quality of the results indicates that less data probably could have been used without serious deterioration of the results. The 96 parallel raw data profiles were taken .133 inches (111 ft) apart while the 3-D migrated profiles were reconstructed at intervals of .532 inches (444 ft).

It is most instructive to examine each of Figures 12 to 14, comparing the raw data, 2-D migrated data, and 3-D migrated data with the true cross-section (Figure 11). Here, the central cross-section (no. 7 in Figure 11) and the corresponding data of Figure 13 are singled out for specific discussion.

Several statements can be made concerning the results shown in Figure 13.

(1) The raw data are extremely difficult to interpret due to diffraction and sideswipe events. Of course, when one starts with a knowledge of the model geometry, it is easy to account for all the arrivals in the raw data.

(2) True structures in the profile plane can be

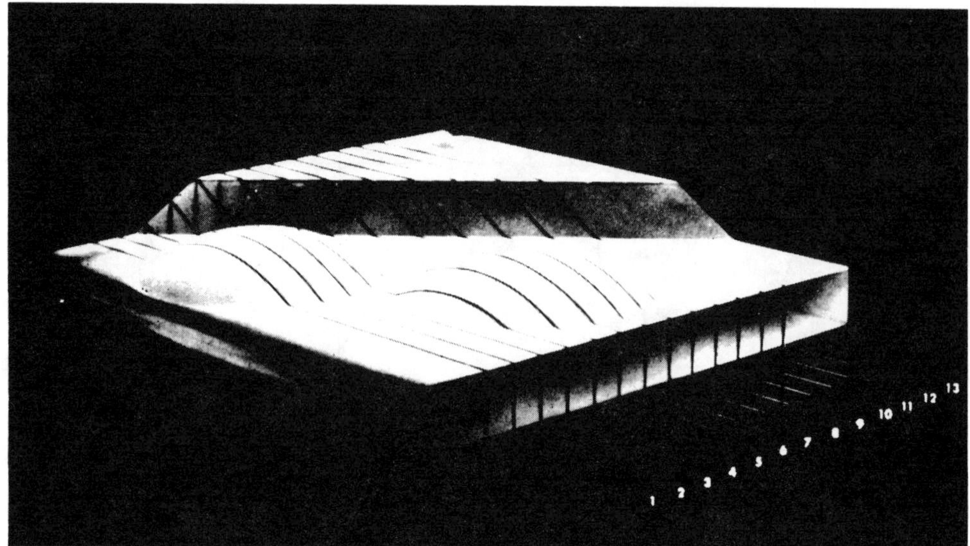


FIG. 11. Photograph of 3-D model.

absent in the raw data due to a component of dip perpendicular to the profile plane (blind structures). For example, no reflected energy from the portion of the fault which lies in the plane of the profile is recorded in the raw data profile.

(3) Conventional 2-D migration of the raw data profile does not eliminate sideswipe events or enhance blind structures. For example, the central hump in Figure 13b is a sideswipe event and has not been removed by conventional migration. This sideswipe causes an ambiguity in lateral correlation. (One could even go so far in this example as to assume the horizontal feature under the hump to represent a gas-fluid contact under an anticlinal trap.) Notice also that arrivals from the correct fault plane position are absent. Correlation across this blind zone is difficult.

(4) Two-D migration produced an increase in the background noise. This is probably a result of using the simplified integration scheme represented by the sum in equation (1).

(5) Three-D migration eliminated sideswipes and brought out blind structures. The resultant profile, shown in Figure 13c, is devoid of any interpretational ambiguities caused by these phenomena.

(6) The background noise created by the 3-D migration is similar to that of the 2-D migration and can probably be reduced by use of more sophisticated numerical techniques.

Similar statements could be made concerning the cross-sections displayed in Figures 12 and 14.

### STRUCTURE MAPS FROM 2-D AND 3-D RESULTS

The upper surface of the model was mapped by first using a more conventional 2-D shooting and migration plan and then using the 3-D migration results. These maps are shown, respectively, in Figures 15 and 16. Contour values used to construct the maps were taken from every fourth shot point as this interval is equal to the distance between the processed profiles. The location of layer terminations and the fault-plane boundaries were read (to the nearest shotpoint) from the profiles.

Several statements can be made concerning the mapping of these experimental data:

(1) Structure could not be determined from the raw data profiles; migration was necessary.

(2) The structural map constructed from the 3-D migrated profiles resulted in a true and precise picture of the upper surface of the model (compare Figure 16 with Figure 10).

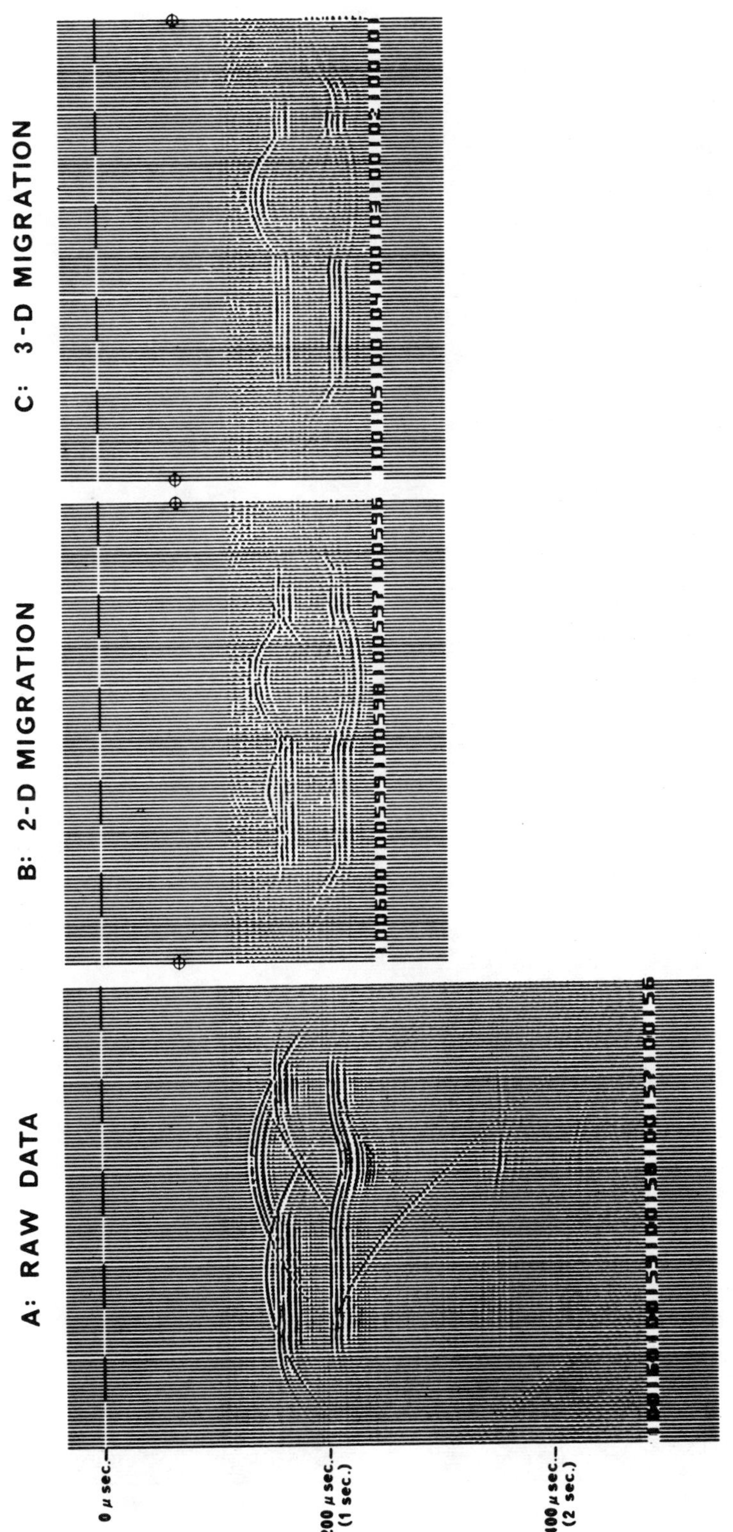


FIG. 12. Results for cross-section no. 3.

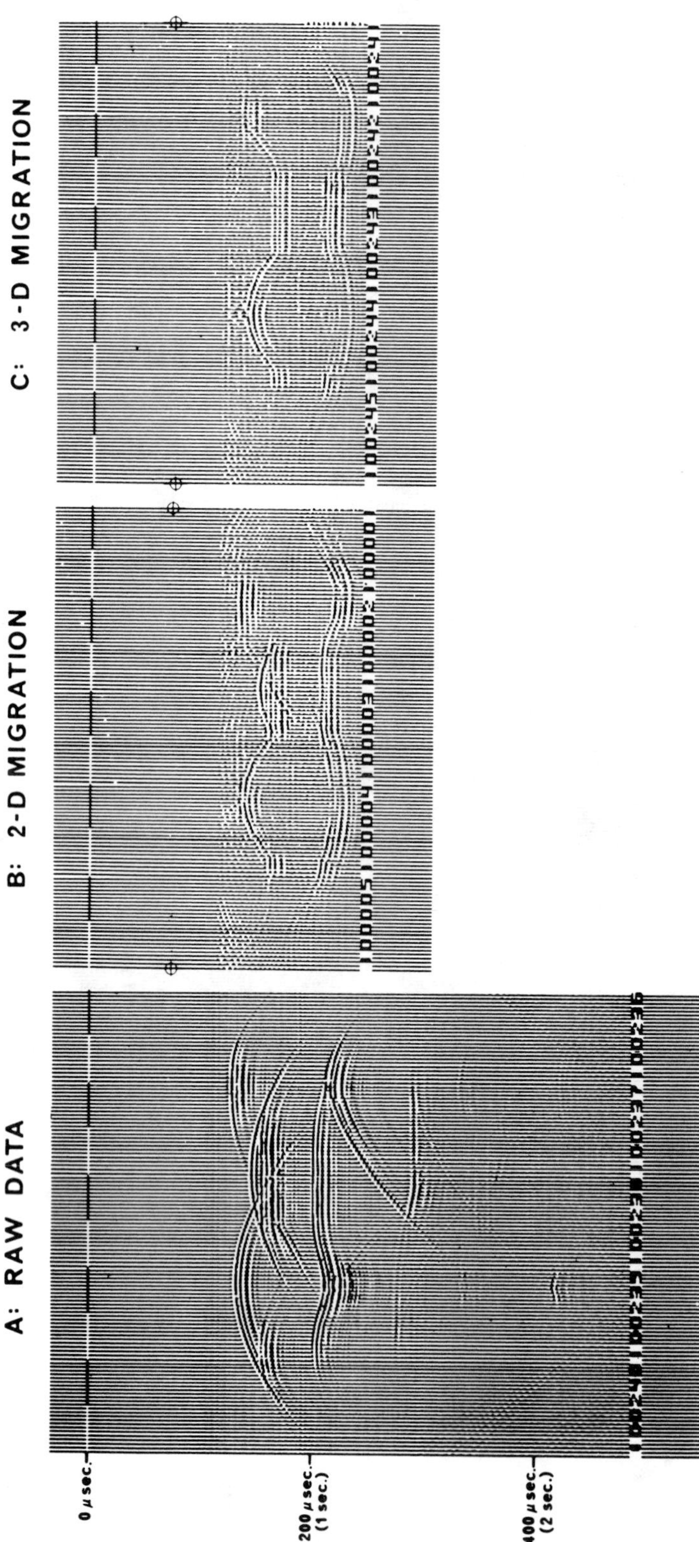


FIG. 13. Results for cross-section no. 7.

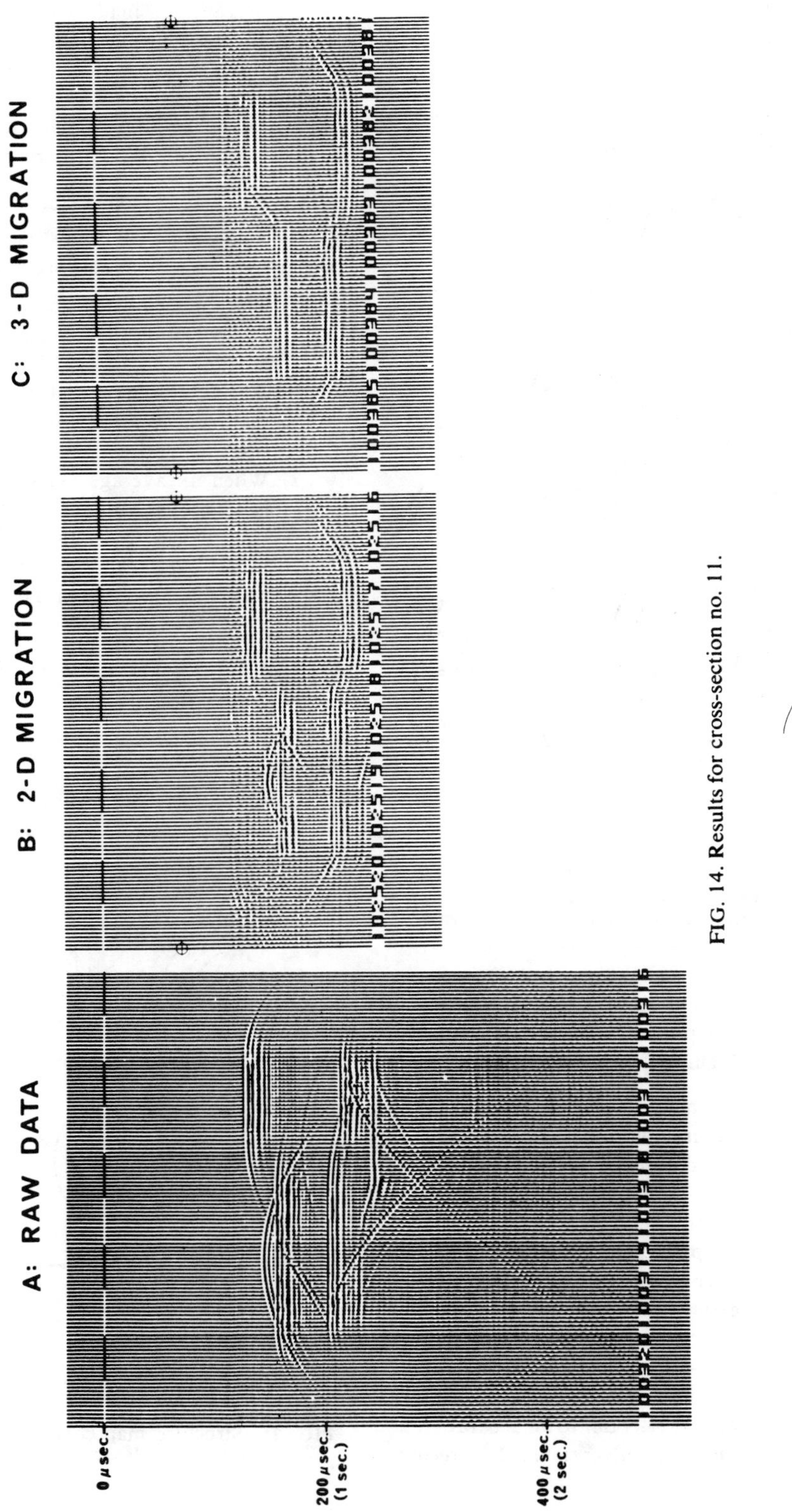


FIG. 14. Results for cross-section no. 11.

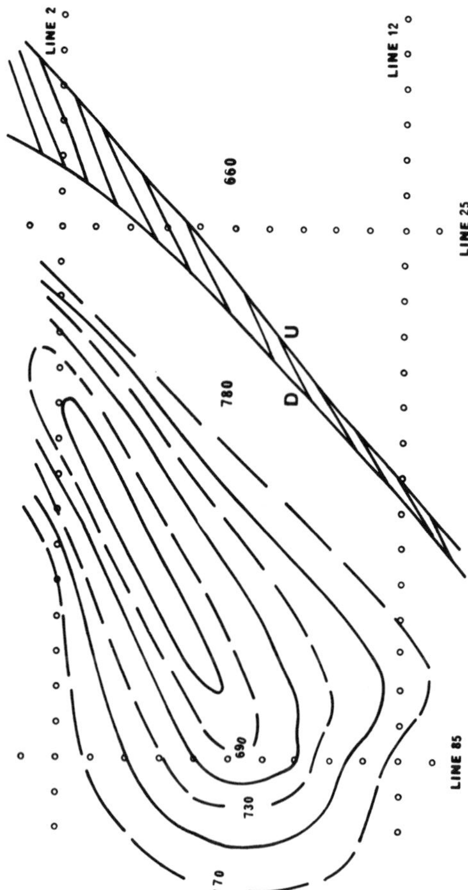


FIG. 15. Structure map constructed from migration of conventional 2-D profiles (contours in msec).

(3) The structural map made from the 2-D migrated profiles is distorted in shape (compare Figure 15 with Figure 10).

With dense enough 2-D coverage and 2-D migration, structural distortions in the 2-D data can be corrected through careful interpretation. As the interpreter develops his map, he can detect which events are sideswipes on individual 2-D profiles and apply appropriate areal displacements. In other words, this experiment does not demonstrate that 3-D migration is essential for delineating the features of this model, although it clearly shows that at least 2-D migration is indispensable. Furthermore, the gathering of dense areal data is essential to an accurate resolution of the model; conventional 2-D recording would have been inadequate. The advantages of 3-D processing will probably be even more evident in layered cases where greater correlation ambiguities exist. The 3-D process will also allow us to image features which are some distance removed from our data-gathering area.

## CONCLUSIONS

The following conclusions can be drawn from the results and discussion presented above:

(1) The migration method of processing seismic-reflection data can be extended to handle 3-D exploration problems with arbitrary source and receiver arrays. A mathematical description is given by equations (1) and (2) above.

(2) When an average velocity-depth function is used over an entire section, false structures caused by lateral velocity variations in the overburden are not eliminated but, rather, are brought into focus by migration.

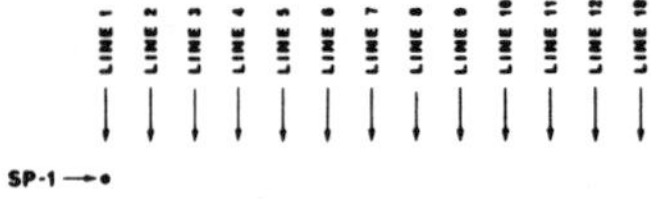


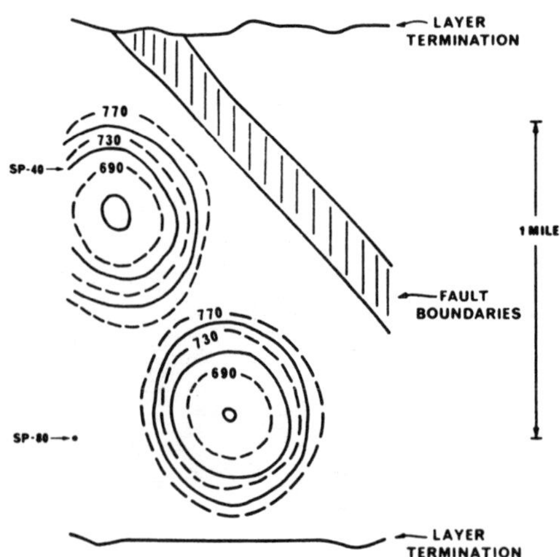


FIG. 16. Structure map constructed from 3-D migrated profiles (contours in msec).

Subsequent trace-by-trace, time-to-depth conversion will eliminate these pseudo-structures but a rather detailed knowledge of the section velocity is required.

(3) When single coverage reflection profiles over 2-D models are processed by conventional migration, an exact reproduction of a true vertical-time section of the model results. The background noise is, however, increased. It has been shown that these noise problems can be eliminated by better numerical techniques (Gardner et al, 1973).

(4) Single coverage profiles recorded over 3-D models are almost totally uninterpretable unless a migration process is applied.

(5) When single coverage reflection profiles over 3-D models are processed by conventional 2-D migration, correlation ambiguities may arise due to sideswipes and blind structures.

(6) Structure maps constructed from a conventional survey with profiles independently processed by 2-D migration show distortions of the structures. However, the distortions can be reduced by increased coverage and careful interpretive techniques.

(7) Simultaneous 3-D migration of a sequence of parallel reflection profiles eliminates the correlation ambiguities caused by side-swipes and blind structures.

(8) Structure maps constructed from the 3-D results give a true and precise picture of the model.

## REFERENCES

Claerbout, Jon F., and Doherty, Stephen M., 1972, Downward continuation of moveout corrected seismograms: Geophysics, v. 37, p. 741–768.

Gardner, G. H. F., French, W. S., and Matzuk, T., 1973, Elements of migration and velocity analysis: Presented at the 43rd Annual International SEG Meeting, October 25, 1973, Mexico City.

Maginness, M. G., 1972, The reconstruction of elastic wave fields from measurements over a transducer array: J. of Sound and Vibration, v. 20, no. 2, p. 219–240.

Morse, P. M., and Ingard, K. U., 1968, Theoretical acoustics: New York, McGraw-Hill Book Co., Inc.

Schneider, William A., 1971, Developments in seismic data processing and analysis (1968–1970): Geophysics, v. 36, p. 1043–1073.

Taner, M. Turhan, and Koehler, Fulton, 1969, Velocity spectra—digital computer derivation and applications of velocity functions: Geophysics, v. 34, p. 859–881.

# Inversion of reflection traveltimes and amplitudes

C. Macdonald*, P. M. Davis*, and D. D. Jackson*

ABSTRACT

A Bayesian nonlinear inversion scheme is used to invert traveltime and amplitude data from supercritical reflections. Reflection amplitudes are determined using a spherical-wave model, where the amplitude of a reflected spherical wave is given by a Sommerfeld integral. This integral is evaluated numerically by integrating over a contour in the complex plane. The spherical-wave model differs from plane-wave models in the regions around the critical points for $P$ and $S$ refraction. In these regions, the spherical-wave amplitudes are sensitive to frequency with a dependance on the ratio $H/\lambda$, between the reflector depth $H$ and the seismic wavelength $\lambda$.

The accuracies of the inversion scheme and the spherical reflection model are tested using control data from the University of Houston physical-model tank. The solution obtained by inversion agrees with the known solution to within 3 percent for all parameters and the amplitudes predicted by the spherical-wave model agree remarkably well with the observed data.

## INTRODUCTION

The seismic inversion problem has received much attention in recent years, and many different methods of solution have been developed. Most of these methods differ either in the data set used for the inversion or in the parameters to be estimated by the inversion. Inversion for velocity only may use traveltime data to solve for one-dimensional (1-D) velocity profiles (Gibson et al., 1979), or two-dimensional (2-D) profiles (Bishop et al., 1985). Other velocity inversion procedures estimate velocity profiles by operating directly on the observed data (Bleistein and Cohen, 1982; Raz, 1982; Carrion and Kuo, 1984; Carrion et al., 1984). One-dimensional impedance profiles can be recovered using normal-incidence seismic traces (Bube and Burridge, 1983; Cooke and Schneider, 1983; Bamberger et al., 1982; Oldenburg et al., 1984). There are also methods that solve for separate velocity and density profiles. Iterative nonlinear inversion methods have been proposed by McAulay (1985) who reconstructs the reflectivity of a stack of plane layers, and by Tarantola (1984a) who models the full wave field. There are also noniterative solutions using the WKBJ approximation (Eiges and Raz, 1985) as well as linear approximations to the wave equation (Tarantola, 1984b; Ikelle et al., 1986) and the Riccati equation (Bregman et al., 1985; Carazzone, 1986). These references cover only a selected portion of the literature on seismic inversion.

For an inversion scheme to be stable, it must approximate true earth structures using simplified models. Otherwise, too many parameters must be solved for in the inversion, which will introduce nonuniqueness and make solution of the problem difficult. It is sensible to start with the most simple models and determine if the inverse problem is well posed. The complexity of the structural model can then be increased until a balance between inversion stability and structural resolution is achieved. As a first attempt at the problem, we consider a model consisting of plane horizontal layers.

Synthetic data are used to find a formulation of the inverse problem which leads to a solution as efficiently as possible. In many cases, nonlinear inverse problems are almost linear (Jackson and Matsu'ura, 1985) and can be solved effectively using iterative, generalized linear inversion techniques (Tarantola and Valette, 1982). We therefore choose a formulation of the seismic problem that exhibits a high degree of linearity. Any iterative nonlinear inversion requires calculation of a large number of model functions, so our synthetic seismograms must be computationally efficient. Having obtained a suitable formulation of the problem, the accuracy of our inversion scheme is tested using control data from the University of Houston physical-model tank. These data consist of a common-depth-point (CDP) gather recorded over a set of three horizontal plane layers. Because the true physical structure has the same form as our model structure, our synthetic seismogram should be able to predict the observed data accurately and our inversion solution should match the known solution.

Since the control data contain supercritical reflections, we require a synthetic modeling algorithm that is accurate in the critical region. In the critical region asymptotic ray theory breaks down and is therefore unsuitable for our purposes.

Manuscript received by the Editor March 3, 1986; revised manuscript received September 5, 1986.
*Department of Earth and Space Sciences, University of California at Los Angeles, Los Angeles, CA 90024.


Reprinted from Geophysics, 52, 606-617

Recently, Winterstein and Hanten (1985) observed critical $P$-wave reflections and some of the typical behavior of the supercritical amplitudes. They observed a shift in the position of the point of maximum amplitude to an offset larger than the critical offset predicted by zero-order asymptotic ray theory and also observed that the maximum was smaller than the ray-theory amplitude. These observations can be explained using the theory for the reflection of spherical waves at plane interfaces. The amplitude of a reflected spherical wave can be determined by evaluating a Sommerfeld integral along a contour in the complex plane (Červený and Ravindra, 1971). The resulting amplitudes are frequency-dependent, with the infinite-frequency values corresponding to the geometrical ray amplitudes. For finite frequencies the difference between ray amplitudes and spherical wave amplitudes may be large.

We show that when reflections from plane horizontal layers are modeled using traveltime and amplitude data, the inverse problem is fairly linear and we may obtain a unique solution by iterative linear inversion. We describe a model for calculating frequency-dependant reflection amplitudes and then show, using the physical-model data, that both the inversion and the synthetic seismic model are accurate when applied to data recorded over plane layers.

## SOLUTION OF THE INVERSE PROBLEM

Consider the general linear inverse problem given by the matrix eqation

$$\mathbf{y} = \underset{\sim}{\mathbf{A}}\boldsymbol{\chi} + \boldsymbol{\varepsilon}, \tag{1}$$

where $\mathbf{y}$ are data, $\underset{\sim}{\mathbf{A}}$ is a linear operator that operates on a set of parameters $\boldsymbol{\chi}$, and $\boldsymbol{\varepsilon}$ are random errors. Given an estimate $\mathbf{x}$ of the true solution $\boldsymbol{\chi}$, we define the data residuals by $\mathbf{e} = \mathbf{y} - \underset{\sim}{\mathbf{A}}\mathbf{x}$. If we know the probability density function $p(\boldsymbol{\varepsilon})$ of the random errors $\boldsymbol{\varepsilon}$, then the likelihood of $\mathbf{x}$ as the solution to the inverse problem is given by $p(\mathbf{e})$ which represents the probability of obtaining a set of random numbers $\mathbf{e}$ from a distribution with density function $p$. The maximum-likelihood solution is the estimate $\mathbf{x}$ of $\boldsymbol{\chi}$ which maximizes the probability function $p(\mathbf{e})$. If we assume that the errors $\boldsymbol{\varepsilon}$ are Gaussian with zero mean and covariance matrix $\underset{\sim}{\mathbf{E}}$, then the function $p(\mathbf{e})$ is given by $p(\mathbf{e}) = a \exp(-S^2)$, where $a$ is a normalization constant and $S^2 = \mathbf{e}^t\underset{\sim}{\mathbf{E}}^{-1}\mathbf{e}$ is the weighted residual sum of squares. The maximum-likelihood problem now reduces to the weighted least-squares procedure of minimizing $S^2$. If $\Delta\mathbf{x} = \boldsymbol{\chi} - \mathbf{x}$, then the errors $\mathbf{e}$ are given by $\mathbf{e} = \underset{\sim}{\mathbf{A}}\Delta\mathbf{x} + \boldsymbol{\varepsilon}$ and the expression can be expanded for $S^2$ to get

$$S^2 = \Delta\mathbf{x}^t\underset{\sim}{\mathbf{A}}^t\underset{\sim}{\mathbf{E}}^{-1}\underset{\sim}{\mathbf{A}}\Delta\mathbf{x} + 2\Delta\mathbf{x}^t\underset{\sim}{\mathbf{A}}^t\underset{\sim}{\mathbf{E}}^{-1}\boldsymbol{\varepsilon} + \boldsymbol{\varepsilon}^t\underset{\sim}{\mathbf{E}}^{-1}\boldsymbol{\varepsilon}. \tag{2}$$

Because the errors $\boldsymbol{\varepsilon}$ are Gaussian with zero mean and covariance matrix $\underset{\sim}{\mathbf{E}}$, the second term in the expansion will have an expectation value of zero and the third term will have an expectation value of $N$, the number of data. The first term is a quadratic form which represents modeling errors alone. Contours of constant $S^2$ in any 2-D plane through parameter space $\mathbf{x}$ will be elliptical.

In the case of a nonlinear inverse problem $\mathbf{y} = \mathbf{f}(\boldsymbol{\chi}) + \boldsymbol{\varepsilon}$, where $\mathbf{f}$ is a nonlinear function, we can define the residual sum of squares $S^2 = \mathbf{e}^t\underset{\sim}{\mathbf{E}}^{-1}\mathbf{e}$, where $\mathbf{e} = \mathbf{y} - \mathbf{f}(\mathbf{x})$. In general the residual function $S^2$ will be nonquadratic with irregular contours, local minima, and other undesirable features. Given an irregular residual function, it may be very difficult to determine the minimum corresponding to the maximum-likelihood solution. There are many techniques for solving this problem (Bevington, 1969), such as grid search methods and the steepest descent method. These methods are generally time-consuming, and if the residual function is highly irregular they may fail. Other methods applicable when the residual function is smooth, such as generalized linear inversion and Marquardt inversion, use linear approximations to the nonlinear function $\mathbf{f}$. We can approximate the function $\mathbf{f}$ about a point $\mathbf{x}_0$ using a Taylor series expansion

$$\mathbf{f}(\mathbf{x}) = \mathbf{f}(\mathbf{x}_0) + \underset{\sim}{\mathbf{F}}\Delta\mathbf{x}, \tag{3}$$

where $\Delta\mathbf{x} = \mathbf{x} - \mathbf{x}_0$. The matrix $\underset{\sim}{\mathbf{F}}$ is given by

$$F_{ij} = \frac{\partial f_i}{\partial x_j}, \tag{4}$$

with the partial derivatives evaluated at the point $\mathbf{x} = \mathbf{x}_0$.

The inverse problem can now be expressed in a linear form

$$\Delta\mathbf{y} = \underset{\sim}{\mathbf{F}}\Delta\mathbf{x} + \boldsymbol{\varepsilon}, \tag{5}$$

where $\Delta\mathbf{y} = \mathbf{y} - \mathbf{f}(\mathbf{x}_0)$. Equation (5) can be solved iteratively to determine a solution $\mathbf{x}$. In theory, this technique can be applied to any nonlinear function. However, unless the linear approximation is valid in a large neighborhood of the point $\mathbf{x}_0$, the inversion will be unstable. In many cases nonlinear functions can be represented by linear functions in regions around $\mathbf{x}_0$ sufficiently large to make the iterative linear inversion effective. If a problem can be linearized, then the inversion can be performed efficiently and much faster than the more general nonlinear inversion techniques. It is therefore sensible to attempt to formulate the nonlinear problem in a form that can be linearized. We can investigate the linearity of a function $\mathbf{f}$ by plotting contours of constant $S^2$ on 2-D planes through parameter space. If the linear approximation of the function is valid, then such contours should be elliptical in a region around the solution. If the function $S^2$ is irregular, then the problem is highly nonlinear and will be difficult to solve.

In many cases the solution to the inverse problem is nonunique, due either to the inability of the data to resolve the parameters fully, or to an inadequate model function or excessive noise levels. One method of stabilizing the problem is the Bayesian inversion technique, as described by Jackson and Matsu'ura (1985); we do not discuss it in detail here. The basic idea of the method is to make prior estimates of the parameters and assign uncertainties to these estimates. By including the estimates and uncertainties as data, we effectively constrain the solution to lie within a few standard deviations of the a priori estimate. However, the constraints are not strict and may be violated if the data require a solution outside of the a priori confidence region. The residual sum of squares will then be the sum of two terms: one term with strictly elliptical contours representing departures from the prior model, and the other with possibly irregular contours representing residuals to the seismic data. For strictly linear problems, the residual sum of squares is a quadratic function of the parameters. If the prior uncertainties are small, then the seismic data will be effectively linear functions within the allowed range of parameter variation, and the error space will be approximately quadratic. However, in order to obtain a residual function with a quadratic form when **f** is highly nonlinear, we would have to know the solution very accurately a priori and inversion would be pointless. The technique is most effective for problems which are pseudolinear where harsh constraints on the solution are not required.

## LINEARITY OF SEISMIC INVERSION

The seismic inversion problem requires the estimation of velocities and densities at depth from reflection data recorded at the surface. The nonlinear model function is a synthetic seismogram. Consider two formulations of the inverse problem. In the first case, we model the full CDP gather using the observed seismic traces as input data. In the second case, we identify reflections on the CDP gather and determine the arrival time and amplitude of the reflected wave at each trace. These times and amplitudes are then used as input data for the inversion. Since the model is plane-layered, we parameterize the medium in terms of the $P$ and $S$ velocities, the density, the attenuation coefficient, and the thickness of each layer.

We investigated the linearity of the problem in each of the above formulations using synthetic data. Geometrical ray theory was used to determine the arrival times and amplitudes of waves reflected from the interfaces of a simple three-layer elastic model. Multiple reflections were not included in the synthetic data. Using a model wavelet, we constructed a synthetic CDP gather from the time and amplitude data. For each formulation of the problem, we constructed contour maps of the residual function $S^2$ by varying two parameters over a grid centered on the known solution and evaluating the residual function at each point. All of the other parameters were held fixed at their known solution.

Figure 1 shows sample contour plots for each case. The variable parameters are the $P$-wave velocities of layers 2 and 3. Figure 1a shows the typical form of the residual function for the case when the data are the observed seismic traces. The problem is clearly highly nonlinear with distorted contours and many local minima. Most of the local minima are due to the oscillatory nature of the wavelet; however, the problem is still highly nonlinear if the seismic pulse is a delta function. It is clear that this formulation of the problem is not suitable for linearization, and it will be difficult to solve by general nonlinear techniques. If we include multiple reflections, the nonlinearity of the problem is even more severe. Figure 1b shows a typical plot obtained for the case when the data are traveltimes and amplitudes. The contours are elliptical, indicating linearity of the model function. Sections for all combinations of parameters show similar form, suggesting that we can approximate the synthetic function using a numerical Taylor expansion. The linearity also implies that all parameters can be uniquely resolved. Although we used a geometrical ray method as a model function, we expect these results to be similar for any synthetic seismogram. We therefore choose to solve the inverse problem using traveltime and amplitude data.

## REFLECTION OF SPHERICAL WAVES

Having chosen a method of solution, we must now find a synthetic model that can accurately predict the observed data.

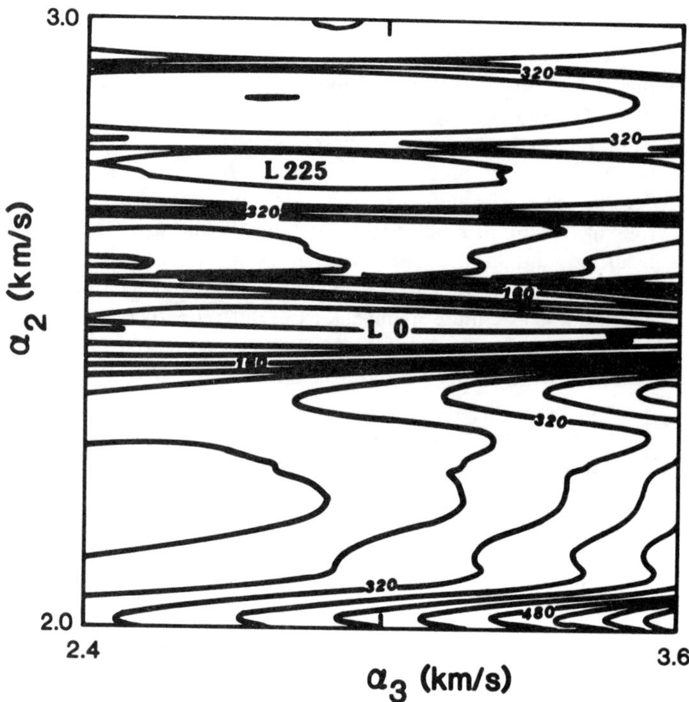


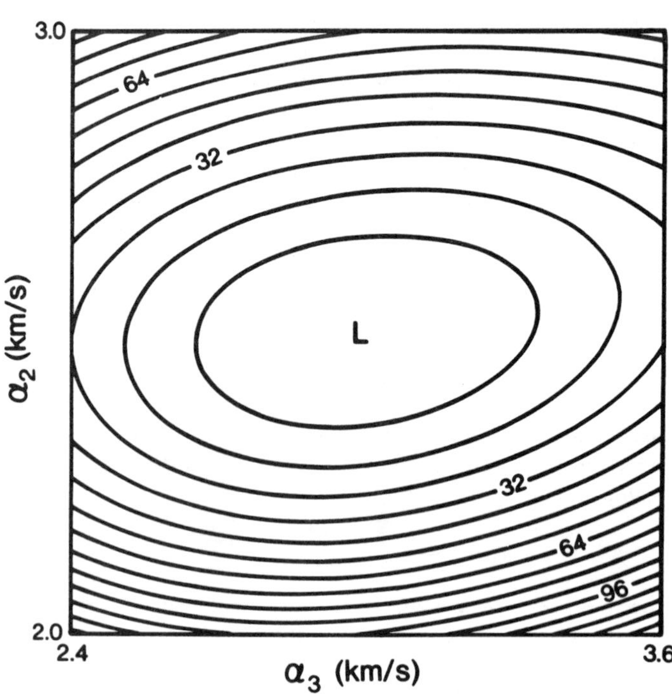


FIG. 1. Contour plots of residual function $S^2$ for a three-layer model. Variable parameters are $\alpha_2$ and $\alpha_3$. All other parameters are fixed at their true values. (a) Input data are seismic traces. The distorted contours indicate nonlinearity. (b) Input data are times and amplitudes. The contours are elliptical, indicating linearity.

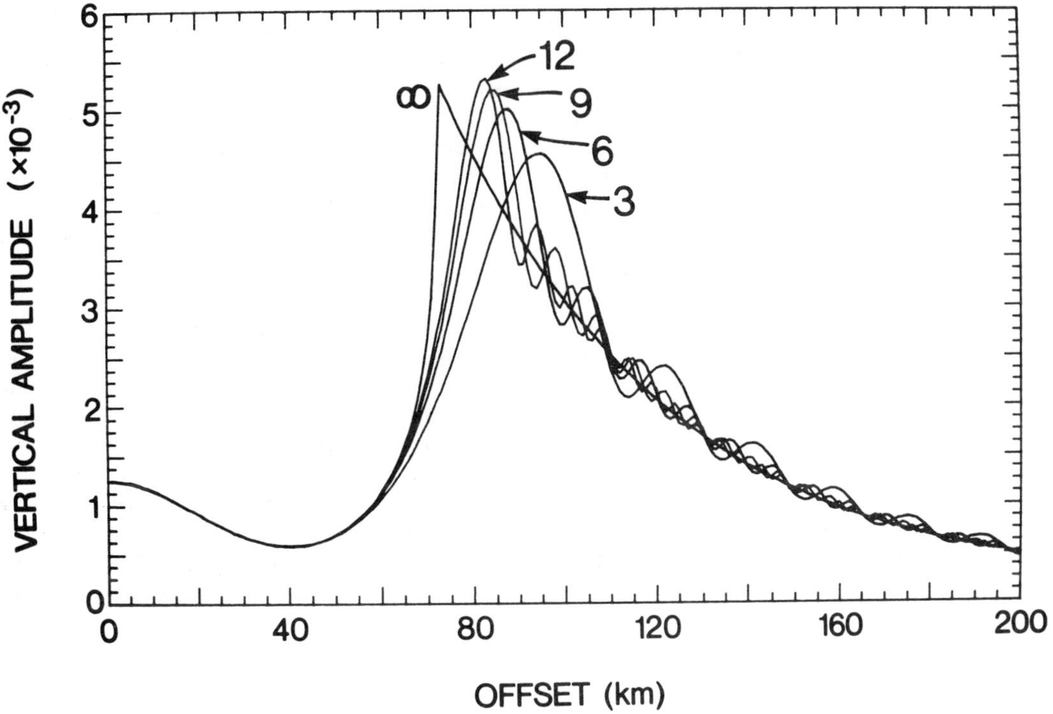


FIG. 2. Vertical component of displacement for spherical waves with frequencies of ∞, 3, 6, 9, and 12 Hz reflected off the third interface of a four-layer crustal model. The *P*-wave velocities in layers 1 to 4 are 5.4, 6.0, 6.75, and 8 km/s, respectively. Each layer has a velocity ratio ($\alpha/\beta$) of $\sqrt{3}$, density of 1 g/cm$^3$, and thickness of 10 km.

To model supercritical reflections, we cannot use plane-wave approximations because they are invalid in the critical region. Thus we consider a model for spherical waves reflected off plane interfaces. Since this model has appeared in the literature before, we only give a brief overview of the mathematics here. The case of a spherical wave reflected off the interface between two elastic half-spaces is dealt with by Červený (1966), Červený (1967b), and Červený and Ravindra (1971). Extension of the method to multilayered media is described by Brekhovskikh (1960) and Červený (1967a).

*P*- and *S*-wave potentials in each layer are specified using the Sommerfeld integral representation. Applying the boundary conditions of continuity of stress and displacement at the top and bottom of each layer determines solutions for the potentials of waves reflected off each interface. Differentiating these potentials yields the following integral equations for the horizontal and vertical displacements $u$ and $w$:

$$u = -\frac{k}{2}\exp(-i\omega t)\int_{-\infty}^{\infty} A_0(q)\sigma(q)H_1^1(krq) \times \exp\left[ikB(q)\right]q^2(1-q^2)^{-1/2}\,dq,$$

and

$$w = \frac{ik}{2}\exp(-i\omega t)\int_{-\infty}^{\infty} A_0(q)\sigma(q)H_0^1(krq) \times \exp\left[ikB(q)\right]q\,dq, \tag{6}$$

where

$$B(q) = \sum_{i=1}^{n} 2h_i\sqrt{(n_i^2 - q^2)}, \tag{7}$$

and $n_i = \alpha_1/\alpha_i$. The above equations give the amplitudes for a wave with frequency $\omega$ and wavenumber $k$ reflected off the $n$th interface in a medium consisting of layers with *P* and *S* velocities $\alpha_i$ and $\beta_i$, density $\rho_i$, and thickness $h_i$. The functions $H_0^1$ and $H_1^1$ are Hankel functions of the first kind of orders 0 and 1, respectively. $A_0(q)$ represents the plane-wave reflection coefficient for interface $n$ and $\sigma(q)$ is the product of plane-wave transmission coefficients for the overlying interfaces (Červený and Ravindra, 1971; Aki and Richards, 1980). The coefficient $A_0$ contains branch points corresponding to the critical refraction of both *P*- and *S*-waves, and in order to evaluate the integrals it is necessary to integrate along a contour in the complex plane which bypasses these branch points. Červený (1967b) describes a modified steepest-descent contour which is suitable for evaluating the integrals.

The resulting reflection amplitudes depend upon the ratio $H/\lambda$ between the radius of curvature $H$ of the incident wave and the seismic wavelength $\lambda$. For large $H/\lambda$ (high frequencies), spherical waves behave as plane waves. For small $H/\lambda$ (low frequencies), the differences between the amplitudes of spherical and plane waves in the critical regions are significant. As frequency decreases, the position of the point of maximum amplitude moves to larger offsets and the magnitude of the maximum decreases. Interference between the reflected and refracted waves at the interface causes these effects. Outside the critical regions at either small or large incident angles the spherical and plane-wave solutions are identical.

Figure 2 illustrates the frequency-dependant nature of the reflection amplitudes for a wave reflected off the Moho in a simple crustal model. The model consists of three layers with $P$ velocities of 5.4, 6.0, and 6.75 km/s each with a thickness of 10 km overlying a half-space with a $P$ velocity of 8.0 km/s. All media have a $P$-to-$S$ velocity ratio of $\sqrt{3}$ and density of 1 g/cm$^3$. Waves with frequencies of $\infty$, 12, 9, 6, and 3 Hz are shown. The shift of the position of maximum amplitude to larger offsets for low frequencies is clearly shown. The frequencies used are common for deep reflection studies and show that the frequency-dependent effects are significant and should be included when modeling crustal reflections. The oscillations of the amplitude curves about the plane-wave curve after the critical point are due to interference effects between the reflected waves and head waves. Since we are considering a single frequency, this interference continues to very large offsets. If we were considering the reflection of a seismic pulse rather than a single frequency, the interference would stop once the head wave and reflection had separated by more than one pulse width. We chose to use the single-frequency approximation because the increased accuracy obtained by including the effect of a multifrequency pulse does not justify the large increase in computation required.

The effects of attenuation and source anisotropy are easily included as simple multiplicative corrections to the amplitudes (Červený, 1967a). The effect of a free surface can also be included using coefficients of conversion (Červený and Ravindra, 1971). Traveltimes of the reflected waves are calculated using ray-tracing techniques.

## CONTROL DATA

The theoretical results described in the preceding sections were tested using control data from the University of Houston physical-model tank. A simple structure consisting of three plane layers was formed by immersing a sheet of Plexiglas in water. Source and receiver transducers were then used to record a CDP gather. Figure 3 illustrates the experimental setup of the physical-model tank. These data give us an opportunity to test the theoretical results against data equivalent

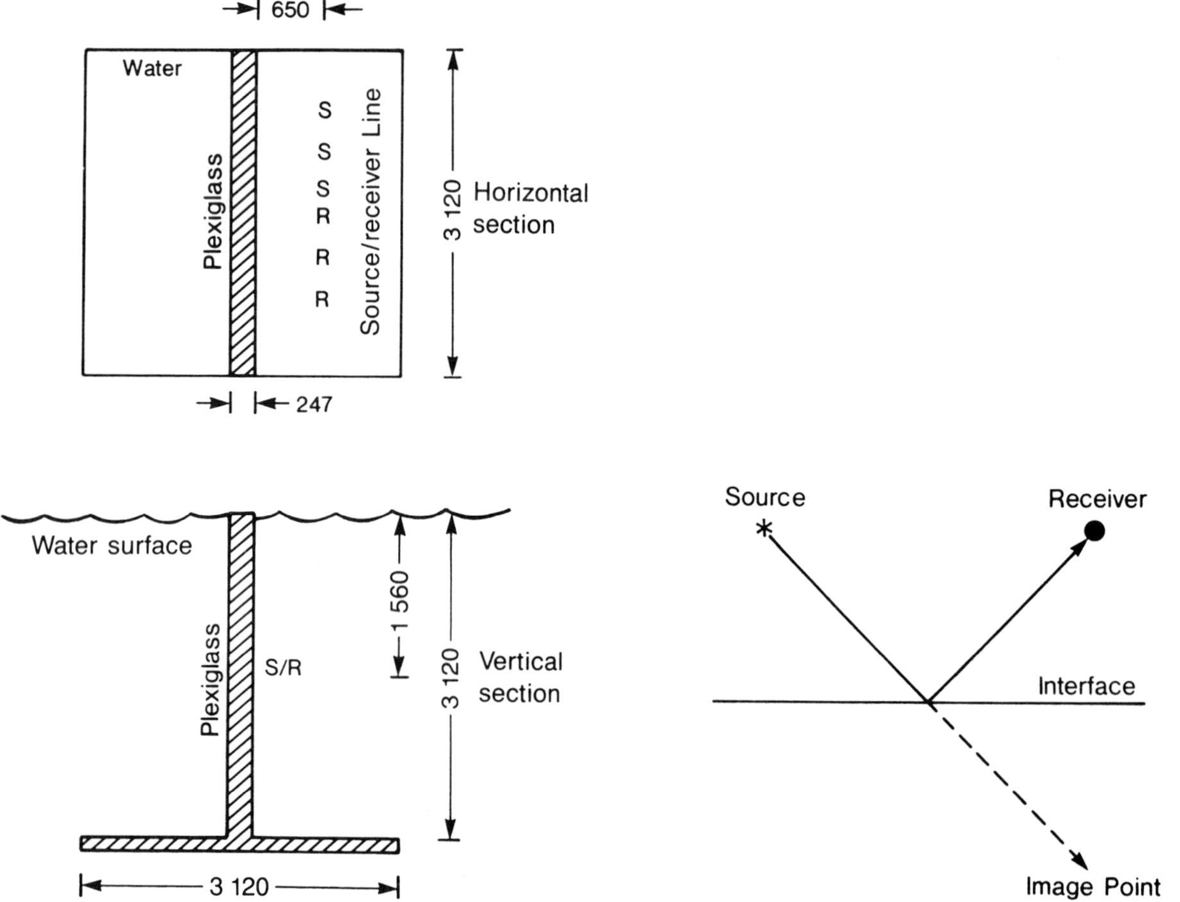


FIG. 3. University of Houston physical-model tank. (a) Horizontal and vertical sections through the Plexiglas model showing the location of the source-receiver line with respect to the Plexiglas. The Plexiglas sheet is immersed in water so that a three-layer structure of water-Plexiglas-water is formed. All dimensions are in scaled meters. The scaling factor converts 1 inch in the physical model into 130 m on the scaled model. (b) Location of image points for recording data set 2. The Plexiglas sheet is removed to record the data set of the image points.

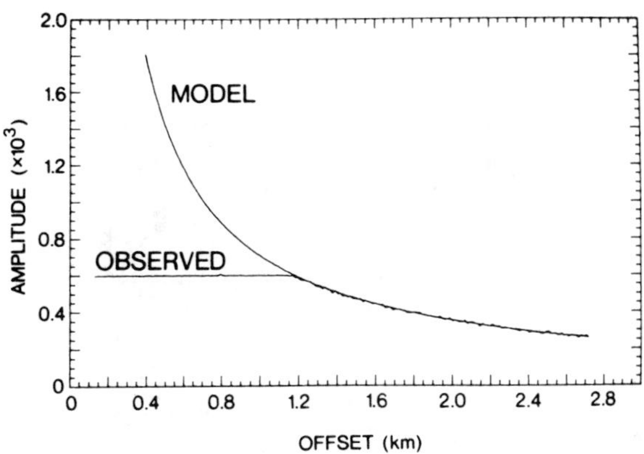


FIG. 5. Amplitude of direct wave and theoretical 1/distance spreading curve. The observed data curve is flat initially due to clipping by the instrumentation.

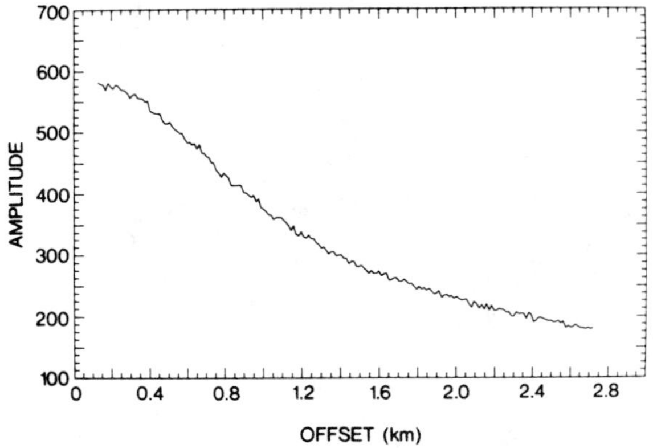


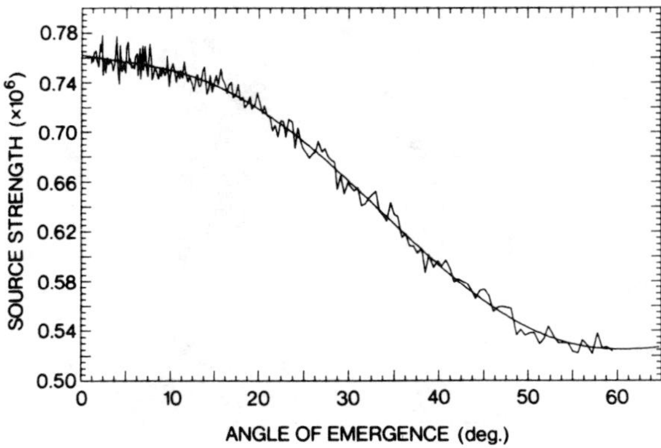


FIG. 6. (a) Amplitude of the image point arrivals $P_i$. (b) Source anisotropy curve determined by correcting the $P_i$ amplitudes for spherical divergence. The smooth curve is a polynomial fit to the source function used as a source model in the inversion.

to field data but with a priori knowledge of the true structure and easily identifiable reflections.

Figure 4 shows the two data sets recorded in the physical-model tank. Data set 1 contains three arrivals of interest, $P$, $PP$, and $PPPP$. The $P$-wave is the direct arrival, the $PP$ wave is a $P$-wave reflection off the top surface of the Plexiglas, and the $PPPP$ wave is a $P$-wave reflection off the back surface of the Plexiglas. Arrivals $PSPP$ and $PSSP$ are reflections of mode-converted $S$-waves. We are not considering $S$-wave reflections, so we neglect these arrivals. Data set 2 has one interesting arrival, $P_i$. This data set was recorded with the Plexiglas removed and the receivers positioned at the image points for the $PP$ arrivals, as illustrated in Figure 3b. The $P_i$ arrival is used to estimate any angular dependance of the source. Since each arrival is a direct wave from the source at a known angle of emergence, we can correct the amplitudes of the arrivals for spherical divergence to get an estimate of the source strength for different angles of emergence. The arrival times and amplitudes of the $P_i$ and $P$-waves were easily determined by hand picking.

Figure 5 shows the amplitude of the direct wave for data set 1 compared with a theoretical 1/distance spreading curve. For small offsets the observed curve is flat due to clipping of the arrivals, but for larger offsets the fit is very good, showing that the spreading correction of 1/distance is valid. We can therefore use this spreading correction to determine the source function from the $P_i$ arrival. Figure 6 shows the observed $P_i$ amplitudes and the estimated angular dependence of source strength. The solid curve is a polynomial fit to the source function which we use to calculate synthetic amplitudes.

For a given trace the arrival time of the $P_i$ wavelet in data set 2 should be almost the same as the arrival time of the $PP$ wave in data set 1. Amplitude and phase differences between the two pulses will be entirely due to the effect of reflection at the water-Plexiglas interface. We can therefore determine the reflection coefficient at each trace by comparing the $P_i$ and $PP$ arrivals. Using a phase-shifted deconvolution technique similar to that of Levy and Oldenburg (1982), we deconvolved windows in data set 1 containing the $PP$ arrival using the $P_i$ arrival at the same trace in data set 2 as a model pulse. The amplitude of the $PP$ wave relative to the $P_i$ wave gives the reflection coefficient. This technique gives a good estimate of the reflection coefficient, because it accounts for phase shifts in the wavelet and minimizes the effects of noise. Using the known start time of the window and the arrival time of the pulse within the window determined by deconvolution, we obtain the arrival time of the $PP$ wave. For the purpose of inversion, we convert the $PP$ reflection coefficients to reflection amplitudes by multiplying the known $P_i$ amplitudes by the corresponding reflection coefficient.

The $PPPP$-wave arrivals were more difficult to determine due to their small amplitudes at large offsets. Arrival times and amplitudes were only determined for the first 60 traces because the amplitudes were too low for reliable estimates to be made at larger offsets. A method similar to that used for the $PP$ wave was used, with the pulse arriving at trace 1 being used for deconvolution. The window was determined using an approximate normal-moveout curve.

Arrival times for the direct arrival ($P$-wave) were very accurately determined by hand picking. The arrival times obtained fit a straight line with 90 percent of all time residuals being

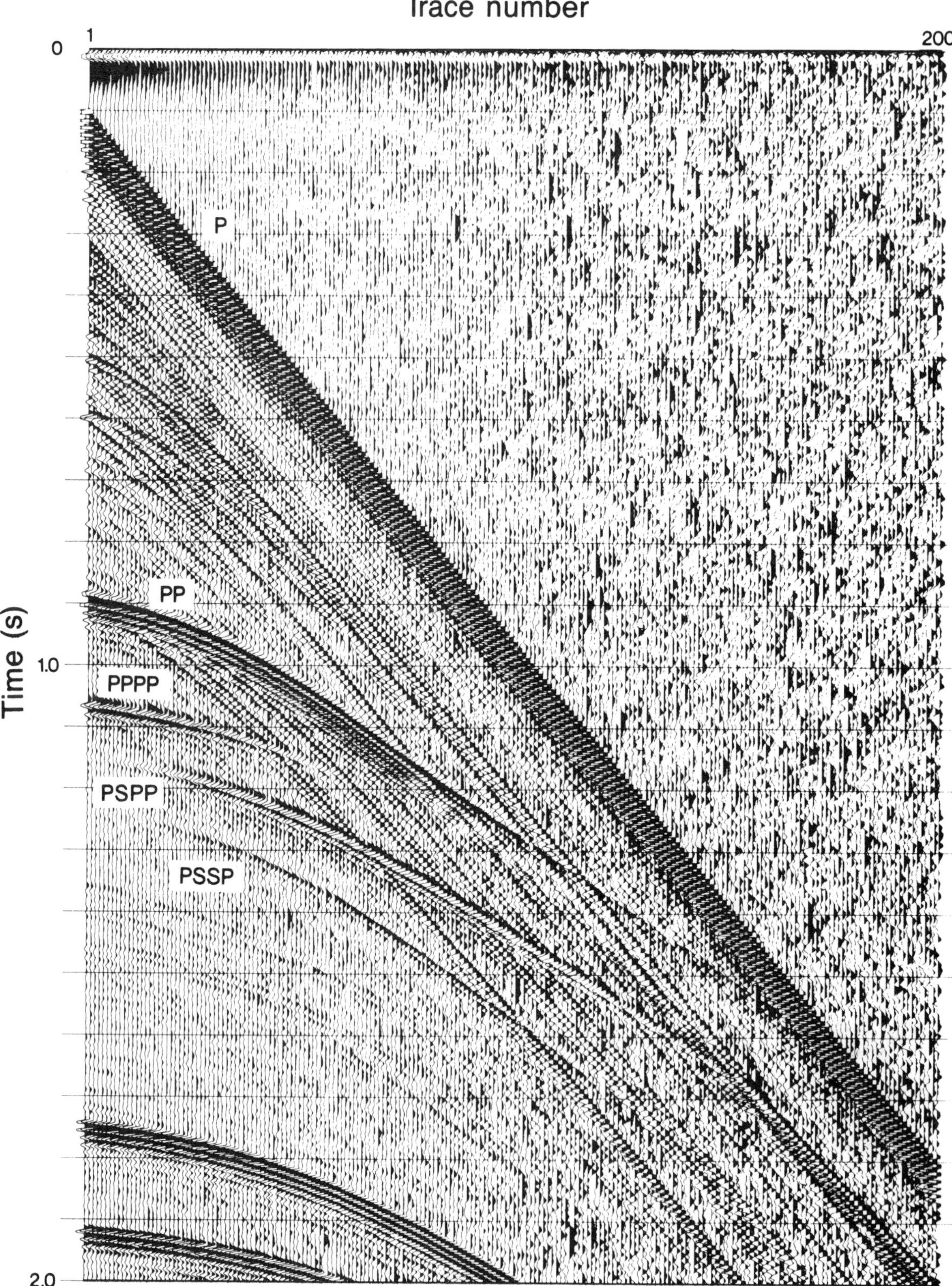


FIG. 4a. Physical-model data. (a) Reflection data set showing arrivals *P* (direct wave), *PP* (reflection off the front of the Plexiglas), *PPPP* (reflection off the back of the Plexiglas), and *PSPP*, *PSSP* (mode-converted *S*-waves).

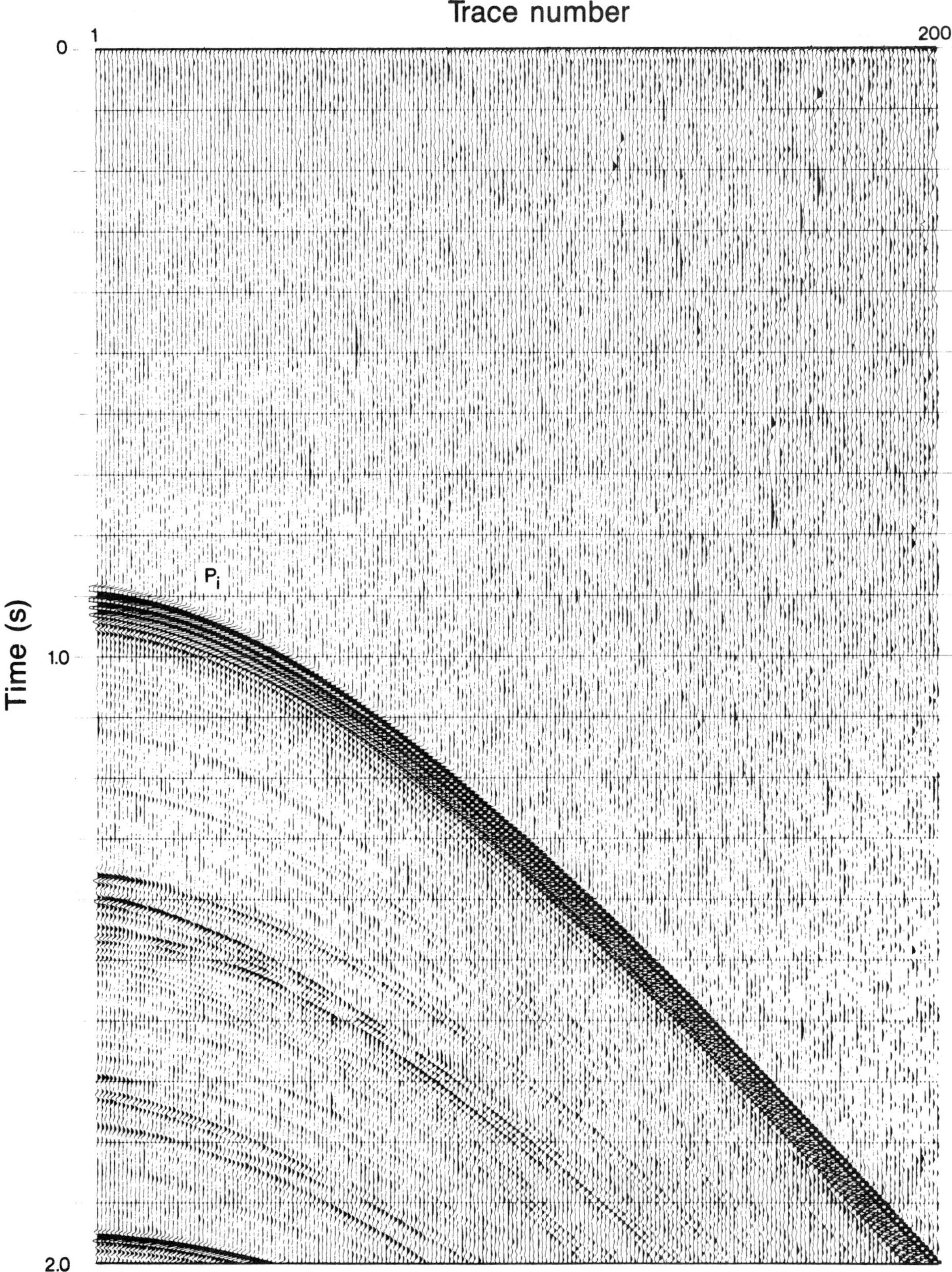


FIG. 4b. Physical-model data. Image point data set showing image point arrivals $P_i$.

less than 1 scaled millisecond. However, this line did not pass through the origin but had an intercept of 12.5 ms (scaled), an error which could have been caused either by a delay induced by the electronic timing mechanism or by an error in the measurement of the inner offset of the array. The dimensions of the source and receiver in the tank are large compared with the inner offset, so it is not unreasonable to expect such a measurement error. However, the solutions obtained by inversion did not differ significantly for the two treatments of the anomaly, so we assumed timing errors and simply subtracted 12.5 ms from all arrival times.

## INVERSION OF THE PHYSICAL MODELING DATA

The procedures described in the previous section determined the arrival times and amplitudes of the *P* and *PP* waves at each of the 200 traces. The arrival times and amplitudes of the *PPPP* wave at the first 60 traces were also determined. These data were used in a Bayesian inversion scheme to estimate the elastic parameters of the physical model. The forward model assumed a structure consisting of three horizontal plane layers characterized by compressional and shear velocities $\alpha_i$ and $\beta_i$, density $\rho_i$, thickness $h_i$, and attenuation coefficient $a_i$. The traveltimes and amplitudes of the direct wave and the *P*-wave reflections off the two interfaces were calculated using the spherical-wave model. The anisotropy curve from the source that was previously determined was included in the forward model. In keeping with the Bayesian philosophy, we used all available knowledge to help simplify the problem. We therefore assumed that layers 1 and 3 were water layers and set $\alpha_1 = \alpha_3$, $\beta_1 = \beta_3 = 0$, $a_1 = 0$, and $\rho_1 = \rho_3 = 1$. The density of water $\rho_1$ was set to 1 g/cm$^3$ because we can only determine relative density and not absolute density. Having made the above assumptions, we are left with eight unknowns to determine by inversion: $\alpha_w$, $\alpha_p$, $\beta_p$, $\rho_p$, $a_p$, $h_w$, $h_p$, and $\lambda$. The subscript $w$ refers to water and $p$ refers to Plexiglas. The distance from the source-receiver line to the Plexiglas is denoted $h_w$ and $\lambda$ is the seismic wavelength.

The inversion was performed by numerically expanding the model function in a Taylor series and iteratively solving the linearized equations until convergence was obtained. In order to test the stability of the inversion, we chose a prior solution that was up to 10 percent away from the true solution. The standard deviations of the prior solution are large so that the solution is not strongly constrained, and we can determine if the data are sufficient to resolve the parameters.

The inversion converged after only five iterations, to the solution given in Table 1. The prior solution and standard deviations are also shown in Table 1. The differences between the prior and final estimates show that the data are sufficient to resolve the parameters without the need for strong Bayesian constraints. The standard deviations of the final solution

**Table 1. Comparison between the known parameters of the physical model and those estimated by inversion.**

| Parameter | Prior solution | Prior std.dev. | Final solution | Final std.dev. | True solution |
|---|---|---|---|---|---|
| $\alpha_w$ (km/s) | 1.4 | 0.5 | 1.524 | 0.0001 | 1.524 |
| $\alpha_p$ (km/s) | 2.95 | 0.5 | 2.813 | 0.006 | 2.828 |
| $\beta_p$ (km/s) | 1.1 | 0.5 | 1.442 | 0.003 | 1.34 |
| $\rho_w$ (g/cm$^3$) | 1.0 | 0.0 | 1.0 | 0.0 | 1.0 |
| $\rho_p$ (g/cm$^3$) | 1.5 | 0.5 | 1.128 | 0.006 | 1.17 |
| $h_1$ (m) | 600.0 | 80.0 | 661.8 | 0.25 | 650.0 |
| $h_2$ (m) | 290.0 | 80.0 | 242.8 | 0.9 | 247.0 |
| $a_p$ (1/m) | 0.000 1 | 0.005 | 0.000 74 | 0.000 05 | — |
| $Q_p$ | — | — | 173 | — | 168 |
| $\lambda$ (m) | 24.0 | 1.0 | 24.2 | 0.08 | — |
| f (Hz) | — | — | 62 | — | 60 |

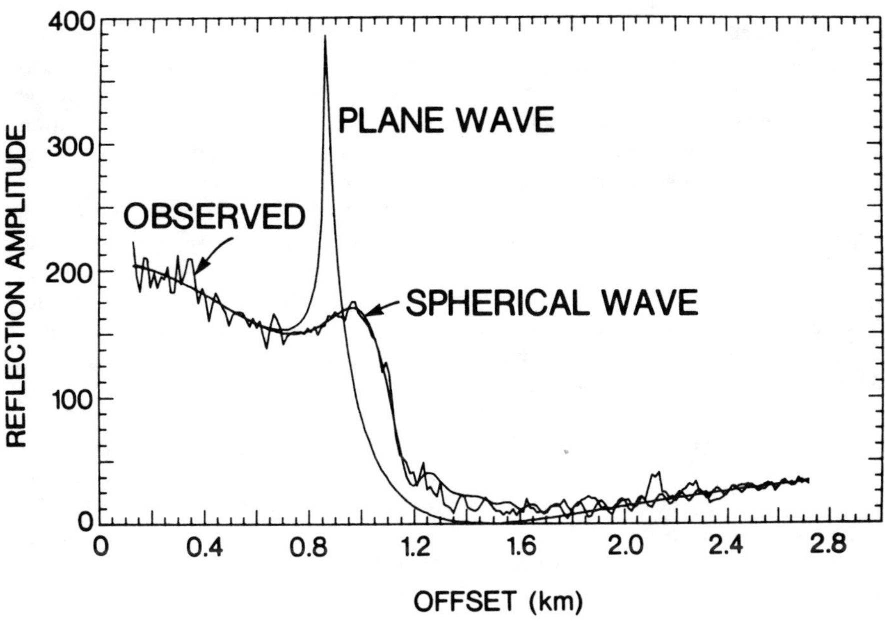


FIG. 7. Fit of the spherical and plane-wave models to the *PP*-wave amplitudes. The plane-wave solution has a sharp peak at the critical offset near 0.85 km. The smooth curve through the data is the spherical-wave solution which matches the observed data at all offsets. Both observed data and the spherical model have a maximum near 1.0 km.

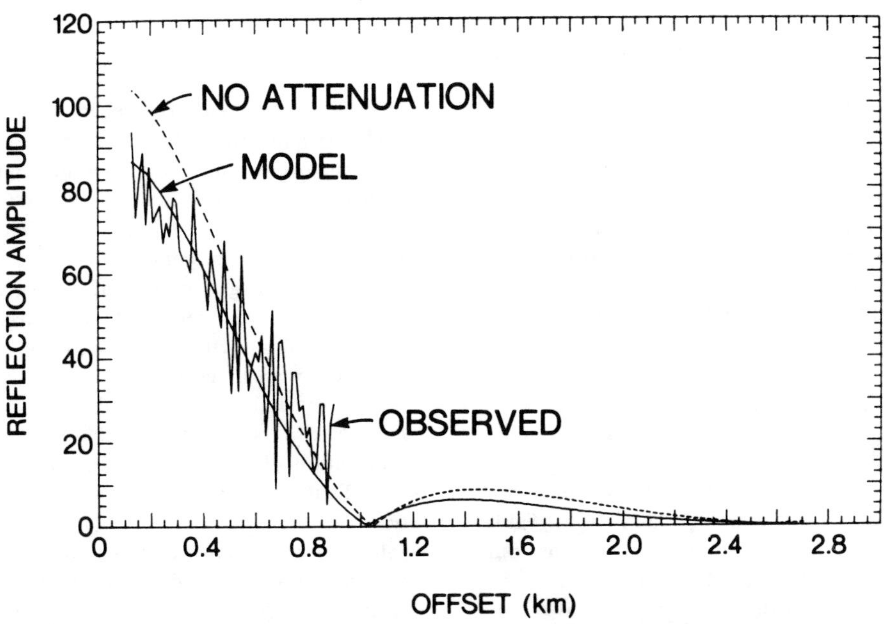


FIG. 8. Fit of inversion solution to *PPPP*-wave amplitude. The solid curve through the data is the solution including attenuation. The significance of attenuation is illustrated by comparison with the dashed line which is the amplitude curve obtained neglecting attenuation.

are determined under the assumption that the problem is fully linear. Because the model function is not fully linear, it is not possible to determine the confidence regions of these uncertainties. However, the relative stability of the parameters is illustrated by the standard deviations given. The solution obtained by inversion can be compared with the true solution (also shown in Table 1). With the exception of $\beta_p$ and $a_p$, the true values of all the parameters are known from laboratory measurements.

In order to obtain spatial dimensions comparable to real earth structures, we performed a temporal scaling that converted the sample interval from its true value of 0.2 μs to a scaled value of 1 ms. The spatial dimensions were then scaled so that the velocity of water determined by the direct wave arrival times was close to 1.5 km/s. With this scaling, 1 inch in the physical-model tank corresponds to 130 m in the scaled model, and the seismic array has an inner offset of 130 m and a trace interval of 13 m.

Because we have used different scaling factors for time and distance, the velocities estimated by inversion will be scaled with respect to the true velocities. Our estimate of $\alpha_w$ was very accurately determined by the direct-wave traveltimes, so we assume that it is reasonable to scale the true velocities by a factor which will give agreement between the true and estimated values of $\alpha_w$. The true velocities listed in Table 1 have been scaled as described above. The solution obtained by inversion agrees remarkably well with the known solution, with all parameter estimates within 3 percent of the true values. The error of approximately 10 m in $h_w$ is related to the timing anomaly of 12.5 ms mentioned previously. By increasing the inner offset to 149 m, this error can be reduced; but since our estimate of $h_w$ is within 2 percent of the expected value, the result obtained is adequate. The frequency corresponding to the wavelength estimated by inversion compares well with the frequency estimated to be dominant in the observed wavelets.

The true values of the *S*-wave velocity and *P*-wave attenuation coefficient are unknown for the Plexiglas used in the physical model. However, Press and Healy (1957) experimentally determined the velocity and attenuation coefficient of *S*-waves in Plexiglas. Their estimate of the attenuation coefficient for shear waves in Plexiglas was used to estimate the attenuation coefficient for *P*-waves using the theoretical result $a_p/a_s \sim 3.6$ (Stacey, 1977), relating *P*-wave and *S*-wave attenuation coefficients. The values of $a_p$ determined by inversion and estimated from Press and Healy's results were converted to quality factor $Q$ using the relationship $Q = \omega/2a\alpha$ between frequency, *P* velocity, $Q$, and attenuation coefficient. $Q$ is independent of the scaling factor, so converting the attenuation coefficients into $Q$ values gives parameters which can be directly compared. These values of $\beta_p$ and $Q_p$ are listed in Table 1 as true values, but they are included only to show that our estimates are not incompatible with other estimates. The discrepancy of 8 percent between the *S* velocity estimates may reflect true differences in the material properties of the Plexiglas used by Press and Healy and the Plexiglas used in the present physical model.

The synthetic amplitude curves generated using the inversion solution show good agreement between theory and observations. Figure 7 shows the fit of the *PP*-wave amplitude data to the spherical-wave model and also to elastic plane-wave amplitudes determined using the same parameter values. The spherical model accurately predicts both the amplitude of the maximum and its location. The plane-wave solution has a very sharp maximum at the critical point (some 200 m before the observed maximum), and is clearly a poor model of the data in the critical region. The spherical-wave model fits the data well at all offsets. The oscillation of the spherical-wave solution after the critical point is small and lies within the noise level of the data.

Figure 8 shows the fit to the *PPPP*-wave amplitudes. Again a very accurate fit was found for all trace amplitudes. The theoretical curve predicts very low amplitude for the later traces, which explains our inability to identify these arrivals on the CDP gather. The dashed line in Figure 8 shows the reflection amplitude with the attenuation effects removed. The difference between the two curves clearly indicates that attenuation is an important factor. Because the refractive index of this interface is negative (Plexiglas overlying water), there is no critical point. We could therefore have used a plane-wave solution to model this reflection.

The use of supercritical data provides stability of the inverse problem and supplies vital extra information, as illustrated by the results of amplitude-only inversion. Using a Marquardt inversion scheme (Bevington, 1969), we use the *PP*-wave amplitudes to solve for the parameters $\alpha_w$, $\alpha_p$, $\beta_p h_w$, and $\lambda$. The solution differed only slightly from the solution obtained using both traveltime and amplitude data. Marquardt inversion does not require the inclusion of prior data, so the supercritical reflection data are sufficient to resolve all the parameters. It is remarkable that using only amplitude data we can resolve parameters such as depth of interface and frequency. This illustrates the great advantages gained by recording supercritical reflections.

## CONCLUSIONS

The Bayesian inversion scheme for inverting reflection traveltimes and amplitudes to determine the best-fitting plane-layer structure is stable. A unique minimum is obtained in the residual sum-of-squares function without the need to impose strong constraints on the parameters. The problem is clearly linear to a good approximation and the synthetic model can be linearized using a numerical Taylor series expansion. A unique solution is then easily obtained using an iterative linear inversion method. The problem is highly nonlinear when the observed seismic traces are used as input data and will be very difficult to solve. The linearity of the inversion has, of course, been obtained at the expense of requiring the identification of arrival times and amplitudes. This is often very difficult and sometimes impossible when reflections interfere.

The physical modeling data show that when reflection traveltimes and amplitudes can be determined, the inversion scheme can efficiently determine a good estimate of the elastic parameters. The data also show that the spherical-wave model can accurately predict supercritical reflection amplitudes.

A major advantage of our inversion method is its simplicity. Since we only model selected parts of the seismic section, our solution is not affected by features in the data which we are unable to predict. If we try to model the data in trace form, we must be able to account for all features in the data. Even the most sophisticated synthetic seismograms are unable to account for all features observed in seismic data, and in any case the large amount of CPU time required for computation renders them unsuitable for inversion purposes.

In regions where the plane-layer model is a good approximation, we expect our inversion to perform well. The requirement for a plane-layered structure obviously limits the applicability of the inversion, so we propose to extend the method to cope with more complex structures.

## ACKNOWLEDGMENTS

We wish to thank Texaco U.S.A. Inc. and Chevron Oil Field Research Company for supporting this research and the Allied Geophysical Laboratories at the University of Houston for supplying the physical-model data.

## REFERENCES

Aki, K., and Richards, P. G., 1980, Quantitative seismology, theory and methods: W. H. Freeman and Co.

Bamberger, A., Chavent, G., Hemons, Ch., and Lailly, P., 1982, Inversion of normal incidence seismograms: Geophysics, **47**, 757–770.

Bevington, P. R., 1969, Data reduction and error analysis for the physical sciences: McGraw-Hill Book Co.

Bishop, T. N., Bube, K. P., Cutler, R. T., Langan, R. T., Love, P. L., Resnick, J. R., Shuey, R. T., Spindler, D. A., and Wyld, H. W., 1985, Tomographic determination of velocity and depth in laterally varying media: Geophysics, **50**, 903–923.

Bleistein, N., and Cohen, J. K., 1982, The velocity inversion problem—Present status, new directions: Geophysics, **47**, 1497–1511.

Bregman, N. D., Chapman, C. H., and Bailey, R. C., 1985, A noniterative procedure for inverting plane-wave reflection data at several angles of incidence using the Riccati equation: Geophys. Prosp., **33**, 185–200.

Brekhovskikh, L. M., 1960, Waves in layered media: Academic Press.

Bube, K. P., and Burridge, R., 1983, The one-dimensional inverse problem of reflection seismology: Soc. Industr. Appl. Math. review, **25**, 497–559.

Carazzone, J. J., 1986, Inversion of *P-SV* seismic data: Geophysics, **51**, 1056–1068.

Carrion, P. M., and Kuo, J. T., 1984, A method for computation of velocity profiles by inversion of large-offset records: Geophysics, **49**, 1249–1258.

Carrion, P. M., Kuo, J. T., and Stoffa, P. L., 1984, Inversion method in the slant stack domain using amplitudes of reflection arrivals: Geophys. Prosp., **32**, 375–391.

Červený, V., 1966, The dynamic properties of reflected and head waves around the critical point: Geophys. Sb., **13**, 135–245.

——— 1967a, The theory of reflected and head waves in the case of layered overburden: Geophys. Sb., **14**, 105–179.

——— 1967b, The amplitude-distance curves for waves reflected at a plane interface for different frequency ranges: Geophys. J. Roy. Astr. Soc., **13**, 187–196.

Červený, V., and Ravindra, R., 1971, Theory of seismic head waves: Univ. of Toronto Press.

Cooke, D. A., and Schneider, W. A., 1983, Generalized linear inversion of reflection seismic data: Geophysics, **48**, 665–676.

Eiges, R., and Raz, S., 1985, Inversion from finite offset data and simultaneous reconstruction of the velocity and density profiles: Geophys. Prosp., **33**, 339–358.

Gibson, B. S., Odegard, M. E., and Sutton, G. H., 1979, Nonlinear least-squares inversion of traveltime data for a linear velocity velocity-depth relationship: Geophysics, **44**, 185–194.

Ikelle, L. T., Diet, J. P., and Tarantola, A., 1986, Linearized inversion of multioffset seismic reflection data in the $\omega$-$k$ domain: Geophysics, **51**, 1266–1276.

Jackson, D. D., and Matsu'ura, M., 1985, A Bayesian approach to nonlinear inversion: J. Geophys. Res., **90**, 581–591.

Levy, S., and Oldenburg, D. W., 1982, The deconvolution of phase shifted wavelets: Geophysics, **47**, 1285–1294.

McAulay, A. D., 1985, Prestack inversion with plane-layer point-source modeling: Geophysics, **50**, 77–89.

Oldenburg, D. W., Levy, S., and Stinson, K., 1984, Root-mean-square velocities and recovery of the acoustic impedance: Geophysics, **49**, 1653–1663.

Press, F., and Healy, J., 1957, Absorption in low-loss media: J. Appl. Phys., **28**, 1323–1325.

Raz, S., 1982, A procedure for the multidimensional inversion of seismic data: Geophysics, **47**, 1422–1430.

Stacey, F. D., 1977, Physics of the Earth: John Wiley and Sons.

Tarantola, A., 1984a, Inversion of seismic reflection data in the acoustic approximation: Geophysics, **49**, 1259–1266.

——— 1984b, Linearized inversion of seismic reflection data: Geophys. Prosp., **32**, 998–1015.

Tarantola, A., and Valette, B., 1982, Generalized nonlinear inverse problems solved using the least-squares criterion: Rev. Geophys. Space Phys., **10**, 251–285.

Winterstein, D. F., and Hanten, J. B., 1985, Supercritical reflections observed in *P*- and *S*-wave data: Geophysics, **50**, 185–195.

# CONVOLUTIONAL BACK-PROJECTION IMAGING OF PHYSICAL MODELS WITH CROSSHOLE SEISMIC DATA[1]

R. J. R. EAST[2], M. H. WORTHINGTON[2] and N. R. GOULTY[3]

## ABSTRACT

EAST, R.J.R., WORTHINGTON, M.H. and GOULTY, N.R. 1988. Convolutional back-projection imaging of physical models with crosshole seismic data. *Geophysical Prospecting* **36**, 139–148.

The development of crosshole seismic tomography as an imaging method for the subsurface has been hampered by the scarcity of real data. For boreholes in excess of a few hundred metres depth, crosshole seismic data acquisition is still a poorly developed and expensive technology. A partial solution to this relative lack of data has been achieved by the use of an ultrasonic seismic modelling system. Such ultrasonic data, obtained in the laboratory from physical models, provide a useful test of crosshole imaging software.

In particular, ultrasonic data have been used to test the efficacy of a convolutional back-projection algorithm, designed for crosshole imaging. The algorithm is described and shown to be less susceptible to noise contamination than a Simultaneous Iterative Reconstruction Technique (SIRT) algorithm, and much more computationally efficient.

## INTRODUCTION

Despite the large number of papers concerned with crosshole tomography that have been published in recent years, few field experiments have been reported. Recent papers containing some discussion of real data have been written by Paulsson and King (1980), Wong, Hurley and West (1983), Ivansson (1985), Peterson, Paulsson and McEvilly (1985), Gustavsson, Ivansson, Moren and Pihl (1986) and Cottin, Deletie, Jacquet-Fracillon, Lakshmanan, Lemoine and Sanchez (1986). The simple

[1] Received January 1987, revision accepted August 1987.

[2] Department of Geology, Imperial College of Science and Technology, Prince Consort Road, London SW7 2BP, U.K.

[3] Department of Geological Sciences, University of Durham, South Road, Durham DH1 3EL, U.K.

Reprinted from Geophysical Prospecting, 36, 139-148 with permission from European Association of Exploration Geophysicists and Blackwell Scientific Publications Ltd.

explanation for this relative scarcity is that, in most localities, crosshole seismic tomography data acquisition is difficult and expensive. There is little doubt that crosshole seismic exploration would develop more rapidly if a greater number of field experiments were carried out. Equally more field data will be recorded when the efficacy of the method becomes established. Our partial solution to this predicament is to generate data by physical modelling with the ultrasonic seismic modelling system at Durham University. These data are a more satisfactory test of imaging software than computer-generated synthetic data plus artificial noise.

The resolution in a velocity image which may be achieved from a crosshole seismic survey is dependent upon the borehole locations, the number and directional distribution of source/receiver raypaths and the structure that is being imaged. We assume that the wavelength of the seismic signals is small compared to the size of the structures through which the energy is propagating, and consequently a ray optic approach to seismic modelling can be applied. Otherwise the resolution would also be a function of signal frequency.

The main requirement is for the region of interest to be crossed by a high density of raypaths with a large range of orientations (Dines and Lytle 1979). This is not always a sufficient condition to ensure good resolution throughout the imaged region. A complex geological structure with large velocity contrasts can result in 'shadow zones' where ray penetration is poor even though many raypaths have been recorded. Poor imaging can also result from misinterpretation of the recorded wavefield; for example, identifying a first arrival head wave from a buried interface as a direct wave.

In any event, using inversion schemes such as generalized matrix inversion or iterative back-projection algorithms like ART or SIRT (Dines and Lytle 1979), the velocity field is defined as a mesh of cells and in most practical applications, a large number of cells are required. For example, if a resolution of 1 m × 1 m was thought to be achievable within a 100 m × 50 m region, then there would be 5000 unknowns (cell velocities) and an equation coefficient matrix of the order of 5000 × 5000.

It is therefore desirable that the fastest possible tomographic inversion algorithms are developed, so that the limiting factor in image clarity will always be the data acquisition rather than the data processing hardware.

The convolutional back-projection algorithm, which is based upon the Radon transform (Radon 1917), is highly efficient. It was first conceived as an imaging technique for radio astronomical purposes (Bracewell 1956). Shepp and Logan (1974) and Ramachandran and Lakshiminarayanan (1971) describe its early application in the medical sciences. More recent references, selected from an extensive literature, are Robinson (1982), Deans (1983) and Durani and Bisset (1984).

In this paper, a simple convolutional back-projection algorithm for crosshole velocity imaging is described. The method is similar to the more familiar slant stack $\tau$-$p$ inversion technique (Phinney, Chowdhury and Frazer 1981). The major limitation of the method, which is solely a straight ray inversion technique, is that the source line must be parallel to the receiver line. However, velocity images can then be obtained with great computational efficiency.

*Convolutional back-projection (the Radon transform)*

The forward Radon transform of a two-dimensional (2D) function $U(x, z)$ is defined by

$$U_R(p, z_s) = \int_{-\infty}^{\infty} U(x, z_s + px)\, dx.$$

The transform is a function of two variables, $z_s$ and $p$. The relation $z = z_s + px$ in the transform represents a line through the original function $U(x, z)$ with an intercept on the $z$-axis of $z_s$ and a slope of $p$. The transform essentially performs the mapping of a line to a point and vice versa.

In the application considered here, $U(x, z)$ represents a seismic slowness field, bounded by the earth's surface and two vertical boreholes (a shot hole at $x = 0$ and a receiver hole at $x = x_{max}$). A line defined by $z = z_s + px$ represents a straight raypath from a source at depth $z_s$ with an angle to the horizontal of $\tan^{-1} p$. The traveltime of this ray is the line integral

$$t(p, z_s) = \sqrt{(1 + p^2)} \int_0^{x_{max}} U(x, z_s + px)\, dx,$$

which is a Radon transform scaled by the factor $\sqrt{(1 + p^2)}$.

Figure 1 depicts a traveltime data spread in $p$–$z_s$ space obtained when only hole to hole raypaths are recorded. $p$–$z_s$ space is only partially covered due to the limited range of ray angles that can be achieved. Hence, only a partial or blurred reconstruction of the slowness field can be obtained.

An estimate of the original slowness field is obtained by the inverse Radon transform of the scaled traveltimes

$$U(x, z) = \frac{1}{2\pi} \int_{p_{min}}^{p_{max}} [t'(p, z_s - px)*(F^{-1}|k|)]\, dp,$$

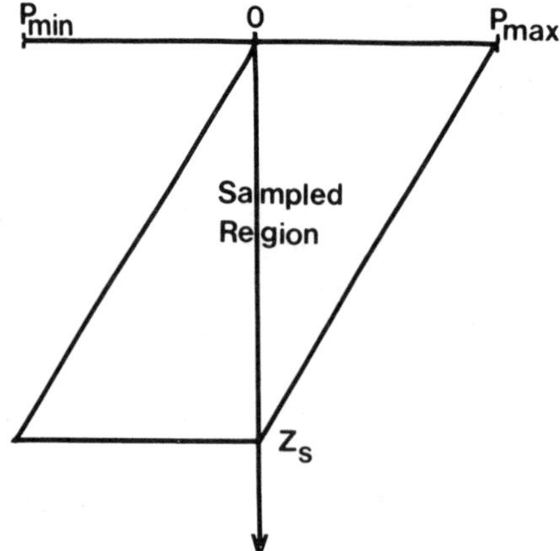


FIG. 1. Sampled region of $p$–$z_s$ space in a typical crosshole experiment.

where $t' = t/\sqrt{(1 + p^2)}$, the asterisk denotes convolution in the spatial domain, $F^{-1}|k|$ is the inverse Fourier transform of the symmetrical ramp function $|k|$ in the wavenumber domain, and $k$ is the angular wavenumber. This expression is similar to the forward transform and performs essentially the same mapping; a line to a point and vice versa. Phinney *et al.* (1981) provide a succinct derivation of the inverse transform.

The convolution operation is equivalent to taking the derivative with respect to $z$ of the Hilbert transform of the data. Alternatively, the Fourier transformed data in the $p$, $k$ domain is multiplied by a symmetric ramp function. The effect of this differentiation, or wavenumber-domain multiplication, is to remove the average slowness level from the inversion so that one is, in fact, imaging a 'slowness difference' field with respect to the average slowness removed. Before any sense can be properly made of the output field, the average must be restored.

Improvement in the quality of the image is possible during this filtering stage through use of the class of generalized omega filters as described by Kwoh, Reed and Truong (1977). The generalized filter is an approximation to the $|k|$ ramp and is described by two parameters, the cut-off wavenumber $k_c$ and the damping factor. The filters closely approximate the $|k|$ ramp at low wavenumbers and then, on approaching the specified cut-off wavenumber $k_c$, smoothly fall to zero according to the damping factor. The values of these parameters are determined by pre-inversion analysis of the wavenumber content of the data and some trial-and-error.

### *The algorithm*

Once the traveltime data have been stored in $p$–$z_s$ space, where $z_s$ represents source depth and the $p$ gradient of the straight line from source to receiver, the inversion is obtained firstly by filtering the data according to the $|k|$ ramp and secondly by calculating the line integrals.

Before transforming the data from $p$–$z_s$ space to $p$–$k$ space and employing the generalized omega filter as described above, abrupt terminations at the edges of data which would result in manifestations of Gibb's phenomenon must be removed. This is achieved by smoothly ramping the missing data samples (for constant $p$) from the last measured sample back to the first measured sample over the length of the Fourier transform. Any gaps in the data set are filled by linear interpolation between its two measured samples on either side. This interpolation represents the creation of traveltime data for raypaths that were not actually recorded, and the inclusion of these interpolated data in the inversion will inevitably result in some smoothing of the slowness image. However, the technique used does have the effect of preserving the average slowness level contained in the recorded data.

Each inverse Radon transform line integral performed is an approximation under the Simpson 1/3 rule. It is composed of interpolated samples at regular intervals along the particular line through $p$–$z_s$ space. The sample interval is made small enough to avoid spatial aliasing. The interpolation procedure used is to take a two-sample wide box around the interpolation point and then to use a 2D sinc interpolating routine for all data points that lie within that box. This means that, for the majority of interpolations, sixteen data points are used. The sinc function is truncated after two samples and linearly tapered to zero over this interval to improve its frequency spectrum.

### *Modelling experiment*

An ultrasonic seismic modelling system, designed for simulation of marine seismic surveys and described by Sharp *et al.* in a paper presented at the 47th EAEG meeting, Budapest, 1985, was converted to a crosshole mode by submerging the receiver beneath the model to be imaged. The piezoelectric sources and receivers were positioned close to the top and bottom surface of the epoxy resin model and the whole assembly was submerged in water.

The chosen model was a wedge-shaped low velocity zone (2810 m/s) embedded within a 5 cm thick block of higher velocity background (2920 m/s) (Fig. 2). A

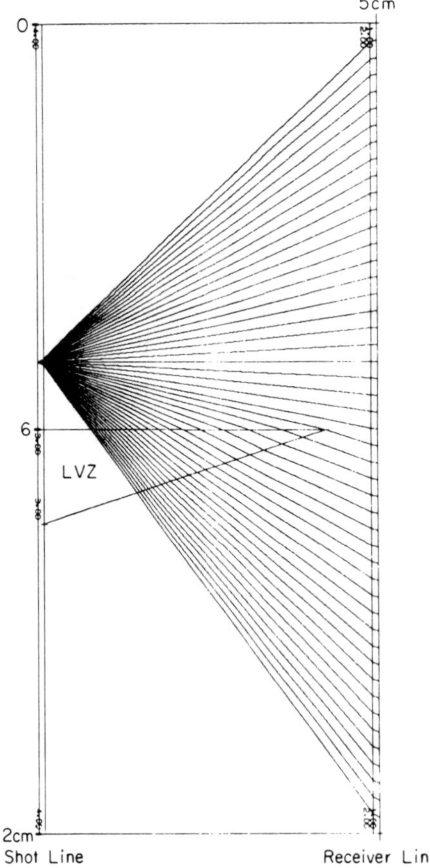


FIG. 2. Wedge model with water layers used for raytracing generation of synthetic data. Shot shown at a depth of 5 cm. LVZ, low velocity zone.

section through the block, 12 cm long, was imaged with source and receiver positions spaced at intervals of 2.5 mm. This gave 2401 raypaths in total. The response recorded by the receiver was stored on magnetic tape with a sampling interval of 0.25 $\mu$s. The source had a dominant frequency of approximately 400 kHz and wavelengths were about 7.5 mm. The model was thus set up to simulate a crosshole survey between two parallel boreholes, one running either side of the block. The model would scale to two parallel wells separated by some 100 m with an imaging frequency of around 200 Hz. By positioning the wedge at the centre of the source and receiver arrays, we ensured that the anomaly was well supported in $p$–$z_s$ space. If the wedge had been located nearer either end of the imaged region, then support, and consequently the resolution of the image, would inevitably have been reduced.

The effect of very thin water layers between the source and receiver transducers and the model is discernible in the modelling. The effect of these low-velocity water layers is merely to focus the source onto the top surface of the block and to mitigate the effect of its directional nature. The other effect, which must be taken into account when examining the inversions, is that the algorithm has insufficient information to resolve these layers and will therefore tend to underestimate slightly the background velocity which is removed during the $|k|$ filtering.

Figure 3 shows a gather of some of the data recorded at a single receiver position. Figure 4 shows data subsequently rearranged into a constant source-receiver offset gather, for zero-offset. The first arrivals in this type of gather all align well and the position of the wedge is apparent where the arrivals come in a little later. It was from such displays that the traveltimes were picked.

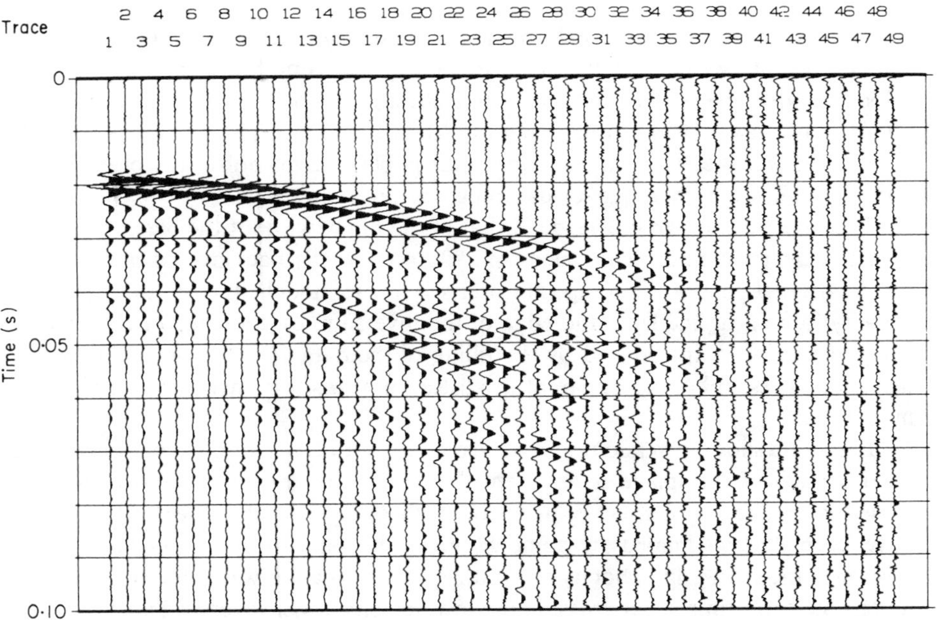


FIG. 3. Ultrasonic data: common receiver point gather.

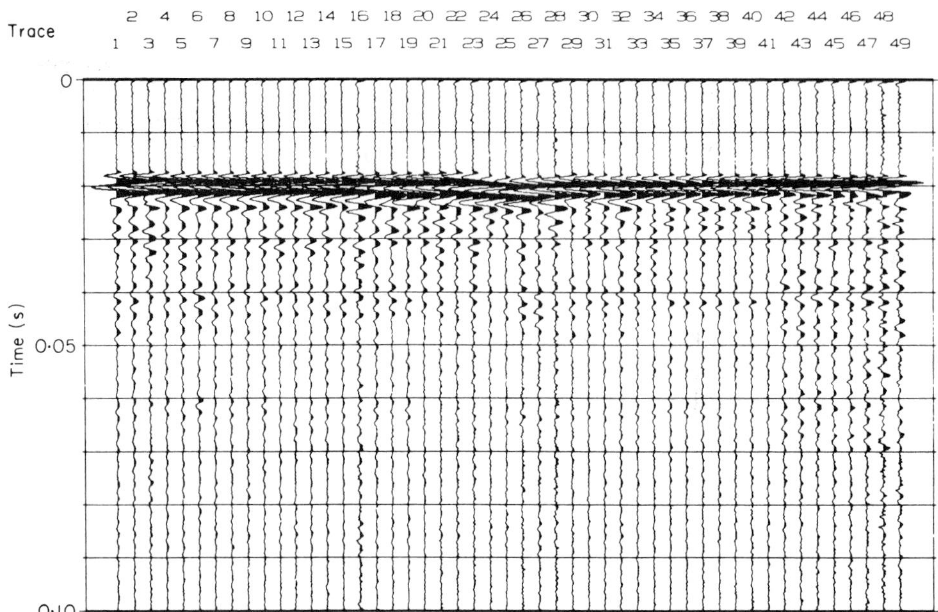


FIG. 4. Ultrasonic data: common source/receiver offset gather. In this gather the offset was zero, i.e. source and receiver were at the same depth for each trace.

## RESULTS

Two synthetic studies were performed, one modelling solely the block containing the low velocity wedge and the other modelling the same block, but with the low-velocity water layers also included. Traveltimes for both synthetic studies were obtained by forward modelling with a ray-tracing package. The omega filter, used in both cases, was the $|k|$ ramp function. This is because no noise was present in the system and hence high frequency damping was not required.

The inversion shown in Fig. 5a is that of the block without the water layers. The wedge is clearly defined. The values of 2920 m/s for the background and 2810 m/s for the deepest part of the wedge are precisely those modelled. Note the small high-velocity lobes near the left-hand edge of the inversion, on the edge of the anomaly. These are artefacts introduced by the filling of the data space prior to filtering and inversion. These lobes are apparent in all Radon transform inversions of the wedge model.

Figure 5b shows inversion of ray-traced data with the water layers. Only a slight deterioration of the image is apparent due to the algorithm's inability to resolve these layers. The major effect of inclusion of these layers is that the model velocities have been underestimated by some 80 m/s. The background velocity is around 2840 m/s and the wedge velocity around 2730 m/s. These lower velocities are clearly due to the extra time spent crossing the low-velocity water layers. This effect should be taken into account when interpreting the inversions of the real data.

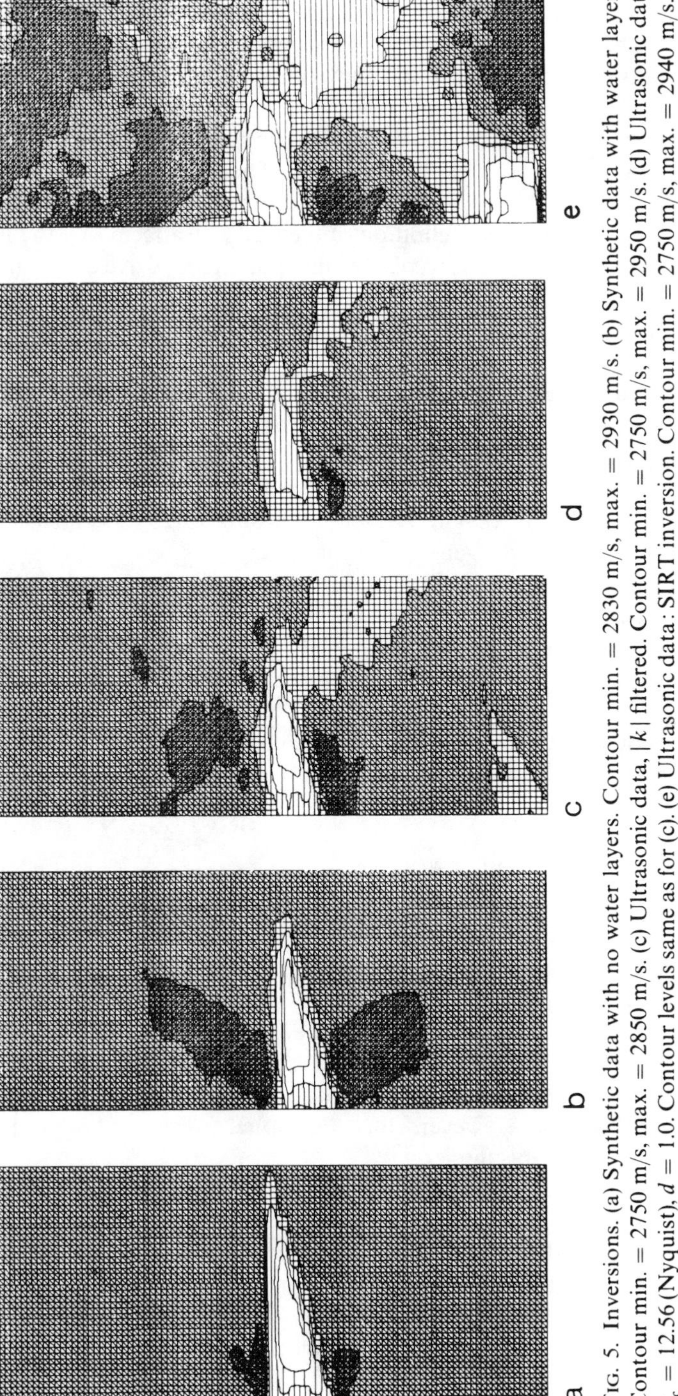


FIG. 5. Inversions. (a) Synthetic data with no water layers. Contour min. = 2830 m/s, max. = 2930 m/s. (b) Synthetic data with water layers. Contour min. = 2750 m/s, max. = 2850 m/s. (c) Ultrasonic data, $|k|$ filtered. Contour min. = 2750 m/s, max. = 2950 m/s. (d) Ultrasonic data. $k_c = 12.56$ (Nyquist), $d = 1.0$. Contour levels same as for (c). (e) Ultrasonic data: SIRT inversion. Contour min. = 2750 m/s, max. = 2940 m/s.

Inversion of the ultrasonic data was tried with various $|k|$ filters. This was to try to suppress noise contamination of the image. High attenuation of the signal during its passage through the block for the longer raypaths meant that high angle rays could not easily be picked, and thus only data for ray angles within the range $\pm 40°$ was used.

Figure 5c shows inversion of the ultrasonic data with no noise suppression (i.e. a simple $|k|$ ramp was used to filter the data). The wedge structure is apparent in the middle of a general background velocity of around 2920 m/s. The velocity of the wedge is around 2730 m/s. Definition of the wedge is affected by the noise contained in the system, and the effect of the finite imaging wavelength (about 7.5 mm). Another factor which also reduced the definition of the image is that none of the high angle raypaths was used in the inversion due to the inability to pick them accurately.

Figure 5d depicts the inversion with a generalized omega filter designed to reduce the noise level. The cut-off frequency $k_c$ was 12.56 (Nyquist) and the damping factor 1.0. This is noticeably a cleaner image than before. The wedge shape is clearer and also much smoother. The background velocity is much the same as before, 2920 m/s, but the wedge has a higher velocity of around 2825 m/s. This is an effect of the smoothing inherent in the generalized filter.

The inversions of the synthetic data showed that underestimation of the background velocity by some 80 m/s was caused by the presence of the water layers in the models. It is suggested that the actual velocities of the epoxy resins making up the tank model are approximately 3000 m/s for the background and 2810 m/s for the wedge.

The amount of CPU time taken by each of these inversions was around 90 s on a VAX 11/780.

## Discussion and Conclusions

Crosshole seismic modelling experiments have proved to be simple and effective in the development of tomographic imaging software. The main advantage of physical modelling data over computer-generated synthetic seismograms is that the effects of noise or uncertainty can be considered more realistically. Noise in the final velocity image will be the result of traveltime errors and the finite bandwidth of the signal.

Time-picking error was estimated to be $\pm 0{\cdot}25$ $\mu$s which corresponds to a velocity error of $\pm 40$ m/s. The receiver, set beneath the block in the tank, was moved by hand. Consequently, position errors could have been up to $\pm 1$ mm for a receiver interval of 2.5 mm. However, this positional uncertainty produces negligible traveltime errors for this model.

A major cause of the reduction in clarity of the images in Figs 5c and 5d compared to the image in Fig. 5b is the finite bandwidth of the real data. The distance between source and receiver was approximately seven wavelengths. Thus, some of the blurring in the image of the sharp end of the wedge will be due to diffraction of the wavefield.

Finally, a multiplicative SIRT type inversion of the tank data, with a cell size of (0.25 cm $\times$ 0.25 cm), is shown in Fig. 5e for comparison. The Radon transform image is less adversely influenced by noise in the traveltime estimates. This is most likely to be due to the interpolation in the method that occurs prior to, and as an essential feature of, the inverse Radon transform integration. Another significant difference between the two methods is that the SIRT inversion took about 15 min of CPU time compared to 90 s for the convolutional back-projection.

The condition that the boreholes must be coplanar can be relaxed to some degree by assuming a velocity model for the structure close to the holes and extrapolating the rays on to two arbitrary parallel planes. Ray positions and traveltimes must then be interpolated to obtain an even sampling of apparent source and apparent receiver positions. Such an operation will be prone to error due to interpolation error and velocity uncertainty.

## References


Bracewell, R.N. 1956. Strip integration in radio-astronomy. *Australian Journal of Physics* **9**, 198–217.

Cottin, J-F., Deletie, P., Jacquet-Fracillon, H., Lakshmanan, J., Lemoine, Y. and Sanchez, M. 1986. Curved ray seismic tomography: application to the Grand Etang Dam (Reunion Island). *First Break* **4**, (7), 25–33.

Deans, S.R. 1983. *The Radon Transform and Some of its Applications.* Wiley-Interscience.

Dines, K.A. and Lytle, R.J. 1979. Computerised geophysical tomography. *Proceedings of the IEEE* **67**, 1065–1073.

Durani, T.S. and Bisset, D. 1984. The Radon transform and its properties. *Geophysics* **49**, 1180–1187.

Gustavsson, M., Ivansson, S., Moren, P. and Pihl, J. 1986. Seismic borehole tomography: measurement system and field studies. *Proceedings of the IEEE* **74**, (2).

Ivansson, S. 1985. A study of methods for tomographic velocity estimation in the presence of low velocity zones. *Geophysics* **50**, 969–988.

Kwoh, Y.S., Reed, I.S. and Truong, T.K. 1977. A generalised Iwl-filter for 3-D reconstruction. *IEEE Transactions on Nuclear Science* **NS-24**, (5), 1990–1995.

Paulsson, B.N. and King, M.S. 1980. Between-hole acoustic surveying and monitoring of a granitic rock mass. *International Journal of Rock Mechanics, Mining Sciences and Geomechanics Abstracts* **17**, 371–376.

Peterson, J., Paulsson, B. and McEvilly, T. 1985. Applications of algebraic reconstruction techniques to crosshole seismic data. *Geophysics* **50**, 1566–1580.

Phinney, R.A., Chowdhury, K.R. and Frazer, L.N. 1981. Transformation and analysis of record sections. *Journal of Geophysical Research* **86**, (B1), 359–377.

Radon, J. 1917. On the determination of functions from their integrals along certain manifolds. *Berichte der Sachsischen Akademie der Wissenschaft* **69**, 262–277.

Ramachandran, G.N. and Lakshiminarayanan, A.V. 1971. Three-dimensional reconstruction from radiographic and electron micrographic application of convolutions instead of Fourier transforms. *Proceedings of the National Academy of Sciences (USA)* **68**, 2236–2240.

Robinson, E.A. 1982. Spectral approach to geophysical inversion by Lorentz, Fourier and Radon transforms. *Proceedings of the IEEE* **70**, (9), 1039–1053.

Shepp, L.A. and Logan, B.F. 1974. The Fourier reconstruction of a head section. *IEEE Transactions on Nuclear Science* **N2-21**, 21–42.

Wong, J., Hurley, P. and West, G. 1983. Crosshole seismology and seismic imaging in crystalline rocks. *Geophysics* **50**, 686–689.

# A three-dimensional perspective on two-dimensional dip moveout

David Forel* and Gerald H. F. Gardner‡


## ABSTRACT

Prestack migration in a constant-velocity medium spreads an impulse on any trace over an ellipsoidal surface with foci at the source and receiver positions for that trace. The same ellipsoid can be obtained by migrating a family of zero-offset traces placed along the line segment from the source to the receiver. The spheres generated by migrating the zero-offset impulses are arranged to be tangent to the ellipsoid. The resulting nonstandard moveout equation is equivalent to two consecutive moveouts, the first requiring no knowledge of velocity and the second being standard normal moveout (NMO). The first of these is referred to as dip moveout (DMO).

Because this DMO-NMO algorithm converts any trace to an equivalent set of zero-offset traces, it can be applied to any ensemble of traces no matter what the variations in azimuth and offset may be. In particular, this three-dimensional perspective on DMO can be used with multifold inline data. Then it becomes clear that velocity-independent DMO operates on radial-trace profiles and not on constant-offset profiles. Inline data over a three-dimensional subsurface will be properly stacked by using DMO followed by NMO.


## INTRODUCTION

The concept of a common-midpoint (CMP) gather loses its simplicity when extended from two-dimensional (2-D) to three-dimensional (3-D) data because, in general, both offset and azimuth vary in the component traces. As Levin (1971) showed, the normal-moveout (NMO) velocity for a CMP gather over a dipping plane depends upon both the strike and dip angles, as well as on the velocity in the medium. A second difficulty with many 3-D surveys is that the data acquisition is irregular and sparse. Theoretically, both the shot and receiver could be placed anywhere, sampling four independent variables, $(x_1, y_1)$ for the shot and $(x_2, y_2)$ for the receiver, but in practice only three variables may be sampled, as in a marine survey consisting of parallel 2-D lines. The four variables may also be sparsely sampled as in many crossed-array surveys on land.

It was pointed out by French et al. (1984) that these problems can be taken in stride by using the concept of dip moveout. In French's method, the dip of the reflector is estimated first, and the data are adjusted according to the dip.

In this paper, the procedure is made independent of interpretation and independent of velocity. Any irregular data can be mapped to a regular data set using velocity-independent processing. We show that the equations derived for the 3-D case reduce to an algorithm in which radial profiles are migrated for the special case of 2-D multifold lines.

## OVERVIEW OF 3-D DMO

Let $(x, y, z)$ be a Cartesian coordinate system with the $z$-axis pointing vertically downward into the earth. The 3-D survey is conducted in the plane $z = 0$, and the traces can have any distribution of offsets and azimuths. For any trace, the distance from the source location to the receiver location is denoted by $2h$, and $t$ refers to the measured arrival time.

Dip moveout (DMO) is viewed here as an operation on the data which brings them from the initial irregular configuration to a regular arrangement, without requiring any use of velocity. The regular arrangement consists of a uniform grid, as shown in Figure 1, ruled on the $z = 0$ plane with a spacing suitable for subsequent migration processes. The traces in the vicinity of each grid point are gathered to form a 2-D offset-time array with an offset axis $k$ and a time axis $t_1$, which are digitized at suitable sampling intervals. DMO can be viewed as a replacement process in which each trace recorded on the $z = 0$ plane makes a contribution to a number of corresponding arrays.

The replacement process can be carried out one trace at a time. First, digitize the straight line segment RS at equally spaced sampling points and locate the nearest grid point for each sampling point. Let the sampling point be at a distance $b$ from the midpoint M of RS. Then the corresponding offset $k$

Manuscript received by the Editor December 16, 1986; revised manuscript recieved October 26, 1987.
*Formerly Allied Geophysical Laboratories, University of Houston; presently Downhole Seismic Services, 3600 Briarpark, Houston, TX 77252.
‡ Allied Geophysical Laboratories, University of Houston, 4800 Calhoun Road, Houston, TX 77004.


Reprinted from Geophysics, 53, 604-610

in the array for that grid point is defined as

$$k^2 = h^2 - b^2. \tag{1}$$

The original trace is added to column $k$ after a uniform contraction of its time axis. That is, time $t_1$ for column $k$ corresponds with time $t$ on the original trace, where

$$t_1 = t\,\frac{k}{h}. \tag{2}$$

Equations (1) and (2) are derived in the Appendix and describe the kinematics of DMO.

Note that each trace contributes only to the arrays at grid points that lie between the source and receiver locations. The larger the offset, the more arrays that receive a contribution.

As shown in the Appendix, the main feature of the arrays constructed in this fashion is that the arrival time for reflections and diffractions is given by

$$t_1^2 = t_0^2 + \frac{4k^2}{V^2}, \tag{3}$$

where $t_0$ is the zero-offset traveltime and $V$ is the velocity of propagation in the medium. Since equation (3) is simply the standard NMO equation, each array can be analyzed using standard velocity analysis methods. A stacking velocity function can be picked and the data stacked into a zero-offset trace at each grid point. Since the velocity $V$ is not dependent upon dip, it is also a suitable velocity function for performing 3-D migration on the stacked data.

## IMPULSE RESPONSE

The DMO transformation defined by equations (1) and (2) can be expressed in geometrical terms by looking at its impulse response. An impulse in the raw data is a point in the five-dimensional space $(x_1, y_1, x_2, y_2, t)$. The response is a curve in the four-dimensional space $(x, y, k, t_1)$. The shape of this curve is determined by the condition that the output grid point $(x, y)$ lies on the shot-receiver line and by equations (1) and (2). The determining equations can be written in terms of $x_1, y_1, x_2, y_2$, and $t$:

$$\frac{x - x_1}{x_2 - x_1} = \frac{y - y_1}{y_2 - y_1}, \tag{4}$$

$$k^2 = \tfrac{1}{4}(x_2 - x_1)^2 + \tfrac{1}{4}(y_2 - y_1)^2 - \left(x - \frac{x_1 + x_2}{2}\right)^2 - \left(y - \frac{y_1 + y_2}{2}\right)^2, \tag{5}$$

and

$$\frac{t_1}{k} = \frac{4t}{\left[(x_2 - x_1)^2 + (y_2 - y_1)^2\right]^{1/2}}. \tag{6}$$

Equations (4) and (6) are linear in $x$, $y$, $k$, and $t_1$ and define a 2-D subspace (a plane) in the four-dimensional space. Equation (5) is quadratic in $x$, $y$, and $k$ and its intersection with the 2-D space is a curve (an ellipse) which is the impulse response curve.

Note that the four-dimensional space $(x, y, k, t_1)$ is an abstraction; the dimension $k$ has to be thought of as orthogonal to the physical coordinates $x$, $y$, $z$, and $t$. Hence, it would be misleading to think of this impulse response as a curve which lies in physical space.

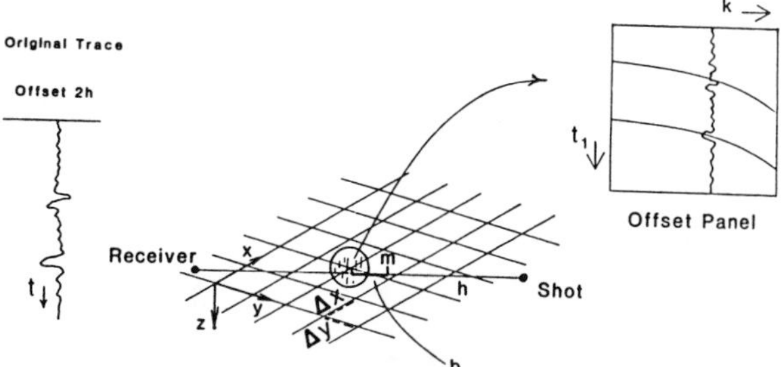


FIG. 1. The 3-D survey is conducted in any orientation on the $z = 0$ plane over a medium of constant velocity $V$. The shot and receiver have offset $2h$; the midpoint is $m$; and events have arrival times $t$. With $h$ fixed, individual traces are mapped to matrices with axes $(k, t_1)$ with equations (1) and (2). The velocity of the subsurface is found by applying equation (3), standard NMO, to a $(k, t_1)$ matrix.

**Table 1. Coordinates specifying the points A and B which define the six line diffractors used to generate 3-D synthetic data.**

| $x_a$ | $y_a$ | $z_a$ | $x_b$ | $y_b$ | $z_b$ |
|---|---|---|---|---|---|
| 166 | −224 | 216 | 80 | 268 | 216 |
| 400 | −354 | 522 | −237 | 414 | 463 |
| 1058 | −52 | −42 | 66 | 702 | 792 |
| 1322 | 457 | 209 | −337 | 370 | 1323 |
| 1698 | 1037 | −96 | −467 | 430 | 1703 |
| 1221 | 1516 | 904 | −708 | −628 | 1915 |

## A 3-D EXAMPLE

For 2-D data acquisition in which the source S and receiver R lie on a straight line, a gather for the construction of a DMO array at location A consists of all traces for which A lies between R and S. For 3-D data, the gather may consist of the same number of traces but with a variable direction from S to R. An example is given here to demonstrate that, for random variations in direction, the $(k, t_1)$ array has the same arrival-time properties as for 2-D data.

To emphasize that the subsurface can be three-dimensional, the synthetic data were constructed by calculating reflection times from six straight-line diffractors. Each line was defined by specifying the coordinates of two points on it $(x_a, y_a, z_a)$ and $(x_b, y_b, z_b)$, which are listed in Table 1. The DMO array was constructed at location (0, 0, 0). The six lines point in a variety of directions but are positioned so that the zero-offset reflection times at (0, 0, 0) are equally spaced at intervals of 0.2 s. The wavelet shape was defined by the function $t\,e^{-\alpha t^2}$ with $\alpha = 5000$, and the wavelet's amplitude was made inversely proportional to the reflection time. The wavelet was positioned so that its zero crossing point occurred at the reflection time.

The DMO gather was constructed by specifying R and S using a uniform random number generator. The coordinates were $[(b \pm h)\cos\theta, (b \pm h)\sin\theta, 0]$, with $h$ taken at 25 m increments from 0 to 1000 m and $b$ taken in 25 m increments from $-h$ to $+h$ for each value of $h$. The angle $\theta$ was picked randomly for each trace to have values between $-\pi/2$ and $+\pi/2$. This procedure defines a gather of about 1600 traces for which (0, 0, 0) lies between R and S.

For each position of R and S, the arrival time to any straight line was obtained using

$$t = \frac{\sqrt{(\mathrm{RN} + \mathrm{SM})^2 + \mathrm{NM}^2}}{V}, \tag{7}$$

where RN and SM are perpendiculars from R and S onto the line. The velocity of propagation $V$ was chosen to be 2500 m/s.

A common-midpoint gather from the data set is shown in Figure 2. For small offsets, the random variation in $\theta$ has a negligible effect, but as RS increases, the arrival times become erratic because of the variations in the azimuths.

The $(k, t_1)$ array constructed from the DMO gather by use of equations (1) and (2) had 40 columns with an increment in $k$ of 25 m starting from $k = 0$. Each column was the sum of all the traces for which $k$ given by equation (1) was within $\pm 25$ m of the $k$ value for that column. The amplitude at time $t_1$ for each trace was obtained by interpolating at time $t_1 h/k$ using a four-point Lagrange interpolation. Each synthetic trace was added until this time exceeded $2h^2/Vb$. As shown in the Appendix, this includes all dip angles up to 90°. A scaling ramp proportional to time was used to correct for spherical divergence.

The $(k, t_1)$ array obtained had six events with hyperbolic

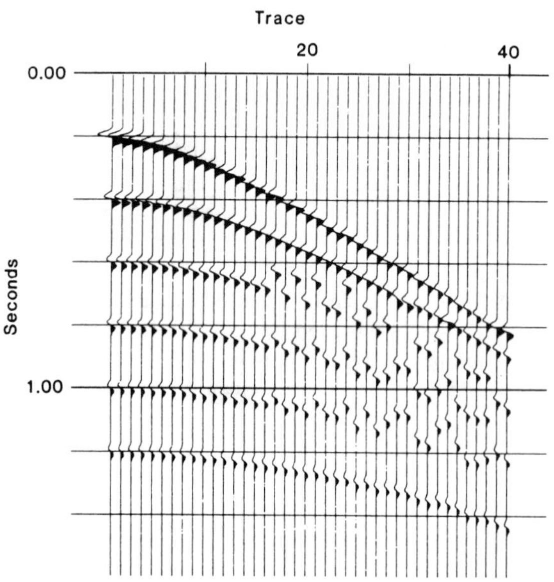


FIG. 2. The CMP gather over the six line diffractors defined in Table 1. For each offset the direction was randomly chosen between $-\pi/2$ and $+\pi/2$. The arrival times lose coherence as the offset increases, in proportion to the dip angle of the normal-incident ray. The velocity of propagation was 2500 m/s.

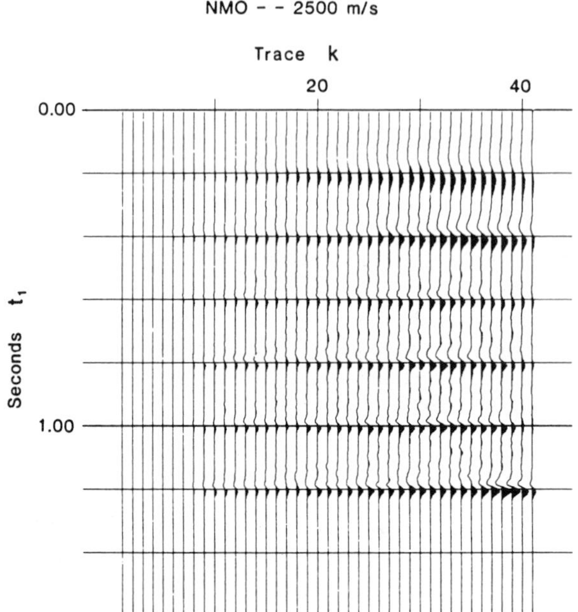


FIG. 3. The $(k, t_1)$ array using data with random azimuthal directions for each offset. An NMO correction for 2500 m/s has been applied. All six events have been flattened, indicating that 3-D data over 3-D structures can be reduced to a standard hyperbolic moveout form without using velocity.

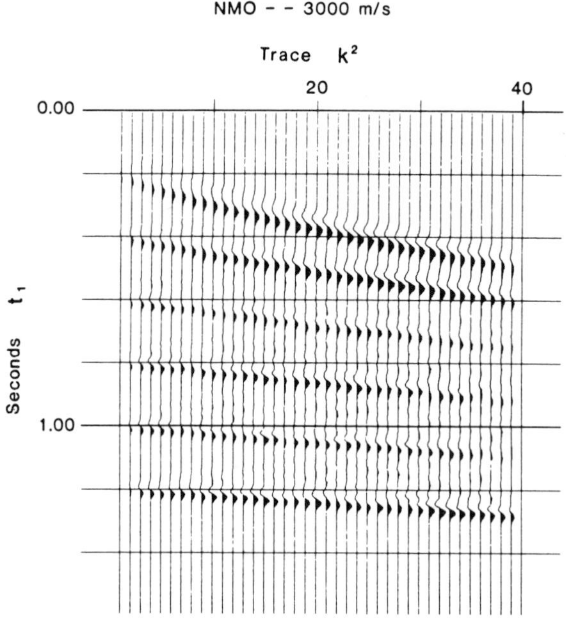


FIG. 4. By plotting $k^2$ instead of $k$, a more even distribution of amplitude is obtained than in Figure 3. In this case, an NMO correction for 3000 m/s was applied; the residual moveout is almost linear.

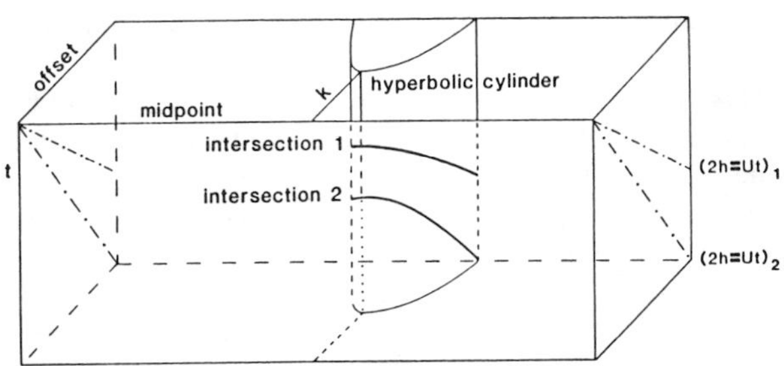


FIG. 5. The 2-D data cube of midpoint, half-offset, and time $(m, k, t_1)$. When an offset $2k$ and time $t_1$ are picked and held fixed in the cube after DMO, equation (1) confines the mapping to a hyperbolic cylinder and equation (2) confines the mapping to a plane of slope $U = 2h/t$, where $U = 2k/t_1$. The intersection of these surfaces is a hyperbola. The figure shows the results for two values of $U$.

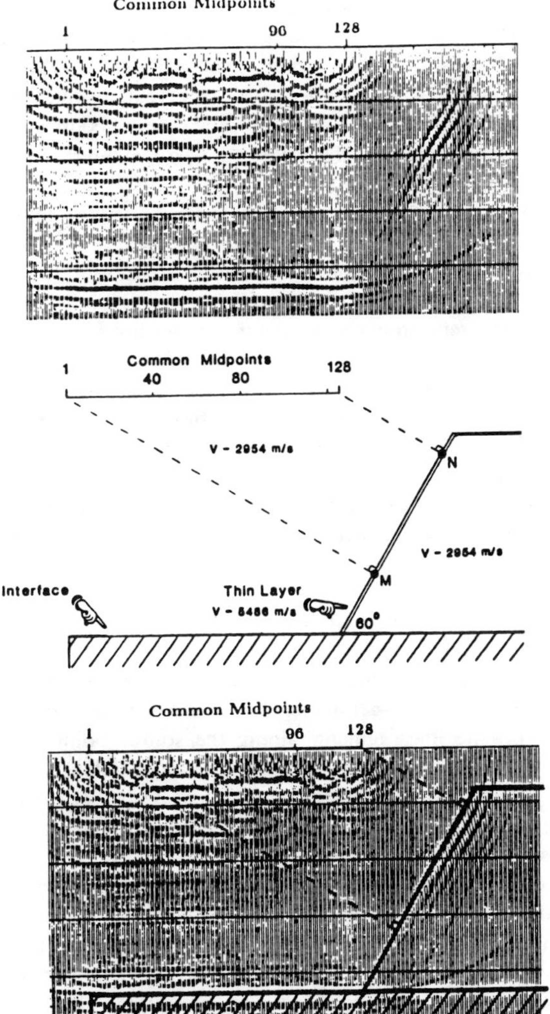


FIG. 6. (Top) After DMO, the section was stacked using a velocity of 3000 m/s and was migrated using a constant velocity of 3000 m/s. Reflections are not seen from the lower part of the ramp. (Middle) Points M and N are the zero-offset reflection points on the ramp from midpoints 1 and 128, respectively. The lower portion of the ramp cannot be imaged because point M occurs partway up the ramp. The upper part of the ramp is well sampled because the reflection point of a CMP moves updip as offset increases. (Bottom) The model diagram is overlaid onto the migrated section.

moveout. To make the amount of moveout clear, an NMO correction for a constant velocity of 2500 m/s was applied with the result shown in Figure 3. All six events are flat in spite of the dip and strike of the reflecting lines and the random variation in the azimuth of each offset. A ramp proportional to time was applied to equalize amplitudes of the events. However, it can be seen that although a stack over these traces will reproduce the timing of the events on the zero-offset trace of Figure 2, the amplitude and phase spectra have undergone a distortion that varies with time and offset.

Note that a more even distribution of amplitude is obtained if $k^2$ is used instead of $k$. Figure 4 illustrates this and also shows that a constant-velocity NMO correction using 3000 m/s leaves a residual moveout which is almost linear.

## THE SPECIAL CASE OF 2-D LINES

Equations (1) and (2) may not be familiar as the basis for DMO, but in the special case of 2-D data they reduce to a known form. Assume, for every trace, that R and S lie on the same straight line over the 3-D subsurface. In this case, the $(k, t_1)$ arrays can be viewed as being perpendicular to the line RS, making the ensemble of arrays into a cube with coordinates $(x, k, t_1)$, as shown in Figure 5. The original data also form a cube $(m, h, t)$. Any data point in the output cube corresponds to a fixed value of $k$, $t_1$ and the grid point $x$. The original data that are combined to form the amplitude at this

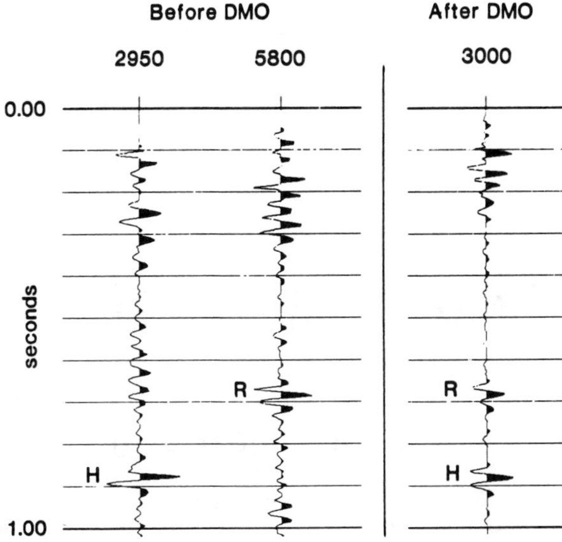


FIG. 7. The medium has a propagation velocity of 2954 m/s. At CMP 96, before DMO, the stacking velocity for the horizontal interface H is 2950 m/s and the ramp R has a stacking velocity of 5800 m/s. After DMO both reflectors have a stacking velocity of 3000 m/s.

point come from values of $h$, $t$, and $b$ which satisfy equations (1) and (2).

Equation (2) is a linear relation between $t$ and $h$ and confines the data points $(m, h, t)$ to the radial plane that passes through the $m$-axis of the original data cube and the point $(0, k, t_1)$.

Equation (1) is quadratic in $h$ and $b$ and confines the data points to the surface of a hyperbolic cylinder in the original data cube which has its axis parallel to the $t$-axis and in the zero-offset plane. Hence, the data values that are summed to form a single data value in the output cube lie on the intersection of these surfaces, which is a hyperbola in a radial plane of the input cube.

In brief, DMO for 2-D lines is equivalent to constant-velocity migration with velocity $U$ of the data that lie in the radial plane $2h = Ut$, a result that was pointed out by Ottolini (1982). Such migration with velocity $U$ gives a very efficient algorithm for DMO in two dimensions.

## A 2-D EXAMPLE

Data were collected in the physical modeling tank at the Allied Geophysical Laboratories for a model which contained a horizontal interface and a thin-layer ramp inclined at 60 degrees, as shown in the middle of Figure 6. Ten offset traces were acquired for each of the 128 CMP gathers. The offset for the first trace was 304.8 m, and the source-receiver pairs were set at increments of 304.8 m. The midpoints were 12.1 m apart. Collection was carried out in the dip direction at a height of

$t$ at offset $= 2h$

DMO

$$k^2 = h^2 - b^2 \tag{1}$$

$$t_1 = t\frac{k}{h} \tag{2}$$

$t_1$ at offset $= 2k$

NMO

$$t_1^2 = t_0^2 + \frac{4k^2}{V^2} \tag{3}$$

$t_0$ at offset $= 0$

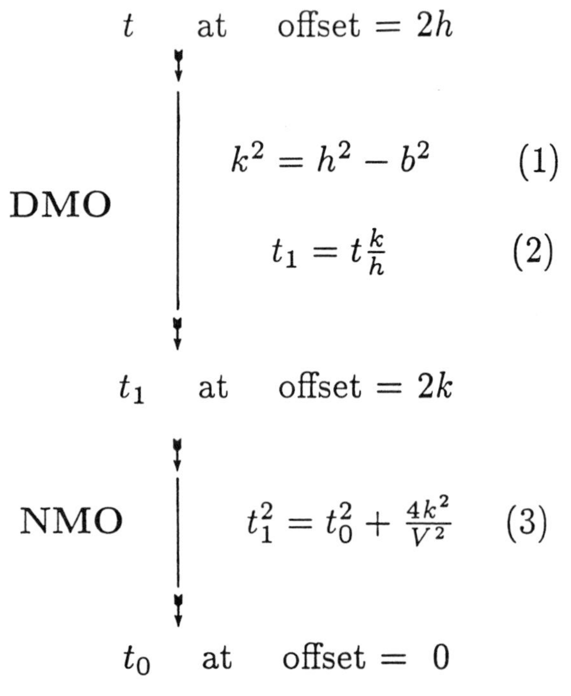

FIG. 8. Summary—The source and receiver have offset $2h$ and data are recorded at time $t$. To apply DMO to a single trace, replace the trace by a family of traces at distances $b$ from the midpoint along the source-receiver segment and having offset axis $k$ and time axis $t_1$. After all traces have been replaced, each $(k, t_1)$ matrix can be analyzed with the NMO equation to get the subsurface velocity and stacked to get zero-offset traces.

1219 m above the horizontal surface. The medium had a propagation velocity of 2954 m/s.

This model was used to demonstrate that after DMO both the horizontal and inclined reflectors can be stacked with the same velocity. The algorithm used for DMO was based on equations (1) and (2) applied to the inline data by operating on radial trace profiles. This approach takes advantage of the regular spacing for the data. Had the data been collected with irregular placement of the sources and receivers, equations (1) and (2) would have been used directly.

DMO was applied by interpolating 24 radial traces in each CMP gather using velocities $U$ from 106 m/s to 2544 m/s in steps of 106 m/s in the equation $2h = Ut$. The radial trace profile for velocity $U$ was migrated with velocity $U$ using an $f$-$k$ constant-velocity algorithm. CMP gathers were interpolated from the migrated radial profiles.

Using DISCO software, a suite of constant-velocity NMO corrections was applied to selected CMP gathers before and after DMO to determine the best stacking velocity. Figure 7 shows the results of analyzing CMP 96. Before DMO, the horizontal interface H shows a stacking velocity of 2950 m/s and the ramp R shows a stacking velocity of 5800 m/s. Figure 7 also shows that after DMO, both events exhibit a stacking velocity of 3000 m/s.

The top of Figure 6 shows the section after the following processing sequence: DMO, constant-velocity (3000 m/s) stack, and constant-velocity (3000 m/s) migration. The lower part of the ramp is poorly imaged for two reasons. First, no zero-offset reflection points occur on the lower part of the ramp. The other reason is that as the offset increases for any particular CMP gather, the reflection point moves updip. Therefore, while little or no information was recorded from the lower part of the ramp, the upper part was sampled by increasing numbers of finite offsets.

The bottom of Figure 6 shows an overlay of the true reflector positions on the data resulting from DMO, then migration.

## DISCUSSION AND SUMMARY

In a constant-velocity medium, any trace can be replaced in two steps by a one-parameter family of zero-offset traces located on the line segment joining the source point and the receiver point: in the first step, which is independent of velocity, intermediate offsets are introduced; in the second, zero-offset is obtained by standard velocity analysis and stack. The steps are indicated in Figure 8.

Except for induced amplitude and phase distortions, the application of DMO, NMO, and stack followed by migration is equivalent to prestack migration of any data set. Since the above analyses and developments are strictly geometrical, the conclusions are only applicable to traveltime. The effect of DMO-NMO on waveshaping has not been considered.

## REFERENCES

French, W. S., Perkins, W. T., and Zoll, R. M., 1984, Partial migration via true CRP stacking: 54th Ann. Internat. Mtg., Soc. Explor. Geophys., Expanded Abstracts, 799–802.

Levin, F. K., 1971, Apparent velocity from dipping interface reflections: Geophysics, **36**, 510–516.

Ottolini, R. A., 1982, Migration of reflection seismic data in angle-midpoint coordinates: PhD. thesis, Stanford Univ.

## APPENDIX

An event at time $t$ on a trace with offset $2h$ is migrated by spreading a wavelet over the ellipsoidal surface at a distance $Vt$ from the shot and receiver locations, as indicated in Figure A-1. This wavefront surface can be regarded as the sum of spherical wavefronts with centers on the line SR joining the shot and receiver locations.

Figure A-2 shows a typical cross-section through the line SR. The radius $R$ of the circle which has its center at $(b, 0)$ and which touches the ellipse can be expressed in terms of $A$, $B$, and $b$ according to equation (A-8).

The equation of the elliptical wavefront is

$$\frac{x^2}{A^2} + \frac{z^2}{B^2} = 1, \qquad \text{(A-1)}$$

where the major and minor axes are given by

$$A^2 = \frac{V^2t^2}{4} \qquad \text{(A-2)}$$

and

$$B^2 = A^2 - h^2, \qquad \text{(A-3)}$$

and where $2h$ is the offset distance from the source to the receiver and $t$ is the total traveltime.

A circle with radius $R$ and center at $(b, 0)$ is given by

$$(x - b)^2 + z^2 = R^2. \qquad \text{(A-4)}$$

When the circle is tangent to the ellipse, then both equation (A-1) and equation (A-4) have the same tangent at $(x, z)$. Let $p = dz/dx$ be the common tangent. Differentiating equations (A-1) and (A-4) with respect to $x$ yields

$$\frac{x}{A^2} + \frac{zp}{B^2} = 0 \qquad \text{(A-5)}$$

and

$$(x - b) + zp = 0. \qquad \text{(A-6)}$$

Elimination of $p$ from equations (A-5) and (A-6) gives

$$b = x\left(1 - \frac{B^2}{A^2}\right). \qquad \text{(A-7)}$$

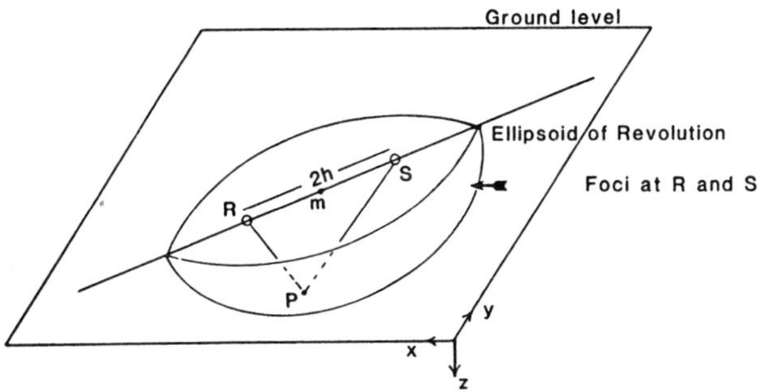


FIG. A-1. A source-receiver pair has finite offset $2h$ and midpoint m. An event recorded at time $t$ over a medium of constant velocity $V$ may have come from any point P on an ellipsoid of revolution whose axis of rotation is the line joining the source and receiver.

This yields an expression for $x$. Multiply equation (A-4) by $1/B^2$ and subtract equation (A-1) to eliminate $z$. Substituting the expression for $x$ into the result will finally yield

$$R^2 = B^2\left(\frac{A^2 - B^2 - b^2}{A^2 - B^2}\right). \qquad \text{(A-8)}$$

Figure A-3 shows an example of circles drawn using equation (A-8) for $A = 1600$ and $B = 884$ m. Note that the circles touch the ellipse as $b$ increases until the contact point reaches ground level, and then the circles shrink into the foci.

### Time relations

Recall that $R$ is the radius of a sphere tangent to the ellipsoid whose center is at a distance $b$ from the midpoint $m$ along the source-receiver line. To return to the velocity discussion, equations (A-2) and (A-3) substituted for $A$ and $B$ in equation (A-8) yield

$$R^2 = \left[\left(\frac{Vt}{2}\right)^2 - h^2\right]\left(1 - \frac{b^2}{h^2}\right). \qquad \text{(A-9)}$$

Since the medium has constant velocity $V$, $R^2 = V^2t_0^2/4$, and hence

$$t_0^2 = \left(t^2 - \frac{4h^2}{V^2}\right)\left(1 - \frac{b^2}{h^2}\right). \qquad \text{(A-10)}$$

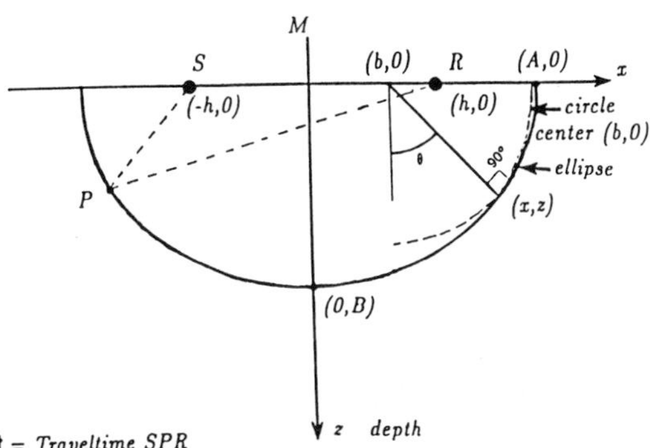


FIG. A-2. The ellipsoidal surface can be regarded as the envelope of a family of spheres with centers at $(b, 0)$ along the segment RS. The distance between the source and the receiver is $2h$ and the midpoint is M. The point of tangency is $(x, z)$ and $\theta$ is the corresponding inclination of the tangent.

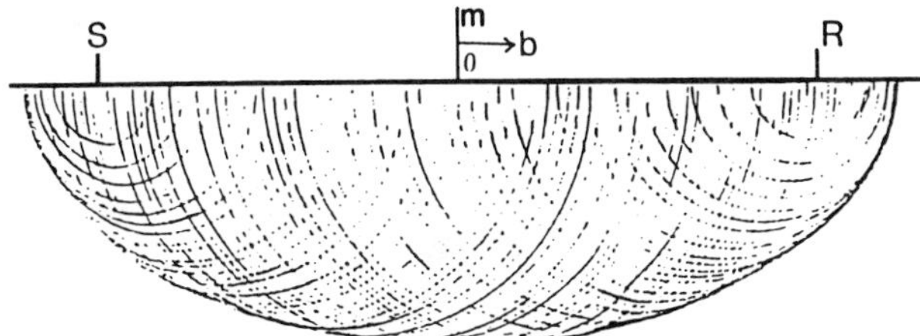


FIG. A-3. An ellipsoidal wavefront created as the envelope of circular wavefronts. At a distance $b$ from m the radius is given by $R^2 = B^2(A^2 - B^2 - b^2)/(A^2 - B^2)$, where $A$ and $B$ are the major and minor axes of the ellipse. The figure shows circles at equal increments in $b$. For better sampling, unequal increments in $b$ could be used.

That is, to calculate $t_0$, the time on the zero-offset trace corresponding to $t$ on the original trace, it is necessary to know the velocity $V$. A solution to this difficulty is to construct a trace with offset $2k$ instead of zero and with arrival time $t_1$ where

$$t_1^2 = t_0^2 + \frac{4k^2}{V^2}. \quad \text{(A-11)}$$

It is important to see that equation (A-11) is the standard NMO correction for offset $2k$. By rearranging equations (A-9), (A-10), and (A-11), we get

$$t_1^2 = t^2\left(1 - \frac{b^2}{h^2}\right) + \frac{4}{V^2}\left(k^2 + b^2 - h^2\right). \quad \text{(A-12)}$$

Choosing $k$ so that

$$k^2 = h^2 - b^2, \quad \text{(A-13)}$$

then

$$t_1^2 = t^2\left(1 - \frac{b^2}{h^2}\right), \quad \text{(A-14)}$$

that is,

$$t_1 = t\,\frac{k}{h}. \quad \text{(A-15)}$$

Equations (A-13) and (A-15), which define the process of DMO, are independent of velocity. The NMO equation (A-11) for the $(k, t_1)$ matrix yields the velocity of the subsurface $V$. Equations (A-13), (A-15), and (A-11) are the same as equations (1), (2), and (3).

As $b$ increases, the point of tangency $(x, y)$ moves up the ellipse toward ground level. The angle of inclination $\theta$ is equal to 90° when $x = A$, and the corresponding value of $b$ can be obtained from equation (A-7). Thus, expressed in terms of $h$ and $t$, $\theta = 90°$ when

$$t = 2h^2/Vb. \quad \text{(A-16)}$$

Amplitudes at times later than that given by equation (A-16) do not contribute to the formation of the ellipsoidal envelope; they may be omitted from the processing without loss.

# Ultrasonic laboratory tests of geophysical tomographic reconstruction

Tien-when Lo*, M. Nafi Toksöz*, Shao-hui Xu*, and Ru-Shan Wu*

## ABSTRACT

In this study, we test geophysical ray tomography and geophysical diffraction tomography by scaled model ultrasonics experiments. First, we compare the performance of these two methods under limited view-angle conditions. Second, we compare the adaptabilities of these two methods to objects of various sizes and acoustic properties. Finally, for diffraction tomography, we compare the Born and Rytov approximations based on the induced image distortion by using these two approximation methods. Our experimental results indicate the following: (1) When the scattered field can be obtained, geophysical diffraction tomography is in general superior to ray tomography because diffraction tomography is less sensitive to the limited view-angle problem and can image small objects of size comparable to a wavelength. (2) The advantage of using ray tomography is that reconstruction can be done using the first arrivals only, the most easily measurable quantity; and there is no restriction on the properties of the object being imaged. (3) For geophysical diffraction tomography, the Rytov approximation is valid over a wider frequency range than the Born approximation in the cross-borehole experiment. In the VSP and the surface reflection tomography experiments, no substantial difference between the Born and Rytov approximations is observed.

## INTRODUCTION

Geophysical tomography has become an important research topic because it is capable of determining subsurface structures in three-dimensional (3-D) space from surface, borehole, and cross-borehole data. In this study we conduct a set of ultrasonic model experiments to image objects using source-receiver geometries analogous to those for surface reflection, vertical seismic profiling (VSP), and cross-borehole measurements. Using these laboratory data, we reconstruct the objects using different tomographic reconstruction algorithms. This permits us to determine the relative performance of different reconstruction algorithms for each geometry.

Tomographic methods can be classified as ray tomography (based on the ray equation) and diffraction tomography (based on the wave equation). Ray tomography has three types of reconstruction algorithms: the series expansion algorithm, the direct Fourier transform algorithm, and the filtered backprojection algorithm. Kak (1985) gives an overview of these three algorithms. Ray tomography works well when the interaction between the illuminating energy and the object under investigation can be successfully described by the ray equation. When the size of the object is comparable to the wavelength of the illuminating waves, diffraction and scattering become the dominant processes. In such cases the system may be described by the wave equation, instead of the ray equation, as proposed by Mueller et al. (1979, 1980). Diffraction tomography also has three types of reconstruction algorithms: the series expansion algorithm, the direct Fourier transform algorithm, and the filtered backpropagation algorithm (Kak, 1985). Using numerical examples, Pan and Kak (1983) compared the direct Fourier transform algorithm and the filtered backpropagation algorithm for diffraction tomography. The series expansion algorithm for the diffraction tomography is described by Mohammad-Djafari and Demoment (1986). This algorithm is similar to the series expansion algorithm for ray tomography except that the forward problem of the series expansion diffraction tomography is a matrix equation relating the object function and the scattered field and the components of this matrix are the Green's functions.

Most geophysical tomographic studies published so far use the ray tomography methods (Dines and Lytle, 1979; McMechan, 1983; Menke, 1984; Bishop et al., 1985; Ivansson, 1985, 1986; Peterson et al., 1985; Chiu et al., 1986; Cottin et al., 1986; Gustavsson et al., 1986; Ramirez, 1986). When ray tomography is used for geophysical applications, there are two inherent problems: (1) "rays" only propagate through a limited portion of the object due to the available source-receiver configurations, and (2) high attenuation in earth materials forces us to illuminate with waves with wavelengths comparable to the size of the subsurface inhomogeneities, resulting in diffraction and scattering which cannot be handled conveniently by the ray equation. To attack these two problems,

Manuscript received by the Editor January 5, 1987; revised manuscript received November 30, 1987.
*Department of Earth, Atmospheric and Planetary Sciences, Massachusetts Institute of Technology, 42 Carleton Street, Cambridge, MA 02142.


Reprinted from Geophysics, 53, 947-956

Devaney (1982, 1984), Esmersoy (1986), and Wu and Toksöz (1987) applied the diffraction tomography techniques to geophysical problems and formulated the filtered backpropagation reconstruction algorithm for geophysical diffraction tomography. These studies were tested using forward theoretical models or numerical data, which were free of noise and satisfied the assumptions.

In this study, we use ultrasonic laboratory data to test the relative performance of geophysical diffraction tomography and ray tomography under three limited view-angle source-receiver configurations: cross-borehole, VSP, and surface reflection.

The filtered backpropagation reconstruction algorithm used in diffraction tomography is derived using the Born or the Rytov approximations. The Born approximation results in a linear mapping between the object function and the complex (amplitude and phase) scattered wave field. The Rytov approximation results in a linear mapping between the object function and the complex phase difference between the total field and the incident field. It should be noted that, although the reconstruction formulas using the Born and the Rytov approximations are similar, the physical assumptions behind these two approximation methods are quite different. In this paper, we compare the diffraction tomography reconstructions using both approximation methods. Kaveh et al. (1981) conducted a laboratory experiment comparing the effects of the Born and Rytov approximations on diffraction tomography. In their study, the object was evenly illuminated from all directions. This paper, on the other hand, studies the effects of the Born and Rytov approximations when the object is illuminated only from limited view angles.

## THE PRINCIPLES OF DIFFRACTION TOMOGRAPHY AND RAY TOMOGRAPHY

In this section we review briefly the filtered backpropagation reconstruction algorithm for diffraction tomography and the algebraic reconstruction technique (ART) for ray tomography used in this paper. More complete discussions of these methods are given by Dines and Lytle (1979), Kak (1985), and Wu and Toksöz (1987).

### Diffraction tomography

For the theoretical derivations and experiments, we use a two-dimensional (2-D) geometry for the source-receiver array and the object as shown in Figure 1. An object with varying wave velocity $C(\mathbf{r})$, where $\mathbf{r}$ is the position vector, is situated in the uniform background medium with velocity $C_0$. From the first Born approximation, we obtain the basic equation for the scattering problem,

$$U_s(\mathbf{r}_g, \mathbf{r}_s) = -k_0^2 \int_V O(\mathbf{r}) G(\mathbf{r}, \mathbf{r}_s) G(\mathbf{r}_g, \mathbf{r})\, d\mathbf{r}, \qquad (1)$$

where subscripts $g$ and $s$ refer to geophone and source, respectively, $U_s(\mathbf{r}_g, \mathbf{r}_s)$ is the scattered field measured at position $\mathbf{r}_g$ when the point source is at position $\mathbf{r}_s$, and $O(\mathbf{r})$ is the object function defined as

$$O(\mathbf{r}) = 1 - \frac{C_0^2}{C^2(\mathbf{r})}. \qquad (2)$$

$G$ is the 2-D Green's function for the background medium

$$G(\mathbf{r}, \mathbf{r}') = \frac{i}{4} H_0^{(1)}(k_0 |\mathbf{r} - \mathbf{r}'|), \qquad (3)$$

where $H_0^{(1)}$ is the zero-order Hankel function of the first kind, and $k_0 = \omega / C_0$ is the wavenumber in the background medium. Wu and Toksöz (1987) derived the filtered backpropagation reconstruction formula to invert equation (1):

$$O(\mathbf{r}) = \frac{1}{(2\pi)^2} \iint dk_s\, dk_g\, J(K_x, K_z | k_s, k_g) \frac{4\gamma_s \gamma_g}{k_0^2} \tilde{U}_s(k_g, k_s) \times \exp\left[-i(\gamma_s d_s + \gamma_g d_g)\right] \exp\left[i(K_x x + K_z z)\right], \qquad (4)$$

where $\tilde{U}_s(k_g, k_s)$ is the 2-D Fourier transform of $U_s(\mathbf{r}_g, \mathbf{r}_s)$, $d_s$ and $d_g$ are the distances from the origin to the source line and the geophone line, $k_s$ and $k_g$ are the wavenumbers along the source line and the geophone line, $\gamma_s = \sqrt{k_0^2 - k_s^2}$, $\gamma_g = \sqrt{k_0^2 - k_g^2}$, and $J(K_x, K_z | k_s, k_g)$ is the Jacobian of coordinate transformation between the $(K_x, K_z)$ coordinate system, and the $(k_g, k_s)$ coordinate system. Equation (4) is the general reconstruction formula. The reconstruction formulas for the cross-borehole, VSP, and surface reflection configurations can be obtained by substituting the $J(K_x, K_z | k_s, k_g)$, $K_x$, and $K_z$ for each configuration into equation (4). In Table 1, we list the reconstruction formulas and the $J(K_x, K_z | k_s, k_g)$, $K_x$, and $K_z$ for the cross-borehole, VSP, and surface reflection configurations.

For the case of reconstruction based on the Rytov approximation, all the reconstruction formulas are the same except that $\tilde{U}_s(k_g, k_s)$ is replaced by $\tilde{\Phi}(k_g, k_s)$, which is the 2-D Fourier transform of the complex phase function $\Phi(\mathbf{r}_g, \mathbf{r}_s)$ defined

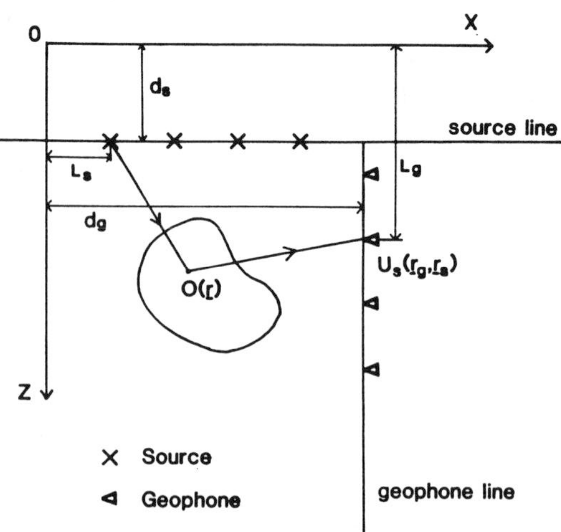


FIG. 1. 2-D geometry used in the derivation of the filtered backpropagation algorithm. $O(\mathbf{r})$ is the object function. $U_s(\mathbf{r}_g, \mathbf{r}_s)$ is the measured scattered wave field. The perpendicular distance from the origin to the source line and the geophone line are $d_s$ and $d_g$, respectively. $\ell_s$ is the distance of the source along the source line. $\ell_g$ is the distance of the geophone along the geophone line.

by

$$\Phi(\mathbf{r}_g, \mathbf{r}_s) = U_i(\mathbf{r}_g, \mathbf{r}_s) \log \frac{U_i(\mathbf{r}_g, \mathbf{r}_s) + U_s(\mathbf{r}_g, \mathbf{r}_s)}{U_i(\mathbf{r}_g, \mathbf{r}_s)}$$
$$= U_i(\mathbf{r}_g, \mathbf{r}_s)\phi_d(\mathbf{r}_g, \mathbf{r}_s), \quad (5)$$

where $U_i(\mathbf{r}_g, \mathbf{r}_s)$ is the incident field and $\phi_d(\mathbf{r}_g, \mathbf{r}_s)$ is the complex phase difference between the total field and the incident field.

### Ray tomography

When we use the algebraic reconstruction technique (ART), the imaging area is first divided into $N$ pixels. Let $f_j$ be the average of a certain physical parameter (such as sonic wave slowness) inside the $j$th pixel and $P_i$ be the line integral of that parameter along the $i$th ray. Then for an imaging system with $N$ pixels and $M$ measurements, the forward problem is

$$P_i = \sum_{j=1}^{N} S_{ij} f_j, \qquad i = 1, 2, \ldots, M, \quad (6)$$

where $S_{ij}$ is the length of the segment of the $i$th ray intersecting the $j$th pixel. To invert $f_j$, the algebraic reconstruction technique uses an iterative approach. It starts with an initial estimate of $f_j$, denoted by $\hat{f}_j$. From this initial estimate, the estimated line integral can be calculated by

$$\hat{P}_i = \sum_{j=1}^{N} S_{ij} \hat{f}_j. \quad (7)$$

The iteration algorithm updates the estimate $\hat{f}_j$ by the recurrence formula

$$\hat{f}_j^{\text{new}} = \hat{f}_j^{\text{old}} + \Delta f_{ij}$$
$$= \hat{f}_j^{\text{old}} + S_{ij} \frac{P_i - \sum_{j=1}^{N} S_{ij} \hat{f}_j^{\text{old}}}{\sum_{j=1}^{N} (S_{ij})^2}. \quad (8)$$

$\Delta f_{ij}$ is the correction on $\hat{f}_j^{\text{old}}$ after examining the $i$th ray. It is the least-squares solution of the following equation:

$$\Delta P_i = P_i - \hat{P}_i$$
$$= \sum_{j=1}^{N} S_{ij}(f_j - \hat{f}_j)$$
$$= \sum_{j=1}^{N} S_{ij} \Delta f_{ij}. \quad (9)$$

Equation (8) keeps updating $\hat{f}_j$ until the difference between the measured and estimated line integrals is smaller than a prespecified threshold. The convergence of this algorithm can be visualized in the vector space (Kak, 1985) and a rigorous proof of the convergence was given by Tanabe (1971).

$S_{ij}$ in equation (8) is calculated by ray tracing. In our experiment, we use a 2-D ray-tracing algorithm similar to the algorithm described by Anderson and Kak (1982). This algorithm is derived by first expressing the position vector of the ray $\mathbf{r}$ in a Taylor series and discarding the third and higher order

**Table 1. Filtered back propagation reconstruction formulas for the cross-borehole, VSP, and surface reflection experiments.**

| | Reconstruction formula | Coordinate transformation |
|---|---|---|
| Cross-borehole | $O(x, z) = \frac{1}{\pi}\int_{-k_0}^{k_0} dk_s \exp(ik_s z - i\gamma_s x) O_1(x, z, k_s)$ | $K_x = \gamma_g - \gamma_s, \quad K_z = k_g + k_s$ |
| | $O_1(x, z, k_s) = \frac{1}{\pi}\int_{-k_0}^{k_0} dk_g \exp\left[ik_g z - i\gamma_g(x_h - x)\right] D(k_g, k_s)$ | $J(K_x, K_z \mid k_s, k_g) = \frac{\lvert k_s\gamma_g + k_g\gamma_s\rvert}{\gamma_g\gamma_s}$ |
| | $D(k_g, k_s) = \tilde{U}(k_g, k_s)\frac{\lvert k_s\gamma_g + k_g\gamma_s\rvert}{k_0^2}$ | |
| VSP | $O(x, z) = \frac{1}{\pi}\int_{-k_0}^{k_0} dk_s \exp(ik_s x - i\gamma_s z) O_1(x, z, k_s)$ | $K_x = k_s + \gamma_g, \quad K_z = k_g - \gamma_s$ |
| | $O_1(x, z, k_s) = \frac{1}{\pi}\int_{-k_0}^{k_0} dk_g \exp\left[ik_g z - i\gamma_g(x_h - x)\right] D(k_g, k_s)$ | $J(K_x, K_z \mid k_s, k_g) = \frac{\lvert k_g k_s + \gamma_g\gamma_s\rvert}{\gamma_g\gamma_s}$ |
| | $D(k_g, k_s) = \tilde{U}(k_g, k_s)\frac{\lvert k_g k_s + \gamma_g\gamma_s\rvert}{k_0^2}$ | |
| Surface reflection | $O(x, z) = \frac{1}{\pi}\int_{-k_0}^{k_0} dk_s \exp\left[ik_s x - i\gamma_s(z - z_s)\right] O_1(x, z, k_s)$ | $K_x = k_g + k_s, \quad K_z = -(\gamma_g + \gamma_s)$ |
| | $O_1(x, z, k_s) = \frac{1}{\pi}\int_{-k_0}^{k_0} dk_g \exp\left[ik_g x - i\gamma_g(z - z_g)\right] D(k_g, k_s)$ | $J(K_x, K_z \mid k_s, k_g) = \frac{\lvert k_s\gamma_g - k_g\gamma_s\rvert}{\gamma_g\gamma_s}$ |
| | $D(k_g, k_s) = \tilde{U}(k_g, k_s)\frac{\lvert k_s\gamma_g - k_g\gamma_s\rvert}{k_0^2}$ | |

terms

$$\mathbf{r}(\ell + \Delta\ell) = \mathbf{r}(\ell) + \frac{d\mathbf{r}}{d\ell}\Delta\ell + \frac{1}{2}\frac{d^2\mathbf{r}}{d\ell^2}(\Delta\ell)^2, \qquad (10)$$

where $\ell$ is the distance along the ray and $\Delta\ell$ is the step size. We then write the ray equation as

$$\frac{d}{d\ell}\left(n\frac{d\mathbf{r}}{d\ell}\right) = \nabla n, \qquad (11)$$

where $n$ is the refractive index. Substituting equation (11) into equation (10), we obtain the following expression for the ray which can be implemented directly as a ray-tracing algorithm:

$$\mathbf{r}(\ell + \Delta\ell) = \mathbf{r}(\ell) + \frac{d\mathbf{r}}{d\ell}\Delta\ell + \frac{1}{2n}\left[\nabla n - \left(\nabla n \cdot \frac{d\mathbf{r}}{d\ell}\right)\frac{d\mathbf{r}}{d\ell}\right](\Delta\ell)^2. \qquad (12)$$

In equation 11, $d\mathbf{r}/d\ell$ is the unit vector tangent to the ray, and $\nabla n$ is calculated by central-difference approximation. The refractive index at an arbitrary position is approximated by bilinear interpolation using the four nearest grid values. We use the shooting method to solve the ray-linking problem, and the launching angle is determined by Newton's method.

The simultaneous iterative reconstruction technique (SIRT) is very similar to the algebraic reconstruction technique except that the correction on each pixel at each iteration is determined from all the measured data simultaneously (Gilbert, 1972). In this study, we use the simultaneous iterative reconstruction technique.

## ULTRASONIC EXPERIMENTS

Ultrasonics experiments simulating geophysical tomography are carried out in a modeling tank. This tank is 100 cm × 60 cm × 50 cm in dimension and is equipped with microcomputer-based control and data acquisition systems. Water is used as a constant-velocity background medium. Objects of various sizes and acoustic properties are used as targets to be imaged. We use one broad-band hydrophone for the source (LC-34) and another for the receiver (ITC-1089D). The frequency range in our experiments is 10 kHz to 200 kHz. Each hydrophone can be moved independently in 3-D space by three step motors. Each step is 0.064 mm. The translation scanning scheme of the hydrophones is controlled by a SLO-SYN step motor controller. The electrical input for the source hydrophone comes from a Panametrics 5055PR pulser. The received signals are filtered by a Krohn-Hite 3202R filter, amplified by a Panametrics 5660B preamplifier, and digitized by a Data Precision DATA6000 digital oscilloscope. The digital oscilloscope and the step motor controller are interfaced with the IBM microcomputer by an IEEE-488 interface bus. Digitized data are transmitted to a VAX 11/780 computer for image reconstruction. Images are displayed on a Comtal image processor. A block diagram of the laboratory setup is shown in Figure 2.

The source and receiver configurations used in our experiments to simulate VSP, cross-borehole, and surface reflection configurations are shown in Figure 3. The object used in the tomography experiments is a gelatin cylinder 90 mm in diameter. The $P$-wave velocity and density of this gelatin cylinder are 1.55 km/s and 1.24 g/cm$^3$, respectively. The difference in $P$-wave velocity between the object and background medium is only 4 percent, which is sufficient to satisfy the constraints for using the Born and Rytov approximations.

Since we use a point source and a point receiver and there is no velocity variation along the axial direction of the gelatin cylinder, which is perpendicular to the source-receiver plane,

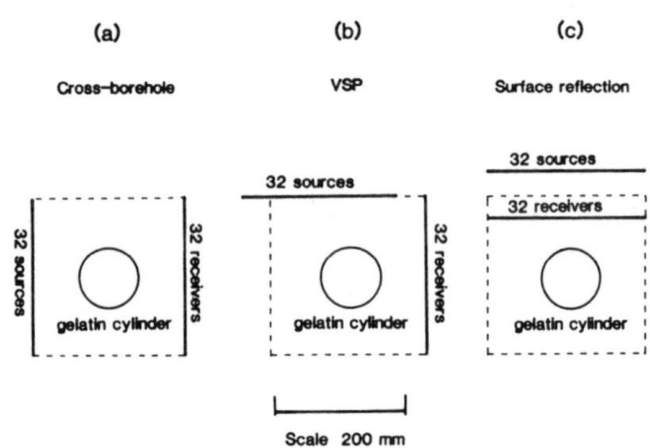


FIG. 3. Top view of the actual layout of the tomography experiments. (a) Cross-borehole experiment. (b) VSP experiment. (c) Surface reflection experiment.

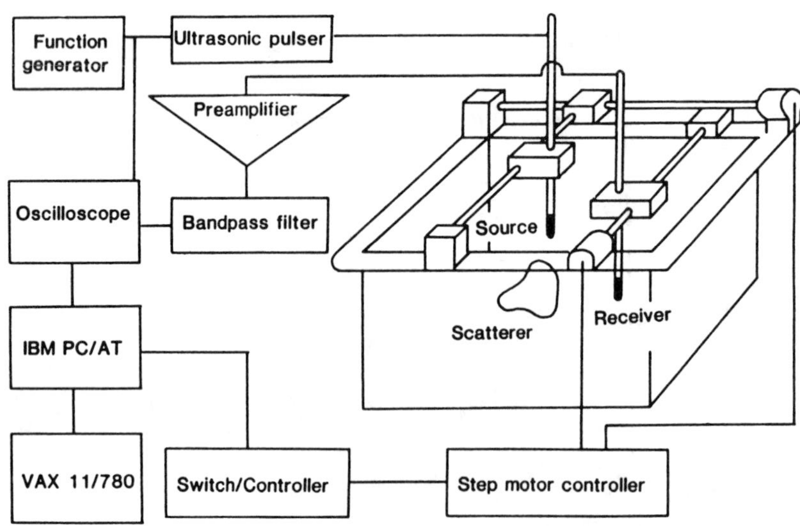


FIG. 2. Block diagram of the microcomputer-based ultrasonic imaging system.

our experiments are "two-and-half-dimensional" (2 1/2-D) experiments (Esmersoy, 1986). In the far field, the 2 1/2-D scattering problem is very similar to the 2-D scattering problem (Esmersoy, 1986). This is why the 2-D reconstruction algorithms listed in Table 1 can be applied to our experiments.

For the diffraction tomography experiments, objects are reconstructed by applying the filtered back-propagation algorithms listed in Table 1 to the scattered wave field induced by the object. We use a dual-experiment method to measure the scattered wave field. First, we put the object inside the water tank, scan the source and the receiver around it, and measure the total wave field. Then, we remove the object and repeat the same scanning procedure to obtain the incident wave field. The difference between these two sets of data is the scattered wave field due to the object. This dual-experiment method also helps eliminate interference from the experimental setup such as the walls of the tank and the hydrophone itself.

For the ray tomography experiments, we use the same data set collected in the diffraction tomography experiments. The traveltimes of the first-arrival *P* waves of the total field waveforms are used to obtain the velocity of the object by the simultaneous iterative reconstruction technique described above.

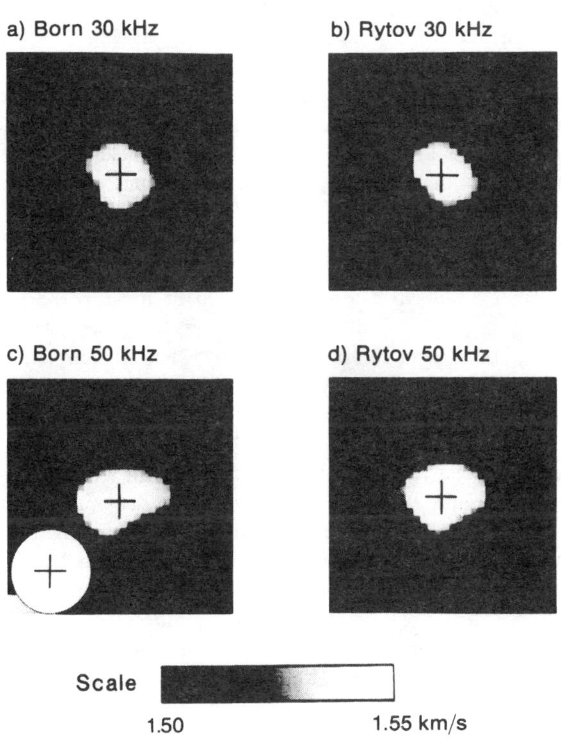


FIG. 5. Images reconstructed by the filtered backpropagation algorithm in the cross-borehole experiment. The gelatin cylinder should be centered at the cross with the size and shape as shown by the white circle at the lower left corner of the figure. (a) Reconstruction with the Born approximation at 30 kHz. (b) Reconstruction with the Rytov approximation at 30 kHz. (c) Reconstruction with the Born approximation at 50 kHz. (d) Reconstruction with the Rytov approximation at 50 kHz.

## EXPERIMENTAL RESULTS

### Cross-borehole tomography experiment

The layout of this experiment is shown in Figure 3a. The source hydrophone is activated at 32 equally spaced positions along a source line, simulating 32 sources in one borehole. The receiver hydrophone records waveforms at 32 equally spaced positions along a receiver line, simulating 32 receivers in another borehole. The gelatin cylinder is placed between the source line and the receiver line. The ultrasonic wave field generated at each source position is measured at 32 receiver positions and therefore 32 × 32 waveforms are recorded for measuring the incident wave field and the same number of data are recorded for measuring the total wave field. The

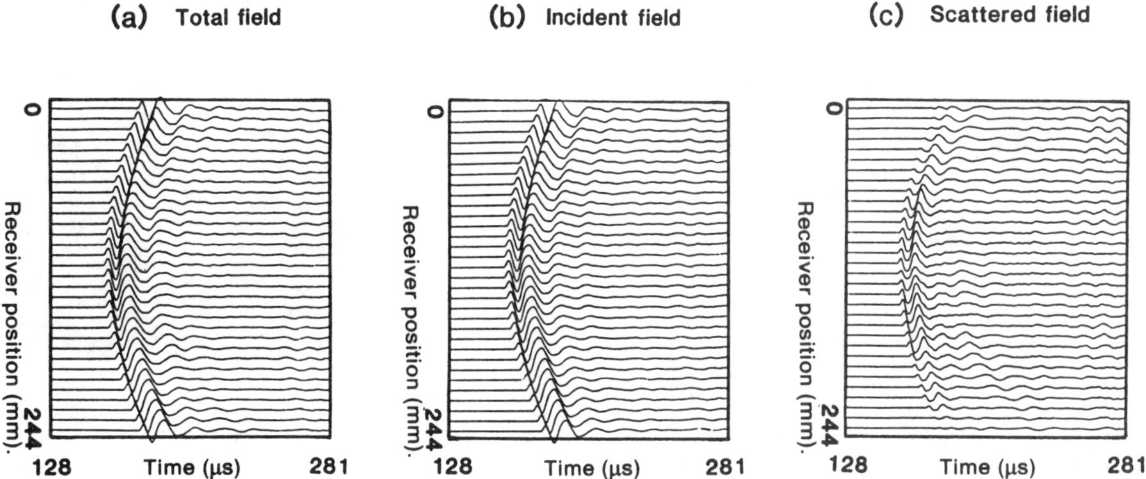


FIG. 4. Examples of waveforms recorded in the cross-borehole experiment. (a) Total field waveforms (measured with the cylinder between source and receiver). (b) Incident field waveforms measured with cylinder removed. (c) Scattered field waveforms [(a)–(b)].

dominant frequency of our signals is 50 kHz, corresponding to a wavelength of 30 mm in water. The sampling interval along the source line and the receiver line is 7.62 mm. Waveforms are digitized at a sampling interval of 300 ns. Figure 4 shows 32 waveforms recorded at 32 receiver positions with the source hydrophone at the middle of the source line. Figure 4a shows the total field waveforms; Figure 4b, the incident field waveforms; and Figure 4c, the scattered field waveforms. Taking the Fourier transform of the waveforms, we obtain the magnitude and phase of the total field, the incident field, and the scattered field at various frequencies. We use Tribolet's algorithm (Tribolet, 1977) with the computation efficiency improvement made by Bonzanigo (1978) to unwrap the phase data. Using these data, we reconstruct the object with the filtered backpropagation algorithm. Reconstructions with both the Born and the Rytov approximations are calculated. Figure 5 shows the reconstructions with the Born and Rytov approximations at 30 kHz and 50 kHz. At the lower frequency, images reconstructed by either the Born or the Rytov approximations are about the same quality (compare Figures 5a and 5b), whereas at the higher frequency, the image reconstructed by the Rytov approximation is less distorted than the one reconstructed by the Born approximation (compare Figures 5c and 5d). The wavelength independence of the Rytov approximation observed in this experiment is consistent with the results of the numerical study by Slaney et al. (1984). In

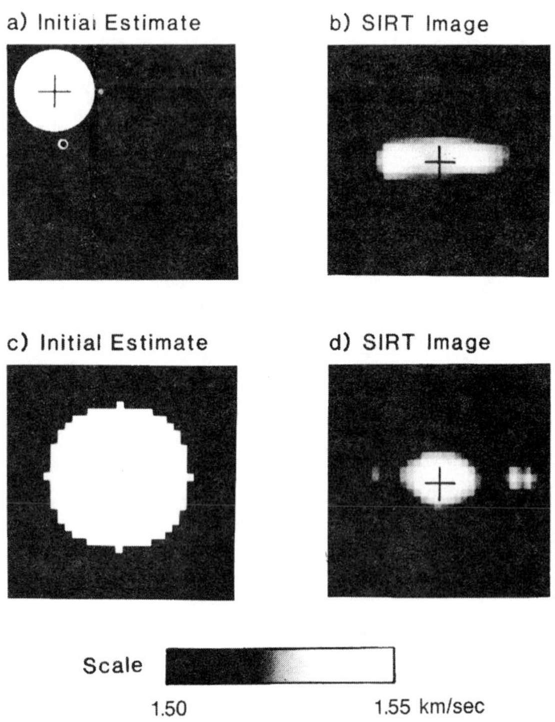


FIG. 6. Images reconstructed by the simultaneous iterative reconstruction technique in the cross-borehole ray tomography experiment. The gelatin cylinder should be centered at the cross with the size and shape as shown by the circle at the upper left corner of the figure. (a) An initial estimate assuming no information about the object function. (b) Reconstruction after twenty iterations with (a) as the initial estimate. (c) An initial estimate with some a priori information about the object. (d) Reconstruction after twenty iterations with (c) as the initial estimate.

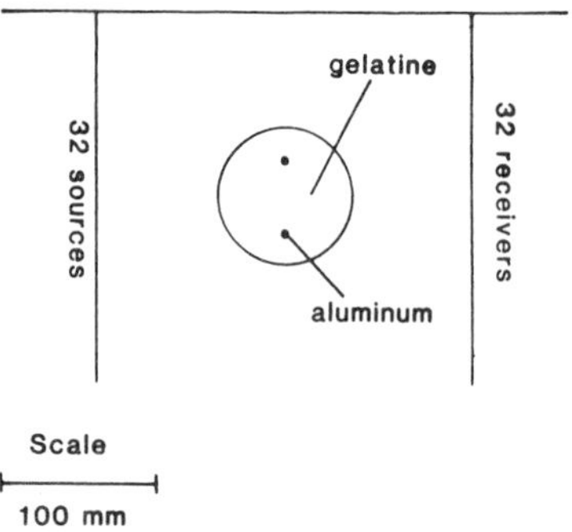


FIG. 7. Experimental configuration of the cross-borehole experiment with a more complex object. The object is a gelatin cylinder with two aluminum rods inside. The dominant source frequency is 30 kHz.

(a) Diffraction tomography

(b) Ray tomography

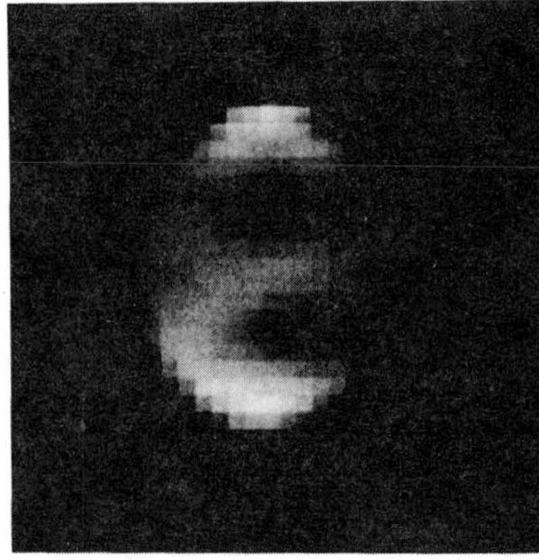

FIG. 8. Images reconstructed by the filtered backpropagation reconstruction algorithm and the simultaneous iterative reconstruction technique for a gelatin cylinder with two aluminum rods in the cross-borehole experiment. (a) Image reconstructed by the filtered backpropagation algorithm with the Born approximation. (b) Image reconstructed by the simultaneous iterative reconstruction technique.

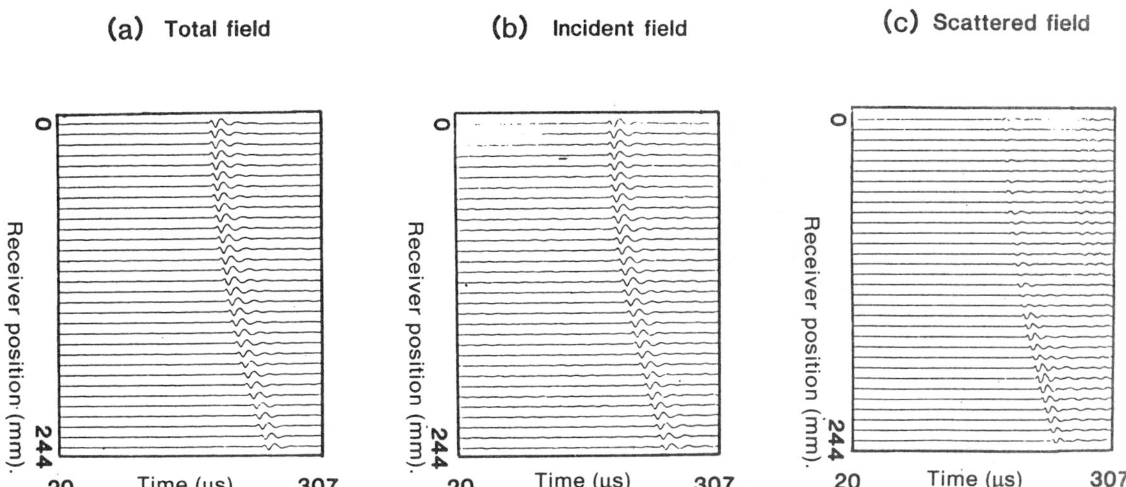


FIG. 9. Examples of waveforms recorded in the VSP experiment. (a) Total field waveforms. (b) Incident field waveforms. (c) Scattered field waveforms [(a)–(b)].

their work, they demonstrated that the validity of the Rytov approximation is judged by the phase change per wavelength, not by the total phase change. Therefore, as long as the velocity contrast between the object and the surrounding medium is small enough [less than a few percent, as suggested by Slaney et al. (1984)], the Rytov approximation is valid without constraints on the size of object. The Born approximation, however, requires that the scattered field be small. This will be violated when the size of the weak inhomogeneity becomes large. Also, note that in the cross-borehole configuration, the information coverage in the frequency domain is poor in the horizontal direction (see Figure 4 in Wu and Toksöz, 1987). This is consistent with the poor horizontal resolution in the images reconstructed in our cross-borehole experiments.

The traveltimes of the first-arrival *P* waves of the total field waveforms are also measured in the cross-borehole experiment to reconstruct the gelatin cylinder by the simultaneous iterative reconstruction technique. The images reconstructed are shown in Figure 6. Using Figure 6a, an initial estimate assuming no information about the object is available, results in Figure 6b as the reconstruction after 20 iterations. Using Figure 6c, another initial estimate circular in shape but with a radius twice the radius of the true object, results in Figure 6d as the corresponding reconstruction after 20 iterations. Comparing Figures 6b and 6d, we note that ray tomography using the simultaneous iterative reconstruction technique improves significantly if the initial model approximates the true object. Further iterations did not improve the images significantly in either case.

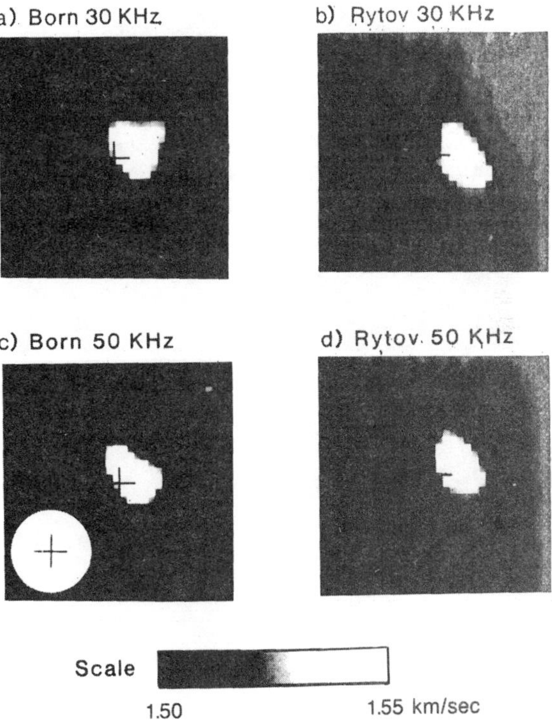


FIG. 10. Images reconstructed by the filtered backpropagation algorithm in the VSP experiment. The gelatin cylinder should be centered at the cross with the size and shape as shown by the white circle at the lower left corner of the figure. (a) Reconstruction with the Born approximation at 30 kHz. (b) Reconstruction with the Rytov approximation at 30 kHz. (c) Reconstruction with the Born approximation at 50 kHz. (d) Reconstruction with the Rytov approximation at 50 kHz.

### Cross-borehole experiment with a more complex object

To investigate further the relative performance of diffraction tomography and ray tomography, we ran another cross-borehole experiment with a more complex object. This object is a gelatin cylinder with two aluminum rods inside as shown in Figure 7. The image reconstructed by diffraction tomography based on the Born approximation is shown in Figure 8a. Both the gelatin cylinder and the two aluminum rods are

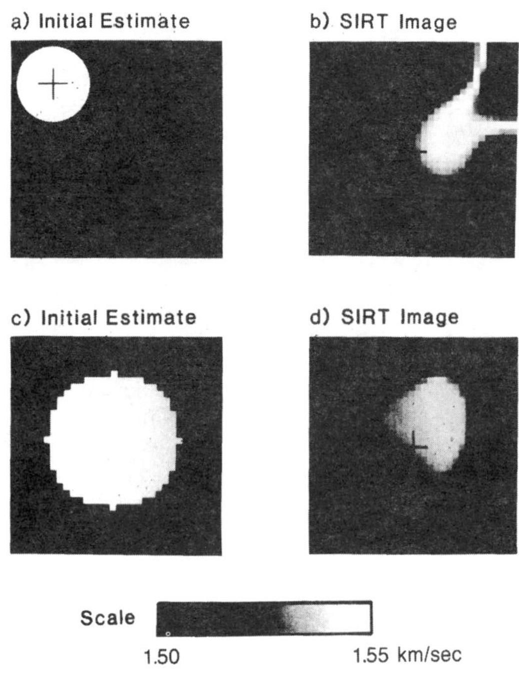


FIG. 11. Images reconstructed by the simultaneous iterative reconstruction technique in the VSP ray tomography experiment. The gelatin cylinder should be centered at the cross with the size and shape as shown by the circle at the upper left corner of the figure. (a) An initial estimate assuming no information about the object function. (b) Reconstruction of the object after 100 iterations with (a) as the initial estimate. (c) An initial estimate with some a priori information about the object. (d) Reconstruction of the object after 100 iterations with (c) as the initial estimate.

successfully reconstructed. The same data are also inverted by ray tomography with the simultaneous iterative reconstruction technique. With an initial estimate such as Figure 6c, the ray tomography reconstruction after 20 iterations is shown in Figure 8b. The gelatin cylinder is reasonably well reconstructed, but the two aluminum rods are poorly reconstructed. This experiment demonstrates that when the size of the object is comparable to the wavelength, diffraction tomography with the filtered backpropagation reconstruction algorithm can reconstruct the object better than ray tomography.

### VSP tomography experiment

In this experiment, the source hydrophone is activated at 32 equally spaced positions along the source line, simulating 32 sources arranged in a straight line on the surface. The receiver hydrophone records waveforms at 32 equally spaced positions along a straight line perpendicular to the source line, simulating 32 receivers in a borehole. Examples of the waveforms recorded with a gelatin cylinder as the scatterer are shown in Figure 9. Figure 10 shows the filtered backpropagation reconstructions at 30 kHz and 50 kHz with the Born and Rytov approximations. All four examples in Figure 10 reconstruct the upper right portion of the object better than the lower left portion. This distortion can be explained by the uneven spectral coverage of the VSP diffraction tomography (Figure 3b in Wu and Toksöz, 1987).

Similar to the cross-borehole ray tomography experiment, the traveltimes of the first-arrival $P$ waves of the total field waveforms are used to reconstruct the object by the simultaneous iterative reconstruction technique. The results are shown in Figure 11. Figures 11a and 11c are initial guess images. Figures 11b and 11d are the corresponding reconstructions after 100 iterations. Comparing Figure 10 and Figure 11, it is apparent that diffraction tomography is less sensitive to the limited view-angle problem than ray tomography.

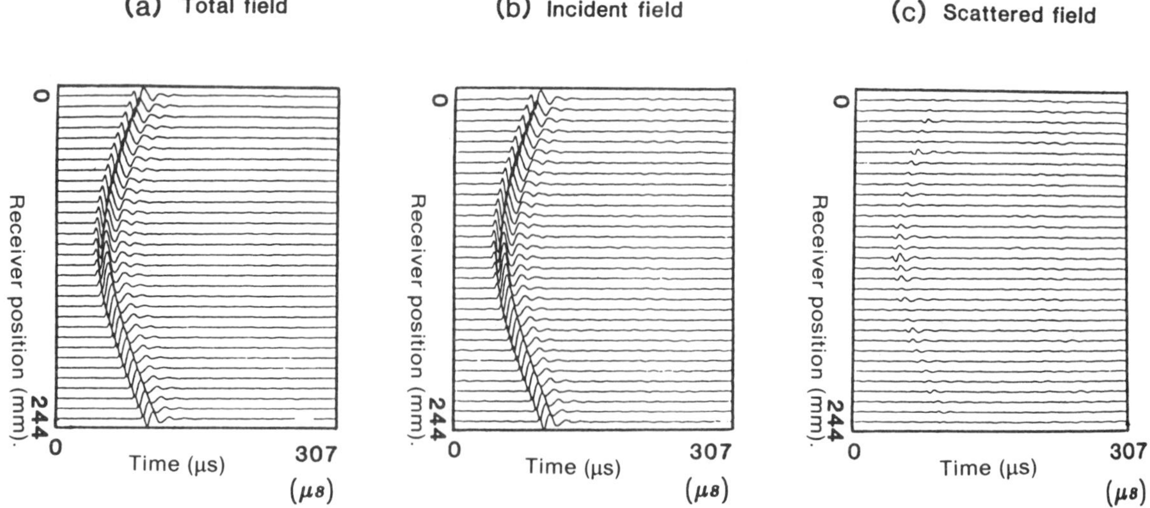


FIG. 12. Examples of waveforms recorded in the surface reflection experiment. (a) Total field waveforms. (b) Incident field waveforms. (c) Scattered field waveforms.

**Surface reflection tomography experiment**

The layout of this experiment is shown in Figure 3c. The source hydrophone scans along a source line, simulating 32 sources on the surface. The receiver hydrophone scans along another line parallel to the source line, simulating 32 receivers also on the surface. Figure 12 is an example of the waveforms recorded in this surface reflection experiment with the source hydrophone situated in the middle of the source line, where Figure 12a is the total field, Figure 12b is the incident field, and Figure 12c is the scattered field. Since the acoustic impedance contrast between the water and the gelatin cylinder is very small, the back-scattered wave field in this experiment is very weak. This reduces the signal-to-noise ratio and makes our reconstruction very noisy. Images reconstructed by the filtered backpropagation algorithm are shown in Figure 13. Figures 13a and 13b are the reconstructions at 30 kHz based on the Born and the Rytov approximations. Figures 13c and 13d are the reconstructions at 50 kHz based on the Born and the Rytov approximations. In this example, the Born approximation performs as well as the Rytov approximation. As discussed by Kaveh et al. (1981) and Wu and Aki (1985), the Born approximation performs well for back scattering. In the surface reflection experiment, the dominant forward scattering component which is disturbing for the Born approximation is not received by the receiver. The input to the reconstruction algorithm is the relatively weak back scattering component of the scattered wave field and this may be one of the reasons why the Born approximation works in this case.

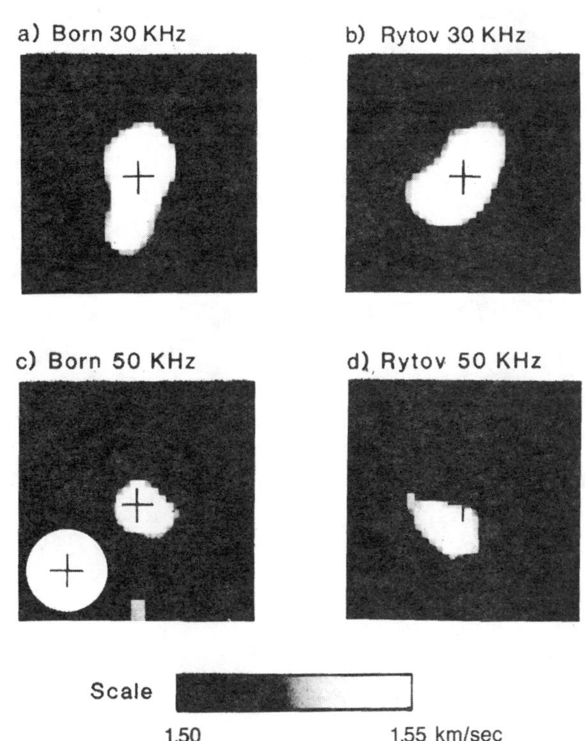


FIG. 13. Images reconstructed by the filtered backpropagation algorithm in the surface reflection experiment. The gelatin cylinder should be centered at the cross with the size and shape as shown by the white circle at the left lower corner of the figure. (a) Reconstruction with the Born approximation at 30 kHz. (b) Reconstruction with the Rytov approximation at 30 kHz. (c) Reconstruction with the Born approximation at 50 kHz. (d) Reconstruction with the Rytov approximation at 50 kHz.

## CONCLUSIONS

Both diffraction tomography and ray tomography can be used for subsurface imaging. These two methods perform differently depending on the source-receiver configuration and the size and properties of the object. When the source-receiver configuration is such that the illuminating waves are directly transmitted through only part of the object, such as in the VSP and surface reflection experiments, diffraction tomography is superior to ray tomography. If the object is uniformly illuminated, as is the case with the cross-borehole experiment, the size and the properties of the object determine the best reconstruction algorithm. In the cross-borehole configuration, if the size of the object is comparable to the wavelength, diffraction tomography is better than ray tomography. If the size of the object is much larger than the wavelength, ray tomography may perform as well as diffraction tomography. These conclusions are based on our laboratory setup where we can separate the scattered wave field by measuring the background field. In field applications, this arrangement may be possible only in enhanced recovery processing, fracturing, or other cases where "before" and "after" imaging can be made.

Two factors closely related to the fidelity of geophysical diffraction tomography are also examined in this paper: (1) source-receiver configuration and (2) approximation methods. Among the three source-receiver configurations tested in this study, the cross-borehole configuration gives the best result. Images reconstructed with the surface reflection configuration have very strong background noise. A VSP configuration images the quadrant of the object facing the source line and the receiver line better than the opposite quadrant of the object.

The Born and Rytov approximations are also compared based on our experimental data. In this paper, we discuss only reconstructed images using data whose wavelengths are 1/3 (50 kHz examples) or 1/1.8 (30 kHz examples) of the diameter of the object (gelatin cylinder). This makes it difficult to decide whether or not the Rytov approximation is superior to the Born approximation (Chernov, 1960; Kaveh et al., 1981; Slaney et al., 1984; Zapalowski et al., 1985) when the size of the weak inhomogeneity is much larger than the wavelength. Our experimental results suggest that for the cross-hole experiments, the Rytov approximation has a wider range of validity than the Born approximation. This is expected, since the Rytov approximation is better suited to the transmission (forward scattering) experiment than the Born approximation. For the VSP and surface reflection tomography experiments, no substantial difference between the Born and Rytov approximations is observed in this study.

## ACKNOWLEDGMENTS

This study is jointly supported by the Full Waveform Acoustic Logging Consortium and the Vertical Seismic Profiling-Reservoir Delineation Consortium at the Earth Resources Laboratory, Massachusetts Institute of Technology.

## REFERENCES


Anderson, A. H., and Kak, A. C., 1982, Digital ray tracing in two-dimensional refractive fields: J. Acoust. Soc. Am., **72**, 1593–1606.

Bishop, T. N., Bube, K. P., Cutler, R. T., Langan, R. T., Love, P. L., Resnick, J. R., Shuey, R. T., Spindler, D. A., and Wyld, H. W., 1985, Tomographic determination of velocity and depth in laterally varying media: Geophysics, **50**, 903–923.

Bonzanigo, F., 1978, An improvement of Tribolet's phase unwrapping algorithm: Inst. Electr. Electron. Eng. Trans., **ASSP-26**, 104–105.

Chernov, L. A., 1960, Wave propagation in a random medium: McGraw-Hill Book Co.

Chiu, S. K., Kanasewich, E. R., and Phadke, S., 1986, Three-dimensional determination of structure and velocity by seismic tomography: Geophysics, **51**, 1559–1571.

Cottin, J. F., Deletie, P., Francillon, H. J., Lakshmanan, J., Lemoine, Y., and Sanchez, M., 1986, Curved ray seismic tomography: application to the Grand Dam (Reunion Island): First Break, **4**, no. 2, 25–30.

Devaney, A. J., 1982, A filtered back propagation algorithm for diffraction tomography: Ultrasonic Imag., **4**, 336–350.

——— 1984, Geophysical diffraction tomography: Inst. Electr. Electron. Eng., Trans., **GE-22**, 3–13.

Dines, K. A., and Lytle, R. J., 1979, Computerized geophysical tomography: Proc. Inst. Electr. Electron. Eng., **67**, 1065–1073.

Esmersoy, C., 1986, The backpropagated field approach to multidimensional velocity inversion: Ph.D. thesis, Massachusetts Inst. Tech.

Gilbert, P., 1972, Iterative methods for the three-dimensional reconstruction of an object from projections: J. Theor. Biol., **36**, 105–117.

Gustavsson, M., Ivansson, S., Moren, P., and Pihl, J., 1986, Seismic borehole tomography—measurement system and field studies: Proc. Inst. Electr. Electron. Eng., **74**, 339–346.

Ivansson, S., 1985, A study of methods for tomographic velocity estimation in the presence of low-velocity zones: Geophysics, **50**, 969–988.

——— 1986, Seismic borehole tomography—theory and computational methods: Proc. Inst. Electr. Electron. Eng., **74**, 328–338.

Kak, A. C., 1985, Tomographic imaging with diffracting and non-diffracting sources, *in* Haykin, S., Ed., Array signal processing, 351–428.

Kaveh, M., Soumekh, M., and Muller, R. K., 1981, A comparison of Born and Rytov approximation in acoustic tomography, *in* Powers, J. P., Ed., Acoustical imaging, **11**, 325–335.

McMechan, G. A., 1983, Seismic tomography in boreholes: Geophys. J. Roy. Astr. Soc., **74**, 601–612.

Menke, W., 1984, The resolving power of cross-borehole tomography: Geophys. Res. Lett., **11**, 105–108.

Mohammad-Djafari, A., and Demoment, G., 1986, Maximum entropy diffraction tomography: Proc., Inst. Electr. Electron. Eng. Internat. Conf. on Acoustics, Speech and Signal Processing, **86**, Tokyo, 1749–1752.

Mueller, R. K., Kaveh, M., and Wade, G., 1979, Reconstructive tomography and applications to ultrasonics: Proc. Inst. Electr. Electron. Eng., **67**, 567–587.

Mueller, R. K., Kaveh, M., and Inverson, R. D., 1980, A new approach to acoustic tomography using diffraction techniques, *in* Metherell, A. F., Ed., Acoustical imaging, **8**, 615–628.

Pan, S. X., and Kak, A. C., 1983, A computational study of reconstruction algorithms for diffraction tomography: interpolation versus filtered backpropagation: Inst. Electr. Electron. Eng., Trans., **ASSP-31**, 1262–1275.

Peterson, J. E., Paulsson, B. N. P., and McEvilly, T. V., 1985, Applications of algebraic reconstruction techniques to crosshole seismic data: Geophysics, **50**, 1566–1580.

Ramirez, A. L., 1986, Recent experiments using geophysical tomography in fractured granite: Proc. Inst. Electr. Electron. Eng., **74**, 347–452.

Slaney, M., Kak, A. C., and Larsen, L., 1984, Limitations of imaging with first-order diffraction tomography: Inst. Electr. Electron. Eng. Trans., **MTT-32**, 860–874.

Tanabe, K., 1971, Projection method for solving a singular system of linear equations and its applications: Numer. Math., **17**, 203–214.

Tribolet, J. M., 1977, A new phase unwrapping algorithm: Inst. Electr. Electron. Eng. Trans., **ASSP-25**, 170–177.

Wu, R. S., and Aki, K., 1985, Scattering characteristics of elastic waves by an elastic heterogeneity: Geophysics, **50**, 582–595.

Wu, R. S., and Toksöz, M. N., 1987, Diffraction tomography and multisource holography applied to seismic imaging: Geophysics, **52**, 11–25.

Zapalowski, L., Leeman, S., and Fiddy, M. A., 1985, Image reconstruction fidelity using the Born and Rytov approximations, *in* Berkhout, A. J., Ridder, J., and van der Wal, L. F., Eds., Acoustical Imaging, **14**, 295–304.

# Anisotropic Wave Propagation and Zero-offset Migration

**N. F. Uren**
*Curtin University of Technology*
*Western Australia*

**G. H. F. Gardner and J. A. McDonald**
*University of Houston*
*U.S.A.*

## Abstract

In reflection seismology, migration may be defined as the transformation of apparent reflector positions to their true positions. In practice this usually means that diffractions are collapsed and reflector dip angles are changed. When a medium is anisotropic, and the axes of velocity symmetry are tilted with respect to the horizontal, horizontal reflection events migrate laterally as well.

The lateral migration of reflections from horizontal layers was studied by constructing an anisotropic scale model representing a medium in which the bedding is inclined to the surface. This situation may be classified as one of transverse isotropy with a tilted axis of symmetry.

The elastic parameters of the anisotropic model were recovered by P- and SH-wave transmission measurements carried out to simulate a walk-away VSP. Numerical modelling of a P-wave CMP gather above a horizontal reflector in such a medium indicated that there would be an asymmetrical distribution of reflection points. The zero-offset reflection point was displaced laterally by a distance equal to more than 20% of the depth, and the spread of reflection points was 15% of the depth. The NMO velocity was found to be a function of offset.

A zero-offset reflection survey was carried out on the model. The P-wave data were migrated with an anisotropic migration algorithm which may be applied to wavefronts of any shape. On the resulting section, the horizontal reflection features are moved laterally by a distance equal to 20% of the depth back to their correct positions.

The lateral displacement of reflections from a horizontal reflector beneath a medium with tilted anisotropic velocity characteristics has significance for the selection of drill locations when such a situation is encountered in the field.

Key words: anisotropic, migration, modelling

## Introduction

In an elastically anisotropic medium, seismic wave velocity is a function of direction. In rocks, anisotropy is often caused by inhomogeneities such as fine layering, crystal structure, grain orientation and fractures. Stress can also cause anisotropy. This paper assumes that the inhomogeneities causing anisotropy are much smaller than the wavelength of the seismic waves involved.

Hooke's Law is normally the starting point for the development of anisotropic wave equations.

$$\sigma_{ij} = C_{ijkl}\epsilon_{kl}, \tag{1}$$

where $\sigma_{ij}$ and $\epsilon_{kl}$ are stress and strain components, and $C_{ijkl}$ is the elastic modulus tensor. A reduced notation compresses the subscripts *ij* to $\alpha$ and *kl* to $\beta$, as seen in Thomsen (1986). Transverse isotropy is the simplest form of anisotropy where there are five independent elastic constants. Wave velocity in a transversely isotropic medium depends only on the angle from a single symmetry axis direction. In sedimentary rocks, the symmetry axis is normal to the bedding planes.

In anisotropic media, rays are usually not normal to the wavefronts, and two velocities must be used for any one wavefront. The *ray velocity* (also called group velocity) is the ratio of the distance travelled to the time taken by a wave starting from one point in a medium and arriving at a second. This is the velocity which one obtains from travel time measurements. The *phase velocity* (plane wave velocity) is the rate of progression of a wave in a direction normal to its wavefront, and this is the velocity which one normally derives theoretically from the anisotropic form of Hooke's law.

Thomsen (1986) gives explicit expressions for plane compressional and shear wave velocities in transversely isotropic media in terms of five elastic constants which are physically more meaningful than the constants of the elastic matrix $C_{\alpha\beta}$.

## Physical Modelling

A very convenient way to demonstrate anisotropy of the type found in dipping sediments is to use physical modelling. A series of scale model experiments was conducted at the Seismic Acoustics Laboratory of the University of Houston. Phenolite, a paper-reinforced plastic, was chosen as the anisotropic medium. The random orientation of wood fibres in paper, combined with the layering effect of many thin sheets, produces transverse isotropy with the symmetry axis normal to the laminations. The sample used had approximately one hundred sheets of paper per inch (2.5 cm) of thickness. Since the wavelengths used in the modelling system were of the order of three millimetres, the long wavelength approximation was satisfied. On this scale, the phenolite was effectively homogeneous to the seismic waves used in the experiments.

Reprinted from Exploration Geophysics, 22, 405-410 with permission from Australian Society of Exploration Geophysicists

A sheet of phenolite was machined and glued as shown in Figure 1. The individual sheets of paper in the phenolite were not visible to the eye, but they were parallel to the glued joints. The upper and lower surfaces are parallel and 2.5 inches (6.4 cm) apart. The distance scale factor used was 1.0 inch equals 1000 feet (1:12,000), so the model represented an anisotropic layer 2500 feet (768 metres) thick. To enable direct comparisons with field situations, distances and velocities will be quoted in full size, rather than in model dimensions.

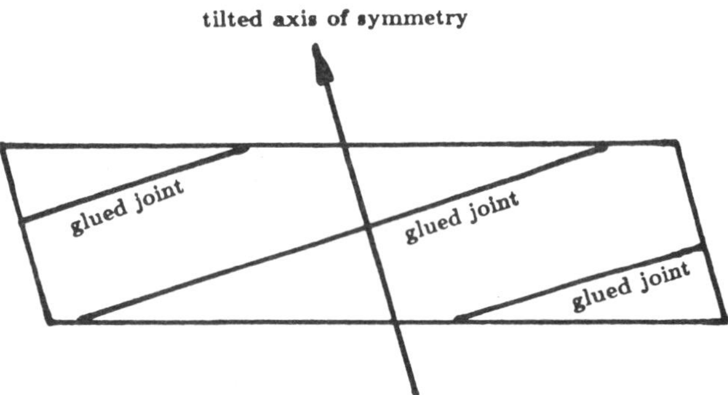


**FIGURE 1**
**Phenolite dipping-layer model.**
**Paper laminations run parallel to the glued joints. Thus the axis of symmetry is tilted with respect to the upper surface of the model. The lower surface represents a horizontal reflector which is 2,500 feet below the upper surface.**

For these reasons, when scale model reflection profiling was carried out subsequently, zero-offset data only were selected for migration. A method for migrating anisotropic zero-offset data was available (Uren, 1989).

TABLE 1

EXPERIMENTALLY DETERMINED ELASTIC PARAMETERS

$\alpha_0$ = 21,100 feet/sec (6,430 m/s)

$\beta_0$ = 10,500 feet/sec (3,200 m/s)

$\epsilon = 0.060$

$\gamma = 0.060$

$\delta^* = 0.25$

Dip angle = $14^0$

After Thomsen (1986)

## Velocity Function Determination

After the dipping-layer model had been constructed, its velocity characteristics were determined by transmission tests similar to what might be achieved with a walk-away VSP. A piezoelectric transducer was placed beneath the model, and a piezoelectric receiver was traversed across the top of the model to record P- and SH-wave travel times at the range of raypath directions permitted by the dimensions of the model.

P- and SH-wave velocities computed from the ratio of raypath lengths to travel times were plotted on polar plots as shown in Figure 2. Thomsen's exact equations (Thomsen, 1986) were used to determine five elastic constants, $\alpha_0$, $\beta_0$, $\epsilon$, $\gamma$ and $\delta^*$ by curve fitting these theoretical relationships to the observed velocity functions. A summary is given in Table 1.

## Numerical Modelling

P-wave raypaths for a common midpoint gather over a numerical model representative of the phenolite model were computed using the experimentally determined elastic parameters. Fermat's principle of least time was used to define the raypaths in the anisotropic medium. As can be seen in Figure 3, offsets range from zero to 8,000 feet (0 to 2428 m). Reflection points are spread over a range of 350 feet (107 m), while the zero offset reflection occurs 550 feet (168 m) to one side of the point on the reflector vertically below the midpoint position. The lateral displacement is more than 20% of the depth, and the spread of reflection points is about 15% of the depth. The moveout curve was found to be non-hyperbolic.

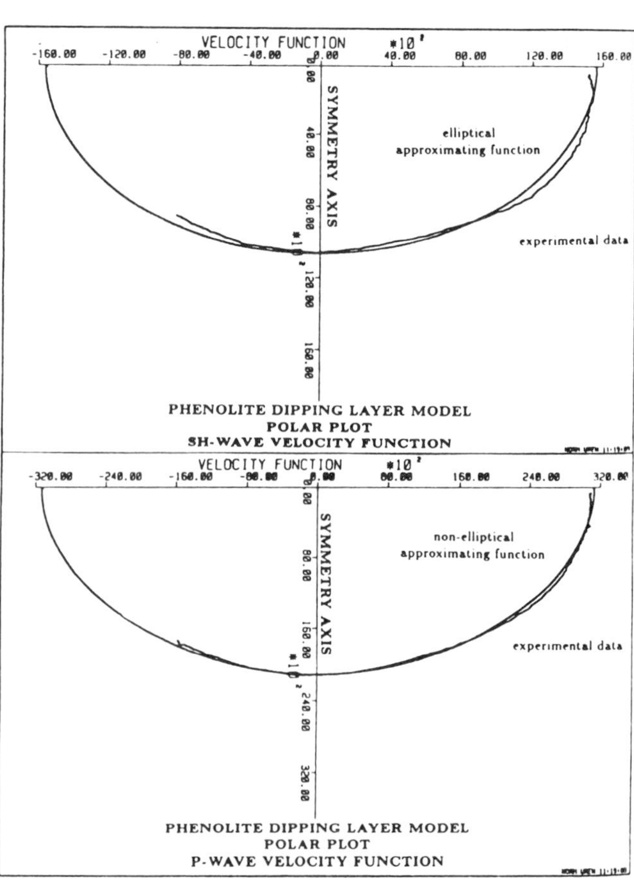


**FIGURE 2**
**Polar plot of P and SH ray velocities.**
**The best fitting theoretical curves (assuming transverse isotropy) are plotted for comparison with the experimentally determined velocity functions.**

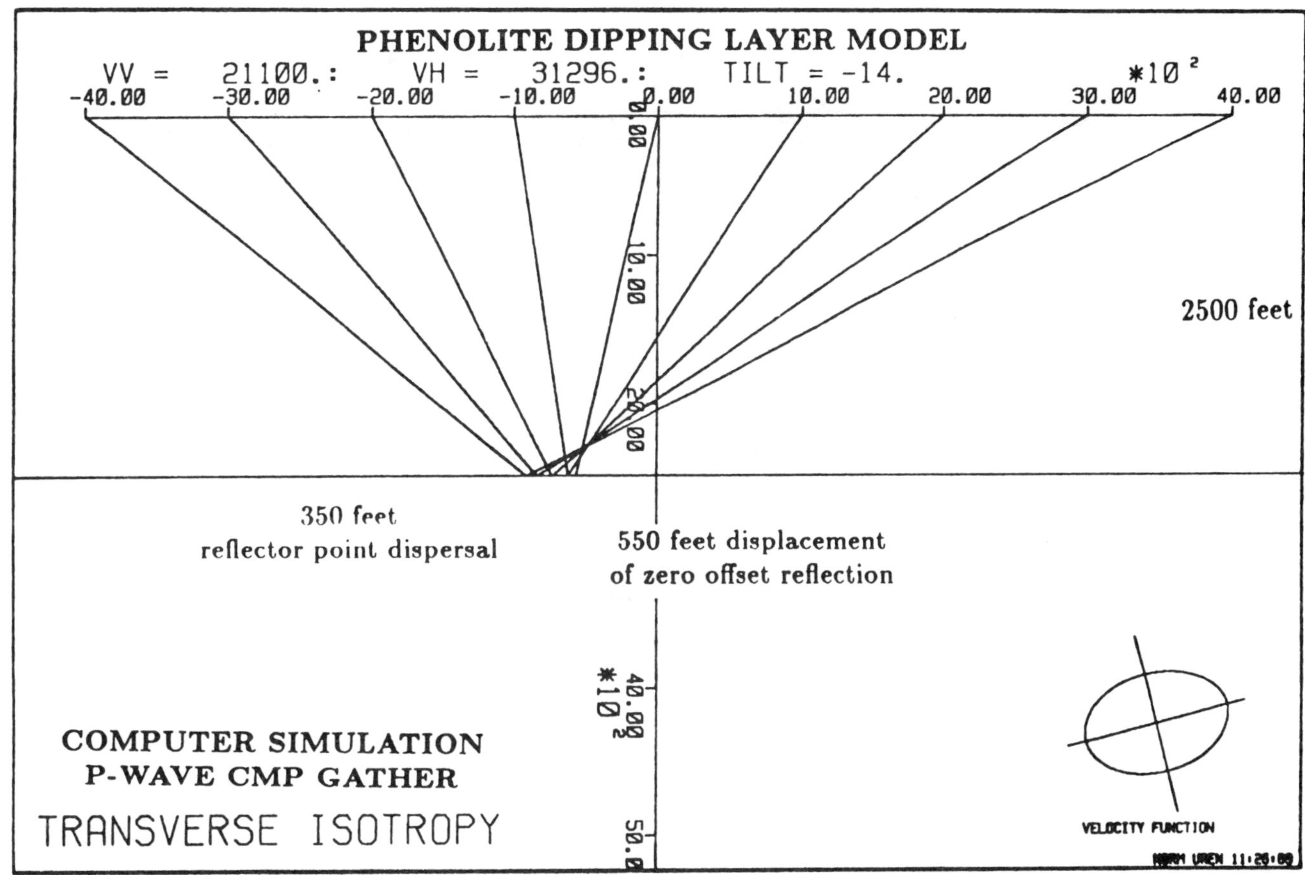


**FIGURE 3**
**Numerical modelling of P-wave raypaths in the phenolite model.**
**The experimental P-wave velocity function was used together with Fermat's principle to define the raypaths. The lateral displacement of the zero-offset reflection is more than 20% of the depth, and the spread of reflection points is about 15%.**

## Anisotropic Zero Offset Migration

An anisotropic form of the migrator's equation may be derived using the exploding reflector model (Uren *et al.,* 1989a). The anisotropic dispersion relation may be written

$$K^2 + L^2 = \frac{4}{v(\alpha)^2}.F^2, \quad (2)$$

where K and L are the horizontal and vertical wavenumbers, F is the frequency, and the anisotropic phase velocity $v(\alpha)$ is a function of *(K/L).*

The expression for the migrated anisotropic section *w(x,z)* is then

$$w(x,z) = \int\int E(K,L).\frac{dF}{dL}.e^{i2\pi(Kx+Lz)}dK.dL, \quad (3)$$

where *E(K,L)* is derived from the Fourier transform of the original time section.

## Reflection Profiling

A rectangular slot was cut into the base of the model, and a zero-offset P-wave reflection profile was recorded at 100 feet (30.5 m) station intervals. The purpose of the slot was to reveal the lateral displacements of zero offset reflections on a horizontal reflector, as predicted by numerical modelling. In order to achieve the zero-offset configuration, the same transducer was used as both source and receiver (Uren *et al.,* 1989b).

Figure 4 shows the unmigrated section in which the glued joints are clearly visible. Each trace is a ten-fold vertical stack. No other processing was used. High amplitude transducer noise was recorded at early times, but this was ignored as being irrelevant to the purpose of the experiment. The slotted section at the base of the model is clearly seen. The diffractions on the right-hand side of the slot are more prominent than those on the left. This is assumed to be due to the oblique nature of the reflection raypaths. An outline of the actual reflecting surface is also shown in Figure 4 at the same scale. The unmigrated section is laterally displaced from the true position of the reflecting surface, although the horizontal reflectors are at their correct depths.

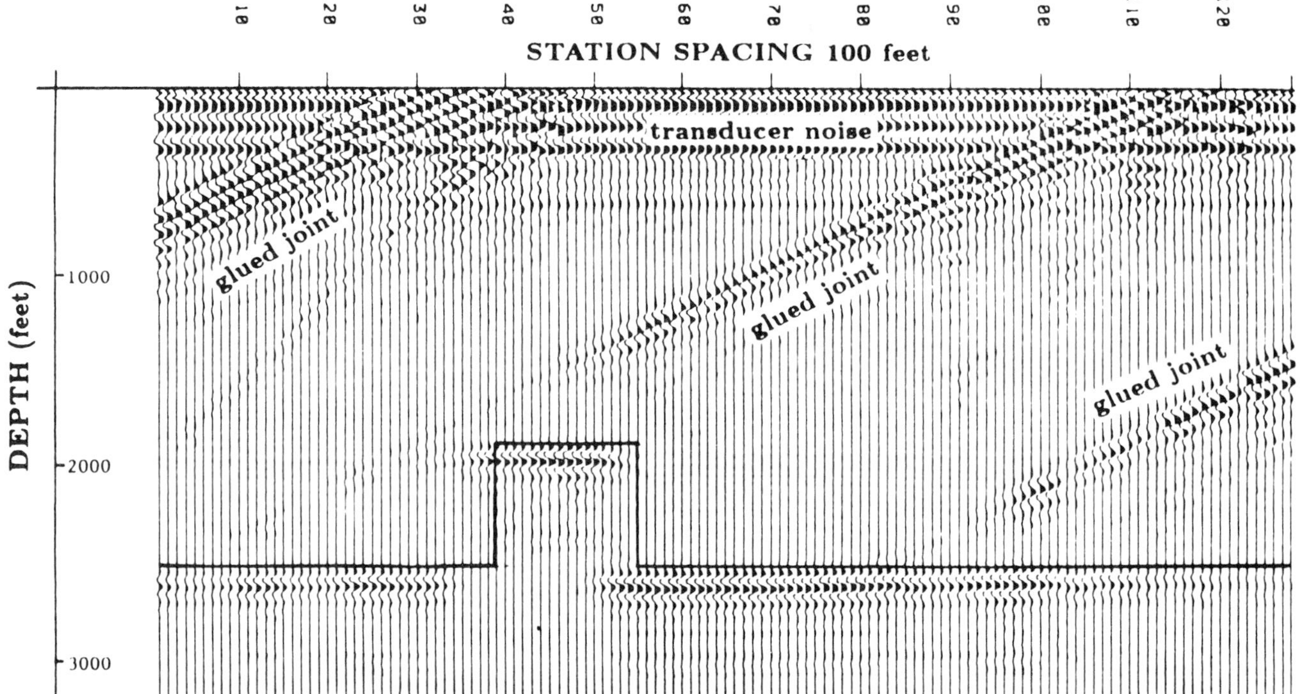


FIGURE 4
Comparison of unmigrated and true sections.
A line diagram of the actual cross section overlies the unmigrated depth section. The lateral displacement of 500 feet agrees with the theoretically predicted displacement within the limit of one trace spacing.

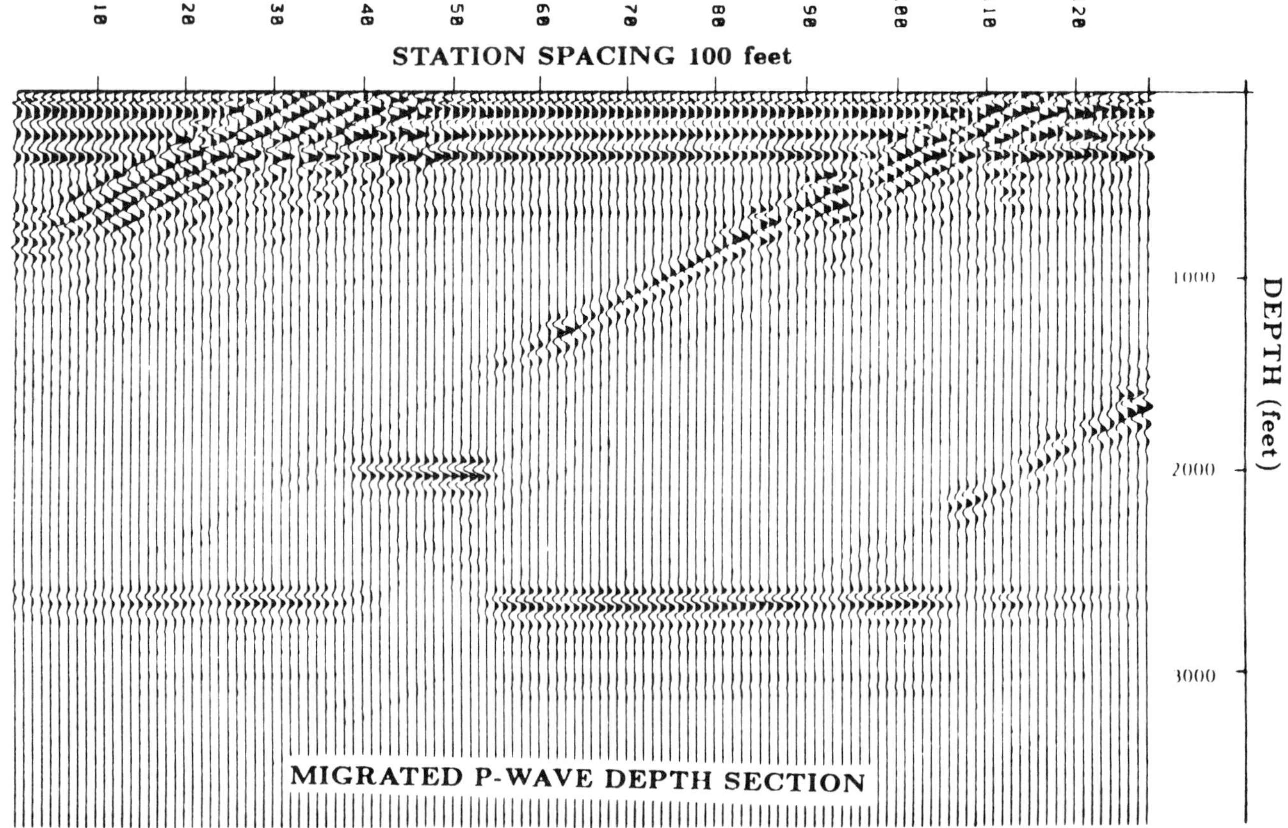


FIGURE 5
Migrated P-wave depth section.
An anisotropic F-K domain algorithm was used. Diffractions are collapsed, but horizontal reflectors do not seem to be affected. The image of the slotted version is migrated back to its correct position.

The data were then migrated using anisotropic F-K migration (Uren *et al.*, 1989a), and the result is shown in Figure 5. A close comparison of Figures 4 and 5 shows that the slotted section has migrated laterally by five traces (500 feet, or 152 m). The correct position of the reflector is 550 feet (168 m) to the left of the unmigrated image. This is considered to be verification of the predicted lateral migration within the limits of accuracy of one station spacing. A dim area to the left of the slotted reflector may be seen on the migrated section in Figure 5. This is due to the lack of illumination of the reflector in that region by obliquely travelling rays. Curved events appear at the base of the slotted section. These are presumed due to the velocity anomaly caused by the air in the slotted part of the model.

Anisotropic P-wave migration of physical model data shifted reflection events by the amounts predicted by numerical modelling of the experimental situation.

## Conclusions

1. Sufficient travel time information was obtained from a simulated walk-away VSP to recover the velocity functions and the five elastic parameters which characterize an anisotropic layer above a reflecting surface in a physical model.
2. It was predicted numerically that P-wave reflections would not occur directly below the common midpoint above a horizontal reflector in a medium which had tilted transverse isotropy.
3. The correct lateral migration of recorded P-wave reflection features from a horizontal reflector was achieved using anisotropic migration.
4. The lateral distances through which reflection features were moved by the anisotropic migration process were an appreciable proportion of the depth to the reflector. This suggests that considerable drill location error could result if anisotropy were ignored when a target is overlain by dipping anisotropic beds.

## Acknowledgements

The authors wish to acknowledge the use of the data acquisition and processing facilities at Allied Geophysical Laboratories at the University of Houston, and the assistance given by the manager of the Seismic Accoustics Laboratory, Dr. K.K. Sekharan.

## References

Thomsen, L., (1986). 'Weak elastic anisotropy'. *Geophysics,* **51**, 1954–1966.

Uren, N. F., (1989). 'Processing of seismic data in the presence of elliptical anisotropy'. Ph.D. thesis, Department of Geosciences, University of Houston, University Park, Houston, Texas.

Uren, N. F., Gardner, G. H. F., and McDonald, J. A., (1989a). 'The anisotropic migrator's equation'. 59th Ann. Internat. Mtg., Soc. Expl. Geophys., Expanded Abstracts, 1184–1186.

Uren, N. F., Gardner, G. H. F., and McDonald, J. A., (1989b). 'Zero offset seismic migration surveys using an anisotropic physical model'. 59th Ann. Internat. Mtg., Soc. Expl. Geophys., Expanded Abstracts, 1044–1046.

# Chapter 5 Wave Propagation In Complex Media

For simple geometries, analytic solutions may be found. For more complicated systems, experiments (numerical or physical) must be performed; and the results, interpreted. The term interpretation is appropriate because, while the goal is often a general insight into wave behavior, the experiment actually performed is just one particular realization of an almost infinite number of possible experiments that could have been done instead.

The results of the experiment apply only to that particular set of experimental conditions. It is in the mind of the interpreter that the general qualities of the experiment are ascertained.

+ Levin and Robinson (1969) attached randomly placed scatterers to the surface of a homogenous block of concrete. The scattered wavefield was interpreted in terms of the correlation distance, i.e., how far apart two detectors can be before knowledge of the wavefield at one detector tells us nothing about the wavefield at the other detector. An example of a real-world situation similar to this is the scattering of Rayleigh waves by near-surface solution features in West Texas.

+ Plona (1980) constructed a synthetic porous rock from sintered glass beads and observed a slow compressional wave predicted by Biot's theory of wave propagation in porous media. This paper has some possible applications both to well logging and the seismoelectric effect.

+ McDonald et al. (1981) constructed a model of a reef to investigate the effect of 3-D spatial sampling on reservoir delineation. This paper was directly inspired by the Michigan reef plays popular in the late 1970s and early 1980s.

+ Tatham et al. (1983) performed a simulated marine shear-wave seismic survey by exploiting double-mode conversions. This model was inspired by the hard carbonate sea-bottom of offshore western Florida. The experiment convinced Geosource's management to perform an actual double-mode conversion survey, whose results were published the next year (Geophysics, 1984, **49**, 493-508).

+ Purnell (1992) has found one way to get around "no-data" zones as a result of high-impedance layers. He imaged targets by acquiring the rays which correspond to converted shear-waves in the high-impedance layer, but are *P*-waves above and below that layer. This is relevant to imaging beneath massive carbonates, evaporites, or basalt flows.

# SCATTERING BY A RANDOM FIELD OF SURFACE SCATTERERS†

FRANKLYN K. LEVIN* AND DONALD J. ROBINSON*

A study of elastic wave scattering by a random distribution of surface scatterers has been carried out with a three-dimensional seismic model. The model consisted of 143 round brass rods cemented to the surface of a large concrete slab. By recording the signals from each detector position with and without scatterers present, we looked at the scattered energy in the absence of the direct signal. We found that (1) the separation for which signals from two detectors ceased to resemble each other (coherence distance) was about 0.3 to 0.4 wavelength of predominant coherent noise, (2) the coherence distance is independent of scatterer size for scatterers with diameters in the range 0.05 to 0.4 wavelength, (3) the amount of scattered energy increases with scatterer size, and (4) all components of motion appear in the scattered energy with about the same average amplitude.

## INTRODUCTION

Since the invention of two-dimensional seismic models by Oliver, Press, and Ewing (1954), most model investigations have been made with thin plates. Two-dimensional models are inexpensive and easy to assemble, to handle, and to change. However, they cannot be used when the problem being studied inherently requires three dimensions. Scattering problems are of this type. In a two-dimensional model, scatterers are infinitely long cylinders; to simulate geology likely to be encountered in the field, scatterers should have a finite size. Consequently, studies of scattering require a three-dimensional model.

There have been few previous model studies of elastic wave scattering. In his note on the nature of a seismogram, Tatel (1954) remarked that when the surface of a small steel block was punctured with many small holes, the resulting model seismogram showed several oscillations between the dilatational and Rayleigh wave arrivals. A number of experimentalists have looked at individual scatterers of various shapes (Miller et al 1967), but none at systems involving a large number of scatterers. In view of the absence of theoretical guidance, their reluctance to undertake the experiments was probably wise. Nonetheless, the field geophysicist rarely sees waves returning from individual obstacles. He operates in areas where boulders, solution cavities, and variations in local geology surround him. Consequently, we decided to examine a system that included a large number of scatterers. We realized that our results were likely to have inexplicable features but believed that, as a start, empirical information was better than no information at all.

## THE THREE-DIMENSIONAL SEISMIC MODEL

Our model consisted of a three-ft by three-ft by nine-inch slab of concrete over a three-ft by three-ft by one-inch slab of marble (Figure 1). Constructing three-dimensional models that are relatively homogeneous and free of cracks is notoriously difficult. In our case, the concrete was a mixture of three parts of cement with one part of riverbed sand. The sand was screened to ensure a grain diameter of less than one-tenth of the shortest wavelength having appreciable amplitude in the incident pulses. To avoid cracking when the concrete set, a special cement developed for dams was used and the slab was immersed in water for several weeks after it had been poured. Homogeneity resulted from the mixture's being vibrated until bubbles ceased to rise. Figure 2, a record

† Presented at the 37th Annual International Meeting of the Society of Exploration Geophysicists, October 31, 1967. Manuscript received by the Editor November 30, 1967.

* Esso Production Research Company, Houston, Texas.

Reprinted from Geophysics, 34. 170-179

FIG. 1. The three-dimensional seismic model with scatterers in place.

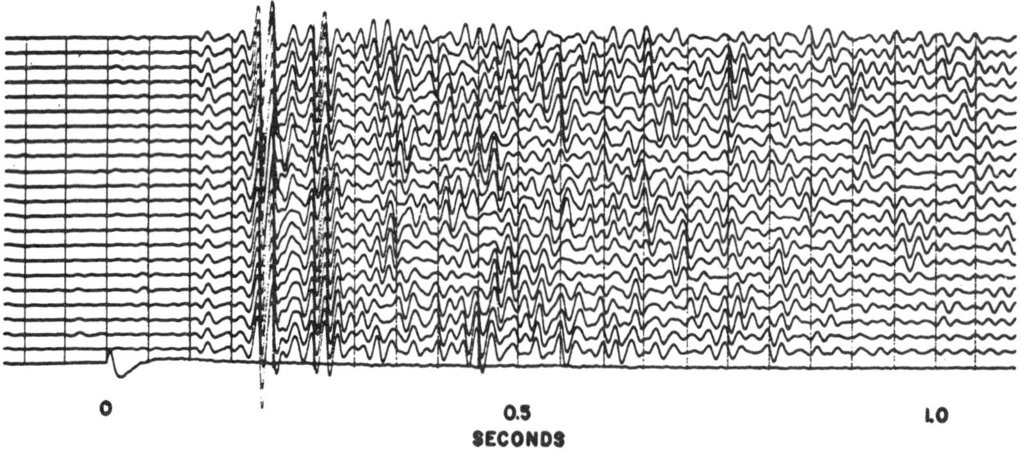


FIG. 2. Signals recorded by moving a detector sensitive to vertical motion on an arc of a circle centered at the source. The radius of the circle was 15 inches and a thousandfold time reduction was used in recording.

section made with a detector at successive positions on an arc of circle around the source, illustrates the degree of homogeneity achieved.

The scatterers were $\frac{1}{2}$-inch long round brass rods cut from $\frac{1}{8}$-, $\frac{1}{4}$-, $\frac{1}{2}$-, $\frac{3}{4}$-, and 1-inch diameter brass stock. We cemented the rods to the concrete surface with phenyl salicylate (salol), a crystalline material that melts at 43°C. To ensure that the scatterers were at random positions, 143 pairs of coordinates were selected from a table of random numbers. The resulting pattern is shown in Figure 3. An observer might object to the obvious holes in the pattern; however, the temptation to fill them had to be resisted. Only if we had repeated our experiments with a large number of patterns, each selected as the one we actually used, could we have been certain our results were statistically valid for a random distribution of surface scatterers. Because the model equipment was needed in other research programs, the added experiments were not performed. In a final step, scatterer coordinates on the model were determined by reducing the pattern of Figure 3 to a 15-inch by 15-inch square.

Elastic pulses were introduced into the model with a one-inch diameter PZT4 disk cemented to the concrete surface. To reduce surface waves, the source holder was surrounded with modeling clay

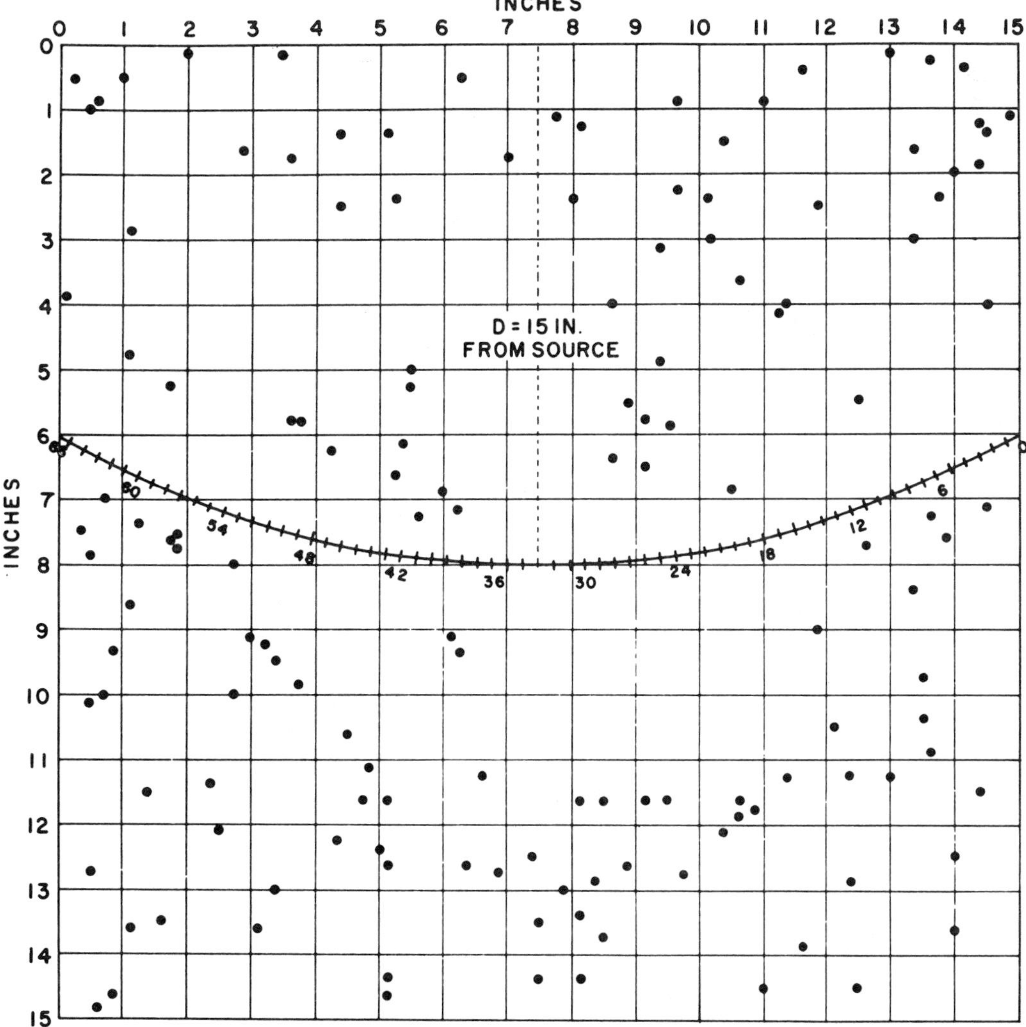


FIG. 3. Positions of scatterers used in the model study.

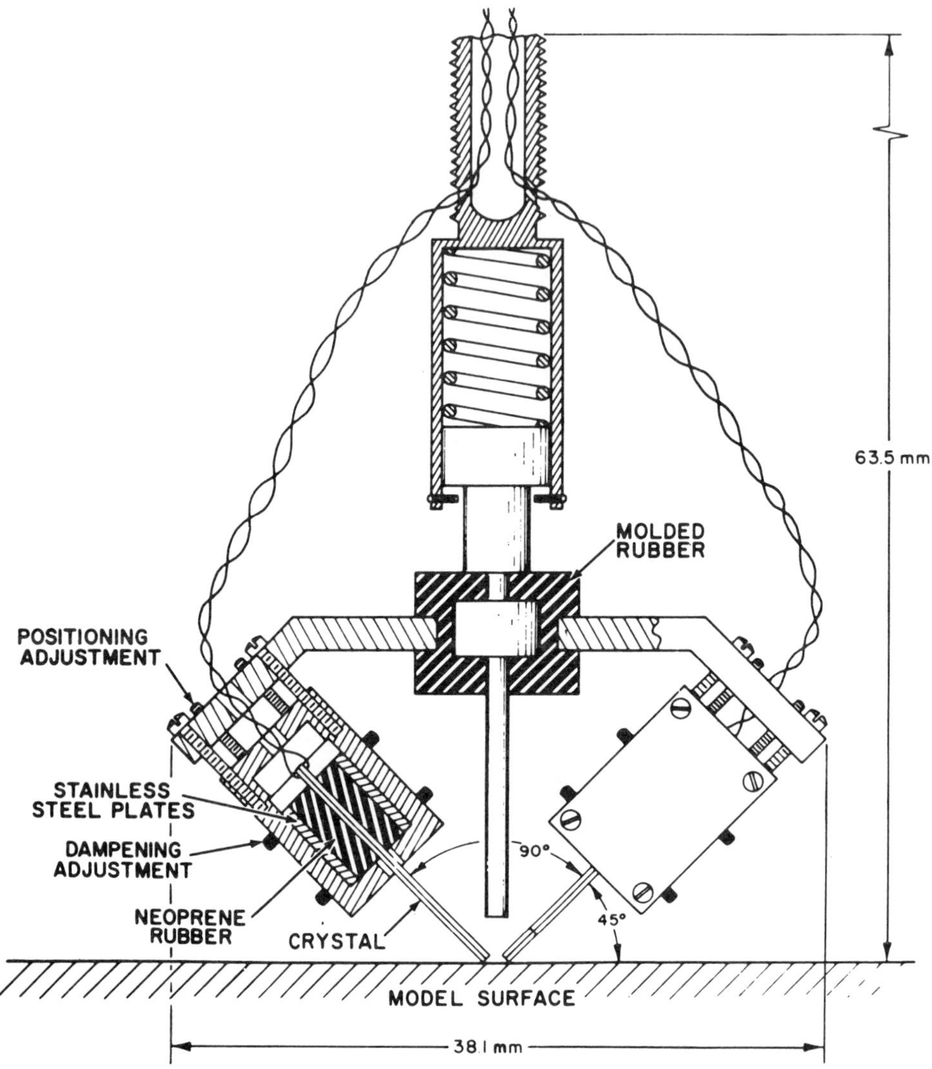


FIG. 4. The bidirectional detector.

pressed against the concrete, and the entire assembly of source and clay was kept under silicone oil.

Theoretical studies of Dunkin (1966) indicated that to characterize scattering processes completely, both horizontal and vertical components of motion should be measured. At the time our model investigations started, no detector with sensitivity and frequency response identical for horizontal and vertical motion was known. We developed a detector that had identical vertical and horizontal responses, but it proved to be too insensitive for our work. Tests with a single one-inch diameter rod showed that the unit (called the bidirectional detector) could pick up scattered events near the rod, but that the far field signals were smaller than the system noise. Figure 4 is a schematic drawing of the bidirectional detector. Its characteristics have been described by Robinson (1967).

When the bidirectional detector was found to lack the sensitivity required for scattering investigations, we fell back on more conventional seismic model detectors. Horizontal motion was measured with a bender bimorph; vertical motion, with a thickness expander crystal. Bender bi-

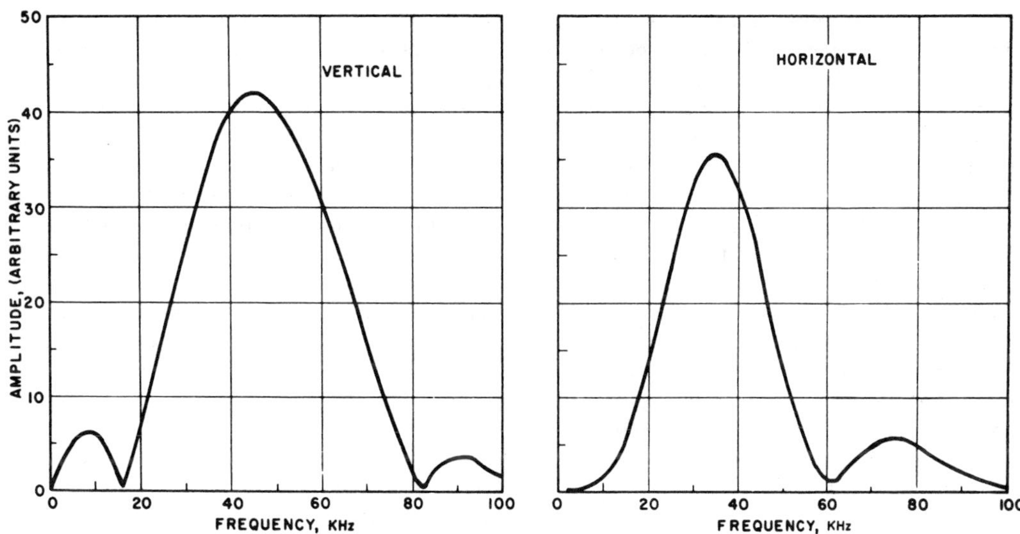


FIG. 5. Spectra of the direct surface wave as detected by the vertical and horizontal motion transducers.

morphs respond best to low frequencies (<50 kHz), while the thickness expander we used had appreciable response to 70 kHz. As best we could tell, the bulk of the scattered energy consisted of Rayleigh waves. Figure 5 shows the spectrum of the direct Rayleigh wave as detected by the vertical and horizontal motion detectors. The vertical motion spectrum is peaked around 45 kHz; the horizontal motion spectrum, at 35 kHz. Spectral responses of the two types of detectors were not simply related, and we cannot be more quantitative.

In an attempt to eliminate the effect of vertical and horizontal transducers with different frequency responses, the outputs of the detectors were passed through a 20 to 30 kHz filter. Although the resulting spectra still did not look much alike (Figure 6), the peak of the vertical

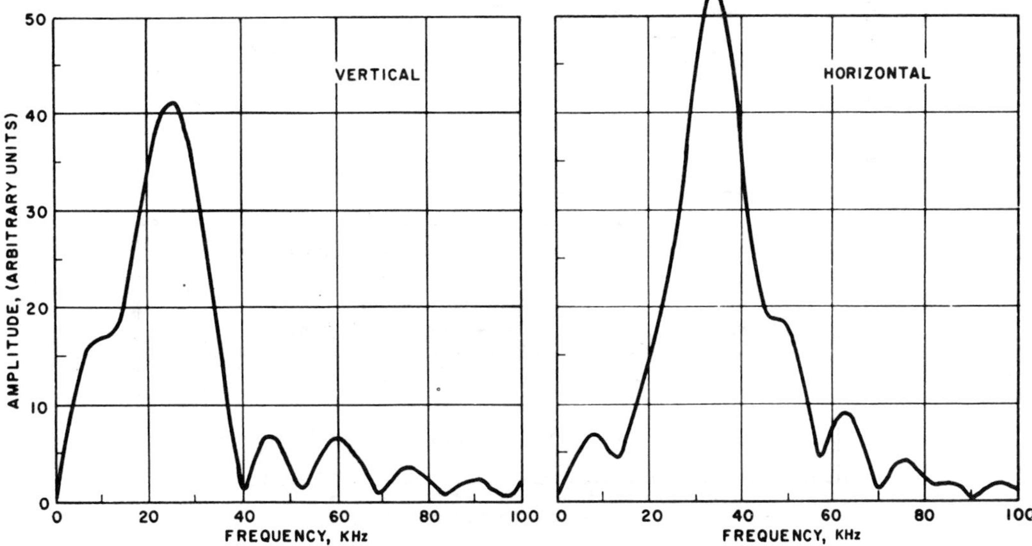


FIG. 6. Same as Figure 5 but with the signals passed through a 20-30 kHz band-pass filter.

motion spectrum shifted from 45 kHz to 25 kHz. The peak of the horizontal motion spectrum did not change appreciably. In retrospect, we believe that instead of trying to relate signals from two such different detectors we should have concentrated on improving the bidirectional detector. Amplification at the individual benders of the unit probably would have given us electrical signals strong enough to override the noise.

The model circuitry included a frequency translator that permitted us to record signals on magnetic tape. The translator, which divides times by a factor of one thousand, was used for all quantitative work. Thus, the abscissas of Figures 5 and 6 are given in the kHz of the actual detected pulses, but the spectra were computed for the time-stretched signals from the translator.

Scattered energy leaves a scatterer in all directions; consequently, even under ideal conditions, the amplitude of a scattered wave is small compared with that of an incident wave. Detecting the scattered energy in the presence of the large incident arrivals is difficult. Our solution to this problem was facilitated by our ability to record signals from the model on magnetic tape. Signals from each detector position were recorded twice: once with no scatterers on the model, once with the scatterers cemented in place. Scatterer-absent and scatterer-present signals were fed into the differencing circuit, and the difference was recorded. The result was a tape representing the effect of the scatterers alone.

Unfortunately, the largest signals on the difference tape resulted from the incident wave's loss of energy: the scattered events we wished to study generally had lower amplitudes. In an attempt to

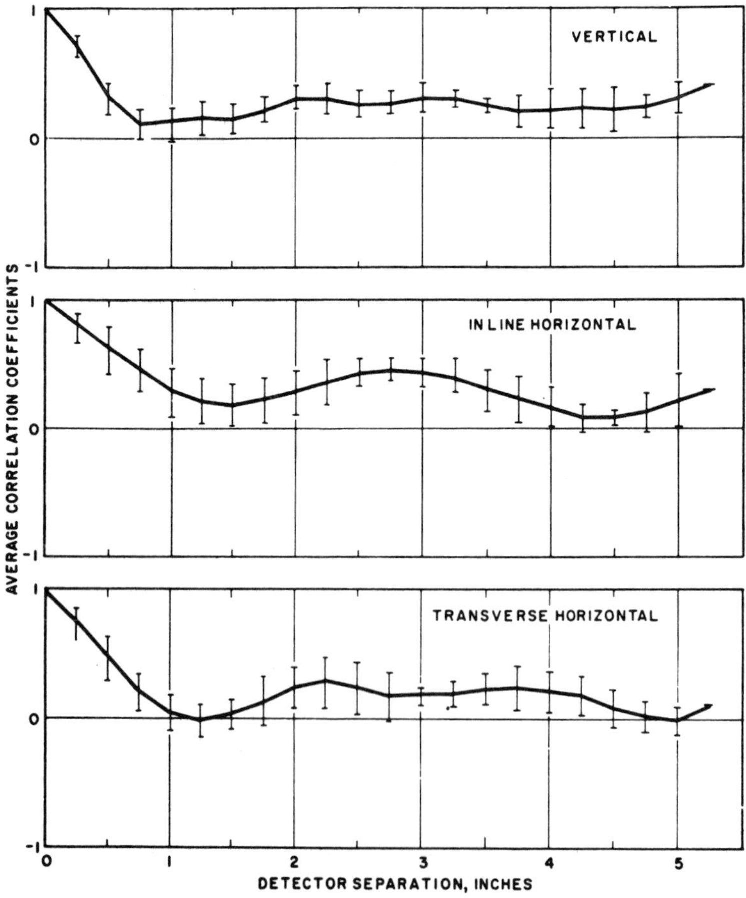


FIG. 7. Correlation coefficient versus detector separation for a distribution of $\frac{1}{8}$-inch diameter rods. The bars indicate standard deviations.

avoid or at least minimize this difficulty, we moved the detector along an arc of a circle centered on the source. The circle had a radius of 15 inches and passed through the square that contained the scatterers (Figure 3). We hoped that the contribution from the incident wave would be common to all positions on the arc so that when the signals from different positions were crosscorrelated, at worst we would get a dc shift in the correlation curve. To a rather good approximation this seemed to be so.

### MODEL EXPERIMENTS

Several different types of experiments were carried out with this seismic model. In some, we simply measured the amplitude of the scattered energy as the size of the scatterers changed, but the distribution of scatterers was maintained. However, most of the experiments involved the determination of a quantity we have called the correlation coefficient.

If $f_j(t)$ is the signal from a detector at position $\mathbf{x}_j$ and $f_k(t)$, the signal from a detector at $\mathbf{x}_k$, we define the correlation coefficient as

$$\rho_{jk} = \int_{-T/2}^{T/2} f_j(t) f_k(t) dt \Bigg/ \left\{ \int_{-T/2}^{T/2} f_j^2(t) dt \int_{-T/2}^{T/2} f_k^2(t) dt \right\}^{1/2} . \quad (1)$$

In the absence of coherent signals, as $|\mathbf{x}_j - \mathbf{x}_k|$ increases, $\rho_{jk}$ decreases from unity, when $|\mathbf{x}_j - \mathbf{x}_k| = 0$, to zero when $f_j(t)$ and $f_k(t)$ become uncorrelated. The separation $|\mathbf{x}_j - \mathbf{x}_k|$ at which $\rho_{jk}$ becomes zero we have called the coherence distance. Our signals were mixtures of coherent and random noise. While we could not separate the two types of noise, by selecting detector positions on an arc of a circle centered on the source, we ensured that the coherent arrivals were common to all detectors. In this case, the correlation coefficient for

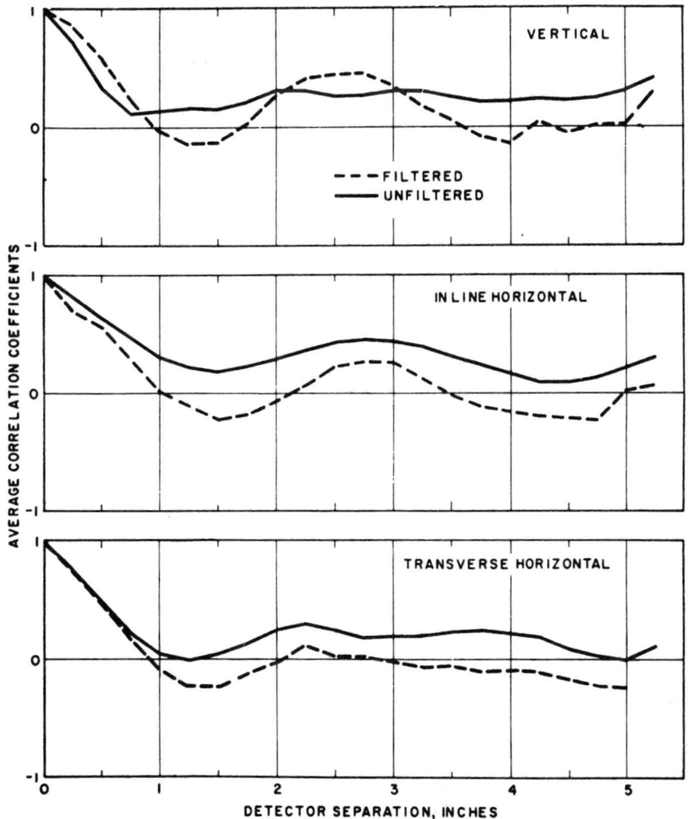


FIG. 8. Same as Figure 7 but for signals both unfiltered and filtered through a 20–30 kHz filter.

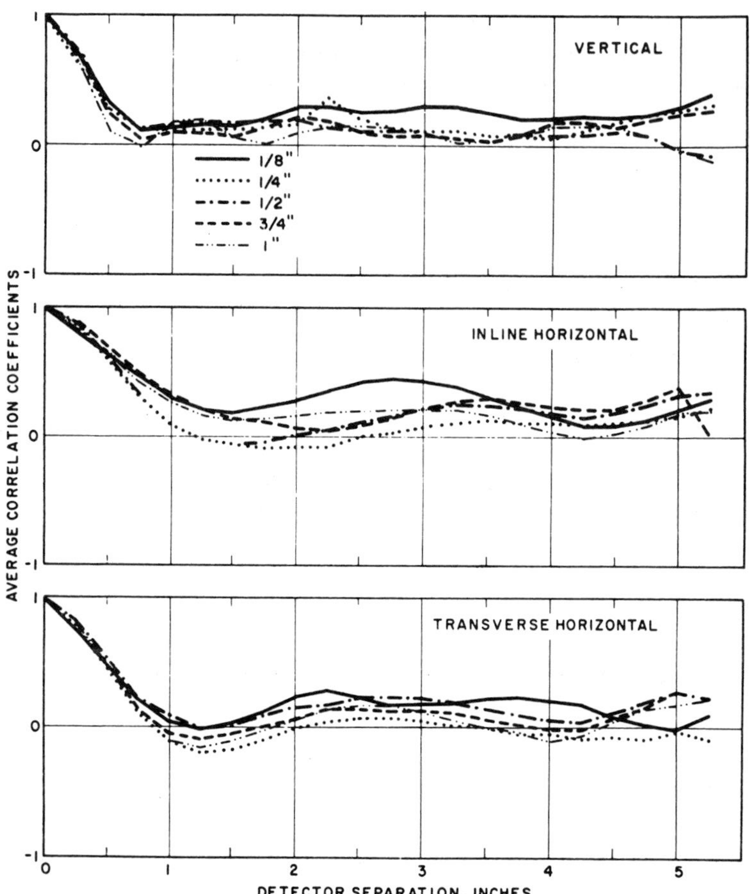


Fig. 9. Correlation coefficient versus detector separation for identical distributions of rods with diameters of $\frac{1}{8}$, $\frac{1}{4}$, $\frac{1}{2}$, $\frac{3}{4}$, and 1 inch.

widely separated detectors is a constant different from zero, and the coherence distance is that smallest separation at which the correlation coefficient takes on the constant value. Detector separations were measured along the arc.

$\rho_{jk}$ was computed with an analog correlator. To minimize the effect of signals common to all positions, we gated in the 500 ms section of the trace that followed the last obvious coherent arrival. As is clear from an examination of Figures 7 and 9, this procedure did not eliminate all contributions from incident waves.

Three components of motion were detected: vertical (motion perpendicular to the surface), in-line horizontal (motion along the line connecting the source and detector), and transverse horizontal (motion perpendicular to the line connecting the source and detector). First, the correlation coefficient was determined for a distribution of $\frac{1}{8}$-inch diameter rods (Figure 7). The coherence distance was 0.7 inch for the vertical component, and 1.0 inch for the horizontal components. Neither of these values could be fixed very closely. The discrepancy between the two coherence distances probably is not real: it almost certainly has its origin in our inability to use transducers with identical frequency responses to detect horizontal and vertical components of motion.

It will be convenient to express distances in terms of wavelengths, where wavelength is the ratio of velocity and frequency. For a continuous frequency spectrum, wavelength really is not a legitimate term. However, the spectra shown in Figure 5 are peaked, and wavelength we defined as the velocity of the Rayleigh wave (8150 ft/sec) divided by the frequency at the peak of the spec-

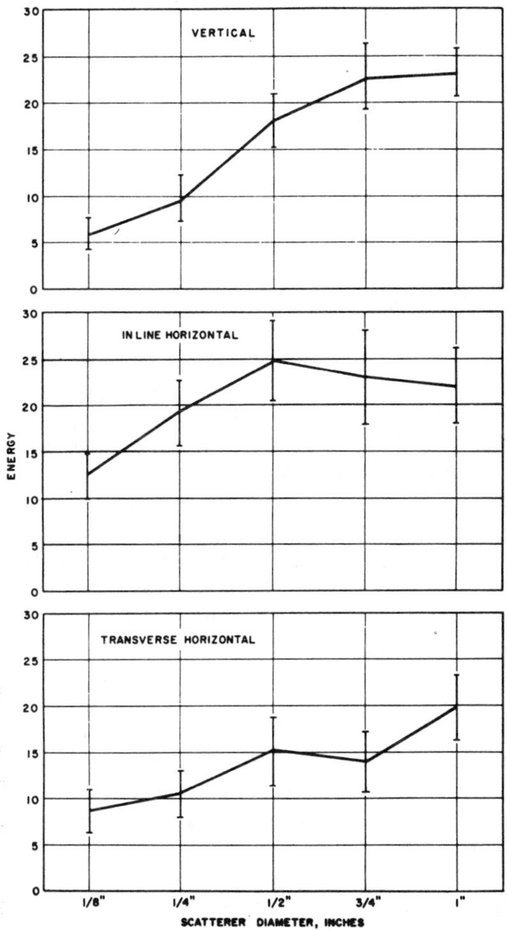


FIG. 10. Energy (integrated squared amplitude) versus scatterer diameter.

trum. This definition gave a wavelength of 2.2 inches for the vertical component of motion and 2.8 inches for the horizontal component, which leads to coherence distances of about 0.32 wavelength and 0.36 wavelength, respectively, for the $\frac{1}{8}$-inch rods. The average distance between the scatterers was 1.25 inches, which is 0.55 wavelength for the vertical and 0.45 wavelength for the horizontal component of motion.

To reduce the effect of different transducer responses, we repeated our measurements, using the 20 to 30 kHz filter mentioned previously. Except for an inexplicable shift in the base line, the coefficient plots (Figure 8) did not differ from the curves for unfiltered data. Corresponding to the spectral shift to lower frequencies, the coherence distance for the vertical motion increased. Because of the sinusoidal nature of the vertical correlation coefficient curve, it was hard to fix an exact value for the coherence distance, but 0.9 inch looks reasonable. This is about 0.23 wavelength.

The scatterers used in the initial experiments were rods $\frac{1}{8}$-inch in diameter, equivalent to 0.045 wavelength. (We have normalized with horizontal motion wavelength of 2.8 inches.) Keeping the pattern of scatterers the same (except where overlaps occurred), we repeated our experiments with rods having diameters of $\frac{1}{4}$, $\frac{1}{2}$, $\frac{3}{4}$, and 1 inch, i.e., 0.089, 0.18, 0.27, and 0.36 wavelengths. Within a rather sizable experimental variation, the coherence distance was independent of scat-

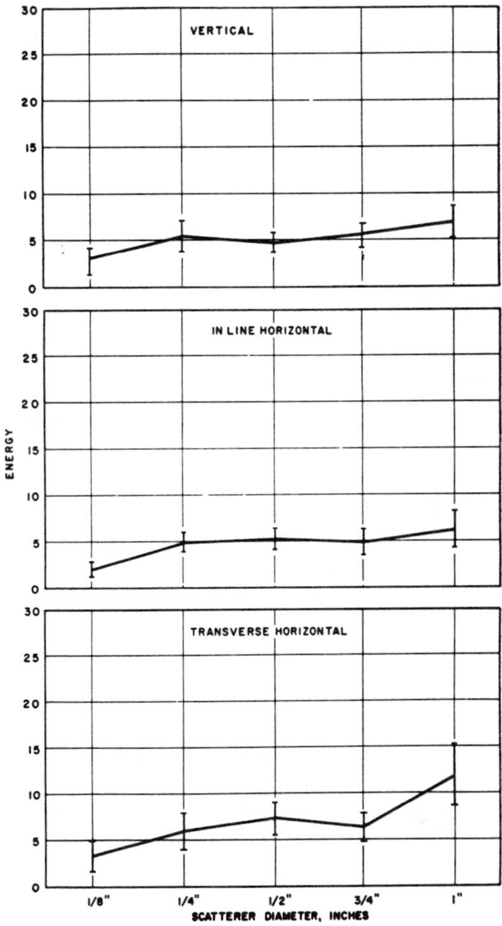


FIG. 11. Same as Figure 10 for signals passed through a 20–30 kHz filter.

terer size (Figure 9). This result was unexpected, since even the smallest scatterers had a greater diameter-to-wavelength ratio than those treated by "small scatterer" theory.

Whatever the behavior of the correlation coefficients, we anticipated that the amount of scattered energy would increase with scatterer diameter: a large rod should intercept a bigger fraction of a wavefront than should a small rod. For the unfiltered data, the anticipated monotonic increase occurred (Figure 10) for the vertical and transverse horizontal components, but not for the in-line horizontal component. When the data were filtered to reduce the effect of transducer characteristics and to obtain greater average wavelengths, the amount of scattered energy that was recorded dropped drastically (Figure 11). In this case, plots of scattered energy against scatterer diameter were similar for all three components of motion. Although scattered energy increased with scatterer diameter, except for the smallest scatterers, the measured increase was not large. Both the filtered and unfiltered curves indicate that no one component of motion was excited in preference to the other two.

## CONCLUSIONS

From the results discussed above, we may draw the following conclusions. For scattering of the type simulated with the model of Figure 1, we find that:

(1) For all three components of motion, the coherence distance is about 0.3 to 0.4 wavelength of the predominant coherent noise.
(2) The coherence distance is independent of the average scatterer size, at least for scatterers having diameters in the range 0.05 to 0.4 wavelength.
(3) The amount of scattered energy usually increases with scatterer size.
(4) All three components of motion appear in the scattered energy with about the same average amplitude.

Of these conclusions, (3) and (4) might have been predicted on the basis of physical reasoning, (1) seems reasonable in retrospect, and (2) was unexpected.

In several ways, our experiments were incomplete and unsatisfactory. By attenuating the Rayleigh wave, we made it impossible to look at the scattering of surface waves in the absence of body waves. Our use of transducers with nonidentical responses prevented our comparing correlation coefficients for different components of motion. Most serious of all, we made measurements for only one distribution of scatterers. This dependence on one distribution introduced deterministic features into what was really a statistical investigation.

In his theoretical investigation of a compressional wave traveling through a medium whose properties vary randomly around average values, Dunkin (1966) found that considering only zero time delay is not adequate. For certain combinations of horizontal and vertical components, correlation maxima can occur at a time delay between the signals that is appreciable. Due to equipment limitations we could not delay one signal relative to another. Also, we did not crosscorrelate vertical and horizontal components but only vertical with vertical, in-line horizontal with in-line horizontal, and transverse horizontal with transverse horizontal. The signals are on tape, and the indicated crosscorrelations can be carried out in the future. Probably the data should be transcribed to digital tape for processing with a computer.

In spite of the limitations of our work, the data should be useful in checking theories of elastic wave scattering. The experimental difficulties we encountered were not trivial; consequently, we do not anticipate that model studies of scattering will be common.

## REFERENCES

Dunkin, J. W., 1969, Scattering of a transient, spherical P wave by a randomly inhomogeneous, elastic half-space: Geophysics, v. 34, no. 3 (in press).

Miller, M. K., Linville, A. F., and Harris, H. K., 1967, Scattering from small surface irregularities: Trans. AGU, v. 48, p. 200.

Oliver, J., Press, F., and Ewing, M., 1954, Two-dimensional model seismology: Geophysics, v. 19, p. 202–219.

Robinson, D. J., 1967, Bidirectional detector of ultrasonic particle motion in solids: Rev. Sci. Inst., v. 38, p. 813–814.

Tatel, H. B., 1954, Note on the nature of a seismogram —II: J. Geophys. Res., v. 59, p. 289–294.

# Observation of a second bulk compressional wave in a porous medium at ultrasonic frequencies

Thomas J. Plona
*Schlumberger-Doll Research, Ridgefield, Connecticut*

(Received 8 November 1979; accepted for publication 7 December 1979)

A second bulk compressional wave has been observed in a water-saturated porous solid composed of sintered glass spheres using an ultrasonic mode conversion technique. The speed of this second compressional wave was measured to be 1040 m/sec in a sample with 18.5% porosity. The theory of Biot, which predicts two bulk compressional waves in porous media, provides a qualitative explanation of the observations. To the author's knowledge, this type of bulk wave has not been observed at ultrasonic frequencies.

PACS numbers: 43.35.Cg, 43.20.Hq, 03.40.Kf, 91.60.Ba

It is the purpose of this letter to describe the experimental observation of a second bulk compressional wave which propagates at speeds approximately 25% of the speed of the normal bulk compressional wave in a fluid-saturated porous medium. This observation has been made using an ultrasonic immersion technique and a fluid-saturated porous medium consisting of water and sintered glass spheres. The excitation of the slow wave in the solid is shown to be consistent with the principles of mode conversion and refraction at plane liquid/solid interfaces. To the author's knowledge, this type of bulk wave has not been observed at ultrasonic frequencies. The existence of a similar low velocity compressional wave in an isotropic fluid-saturated porous media was predicted by Biot.[1]

An ultrasonic immersion technique was used to generate the bulk modes in a solid plate based on the concept of mode conversion and refraction at liquid-solid interfaces. Using this technique, which is similar to that of Smith[2] and Hartman and Jarzynski,[3] sound speeds of the normal compressional wave, the shear wave, and a slow compressional wave were determined. The geometry is depicted in Fig. 1. The pulsed ultrasonic system incorporated commercially available pulser/receiver electronics and nondestructive testing type broadband transducers. The transducers used included a pair of 28.6-mm-diam 500-kHz transducers, and a pair of 25.4-mm-diam 2.25-MHz transducers. Total acoustic path lengths between transmitter/receiver were typically 130 $\mu$sec in water. The velocities were determined from the measured arrival times of the appropriate pulses on an oscilloscope and a corresponding measurement of the angle of incidence. Uncertainty in the measured speeds was about 3%.

The solid samples used were water-saturated disks composed of sintered glass spheres. Aggregates of solid spheres of diameters between 0.21 and 0.29 mm were sintered at temperatures between 700 and 740 °C. Four samples of varying porosity (given in Table I) were fabricated by controlling the maximum oven temperature and the total time at that temperature. The samples were ground to achieve plane parallel faces, and water-saturated using a vacuum impregnation technique. Sample thickness varied between 14 and 21 mm and sample diameters between 90 and 100 mm. The anisotropy of the samples was tested by the measurement of the fast compressional speed in three perpendicular directions on one of the samples. The speeds were identical to within experimental uncertainty. The bulk glass had a density of 2.48 g/cm$^3$, compressional speed of 5.69 km/sec, and a shear speed of 3.46 km/sec as calculated from the manufacturer's modulus data.

The existence of a second bulk compressional wave in a porous solid can be demonstrated by studying Fig. 2. This shows the received pulses recorded after propagating through the water/porous solid/water system at four different angles of incidence. These results were obtained using sample 1 and 2.25 MHz broadband transmitting and receiving transducers. For the case of normal incidence, only

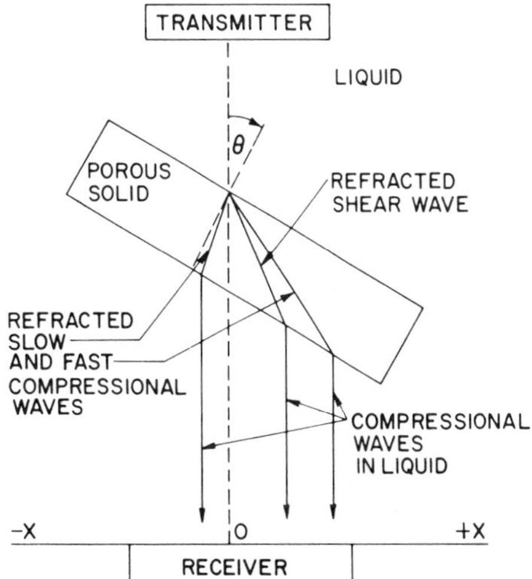


FIG. 1. Schematic drawing of mode conversion and refraction at liquid/porous solid/liquid interfaces. Reflected waves are not shown.

Reprinted from Applied Physics Letters, 36, 259-261 with permission from American Institute of Physics

TABLE I. Measured sound speeds, porosities, and slow wave to fast wave amplitude ratio at normal incidence for water-saturated sintered glass samples.

| Sample | Porosity (%) | Fast compressional speed (km/sec) | Shear speed (km/sec) | Slow compressional speed (km/sec) | Amplitude ratio ($\theta = 0$) |
|---|---|---|---|---|---|
| 1 | 28.3 | 4.05 | 2.37 | 1.04 | 0.38 |
| 2 | 25.8 | 4.18 | 2.50 | 1.00 | 0.26 |
| 3 | 18.5 | 4.84 | 2.93 | 0.82 | 0.21 |
| 4 | 7.5 | 5.50 | 3.31 | ... | ... |

compressional waves are generated in an isotropic solid. One anticipates, therefore, an echo pattern consisting of a direct, fast compressional, and then, equispaced in time, a set of multiple reflections. Pulses $A$, $C$, $E$ and $G$ in Fig. 2(a) are these arrivals. However, pulses $D$ and $F$ are additional arrivals not observed in nonporous solids. Pulse $D$ represents an additional bulk wave traveling through the porous solid at a velocity much less than the normal compressional wave. Pulse $F$ arrives later than pulse $D$ by a time corresponding to the time difference between pulse $A$ and $C$. Thus pulse $F$ (which is also inverted with respect to pulse $D$) corresponds to a twice-reflected path in the solid where the arrival travels once through the solid at the slow speed and twice at the fast compressional speed. This is significant since it demonstrates that mode conversion between this additional bulk wave and the normal compressional wave occurs at the boundaries. Therefore, the existence of the first additional arrival, pulse $D$, and the mode converted arrival, pulse $F$, *both at normal incidence*, are evidence that this second bulk wave is primarily a compressional vibration.

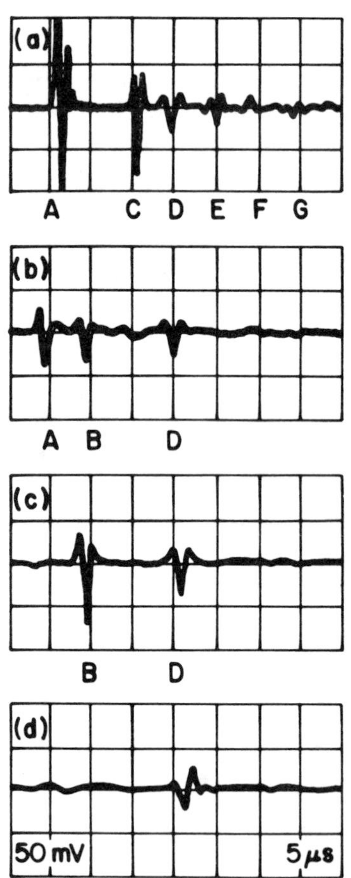


FIG. 2. Received pulses after propagation through water/porous solid/water system at various angles of incidence $\theta$. (a) $\theta = 0°$, (b) $0 < \theta < \theta_c^*$, (c) $\theta_c^* < \theta < \theta_s^*$, (d) $\theta_s^* < \theta < 90°$.

As the angle of incidence $\theta$ is increased from zero to some finite value less than the longitudinal critical angle, i.e., $\theta_c^* = \arcsin$ (speed in water/fast compressional speed in solid), only three pulses are received [Fig. 2(b)]. Pulse $A$ is the usual compressional wave, pulse $B$ is the mode-converted shear wave, and pulse $D$ is the slow speed compressional wave. The fast compressional arrival has shifted toward a shorter arrival time, while the slow arrival has shifted toward a longer time. Based on refraction principles, these time shifts are indicative of the fast and slow compressional wave speeds being faster and slower than the speed in water, respectively. [A similar time shifting result is observed also in Figs. 2(c) and 2(d).]

As the angle of incidence is increased beyond $\theta_c^*$ but less than the shear critical angle, i.e., $\theta_s^* = \arcsin$ (speed in water/shear speed in solid), only the shear arrival (pulse $B$) and the slow compressional (pulse $D$) are observed [Fig. 2(c)]. The fast compressional wave has been critically refracted and no longer propagates through the disk. In this configuration the spatial separation of the ray paths due to refraction as shown in Fig. 1 is most clearly evident. As the receiver was translated in the $+x$ direction, it was observed that the amplitude of the shear arrival increased to a maximum relative to the slow wave amplitude. Conversely, as the receiver was translated in the $-x$ direction, the slow wave amplitude was observed to increase to a maximum relative to the shear. This beam shift to the negative side of the transmitter-receiver axis is further indication that the speed of the slow wave is less than that of water.

Finally, the received signal when $\theta > \theta_s^*$ is shown in Fig. 2(d). In this case only the slow compressional arrival is observed since the fast compressional and shear wave have both been critically refracted. A critical angle for the slow wave was neither observed nor expected.

Velocity measurements were made using the 500 kHz transducer pair. At the lower frequencies, dispersion and attenuation were less severe and pulse distortion was minimal. Measured sound speeds averaged from data taken at several different angles of incidence are given in Table I. Also included is the ratio of the direct slow wave peak amplitude to the direct fast wave peak amplitude at normal incidence.

In these experiments, the observed "extra" arrival reflects and refracts in a manner consistent with a low-speed compressional wave in the porous solid. The identification of this observed "extra" arrival as a bulk wave and not a miscellaneous echo or scattering path was made by systematically perturbing (via translations) all combinations of the source, receiver, and sample relative positions and using short time pulses. If the "extra" arrival was not a bulk wave, its timing relationship to the known shear and compressional bulk waves would have to be dependent on such source, receiver, or sample translation. This type of dependency was never observed. In addition, this type of "extra" arrival was never observed in samples of similar dimensions made from such nonporous materials as aluminum, stainless steel, or Plexiglas.

The data in Table I indicate that both the fast compressional and shear speed increase with decreasing porosity while the slow wave speed decreases. The slow wave speed at a porosity of 7.5%, however, was not measured since the appropriate pulse was too small to uniquely identify. The amplitude ratio data is consistent with this fact since as porosity decreases, the slow wave amplitude also decreases. Thus, the largest slow wave amplitude was observed at the largest porosity (i.e., sample 1, porosity = 28.3%).

A complete identification of the mechanism responsible for this observed slow compressional wave is still under investigation. The Biot porous media theory which predicts the existence of a similar slow compressional wave seems to provide the most reasonable explanation. However, a more thorough comparison of the relationship between Biot's theory and the properties of this observed slow compressional wave is necessary. Currently, measurements of wave speeds and attenuations for all three waves are being made on a larger suite of samples.

If one assumes a Biot model, then the conditions that favor a large amplitude second compressional wave are those of high acoustic frequency and a large fluid permeability in the sample. These conditions are met in this investigation where the ultrasonic frequencies are in the low MHz range, and the fluid permeability is approximately 1 $\mu m^2$. This work has applications in wave propagation studies in various types of porous media such as marine sediments[4] and porous rocks.[5]

In summary, a second bulk compressional wave in a water-saturated sintered glass porous media has been observed using a mode conversion technique. This slow compressional wave, which to the author's knowledge has not been directly observed before, has a velocity less than that of the usually compressional and shear wave in the solid and also less than that of water. This observed wave can be tentatively identified with the slow speed compressional wave predicted by Biot.

The author would like to thank A.J. Devaney and R. Johnson of Schlumberger-Doll Research for their helpful discussions.

[1]M.A. Biot, J. Acoust. Soc. Am. **28**, 168 (1956).
[2]R.E. Smith, J. Appl. Phys. **43**, 2555 (1972).
[3]B. Hartmann, J. Jarzynski, J. Acoust. Soc. Am. **56**, 1469 (1974).
[4]R.D. Stoll, *Physics of Sound in Marine Sediments*, edited by L. Hampton (Plenum, New York, 1974), p. 19.
[5]N.C. Dutta and H. Ode, Geophysics **14**, 1777(1979).

# Areal seismic methods for determining the extent of acoustic discontinuities

John A. McDonald,* G. H. F. Gardner,* and J. S. Kotcher*

The collection of areal seismic reflection data is becoming fairly routine. It is now generally realized that the solution of three-dimensional (3-D) structural problems is only possible when the target has been adequately sampled and the data have been correctly migrated to produce an accurate image of the subsurface. However, many of our exploration prospects are associated with lithological changes or stratigraphic features rather than structural features. We show how areal seismic techniques can provide an added dimension in determining the extent of acoustic discontinuities in areas where the strata are generally flat.

Seismic reflection data were collected in a water tank over a flat plate containing an acoustic discontinuity of irregular shape and small size. The data were migrated to create seismic sections through the discontinuity, and sufficient sections were produced to define the areal extent of the anomaly. The resolution on the seismic sections at the change in acoustic properties was improved by increasing the migration aperture.

Similar experiments were carried out in the field in an area over essentially flat lying layers and a pinnacle reef. A common-depth-point (CDP) stacked section had shown that the seismic expression of the reef was the classical degradation of the reflection from the depth of the reef in the seismic section. The diagnostic loss of reflectivity was also seen on the sections, constructed from the areal data that intersected the anomaly. A composite interpretation based on the migrated areal data allowed the areal extent of the reef to be determined. When two known producing wells were located on the map, they were found to lie within the discontinuity in reflectivity. In addition, an interpretation based on a CDP section from a line crossing the anomaly showed that the anomaly in the vertical plane of the line coincided with the interpretation based on areal data.

## INTRODUCTION

Seismic reflection data have been used for many years to aid and evaluate our interpretations of geologic structure. The solution of structural problems with seismic data makes use of only a little of the information contained in the seismic signal. We will demonstrate that some of this other information in the reflection data may be applied to the study of *areas* without appreciable structure in the subsurface.

The particular problem we studied was that of the detection of pinnacle reefs in carbonate sections. Pinnacle reefs of sufficient porosity have high hydrocarbon bearing potential but have no apparent structural expression on conventional seismic sections. However, we will show that it is possible, by unusual data collection and processing techniques, to describe the areal extent of such a reef by seismic methods. The technique also will be demonstrated by the use of physical models.

The nonstructural seismic effect of reefs has been recognized for a long time. For instance, Van Siclen (1957) realized that reefs could be mapped areally from the false structure on seismic sections. Because of higher velocities in the limestone reef than in the surrounding shale sheath, a prominent subreef reflector appeared to be at a shallower depth beneath the reef than elsewhere. More recently, Klose and Holland (1975) showed an example of a more difficult problem in which the overlying sediments do not drape over the reef. Klose and Holland pointed out that reflection amplitude, frequency changes, phase reversals, and velocity changes due to differences in lithology should be used as diagnostic tools to define reefs. Bubb and Hatlelid (1977) illustrated many of the ways carbonate buildups may be recognized on seismic sections.

We show how some of the properties of the seismic signal may be measured, and this information then used to interpret changes in an anhydrite-carbonate sequence.

## GEOLOGIC BACKGROUND TO REEF DEVELOPMENT

Some knowledge of the depositional environment during reef development is necessary in planning a geophysical experiment to determine the areal extent of reefs. Our field techniques were developed in the northern Michigan reef trend, and we will confine our geologic descriptions to that area.

The northern Michigan reefs form a barrier trending southwest to northeast. Inside the barrier, to the south, is a series of pinnacle reefs trending in the same direction. The trend is about 160 miles long and about 10 miles wide. The reefs are of the Silurian period; the nomenclature of the formations surrounding them, shown in Figure 1, is adapted from Mesolella et al (1974). The base of the section of interest in this study is the Niagara nonreef carbonates. The transition to the A-1 evaporite, a halite interbedded with anhydrite and dolomite, takes place in an interval of 5–10 ft. The evaporite is overlain with the basal A-1 carbonate, a shaly limestone. Next in sequence is the A-2 evaporite which is similar to

Presented in part at the 48th Annual International SEG Meeting November 1, 1978, in San Francisco. Manuscript received by the Editor May 29, 1979; revised manuscript received April 21, 1980.
*Gulf Research and Development Company, P.O. Drawer 2038, Pittsburgh, PA 15230.
0016-8033/81/0101-0002$03.00. 

Reprinted from Geophysics, 46, 2-16

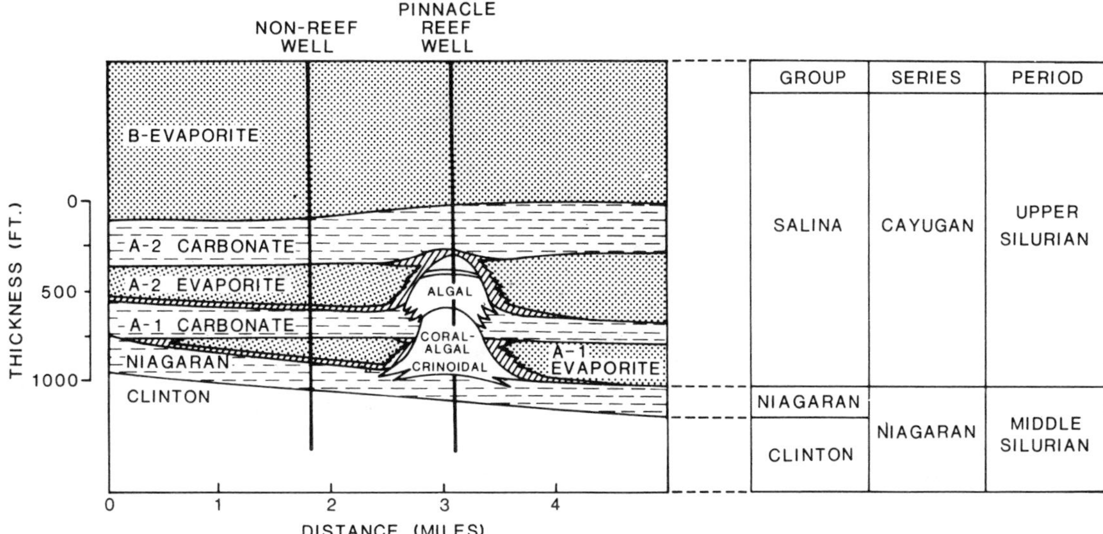


FIG. 1. Cross-section of the depositional sequence and of reef formation in the upper Silurian period in Michigan (after Mesolella et al, 1974).

the A-1 evaporite. At the top of the section is the A-2 carbonate, a coarser grained version of the A-1. In the northern reef trend, this sequence is 500–750 ft thick.

Mesolella et al (1974) performed an exhaustive study of well data from depositional environments both on and off the reef trend and have proposed an interesting model for deposition. In general, the pinnacle reefs have, in ascending order, a crinoidal zone which is separated by carbonates from a coral-algal zone, an algal zone, and a laminar stromatoporoidal and stromatolitic zone. Carbonates also separate the two upper zones. Vuggy porosity in the productive zone was developed by coral skeletons selectively dissolving from the matrix. Mesolella et al (1974) postulated a depositional sequence in which the Niagara, the A-1, and the A-2 carbonates were laid down during high sea levels, and, at low sea levels, the A-1 and A-2 evaporites were deposited. This hypothesis requires exposure of the reefs to the atmosphere during their evolution. It is supported by core data which show limestone of the coral-algal zone to be discolored, probably due to exposure to the atmosphere. In addition, large fractures and large vugs in the carbonates indicate leaching during diagenesis.

The Niagara and A-1 carbonates are thickest around the margins of the Michigan basin where they are several hundred feet thick. The A-2 carbonate, unlike the A-1, is thinnest at the basin margin. Mesolella et al (1974) suggested that this thinning is an indication that physical factors dominated biological factors during this period of carbonate deposition.

The sedimentary section appears to be without structure and is merely a sequence of horizontal layers. On a seismic section the reefs manifest themselves as a degradation of the seismic reflection. Reflections disappear when a seismic line passes from a nonreef area, with its succession of carbonates and evaporites, to a reef. There are no strong acoustic discontinuities in the conglomerate of corals and fossilized algae. Good examples of reef signatures on seismic sections have been given by McClintock (1976), and similar sections will be given later in this paper.

## LITHOLOGIC MODEL

The development of the areal seismic data processing techniques for reef detection was considerably aided by making use of data collected in a model experiment. In this section we describe the model, the results, and the conclusions arrived at after the data had been processed.

A model was designed which presented no structural relief but which contained variations in acoustic properties. Readily available materials with known properties were used. The model and its physical properties are shown in Figure 2. The basic unit was a plate of Plexiglas in which was cut an irregular shaped hole; the hole was then filled with two layers of different kinds of silicone rubber. The model was immersed in water with the upper surface horizontal.

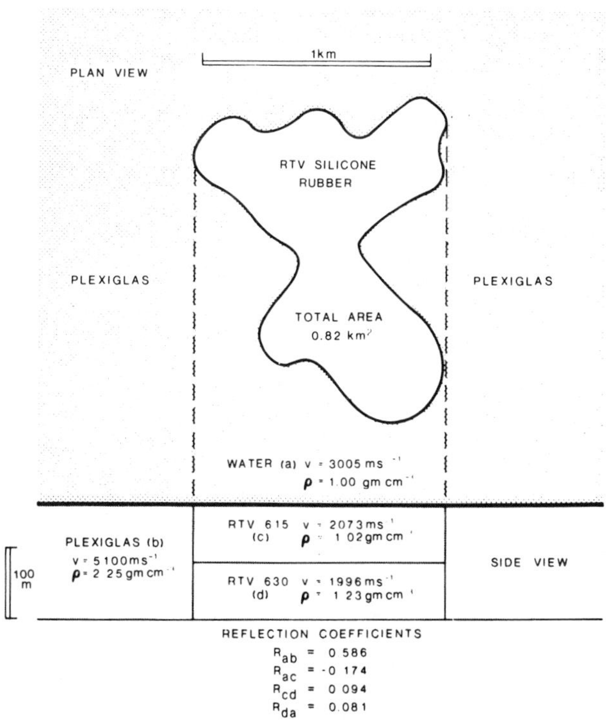


FIG. 2. Plan and side view of the model showing physical properties of the materials used.

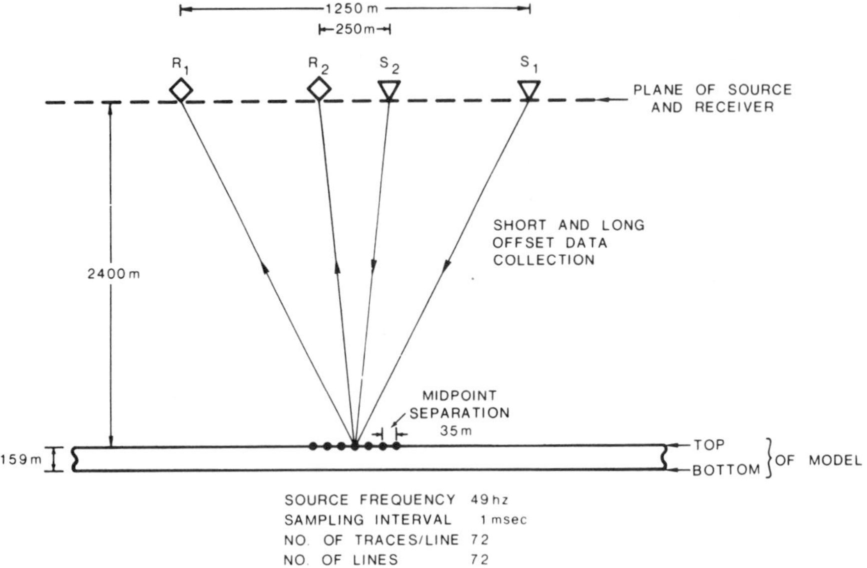


FIG. 3. Pertinent model dimensions and parameters, $R_n$-receiver, $S_n$-source.

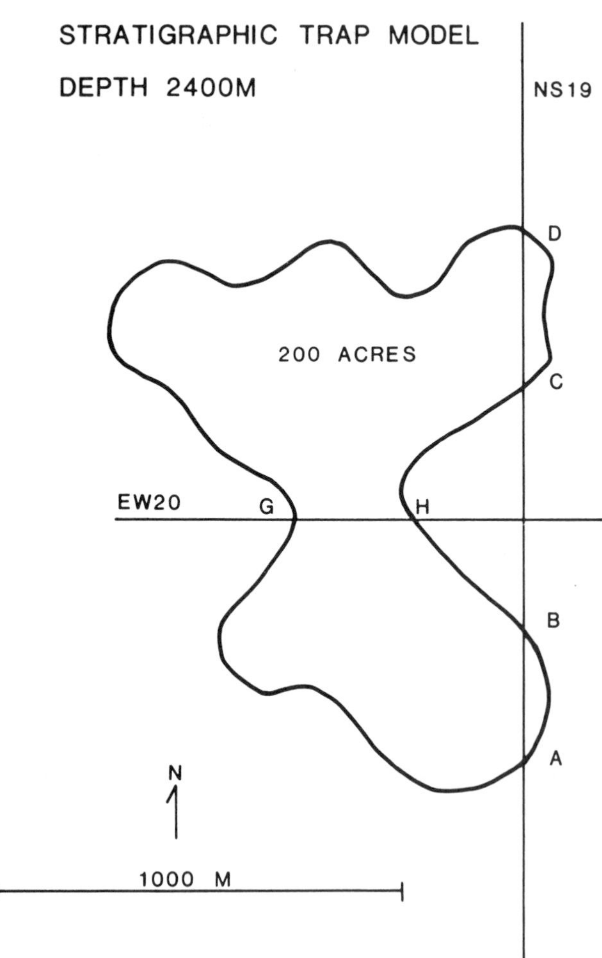


FIG. 4. Location of two lines (EW20 and NS19) for which seismic sections were produced by 3-D migration of the data over the model.

The experimental data collection was similar to that described by French (1974) in which a model was suspended in a tank and water was used as the medium to transmit ultrasonic pulses generated and detected by piezoelectric transducers tracking above the model. Pertinent dimensions and parameters of the model are shown in Figure 3. Data were collected with a source-to-receiver separation of 250 m and also of 1250 m in order that estimates could be made of velocity. Seventy-two lines were traversed over the model, each with 72 recordings; both source increment and line spacing were 35 m. In this manner, an area of 2485 × 2485 m was sampled twice by 5184 seismic traces with a spatial separation of 35 m.

## PROCESSING THE MODEL DATA

The aim of the experiment was to define the limits of the acoustic discontinuity, the shape of which, in this case, was already known. To facilitate data handling, a Cartesian coordinate system was set up which encompassed the area covered by the seismic survey. Thereby each source and each receiver position was given coordinates. Each of the 5184 traces was then identified by the coordinates of the associated source and receiver positions. These four numbers were written in a header for each trace and uniquely defined the trace for processing.

The seismic traces were not subjected to any deconvolution or filtering techniques.

A computer program was written to construct seismic sections by three-dimensional (3-D) migration of the areal data into a single vertical plane. A representation of a seismic section was constructed by (1) differentiating each input trace, (2) shifting it according to the traveltime formula given below, (3) weighting it according to the apodization formula given below, and (4) summing it into the output trace. This set of operations was discussed by Gardner et al (1974) and directly implements Kirchhoff's integral solution of the wave equation.

The apodization formula can be explained by analogy with an aperture in an optical system. In this application, a circular aperture was specified at each reconstruction point of the vertical section, the radius of which increased directly proportional to time. Only data within the aperture were used. That is to say, any input trace had a midpoint for its source and receiver positions; at some time along this trace the distance from the midpoint to the reconstruction point became less than the aperture radius at this time; from then on the input trace contributed to the output with a weight dependent upon the relative distance of the midpoint from the reconstruction point and the radius of the aperture. The best weighting function to use depends on the particular effect sought. In general, any weighting which monotonically decreases to zero at the aperture boundary is suitable. In the present application, the weight varied linearly at each radius with the area enclosed by a circle of that radius and vanished at the boundary of the aperture.

A time $T$ on any input trace is associated with a time $T_0$ on the output trace by the time taken by a pulse to travel from the source to $T_0$ and back to the receiver. The formula used was a straightforward generalization of Dix's formula (Gardner et al, 1974)

$$T = \sqrt{\left(\frac{T_0}{2}\right)^2 + \left(\frac{D_1}{V_0}\right)^2} + \sqrt{\left(\frac{T_0}{2}\right)^2 + \left(\frac{D_2}{V_0}\right)^2} \quad (1)$$

where

$D_1$ = distance from reconstruction point to source,
$D_2$ = distance from reconstruction point to receiver, and
$V_0$ = root-mean-square velocity at time $T_0$.

Migration was accomplished by weighting the value of the amplitude of the derivative trace at time $T$ and adding it to the output trace at time $T_0$.

The square roots in equation (1) can be calculated approximately but very satisfactorily, by using Newton's method of iteration and a precalculated table. The table was designed to cover the range of values in $T$, $D$, and $V$ which commonly occur. It can be shown that the estimated values have their maximum errors about half way between tabulated values and, furthermore, the biggest error occurs for the smallest value of estimation. For example, at a depth of $T$ = 1000 msec the error in the estimated traveltime is less than 0.015 msec.

The areal data were used to produce seismic sections for the lines shown in Figure 4. In the case of the east-west line (EW20), two versions were produced: one with a circular aperture of 100 m and one with an aperture of 500 m. An important parameter in 3-D migration is the aperture radius at a given time down the section. The larger the radius the better the resolving power of the migration for neighboring points (Gardner and Kotcher, 1977). This is illustrated in Figure 5 where two versions of the EW20 seismic section are compared. The smaller aperture (migration radius 100 m), which contained about 25 traces, results in a more gradual change in amplitude across the acoustic discontinuity. The

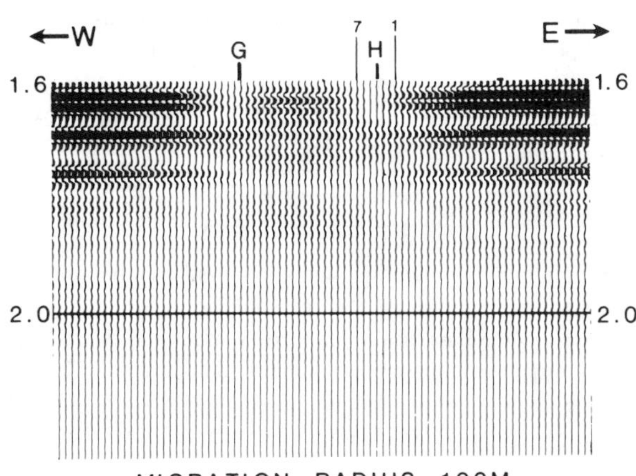


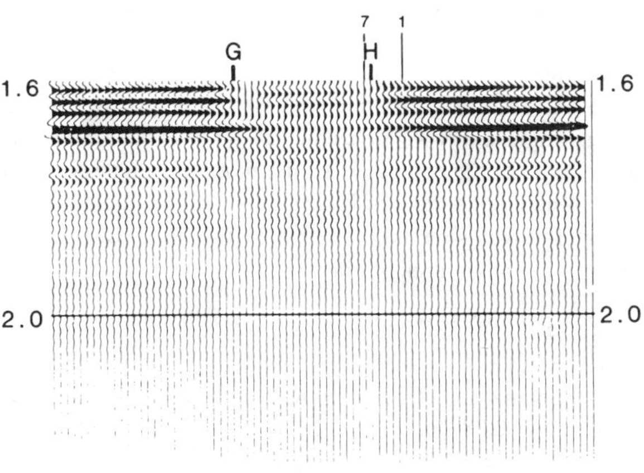


FIG. 5. The seismic section for line EW 20 (Figure 4) migrated with two different radii. The small numbers indicate the traces shown in detail in Figure 6.

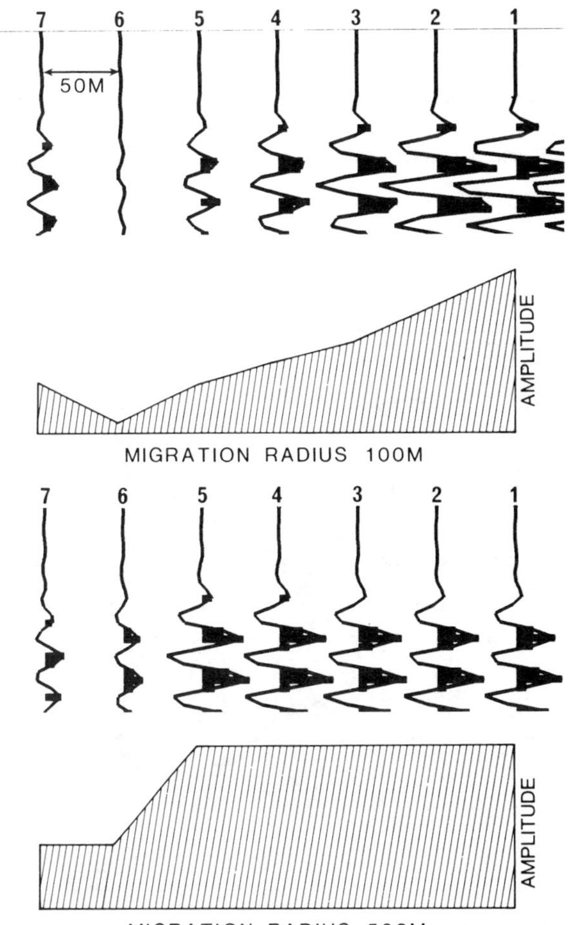


FIG. 6. Detail from Figure 5 showing the transition from reflections from Plexiglas to reflections from silicone rubber. The hachured zones indicate the change in amplitude from trace to trace.

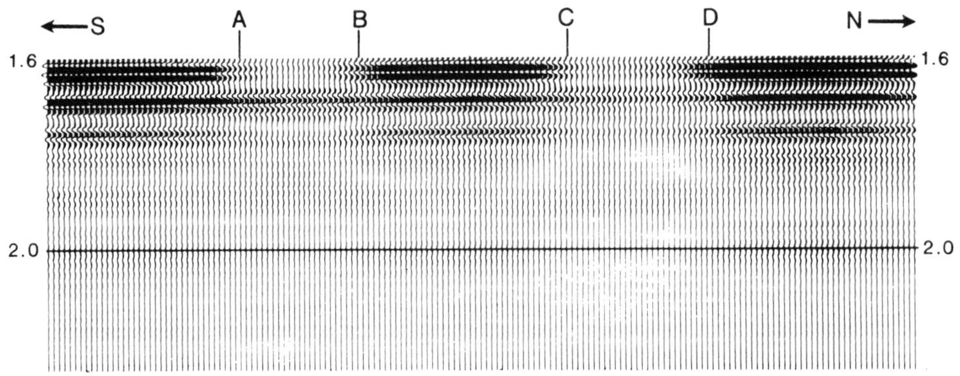


FIG. 7. The seismic section for line NS19 (Figure 4). This line crosses the Plexiglas-rubber boundary four times.

detail at the discontinuity in Figure 5 is shown in Figure 6. The larger aperture (migration radius 500 m), which contained about 625 traces in each output trace, results in a much more abrupt change in amplitude. The choice of aperture size is largely a compromise between low resolution with shorter computation time and high resolution with longer computer time.

A second example of a seismic section is shown in Figure 7 which has been constructed on a north-south line intersecting the discontinuity four times. From a series of parallel seismic sections, either east-west or north-south, or using both, the location of the boundary can be detected and the boundary drawn. This gives a fairly good representation of the areal extent of the low-reflectivity zone. However, a much more direct and less arduous method is to reconstruct a horizontal seismic section directly from the input data. To increase the information content of the horizontal section, the Hilbert transform of the data was constructed simultaneously, and from this the power envelope and the phase angle were deduced (Taner and Sheriff, 1977; Taner et al, 1979).

The calculation for a horizontal seismic section was performed by setting up a grid of points, spaced 35 × 35 m, and for each point differentiating the input trace amplitude, calculating the Hilbert transform, and time shifting according to the traveltime formula. Trace amplitudes then were weighted according to the apodization function, and the input trace amplitude was placed in one grid-point array and its Hilbert transform placed in another. From the resultant trace grid values $A$ and their Hilbert transform values $B$, the amplitude and phase angle were constructed by the formulas

$$\text{power} = (A^2 + B^2)^{1/2},$$
$$\text{phase angle} = \tan^{-1}(B/A).$$

These gridded values of power and phase can be presented graphically in several ways. One example is shown in Figure 8 which displays the data at a time of 1636 msec, corresponding to a horizon just below the surface of the model shown in Figure 3. The figure shows contour lines of constant phase angle, and the size of the plus signs at the grid points are proportional to the power. The contour lines clearly outline the area of low reflectivity (silicone rubber) shown in Figure 2; this is because the phase angle changes rapidly through 180 degrees across the boundary. As was seen in Figures 5 and 6, the change in size of the pluses indicates a rapid drop in power at the boundary.

A second mode of representation is shown in Figure 9. Here the phase angle and power calculated above have been interpolated to a fine grid and then output to a color plotter. Both these displays give reasonable pictures of Figure 2.

In the following we will show how this technique has been applied to a real geologic problem in the field.

HORIZONTAL SEISMIC SECTION AT 1636 MS

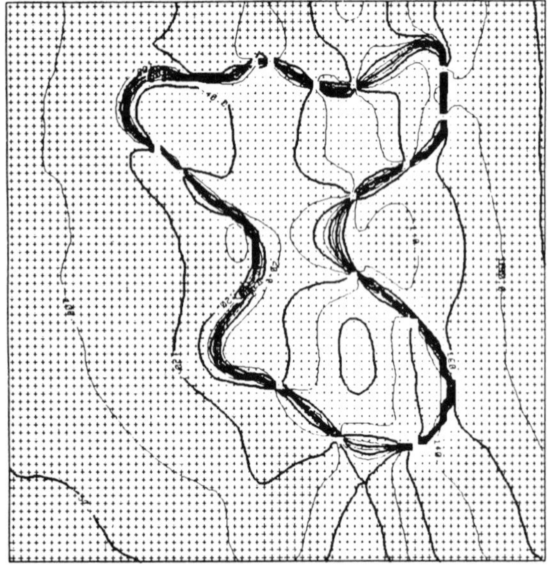

FIG. 8. A migrated horizontal seismic section at 1636 msec corresponding to a horizon just below the surface of the model shown in Figures 2 and 3. Phase contours are represented by lines at intervals of 20 degrees; the size of the pluses indicates the amplitude.

## PLANNING THE DATA COLLECTION

It was decided to design a field experiment in which the area covering a pinnacle reef could be adequately sampled seismically, and then these data could be processed to produce an areal representation of the reef in a manner similar to that used for the model data. First, it should be pointed out why conventional seismic line spacings are inadequate for this kind of experiment and areal methods are efficient.

Acceptable producing reefs in the northern Michigan trend seem

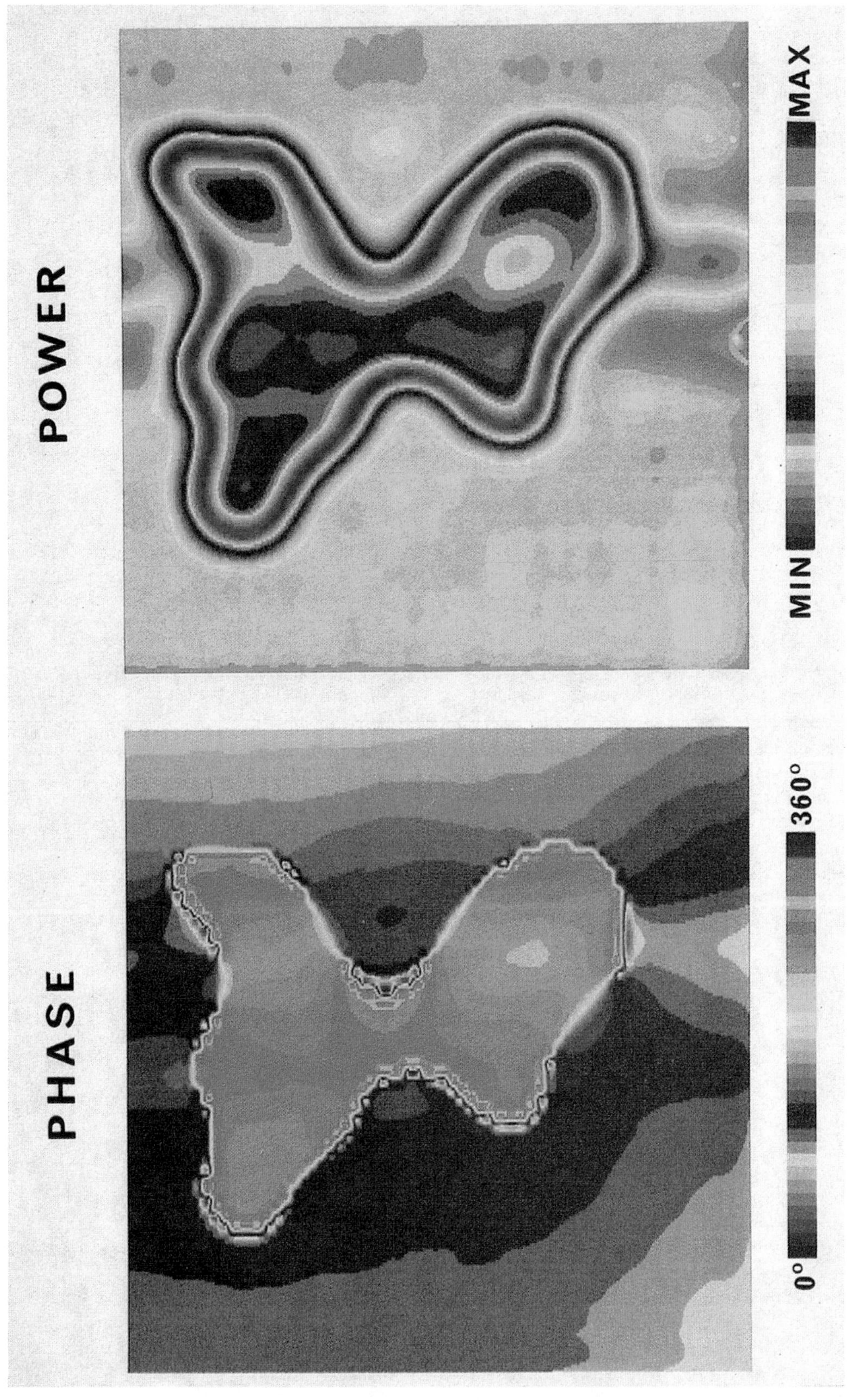


FIG. 9. Color plots of the data shown in Figure 8. In this case, the phase and power have been separated.

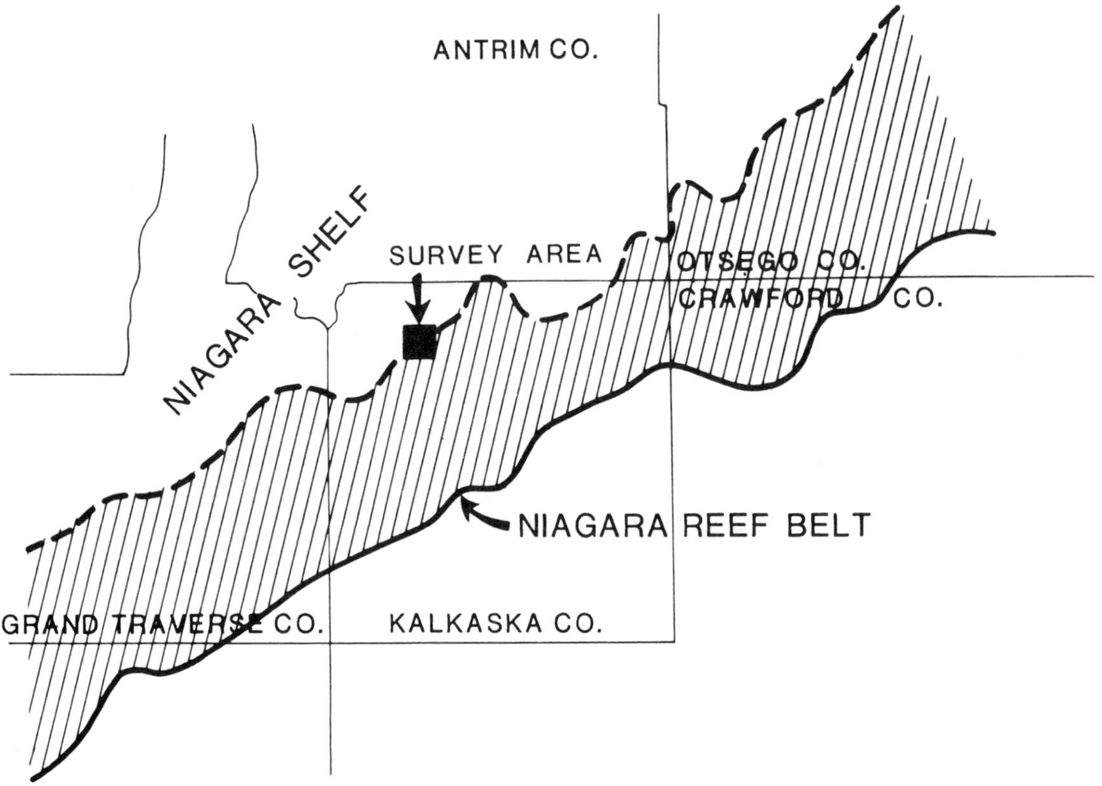


FIG. 10. Index map showing the location of the survey in the Niagara reef trend (after Mesolella et al, 1974).

to vary in area from 60 to 600 acres, with many around 160 acres in area. If we assume that reef horizontal cross-section is circular, the diameter for an area of 60 acres is 1824 ft; for an area of 160 acres, it is 2979 ft. In order to be certain that a network of lines on a grid separation $l$ crosses at least some portion of a reef of radius $r$, it can be shown (Chon, 1976) that

$$\frac{4r}{l}\left[1 - \frac{r}{l}\right]$$

must be unity, which occurs when $r = l/2$. In other words, the separation of lines in the grid must be equal to, or less than, the diameter of the reef. In order to detect all the small 60-acre reefs, the lines would have to be not more than one-third mile apart and, even for the average reef, half-mile spacings would be necessary. In addition, such a plan would only ensure that all reefs were crossed at least once; it would not define the areal extent of any reef. However, in northern Michigan, environmental and other requirements do not generally permit lines at one-third or even one-half mile spacings. The preferred locations on seismic lines is on section line roads for which the grid interval would be 5280 ft. The ratio $r/l$ then becomes 0.17 for 60-acre reefs and 0.28 for 160-acre reefs, reducing the probabilities of their discovery to 55 and 82 percent, respectively. Therefore, in order to detect all reefs *and* to determine their areal extent, it is necessary to scan the surface point-by-point regarding each subsurface point as discrete. Space, as well as time, has to be sampled correctly to avoid undersampling and spatial aliasing. To obtain the correct spatial sampling, we need to use the Shannon theorem

$$\Delta x = \frac{V}{2f},$$

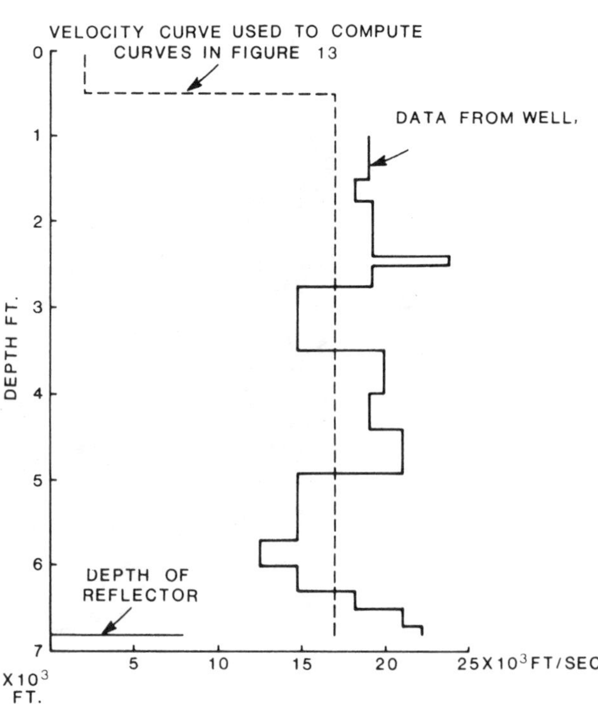


FIG. 11. Interval velocity versus depth from a well about one mile from the survey area to the depth of the A-2 carbonate. The broken line shows the simplified curve used to calculate the traveltime curves given in Figure 13.

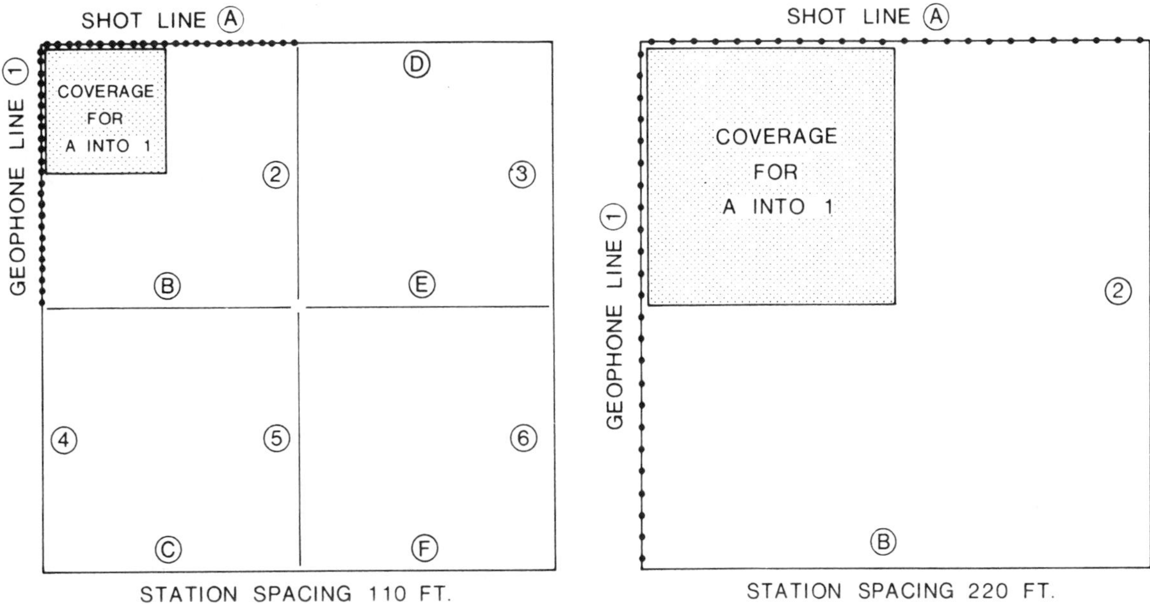


FIG. 12. Two methods of areally sampling a one-mile section with a 24-channel recording system. Left: each shot line occupied twice, total 384 shots. Shot lines B and E not on section lines. Total traces generated, 9216. Right: each shot line occupied twice, total 96 shots. Total traces generated, 2304.

where $\Delta x$ is the maximum distance between sample points when a seismic wave of frequency $f$ is transmitted through a geologic section of velocity $v$.

A survey area was chosen in the northern portion of the southern peninsula of Michigan, Kalkaska County (Figure 10). Smoothed interval velocities from a nearby well plotted as a function of depth are shown in Figure 11. It can be seen that the velocities are high to the target Niagaran (depth of reflector) but that there are two low-velocity zones. To estimate the required spatial sampling interval $\Delta x$, we assumed the simplified velocity structure shown by the dashed line in Figure 11. Another factor was the number of recording channels available in the seismic system, which in this experiment was 24.

There are only two ways reasonably to sample a square-mile section with a 24-channel system as illustrated in Figure 12. The seemingly preferable arrangement with the denser sampling (left side of figure) requires six positionings of the geophones and 384 shot holes. In addition, 96 of these shot holes do not lie on section lines (shot lines B and E). The only alternative (right side of figure) requires two positionings of the geophones and 96 shot holes. The

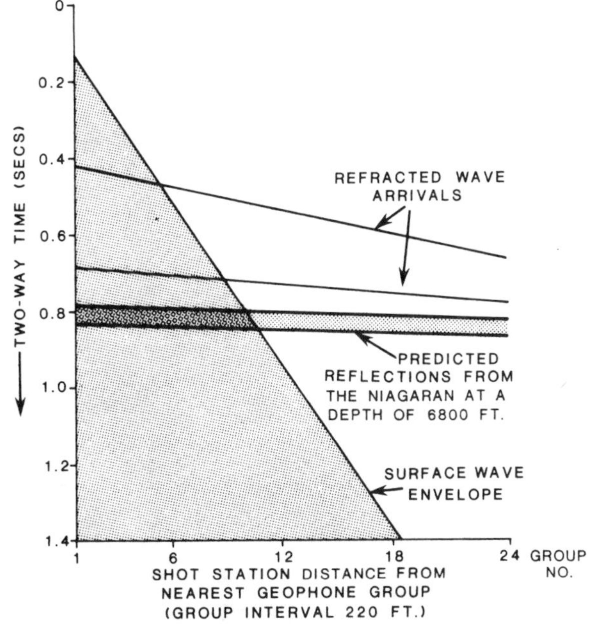


FIG. 13. Traveltime curves calculated from the simplified curve shown in Figure 11. The traveltime is plotted as a function of the geophone groups number; Group 1 being at the section corner, Group 24 being 5170 ft from the shot line.

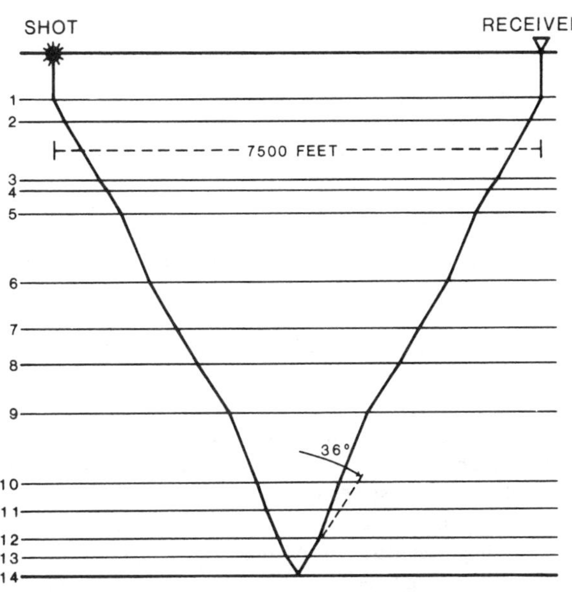


FIG. 14. Calculated raypath for the longest offset to be expected in the areal survey. This shows that the angle of incidence at the A-2 carbonate (reflector 14) is 36 degrees.

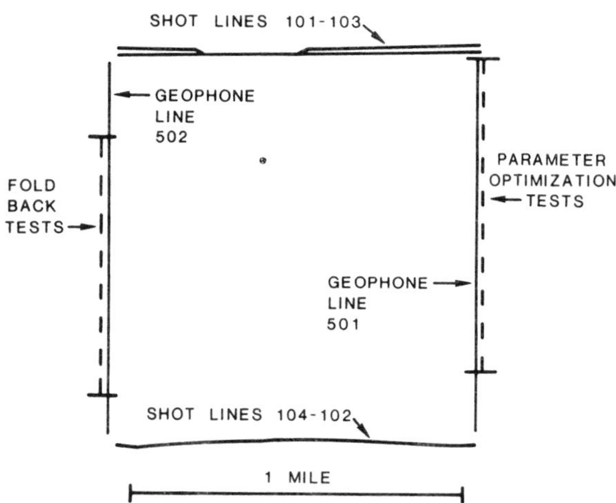


FIG. 15. The layout used for carrying out the fold-back and parameter optimization tests. Also shown are the lines used for the areal survey.

220-ft station spacing dictated by the latter choice meant that $\Delta x = 110$ ft and the Nyquist frequency was about 80 Hz. Preliminary indications were that reflections from the Niagaran were at 30–40 Hz.

The averaged velocity curve (dashed line) shown in Figure 11 was used to produce the simple traveltime curves shown in Figure 13 for surface waves, for refractions along the upper surface of the higher velocity sequence, and for the predicted reflections from the Niagaran. It can be seen that surface waves precede the reflections on the seismic section on only the first ten groups. Further, the migration process causes any waves of radically different velocity to the reflections to be degraded substantially. Also, the velocity data shown in Figure 11 were used to calculate the raypath for the longest source-receiver offset (7500 ft) we expected to use. The result (Figure 14) indicated that recording across the diagonal of a section would present little problem in obtaining reflections from the Niagaran.

Prior to the field recording of data, parameter optimization tests were carried out at the field location. These tests were designed to determine the most suitable dynamite charge size, type, and depth and to measure the predominate surface wavelengths and phase velocities.

## DETERMINING THE SOURCE PARAMETERS

The two most prominent geologic markers in the geologic section in Kalkaska County are the Dundee limestone at about 0.5 sec (~2500 ft) and the A-2 evaporite at about 1.0 sec (~5000 ft). These markers reflect energy predominantly in the range 25–35 Hz.

Charge depths were limited by environmental requirements to 20 ft or less. Practical limitations imposed by the water table reduced this to 10 ft. Because of the size of the camouflet generated, the largest charge that could be exploded at this depth was 2 lb. No data have been published showing source frequency content in glacial till, but the physical properties of clays are similar. O'Brien (1969) showed that a reduction in charge weight significantly increases the high-frequency content of the source signature. O'Brien measured the energy spectra 100 ft from charges placed 50 ft deep in a clay bed. He showed that for 3.4 lb, the peak was at 22 Hz, and for 0.68 lb, it was at 40 Hz. For our markers it appeared that a charge weight from 1 to 3 lb would be suitable.

**Table 1. Surface wave observations.**

| | |
|---|---|
| Wavelength, $\lambda$ | 225 ft |
| Frequency, $f$ | 20 Hz |
| Phase velocity, $C_p$ | 4300 ft/sec |
| Group velocity, $C_g$ | 2200 ft/sec |

Because the area of the survey has residences located in it and pipelines crossing it, single charges of 2 lb could not always be exploded. Therefore, experiments were carried out with single smaller charges and with arrays of small charges. No difference was found between the two types of charges for reflected waves from the prominent geologic markers, but there was, of course, a noticeable reduction in the amount of surface wave energy generated when the smaller charges were used. However, as we will show, surface wave noise is rather unimportant in this type of survey and, wherever possible, single charges were used.

As some compensation for attenuation and geometrical spreading, the charge size was increased (to a maximum of 2 lb) as the shotpoint receiver offset increased. For a typical crossed array (Figure 15) of shotpoint line 101 into receiver line 501, 0.25 lb charges were used for SPs 1–6, 0.5 lb for SPs 7–12, 1 lb for SPs 13–18, and 2 lb for SPs 19–24. This range of values is greater than that implied by the cube-root scaling law (O'Brien, 1969) but was justified heuristically. Where larger charges could not be used singly, they were replaced by patterns of smaller charges and the records were vertically stacked during processing.

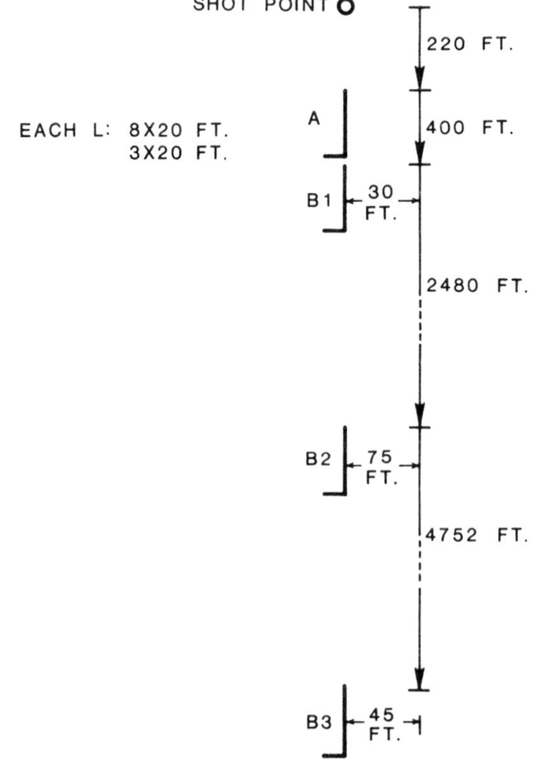


FIG. 16. Detail of the layout for the parameter optimization tests the location of which is shown in Figure 15.

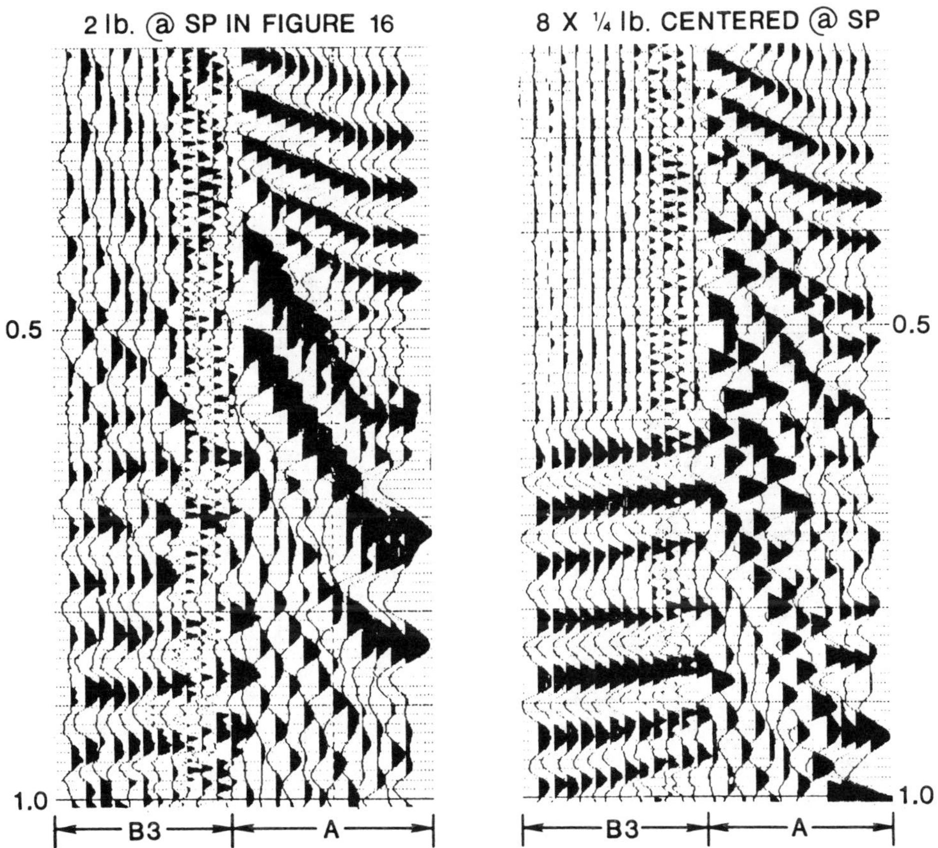


FIG. 17. Comparison between single shot records and records from arrays of shots for the L-shaped geophone arrays A and B3 shown in Figure 16.

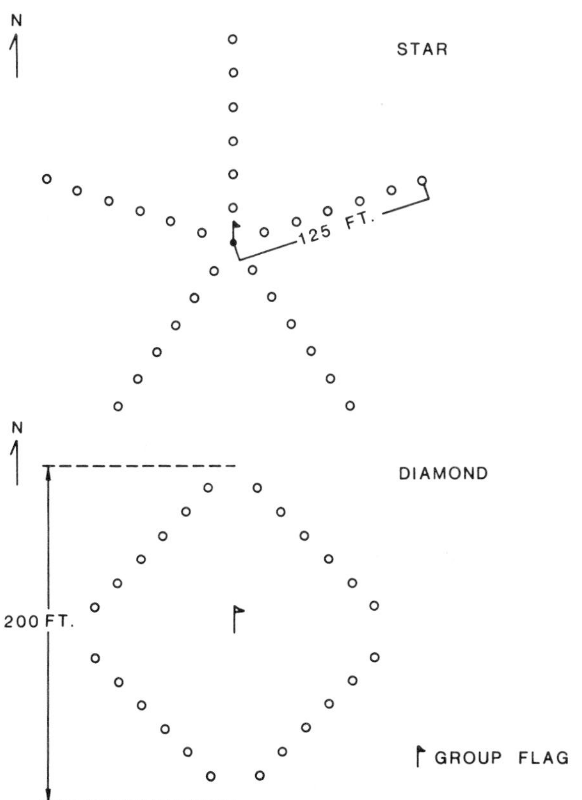


FIG. 18. The two noise-reducing geophone arrays tested.

## DETERMINING THE ARRAY PARAMETERS

Measurements of the surface wave parameters were carried out on the eastern side of the proposed survey area (Figure 15) using L-shaped arrays at different offsets to determine wavelengths and phase and group velocities (Figure 16). The maximum offset of 4752 ft equalled the offset to the first 21 groups in the proposed layout. The results of the surface wave measurements are given in Table 1.

A comparison between a single shot and a shot array record is shown in Figure 17. The single shot used was 2 lb at a depth of 9 ft at the shotpoint shown in Figure 16. The shot array was eight 0.25 lb charges centered about the shotpoint in line with a shot hole separation of 20 ft. The differences in the records are clearly discernible but, as was explained above, a compromise had to be made between single shots and shot patterns.

The purpose of a geophone array is to reduce by as big a factor as possible the amplitude of unwanted energy, usually surface waves. There is always a limit to the number of geophones available, and there is usually a practical limitation in actually placing geophone arrays on the ground. Also, in areal data collection two-dimensional (2-D) arrays should be used. Unless such an array can be placed out on fairly level ground, it is useless. In this experiment, the maximum elevation variation over the mile section was ±8 ft and areal arrays were feasible. The maximum number of geophones available per group was 30; two possible arrays for this number are shown in Figure 18. The normalized response $R_n$ of an array of $n$ geophones to a sine wave of frequency $\omega$ and wavenumber $\kappa$ is given by

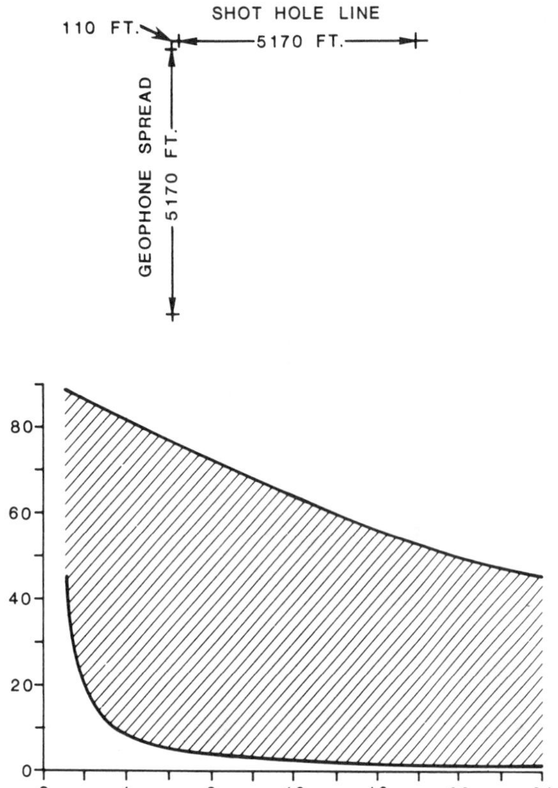


FIG. 19. The range of azimuthal angles for 24 geophone groups at 220-ft intervals in a crossed array layout. For example, the 8th group from the section corner would receive surface waves in an arc from 5 to 72 degrees.

$$R_n = \frac{\sin(n\,\pi\,\kappa\,d)}{\omega\,\sin(\pi\,\kappa\,d)},$$

where $d$ is the separation between the geophones which are assumed to be of equal sensitivity. The length of the array $D$ in the direction of the propagating wave is given by

$$D = (n - 1)d. \tag{2}$$

In areal surveying every geophone group presents a different azimuth to every shotpoint. Figure 19 shows the range of azimuths for each group in this experiment. The response of any array with a finite number of elements (geophones) has an azimuthal dependence. The variation in apparent $n$ as a function of azimuth $\phi$ for the five-pointed star array (Figure 18) is shown in Figure 20. It can be seen that there are ten response maxima, where $n = 30$, in 180 degrees. The apparent length of the array also varies as a function of azimuth; from a minimum of 0.90 $D$ where $\phi = 0$ to 0.938 $D$ where $\phi = 0.05\pi$. It can be shown from equation (2) that an array has to be of length $(\lambda - d)$ to reject totally a wavelength $\lambda$ if it is assumed that the geophones have equal weight. For a large number of geophones, it can also be assumed that the geophone locations sample a particular azimuth evenly in space and the azimuthal values of $D$ may be determined. These range from 0.06 $D$ for $\phi = 0$ to 0.323 $D$ for $\phi = 0.05\pi$. The "length" of an areal array may then be calculated for all values of $\phi$ which will totally reject the wavelength $\lambda$. For a value of $D = 250$ ft, it can show that the longest wavelength rejected is 265 ft along the azimuth $\phi = 0$ for the star array. However, the best response

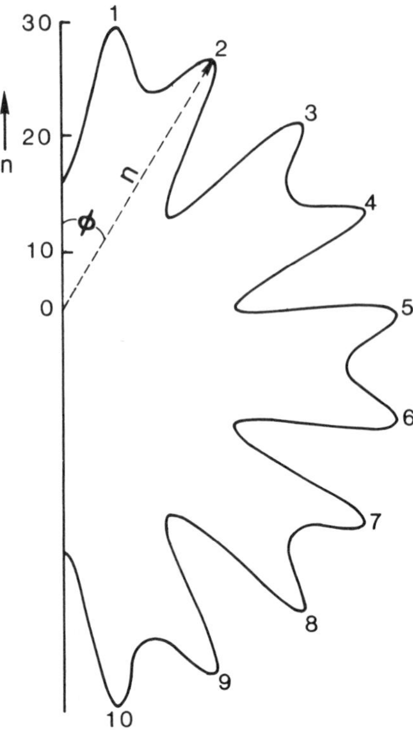


FIG. 20. The variation in the apparent number of array elements, $n$, as a function of azimuth $\phi$, for the 5-pointed star shown in Figure 18.

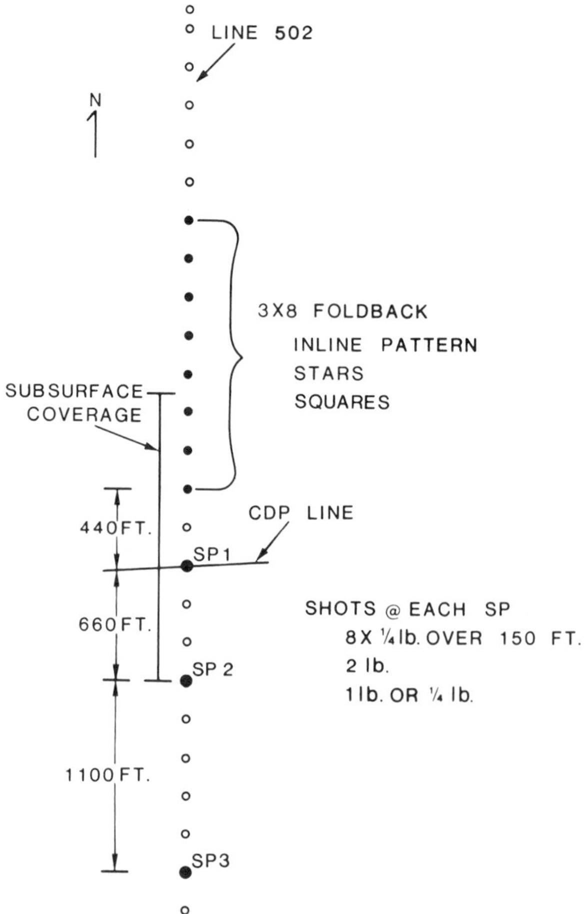


FIG. 21. Detail of the foldback tests shown in Figure 15.

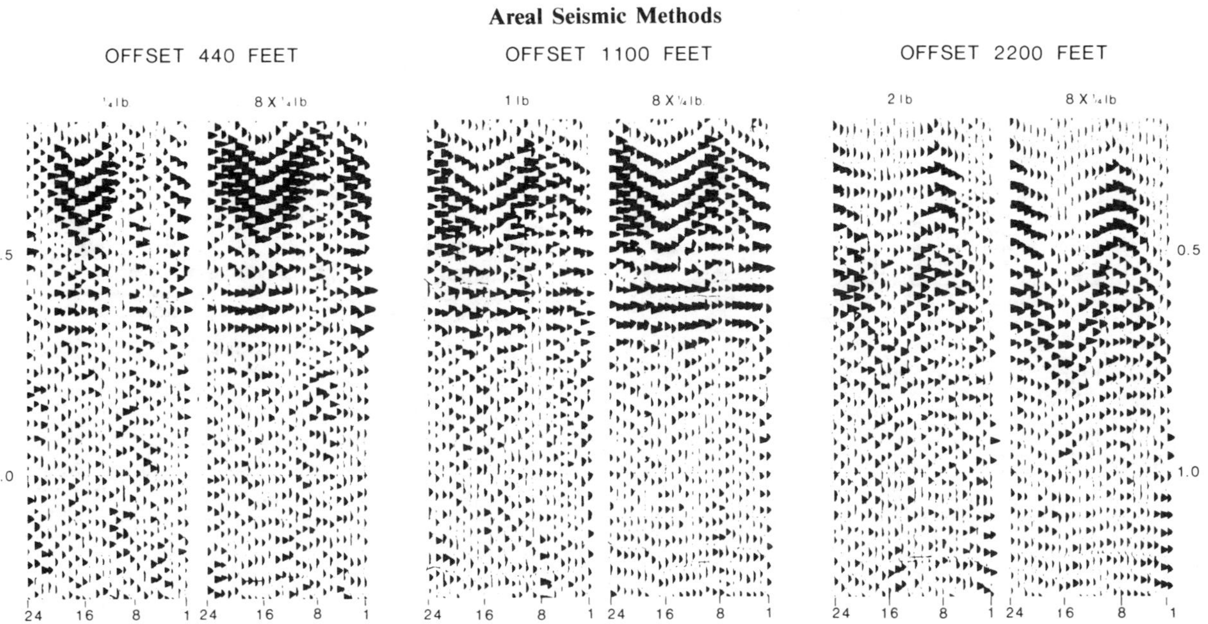


FIG. 22. Comparisons of in-line, star and square geophone arrays for both single shots and shot arrays for the fold-back tests shown in Figure 21.

is in the direction $\phi = 0.05\pi$, and in subsequent increments of $0.1\pi$, which has a rejection band for $\lambda$ from 8.92 to 258.6 ft (78.8 m). Our experimentally measured value of 223 ft given in Table 1 should then be attenuated by a maximum of 16 by the star array.

A similar analysis can be made for the square array (Figure 18). The value of $n$ and, hence, the rejection factor, will vary from 12 to 24. In some directions the square has heavily weighted geophone locations. Because the star array response is less directional, it was preferred in these experiments.

Despite careful experiments to design multidirectional noise reducing arrays, it is probable that the effort is not worthwhile. From the assumed surface wave group velocity used in Figure 13 and confirmed by measurement (Table 1), it can be seen (Figure 13) that beyond shotpoint 10 the surface waves arrive at the receivers later than the reflections from the A-2 carbonate. Thus, geophone arrays at the longer offsets serve little purpose. However, as shots were placed at both ends of the geophone lines, arrays were laid out at all group locations.

As a final check on the shot and geophone array parameters adopted for these experiments, a fold-back test was carried out in which the star arrays, square arrays, and conventional in-line arrays were laid out in coincidence. The in-line arrays had $D = 300$ ft and $n = 24$. Data were recorded from single shots and from shot arrays ($D = 150$ ft, $n = 8$) for each of the step-out locations shown in Figure 21. The results of these comparisons are shown in Figure 22; the range of offsets covered was from 440 to 3740 ft. The data have had the moveout removed and a 20–50 Hz filter applied. As would be expected, the combination of the star geophone array and linear shot array, in general, is the most effective combination. However, at longer offsets the in-line geophone arrays were less effective.

**Table 2. Recording parameters.**

| | |
|---|---|
| Recording system | DDS 620 |
| No. of channels | 24 |
| Geophones | 8 Hz |
| Geophones/group | 30 |
| Trace length | 3 sec |
| Sampling interval | 2 msec |
| Recording filters | out/out |

FIG. 23. The midpoint coverage achieved with the shooting geometry shown in Figure 15.

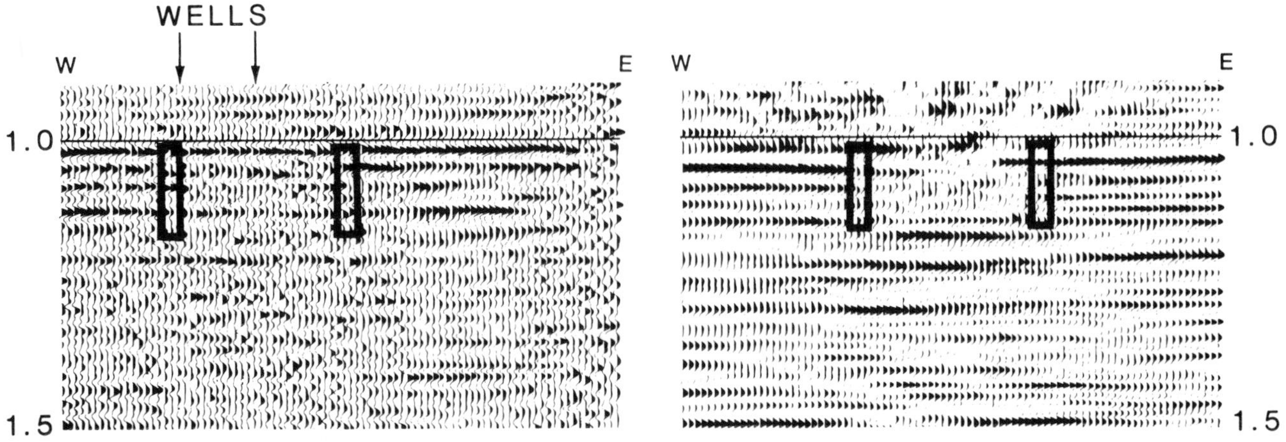


FIG. 24. Comparison between a CDP and a three dimensionally migrated seismic section, line 9EW, both west to east across the anomaly. The locations of these sections are shown in Figure 25. Left: CPD section. Right: Migrated section.

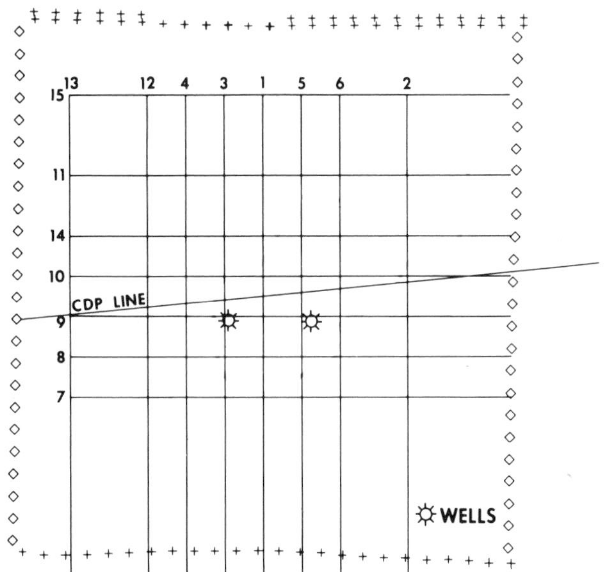


FIG. 25. The locations of the fifteen lines used in the areal interpretation and of the CDP line shown in Figure 24.

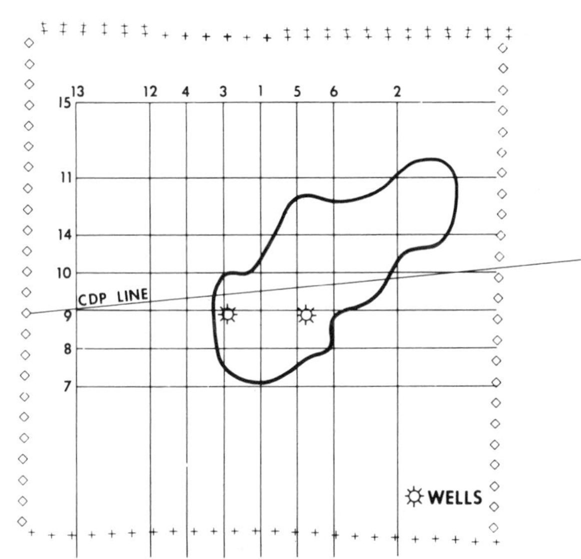


FIG. 27. Same as Figure 25 with the interpretation (zone of low reflectivity) added.

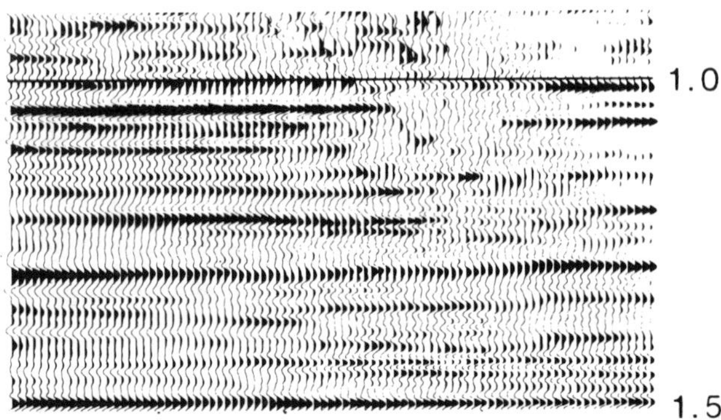


FIG. 26. Two of the seismic sections used in the interpretation without the anomaly picked. Line 6NS evidently crosses the anomaly, line 13NS does not.

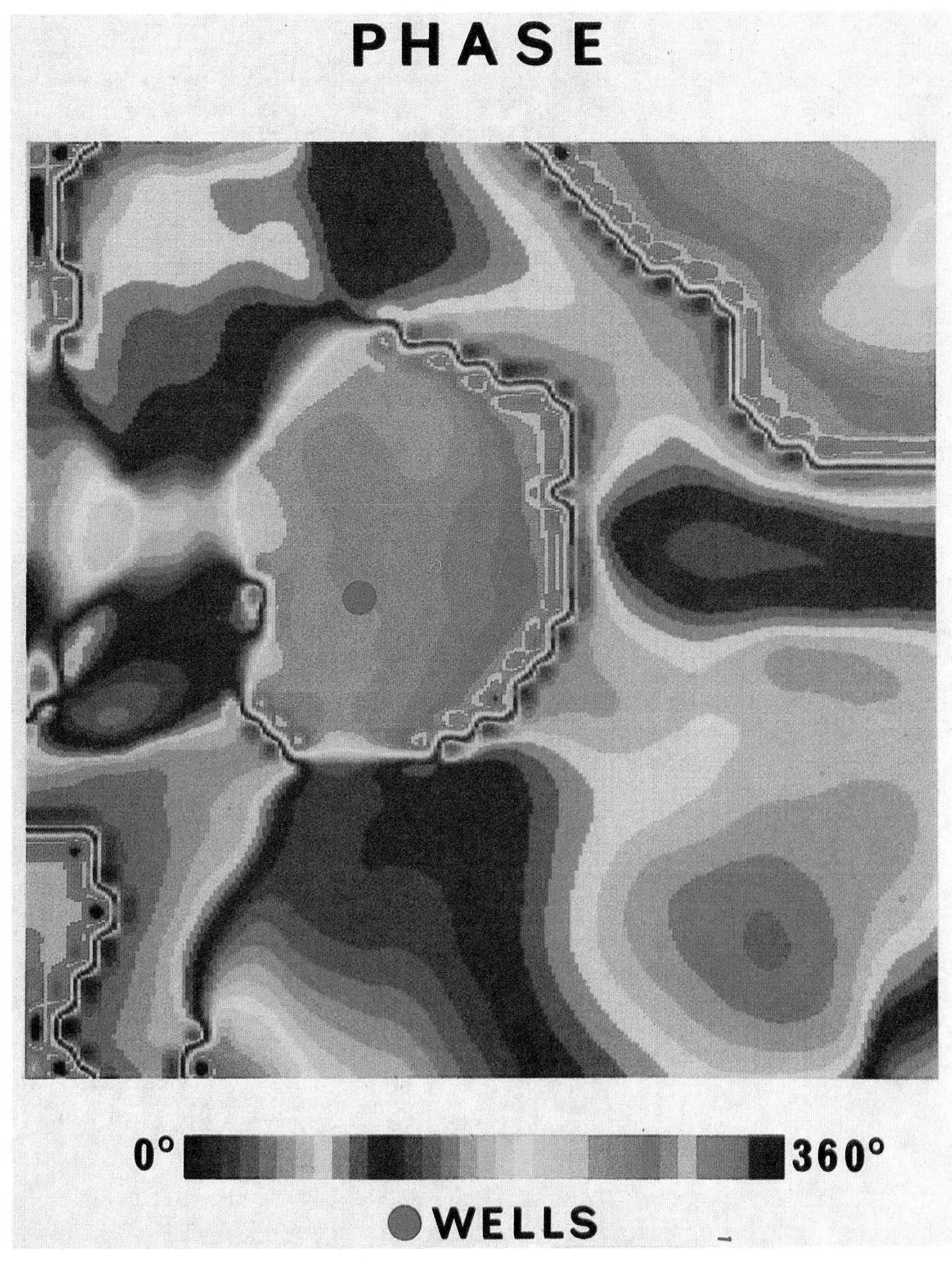


FIG. 28. A horizontal seismic section (instantaneous phase) at the time of the A-2 carbonate showing the anomaly with the well locations.

## FIELD RECORDING AND PRELIMINARY PROCESSING

The field records were obtained by using the geometry shown in Figure 15. Each shot line generated energy into each geophone group line. This geometry covered the section with 2304 traces at a spatial increment of about 110 ft (Figure 23). The recording parameters are given in Table 2.

The first stage in the data processing was to establish the geometry of the experiment. Reliable surveying information had been obtained by using an electronic distance measuring device to establish the locations of the section corners and of other suitable base stations. All station locations (shotpoint and geophone group) were then tied into these bases using a transit. An accurate large-scale map was drawn from the surveying information, and the map was digitized to produce the coordinates of all the stations with respect to an arbitrary origin. These coordinates were recorded on magnetic tape.

The seismic data were demultiplexed, deconvolved, and scaled. The deconvolution operator had 64 points and was used over the time window 0.25–2.75 sec. The data were passed through a zero phase filter with a passband of 20–60 Hz with 24 dB/octave slopes.

Two common preliminary steps were omitted in our processing sequence. These were time corrections for (1) elevation and for (2) residual statics. The first omission is easy to justify; the 96 stations occupied in this survey were at an average elevation of 1078 ft with a maximum deviation of ±8 ft. For the weathering velocity used, this value represented a correction of about ±4 msec. The residual static corrections were omitted because no suitable computer program was available. This may seem to have been a risky procedure, but the results appear to have justified the omission.

## MIGRATION PROCESSING

The programs used to process and display the model data referred to earlier were used to process and display the field data. First, in Figure 24, we show a comparison of part of a CDP seismic section from a line over the target reef with a seismic section reconstructed from the areal seismic data. The velocity function determined from the CDP seismic section was used for the 3-D migration. A discontinuity at the level of the A-2 carbonate, indicative of a reef (Caughlin et al, 1976), is apparent on both seismic sections.

A series of parallel west-east and parallel north-south seismic sections were reconstructed from the areal data, as indicated in Figure 25. All seismic sections were reconstructed with trace spacings of 55 ft producing a total of 13.3 miles of section. The migration aperture used (1000 ft) meant that about 280 traces were migrated into an output trace. A large number of traces were used to smooth over the avoidance of the static time correction and to sharpen the focus at the discontinuity.

Two of the sections generated are shown in Figure 26; one of them is across the discontinuity and one off it. The 15 lines shown in Figure 25 were then interpreted in the same manner using the reef criteria established by Caughlin et al (1976). The extent of the discontinuity and its relationship to the migrated lines in the section are shown in Figure 27; also shown are the locations of the producing wells.

The technique used to generate horizontal displays for the model data (Figure 9) also was used for the real seismic data. Horizontal cuts were made through the data at the horizon which exhibited the anomaly, the A2 carbonate. This display, shown in Figure 28, also gives the well locations. The differences between the interpretations in Figures 27 and 28 should be noted. Figure 28 represents the instantaneous phase at a precise time in the section; Figure 27 is a representation of both phase and amplitude over several tens of milliseconds.

## CONCLUSIONS

We have shown in these experiments that areal seismic techniques can be applied successfully to the study of lithological or stratigraphic problems. Recent advances in instrumentation have increased the number of channels available for simultaneous recording fifty-fold compared to the 24 channels used in these experiments. With these large numbers of channels, areal methods are probably more efficient than conventional CDP methods for sampling the subsurface. We have also shown that many of the noise problems which are inherent to in-line shooting methods can be safely ignored without decreasing the quality of areally collected data. Finally, it is evident that the black-and-white vertical seismic section may no longer provide everything that an interpreter requires in making an evaluation of a prospect. Lithological changes which have a horizontal expression need to be illustrated horizontally, and parameter changes can be most successfully demonstrated in color.

## ACKNOWLEDGMENTS

Many people contributed to the success of this project. R. L. Geyer gave invaluable advice on the design of the field experiments. R. A. Bain of the Mid-Continent office of Gulf Exploration & Production Company gave every encouragement. The experiments were carried out by the Gulf Laboratory research crew which included A. L. Cogley, J. R. Decheck, A. R. Lingenfelter, R. G. Slater, and D. C. Stothart. Backup to the research crew was provided by a Seismograph Service Corporation crew under the management of J. Craft.

The model experiments were carried out by R. I. Morris. Much of the program development was carried out by R. V. Hetrick, and R. I. Morris was in charge of all the data processing. N. J. Kear and S. S. Witt drafted the figures.

## REFERENCES

Bubb, J. N., and Hatlelid, W. G., 1977, Seismic stratigraphy and global changes of sea level, part 10: Seismic recognition of carbonate buildups, *in* Seismic stratigraphy—Applications to hydrocarbon exploration: AAPG memoir 26, p. 185–204.

Caughlin, W. C., Lucia, F. J., and McIver, N. L., 1976, The detection and development of Silurian reefs in northern Michigan: Geophysics, v. 41, p. 646–658.

French, W. S., 1974, Two-dimensional and three-dimensional migration of model experiment reflection profiles: Geophysics, v. 39, p. 265–277.

Gardner, G. H. F., French, W. S., and Matzuk, T., 1974, Elements of migration and velocity analysis: Geophysics, v. 39, p. 811–825.

Gardner, G. H. F., and Kotcher, J. S., 1977, Improvements in velocity analysis, dip analysis, statics, and signal-to-noise ratio by the use of Kirchhoff summation: Geophysics, v. 42, p. 152–153.

Klose, G. W., and Holland, W. G., 1975, Seismic expression of a carbonate-evaporite sequence in northern Alberta, Canada: J. Can. SEG, v. 11, p. 38–47.

McClintock, P. L., 1976, Noise, drift hamper Michigan seismic: Oil and Gas J., v. 74, p. 105–110.

Mesolella, K. J., Robinson, J. D., McCormick, L. M., and Ormiston, A. R., 1974, Cyclic deposition of Silurian carbonates and evaporites in Michigan basin: AAPG Bull., v. 58, p. 34–62.

O'Brien, P. N. S., 1969, Some experiments concerning the primary seismic pulse: Geophys. Prosp., v. 17, p. 511–544.

Taner, M. T., and Sheriff, R. E., 1977, Application of amplitude, frequency, and other attributes to stratigraphic and hydrocarbon determination, *in* Seismic stratigraphy—Applications to hydrocarbon exploration: AAPG memoir 26, p. 301–327.

Taner, M. T., Koehler, F., and Sheriff, R. E., 1979, Complex seismic trace analysis: Geophysics, v. 44, p. 1041–1063.

Van Siclen, De W. C., 1957, Organic reefs of Pennsylvanian age in Haskell County, Texas: Geophysics, v. 22, p. 610–629.

# Seismic shear-wave observations in a physical model experiment

Robert H. Tatham*, Donald V. Goolsbee*, Wulf F. Massell*, and H. Roice Nelson‡

## ABSTRACT

The observation and common-depth-point (CDP) processing of mode-converted shear waves is demonstrated for real data collected in a physical model experiment. The model, submerged in water, represented water depth scaled to 250 ft, the first subsea reflector at 4000 ft, and the last reflector at 7000 ft below the sea floor with a structural wedge at the center. Very efficient mode conversion, from *P* to *SV* and back to *P*, is anticipated for angles of incidence at the liquid-solid interface (sea floor) between 35 and 80 degrees.

The model, constructed of Plexiglas and 3180 resin, will support elastic shear-wave propagation. One anticipated problem, internal reflections from the sides of the model, was solved by tapering the sides of the model to 45 degrees off vertical. The *P* wave reflection coefficient at an interface between Plexiglas and water is 35 percent for vertical incidence, but it diminishes to very nearly zero between 43 and 75 degrees. Thus, by tapering the sides of the model, any undesired internal *P* wave reflections had to undergo at least two reflections at angles of incidence in the low reflection coefficient range for *P* waves.

Data were collected in both an end-on CDP mode, with offsets from 1000 ft to 20,000 ft, and a variety of walk-away experiments with scaled ranges from 1000 ft to 31,000 ft. Processing and analysis of the data confirm the existence of mode-converted shear-wave reflections in a modeled marine environment. In particular, the *S* wave reflections from all interfaces are identified on both the 100 percent gathered records and the final stacked records. These *SV* wave reflections were isolated for stacking by considering those portions of the gathered records, both offset and arrival time, that correspond to optimum angles of incidence. In addition, $\tau$-$p$ processing isolated particular angles of incidence, further confirming the incidence angle-range criterion. Thus, the desired events are unambiguously identified as mode-converted shear waves.

## INTRODUCTION

Recent success in developing seismic shear-wave techniques for petroleum exploration in land environments gives us encouragement to reexamine possible shear-wave sources for the marine environment. Possible methods include placing both sources and receivers on the sea floor with direct generation and recording of seismic shear waves. This represents a direct extension of land techniques and may be applicable to some shallow-water areas where the sediments at the sea floor will support shear-wave propagation. For deep water, or areas with unconsolidated sediments, such direct methods may not be possible. Conventionally generated *P* waves, however, are converted to shear waves (*SV* polarization) when interacting with a discontinuity in elastic parameters, such as the water bottom or base of unconsolidated sediments. Tatham and Stoffa (1976) suggested such a method of recording shear-wave data in a marine environment. Since the publication of that paper, viable shear-wave sources for land operations have evolved and recent attention has been directed to evaluating the interpretational value of *S* wave data. Industry interest in the potential usefulness of shear-wave recording and interpretation is reflected in the report of over 125 crew months (about 6000 miles) of land shear-wave recording during 1980 (Senti, 1981). With this expression of interest in shear-wave recording we are reexamining some of the challenging problems in developing marine shear-wave sources.

## MODE CONVERSION

In general, when *P* wave energy propagating in a solid material encounters an impedance discontinuity, four outgoing seismic waves result: reflected *P* and *S* waves and refracted *P* and *S* waves. If the incident wave is propagating in a water layer, however, there can be no reflected *S* wave. In such a case, the incident *P* wave impinges on the water bottom (or base of unconsolidated sediments) and energy is partitioned into three waves—reflected *P*, refracted *P*, and refracted (mode-converted) *S*. As the angle of incidence *i* increases from vertical, mode conversion to shear waves generally becomes more efficient. Further, as *i* exceeds the *P* wave critical angle, no *P* wave energy is refracted into the

Presented at the 51st Annual International SEG Meeting October 12, 1981, in Los Angeles. Manuscript received by the Editor January 5, 1982; revised manuscript received August 23, 1982.
*Petty-Ray Geophysical, P.O. Box 36306, Houston, TX 77036.
‡Formerly Seismic Acoustics Lab, University of Houston, Houston; presently Landmark Graphics, Houston.


Reprinted from Geophysics, 48, 688-701

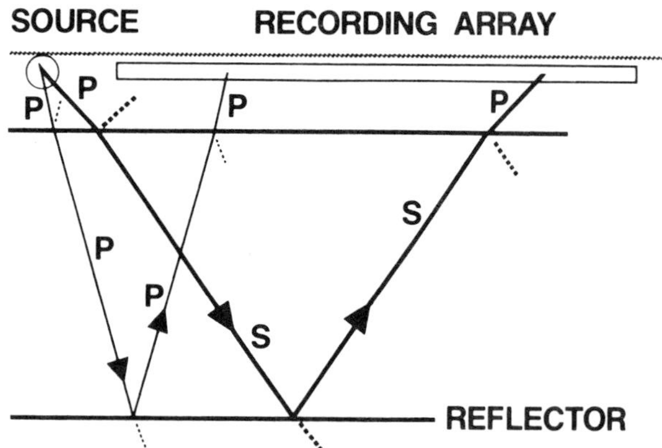


FIG. 1. Geometry of a typical marine seismic recording array, showing a conventionally reflected *P* wave and a mode-converted *PSSP* ray at large angle of incidence being detected at large offset. (After Tatham and Stoffa, 1976.)

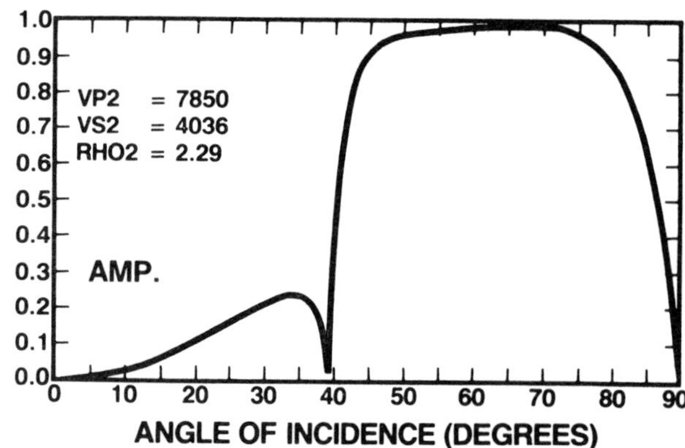


FIG. 2. Amplitude, as a fraction of the original *P* wave amplitude, for a double mode-converted wave (*PSSP*) at all angles of incidence. Note large amplitude of *S* wave, after two mode conversions, at large angles of incidence (from the vertical) beyond the critical angle. The water has a *P* wave velocity of 5000 ft/sec and a density of 1.0 g/cm$^3$. Sediments have a *P* wave velocity of 7850 ft/sec, an *S* wave velocity of 4036 ft/sec, and a density of 2.3 g/cm$^3$.

subsea sediments. For these angles, only two possible avenues of energy propagation are available—the reflected *P* wave and the refracted (mode-converted) *S* wave. For some contrasts in elastic constants most of the incident energy goes into the refracted *S* wave and, for other elastic parameters, most of the energy is reflected as a *P* wave. For subsea sediments with a *P* wave velocity between 6500–9000 ft/sec, a range that includes many shallow sedimentary rocks, the mode conversion is extremely efficient (Tatham et al, 1977).

Figure 1 illustrates the recording geometry for observing mode-converted *S* waves. We assume that the ray of interest leaves the source as a *P* wave, is converted from a *P* wave to an *S* wave (*SV*) at only one interface (such as the water bottom or base of unconsolidated sediments), is reflected from depth as an *S* wave, and then undergoes mode conversion from an *S* wave to a *P* wave upon reentering the upper (water) layer. Keep in mind that mode conversion occurs only for nonvertical angles of incidence. Since mode conversion becomes more efficient as the angle of incidence increases, we would expect to observe mode-converted arrivals primarily on the longer-offset traces of multifold CDP data.

This double mode conversion may, at first thought, appear rather esoteric. It is, however, equivalent to the *SKS* phase of

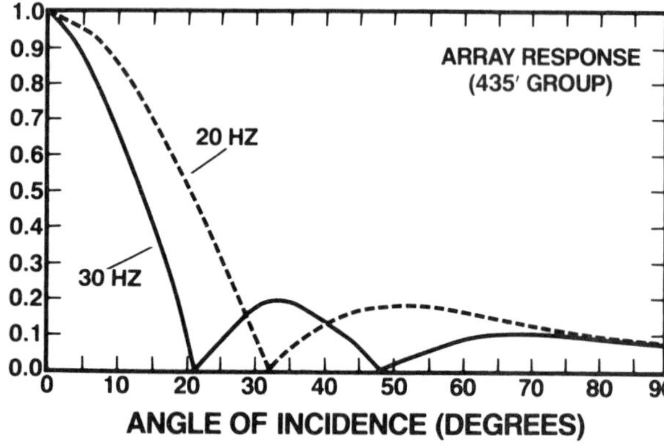


FIG. 3a. Array response, as a function of angle of incidence, for the 435 ft hydrophone arrays employed in recording the data reported by Tatham and Stoffa (1976). Near-surface (water) velocity was assumed to be 5000 ft/sec, and the response is plotted for 20 and 30 Hz signals.

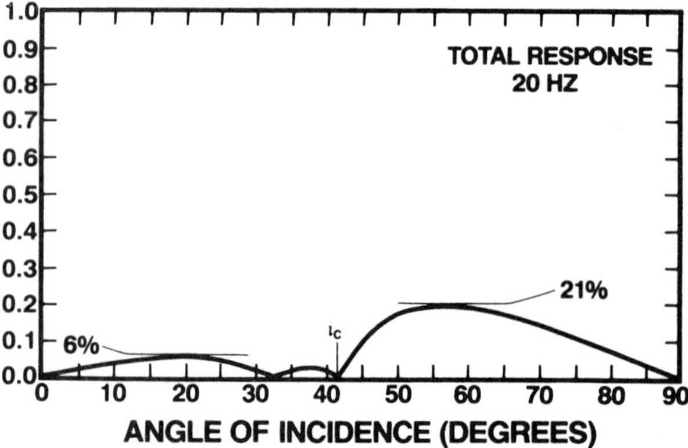


FIG. 3b. Applying mode-conversion coefficients from Figure 2 to the 20 Hz array-response curve in Figure 3a. Note that the maximum anticipated amplitude, accounting for array response, is only about 20 percent of the original *P* wave amplitude.

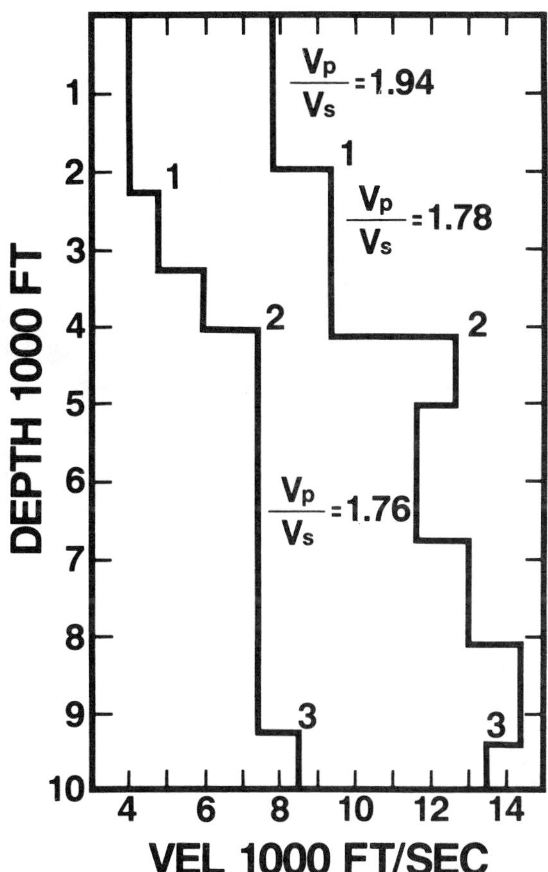


FIG. 4. $P$ and $S$ wave velocity profiles as a function of depth for an area offshore the Florida panhandle. $V_p/V_s$ values are averages taken between the surface and interface 1, and between interfaces 1 and 2, and 2 and 3, respectively. (After Tatham and Stoffa, 1976).

earthquake seismology, except that we also include a reflection. Richter (1958, p. 259) pointed out that the observed *SKS* phase, at the proper ranges, is often large and increases in amplitude with increasing range.

Figure 2 shows the anticipated amplitude (displacement potential) for all angles of incidence, for plane $P$ waves converted to $S$ waves at the water bottom, and with another mode conversion back to $P$ wave upon reentering the water layer. The elastic parameters used for the calculation were determined by velocity analysis of the data considered by Tatham and Stoffa (1976). The amplitude is the fraction of the incident $P$ wave for just the two transmission mode-conversion coefficients, ignoring attenuation and the shear-wave reflection coefficient of the subsurface reflector. Note that for angles of incidence less than the critical angle of $P$ waves (39 degrees), the resulting amplitude ranges from 5 to 20 percent of the incident $P$ wave amplitude. Beyond the critical angle, however, the $S$ wave reflection with two mode conversions has an amplitude nearly equal to the reflection coefficient of the $S$ wave, i.e., most of the energy is mode-converted rather than reflected at the water bottom. Recall that beyond the critical angle no $P$ wave energy penetrates the subsurface.

**Table 1. Model velocities and densities.**

| | (g/cm$^3$) | Actual velocity (ft/sec) | Scaled velocity (ft/sec) |
|---|---|---|---|
| Water | 1.00 | 5000 | 12,000 |
| Plexiglas | 1.17 | $V_p$ = 9000<br>$V_s$ = 4500 | $V_p$ = 21,600<br>$V_s$ = 10,800 |
| 3180 resin | 1.42 | $V_p$ = 8654<br>$V_s$ = 4327 | $V_p$ = 20,770<br>$V_s$ = 10,385 |

## PREVIOUS INVESTIGATION

As mentioned above, the earlier work of Tatham and Stoffa (1976) made suggestions of observations of mode-converted shear waves for some routinely recorded marine data offshore the Florida panhandle. The observed *SV* waves were, at best, weak. The disparity between predicted and observed amplitudes for the shear-wave energy is attributed to the rather long (435 ft) hydrophone groups contained in the recording array. The suppression of the low-velocity events that traveled most of their travel path as *SV* waves is nearly 80 percent. Figure 3a shows the hydrophone array response, for the 435 ft array, adapted to angle of incidence. The total response for shear waves, the array response multiplied by the mode-conversion curves of Figure 2, is given in Figure 3b. Note that the observed events, accounting for source array response and ignoring attenuation, should be at most 21 percent of that predicted by the mode-conversion curves alone. This value is in accordance with the reported observations.

In spite of these rather weak events, both $P$ and $S$ wave velocity profiles were determined and are shown in Figure 4. These interpreted velocity profiles were used as a guide for constructing the physical model.

## PHYSICAL MODEL STUDY

The Seismic Acoustics Laboratory at the Univ. of Houston operates a large water-filled tank for gathering ultrasonic acoustic data over physical models representing earth structures. The 3-D models allow body-wave transmission, and no "plate-wave" considerations enter into the scheme. Model scaling is such that one inch on a model represents 1000 ft of real earth, ultrasonic frequencies scale to seismic frequencies, and velocities scale to 2.4 times the velocity of the model materials. Thus, water has a scaled velocity of about 12,000 ft/sec, and data were typically recorded at a scaled sample interval of 1 msec.

### Model construction

A physical model to test the recording of mode-converted shear-wave data was constructed at the Seismic Acoustics Laboratory. The materials, listed in Table 1, were chosen such that mode-converted shear waves should be observed. The physical dimensions were sufficient to allow a single walk-away spread to a total scaled distance of over 30,000 ft and an end-on marine cable configuration could be simulated with range up to 20,000 ft. The structure, varying only slightly with a short 9½ degree ramp, is unchanging normal to the profile. All four sides are tapered. The first reflecting interface, as shown in Figure 5, is at a scaled depth of 4000 ft (4 inches in the model) with a 1000 ft ramp in the center of the model. The last layer is 1000 ft thick and the total model thickness scales to 7000 ft. The mode-conversion

Table 2. Normal moveout velocities.

| P wave | | S wave | |
|---|---|---|---|
| $t$ (msec) | $V$(scaled) (ft/sec) | $t$ (msec) | $V$(scaled) (ft/sec) |
| 20 | 12,000 | 0 | 5,900 |
| 100 | 19,600 | 40 | 5,900 |
| 200 | 20,400 | 200 | 9,600 |
| 3000 | 21,000 | 400 | 10,095 |
| | | 1000 | 10,250 |
| | | 1800 | 10,500 |
| | | 3000 | 11,000 |

coefficients for Plexiglas and water are shown in Figure 6. These compare quite favorably with the curves developed for offshore Florida. The normal incidence reflection coefficient between the Plexiglas and 3180 resin is about 3.5 percent and about 35 percent between the Plexiglas and water.

One anticipated problem was undesired $P$ wave energy being reflected and scattered from the sides of the model and arriving at the receiver ahead of the desired $S$ wave reflections. Attenuation of such spurious energy was realized by tapering the sides of the model to 45 degrees. This design was selected to exploit low $P$ wave reflection coefficients in a 43–85 degree range, as shown in Figure 7. The $P$ waves are incident upon the Plexiglas-water boundary from within the Plexiglas. As illustrated in Figure 8, at least two reflections with angles of incidence in the 43–80 degree

*(Text continued on p. 701)*

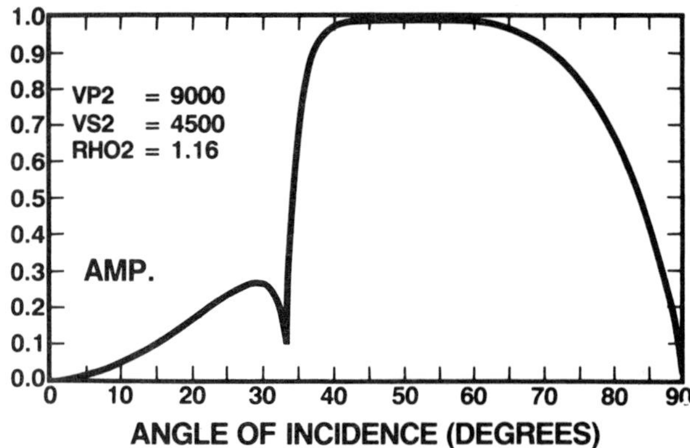


FIG. 6. Mode-conversion coefficients (amplitude) for water-Plexiglas interface. Note similarity to Figure 2. Curve is the same when scaled velocities are substituted.

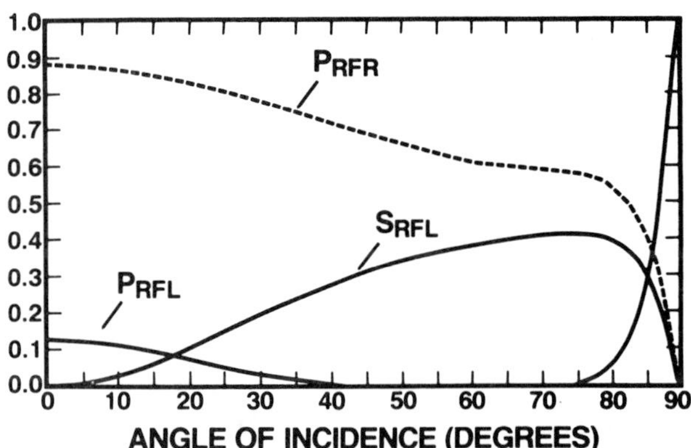


FIG. 7. Reflection coefficients (energy), as a function of angle of incidence, for a $P$ wave incident upon a Plexiglas-water interface. $P$ wave is incident upon the interface from within the Plexiglas. Note the low value of $P$ wave reflection coefficient for angles between 40 degrees and 70 degrees.

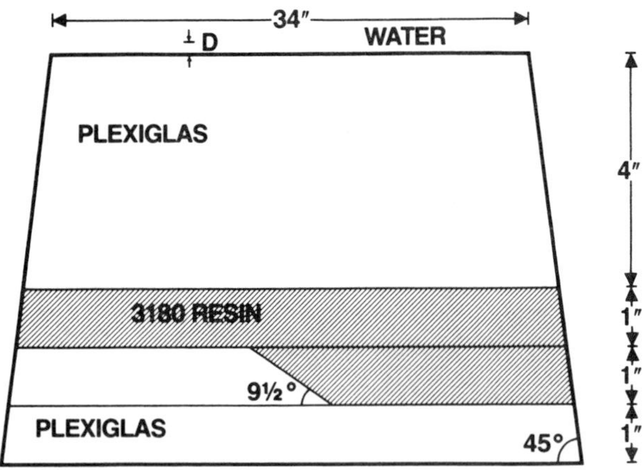


FIG. 5. Cross-section of physical model constructed at the Seismic Acoustics Laboratory at the University of Houston. Vertical exaggeration is four to one. The first layer is 4 inches thick, which scales to 4000 ft (scale factor is 1 inch = 1000 ft.) The 1000 ft (scaled) ramp in the center of the model adds structural interest. The distance $D$ between source and top of model scales to 250 ft. The model is immersed in water to conduct the experiment. Physical properties of water, Plexiglas, and 3180 resin are summarized in Table 1. The model, tapered on all four sides, is 4 inches wide by 34 inches long at the top and 18 inches wide by 48 inches long at the base.

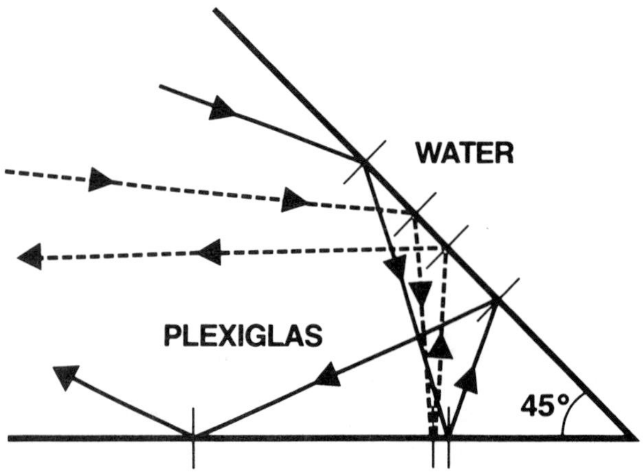


FIG. 8. Two raypaths for a ray traveling toward a 45 degree corner in Plexiglas. Note that for an outgoing ray to leave the corner area toward the upper surface of the model it must have at least two reflections at angles of incidence between 40 and 70 degrees.

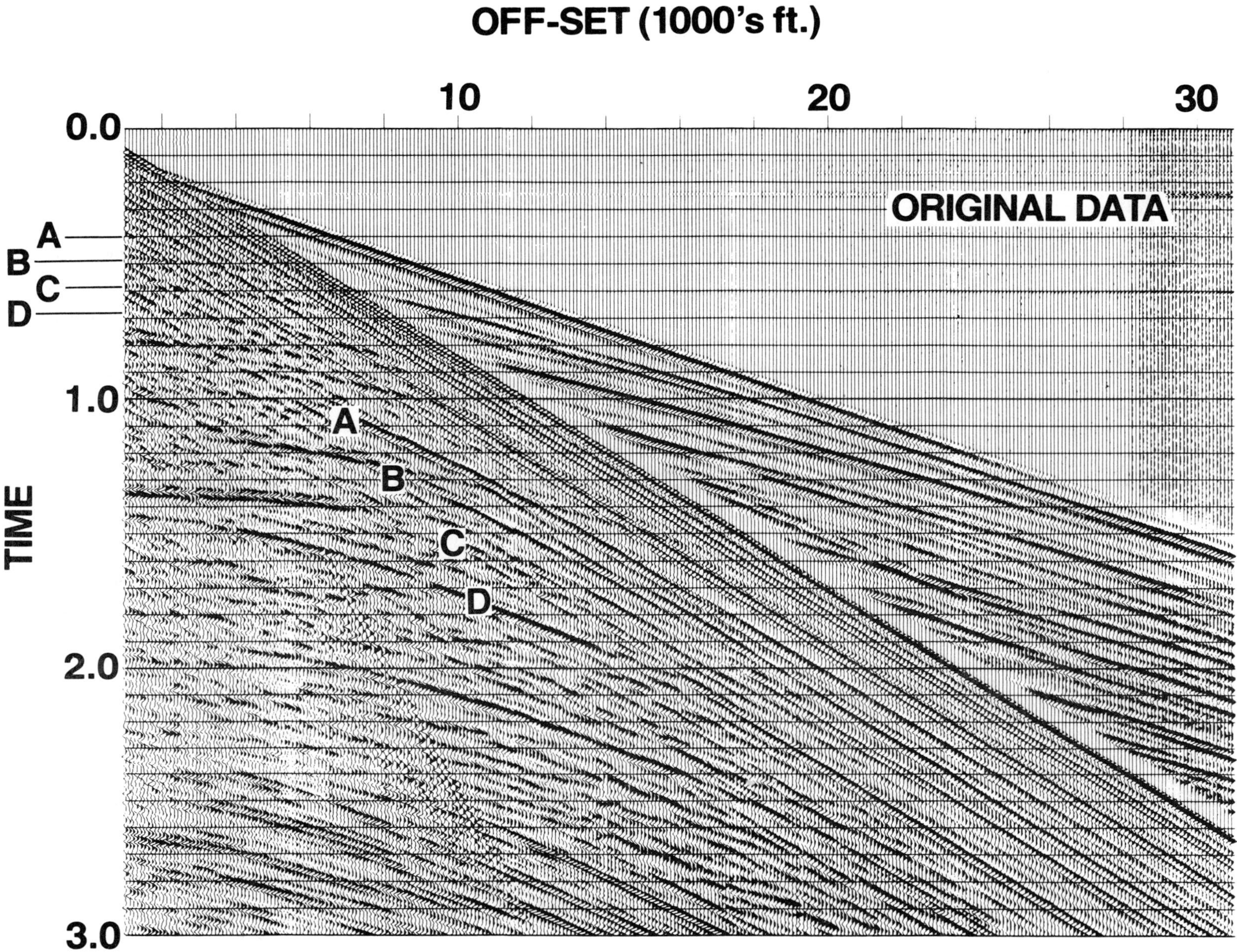


FIG. 9. Data recorded over the physical model. Ranges (offsets) are 1000 ft (scaled) to 31,000 ft. Source and detector ($P$ wave transducers) were 250 ft above the solid interface, and the walk-away was done in 100 ft increments. Note numerous reflection and refraction events. Letters A, B, C, and D approximately locate particular $P$ and $S$ wave reflections.

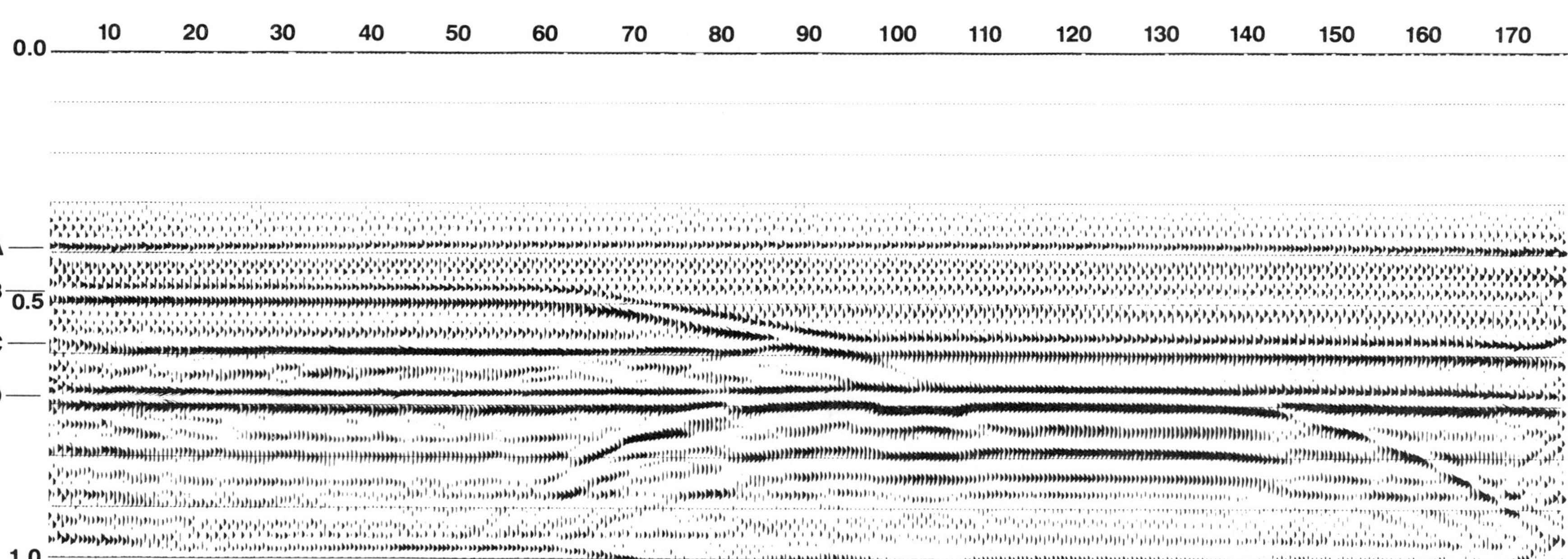


FIG. 10. Conventional *P* wave CDP stack of data collected over the physical model. Ranges of 1000 to 8000 ft were used, and the 200 ft shot spacing yields an 18-fold CDP stack. Note the reflections at the lower Plexiglas-resin interface at 0.4 sec (event A), the resin-Plexiglas interface (including ramp) at about 0.5 sec on the left-side of the section (event B), and the base of the model (Plexiglas-water interface) at about 0.7 sec (event D). No reflection is observed from the top of the model, with depth of 250 ft and minimum hydrophone offset of 1000 ft.

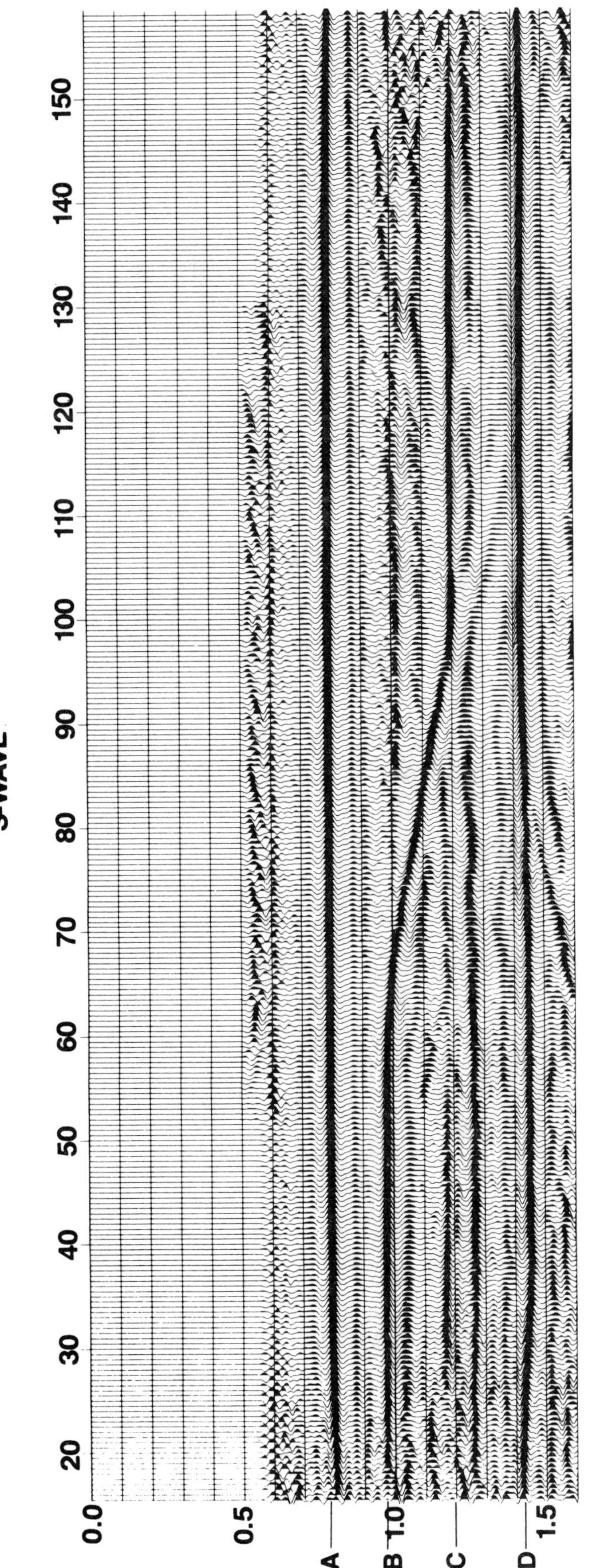


FIG. 11. CDP stack of data collected over the physical model, but applying shear-wave velocities. Ranges of 5600 to 25,000 ft were employed. Since $S$ wave velocity is about one-half the $P$ wave velocity, the time scale of the display is one-half that of Figure 10. Note reflections from Plexiglas-resin interface at about 0.74 sec (event A), resin-Plexiglas interface that includes the ramp (event B), and the base of the model (Plexiglas-water) at 1.35 sec (event D). The observed events correlate very well with the events observed in the $P$ wave data. The 180 degrees phase reversal is consistent with the double mode conversion.

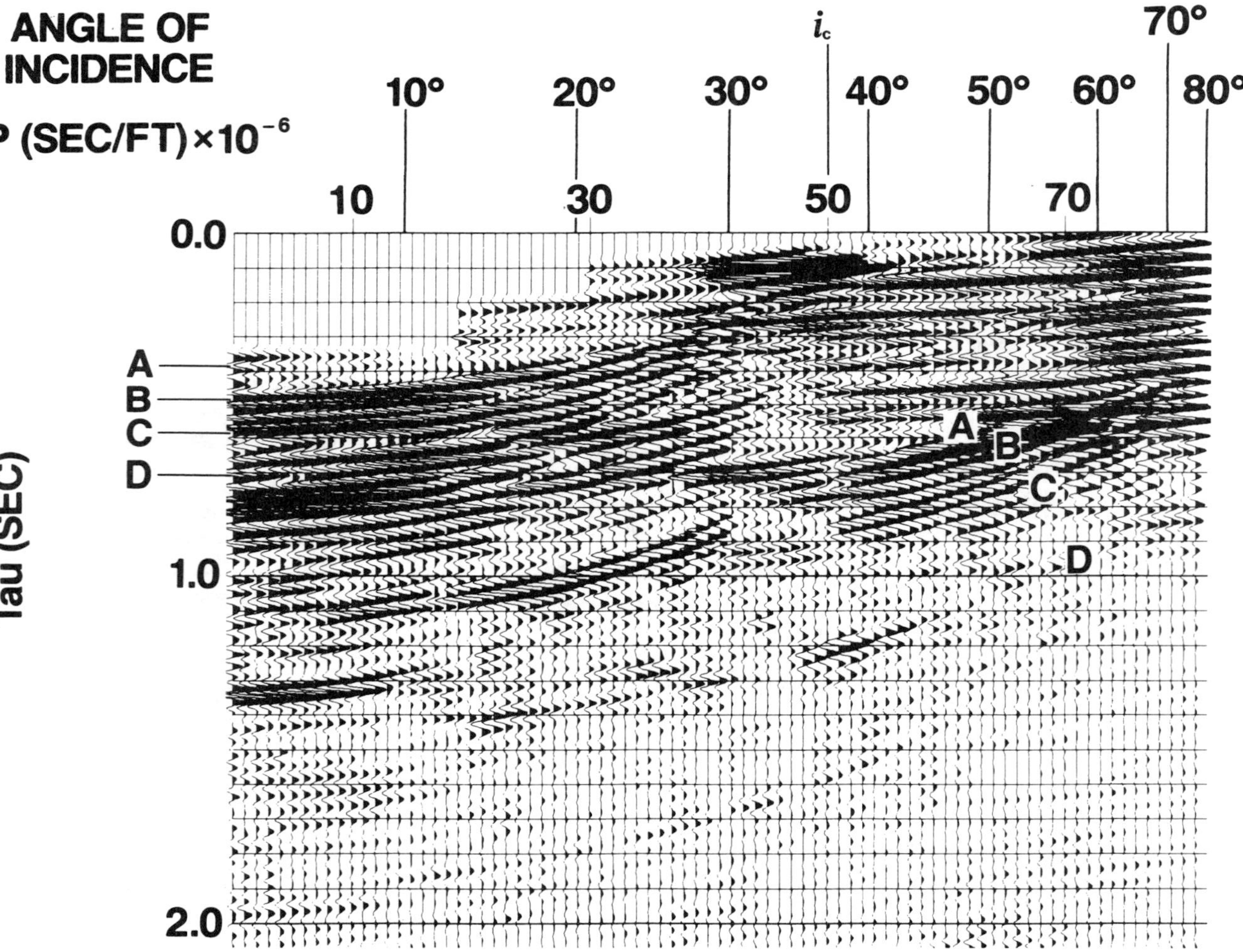


FIG. 12. Tau-$p$ transform (slant-stack) applied to data (single field record) shown in Figure 9. Vertical axis is time, and each trace represents a constant ray parameter $p$. Knowing the water velocity, each ray parameter represents a particular angle of incidence. Note difference in character beyond the critical angle. Reflection hyperbolas appear as ellipses in the tau-$p$ transform. The ellipses beyond the critical angle are shear-wave (mode-converted) reflections. Some muting of noise is applied at low $p$ and low tau values. Letters approximately locate $P$ and $S$ wave reflections shown in Figures 10 and 11.

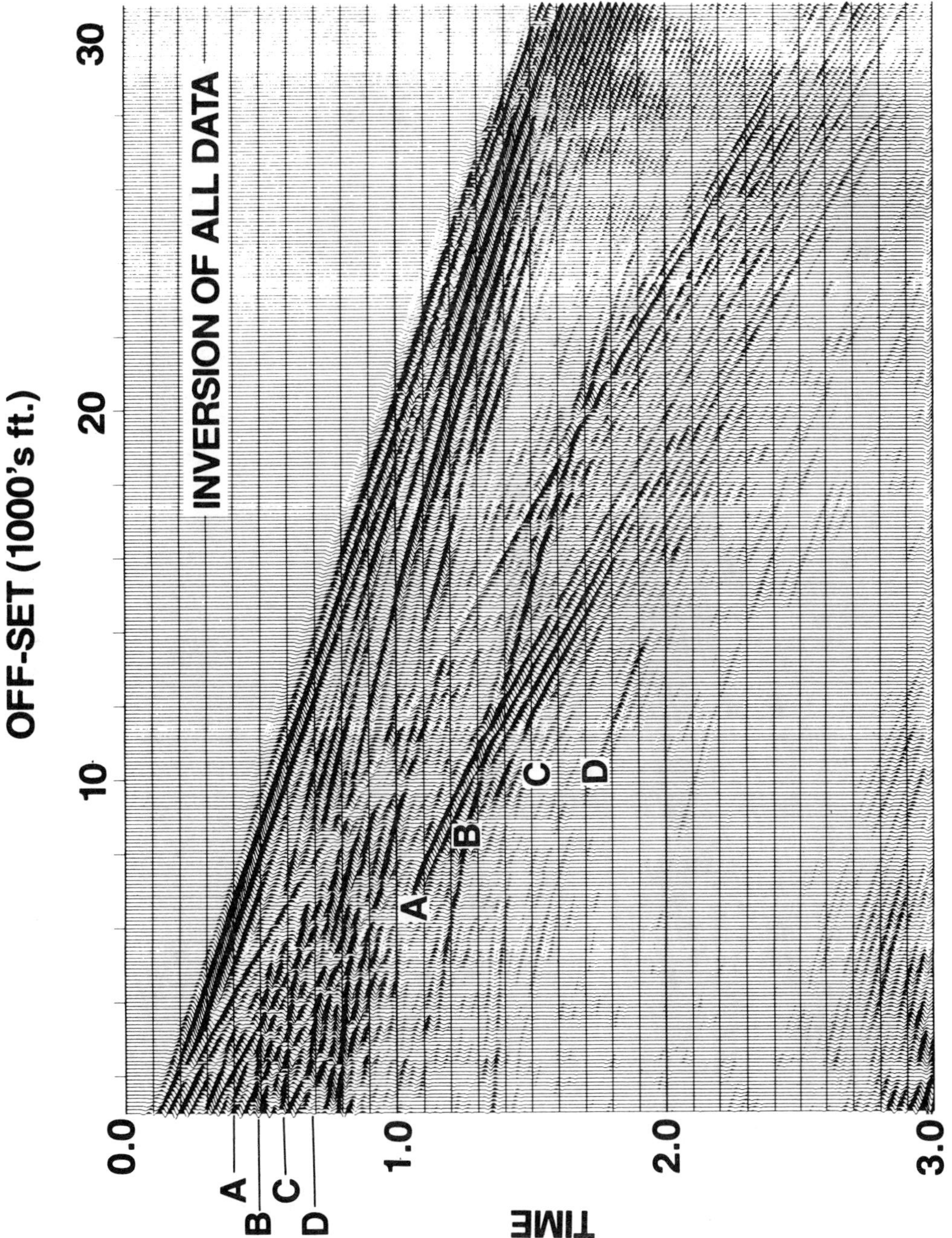


FIG. 13. Inverse transform (slant-stack) applied to data in Figure 12. Note similarity to Figure 9, but with some noise removed. The noise was muted in the tau-$p$ transform space. Also, different play-back scaling was applied. Letters identify approximate location of events considered in Figures 10 through 17.

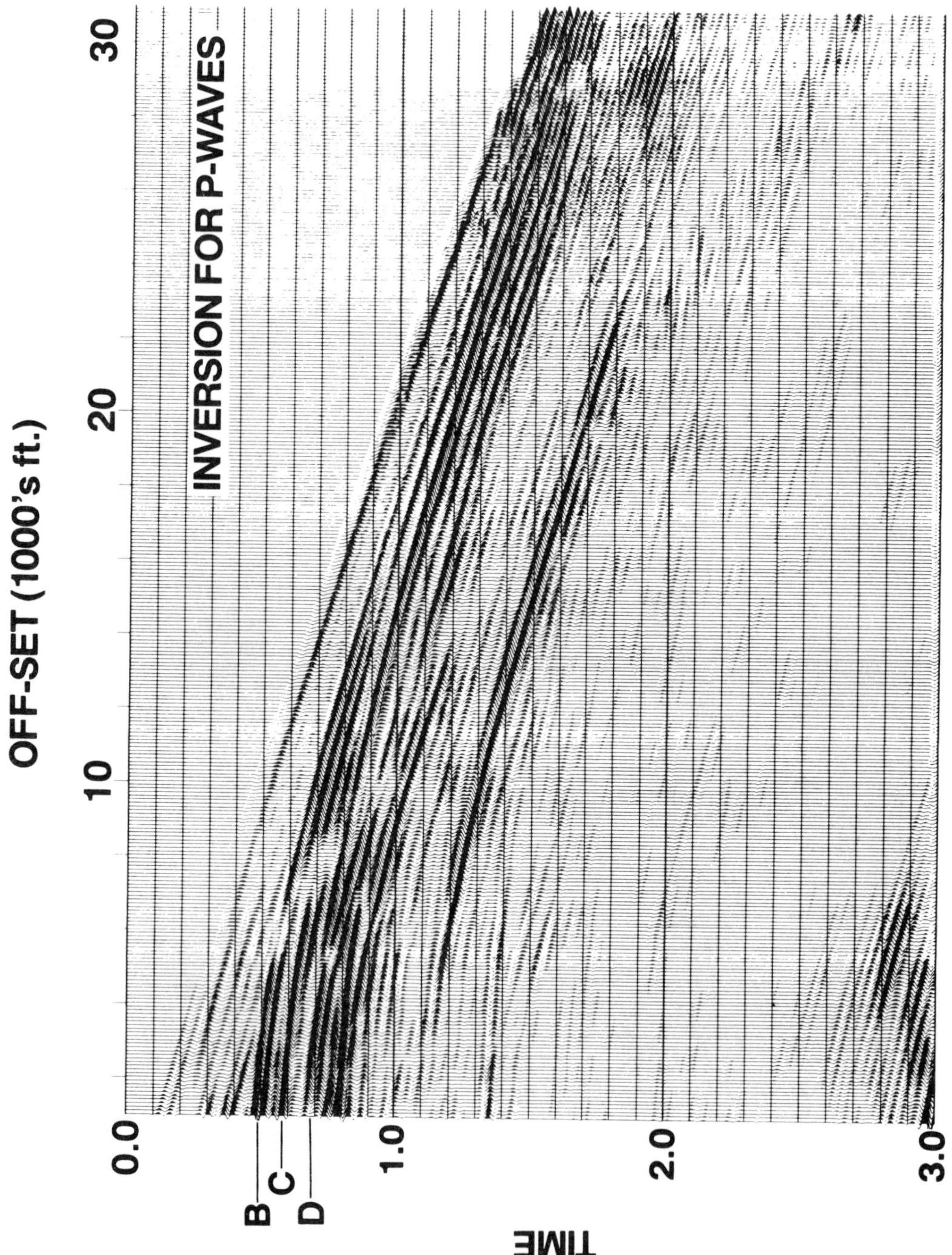


FIG. 14. Inverse transform applied to data in Figure 12, but including only ray parameters at angles of incidence less than the critical angle. Thus, this record is primarily *P* wave energy. Letters identify relevant events.

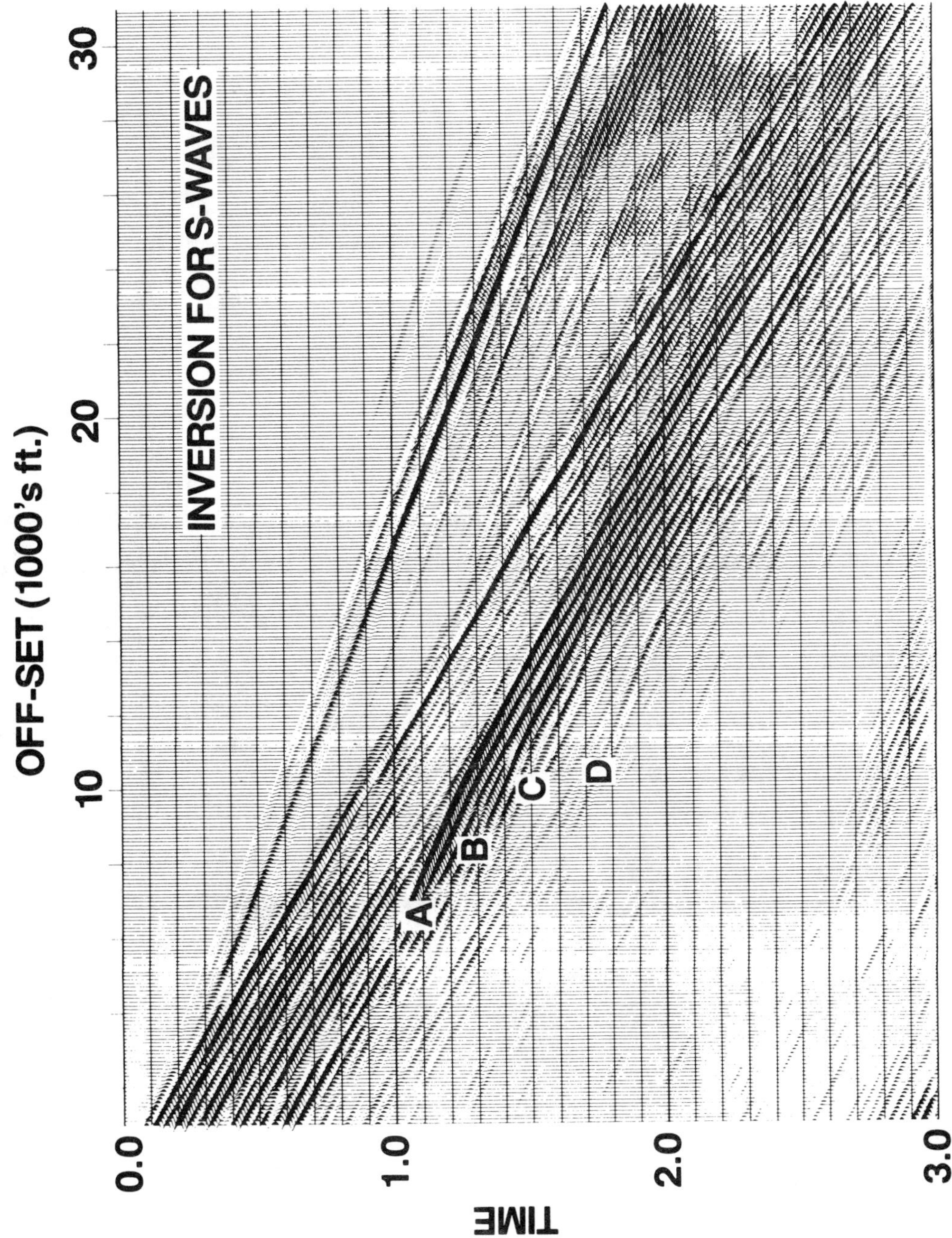


FIG. 15. Inverse transform applied to data in Figure 12, but including ray parameters at angles of incidence between 40 and 75 degrees. Thus, this record is primarily *S* wave energy and low-velocity noise. The *S* wave reflections show curvature. Index letters show approximate locations of *S* wave reflections.

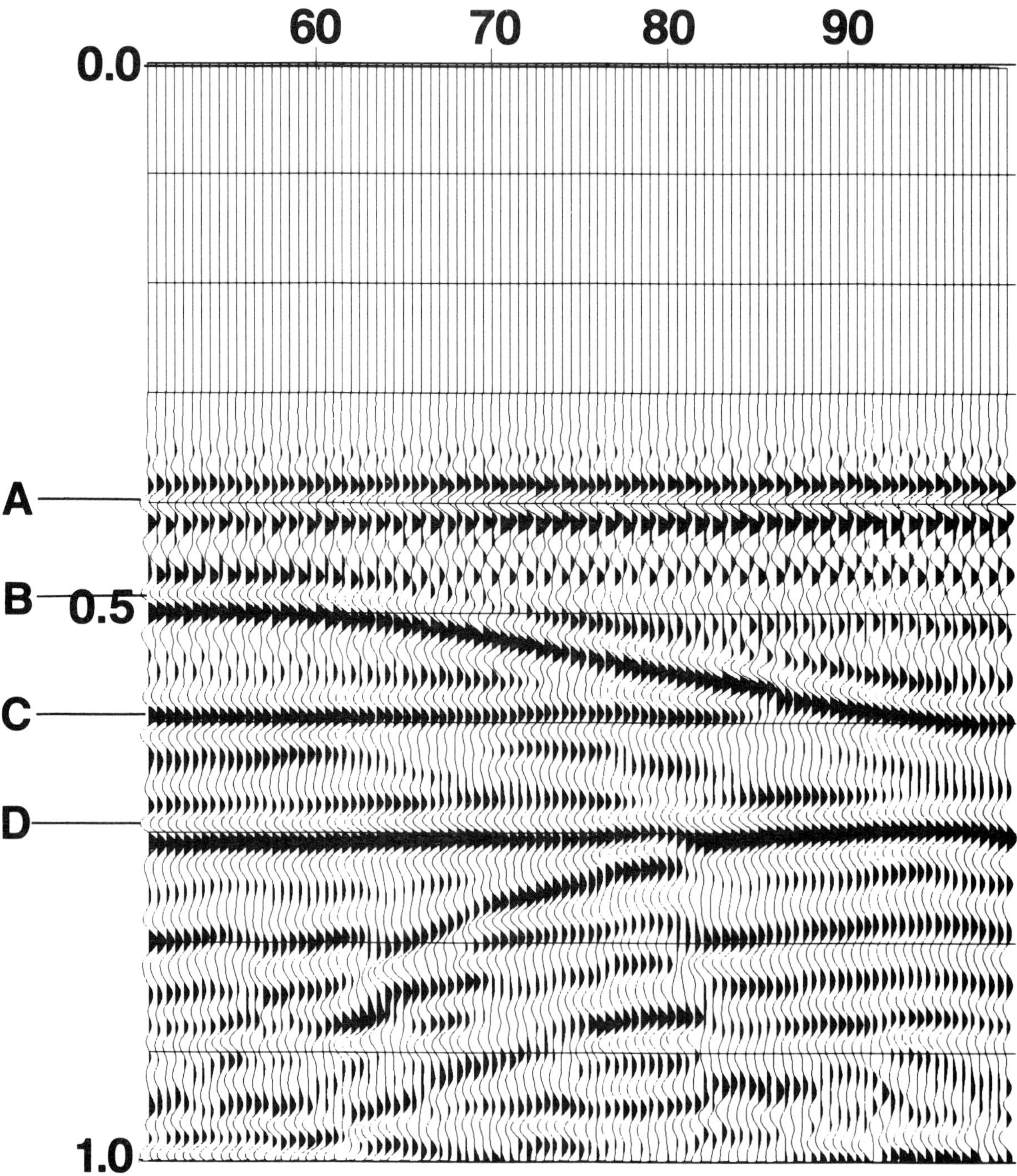


FIG. 16. Stack of CDP gathers, at $P$ wave velocity, after the gathered records were transformed (slant-stack) to tau-$p$ space, and inverse-transformed with angles of incidence less than the critical angle. Only full 48-fold gathers were used, thus the lack of stacked traces from the CDP taper. Comparison with Figure 11 confirms the success of this procedure.

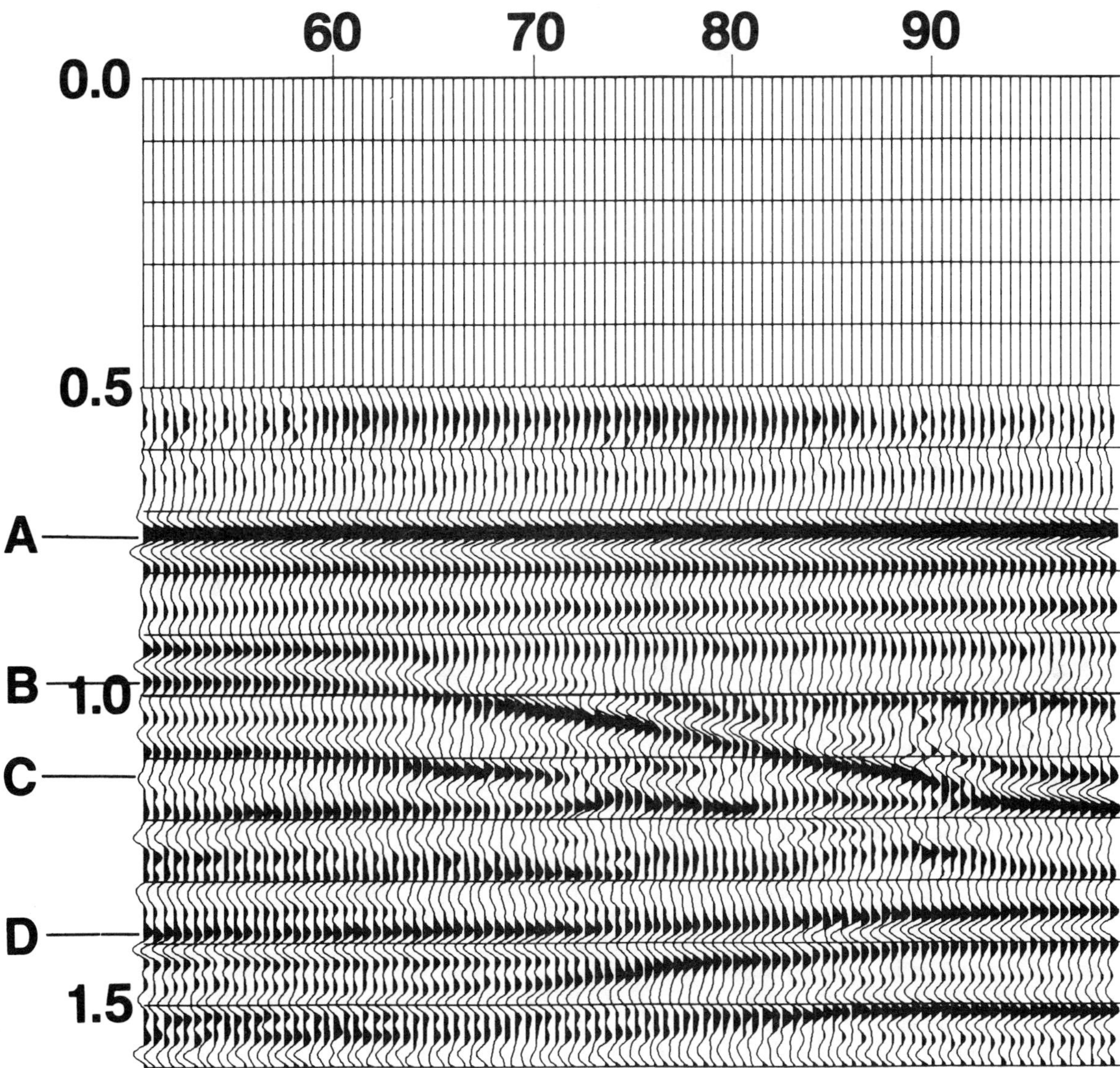


FIG. 17. Stack of CDP gathers, at *S* wave velocities, after the gathered records were transformed (slant-stack) to tau-*p* space, and inverse-transformed with angles of incidence between 40 and 75 degrees. As with the *P* wave stack, only full 48-trace gathers were used, and thus the lack of stacked traces from the CDP taper. Comparison with Figure 11 further confirms that these stacked events represent mode-converted shear waves from angles of incidence beyond the critical angle.

range are required for any ray originating from the surface to return to the surface of the model. Hence, almost no spurious *P* wave energy is seen in the records.

### Test data

Figure 9 shows a walk-away spread with a shot spacing of 100 ft, the near offset of 1000 ft and the longest offset of 31,000 ft. The source and detector were 0.25 inches (250 scaled ft) above the solid Plexiglas interface, thus at an equivalent water depth of 250 ft. Immersion in the water, however, eliminated the usual free surface, and thus we see no waveguide effects resulting from the water layer. Many reflected waves, as well as direct arrivals, are observed and will be discussed below.

In addition to the walk-away spread, a conventional end-on CDP marine line was recorded. Both shot and detector spacing was 200 ft, with ranges from 1000 to 20,000 ft. This led to a maximum CDP coverage of 48 fold, with considerable taper at the ends of the lines.

Figure 10 shows a conventional *P* wave stack, utilizing ranges of 1000 to 8000 ft (18 fold) in the CDP gathers. Table 2 gives normal moveout velocities applied. Note the presence of the ramp at 0.4 to 0.9 sec at the center of the model. Figure 11 shows the stack with shear-wave velocities. Ranges from 5600 ft to a maximum 25,000 ft were considered in the CDP stack. Since *S* wave velocities are about one-half the *P* wave velocities, the *S* wave section is displayed at one-half the time scale of the *P* wave section, and thus the *P* and *S* wave events should nearly correlate. Note the presence of the ramp at 1.00 to 1.40 sec on the *S* wave section, as well as all the reflections observed on the *P* wave section, thus fully confirming the presence of mode-converted shear-wave (*SV*) reflections.

### Tau-*p* processing

Applying a tau-*p* transform (slant stack) to the shot-oriented field data, or CDP gathers, will display the data as a function of angle of incidence (Stoffa et al, 1981). Figure 12 shows a tau-*p* transform applied to the data shown in Figure 9. Note that the reflection hyperbolas of Figure 9 are seen as ellipses in Figure 12. Also note the strong change in character at the critical angle. The ellipses beyond the critical angle are mode-converted shear waves. Individual values of *p* correspond to particular angles of incidence, as indicated in Figure 12.

An inverse transform can be applied to the data of Figure 12, resulting in the original shot-oriented field record, or the original CDP gather.

Figure 13 represents an inverse transform applied to all the data in Figure 12. Comparison with Figure 9 confirms the success of this inversion. Figure 14 is an inversion, to record space, of the data in Figure 12, but using only those ray parameters (0 to $42 \times 10^{-6}$ sec/ft) corresponding to angles of incidence between 0 and 33 degrees. That is, no postcritical reflections are considered. Note the improvement in the overall quality of the *P* wave reflections. Figure 15 is a similar inversion of the data in Figure 12, but with ray parameters restricted (*p* from $55 \times 10^{-6}$ to $80 \times 10^{-6}$ sec/ft) to angles of incidence between 40 and 75 degrees. Note the presence of reflections, with curvature consistent with *S* wave velocities, at the longer ranges.

Further tau-*p* processing was applied to the CDP gathered records constructed from the end-on marine profile. Only those CDP gathered records with offsets of 1000 to 20,000 ft were considered (48 fold). A forward tau-*p* transform was applied to each record and two inverse transforms were then applied. The result was two sets of CDP gathers, one with a range of ray parameter angles of incidence consistent with *P* wave data and another consistent with *S* wave data. That is, the *P* wave gathers represent angles of incidence between vertical and 35 degrees while the *S* wave gathers represent angles of incidence between 40 and 75 degrees.

Figure 16 is a stack of the *P* wave gathers, and Figure 17 is a stack of the *S* wave gathers. Only full 48-fold gathers were used in the stack. Comparison of the data in Figures 16 and 17 with the conventionally stacked data in Figures 10 and 11 adds confidence to the original identification of the *P* and *S* wave reflections and shows some of the potential of tau-*p* processing of seismic reflection data.

## DISCUSSION AND CONCLUSIONS

The observation and CDP processing capability of mode-converted shear waves has been demonstrated for real data collected in a physical model experiment. That is, observation of shear-wave energy in data generated and recorded as *P* waves in a simulated marine environment has been confirmed. The primary difference between the recording geometry and that employed in a typical marine setting was the large offsets (up to 20,000 ft) applied. Examination of the walk-away analysis suggests that shear waves may have been successfully stacked with a maximum range as short as 16,000 ft, but this is still greater than a typical marine configuration.

A significant difference between the model and real marine case is the absence, in the model, of the water layer itself. Thus, no guided waves in the water-layer waveguide were present to contaminate our observations further. The scaled velocities of the model were different from those for the real earth, but the consistency of velocity ratios across interfaces makes this consideration insignificant.

This study suggests that further work should be applied to actual offshore situations. Some operational problems, such as large offsets, can be overcome. Possible anisotropy effects, however, may be difficult. Levin (1979, 1980) suggested that rock anisotropy may have a strong effect on the *SV* polarized shear waves. Actual field experiments will be required to determine fully the extent of this potential problem.

## ACKNOWLEDGMENTS

We wish to thank Petty-Ray Geophysical, Geosource Inc., for encouragement to conduct these experiments and publish the results and the Seismic Acoustics Laboratory, Univ. of Houston, for assistance in conducting the experiments. Dr. C. Walker and Dr. J. Keeney offered critical reviews of the manuscript.

## REFERENCES

Levin, F. K., 1979, Seismic velocities in transversely isotropic media: Geophysics, v. 44, p. 918–936.

——— 1980, Seismic velocities in transversely isotropic media, II: Geophysics, v. 45, p. 3–17.

Richter, C. F., 1958, Elementary seismology: San Francisco, W. H. Freeman and Co.

Senti, R. J., 1981, Special Report, Geophysical Activity in 1980: Geophysics, v. 46, p. 1316–1333.

Stoffa, P. L., Buhl, P., Diebold, J. B., and Friedman, W., 1981, Direct mapping of seismic data to the domain of intercept time and ray parameter, a plane-wave decomposition: Geophysics, v. 46, p. 255–267.

Tatham, R. H., Danbom, S. H., Tyce, R. C., and Omnes, G. C., 1977, Seismic parameters and their estimation: Progress in estimation and interpretation: SEG Cont. Ed. School, The stationary convolutional model of the reflection seismogram and progress in measurement of subsurface seismic parameters: Velocity, density, specific attenuation: Houston, Texas, November 28–29.

Tatham, R. H., and Stoffa, P. L., 1976, $V_p/V_s$—A potential hydrocarbon indicator: Geophysics, v. 41, p. 837–849.

# Imaging beneath a high-velocity layer using converted waves

Guy W. Purnell*

ABSTRACT

High-velocity layers (HVLs) often hinder seismic imaging of deeper reflectors using conventional techniques. A major factor is often the unusual energy partitioning of waves incident at an HVL boundary from lower-velocity material. Using elastic physical modeling, I demonstrate that one effect of this factor is to limit the range of dips beneath an HVL that can be imaged using unconverted $P$-wave arrivals. At the same time, however, partitioning may also result in $P$-waves outside the HVL coupling efficiently with $S$-waves inside. By exploiting some of the waves that convert upon transmission into and/or out of the physical-model HVL, I am able to image a much broader range of underlying dips. This is accomplished by acoustic migration tailored (via the migration velocities used) for selected families of converted-wave arrivals.

## INTRODUCTION

In many areas of the world, the presence of high-velocity layers (HVLs) makes it difficult to seismically image deeper reflectors. One component of the problem is the relatively limited penetration of $P$-waves through HVLs. Another component concerns processing the data to effectively exploit waves (reflections from deeper interfaces) that do penetrate. In general, the class of such waves may include $P$-waves and $S$-waves that experience one or more mode conversions along the path from source to receiver.

This paper addresses the energy partitioning of waves incident at HVL boundaries and demonstrates one way to exploit these mode conversions.

Rock salt, basalt, and carbonate are materials that typically form such problem layers. The $P$-wave penetration problem often seems to occur when they are juxtaposed against much lower velocity materials. In that situation, however, mode conversion of both reflected and transmitted waves may also be very efficient.

Although partitioning may divert an unusually large fraction of incident energy from $P$-to-$P$ transmission, a number of other physical phenomena may also weaken or obscure sub-HVL reflections. For example, $P$-to-$P$ energy transmitted into an HVL is subject to attenuation by wave absorption or by scattering at heterogeneities. Therefore, in a given geologic situation, the relative importances of boundary effects and internal effects are worth considering.

In the case of basalt, some published work (e.g., Fuller et al., 1988, and Pujol et al., 1989) suggests that interior effects (absorption and scattering) may not always play significant roles in weak $P$-wave penetration. On the other hand, the previous two citations and Papworth (1985) each identify strong $S$-wave arrivals associated with $P$-to-$S$ conversion at basalt surfaces encountered in land surveys.

Likewise, in marine crustal refraction work, waves that convert between $P$ and $S$ upon transmission through the top-of-basement interface (usually at basalt or limestone beneath soft sediments) sometimes provide additional useful information. This has motivated studies of the factors that affect their occurrence (e.g., White and Stephen, 1980).

From seismic reflection work, there are few published real-data examples of exploiting waves that convert when transmitted across HVL boundaries. In one exception, Tatham and Goolsbee (1984) present results from an area with a hard-carbonate water bottom.

From these and other examples, it appears that HVL boundary effects are typically important, regardless of whether internal absorption or scattering are significant; furthermore, the effects are magnified as velocity contrasts increase. Consequently, this paper concentrates on the boundary effects. For simplicity, I restrict attention to the case of a smooth HVL boundary, although the effects of roughness on the boundary can sometimes be important.

In this study, I rely heavily on three-dimensional (3-D) "solid" physical-model experiments conducted at the Seis-

Presented at the 60th Annual International Meeting, Society of Exploration Geophysicists. Manuscript received by the Editor August 6, 1991; revised manuscript received February 24, 1992.
*Texaco, E&P Technology Division, P. O. Box 770070, Houston, TX 77215-0070.


Reprinted from Geophysics, 57, 1444-1453

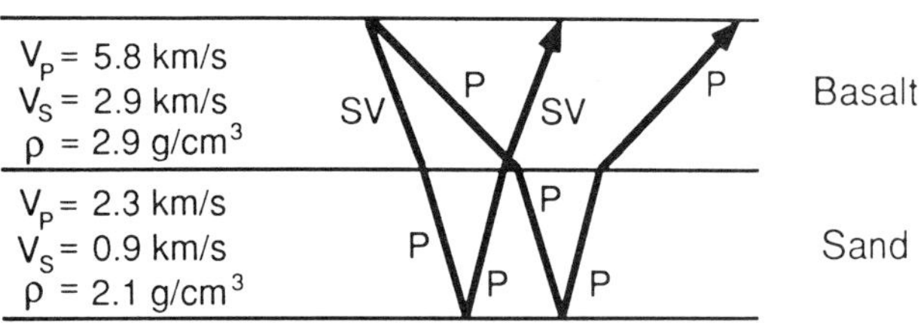


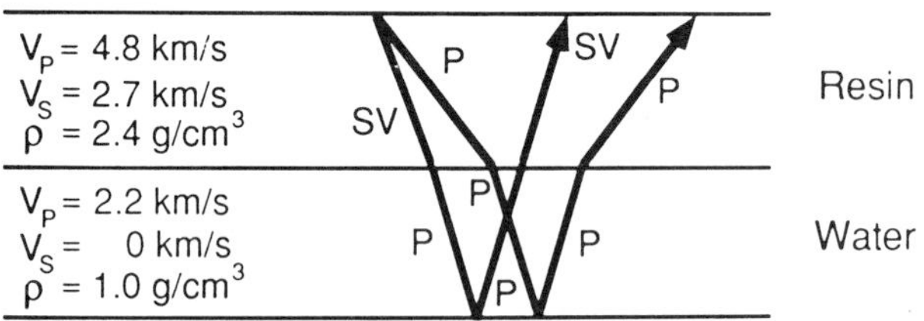


FIG. 1. *PPPP* and *SPPS* raypaths for reflection beneath a surface high-velocity layer . (a) Basalt/sand model. (b) Resin/water model (with scaled velocities).

mic Acoustics Laboratory at the University of Houston. They provided scaled seismic data for studying 3-D elastic wave propagation and testing various data-processing schemes.

## REFLECTION AND TRANSMISSION COEFFICIENTS

### Real-world analog

Consider an idealized model of a surface basalt flow above a layer of unconsolidated sand (Figure 1a). As depicted, it is possible to trace a *P*-wave down to the base-of-sand interface and back up without conversion. The difficulty in exploiting this *PPPP* reflection is that the *P*-to-*P* transmission coefficient for the upcoming reflected *P*-wave, while large at normal incidence, is limited by the small *P*-wave critical angle (23 degrees) at the base of the basalt. This can be seen by inspecting the energy transmission coefficients for a plane *P*-wave incident at the basalt/sand contact from above (black curves in Figure 2a) and from below (black curves in Figure 2c).

Figure 1a also depicts a second possible reflection raypath (*SPPS*) associated with an initial downgoing *S*-wave in the

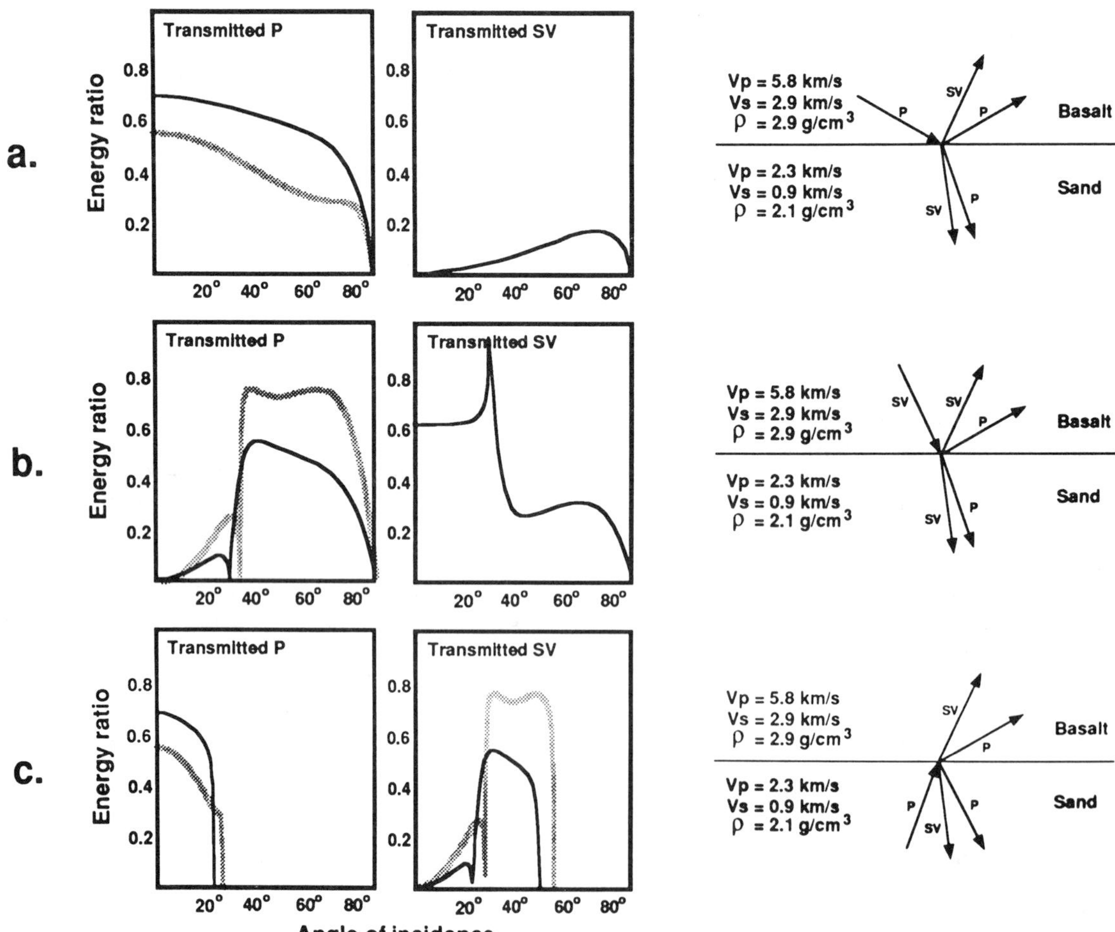


FIG. 2. Plane-wave energy transmission coefficients as functions of incidence angle. (a) *P*-wave incident at basalt/sand contact (black curves) and at resin/water contact (grey curves). (b) *SV*-wave incident at basalt/sand contact (black) and resin/water contact (grey). (c) *P*-wave incident at sand/basalt contact (black) and at water/resin contact (grey).

basalt. Along this path, a wave converts from $S$ to $P$ upon exiting the basalt and from $P$ to $S$ upon reentering the basalt. Reflection at the target is therefore from $P$ to $P$. In most geologic situations, the net energy of waves that follow such paths would be much smaller than that of the normally exploited unconverted $P$-waves. In this situation, however, the close match between $V_S$ within the basalt and $V_P$ outside leads to efficient coupling between $S$-waves within and $P$-waves outside. The match is also apparent in Figure 1a from the minimal ray bending along the *SPPS* path.

Inspecting the energy transmission coefficients (Figures 2b and 2c) for the *SPPS* ray type reveals that the upcoming reflections are limited on the high-angle end by the $S$-wave critical angle (52 degrees) and on the low end by the efficiency of $P$-to-$S$ conversion. This suggests that it may be possible to avoid some of the penetration problem that exists for $P$-waves by using sub-HVL reflections associated with waves that propagate through the HVL in the $S$ mode and elsewhere in the $P$ mode.

If a surface low-velocity layer (LVL) were present, the screening effect at the HVL's upper surface could be quite severe (e.g., Fuller et al., 1988; Purnell et al., 1990) and typically would exceed the effect at the HVL's lower surface. Nevertheless, the effect at the base can still exert a significant angle-dependent weighting effect on upcoming waves and is potentially the major difficulty when the sources and receivers are placed at or below the HVL's upper surface.

If a surface LVL were added to the model in Figure 1a, waves that propagate in the $S$ mode through the HVL and elsewhere in the $P$ mode would carry more energy than waves that follow other conversion paths (Purnell, 1990). Some of the other possible conversion paths have been discussed elsewhere (e.g., Krohn, 1988).

### Physical model

Figure 1b depicts a physical-model analog of the basalt/sand model. (The physical-model velocities are scaled; however, it is the velocity contrast that is important here, not the individual velocity magnitudes). Other than the basalt case, the velocity contrasts and resultant coupling between $P$- and $S$-waves in the physical model also mimic what would be observed in situations involving carbonate or salt juxtaposed against poorly consolidated sediments.

The lower medium (water) in the physical model has a Poisson's ratio of 0.5, corresponding to a fluid. For the purposes of this study, the results of using water as a medium should not be too different, qualitatively, from real-earth situations in which Poisson's ratios of 0.45 or higher occur (e.g., for clays, shales, water-saturated sands or other poorly consolidated sediments). Poisson's ratio for the upper medium (resin) is 0.28, which corresponds to intermediate values for basalt, granite, rock salt, and limestone. Consequently, both $P$- and $S$-waves can propagate within the physical-model HVL, but only $P$-waves can propagate within the lower medium.

The energy transmission coefficients for the physical model are shown as grey curves in Figures 2a–c. Energy partitioning is qualitatively very similar between the basalt/sand model and the physical model, except for the absence of $S$-wave transmission into the lower medium of the physical model. The energy transmission curves (Figures 2b–c, in particular) indicate unusually efficient coupling between $P$- and $S$-waves at the HVL base, even though the materials in the physical model do not closely conform to one of White and Stephens's requirements for near-perfect conversion efficiency, i.e., that the $V_P$ of the lower medium equals the $V_S$ of the HVL.

### Cumulative transmission coefficients

The preceding section dealt with the effects of transmission coefficients, one interface at a time. The product of all the transmission coefficients encountered along a given raypath can be used as an estimated upper bound on the relative strength of a reflection recorded at the surface.

Figure 3 shows the cumulative energy transmission coefficients for the physical model, plotted as functions of the downgoing incidence angle at the resin/water contact. (Here, the base of the water layer is treated as a perfect reflector.) Most of the energy returns to the surface with the *PPPP* arrival or the *SPPS* arrival, which accounts for more than twice as much energy as the former. The singly-converted arrivals, *PPPS* and *SPPP*, carry much less energy than the others.

The *PPPP*, *SPPS*, and *PPPS* arrivals are unaffected by critical angles, insofar as they are observed up to angles approaching 90 degrees. To understand why this is so, consider the *PPPP* and *SPPS* raypaths depicted in Figure 1b. Since $(V_{S(water)} < V_{P(water)} < V_{S(resin)} < V_{P(resin)})$, there are no critical-angle cutoffs for downgoing rays. For upgoing rays, critical-angle limits exist at 28 degrees (for $P$-to-$P$ transmission) and at 58 degrees (for $P$-to-$S$ transmission). Since this is a flat-layered model, *PPPP* and *SPPS* have symmetric down- and upgoing paths; therefore, their critical-angle limits are reached for infinitely large source-receiver offset only. *PPPS* rays strike the HVL base over the same angular range as *PPPP* rays, so they are similarly affected.

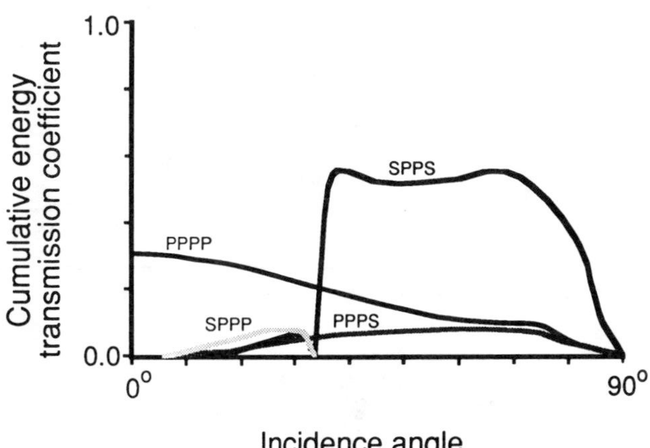


FIG. 3. Cumulative energy transmission coefficients for the resin/water model's *PPPP* and *SPPS* reflection arrivals (depicted in Figure 1b), as well as for the corresponding *SPPP* and *PPPS* reflections.

## PHYSICAL-MODEL DATA ACQUISITION

In the physical-model experiments, ultrasonic piezoelectric transducers served as source and receiver. Since they were coupled directly to the top surface of the HVL, $P$-waves, $S$-waves, and surface waves were directly generated and recorded. For convenience in relating the model data to actual seismic data, I use scale factors of 1:12000 for length and 1:8000 for time; densities are unscaled. After scaling, most of the recorded reflections are in the 5–35 Hz range.

The experimental data consist of shot records simulating a two-dimensional (2-D) seismic line across a surface HVL matching the one depicted in Figure 1b. Figure 4 shows the experiment in cross section, while Table 1 lists the recording parameters for the seismic line. As illustrated in Figure 4, seven dipping-plane reflectors are embedded in the lower medium and oriented such that the seismic line runs in the true dip direction for all of them. Their dips range from 0 to 70 degrees.

## DATA PROCESSING STRATEGY

The goal in processing these data is to obtain the most complete imaging possible for the dipping reflectors. For convenience in discussing the data, I group the converted-wave reflections into two classes: (1) "Class 1," those with symmetric downgoing and upgoing conversion paths and (2) "Class 2," those with asymmetric conversion paths. The only Class-1 converted-wave arrivals in these data are waves that travel in the HVL (both down and up) as $S$-waves and elsewhere as $P$-waves. The Class-2 arrivals here are waves that traverse the HVL in the $P$ mode on the downgoing path and in the $S$ mode on the upgoing path (or vice versa). In data processing, therefore, I am concerned with imaging the dipping reflectors using each of the four types of primary reflection arrivals: $PPPP$, $SPPS$, $SPPP$, and $PPPS$.

Rather than present the results of applying DMO, CMP stack, and acoustic poststack depth migration, I will show the results of acoustic prestack depth migration. This approach has the advantage of demonstrating the same points, while avoiding uncertainty as to the correctness of the CMP stack step (with or without prior DMO). Another advantage of prestack migration is that it can be modified readily to accommodate different velocities for upgoing and downgoing waves in the same layer. This is one way to handle the Class-2 arrivals using acoustic prestack migration, as Dillon et al. (1988) showed for mixed-mode VSP data. Conventional CMP stack followed by poststack migration cannot properly handle the Class-2 arrivals.

The prestack migration used here is shot depth migration implemented with an acoustic phase-shift algorithm. To image the dipping targets using the unconverted $P$-waves, I use the known $P$-wave velocity field. For imaging the Class-1 converted-wave arrivals, I use the known $V_S$ for the HVL and use $V_P$ elsewhere. For imaging the Class-2 arrivals, I use $V_S$ only for the downgoing (or upgoing) path through the HVL and use the appropriate $V_P$ elsewhere.

**Table 1. Recording parameters for 2-D line (in scaled units).**

| | |
|---|---|
| Source interval | 60 m |
| Number of source points ($SP$s) | 76 |
| Receiver interval | 30 m |
| Number of receivers per $SP$ | 152 (maximum) |
| Receiver layout | end-on, shot pulling |
| Near offset | 274 m |
| Far offset | 4804 m |
| CMP fold | 38 (maximum) |

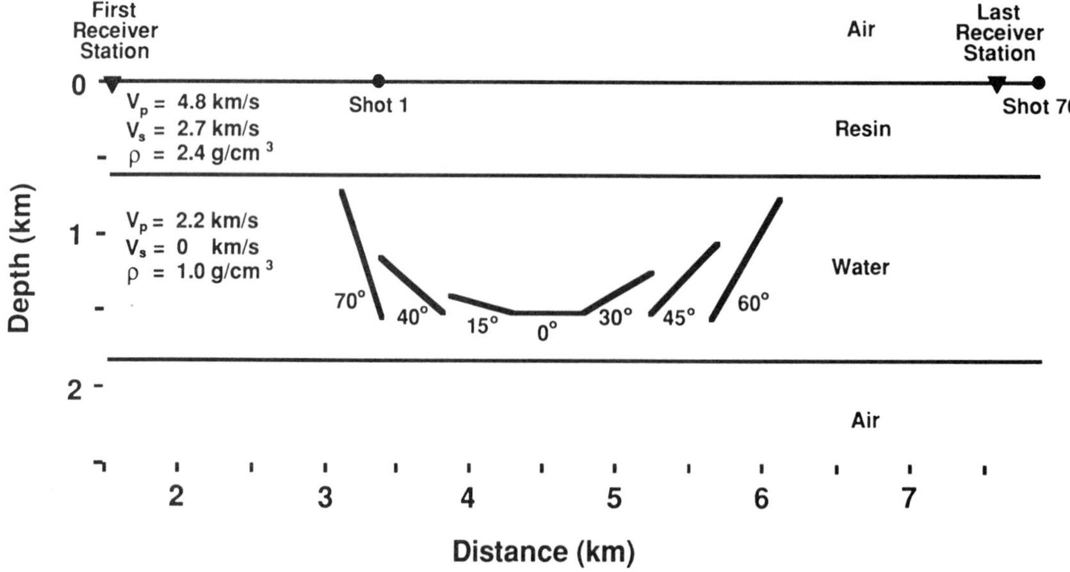


FIG. 4. Cross section of a 3-D physical model containing seven dipping reflectors beneath a surface high-velocity layer. The model is constant in the cross line direction. Also, only part of its 9144-m (scaled) inline extent is shown. The velocities posted are scaled to their real-world equivalents; densities are unscaled.

## DISCUSSION

On a conventional CMP stack (Figure 5) of the seismic line, various reflections from the 0-, 15-, and 30-degree targets are conspicuous.

### Unconverted *P*-wave reflections

A prestack depth migration (Figure 6) for the unconverted *P*-wave reflections (*PPPP*) shows clear images of the 0- and 15-degree target reflectors, but images of the 30-degree and higher-dip reflectors are not present. About 150 m and deeper beneath those images are spurious images produced from converted-wave reflections and multiples migrated with inappropriate velocities. Images of the 30-degree and higher-dip reflectors are not present because, even if a significant amount of energy were reflected there, the upgoing reflected waves (which can only be *P*-waves) encounter a critical angle of 28 degrees for *P*-to-*P* transmission into the HVL.

This can be verified by attempting to trace *PPPP* rays that reflect at an interface having a dip greater than 28 degrees. For normal-incidence reflection at such an interface, a ray traced up will meet the HVL base at the same angle as the reflector dip angle, which exceeds the critical angle overhead; therefore, no normal-incidence raypaths exist connecting the source-receiver plane and any reflector having a dip greater than 28 degrees. Even considering nonnormal-incidence reflection at such an interface, the reflected rays meet the HVL base at angles greater than the critical angle. Figure 7 illustrates this with shot-record ray tracing to a 30-degree plane reflector, analogous to the one in the physical model.

### Class-1 converted-wave reflections

The migration result (Figure 8) for the *SPPS* Class-1 reflections shows clear images over a much wider range of reflector dips than does the unconverted *P*-wave result (Figure 6).

In contrast with the *PPPP* ray tracing, Figure 9 reveals no difficulty in tracing *SPPS* raypaths from the shot to the 30-degree reflector and back up to the receivers. Most of the *SPPS* rays undergo minimal bending. Furthermore, the amount of energy that travels along such *SPPS* arrivals is large because *P*-to-*S* transmission into the HVL from below is quite efficient between the *P*-wave critical angle and the *S*-wave critical angle (Figure 2c). In the physical model, a wave incident beyond the *S*-wave critical angle (58 degrees) does not penetrate, which explains why the 60- and 70-degree reflectors are not imaged in Figure 8. A legitimate image of the 0-degree reflector exists because of arrivals recorded at offsets well away from zero offset.

Unlike the *PPPP* migration, interpretation of the *SPPS* migration is complicated by spurious images at depths above, as well as below, the correct images. The shallower images are from unconverted *P*-wave and Class-2 reflections (*PPPP*, *PPPS*, and *SPPP*); the deeper images are associated with multiples.

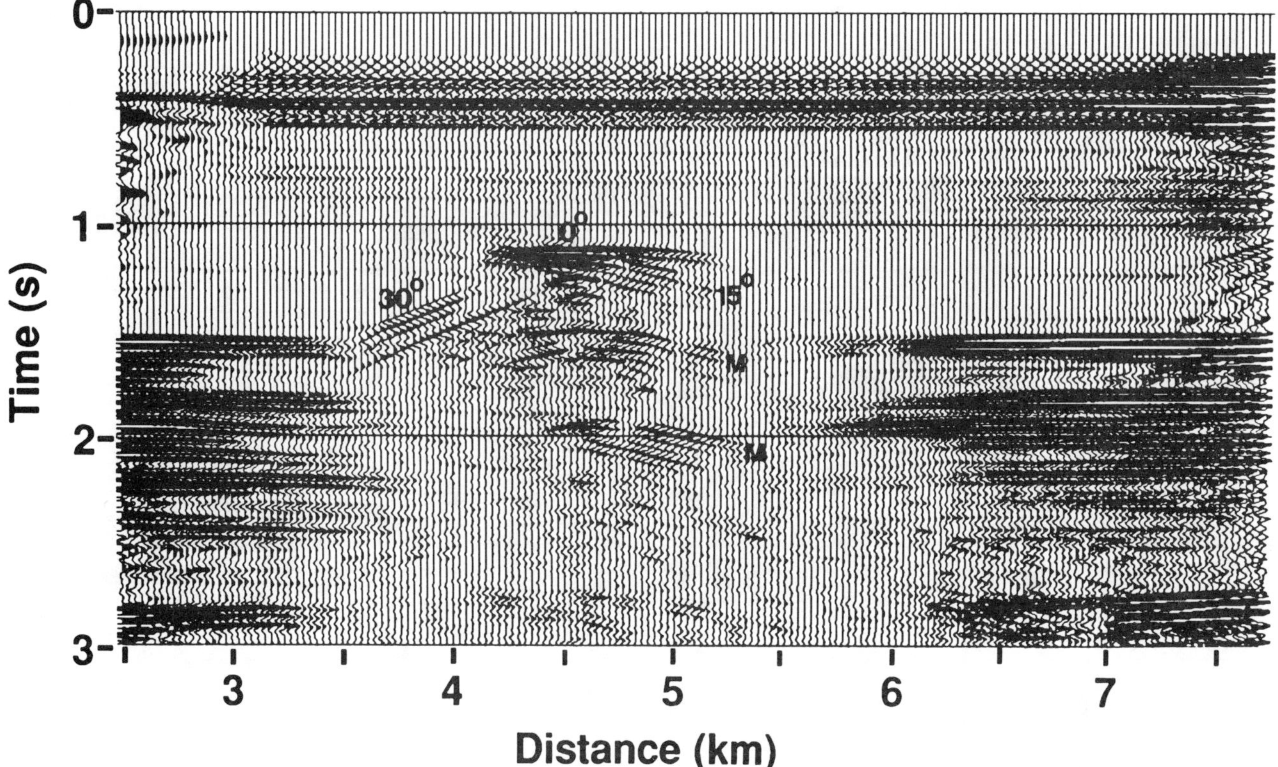


FIG. 5. CMP stack of the 2-D seismic line. Sets of primary reflections from some of the dripping targets are annotated, along with two sets of multiples ("M"). Most of the conspicuous events beneath about 1.6 s are multiples.

**Class-2 converted-wave reflections**

There are two types of Class-2 converted-wave reflections from the dipping targets (*SPPP* and *PPPS*). Figure 10 shows the prestack migration result for the *SPPP* arrivals, which provide good images of the 0- and 15-degree reflectors. Overall, this result does not differ very much from that of migrating for the *PPPP* arrivals. The strong *SP* image of the HVL base results from the efficiency of *S*-to-*P* conversion upon reflection for some incidence angles at the HVL base. The raypath diagram in Figure 11 indicates that it should be possible to record *SPPP* reflections from the 30-degree target when shooting downdip. Boosting the display gain used for Figure 10 reveals that a weak image of that target is indeed present.

Figure 12 shows the prestack migration result for the second Class-2 type (*PPPS*). Here, the 30-degree target is strongly imaged in addition to the 0- and 15-degree reflectors. A strong *PS* image of the HVL base is also present. Figure 13 shows that, when shooting downdip, this ray type illuminates a wider region of the 30-degree reflector than does the *SPPP* type. Comparing Figures 10 and 12, the *PPPS* branch appears to be the more useful of the two Class-2 arrivals.

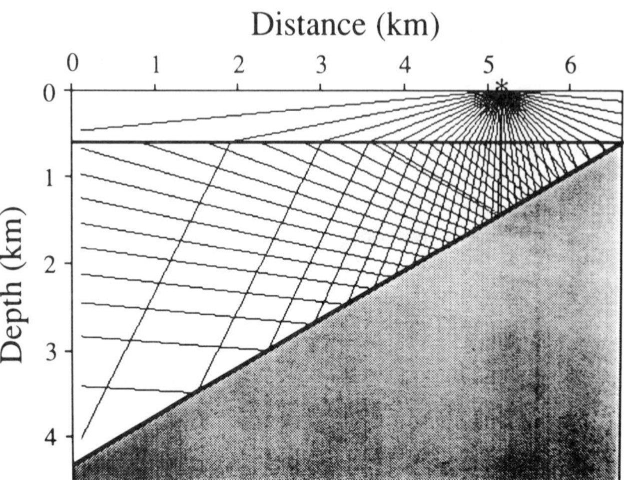


FIG. 7. Attempt to trace *PPPP* reflection raypaths for a 30-degree plane reflector analogous to the one in the physical model. All of the downgoing rays reach the reflector, but none of the upcoming rays return to the surface because they strike the base of the high-velocity layer at angles greater than the 28-degree critical angle for upgoing *P*-to-*P* transmission.

## CONCLUSIONS

The physical-model data demonstrate that the range of reflector dips that are imaged using unconverted *P*-wave arrivals can be greatly restricted when an overlying HVL is present. Furthermore, the data demonstrate that, for certain $V_P$ and $V_S$ contrasts, highly efficient coupling can occur

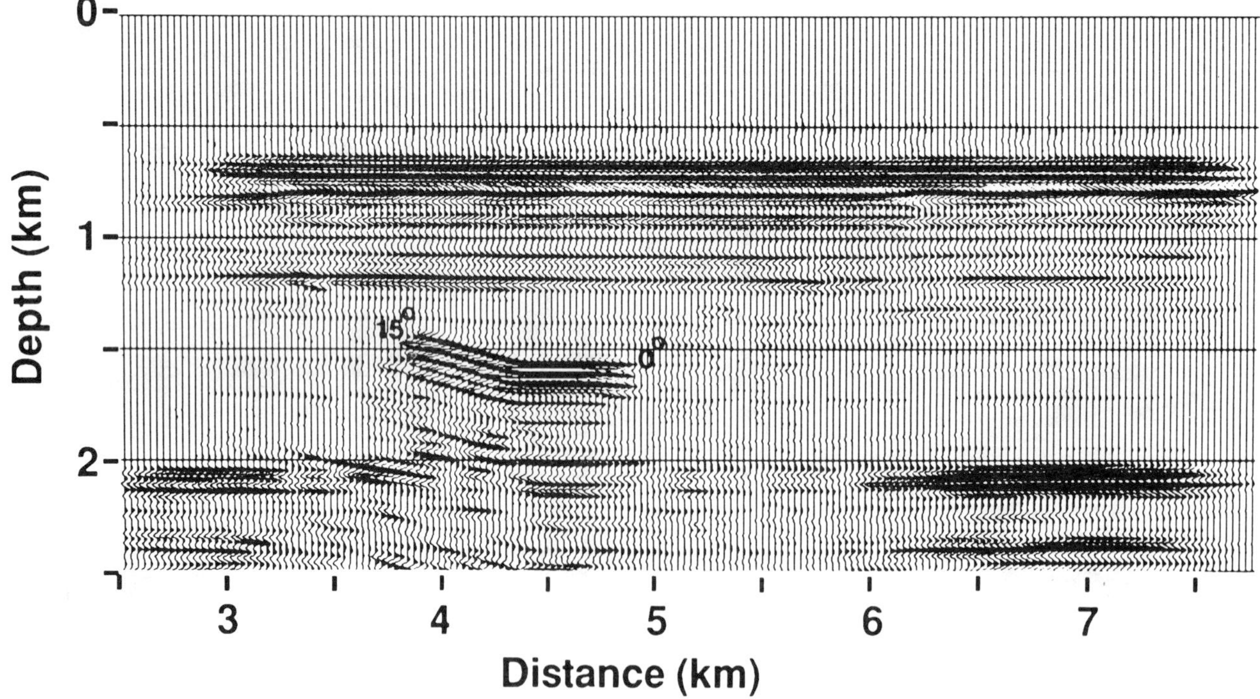


FIG. 6. Prestack depth migration of the 2-D seismic line using the known *P*-wave velocities. Images of only the 0- and 15-degree reflectors are present. About 150 m and deeper beneath those images are spurious images produced from converted-wave reflections and multiples migrated with velocities inappropriate to them. Images of the 30-degree and higher-dip reflectors are not present because of the critical angle problem (illustrated in Figure 7) for *PPPP* reflections.

between $S$-waves within an HVL and $P$-waves outside. Consequently, a sub-HVL reflection that travels through the HVL in the $S$ mode and elsewhere in the $P$ mode can be significantly strong (at the recording surface) relative to the corresponding unconverted $P$-wave arrival.

Data processing that exploits some of these converted waves can image a much greater range of sub-HVL reflector dips than data processing that uses the corresponding unconverted $P$-wave arrivals. This is because of the efficient coupling cited above, and because such waves are less restricted by critical angles.

Acoustic migration is successful in imaging the physical-model reflectors using both symmetric-path and asymmetric-path converted waves. However, the acoustic approach requires selecting a single family of converted waves (via the migration velocities used). The other unselected families are mismigrated and produce artifacts which, in real-world data, could complicate interpretation. If this processing approach is pursued, improved techniques of removing the unselected families of events before migration may be necessary, as well as determining practical means of combining the results after migration.

## ACKNOWLEDGMENTS

The Allied Geophysical Laboratories (University of Houston) and their industrial sponsors provided the physical-modeling facilities. At AGL, I especially thank J. A. McDonald, G. H. F. Gardner, and K. K. Sekharan. At Texaco, I gratefully acknowledge help from J. H. Higginbotham, J. R. Rogers, Y. Shin, D. V. Sukup, and R. H. Tatham. Finally, I thank Texaco, Inc. for its support and for permission to publish this paper.

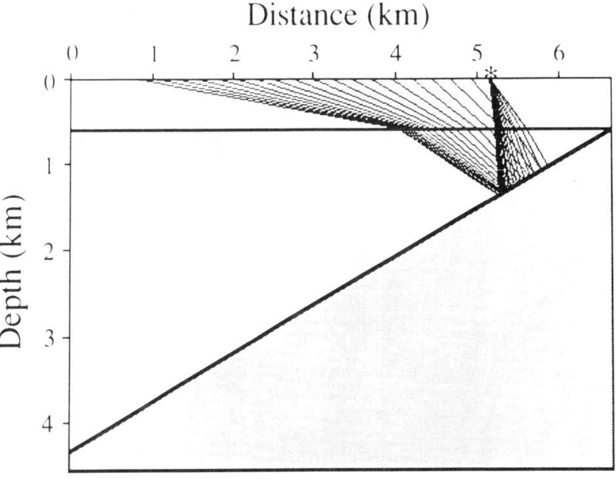


FIG. 9. $SPPS$ rays can be traced with little difficulty from a surface source to the 30-degree plane and back to the surface. Most of the rays undergo little bending.

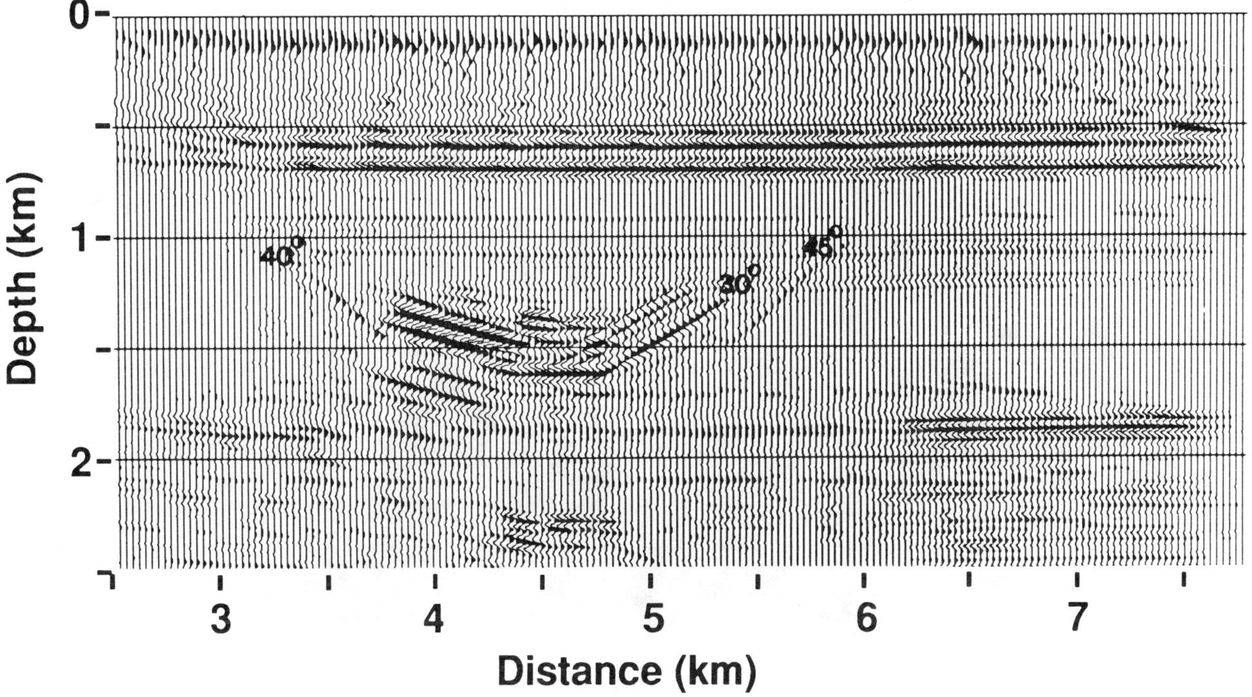


FIG. 8. Prestack depth migration of the 2-D seismic line using $V_S$ in the high-velocity layer and $V_P$ elsewhere. Migrating with these velocities exploits waves that travel in the $S$ mode within the high-velocity layer and in the $P$ mode elsewhere. The higher critical angle for $P$-to-$S$ transmission into the HVL (58 degrees) than for $P$-to-$P$ transmission (28 degrees) enables using $SPPS$ reflections to image all but the 60- and 70-degree targets.

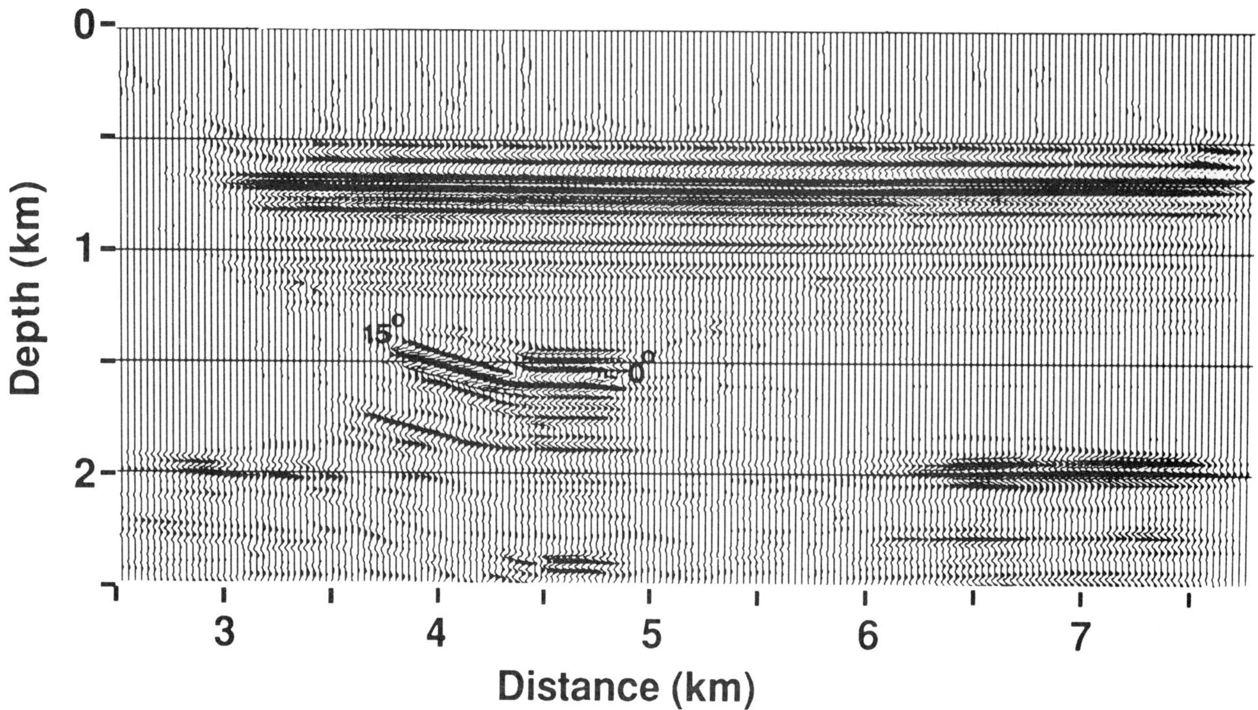


FIG. 10. Prestack depth migration of the 2-D seismic line using $V_S$ only for *downgoing* waves in the high-velocity layer and $V_P$ for all other downgoing and upgoing paths. These velocities are optimal for imaging *SP* reflections from the HVL base and *SPPP* reflections from the dipping targets.

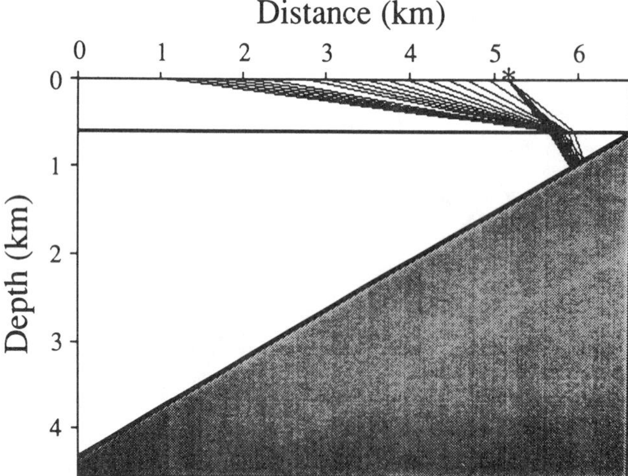


FIG. 11. *SPPP* reflection raypaths for the 30-degree plane.

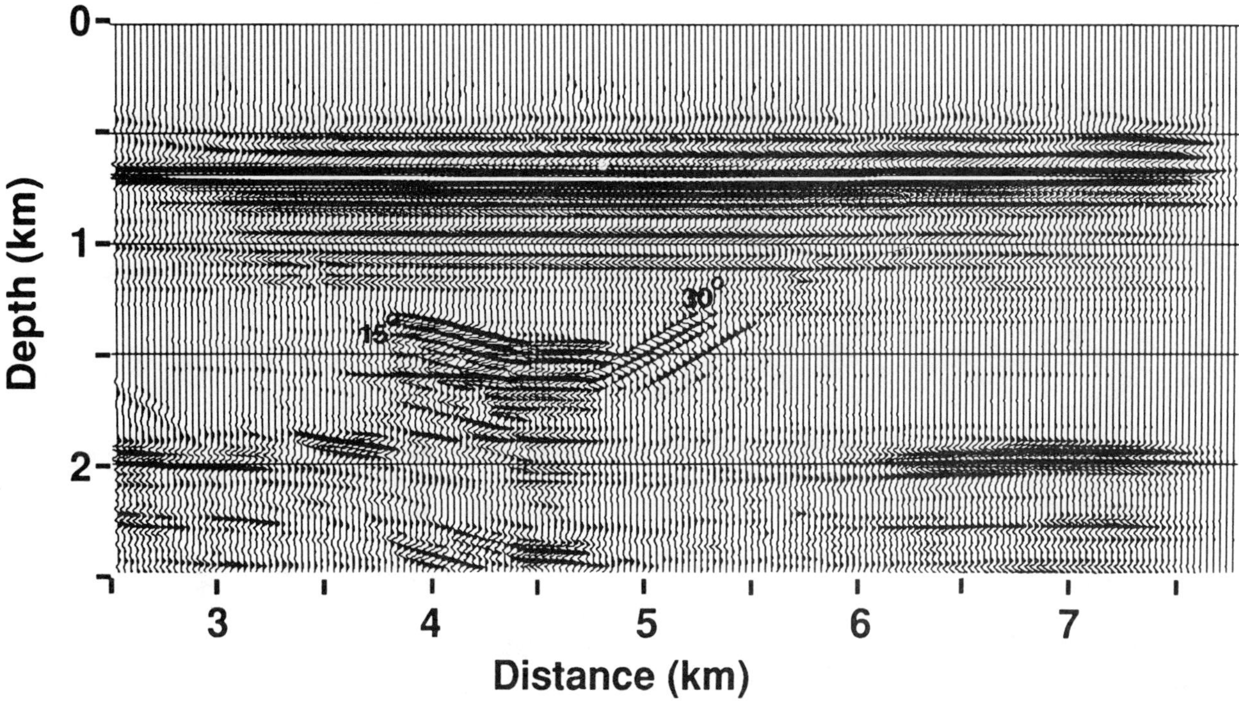


FIG. 12. Prestack depth migration of the 2-D seismic line using $V_S$ only for *upgoing* waves in the high-velocity layer and $V_P$ for all other downgoing and upgoing waves. These velocities are optimal for imaging PS reflections from the HVL base and *PPPS* reflections from the dipping targets. This migration result is better than the *PPPP* and *SPPP* migrations (Figures 6 and 10), in that a strong image of the 30-degree target is formed in the correct location.

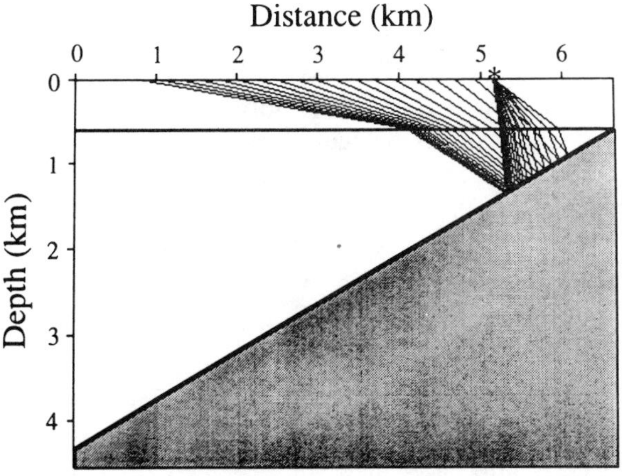


FIG. 13. *PPPS* reflection raypaths for the 30-degree plane.

## REFERENCES


Dillon, P. B., Ahmed, H., and Roberts, T., 1988, Migration of mixed-mode VSP wavefields: Geophys. Prosp., **36**, 825–846.

Fuller, B. N., Pujol, J. M., and Smithson, S. B., 1988, Seismic reflection profiling in the Columbia Plateau: 58th Ann. Internat. Mtg., Soc. Expl. Geophys., Expanded Abstracts, 34–37.

Krohn, C. E., 1988, Computer modeling of *P-SV* waves: 58th Ann. Internat. Mtg., Soc. Expl. Geophys., Expanded Abstracts, 1005–1008.

Papworth, T. J., 1985, Seismic exploration over basalt-covered areas in the U.K.: First Break, **3**, 20–32.

Pujol, J., Fuller, B. N., and Smithson, S. B., 1989, Interpretation of a vertical seismic profile conducted in the Columbia Plateau basalts: Geophysics, **54**, 1258–1266.

Purnell, G. W., 1990, Imaging beneath a high-velocity layer using converted waves: Ph.D. dissertation, Univ. of Houston.

Purnell, G. W., McDonald, J. A., Sekharan, K. K., and Gardner, G. H. F., 1990, Imaging beneath a high-velocity layer using converted waves: 60th Ann. Internat. Mtg., Soc. Expl. Geophys., Expanded Abstracts, 752–755.

Tatham, R. H., and Goolsbee, D. V., 1984, Separation of *S*-wave and *P*-wave reflections offshore western Florida: Geophysics, **49**, 493–508.

White, R. S., and Stephen, R. A., 1980, Compressional to shear-wave conversion in oceanic crust: Geophys. J. Roy. Astr. Soc., **63**, 547–565.

# Chapter 6
# The Wave of the Future

By the early 1980s, virtually all physical modeling research was conducted by, or through, universities or governmental facilities. The majority of this work was 3-D, although a reader interested in modern 2-D modeling may refer to Pant et al. (1988). The current trends, as inferred from the host institutions of scientists presenting physical modeling papers, would indicate that physical modeling is likely to continue to be performed primarily at academic and other not-for-profit research laboratories.

The move to 3-D systems occurred as the electronic components necessary to perform data acquisition became less expensive. A key innovation was the low-cost mini-computer, which revolutionized both management of data acquisition and recording of that same data in digitized form. A PDP-10 (with associated Biomation digitizer) was installed by 1978 at Gulf Research to acquire seismic physical model data, and presumably mini-computers came into similar use at other oil company research facilities at about the same time. By 1980, a system similar to the one at Gulf Research had been installed at the University of Houston. Such computer control for physical modeling facilities has now become almost standard.

This last chapter includes papers that we feel demonstrate some of the flavor of future directions in research involving physical models. This choice is naturally subjective, but it is guided by the notion that physical model experiments should be done primarily on those systems whose response is difficult to obtain by numerical means. Specifically, elastic (as opposed to acoustic) wave propagation experiments fall into this realm. For elastic problems, the simultaneous recording of all three components of motion is the ultimate goal. Currently, most experimenters acquire the three components of motion in three separate experiments, thus making quantitative comparison of amplitudes on the three separate components perilous. In spite of this handicap, a number of interesting results have emerged.

+ Balch et al. (1991) show that the several different scattered wavefronts (PP, PS, SS, SP) can be used as multiple (and hence redundant) constraints on reconstructing images.

+ Brown et al. (1991) investigate the complexities of wave behavior in an orthorhombic medium. (Such media are not difficult to imagine: they might be caused by vertical fractures imposed on a stratigraphically layered sequence.) Especially interesting are the illustrations of the manner in which separate shear-wave wavefronts can merge.

+ Ebrom et al. (1992) demonstrate that three-component data offers some unique possibilities for recognizing out-of-plane energy.

+ Martin et al. (1993) illustrates the use of one of the most interesting new approaches to physical modeling in the last few years. This is the use of lasers, both as source and receiver. Lasers allow one to escape the intrinsic bandwidth limitation of piezoelectric transducers. The clarity of event resolution is well illustrated in their simulated crosshole experiment, wherein many different events may be easily distinguished. Lasers carry their own drawbacks, however, which are also discussed in this paper.

# THE USE OF FORWARD- AND BACK-SCATTERED P-, S- AND CONVERTED WAVES IN CROSS-BOREHOLE IMAGING[1]

A. H. BALCH, H. CHANG, G. S. HOFLAND, K. A. RANZINGER and C. ERDEMIR[2]

## Abstract

BALCH, A.H., CHANG, H., HOFLAND, G.S., RANZINGER, K.A. and ERDEMIR, C. 1991. The use of forward- and back-scattered P-, S- and converted waves in cross-borehole imaging. *Geophysical Prospecting* **39**, 887–913.

The principles of imaging, for example that of prestack migration, can be applied to cross-borehole seismic geometry just as they can to surface seismic configurations. However, when using actual cross-borehole data, a number of difficulties arise that are rarely or never encountered in imaging surface seismic data: discontinuities may reflect or diffract incident seismic waves in any direction. If a discontinuity lies between the lines of sources and receivers, forward-scattered, or interwell, events may be recorded. If a discontinuity lies outside the interwell region, back-scattered, or extra-well, events may be recorded. Many angles of incidence are possible, and all possible reflected modes (P–P, P–S, S–P and S–S) are present, frequently in nearly equal proportions. The planes of the reflectors dip from 0 to $\pm 90°$.

In order to deal with these complexities we first separate propagation modes at the receiver borehole using both polarization and velocity. Next we compensate for phase distortion due to dispersion. Finally, and most importantly, we migrate or image the data in cross-borehole common-source gathers. To do this, a finite-difference solution to the 2D scalar wave equation, using reverse time, for an arbitrary distribution of velocities, is used to project the separated, reflected-diffracted wavefield back into the medium.

There are four reflection modes (P–P, P–S, S–P and S–S), so we can apply four different imaging conditions. The zones outside the boreholes as well as inside the boreholes can be imaged with these conditions.

These operations are repeated for each common-source gather: each common-source gather generates four partial images in each image space. This multiplicity of partial images can be stacked in various combinations to yield a final image of the subsurface.

Our experiments using solid (not fluid) physical models indicate that when these procedures are correctly applied, high quality cross-borehole images can be obtained. These images appear with great clarity even though some of the weak diffractions causing diffraction images may be almost totally obscured by other high-amplitude events on the raw data.

[1] Received July 1990, revision accepted April 1991.

[2] Geophysics Department, Colorado School of Mines, Golden, CO 80401, U.S.A.

## Introduction

The interpretation and processing of cross-borehole seismic data present a number of new challenges and problems in applied seismology. Cross-borehole seismology is being used in diverse areas such as: petroleum reservoir analysis; evaluation of hazards associated with abandoned mines, especially coal-mines; and in military science, especially the detection and neutralization of hostile tunnels.

Cross-borehole data are extremely complex. 'Raw' unmigrated, or unimaged, cross-borehole data are difficult to interpret. The presence of compressional, shear and mixed-mode reflections, the need to record multiple components, the presence of events reflected from 'behind', 'in front' and 'all sides' of the receiver borehole, and the presence of coherent and random noise, all contribute to this complexity. The sheer volume of the raw data is a daunting mass of information for a user to process and interpret, and cross-borehole data associated with even simple geology may require extensive processing in order to produce a representation of the subsurface that can be interpreted with confidence.

The most important processing step is imaging, for which prestack migration can be used. Due to the source–receiver geometry, cross-borehole data plots bear no resemblance to the subsurface geology (in contrast to surface seismic data plots).

Some reflections originate in the region between boreholes (Fig. 1). We call these forward-reflected or interwell events. Others can originate in the region to the left and to the right of both boreholes. We call these back-scattered, back-reflected or extra-well events. The distinction between interwell and extra-well events is important in imaging.

We have chosen to use solutions to the scalar wave equation to implement the migration. The principles of this approach were described by Hu *et al.* (1988) but they only considered migration of pure single (compressional) mode wavefields. Besides migrating multimodes and mixed-mode reflections, our investigation goes beyond the work of Hu *et al.* in two additional ways: we treat both interwell and extra-well reflections and we use 'inverse $Q$-filtering' (Chang 1991) to reduce the effects of dispersion. However, we do not consider the specific imaging technique to be a critical element in this investigation. There are a large variety of techniques applicable to this problem, including those described by Beydoun *et al.* (1989) and Pratt and Worthington (1990).

The cross-borehole data contain a mixture of P–P, P–S, S–P and S–S reflections. These modes have been frequently observed in the field. They are well documented in the Chateaurenard cross-hole study (Delvaux *et al.* 1987). We can migrate these modes one at a time, with scalar wave theory. It is desirable, for optimum imaging, to separate shear and compressional modes at the receiver borehole; this is done on the basis of polarization and apparent velocity. Although mode separation does not completely eliminate the mixed-mode problem (we cannot separate P–P and S–P reflections, for example), it mitigates the problem considerably. Alternative ways of dealing with the multimode problem include that of Beydoun *et al.* (1989), who chose to migrate upgoing S–S and S–P reflections, and Iverson (1986) who attempted to obtain P–P images of horizontal, nearly parallel reflectors by a Wyatt type P–P migration of mixed-mode data.

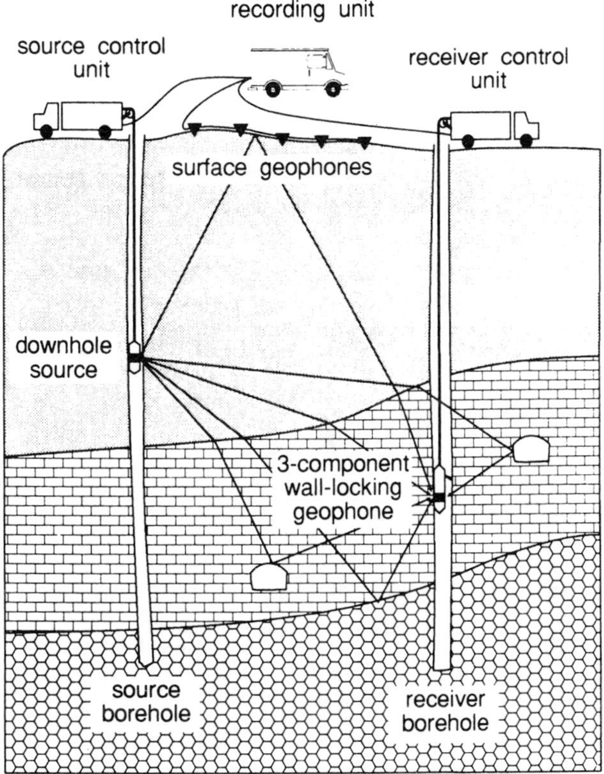


FIG. 1. Schematic diagram of the cross-borehole seismic process. Seismic energy from a source borehole is reflected off discontinuities into a receiver borehole. Reflections may originate from a region between or a region outside the boreholes.

As a result of the mixture of modes and the mixture of interwell and extra-well events, cross-borehole migration produces many artefacts. Since the source–receiver geometry is different for each common-source gather, the artefacts shift position on the various migrated gathers, whereas the correct images remain stationary. By stacking the partial images from each migrated gather, we can then suppress the artefacts and enhance the true images.

Finally, attenuation and dispersion effects are difficult to take into account in imaging. We have used an 'inverse $Q$-filtering' process on the original data to create a new data set that approximates the data we would have obtained if there had been no attenuation and, more important, dispersion. $Q$-deconvolution has enabled the migration process to produce better-defined images.

Our solid physical model data contain all the complexities described above. Our experience with these data indicates that high quality images of the subsurface can be obtained, notwithstanding the complexities.

## The Physical Model

### *Laboratory procedure*

*Physical modelling laboratory.* Figure 2 is a schematic diagram of the 2D, physical, seismic, ultrasound modelling system used to generate the cross-borehole data. A Panametrics 5055 PR pulse generator energizes the source transducer at about 30 times per second and simultaneously sends a synchronization, or zero time, pulse to a Tektronix 11401 digitizing oscilloscope. The source transducer was mounted on the edge of a $\frac{1}{4}$ inch sheet of perspex; it exerts a force normal to the surface to which it is attached.

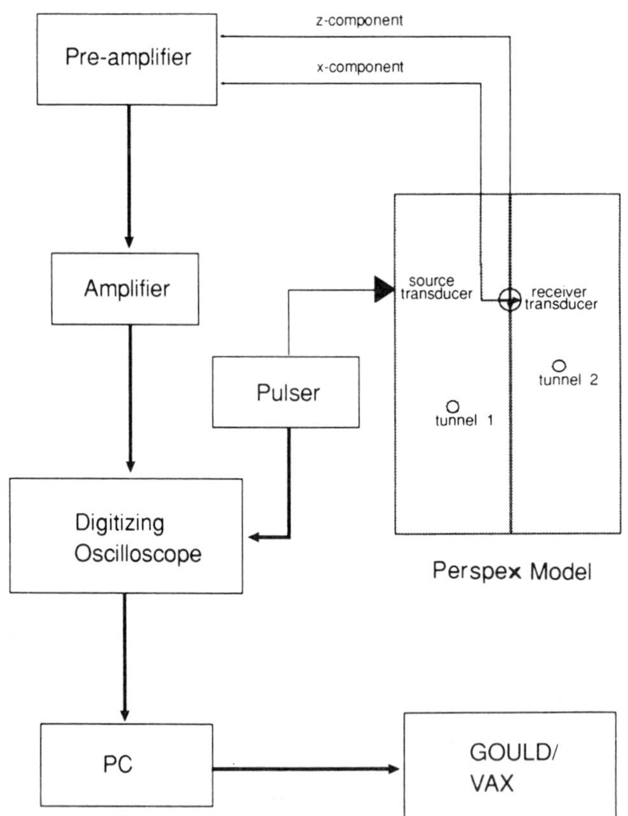


Fig. 2. Scheme of the 2D physical seismic ultrasound modelling system used to generate the cross-borehole model data.

The wavefield generated by the source is sensed by vertical and horizontal ($z$ and $x$) transducers, whose outputs are amplified by Bruel and Kjaer 2635 charge amplifiers. The amplified waveforms are fed to the oscilloscope for display. The system typically is pulsed every 33 ms; the wavefield itself dies out in a few ms. The repetitive wave-form display appears stationary to the observer. The wave forms are usually averaged, or stacked, in real time, by the oscilloscope. The observer typically sees an 8- to 256-fold average, depending on the signal-to-random-noise ratio.

The averaged wave form is downloaded in digital form to a Compaq 386 desktop computer and stored there. The receivers are then moved to a new location and the recording cycle is repeated. When a complete data set, or 'common-source gather', is obtained, it is uploaded to a mainframe computer, converted to SEG-Y format, and stored there either on hard disk or on magnetic tape. From this point onwards the data are treated as actual field recordings.

*Scaling.* We observed a dominant wavelength of about 3 inches in the perspex model. The dominant wavelength in a sample field data set used in tunnel detection was approximately 3 m. We chose to make our model wavelength dimension consistent with the field wavelength dimensions. Thus our scale is such that 1 inch (model) represents approximately 1 m (real earth) and a tunnel 2 m in diameter is represented by a 2 inch hole in the perspex. The rest of our scaling dimensions are determined by this choice, since the earth and perspex velocities are fixed. If the earth medium P-wave velocity is 4500 m/s and the perspex velocity is 7300 ft/s, then 1 earth ms represents 50 model $\mu$s approximately.

*The model.* Figure 3 is a sketch of the cross-borehole model. The model consists of a 4 ft by 6 ft sheet of $\frac{1}{4}$ inch perspex. Two holes, representing tunnels 2 m in diameter, were cut into the sheet. A vertical line of 33 ultrasound source points at 2 inch spacing is located along the left-hand edge. The sources are $\frac{1}{4}$ inch in diameter and produce a force normal to the edge. A vertical line of 256 two-component seismic detectors, at $\frac{1}{4}$ inch spacing, is located approximately in the middle of the sheet. The polarized detectors are sensitive to motion in one direction, parallel to the surface upon which they are placed. The source and receiver lines are considered to be boreholes.

This model provides us with five reflecting/diffracting surfaces: two diffracting holes, or tunnels, two horizontal edges and one vertical edge (see Fig. 4). The horizontal edges can be considered to be flat-lying formation boundaries, the holes can be associated with tunnels, or other subsurface voids, and the vertical edge represents a fault.

This model incorporates both compressional-mode, P, and shear-mode, S, waves. P- and S-modes are generated by the source itself, and both modes are generated every time a reflection occurs. This complicates the model considerably, but it is similar to the cross-hole data that would be recorded in the real earth.

Reflectors in the region between the boreholes scatter energy forward into the receiver line. The vertical edge and the right-hand tunnel scatter energy back into the receiver line. These extra-well reflectors lie outside the region between the boreholes. As we shall subsequently see, this distinction is important when the data are to be imaged.

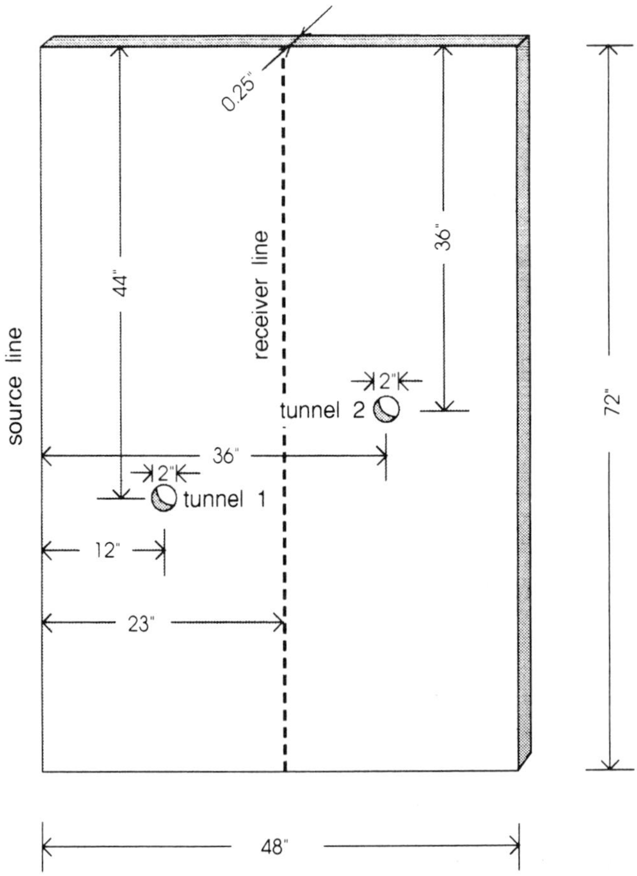


FIG. 3. Sketch of the seismic model, showing the location of the target holes (or tunnels), source and receiver lines, and edge reflectors. Scale: 1 inch represents 1 m, approximately.

## *The laboratory model data*

The model data were acquired in common-source gather sets. That is, the source remained stationary while all 256 receiver positions were occupied by the horizontal and vertical component transducers. One such set, for the vertical $z$-component detector is shown in Fig. 5. The earliest event is the direct P-wave. Figure 6 is a 'magnified' view of a portion of the $z$-component data plotted in Fig. 5. The appearance of polarity reversal on the P-wave direct arrival is due to the fact that the initial displacement is always radially away from the source, which will be up or down depending on the position of the detector relative to the source.

A strong first arrival shear event is clearly shown on both Figs 5 and 7, in addition to the first arrival compressional event. In Fig. 6, a polarity change in the $z$-component of the shear-wave first arrival can be observed. The source exerts a

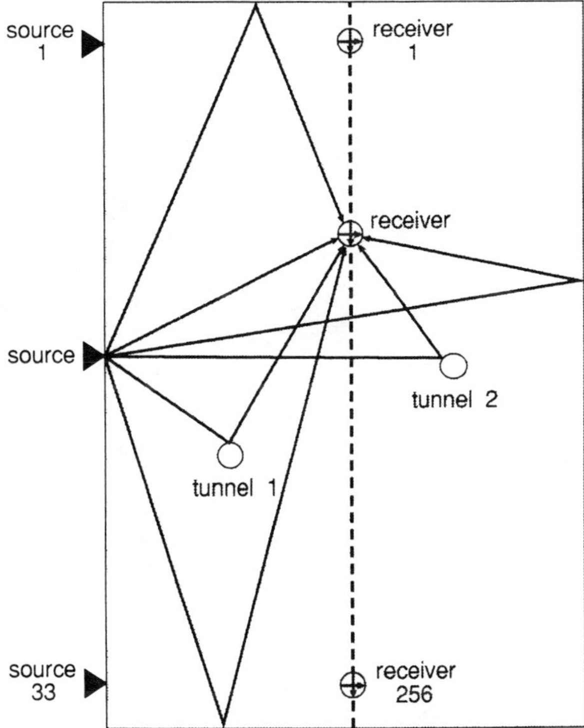


FIG. 4. Typical raypath diagram for the model shown in Fig. 3. Source and receiver lines are indicated. The rays represent all four modes of reflection: P–P, P–S, S–P, S–S. Note that two reflectors and one tunnel are located in the region between the boreholes, while one reflector and one tunnel are located outside the region between the boreholes.

horizontal force on the vertical left-hand edge of the model (normal to the edge). Due to the symmetry of this boundary condition, about a horizontal line passing through the source, the vertical component of motion of the direct shear wave must be in opposite directions above and below the line of symmetry. All possible reflected modes from the top and bottom edges of the model can be identified, and are indicated on Figs 5 and/or 7: PP, PS, SP and SS. A strong reflected P-mode from the vertical right-hand edge is shown in Fig. 7.

Reflections/diffractions from the tunnels can be identified, but they are considerably lower in amplitude, which is consistent with theory and field experience. It is significant that scattered events are observed even though the tunnels are only about two-thirds the dominant P-wavelength. All four modes scattered from the left-hand tunnel, P–P, P–S, S–P and S–S, can be identified. Scattered mode energy from the right-hand tunnel (tunnel 2) can be seen. Little shear-mode energy is incident on the right-hand tunnel. This is because the source produces a normal horizontal force on the vertical edge of the model; most of the shear-mode energy is radiated up or down. Numerous multiply-reflected coherent events appear on the figures.

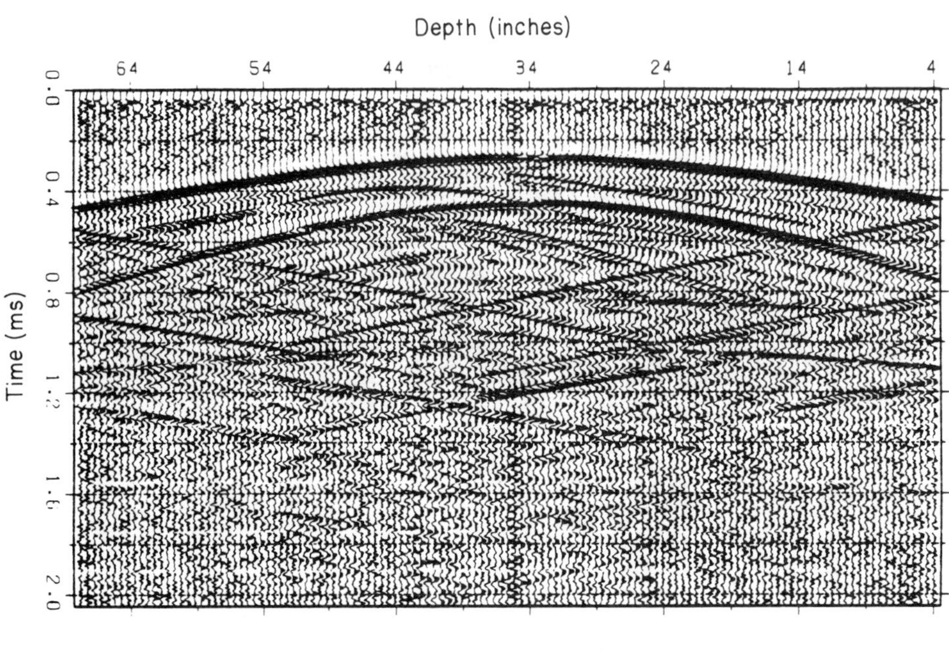


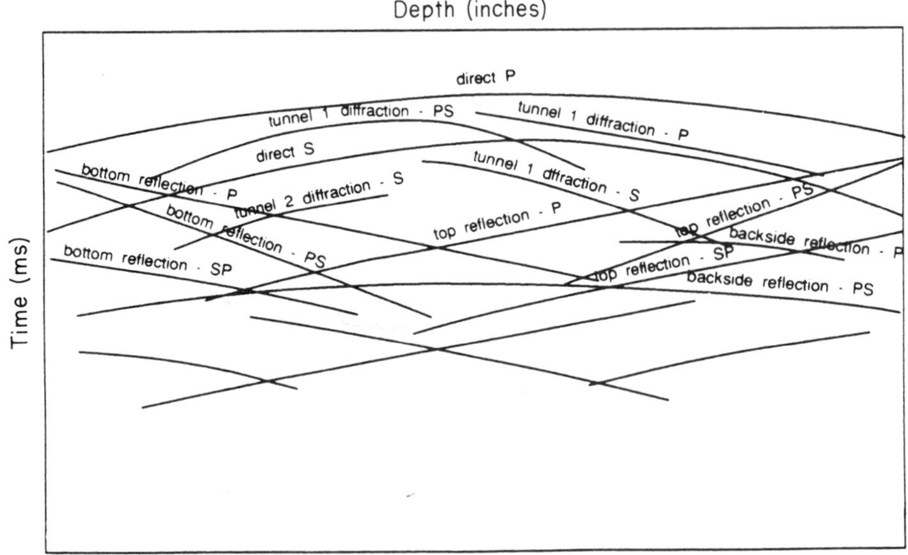


FIG. 5. Common-source gather data set after AGC: $z$- or vertical component receiver. The source was located 34 inches from the top of the model. Detectors or receivers were spaced $\frac{1}{4}$ inch apart in a vertical line, in the middle of the model (Fig. 4). Some of the coherent events are identified in the lower portion of the figure, including several tunnel diffraction modes. 128 traces were plotted.

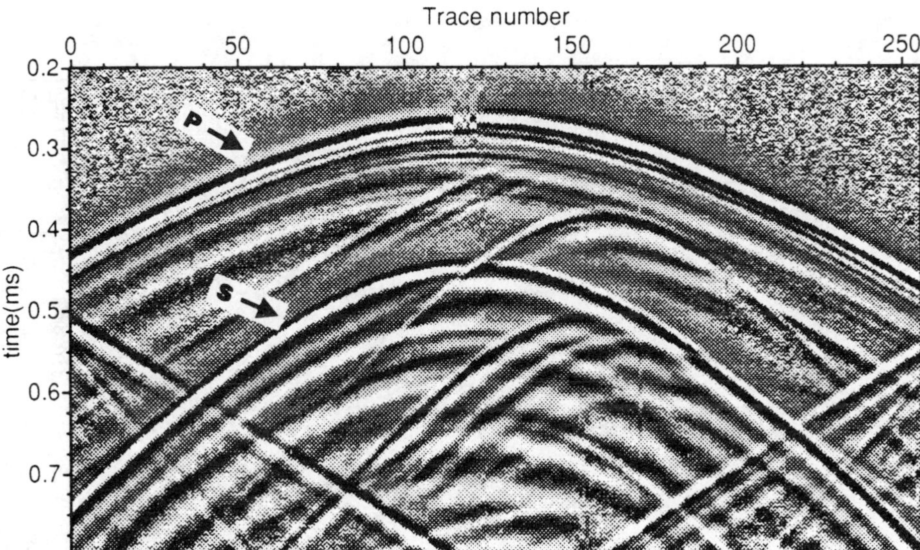


FIG. 6. Enlarged view of the *z*-component data plot in Fig. 5, showing the data from 0.200 to 0.800 ms. Note the *z*-component polarity reversal on the P-wave first arrivals (arrow P) and shear-wave first arrivals (arrow S).

## PROCESSING THE PHYSICAL DATA

### *Sequence of steps*

There were four principal steps in the data processing sequence: mode separation, automatic gain control (AGC), deconvolution, and imaging (or prestack migration). After migration the partial images were combined to yield final stacked images. The processing flow is indicated in Fig. 8. Our main purpose in migrating the data was to delineate the reflecting/diffracting boundaries. We found we could do this effectively with data to which AGC had been applied before stacking. This has the effect of giving approximately equal weight to the images (and noise) on every migrated partial image.

### *Mode separation*

The migration, or imaging, scheme used on these data is described in detail by Hofland (1990). Hofland's software accommodates an arbitrary distribution of velocities, but uses solutions to the scalar wave equation. It is therefore desirable to separate the recorded data into compressional and shear modes before attempting the migration.

Two methods (Dankbaar 1987; Ranzinger 1990) were used for mode separation. Both methods utilize polarization and apparent velocity of the wave to achieve the

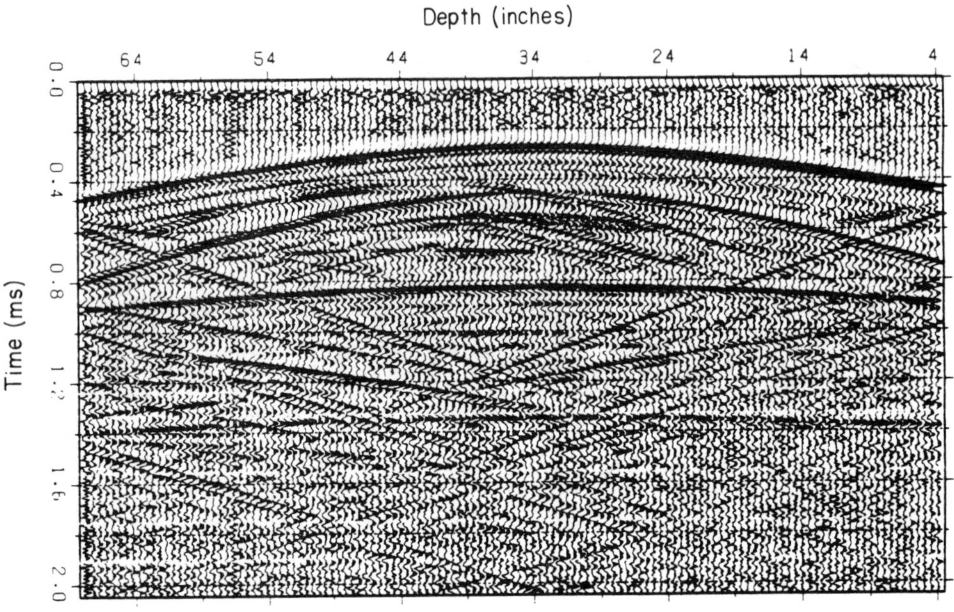


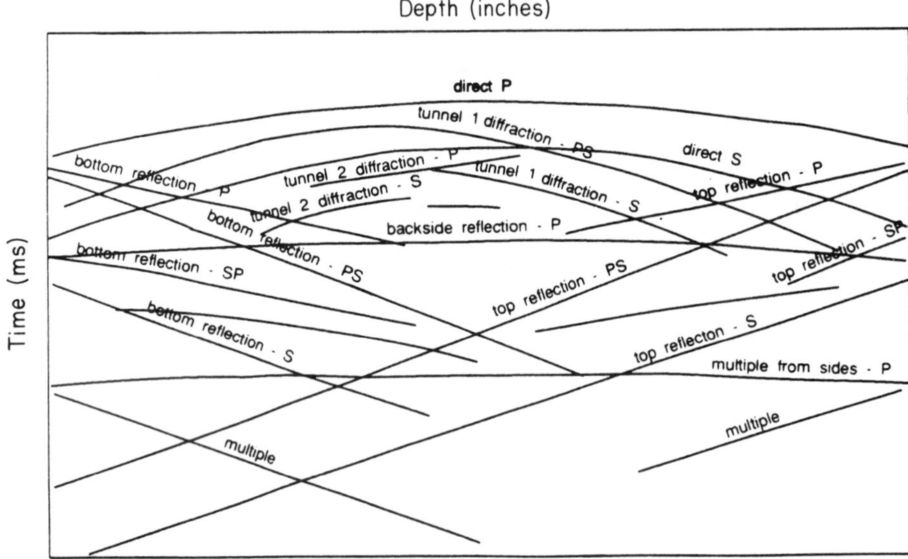


FIG. 7. Common-source gather data set after AGC: $x$- or horizontal component receiver. The source was located 34 inches from the top of the model. Detectors, or receivers, were spaced $\frac{1}{4}$ inch apart in a vertical line, in the middle of the model (Fig. 4). Some of the coherent events are identified in the lower portion of the figure. 128 traces were plotted.

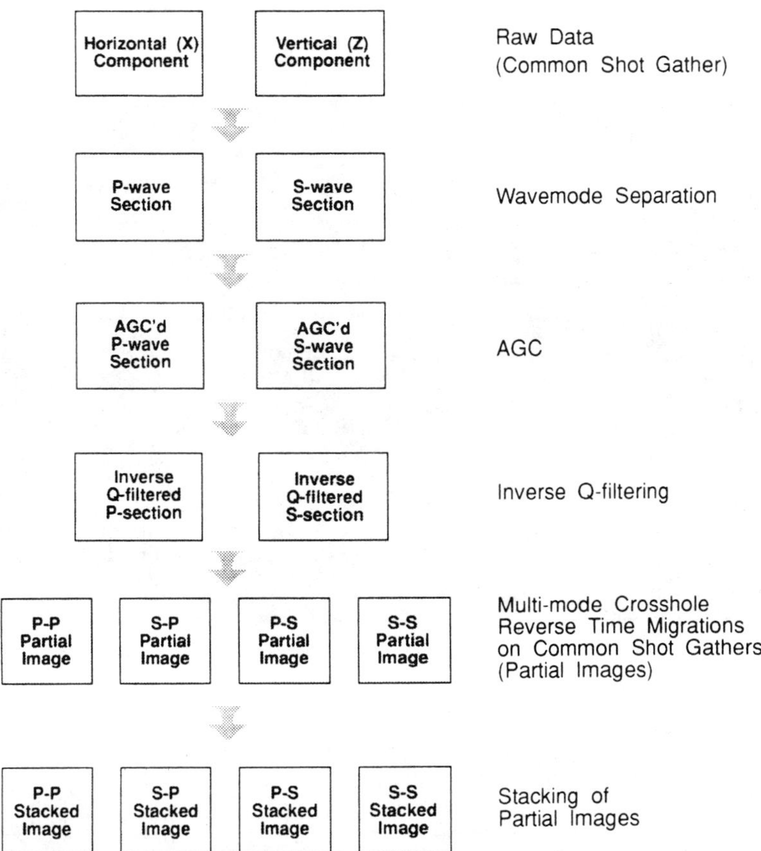


FIG. 8. Cross-borehole data processing flow diagram. The location of the AGC step is optional, but time-variant amplitude control is desirable, somewhere in the sequence.

desired separation. Ranzinger's method explicitly accommodates variable P- and S-mode velocities but is more computer intensive than that of Dankbaar. The velocity in this model was nearly constant, thus Dankbaar's method was used for economy's sake. It should be noted that Dankbaar's formulation depends upon whether the events are assumed to be coming from the left or from the right of the receiver borehole. We must use a formulation appropriate to the space in which we are attempting to image.

Figure 9 shows two common-source gathers after mode separation using Dankbaar's method. The gathers shown are for the space between boreholes. The results have been plotted with AGC, in order to display all the coherent events on a single plot in spite of a considerable variance in actual amplitude. An unfortunate byproduct of this is that the P-wave's first arrival, although greatly suppressed by the mode separation, has been boosted by the AGC on the S-wave display. In Fig. 9 (bottom) this event is plotted with an amplitude far out of proportion to its actual amplitude relative to the shear events shown in the figure.

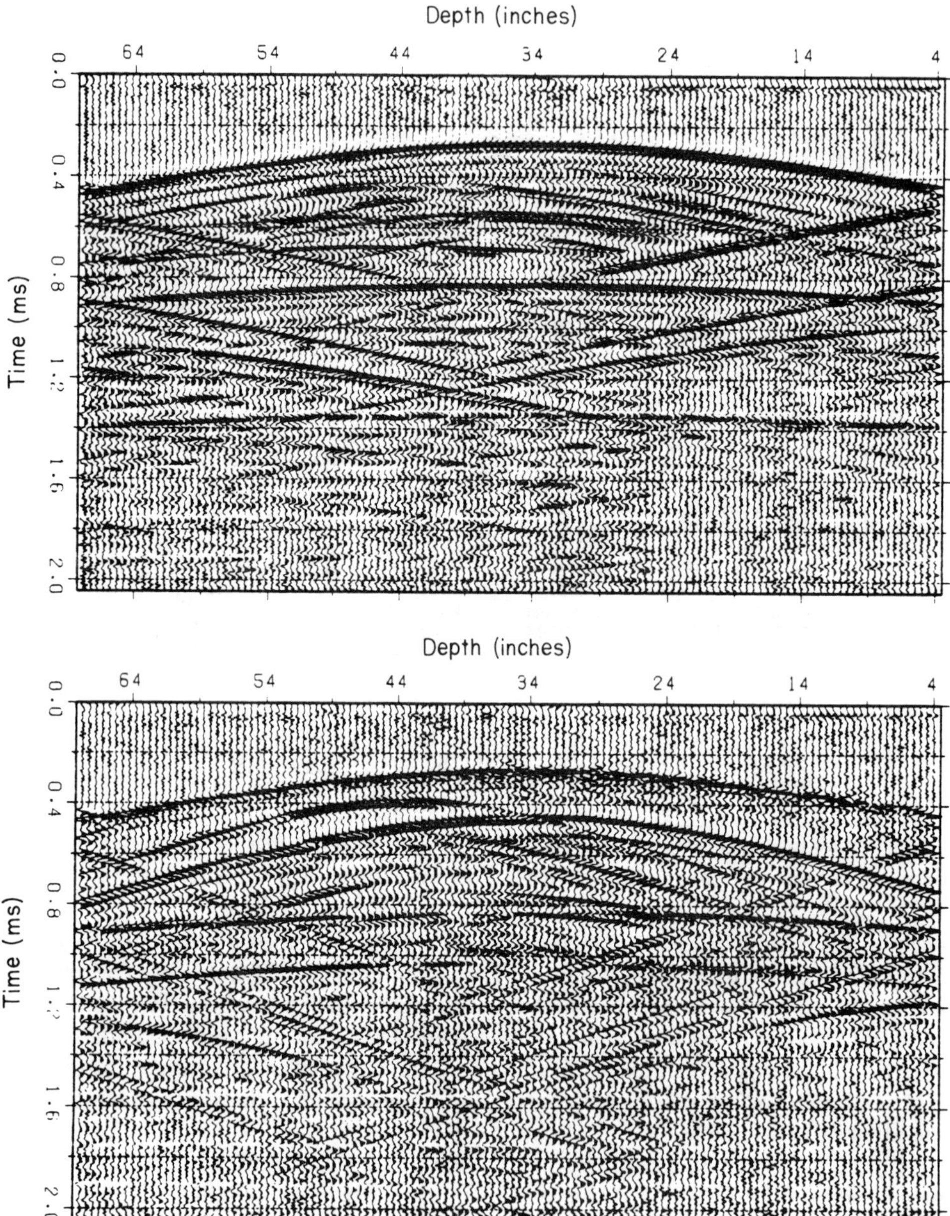


FIG. 9. Mode-separated common-source gathers, plotted with AGC, for the space between boreholes. The upper figure shows primarily P-waves and the lower figure displays predominately S-waves. Dankbaar's (1987) separation method was used to obtain these data from the data shown in Figs 5 and 7. 128 traces were plotted.

All 33 two-component common-source gathers were mode separated; this yielded 132 mode-separated data sets for migration: 33 sets each for the P- and S-wave data sets, between and to the right of the receiver borehole. The reflection history of events arriving at the receiver line is unknown. As a result the P-mode sets contain direct P-arrivals, P–P reflections and S–P converted wave reflections. Similarly, the S-mode set contains direct shear, plus S–S and P–S reflections.

### *Inverse Q-filtering*

Our main purpose in deconvolving the recorded data was to reduce the effects of attenuation and dispersion in the medium. The migration scheme, described in the next section, is based on a non-attenuating, non-dispersive model of the earth. If only one common-source gather were to be imaged, errors due to dispersion would merely tend to 'smear' the images a little, and possibly mislocate them (over- or under-migration). Our strategy is to migrate all 33 of the gathers, and then combine or 'stack' the results. If the various images are mislocated by a different amount, destructive interference may cause a significant deterioration in the image quality.

Figure 10 illustrates inverse $Q$-filtering with a synthetic example. The first trace shows a series of reflection pulses that might be observed in an ideal medium excited by an impulsive seismic source. The second trace shows the effect of dispersion and attenuation. As time increases (i.e. as the travel distance increases) the various frequency components that constituted the original pulses have been attenuated differently and have travelled at different speeds. The dominant frequency of the recorded pulse becomes lower as the travel distance increases and the phase relationships amongst the frequency components change. These effects account for the change in the appearance of the reflected arrival. Since the velocity is frequency-dependent, the 'apparent' velocity (phase velocity) changes with distance travelled. No single migration velocity will be appropriate for all events, i.e. the amount of under- or over-migration will depend on the distance a reflected pulse has travelled.

To compensate for this attenuation–dispersion effect, we have utilized a so-called 'inverse $Q$-filtering' process (Chang 1991). We attempt to create the data set we would have obtained if there had been no attenuation and dispersion. This set is then migrated as if the medium were perfectly elastic. This procedure appears easier than attempting to take these effects into account explicitly during the migration process. The medium is assumed to be linear and have a constant, or frequency-independent $Q$. The attenuation filtering process is assumed to be minimum phase.

The third trace in Fig. 10 shows the result of applying inverse $Q$-phase filtering to the dispersed waveforms that appear on the second trace. Note that phase distortion has been removed and that the 'correct' (i.e. non-dispersive) arrival times have been restored.

We can apply an additional filter to compensate for the progressive attenuative loss of the higher frequencies over time. The fourth trace in Fig. 10 shows the result of this operation. The amplitude of wave forms and the high frequencies have been restored. Because of frequency-domain clipping of the inverse $Q$-amplitude filter, the

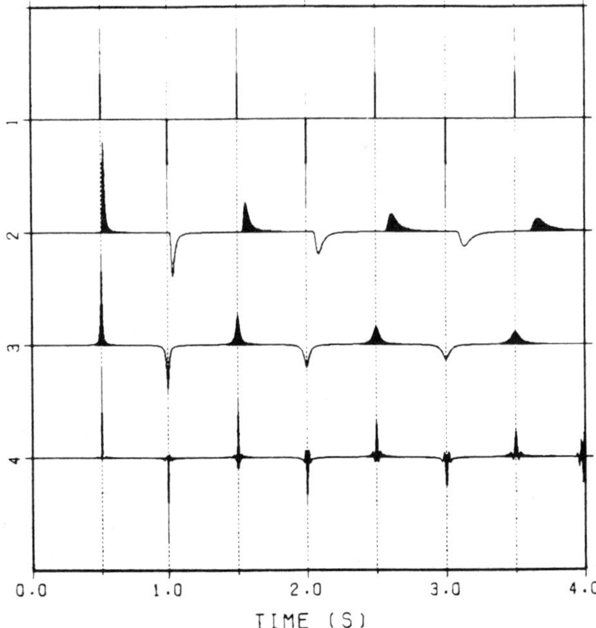


FIG. 10. Synthetic example demonstrating the effect of attenuation and dispersion on reflected waveforms, and how this effect is mitigated by inverse $Q$-filtering. Trace 1 represents an original reflection sequence due to a source impulse in an ideal elastic medium. Trace 2 (after forward $Q$-filtering) shows the effect on the reflected pulses, over space and time, of dispersion and attenuation. In Trace 3 (after inverse $Q$-phase filtering) the reflections have been phase compensated, which restores the arrival times and original symmetry. In Trace 4 (after inverse $Q$-amplitude filtering) the frequency components have been equalized.

later arrivals have not been fully recovered. The principal limitations on the effectiveness of the process have two causes. Firstly, the $Q$ of the medium is never known exactly. Errors in $Q$-estimates limit the resolvability of the process. Secondly, because of computational approximations involved in inverse $Q$-amplitude filtering (Chang 1991), amplitude clipping is needed for stability. Consequently later arrivals are usually not fully recovered. In a real data situation, ambient noise is always present and the noise usually controls the limit of resolution.

### *Imaging the reflections (prestack migration)*

*Imaging principle.* The most important step is to create images of the discontinuities that caused the reflections and diffractions. In order to do this we applied the principles of prestack migration of surface and/or VSP data to the cross-borehole geometry. These principles and procedures have been described elsewhere (Claerbout 1971; Chang and McMechan 1986; Whitmore and Lines 1986; Hofland 1990).

The following procedure was used. Project the reflected/diffracted wavefield back into the medium, i.e. estimate the reflected/diffracted wavefield for all time, everywhere within the medium to be imaged (image space) based on the recordings of the reflected/diffracted wavefield along a portion of the boundary of the medium, and on the medium velocities. Estimate the source wavefield for all time, everywhere within the medium. Apply the imaging principle (Claerbout 1971) that reflecting/diffracting points are located at those places in the medium where the source wavefield and reflected wavefield are time coincident. This principle is illustrated in Fig. 11. The source wavefront encounters a diffractor; a diffracted wave is eventually recorded at the receiver borehole. A typical receiver registers this event at time 16 (Fig. 11). We extrapolate the recorded event(s) backwards in time, and extrapolate the source wavefield forward in time. Note that the diffractor is located where the

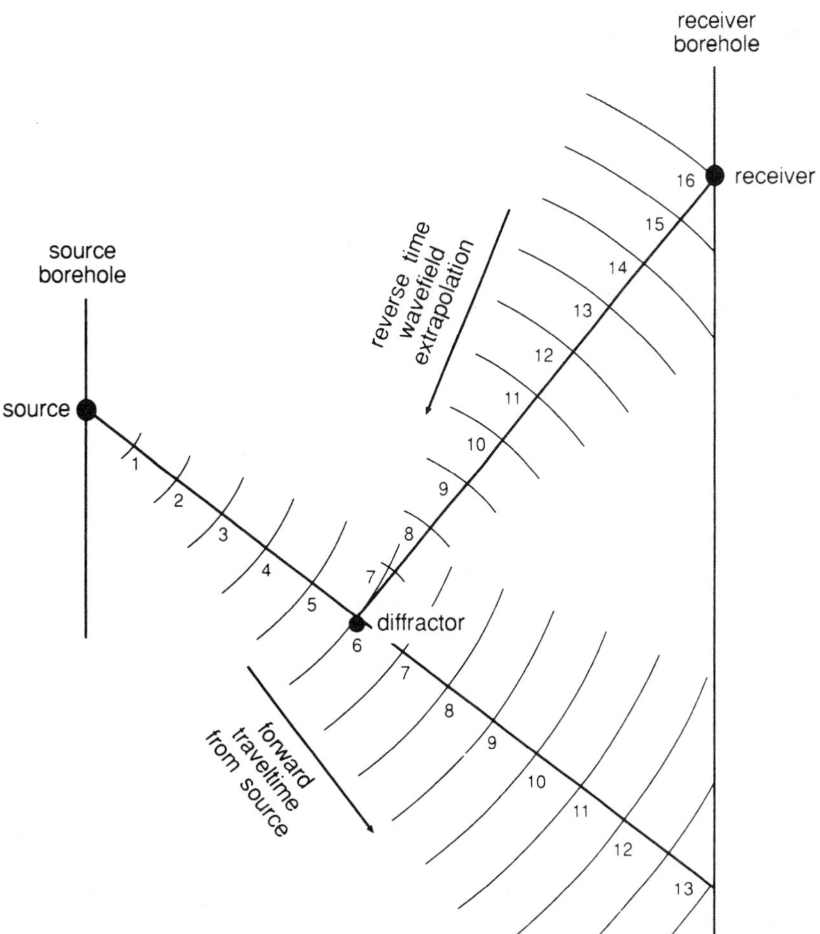


FIG. 11. The imaging principle. The diffracted wavefield and the source wavefield are space–time coincident where the diffractor is located.

source and reflected wavefields are time coincident. If we regard a reflecting surface as a continuum of diffractors, the principle applies equally well to reflectors.

*The reflected wavefield.* We used a finite-difference solution to the scalar wave equation to back project the reflected wavefield into the medium. The wavefield recorded along the line of receivers (treated as a scalar field) was used as a boundary condition.

The finite-difference algorithm was fourth order in space, second order in time (Dablain 1986). To reduce boundary reflections, absorbing boundary conditions (Clayton and Engquist 1977) were applied.

*The source wavefield.* We first calculated the direct traveltime from the source to every point in the medium. Then we used a weighted impulse (delta function) for the source wave form or else a wave form derived from deconvolved direct arrivals observed at the receiver line.

Procedures using Fermat's principle to map wavefronts and raypaths (Ranzinger 1990; Schneider 1990), were used to calculate the traveltimes. For an inhomogeneous medium, these methods are more efficient to implement by computer than by ray tracing.

*Migration or imaging.* The migration principle requires us to extrapolate the reflected wavefield back into the medium, based on observation of the reflected wavefield along the recording line. Our recording line does not enclose the medium, thus we can only partially reconstruct the reflected wavefield. For this reason the image that we shall obtain is incomplete, or distorted. In the terminology of surface data processing, 'migration smiles' result. We call this image a partial image.

We construct partial images from every common-source gather. Those portions of each partial image that correspond to true reflectors in the medium are always located in their proper place. Those which are migration smiles, or other artefacts, are located at different places on each partial image. When the partial images are combined, or 'stacked', the artefacts interfere destructively, leaving a final image that primarily corresponds to the true reflectors.

We choose our source wave velocity separately from our choice of the velocity of the reflected waves. Our choices depend on the mode we wish to image. For example, if we wish to image the SP reflections we would use the S-wave velocity for the source, and the P-wave velocity for the reflected wavefield. The choices are tabulated as follows:

| | $R = V_p$ | $R = V_s$ |
|---|---|---|
| $S = V_p$ | (PP) | (PS) |
| $S = V_s$ | (SP) | (SS) |

where $S$ is the source velocity, $R$ is the reflection velocity, $V_p$ is the P-wave velocity and $V_s$ is the shear-wave velocity.

When we use $V_p$ for our source velocity and $V_s$ for our reflected wave velocity, we will properly image the PS reflection, but we will obtain a false image due to the back-projected SS energy. Similar arguments apply to the other migrations: we image one unwanted reflected mode.

Fortunately the desired images will be properly located, since proper source and reflected wave velocities were used. The extraneous images will be mislocated. The amount of mislocation will vary amongst the various partial images, whereas the desired images will remain stationary at their proper locations. When the partial images are combined, the desired images will reinforce one another, but the false images tend to interfere destructively.

*The image spaces.* It is difficult to distinguish between interwell and extra-well reflections/diffractions at the receiver line. We therefore divide our medium into two image spaces. One space lies to the left of the receiver line, the other lies to the right. First the entire reflected wavefield is projected into the left-hand space. We use the source field in the left-hand space when we apply the imaging condition. The reflector images in the left-hand space will then be properly located. The images that really belong in the right-hand space will appear as extraneous, mislocated images. Each migrated common-source gather will locate the real reflector partial images in the correct positions. Each common-source gather will mislocate the extraneous partial images in a different location. As before, when the partial images are stacked, the real images will interfere constructively and the extraneous ones will interfere destructively. The result is a good quality representation of the real reflecting discontinuities in the left-hand image space.

We see that for every common-source gather, we need to make eight partial images: four mode-types in each of two image spaces. In the most general 2D case, we would have three image spaces: to the left, between and to the right of the boreholes.

*The partial images.* A single synthetic example and four real examples will illustrate the partial image concept discussed above. Figure 12 is a computer-generated synthetic common-source gather with the same geometry as the physical model shown in Fig. 4. A compressional, P, first arrival, a PP diffraction from the left-hand tunnel and PP reflectors from the top, bottom and right-hand edge are shown. In this demonstration example all amplitude losses and mode conversions have been ignored.

Figure 12b is the PP partial image obtained by migrating the data shown in Fig. 12a (in both image spaces).

In the left-hand image space (Fig. 12b), we see a good representation of the top and bottom reflectors. The edges of the images are slightly curved, due to 'smiles'. Only part of the reflecting edge is restored because specular reflection into the receiver array only occurs over part of the reflecting edges. The diffraction from the hole or tunnel is imaged as a short horizontal line at the hole location. Only a relatively small segment of this diffraction could be back-extrapolated. This is due to the limited size of the receiver array and to the need to mute the diffraction where it overlaps and is obscured by the direct arrival. These difficulties can be overcome by stacking of multiple partial images.

In the right-hand image space (Fig. 12b) we see a good quality image of the vertical edge of the model. The curved ends of the images are further examples of 'smiles'. Three false images appear in the right-hand image space. The upper and lower images result from projecting the top and bottom edge reflections into the

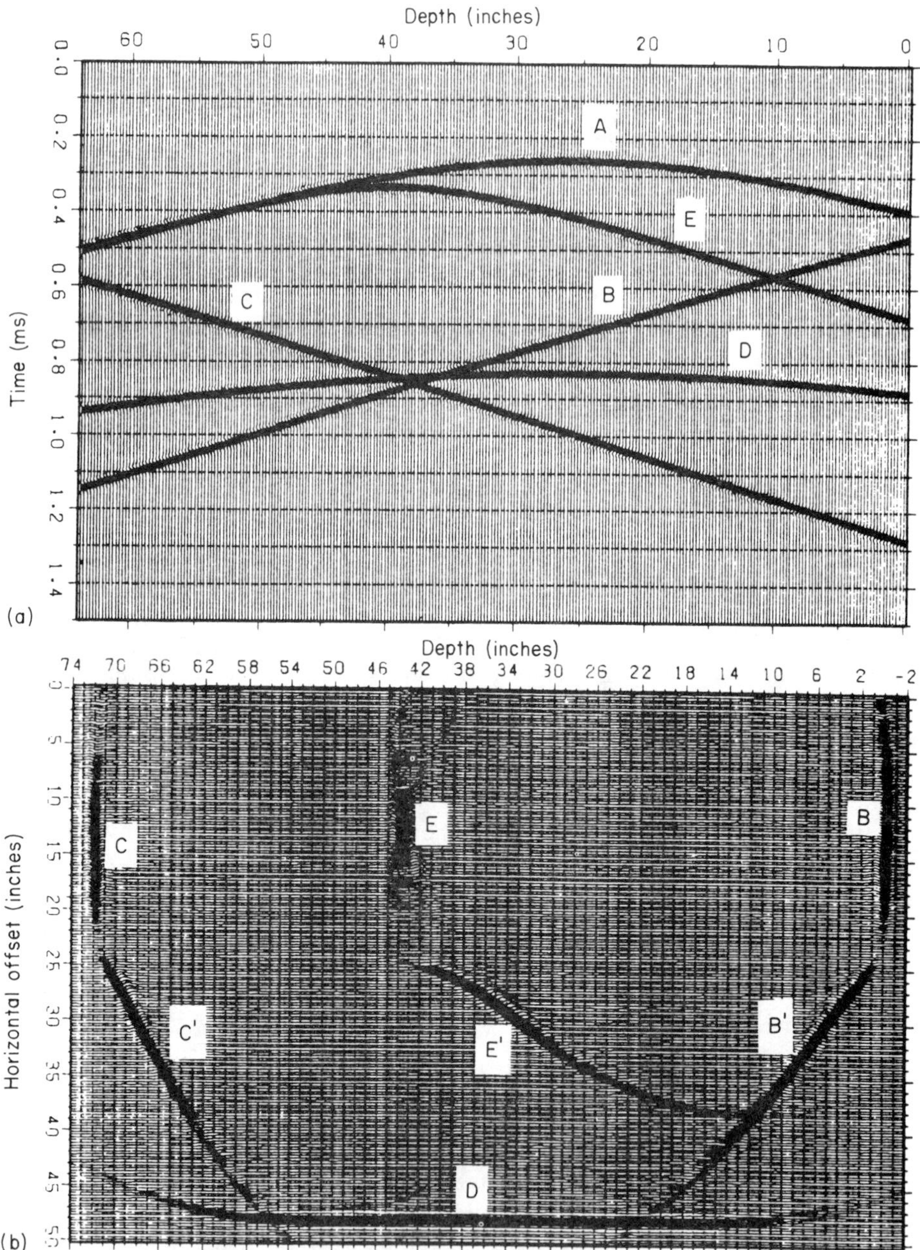


FIG. 12. Synthetic example illustrating the partial image idea. (a) Computer generated synthetic cross-borehole PP common-source gather data set, based on Fig. 3. Events are: A, direct P-arrival; B, PP reflection from top edge; C, PP reflection from bottom edge; D, PP reflection from right-hand vertical edge; E, PP reflection/diffraction from the tunnel. (b) Migrated, imaged version of the synthetic data above. Images are B, B′ top reflector, C, C′ bottom reflector; D, left-hand vertical edge reflector; and E, E′ tunnel 1. Primed images are false images resulting from projections into the incorrect image space.

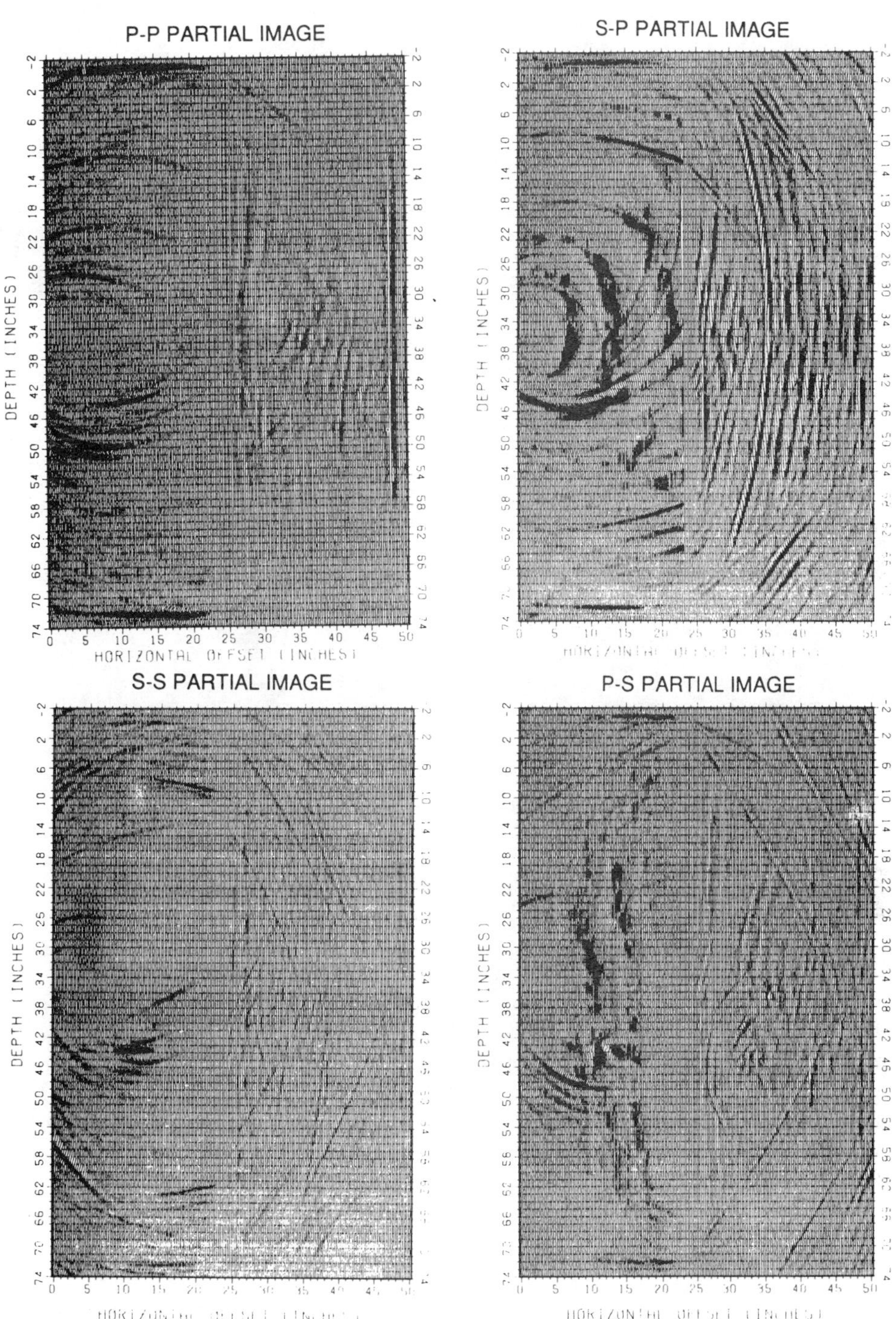


FIG. 13. Set of four partial images resulting from migrating a single common-source gather. The common-source gather input data were obtained from the model shown in Figs 3 and 4. Artefacts are unavoidable, and are due to imaging of wrong modes, and imaging in the wrong image space.

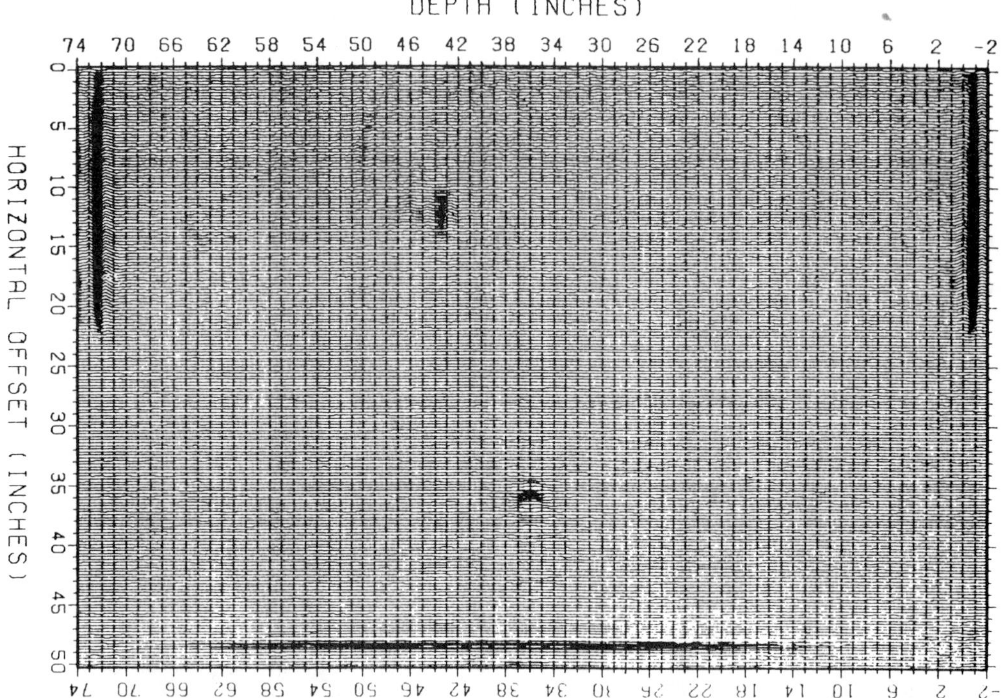


FIG. 14. Final stacked PP image of the model shown in Fig. 3, based on 33 common-source gathers. All three edges and both tunnels can be seen clearly and have the correct size and locations. Edge images are created only along those portions of the edge that reflect source energy into the receiver array.

right-hand image space. As stated before, all events are projected into both spaces. We cannot determine the correct space for projection based on the original data, i.e. the top and bottom reflections in Fig. 12a are equally well accounted for by *either* or *both* of the resultant migrated images shown in Fig. 12b. The curved image in the middle of the right-hand space (Fig. 12b), labelled E′, comes from projecting the tunnel diffraction into the right-hand image space and imaging it there. As before, the diffracted event is equally accounted for by a tunnel on the left or a curved reflector on the right. Although this demonstration example is based on a scalar wave computation, the ambiguity cannot be resolved by measuring all components of motion.

When the source–receiver geometry is changed by recording additional common-source gathers, the true images appear in their proper places on every gather but the false images move.

Partial images based on a common-source gather of the real (physical model) data are shown in Fig. 13. Many more artefacts appear on the real data migrations since all modes and all reflections were present in the real input data.

*The stacked images.* The final stacked PP image is shown in Fig. 14. All five reflectors/diffractors: top, bottom and side edges, plus left- and right-hand tunnels,

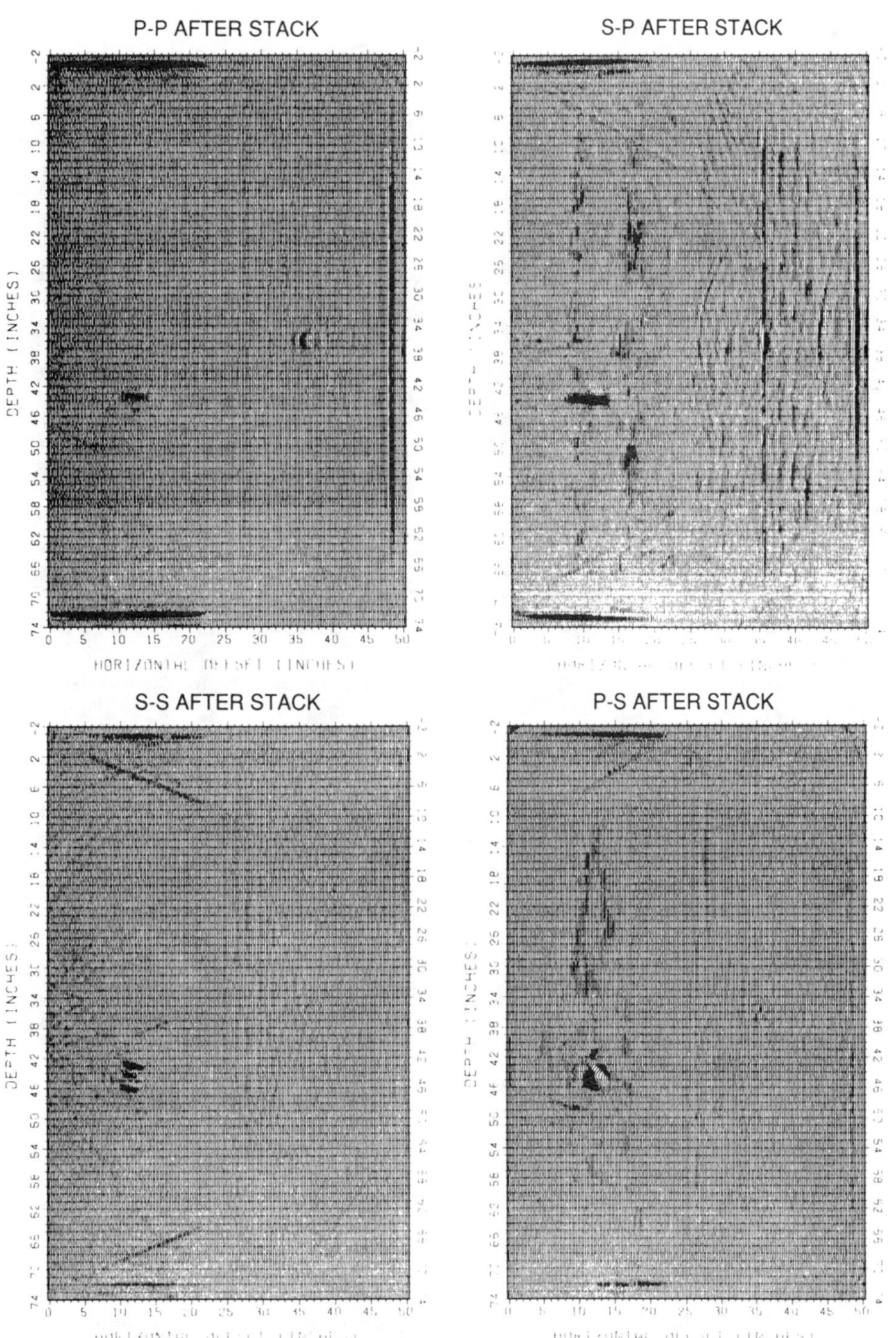


FIG. 15. Final stacked images of the physical model shown in Fig. 3 based on 33 cross-borehole common-source gathers. All four reflected/diffracted modes have been imaged: PP, PS, SS, SP.

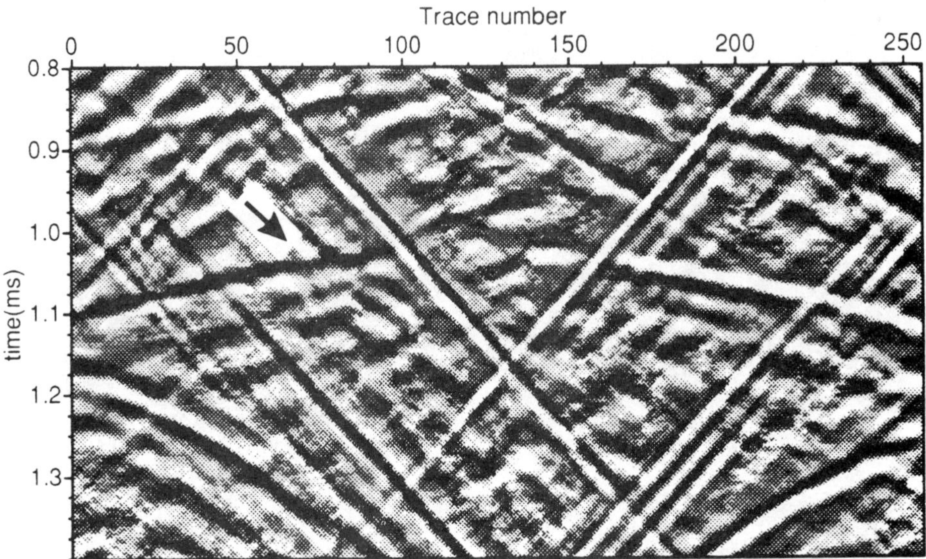


FIG. 16. Enlarged portion of Fig. 5 showing the vertical component recordings from a common-source gather (source 34 inches from the top of the model, opposite receiver 121). The reflection from the right-hand vertical edge of the model is indicated (arrow). Note the polarity reversal as we pass the source level, due to reversal in the sign of the illuminated angle.

appear with remarkable clarity, and at the correct location. The edge images are sharp and clear, but do not extend over the entire length of the true physical edges (Fig. 3). This is because the edges are imaged only where specular reflection into the receiver array occurs. We see no image along those portions of the edges that do not directly reflect energy into the array.

The holes are correctly located and are approximately the same size as the actual holes in the model. The set of four mode migrated stacked images is shown in Fig. 15. All four show the reflecting edges clearly. The PP mode migration shows all five images best. The SS mode migration shows the top and bottom edges and the left-hand tunnel quite well. The right-hand images are missing. This can be explained in part by the fact that the source exerted a (horizontal) force normal to the (vertical) left-hand edge of the model. Very little shear energy was radiated horizontally. Almost no SS energy was reflected into the receiver array from the right side of the model.

The mixed-mode migrations show all five images, but the background noise is high. This is due in part to the low-amplitude of the mixed-mode reflections/diffractions received at the borehole, and in the case of PS images also due to a sign change in the PS reflection due to a sign change in the angle of illumination. We believe that with additional effort, the mixed-mode images could be greatly improved. In any case we have demonstrated that mixed-mode images can be obtained from cross-borehole data, even using scalar wave theory.

### *The sign change problem*

The problem encountered in attempting to image the PS reflections is illustrated in the case of the PS reflection from the right-hand edge. The source exerts a force on the left-hand edge which is normal to that edge. Conditions of symmetry about a line through the source-point normal to the reflector (a horizontal line in this case) require that the vertical component of the reflected PS wave motion be in opposite directions on either side of this line. The phenomenon is shown in Fig. 16, which is an enlarged version of Fig. 5, from 0.800 to 1.400 ms. The polarity change in the $z$-component of the PS reflection from the right-hand edge is apparent. It occurs at the source level, as expected. As a result of this polarity change in the vertical component of motion, caused by the reversal in sign of the illumination angle, there results a reversal in the sign of the reflection on the S-mode separated data set (Fig. 17) and thus we obtain an image whose polarity also depends on the sign of the illumination angle. If we sum or stack all the partial images, we stack images of mixed polarity. This can cause destructive interference and degrade the final image instead of enhancing it.

Figure 18 illustrates this degradation. (a), (b) and (c) show greatly enlarged plots of partial stacks of the PS image of the vertical reflector. (a) was obtained by stacking the partial images from the upper 10 sources, (b) from the middle 13, and (c) from the lower 10 sources. (a) has predominately positive illumination angles, (b) mixed angles, and (c) has negative illumination angles. The sign reversal in the

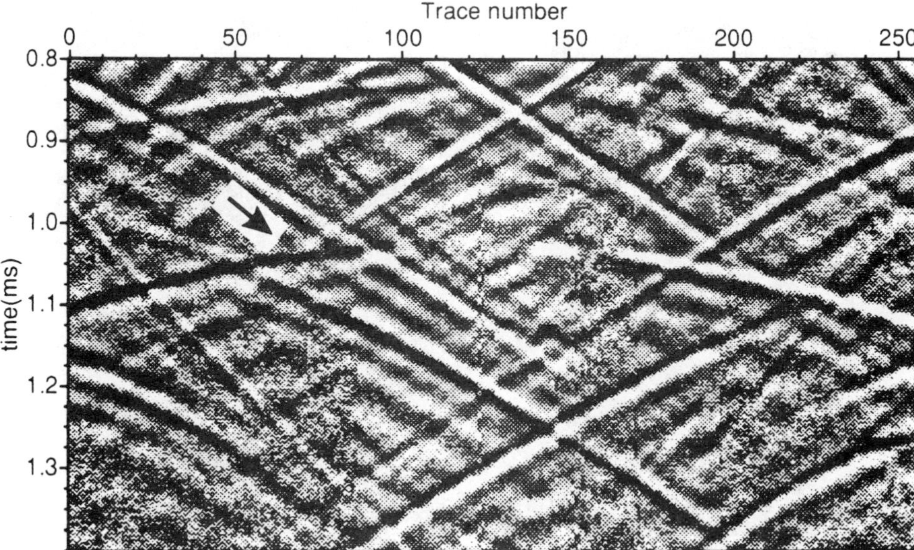


FIG. 17. Enlarged portion of the data of Fig. 9 (bottom), showing S-mode separated data from the common-source gather with sources opposite trace number 121. Note the polarity reversal of the PS reflection from the right-hand vertical edge (arrow) as we pass the source level and the illumination angle is reversed.

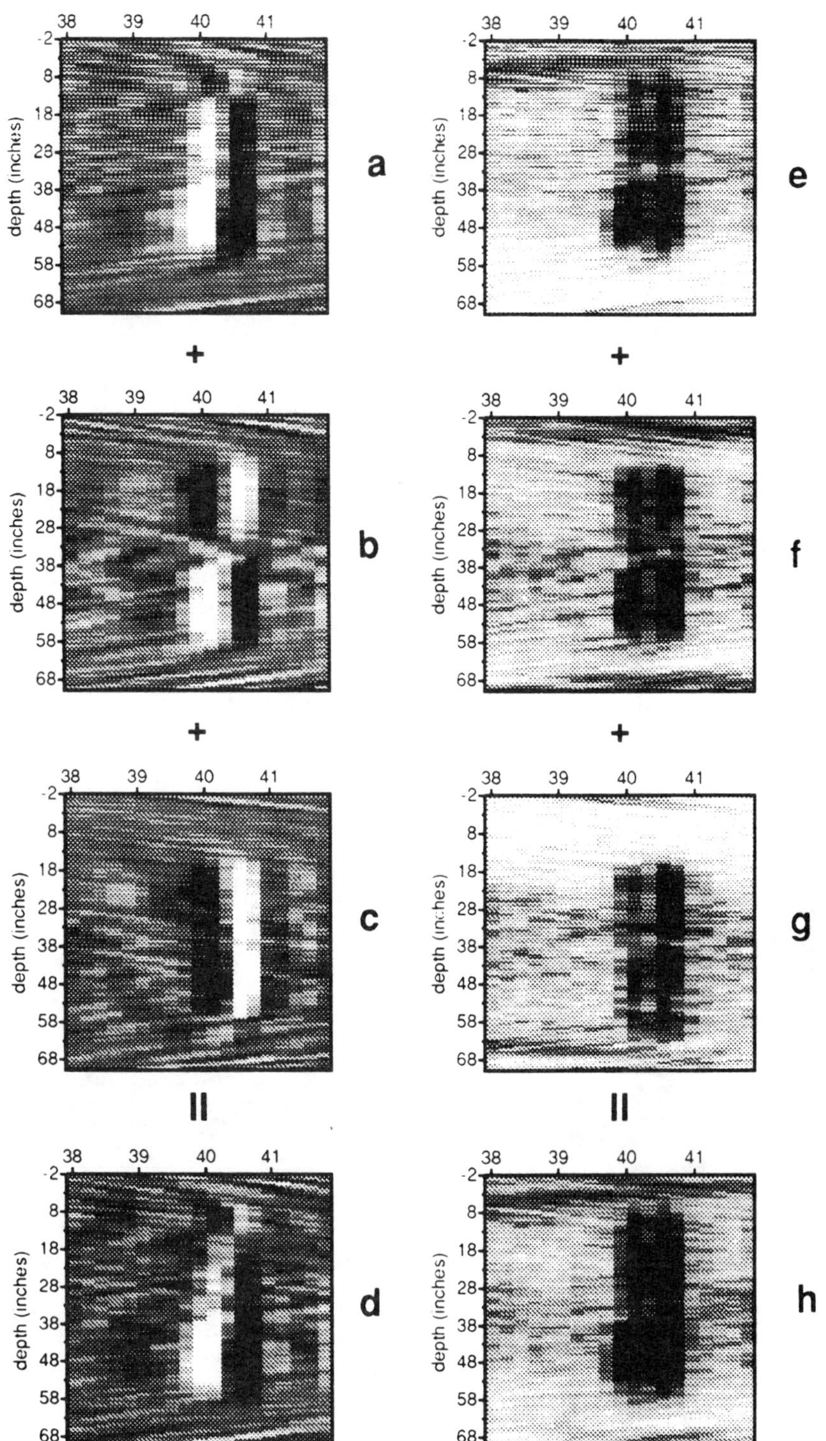
38
39
40
41
depth (inches)
-2
8
18
28
38
48
58
68
a
b
c
d
e
f
g
h
+
=

image is apparent. When these data are summed, or stacked, we obtain (d). As a result of destructive interference, the image quality in (d) is poor.

We see no polarity reversal from the vertical edge SP wave reflection on the P-wave version of the gather associated with the source located opposite receiver location 121. Had the source exerted a tangential force on the left-hand edge, as is usually the case for a shear-wave source, the polarity would have been reversed. But in the present instance, the incident shear wave was due to the application of a force normal to the edge. Under these conditions the polarity is an even function of the incident angle. The analogous case for borehole sources is illustrated by Delvaux *et al.* (1987, Fig. 5). When the partial SP images are summed, constructive interference results which helps to explain the strong vertical edge SP reflector image seen in Fig. 15.

A complete treatment of the problem is beyond the scope of this paper. One partial solution is to stack, or sum, absolute values. This is demonstrated in Fig. 18e, f, g, h. Constructive reinforcement has been obtained, at the expense of resolution and of 'sign of the reflectivity'.

Edge artefacts are especially prominent on the SS mode image. We believe they are mismigrated PS events. The PS reflections from the horizontal edges are quite strong, even stronger than the SS, for certain source positions. These reflection amplitudes vary greatly amongst the common-source gathers. These two factors prevent total cancellation of the image on the stacked version.

On the PS panel, Fig. 15, we see a similar artefact due to uncancelled SS images. Similarly we see uncancelled PP edge images on the SP panel.

## Conclusions

Cross-borehole seismic recordings contain a complicated combination of events. In their raw form these recordings can be difficult to interpret at best. By migrating physical model cross-borehole data, we have seen that these data can be converted into meaningful images of both interwell and extra-well reflecting bodies between two boreholes. The images are much more interpretable geologically than the original recordings.

---

Fig. 18. An example of reflector image sign change due to reversal of the polarity of the illumination angle. (a) is a plot of a stack of 10 partial images of the right-hand edge of the model, associated with sources near the top of the model (predominately positive illumination angle). (b) is associated with the middle 13 sources (mixed illumination angles). (c) is associated with the bottom 10 sources (predominately negative illumination angles). The image in (a) is opposite in sign to that in (c). The image in (b) experiences a sign reversal in the middle of the image (as the illumination angle reverses sign). (d) is a summation of (a), (b) and (c) and illustrates the deterioration in image quality when images of (possibly) opposing signs are summed. In (e), (f), (g) and (h) the same procedure is carried out after the images (a), (b) and (c) have been replaced by their absolute values. Image enhancement was obtained here at the expense of resolution.

The fact that SS and mixed-mode reflections can be imaged is of interest and value. We have the possibility of generating four images from one set of cross-borehole data. These images might be directly combined to yield a further enhanced image. They can also be compared to obtain further information about rock properties. A fluid-filled void, for example, would image quite differently on reflections using shear modes than on reflections involving P-modes.

Mode separation is helpful in reducing the number of image artefacts caused by migrating S-modes at P-mode velocity, and vice versa.

Inverse $Q$-filtering, especially to reduce the effect of dispersion, was helpful in the model study. It should be noted that perspex may be more dispersive than most real earth media.

We were able to obtain good images of rather small reflectors/diffractors using mode separation and stacking without having to solve the vector wave equation.

## Acknowledgements

This work was supported by the U.S. Geological Survey, under USGS-SIP-3343G-01, project 3.1.11, under the direction of Joseph P. Rousseau; the U.S. Bureau of Mines under contract J029 003, represented by Robert Munson; the U.S. Army, contract DAAK 7090 P1147, assisted by Frank Ruskey and Raymond Dennis; and the U.S. Department of Energy contract no. DE-AC22-89BC 14478 supervised by Robert Lemmon.

We thank the contributors for their generous support and encouragement which made this investigation possible.

## References

Beydoun, W.B., Delvaux, J., Mendes, M., Noval, G. and Tarantola, A. 1989. Practical aspects of an elastic migration/inversion of crosshole data for reservoir characterization. *Geophysics* **54**, 1587–1595.

Chang, H. 1991. Multi-mode crosshole reflection imaging of multi-component physical model data. Ph.D. thesis 4016, Colorado School of Mines, Golden, Colorado.

Chang, W.F. and McMechan, G.A. 1986. Reverse time migration of offset vertical seismic profiling data using excitation-time imaging condition. *Geophysics* **51**, 67–68.

Claerbout, J.C. 1971. Toward a unified theory of reflector mapping. *Geophysics* **49**, 467–481

Clayton, R.W. and Engquist, B. 1977. Absorbing boundary conditions for acoustic and elastic wave equations. *Bulletin of the Seismological Society of America* **67**, 1529–1540.

Dablain, M.A. 1986. The application of higher order differencing to the scalar wave equation. *Geophysics* **51**, 54–66.

Dankbaar, J.W.M. 1987. Vertical seismic profiling separation of P- and S-waves. *Geophysical Prospecting* **35**, 803–814.

Delvaux, J., Nicoletts, L., Noval, G. and Dutzer, J.F. 1987. Acquisition techniques in cross hole seismic surveys. *Proceedings SPE, Dallas*, 413–419.

Hofland, G.S. 1990. Multi-mode, multiple offset VSP reverse time migration (imaging) of a complex, physical earth model. M.S. thesis T3707, Colorado School of Mines, Golden, Colorado.

HU, LIANG-ZIE, MCMECHAN, S.A. and HARRIS, J.M. 1988. Acoustic prestack migration of cross-hole data. *Geophysics* **53**, 1015–1024.

IVERSON, W.P. 1986. Crosswell reflection seismology: applications and limitations. 56th SEG meeting, Houston, Expanded Abstracts, 232–235.

PRATT, R.G. and WORTHINGTON, M.H. 1990. Inverse theory applied to multi-source cross-hole tomography. Part 1: Acoustic wave equation methods. *Geophysical Prospecting* **38**, 287–310.

RANZINGER, K.A. 1990. Wave mode separation of multi component, multi offset VSP data acquired from a complex physical earth model: M.S. thesis T3812, Colorado School of Mines, Golden, Colorado.

SCHNEIDER, W.A. 1990. A three-dimensional physical modeling study applying tomographic inversion and seismic migration to the tunnel detection problem. Ph.D. thesis T3866, Colorado School of Mines, Golden, Colorado.

WHITMORE, N.D. and LINES, L.R. 1986. Vertical seismic profiling depth migration of a salt dome flank. *Geophysics* **51**, 1087–1109.

# Scaled physical modelling of anisotropic wave propagation: multioffset profiles over an orthorhombic medium

R. James Brown, Don C. Lawton and Scott P. Cheadle*

*Department of Geology and Geophysics, The University of Calgary, Calgary, Alberta, Canada* T2N 1N4

Accepted 1991 June 19. Received 1991 June 14; in original form 1991 January 18

## SUMMARY

In this paper we report further results of scaled physical modelling experiments in the laboratory in which ultrasonic elastic waves are propagated through an anisotropic medium of orthorhombic symmetry. Whereas our earlier experiments consisted for the most part in sending and receiving on opposite faces of a small cube of phenolic laminate, these new results are from multioffset profiles run parallel and at 45° to principal directions on a larger slab of phenolic.

The variation of NMO velocity with offset (or angle of incidence) has been determined for compressional and transverse shear waves along profiles in the two principal directions on the 3-face (parallel to laminations) of the slab. These observed group velocities differ from the exact theoretical values by a maximum of about 1 per cent or less and also compare favourably with the theoretical velocities calculated from Thomsen's first-order equations, with maximum differences of about 2 per cent. Differences between the observed and theoretical velocities are attributed to some combination of finite transducer size (geometrical or effective path length effects, array attenuation effects, or interference with the otherwise free surface), sample inhomogeneity and/or anelasticity, and experimental error.

The transmission shot gathers acquired for propagation in symmetry planes, and for source–receiver pairs with the same polarization, are similar in form to records acquired over a transversely isotropic medium. The effect of the shear-wave window and the variation of the hyperbolic NMO parameter with offset are clearly seen. Transmission records were also acquired in off-symmetry planes, namely along profiles at 45° to principal directions. On these records, which include all nine possible pairs of source–receiver polarizations, we see clear shear-wave splitting at and near zero offset and more complicated wave effects with increasing offset, such as one or another wave phase dying out. This could be due to cusping of wave surfaces or rapid changes of amplitude and/or polarization with ray direction, possibly as consequences of nearby shear-wave singularities.

**Key words:** anisotropy, NMO velocity, orthorhombic symmetry, physical modelling, shear-wave splitting, singularities.

## INTRODUCTION

Using as a medium the anisotropic phenolic laminate described below and by Cheadle, Brown & Lawton (1991), physical seismic experiments have been continuing at the University of Calgary with scaled-down models (1:5000) and scaled-up frequencies (5000:1). In this earlier work we looked primarily at body waves propagating along principal (or symmetry) directions, corresponding to zero-offset transmission through each of three principal faces (symmetry planes) of the phenolic. The only departure from this simplest case, necessary to enable determination of the nine stiffnesses of the material, was propagation along paths at 45° to each of two principal directions and contained within the principal plane thereby subtended. Such records correspond to relatively far-offset shots (angle of incidence = 45°) along profiles oriented in principal directions.

This paper presents the results of laboratory experiments

* Now at Veritas Seismic Ltd, 200, 615–3rd Ave SW, Calgary, Alberta, Canada T2P 0G6.

Reprinted from Geophysical Journal International, 107, 693-702 with permission from Royal Astronomical Society, Blackwell Scientific Publications Ltd.

in which transmission shot gathers have been recorded for a wide range of offsets and for two fundamentally different profile orientations: (a) along principal directions, for which the sagittal or propagation planes are principal or symmetry planes, and (b) at 45° to two principal directions, for which the sagittal planes are no longer principal planes. For this latter case we are reporting our first experimental results for cases where the propagation paths contain components of all three principal directions. Our earlier experiments (as well as some reported upon in this paper) for propagation within symmetry planes yielded results that are considerably less complex than for the case of an arbitrary direction of propagation. For example, the variation of phase velocity with direction in a symmetry plane of an orthorhombic medium for one particular wave phase is of the same algebraic form as in the transverse-isotropy (TI) case, although the interrelationship of these variations between different wave phases is not, in general, the same as in the TI case.

Experimental results obtained on synthetic materials that are essentially transversely isotropic have been reported by Ebrom *et al.* (1990) and Rathore *et al.* (1990), among others. Similar laboratory experiments using ultrasonic waves on real rock samples have been carried out by Jones & Wang (1981), Lin (1985) and Sayers (1988, 1990), for example. However, we are not aware of any other published results to date of the kind reported here of physical seismic modelling using orthorhombic materials. We document our results in two related areas, the variation of stacking or NMO velocity with offset for cases of sagittal symmetry and the analysis of shot gathers obtained for various combinations of shot and receiver polarization (vertical, radial, transverse) in both symmetry and off-symmetry planes.

## LABORATORY SET-UP

The set-up of the laboratory equipment used in these experiments is very similar to that described by Cheadle *et al.* (1991). Flat-faced piezoelectric transducers are used as sources and receivers, both types having an active element 12.6 mm in diameter. The compressional-wave transducer (Panametrics V103) is sensitive to displacement normal to the contact face, whereas the shear-wave transducer (V153) is sensitive to displacement tangential to the transducer face, in a direction parallel to the cable connector. A source pulse is obtained by driving the source transducer ($P$ or $S$) with a single cycle of a 28 V square wave. The width of the pulse is adjusted manually to produce an approximately symmetrical broadband wavelet with a central frequency of about 600 kHz, corresponding to minimum wavelengths of about 2.5 mm and 5 mm for $SH$ and $qP$, respectively. Data are recorded using a Nicolet digital oscilloscope at a sampling interval of 50 ns. The transducers are coupled to the surface

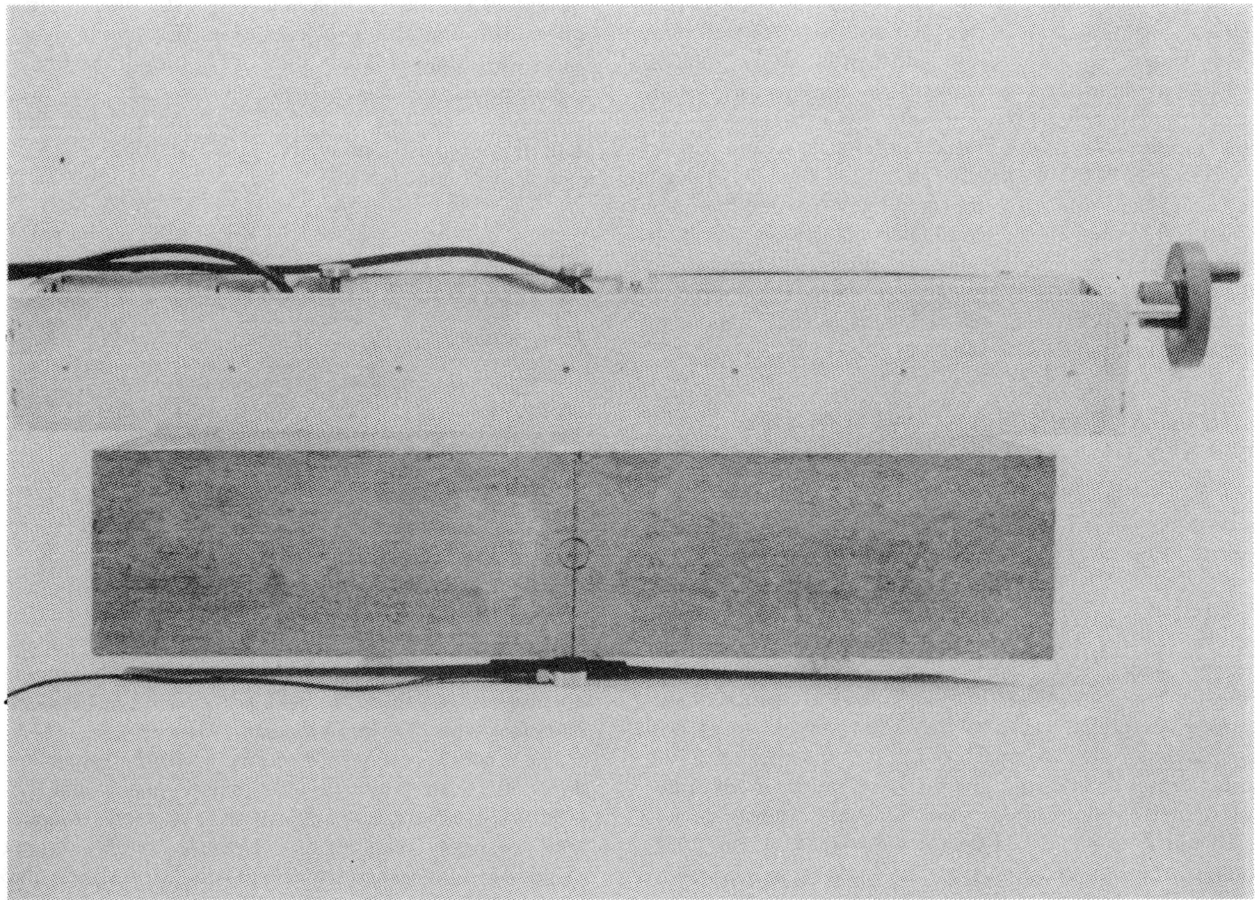

**Figure 1.** Photograph of a slab of Phenolic CE with the experimental arrangement used in transmission experiments, i.e. with source and receiver transducers on opposite surfaces of the slab. The slab thickness is 104 mm (520 m at 1:5000 scaling).

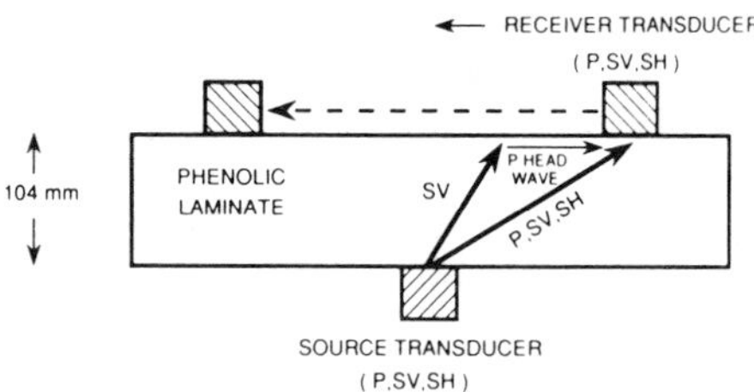


**Figure 2.** Schematic diagram of wave propagation paths and transmission-experiment design as seen in section.

of the model using a viscous pharmaceutical wax and a uniform pressure is applied to each transducer to ensure consistent coupling between the transducers and the model.

Figures 1 and 2 are, respectively, a photograph of the apparatus used for the experiments described in this paper and a schematic diagram of the system. As illustrated in these figures, we carried out transmission experiments in which the source and receiver transducers were on opposite sides of a phenolic slab. This reduces the number of unwanted arrivals (e.g. converted reflections, direct waves, additional head waves) that would accompany reflection experiments, wherein both transducers would be on the same surface. During an experiment, the source transducer was fixed below the slab while the receiver transducer was moved incrementally, on a calibrated worm drive, across the top of the slab (Figs 1 and 2).

The material used, Phenolic CE, is constructed of canvas laminae that are bonded together with a phenolic resin. Each lamina is woven from fine cotton fibres, roughly 0.1 mm in diameter, and consists of a *warp* (the set of straight parallel fibres that are stretched tight) and a *woof* (the parallel set of fibres that are orthogonal to and curl up and down over the warp). The density of fibres is about 8 per cm in the warp and 9 per cm in the woof, and there are about 25 laminae per cm. Phenolic CE is available commercially and is commonly used in applications requiring both electrical insulation and mechanical strength (e.g. in transformers) as well as for machining components such as gears, pulleys and sheaves.

## SHOT GATHERS WITHIN SYMMETRY PLANES OF THE PHENOLIC

The physical modelling capability described in the preceding section and by Cheadle *et al.* (1991) has been applied to generate several shot gathers along different profiles on the 3-face of the phenolic slab (taken as horizontal; i.e. the 3-direction is taken as the vertical). Profiles were initially shot at 0° (the 1-direction) and 90° (the 2-direction) for vertical, radial and transverse shot–receiver polarizations (Fig. 3). Fig. 2 also shows how an *SV*-to-*P* converted phase, termed the local *SP* wave by Booth & Crampin (1985), will be generated at the critical angle. For the corresponding critical offset and beyond, the shear-wave arrivals will suffer interference from this head wave. This critical offset defines the limits of the so called shear-wave window (Evans 1984; Crampin, Evans & Üçer 1985; Booth & Crampin 1985). One might expect the polarization of this local *SP* wave to be within the sagittal plane and perhaps closer to horizontal than vertical. Hence we might expect it to be strongest on radial-receiver records and the weakest on transverse-receiver records. [These conventional (isotropic) expectations should be applied with caution in the anisotropic case, especially for ray paths in off-symmetry planes.]

For the 0° and 90° records (Fig. 3) there is sagittal symmetry (i.e. the sagittal planes are symmetry planes) and the wave phenomena that one observes in the records are similar to what would have been seen for a transversely isotropic medium. This is true also for *SH* propagation in a symmetry plane even though the polarization then is not in this sagittal plane (Crampin & Kirkwood 1981). In Fig. 3, only the V–V record (vertically polarized source and receiver) is shown for the 90° profile as this family of records is very similar in nature to the 0° records. The records in Fig. 3 also show a significant broadening (lower frequency content) of the wavelet with increasing source–receiver offset. This is primarily due to array effects of the source and receiver transducers caused by their finite size and possibly partly due to anelastic attenuation.

The effect of the shear-wave window is seen in Fig. 3 where the local *SP* phase (Fig. 2) (labelled *P* head wave in Fig. 3) arrives before and interferes with the *SV* phase. Note that this head wave does indeed have greater amplitude on the radial–radial (R–R) record (Fig. 3, bottom left) as expected. Thus, outside the shear-wave window (from about −250 to 250 m on this scaled section) the *SP* head wave totally obscures the *SV*. On the V–V sections (Fig. 3, top) the interference is apparently not nearly as great but the picking of accurate *SV* onset times is still made practically impossible. For the T–T case (Fig. 3, bottom right) the shear-wave window is not a factor because of the transverse polarizations of source and receiver and the sagittal symmetry. The fact that these are transmission and not reflection records probably also contributes to their simplicity of appearance.

## THEORY OF VELOCITY VARIATION IN SYMMETRY PLANES

### Introduction

The variation in velocity (group and phase) with direction of propagation is one of the fundamental properties of seismic anisotropy. The actual directional dependence of velocities for various symmetries has been studied by Backus (1965), Crampin (1977, 1981) and Crampin & Kirkwood (1981), for example. The specific dependence of NMO velocity on angle of incidence for the case of transverse isotropy (TI) has been examined by Postma (1955), Daley & Hron (1977), Levin (1979), Berryman (1979), Thomsen (1986) and Uren, Gardner & McDonald (1990), among others. Winterstein (1986) has studied the depth discrepancies resulting from the application of observed NMO (or stacking) velocities, particularly for *SH*, in TI media.

The exact expressions for seismic velocities as functions of direction in an arbitrary off-symmetry plane and for an arbitrary symmetry system can be extremely complicated and not easily manipulated. However, within symmetry planes—for any symmetry system—the expressions simplify to the same algebraic form as those that apply for the TI system (Postma 1955; Crampin & Kirkwood 1981; Crampin 1981). These exact expressions may further be simplified to

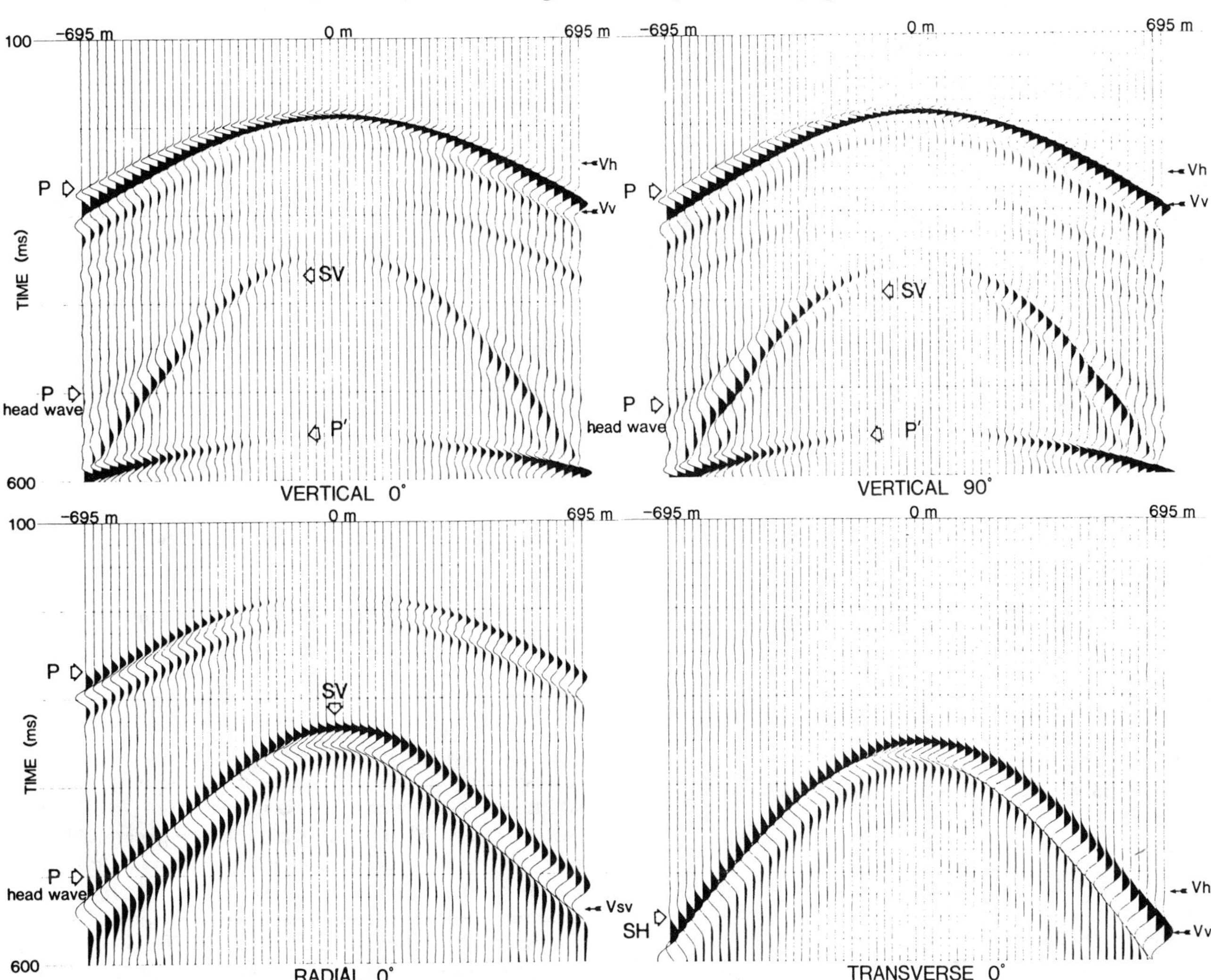


**Figure 3.** Transmission shot gathers recorded across the 3-face of the Phenolic CE slab. Traveltimes and offsets are scaled by 5000:1. The nominal trace spacing is 25 m, although this value varies slightly with position because of transducer finiteness (see text). Top left: profile in the 1-direction (0°) with vertically polarized source and receiver (V–V). Top right: profile in the 2-direction (90°) with vertically polarized source and receiver (V–V). Bottom left: profile in the 1-direction (0°) with radially polarized source and receiver (R–R). Bottom right: profile in the 1-direction (0°) with transversely polarized source and receiver (T–T). Hollow arrows identify various direct and refracted arrivals ($P'$ is the first $P$-wave multiple through the slab). Solid arrows marked $Vv$, $Vh$ and $Vsv$ indicate hypothetical onset times on the far trace assuming the vertical $P$-wave velocity, horizontal $P$-wave velocity and axial $SV$-wave velocity, respectively, to persist isotropically throughout the slab.

approximate expressions that are valid for cases in which the anisotropy is weak (Backus 1965; Crampin 1977; Thomsen 1986). Such simplified approximate expressions can provide reasonable estimates of the velocity variations in symmetry planes for all symmetry systems (Crampin & Kirkwood 1981) and are also important in providing intuition into the physical nature of the directional variation of a particular wave velocity (Thomsen 1986). For most computational purposes, of course, the exact expressions should be used.

## Phase velocity

We have derived the exact expressions for the phase velocities as functions of phase angle (wavefront-normal direction), $v_{SH}(\theta)$ and $v_P(\theta)$, in a vertical symmetry plane of an orthorhombic medium. We assume the 3-direction ($\theta = 0$) to be aligned with the vertical and consider profiles in both the 1- and 2-directions (propagation in the 1–3 and 2–3 planes). Starting with equations (A-13) and (A-14) of Cheadle *et al.* (1991) (see also Musgrave 1970, chapter 9), we obtain for $SH$ and $qP$ waves in the 1–3 plane:

$$v_{SH}^2(\theta) = \frac{C_{44}}{\rho}\cos^2\theta + \frac{C_{66}}{\rho}\sin^2\theta \tag{1a}$$

and

$$v_P^2(\theta) = \frac{1}{2\rho}[C_{33} + C_{55} + (C_{11} - C_{33})\sin^2\theta + D(\theta)] \tag{1b}$$

where

$$\begin{aligned} D^2(\theta) = {} & (C_{33} - C_{55})^2 + 2[2(C_{13} + C_{55})^2 - (C_{33} - C_{55}) \\ & \times (C_{11} + C_{33} - 2C_{55})]\sin^2\theta \\ & + [(C_{11} + C_{33} - 2C_{55})^2 - 4(C_{13} + C_{55})^2]\sin^4\theta. \end{aligned} \tag{1c}$$

For a profile in the 2-direction (propagation in the 2–3

plane), the index changes are: 4→5 (for $SH$), 1→2 and 5→4 (for $P$). The expressions for $v_{SV}(\theta)$ are as given for $v_P(\theta)$ but with a change of sign on $D(\theta)$.

Equations (1), valid in symmetry planes of an orthorhombic medium, are of the same algebraic form as those given by Daley & Hron (1977) and Thomsen (1986), for example, for TI media (and conform to Thomsen's notation). They reduce exactly, of course, to the TI equations upon degeneracy of the orthorhombic system described by: $C_{11}=C_{22}$; $C_{13}=C_{23}$; $C_{44}=C_{55}$ and $C_{12}=C_{11}-C_{66}$. (Going in the opposite direction, however, from the TI to the more general orthorhombic expressions cannot be done in an immediately obvious manner without some additional knowledge).

By utilizing the Thomsen anisotropy parameters $\gamma$, $\delta$ and $\varepsilon$ for each vertical symmetry plane, equations (1) may be written in the form

$$v_{SH}(\theta)=v_{SH}(0)(1+2\gamma\sin^2\theta)^{1/2} \tag{2a}$$

and

$$v_P(\theta)=v_P(0)[1+\varepsilon\sin^2\theta+D^*(\theta)]^{1/2} \tag{2b}$$

where

$$D^*(\theta)=[E^2+2E(2\delta-\varepsilon)\sin^2\theta\cos^2\theta+\varepsilon(2E+\varepsilon)\sin^4\theta]^{1/2} \tag{3a}$$

in which we introduce $E$ (*not* a first-order small quantity):

$$E=\frac{1}{2}\left[1-\frac{v_{SV}^2(0)}{v_P^2(0)}\right] \tag{3b}$$

and where, for the 1–3 plane,

$$\gamma=\frac{C_{66}-C_{44}}{2C_{44}}, \tag{4a}$$

$$\delta=\frac{(C_{13}+C_{55})^2-(C_{33}-C_{55})^2}{2C_{33}(C_{33}-C_{55})}, \tag{4b}$$

$$\varepsilon=\frac{C_{11}-C_{33}}{2C_{33}}, \tag{4c}$$

and

$$E=\frac{C_{33}-C_{55}}{2C_{33}}. \tag{4d}$$

The same changes of index as given above for equations (1) apply for equations (4). Note in the above equations that $v_P(0)$ is the same for both profile directions. However, unlike the TI case, $v_{SH}(0)$ is not the same for both profiles and for a particular profile $v_{SH}(0)\neq v_{SV}(0)$ in general.

## Group velocity

The group velocity in a symmetry plane can be determined from the phase velocity using well-known relationships (e.g. Postma 1955; Backus 1965; Berryman 1979; Crampin 1981; Radovich & Levin 1982):

$$V^2[\phi(\theta)]=v^2(\theta)+\left(\frac{dv}{d\theta}\right)^2 \tag{5a}$$

and

$$\tan\phi=\frac{v\tan\theta+\left(\dfrac{dv}{d\theta}\right)}{v-\left(\dfrac{dv}{d\theta}\right)\tan\theta} \tag{5b}$$

where $\phi$ is group or ray angle. It can also be shown in a straightforward way that

$$\phi-\theta=\tan^{-1}\left[\frac{1}{v}\left(\frac{dv}{d\theta}\right)\right]=\tan^{-1}\left[\frac{1}{V}\left(\frac{dV}{d\phi}\right)\right]. \tag{5c}$$

Applying equation (5a) to equations (1) or (2) gives the following group-velocity functions:

$$V_{SH}^2[\phi(\theta)]=v_{SH}^2(\theta)\left[1+\left(\frac{\gamma\sin 2\theta}{1+2\gamma\sin^2\theta}\right)^2\right] \tag{6a}$$

and

$$V_P^2[\phi(\theta)]=v_P^2(\theta)\left\{1+\left[\frac{\varepsilon\sin 2\theta+(dD^*/d\theta)}{2[1+\varepsilon\sin^2\theta+D^*(\theta)]}\right]^2\right\}. \tag{6b}$$

In computing group velocities from these equations for the purpose of comparing with those that we have observed for various angles of incidence, it is essential to distinguish between $\theta$, the phase angle or slowness direction, and $\phi$, the group angle or angle of incidence of the ray. A computed group velocity must be plotted against the corresponding group angle, $\phi$, and not the phase angle, $\theta$, which may have been inserted into equation (6). Thus we write $v(\theta)$ and $V[\phi(\theta)]$ to denote the (in general different) phase and group velocities at one and the same point on a wavefront, for example at point $A$ (Fig. 4). In addition, however, there exists some neighbouring point $B$ (Fig. 4) on the wavefront where the group-velocity direction $\phi_B$ is equal to $\theta_A$ the phase-velocity direction at $A$. We denote such group velocities as $V(\phi=\theta)$. If one plots the computed group velocities against the related $\theta$ values, and on the same set of axes plots for comparison observed group velocities versus incidence angles, then one is in fact plotting $V(\phi_A)$ versus $\phi_B$. It can be shown that $V[\phi(\theta)]$ and $V(\phi=\theta)$ differ from $v(\theta)$ by second-order quantities that are of equal magnitude but opposite sign; or in other words, $V(\phi=\theta)$ differs from $V[\phi(\theta)]$ by about twice as much as $v(\theta)$ does.

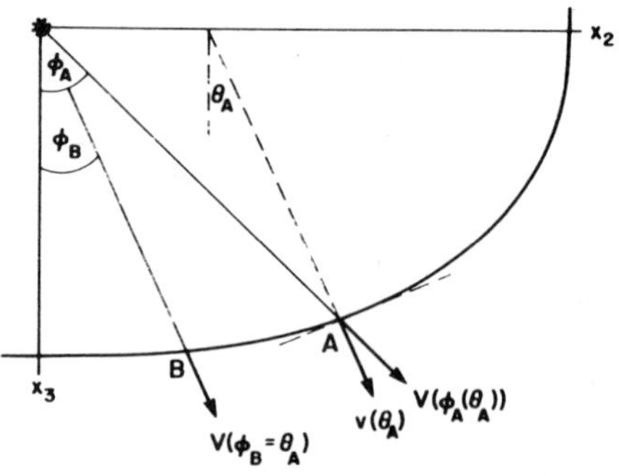


**Figure 4.** Sketch of an anisotropic wavefront showing the phase velocity $v(\theta)$ at $A$, the group velocity $V[\phi(\theta)]$ at $A$, and the group velocity $V(\phi=\theta)$ at $B$.

In view of the above, one should also calculate the group angle, $\phi$, so that one may plot the pair $\{\phi, V[\phi(\theta)]\}$. Applying equation (5b) or (5c) to equations (1) or (2) yields

$$\phi_{SH} = \theta_{SH} + \tan^{-1}\left(\frac{\gamma \sin 2\theta_{SH}}{1 + 2\gamma \sin^2 \theta_{SH}}\right) \tag{7a}$$

and

$$\phi_P = \theta_P + \tan^{-1}\left[\frac{\varepsilon \sin 2\theta_P + (dD^*/d\theta_P)}{2[1 + \varepsilon \sin^2 \theta_P + D^*(\theta_P)]}\right]. \tag{7b}$$

### Series approximations

Equations (2) may be expanded in series form to any desired order, as was done to first order by Thomsen (1986). Since a first-order expansion is not sufficient to resolve any difference between group and phase velocities, we have expanded these expressions to include the second-order terms, in which a difference between these two velocities first appears. For our test material, maximum anisotropies are about 10 per cent (*SH*) and 20 per cent (*P*), so that second-order dimensionless quantities, such as relative differences between group and phase velocities, should be on the order of about 1 to 4 per cent and truncated third- and higher order quantities should be on the order of about 0.1 to 0.8 per cent. Such accuracy is probably acceptable in most but not all situations and, in any case, the second-order expressions are not so much simpler than the exact ones as to warrant their use routinely. We have therefore used the exact expressions in obtaining our results here. Along the way, however, we compared our observed group velocities with the velocities given by Thomsen's (1986) expressions and found typical maximum differences of only about 2 per cent. It might also be mentioned that Thomsen (1986) also gives approximate equations for the anisotropy parameters $\gamma$, $\delta$ and $\varepsilon$. The approximation for $\delta$ in particular can be quite bad (out by a factor of about 2 for our data). As Thomsen (1986) stresses, the linearized approximate expressions are intended to aid one's intuition or understanding of the physics, not for computational purposes.

The three anisotropy parameters of Thomsen (1986) have been discussed and evaluated (exactly) by Cheadle *et al.* (1991) for a cube of Phenolic CE for each of the three symmetry planes. These values have also been determined separately for the larger slab of phenolic used in the profile experiments described in this paper. These are listed in Table 1 along with a similar comparison of the measured velocities.

**Table 1.** Comparison of parameters determined in two different samples of Phenolic CE: the slab used in this study and the cube used by Cheadle *et al.* (1991). The 45° velocities ($V_{44}$ etc. for *qP* or $V_{4\bar{4}}$ etc. for *qSV*) could not be measured accurately for the slab and were taken to be equal to the corresponding cube values. This affects the off-diagonal stiffnesses and the $\delta$ values slightly.

| Parameter | Slab | Cube |
|---|---|---|
| $V_{11}$ (m/s) | 3502 | 3576 |
| $V_{22}$ | 3341 | 3365 |
| $V_{33}$ | 2935 | 2925 |
| $V_{23}$ | 1515 | 1516 |
| $V_{13}$ | 1600 | 1606 |
| $V_{12}$ | 1646 | 1662 |
| $V_{44}$ | 3094 | 3094 |
| $V_{55}$ | 3219 | 3219 |
| $V_{66}$ | 3378 | 3378 |
| $V_{4\bar{4}}$ | 1569 | 1569 |
| $V_{5\bar{5}}$ | 1620 | 1620 |
| $V_{6\bar{6}}$ | 1810 | 1810 |
| $\gamma(2\text{–}3)$ | 0.0292 | 0.0355 |
| $\gamma(1\text{–}3)$ | 0.0902 | 0.1009 |
| $\gamma(1\text{–}2)$ | 0.0577 | 0.0611 |
| $\delta(2\text{–}3)$ | 0.1540 | 0.1656 |
| $\delta(1\text{–}3)$ | 0.1166 | 0.1133 |
| $\delta(1\text{–}2)$ | −0.0975 | −0.1131 |
| $\varepsilon(2\text{–}3)$ | 0.1479 | 0.1617 |
| $\varepsilon(1\text{–}3)$ | 0.2118 | 0.2473 |
| $\varepsilon(1\text{–}2)$ | 0.0494 | 0.0645 |

## COMPARISON OF OBSERVED AND THEORETICAL GROUP VELOCITIES

In order to determine experimentally the variation of these NMO velocities with direction (angle of incidence) a geometrical correction first had to be applied. For any particular source–receiver offset for which a traveltime measurement has been made, an initial path length and angle of incidence are determined from basic trigonometry. In doing this, the source and receiver 'points' are taken to be the centres of the respective transducers. Preliminary calculation showed that the effective path length was in fact shorter than the nominal distance between transducer centres and was in fact close but not exactly equal to the distance between nearest edges of transducers, which constituted a significant difference. (The transducer diameter of 12.6 mm scales up to 63 m at 1:5000). The correction was determined by repeating the shot profile with precisely the same geometry and the same thickness of material (104 mm), except that we used an isotropic medium (namely Plexiglas, which has *P*-wave and *S*-wave velocities of 2740 and 1385 m s$^{-1}$, respectively). The variations in traveltime with offset for this material were entirely due to geometry, with no anisotropic contribution. We were therefore able, on the basis of measured traveltimes and a known (invariant) velocity, to determine an effective path length, and thus also incidence angle, for each nominal source–receiver offset. This empirical correction was calculated and applied independently for each of the *P* and *SH* transducer types.

Figure 5 shows the agreement between the experimentally determined NMO velocities and those calculated from equations (6) and (7). Maximum differences are seen to be about 1 per cent or less. Most of the discrepancy for the *SH* cases (Figs 5c and d) seems to be associated with a curious rapid group-velocity increase observed at very small offsets (<5°) followed by a decrease with offset around 15°. This behaviour was more pronounced before applying the geometrical correction and is likely a residue of a procedure which, although it accounts for the gross ray path geometry, does not include provision for the non-orthogonal orientation of the anisotropic wavefront to the ray. In

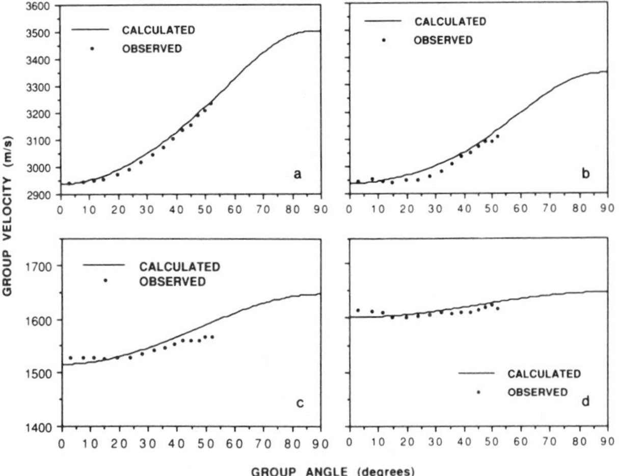


**Figure 5.** Comparison between observed and calculated group velocities versus angle of incidence (group angle): (a) $qP$ group velocity for a profile in the 1-direction; (b) $qP$ group velocity for a profile in the 2-direction; (c) $SH$ group velocity for a profile in the 1-direction; and (d) $SH$ group velocity for a profile in the 2-direction.

addition, our relatively large transducers could be causing problems in at least two other ways: they may be interfering with the free surface in a manner that requires a special free-surface correction and, as can be seen in Fig. 3, they act as arrays and cause high-frequency degradation which increases with offset. Other error sources are measurement uncertainty (±150 ns or, scaled at 5000:1, ±0.75 ms) and the possibility that the phenolic material may have some measurable inhomogeneity and/or anelasticity.

## ANALYSIS OF THE VECTOR WAVEFIELD IN OFF-SYMMETRY PLANES

Using an experimental procedure similar to that described above, profiles were shot on the 3-face at 45° to each of the 1- and 2-axes. These were acquired first with radial (R–R) and then transverse (T–T) shot–receiver polarizations (Fig. 6). In these we observe effects, not seen in the 0° (1-direction) or 90° (2-direction) records, that give a hint of the complexity of wave propagation in off-symmetry planes for an orthorhombic medium. In both of these shot records (Fig. 6), split shear waves are clearly visible on the zero-offset traces (identical on the two sections) and for the near offsets. However, as offset increases, one of the $qS$ arrivals dies out: the $S_1$ on the R–R record and the $S_2$ on the T–T record. This effect could conceivably be caused by cusping on complex wave surfaces, by rotations of polarizations into the plane perpendicular to the particular receiver polarization of each record, or by a relative-amplitude effect wherein the phases in question do not actually die out but simply become very low-amplitude at larger offsets. (Note that each trace is true-relative-amplitude but traces have been balanced relative to one another simply to maintain a roughly constant visible dynamic range.) The wave phenomena we observe could also be due to some combination of these three effects. All of these are related manifestations of the general complexities of shear-wave propagation in arbitrary anisotropic media, complexities which are most pronounced in the neighbourhood of cusps on the wave or group-velocity surfaces or of singularities on the slowness or phase-velocity surfaces.

**Figure 7.** Section of synthetic $qS$-wave surfaces (source at the asterisk) in an off-symmetry plane of an orthorhombic material numerically modelled as Greenhorn shale (Jones & Wang 1981) with superposed cracks (from Dellinger & Etgen 1989).

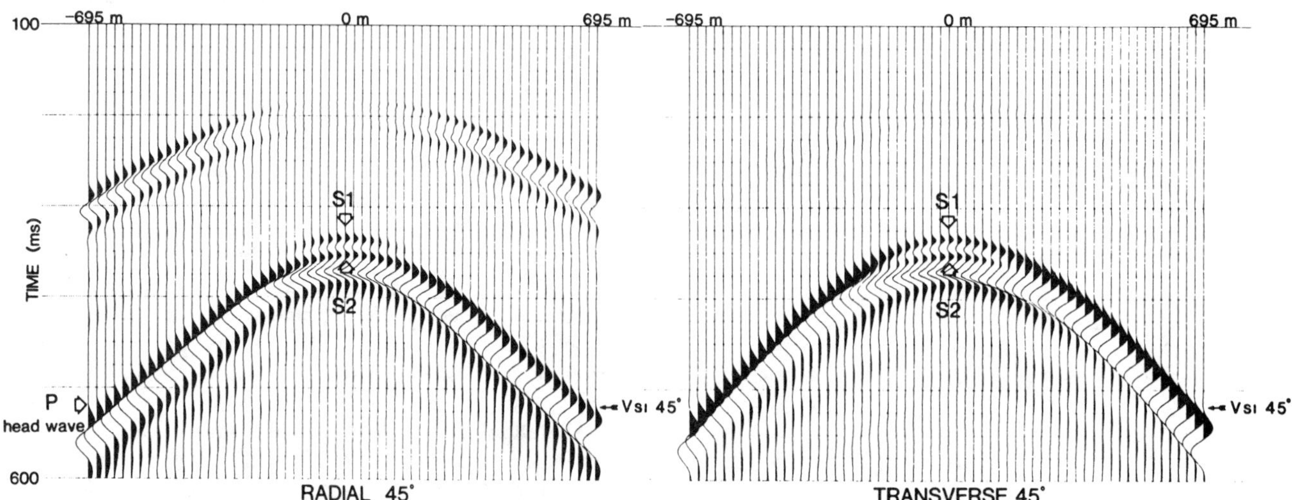


**Figure 6.** Shot gathers across the 3-face of the slab of Phenolic CE along a profile at 45° to the 1- and 2-directions. The records have the same acquisition geometry as those shown in Fig. 3. Left: radially polarized source and receiver (R–R). Right: transversely polarized source and receiver (T–T). In both records, $V_{S1}$ indicates the expected arrival times of the fast shear wave.

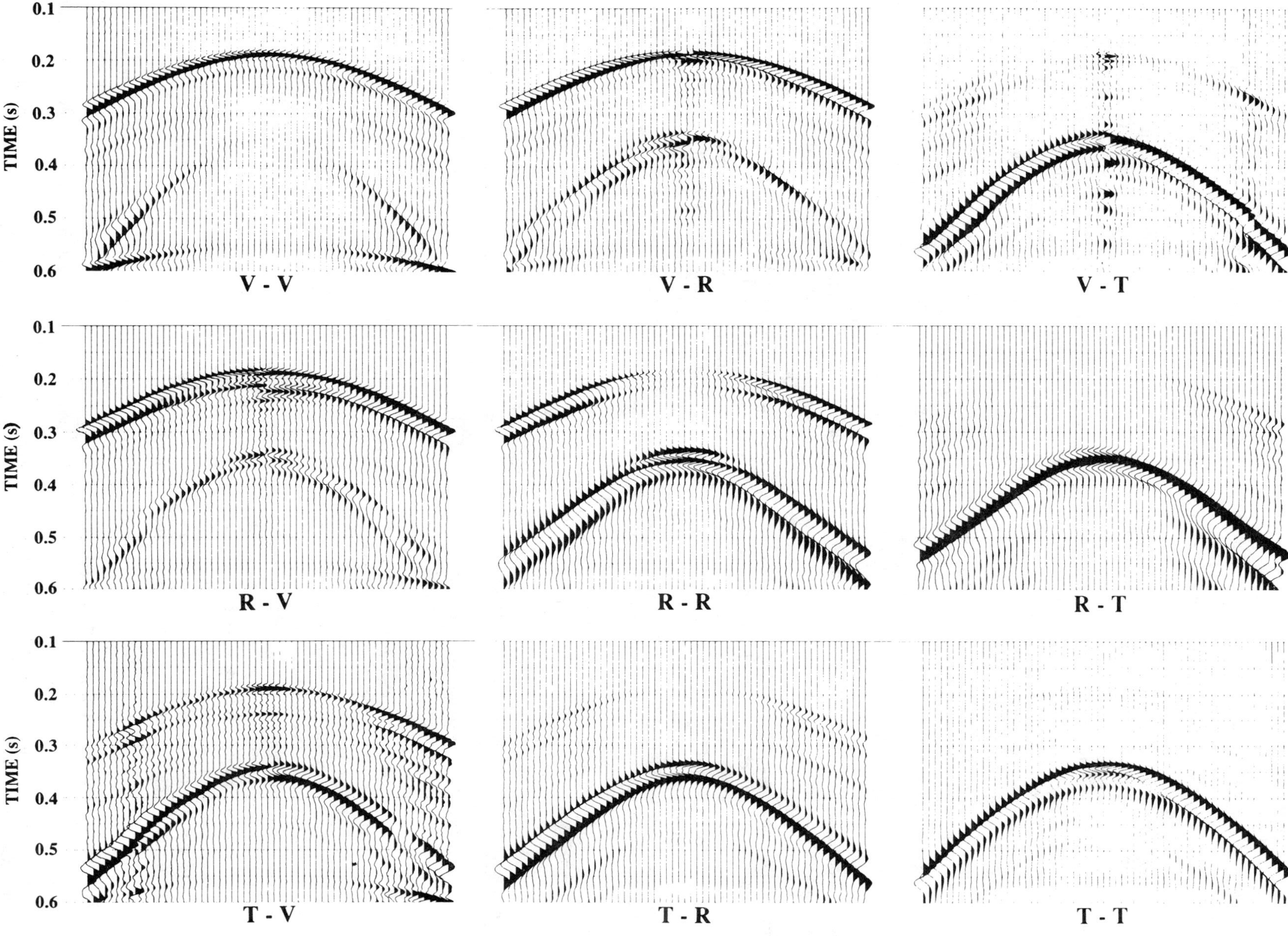


**Figure 8.** Nine-component record for a profile at 45° to the 1- and 2-directions, filtered at 5/10–100/200 Hz (after frequency scaling of 1 : 5000). The source–receiver polarization directions are shown by the label below each record.

Musgrave (1970) discusses these topics in some detail and shows a number of relatively simple theoretical examples of wave and slowness surfaces in symmetry planes. The full extent of the complexities are, however, generally realized in off-symmetry planes. Dellinger & Etgen (1989) show a synthetic example, generated on the basis of a particular orthorhombic medium (namely, cracked Greenhorn shale), of wave surfaces in an off-symmetry plane (Fig. 7). Crampin & Yedlin (1981) and Crampin (1981, 1985, 1989, 1991) discuss the topic of shear-wave singularities and show theoretical examples of the rapid fluctuations in amplitudes and polarizations near such singularities.

In order to provide more information on the polarizations for a particular shot, we decided to record a full 'nine-component' data set. So, for example, instead of just recording with a transversely sensitive transducer in the case of the transverse source (T–T), we recorded also with vertically and radially sensitive transducers (T–V and T–R); and so on for all three source transducers, resulting in nine records for the particular shot location (Fig. 8). It is of ancillary interest to note the manifestation of negative reciprocity, or antireciprocity, in Fig. 8. According to this principle (see e.g. Knopoff & Gangi 1959), interchanging orthogonally polarized transducers, but keeping the shot and receiver locations unchanged, will lead to displacement fields that are equal but of opposite polarity (assuming only an elastic medium and equal shot-impulse magnitudes).

Due to the fact that it can be difficult to replicate the source pulses in different experiments run at intervals of days or more, but not too difficult to maintain a constant source signature for a few hours, we decided to shoot all nine records as one experiment. Comparison of Fig. 6 with the R–R and T–T records of Fig. 8, shows that, although there are some differences in pulse shape and therefore some of the fine details on traces, the overall results in terms of wave phases present, traveltimes, etc., appear to be repeatable between experiments. One can also see indications of split shear waves on the other off-diagonal records of Fig. 8 (V–R and R–V more so than V–T and T–V).

In Fig. 8, the $S_1$ arrival that is seen to die out on the R–R record persists to large offsets on the R–T record. This could be the result of a polarization rotation, perhaps near a singularity or, since we have not displayed true trace-to-trace relative amplitudes, the effect could be one of amplitude scaling. At first glance it appears as if the $S_2$ phase on the T–T record is scarcely visible on the T–R record (Fig. 8) and that the same $S_1$ arrival dominates both records. However, the apparent onset times of these phases on the two records are not the same at far offsets: the $S_1$ seems to arrive later on the T–T record. In fact, the earlier far-trace onset on the T–R (and R–T) record may include some *SP* head-wave energy beyond the shear-wave window, although one might have expected to see this also on the R–R record.

## CONCLUSIONS AND FUTURE DIRECTIONS

We have shown that observed stacking or NMO velocities for *qP*- and *SH*-wave propagation in symmetry planes of our orthorhombic medium conform closely, that is within about 1 per cent, to the theoretical group-velocity variation. The observed *SH* velocities do, however, exhibit some anomalous variation (departure from the expected monotonic variation). We have suggested as possible reasons for this: effects of large transducer size (geometrical or effective path length effects, array attenuation effects, and free-surface contamination), sample inhomogeneity and/or anelasticity, and experimental error. We are pursuing the investigation of these possible explanations.

It has become clear to us in examining the full vector-wavefield records, even for this orthorhombic medium that is only moderately anisotropic, that it is extremely difficult to identify all wave phases and determine intuitively what sorts of propagation effects one is seeing when the sagittal plane is an off-symmetry plane. What we feel is needed then is to compare physical modelling results with synthetic seismograms and other theoretical results (e.g. positions of singularities, wave-surface mapping, etc.) computed by means of seismic anisotropic numerical modelling. Only then, in our opinion, will it be possible to identify many of the complicated and rapidly (spatially) fluctuating effects that occur; for example, around singularities and cusps. Moreover, by making such comparisons with a number of numerical modelling algorithms based on different theoretical schemes, it may well be possible to determine which of those techniques are superior for the tasks at hand. We are now, therefore, pursuing collaborations with a number of groups who have developed these capabilities.

Some of our next laboratory experiments will entail the generation and recording of waves on ray paths close to singular or conical directions in order to observe what sorts of polarization and amplitude fluctuations actually occur and to compare these, and the different patterns of arrivals in different sagittal planes, with those predicted numerically.

## ACKNOWLEDGMENTS

We are most grateful to the sponsors of the CREWES Project (Consortium for Research in Elastic Wave Exploration Seismology) at the University of Calgary for their crucial financial support, and to its director Rob Stewart for his scientific and moral support of this work. Eric Gallant is acknowledged for his stalwart technical assistance as is David Eaton for pointing out the negative reciprocity in the nine-component record set (Fig. 8). We thank Stuart Crampin and Dan Ebrom, as well as an anonymous reviewer, for providing salient and constructive suggestions which have contributed to an improved paper. Finally, John Lovell displayed a blend of saintly patience and friendly persuasion in helping us to meet the publication deadlines.

## REFERENCES

Backus, G. E., 1965. Possible forms of seismic anisotropy of the uppermost mantle under oceans, *J. geophys. Res.*, **70,** 3429–3439.

Berryman, J. G., 1979. Long-wave elastic anisotropy in transversely isotropic media, *Geophysics*, **44,** 896–917.

Booth, D. C. & Crampin, S., 1985. Shear-wave polarization on a curved wavefront at an isotropic free surface, *Geophys. J. R. astr. Soc.*, **83,** 31–45.

Cheadle, S. P., Brown, R. J. & Lawton, D. C., 1991. Orthorhombic anisotropy: A physical seismic modeling study, *Geophysics*, **56**, in press.

Crampin, S., 1977. A review of the effects of anisotropic layering on the propagation of seismic waves, *Geophys. J. R. astr. Soc.*, **49**, 9–27.

Crampin, S., 1981. A review of wave motion in anisotropic and cracked elastic-media, *Wave Motion*, **3**, 343–391.

Crampin, S., 1985. Evaluation of anisotropy by shear-wave splitting, *Geophysics*, **50**, 142–152.

Crampin, S., 1989. Suggestions for a consistent terminology for seismic anisotropy, *Geophys. Prosp.*, **37**, 753–770.

Crampin, S., 1991. Effects of point singularities on shear-wave propagation in sedimentary basins, *Geophys. J. Int.*, this issue.

Crampin, S. & Kirkwood, S. C., 1981. Velocity variations in systems of anisotropic symmetry, *J. Geophys.*, **49**, 35–42.

Crampin, S. & Yedlin, M., 1981. Shear-wave singularities of wave propagation in anisotropic media, *J. Geophys.*, **49**, 43–46.

Crampin, S., Evans, R. & Üçer, S. B., 1985. Analysis of records of local earthquakes: the Turkish Dilatancy Project (TDP1 and TDP2), *Geophys. J. R. astr. Soc.*, **83**, 1–16.

Daley, P. F. & Hron, F., 1977. Reflection and transmission coefficients for transversely isotropic media, *Bull. seism. Soc. Am.*, **67**, 661–675.

Dellinger, J. & Etgen, J., 1989. Wave-type separation in 3-D anisotropic media, *59th Ann. Internat. Mtg, Soc. Expl. Geophys., Expanded Abstracts*, pp. 977–979.

Ebrom, D. A., Tatham, R. H., Sekharan, K. K., McDonald, J. A. & Gardner, G. H. F., 1990. Hyperbolic traveltime analysis of first arrivals in an azimuthally anisotropic medium: A physical modeling study, *Geophysics*, **55**, 185–191.

Evans, R., 1984. Effects of the free surface on shear wavetrains, *Geophys. J. R. astr. Soc.*, **76**, 165–172.

Jones, E. A. & Wang, H. F., 1981. Ultrasonic velocities in Cretaceous shales from the Williston basin, *Geophysics*, **46**, 288–297.

Knopoff, L. & Gangi, A. F., 1959. Seismic reciprocity, *Geophysics*, **24**, 681–691.

Levin, F. K., 1979. Seismic velocities in transversely isotropic media, *Geophysics*, **44**, 918–936.

Lin, W., 1985. Ultrasonic velocities and dynamic elastic moduli of Mesaverde rock, *Lawrence Livermore Nat. Lab. Rep. 20273*, rev. 1.

Musgrave, M. J. P., 1970. *Crystal Acoustics*, Holden-Day, San Francisco.

Postma, G. W., 1955. Wave propagation in a stratified medium, *Geophysics*, **20**, 780–806.

Radovich, B. J. & Levin, F. K., 1982. Instantaneous velocities and reflection times for transversely isotropic solids, *Geophysics*, **47**, 316–322.

Rathore, J. S., Fjaer, E., Holt, R. M. & Renlie, L., 1990. Acoustic anisotropy of synthetics with controlled crack geometries, *4th International Workshop on Seismic Anisotropy, July 2–6, 1990 (Abstracts)*, British Geological Survey, Edinburgh.

Sayers, C. M., 1988. Stress-induced ultrasonic wave velocity anisotropy in fractured rock, *Ultrasonics*, **26**, 311–317.

Sayers, C. M., 1990. Ultrasonic studies of deformation and failure of rocks, in *Elastic Waves and Ultrasonic Nondestructive Evaluation*, eds Datta, S. K., Achenbach, J. D. & Rajapakse, Y. S., Elsevier, Amsterdam.

Thomsen, L., 1986. Weak elastic anisotropy, *Geophysics*, **51**, 1954–1966.

Uren, N. F., Gardner, G. H. F. & McDonald, J. A., 1990. Normal moveout in anisotropic media, *Geophysics*, **55**, 1634–1636.

Winterstein, D. F., 1986. Anisotropy effects in *P*-wave and *SH*-wave stacking velocities contain information on lithology, *Geophysics*, **51**, 661–672.

# Sideswipe in a 3-D, 3-C model survey

*By DAN EBROM and JOHN McDONALD*
*Allied Geophysical Laboratories and Department of Geosciences*
*University of Houston*

*and BOB TATHAM*
*Texaco, Inc.*
*Houston, Texas*

Development geophysics today is heavily dependent upon 3-D surface seismic acquisition. Why do 3-D? One reason, of course, is that a 3-D survey is commonly done at a tighter receiver spacing than most 2-D lines, thus allowing greater in-line spatial resolution. But another reason is that, when energy returns from out of the source-receiver plane, no 2-D data set can unambiguously identify this out-of-the-plane intruder (sideswipe). In order to remove sideswipe from our 2-D lines, we must perform a 3-D migration. Simply put, 3-D problems require 3-D solutions.

A 3-D data set gives us more than simply a tightly spaced set of 2-D lines. By comparing the timing of events from one line to another, we can identify sideswipe even before we go to migration to remove it. As yet, though, we have assumed that all of the events in our data, in plane or not, are *P*-waves.

Now, our geophones record the vertical component of any wave that has traveled past—be it ground roll, air wave, head wave, *P*-wave, or shear wave (*S*-wave). Body waves (*P*-waves or *S*-waves) that travel along nearly horizontal rays (scattered perhaps from a salt diapir) will have a different vertical component of motion than had they been traveling along rays nearly vertical (Figure 1). Neglecting the effects of the weathering layer for a moment, we find that a *P*-wave traveling horizontally will have almost none of its particle motion received by a vertical-component geophone. An *S*-wave traveling horizontally, on the other hand, will have almost all of its motion sensed by the vertical component geophone.

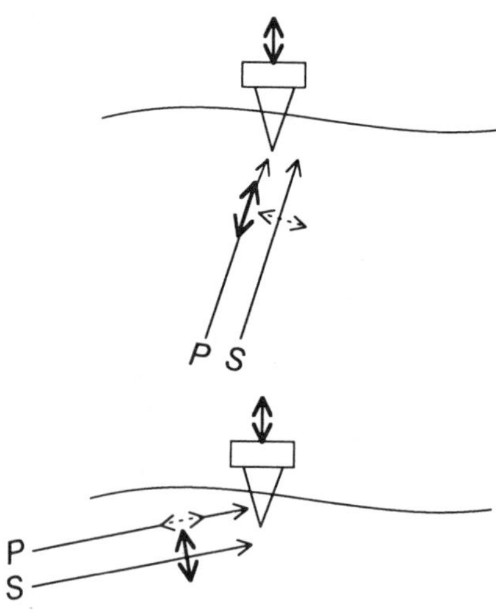


**Figure 1. *P*-waves traveling subvertically are detected by a vertical component geophone, while *S*-waves traveling along the same ray are rejected. But *P*-waves traveling subhorizontally are rejected by a vertical component geophone, while the *S*-waves are detected.**

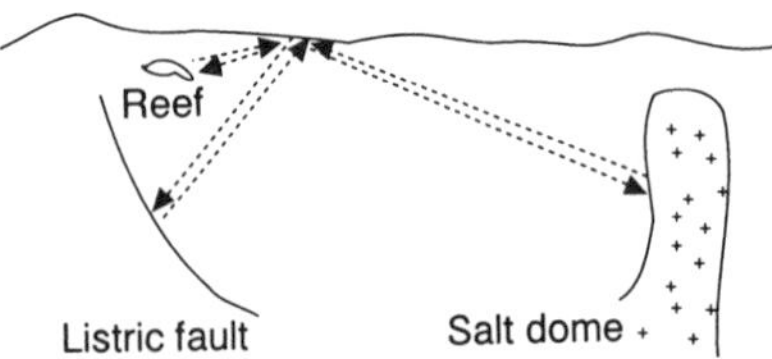


**Figure 2. High angle (nonvertical) returns of seismic energy in an extensional basin like the Gulf of Mexico. Even though the layering is predominantly subhorizontal, many interesting exploration objectives give rise to high angle returns.**

The weathering layer does diminish the effects of angle. By providing a low-velocity zone just beneath the receivers, the weathering layer refracts rays so that they are more nearly vertical. However, the thickness and velocity contrast of the weathering layer is highly variable (for *P*-waves this is frequently due to water table elevation changes), so that the magnitude of this beneficial refraction effect is difficult to predict. And, of course, if we were to place our receiver in a borehole (as for VSP data), the weathering layer effects could be entirely absent.

Our single-component vertical geophone works very well for what it was originally designed to do: pick up the particle motion associated with nearly vertically traveling *P*-waves. As an unintended bonus, it also rejects the particle motion from nearly vertically traveling *S*-waves. This works fine as long as the subsurface is fairly flat and source-receiver offsets are small. But flat geology is not the rule: structural deformations occur in most sorts of basins. These structures can range from diapirs to listric faults to overthrusts. Waves that have been scattered from these high-angle features will not necessarily arrive at the surface along nearly vertical rays (Figure 2).

How then do we record the data set that will give us the most amount of information about the origins of sideswipe? By recording not just in 3-D with single component receivers but in 3-D with three-component (3-C) receivers. Now, any seismic body wave that appears at the surface will have its true particle motion correctly recorded.

Reprinted from The Leading Edge, Vol. 11, No. 11, 45-49

In order to prove out these ideas, we performed a scale-model survey over an anticline, acquiring each of the three components of particle motion. The methodology we used is commonly referred to as "nine-component" data acquisition because three separate source orientations are recorded. (Three source orientations × three receiver components = nine components.)

An ultrasonic investigation of a scaled geologic feature was done in such a way that the wave propagation in the model closely approximated wave propagation in the real earth. The scale factor was 1:10 000 so 1 cm on the model corresponded to 100 m on the earth. The scale model was a dome carved in the bottom of a piece of otherwise homogenous Plexiglas. This geometry simulated an anticline piercing an otherwise flat horizon.

The dome, of course, has a negative reflection coefficient for rays propagating straight downward, since the air on the other side of the interface has an

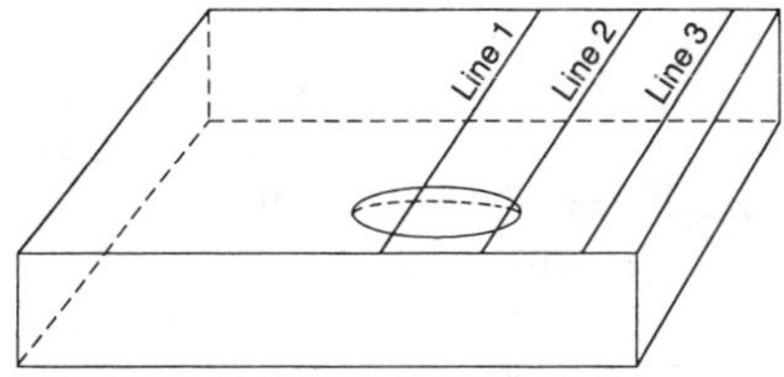


**Figure 3. Perspective view of the physical model. Line 1 runs over the crest of the anticline, line 2 over the flanks, and line 3 misses it entirely.**

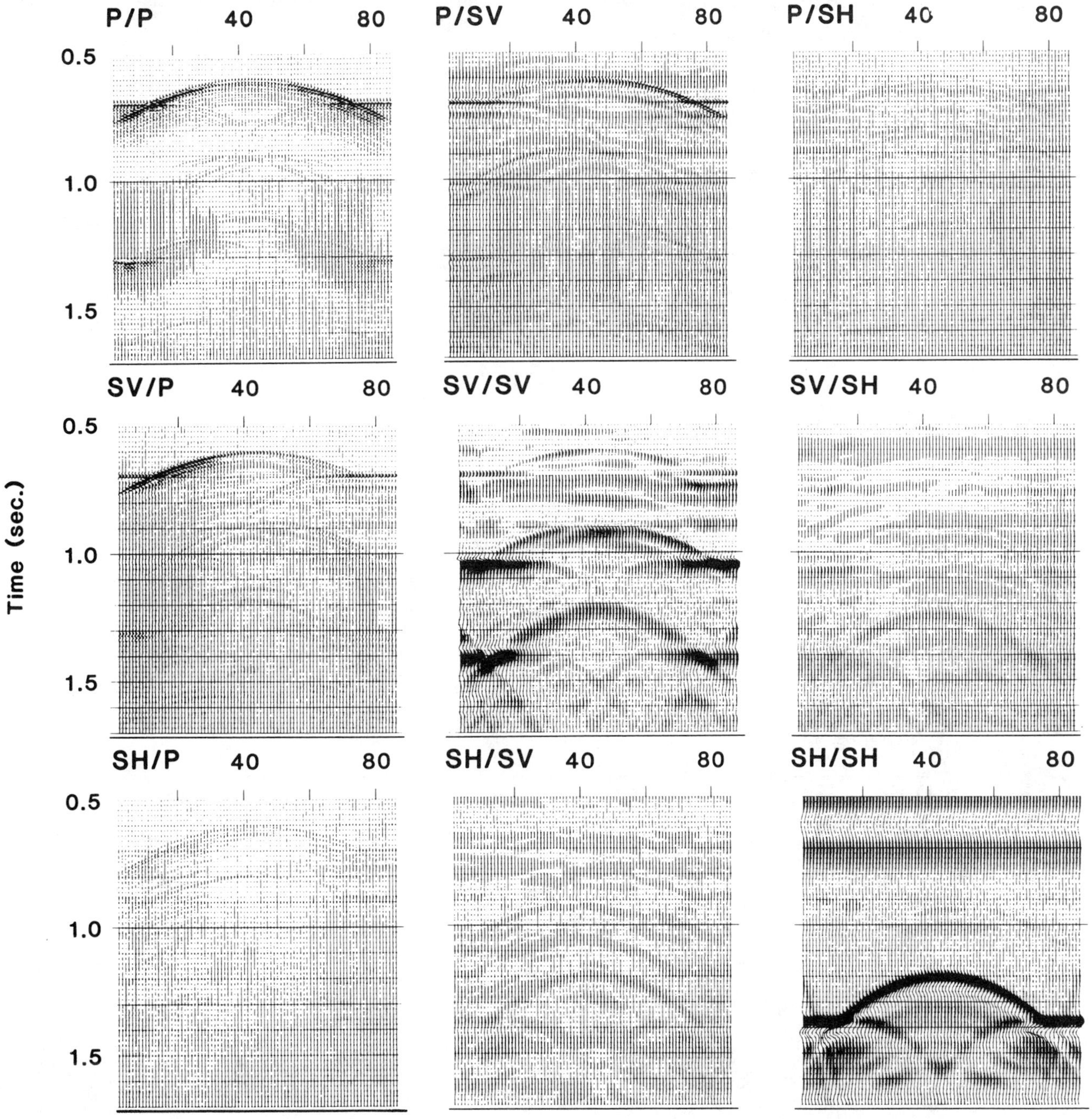

**Figure 4. Nine-component data from line 1 which goes over the center of the dome. Sections are identified by the source polarization, followed by the receiver polarization. The anticline's center is at midpoint 44. Data are single-fold with in-line offset of 760 m and with total length of 2.1 km.**

acoustic impedance much less than that of the Plexiglas. All of the distances given in the following text are scaled, or field, units. (Divide by 10 000 to get unscaled or laboratory units.)

We acquired what might be called a "very sparse 3-D" 3-C data set. That is, the lateral spacing of the 2-D lines was much greater than the in-line receiver spacing. The first line was over the crest of the dome, the second was over its flanks, and the third did not pass over any part of the dome (Figure 3).

The acquisition was single-fold, which is acceptable in model data because of the high signal-to-noise ratio. The source transducer was trailing the receiver transducer by 760 m, i.e., we were pushing our line. Receiver spacing was 25 m. The depth to the top of the dome was 740 m, while the depth to the flat plain was 870 m. The dome had a radius in plan view of 570 m. The velocity in the model was 2780 m/s for *P*-waves and 1410 m/s for *S*-waves. This yields a $V_p/V_s$ ratio of about 2.

The first line's data are shown in Figure 4. Each combination of source and receiver polarizations is shown by a 3 × 3 matrix. Each combination is designated by the notation "source polarization/receiver polarization." Thus, the

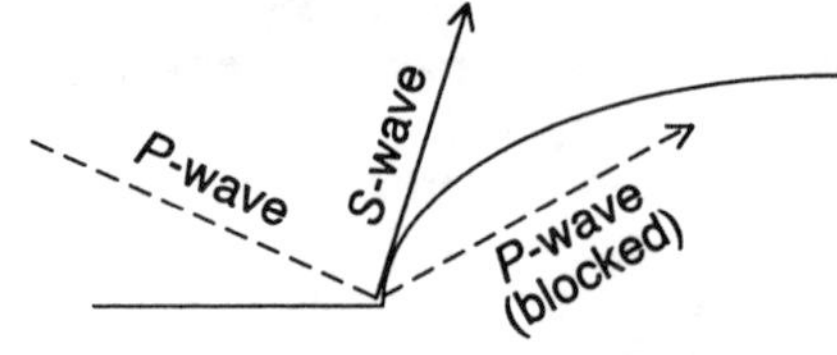


**Figure 5. Mode conversions can get reflections back to the surface in geometries where same-mode reflections would be obstructed. Note how the *P*-to-*S* mode conversion can have an unobstructed reflection point closer to the dome than a *P-P* or (*S-S*) reflection.**

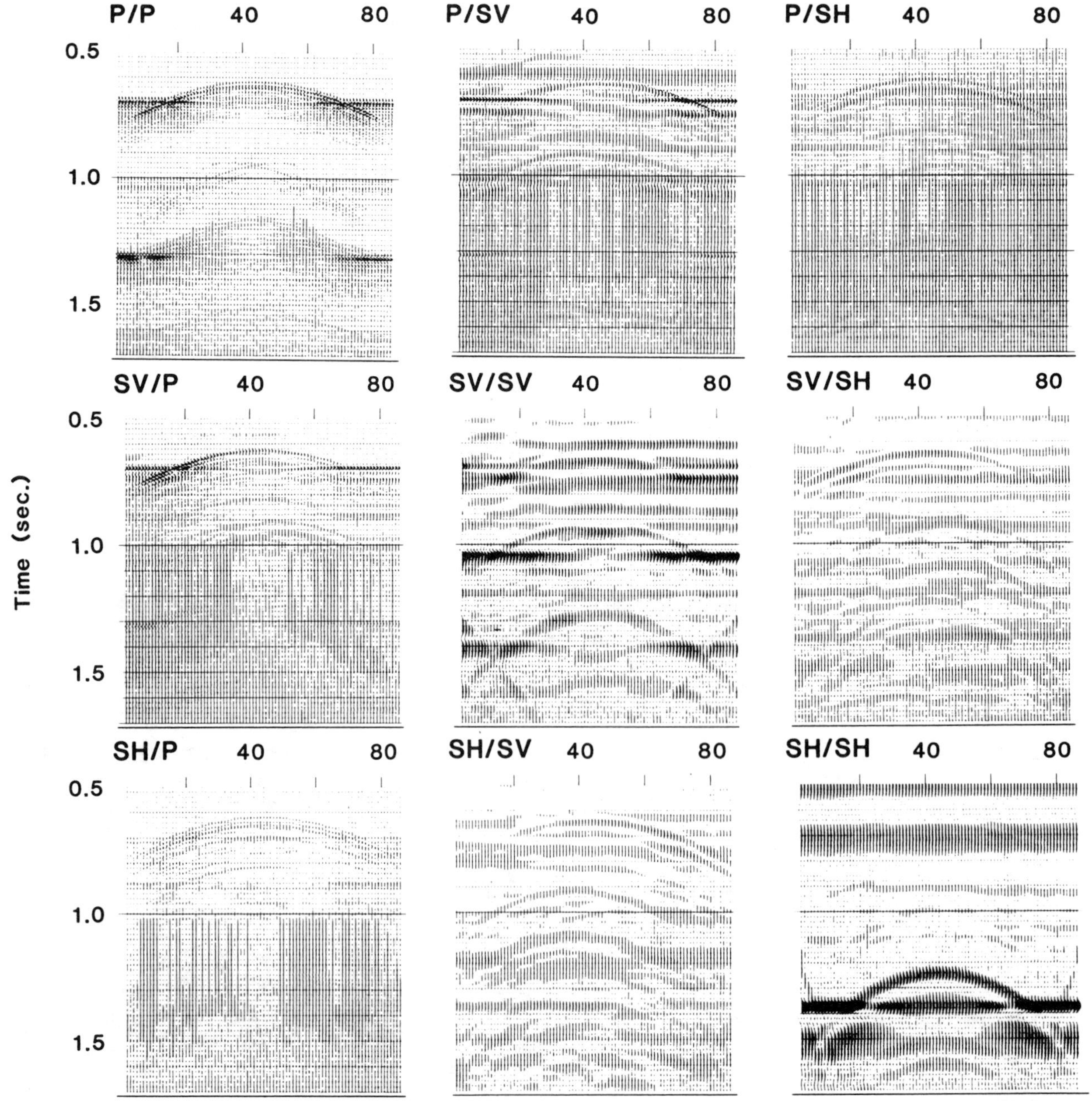

**Figure 6. Nine-component data from line 2 which goes over the flanks of the dome. Identification and parameters as in Figure 4.**

possible combinations are *P/P*, *P/SV*, *P/SH*, *SV/P*, *SV/SV*, *SV/SH*, *SH/P*, *SH/SV*, and *SH/SH*. As this line was over the crest, the common mode source/receiver combinations (*P/P*, *SV/SV*, and *SH/SH*) show the structural feature most clearly. The common mode combinations form the diagonal of the 3 × 3 matrix in Figure 4.

As would be expected, the events visible with the source at one polarization and the receiver at another (e.g., *P* source and *SV* receiver) are also observed when the source and receiver polarizations are exchanged (e.g., *SV* source and *P* receiver). In our display format, this means that figures symmetric about the diagonal should appear similar.

Note that all of these events originate at the same reflector and that they are all primary events (with the exception of a multiple on the *P/P* section). The same reflector can appear on the sections at three separate times depending upon whether it arrives as a *P*-wave reflection (earliest), *S*-wave reflection (latest), or mode conversion (in between *P*-waves and *S*-waves in arrival time).

The zero offset line is expected to show the best rejection of the cross components. This is observed to be the case: the *P/SH*, *SH/P*, *SH/SV*, and *SV/SH* show very low amplitude. (A quick word about amplitudes—the plots in all these figures have been normalized to the same reference amplitude, with no gain control or whole trace balancing.)

The *P/P* acquisition shows the dome,

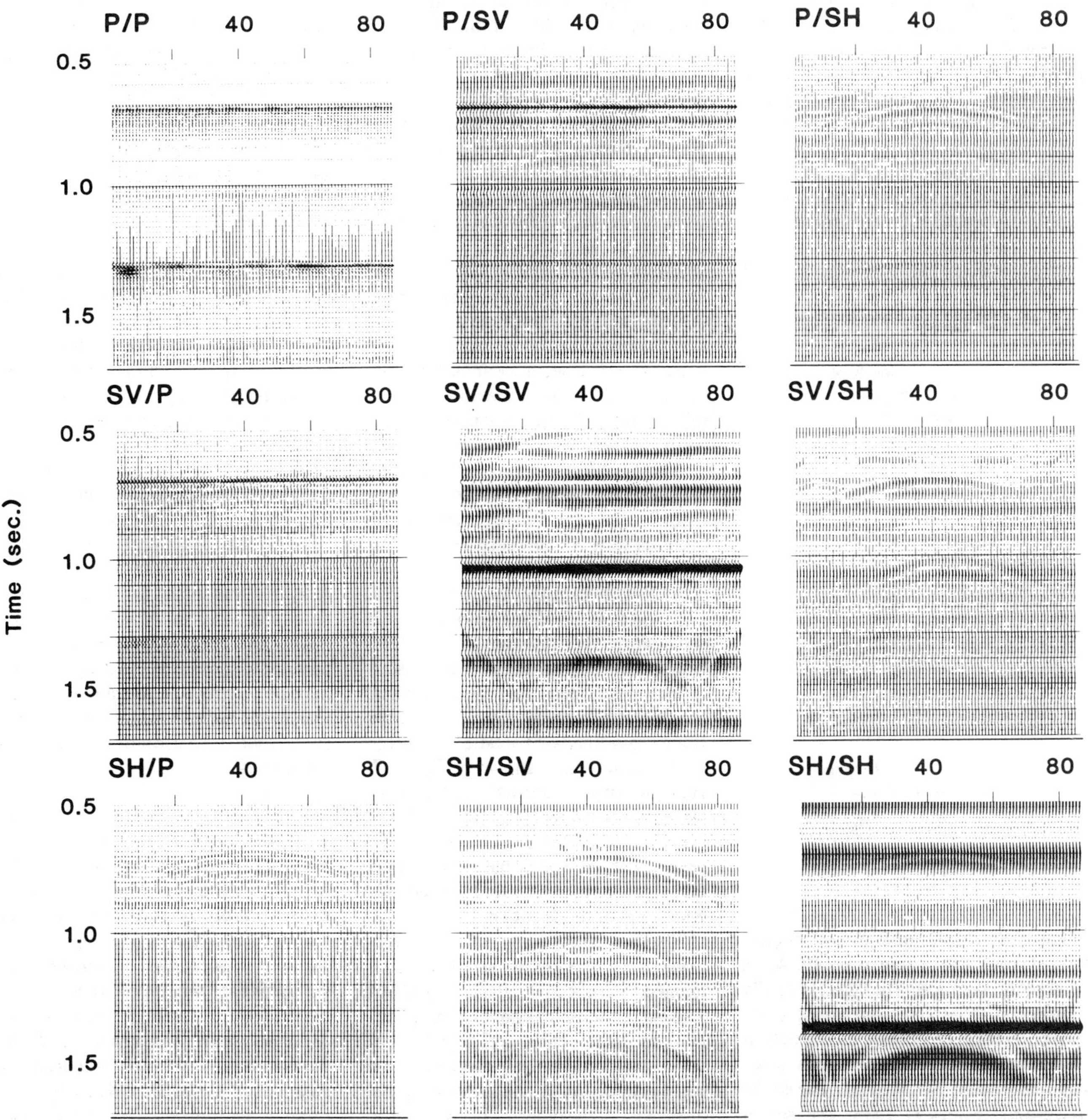


**Figure 7. Nine-component data from line 3. Although this line does not pass over any portion of the dome, out-of-plane dome arrivals can be seen in several of the sections. Identifications and parameters as in Figure 4.**

the horizontal reflector, and a multiple, all at *P*-wave velocity. The *P/SV* and *SV/P* sections show the same events as each other (*P*-wave and mode conversion), but the amplitudes are larger on the origin side of the dome when the *SV* source was trailing and larger on the terminus side of the dome when the *SV* receiver was leading.

The *SV/SV* line shows a *P*-wave reflection at about 600 ms, a mode conversion at about 900 ms, and an *S*-wave event at about 1180 ms. Note that the flat portion of these reflection events extends further into the center of the section for the mode conversion than for the simple *P-P* or *S-S* events. This is a graphic illustration of the asymmetry of raypaths involving mode conversions (Figure 5). The dome occludes those rays which would pass through it. Hence, asymmetric raypaths can return from points on the subsurface which would be occluded for symmetric raypaths. Two diagonal linear events can be seen entering the section at times of about 1300 ms; these are reflections from the sides of the model.

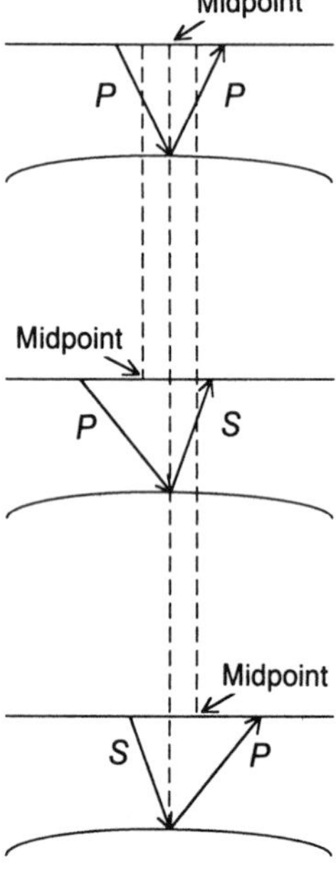


**Figure 8. *P*-to-*S* and *S*-to-*P* reflections from the dome top will have midpoint locations that are laterally displaced from the same-mode midpoint. Hence, the crests of the events for these two mode conversions (midpoints 38 and 50) will appear to be displaced laterally from the same-mode reflection crest (midpoint 44). Compare events at 1000 ms on sections *SV/SH* and *SH/SV* of Figure 7.**

The *SH/SH* section demonstrates the clarity of imaging possible with a shear wave that does not couple (at normal incidence) to *P*-waves. The *SH* polarization discriminates against in-line particle motion, and thus discriminates against in-line *P*-waves. Thus, whereas the *SV/SV* section shows three events, the *SH/SH* section shows but one, making the job of interpretation much easier. The *SH/SH* section shows especially prominently the diffractions from the base of the dome. From before 500 ms down to 900 ms, there is a flat low-frequency (about 5 Hz) event which is due to the direct shear wave. Interestingly, the direct wave is not as noticeable on the *SV/SV* section.

The second line has its matrix of sections shown in Figure 6. This line is over the flanks of the dome, at a lateral offset from the crest of the dome of about two-thirds of the dome radius. On this line, all of the events observed on line 1 can still be found, but the timing has changed. The reflections from the dome are at greater times because the reflection points are no longer directly below the surface acquisition line.

The *SH/SH* section shows the same time for a flat portion of the reflector as before (1340 ms) but the crest of the dome is now 40 ms later than before (1220 ms as opposed to 1180 ms). Note that a *P*-wave event is now noticeable at 620, where it interferes with the direct shear wave. This *P*-wave is now detectable on the *SH/SH* section because it reflects from out of the plane and, thus, has particle motion which possesses some cross-line horizontal component.

The third line is shown in Figure 7. Here the surface acquisition line does not pass over the dome at all, being laterally offset from the crest of the dome by about 1.33 dome radii. Yet the dome still produces events detectable on several of the sections.

Those sections on which the reflections from the dome are least pronounced are the *P/P*, *P/SV*, and *SV/P*. This makes sense in light of these polarizations pairs' enhanced rejection of out-of-the-plane particle motions. Even so, the rejection is not complete, and inspection of the other sections serves to allow the interpreter to more certainly classify these sideswipe events.

Those sections on which the domal events are most pronounced always include an *SH* polarization as either a source or a receiver. On the *SH/SH* section, the dome is present as an *S*-wave event whose crest is at about 1400 ms, interfering with the direct reflection from the flat reflector directly below the acquisition line. On the *P/SH* and *SH/P* sections, the out-of-plane *P*-wave reflection is visible at a time of 680 ms.

The *SH/SV* and *SV/SH* sections contain all three types of arrival from the dome, with a *P*-wave event at 680 ms, a mode conversion at 1000 ms, and an *S*-wave event at 1400 ms. The crests of the mode conversions on the two different sections are displaced laterally in opposite directions. This is due to the fact that an *S*-to-*P* mode conversion and a *P*-to-*S* mode conversion do not have the same reflection point, given the same source and receiver locations (Figure 8).

What does all this tell us? First, that much more is going on in the subsurface than is detected by vertical component geophones. Second, that multicomponent data give us additional information to distinguish in-plane reflectors from sideswipe (out-of-plane energy).

There is ongoing improvement in the total number of channels that can be simultaneously acquired, but hardware is still a limiting factor in land 3-D surveys. When we acquire multicomponent data, we increase the number of channels that must be dedicated to a particular receiver location. Philosophically, and perhaps pragmatically, experiments such as these suggest that it may be worthwhile to trade off spatial resolution in order to get a useful sideswipe tool in the form of multicomponent data. LE

---

*Acknowledgments: We wish to thank Rob Stewart of the University of Calgary for encouraging the publication of these results. Additionally, we would like to thank the corporate sponsors of the Seismic Acoustics Laboratory. The machining of this physical model was done at the Bellaire Texaco Machine Shop, under the supervision of Shorty Turlak. K.K. Sekharan, the technical manager of the Seismic Acoustics Laboratory, was responsible for the data acquisition systems and procedures.*

# Laser Ultrasonics applied to Seismic Physical Modeling

Martin, D.*, Pouet, B.**, and Rasolofosaon, P. N. J.*

*Rock Physics and Physical Modeling Lab., Institut Français du Pétrole (IFP)
**Formerly IFP, now Northwestern University, Evanston, IL

## SUMMARY

In the present state-of-the-art, the "physical modeling" of a seismic experiment in the laboratory on scaled-down models suffers from two major drawbacks. First, coupling problems between the emitter/receiver and the model to be tested usually require that experiments be performed in a water tank, and therefore, only marine seismics can be simulated. Second, it is rather difficult to scale-down the size of the emitters and the receivers with respect to the wavelength.

This paper introduces a relatively new laboratory technique, called "laser ultrasonics," which is widely used in non-destructive evaluation. Laser ultrasonics overcomes the aforementioned drawbacks and is well adapted to seismic prospecting problems. The method, based on photo-acoustic principles, uses a pulsed laser to generate ultrasonic waves and a laser interferometer to detect the resulting vibrations in the tested scaled model.

A general presentation and a critical evaluation of the method are made. The great promise of the technique is illustrated by some of its recent applications to seismic modeling.

In conclusion, in the present state of the technology laser ultrasonics is an outstanding research tool for seismic physical modeling but not well suited to immediate large scale or industrial applications. Nevertheless, considering the continual evolution of the technology mainly in the domains of optics and electronics, "seismic physical modeling" will probably soon be revolutionized by the advent of the photo-acoustic methods.

## INTRODUCTION

The generation and detection of elastic waves by photo-acoustic methods, commonly called "photoacoustics" or more specifically "laser ultrasonics", is now well known and has evolved from science to engineering applications, at least as is reported in Hutchins and Tam (1986) in the Special Issue on the topic in the September 1986 issue on the IEEE Transactions on Ultrasonics, Ferroelectrics, and Frequency Control.

The application of these techniques to seismic physical modeling has recently been developed and reported by Pouet and Rasolofosaon (1990), Pouet (1991), and Martin et al. (1991 and 1992). This paper makes a general presentation and critically evaluates the illustrated method by examining some of its recent applications.

First we propose a brief state-of-the-art of seismic modeling in order to situate the technique in the present context. Second, we explain the reasons leading to the choice of the photo-acoustic methods applied to seismic physical modeling and we present the experimental set-up. Next, we illustrate the great promises of the technique by two types of applications, one related to cross-well seismic modeling, and the other related to reflection seismic modeling and imaging in isotropic and anisotropic media. Last, we present a critical evaluation of the laser technique in view of the present state of the technology and then propose some prospective views.

## SEISMIC MODELING: A STATE-OF-THE-ART

Seismic prospecting experiments can be simulated numerically by computing theoretical models or physically in the laboratory by using scaled-down analogical "physical models". Both methods have in common two main complementary goals, one typically academic and another more applied, which are respectively,

(1) to understand the phenomena of wave propagation in complex contexts (rheology of the media, geometry of propagation...), and
(2) to improve seismic data acquisition, processing, and interpretation in order to obtain the best three-dimensional (3-D) seismic image of the subsurface compatible with the geology and leading to the mapping of hydrocarbon reservoirs.

A brief comparative state-of-the-art of both of these approaches follows.

Concerning numerical techniques, the main advantages are the

(1) high versatility of the method (i.e., no problem to change a physical or a geometrical parameter of whatever part of the model), and
(2) ease of providing at low cost everywhere in the model the particle motion, the pressure, the curl or the divergence of the displacement vector and wavefronts visualization (snapshots) at fixed times.

The main drawbacks are the

(1) the excessive time consumed in true 3-D computation on large models, and
(2) restrictive approximations on the behavior (elasticity, isotropy) and on the geometry (2-D, tabular if 3-D) of the media of propagation.

Concerning "physical modeling", the main advantage is certainly the possibility, at least in principle, of simulating a complete 3-D geometry with media of arbitrary complexity (attenuation, anisotropy...).

The main drawbacks are

(1) low versatility of the method (i.e., changing any parameter of the medium of propagation generally implies the building of a new model),

(2) problem of scaling down in the laboratory the sizes of the emitters and the receivers with respect to the elastic wavelength, and

(3) coupling problems between the model to be tested and the emitters/receivers.

Note that the frontiers of the domains of application of each method are continually moving following the evolution of the computing techniques and the technology of the computers, and the advent of unconventional physical modeling methods such as those based on photo-acoustical techniques.

Sometimes we question whether numerical modeling or physical modeling is the best. However, such a question makes no sense because they are two totally different approaches which are not comparable. In other words, to try to impose a hierarchy between these two approaches is not sensible. From one side, numerical modeling consists in applying mathematical tools in the context of some mathematically formulated physical hypotheses. The purpose is to check all the implications of the initial physical assumptions. For this kind of exercise our physical vision and intuition is often sadly limited, especially in complex situations, thus we use numerical modeling to compensate. The output data are synthetic data which have no physical reality. From the other side, physical modeling consists in doing actual physical experiments in the laboratory. The output data represents a physical reality which is often used to check the validity of theoretical models, and sometimes to find new phenomena which cannot be explained by available theories. In other words physical models can be used to validate numerical algorithms (modeling, migration...) only on limited cases because of economical reasons. Because of its high versatility, a validated numerical method can allow, at low cost and under reasonable conditions, a broader number of tests in order to help our physical vision and intuition. Thus numerical modeling and physical modeling are complementary rather than competing. Clearly numerical modeling will never replace physical modeling and vice-versa, just as neither can replace an actual field experiment. Nevertheless, to understand more and more the physical phenomena involved numerical modeling and physical modeling can both help improve the present techniques, and guide our efforts in the field.

## LASER ULTRASONICS, A BREAKTHROUGH IN SEISMIC MODELING

### Why the laser technique?

Currently, the simulation of seismic experiments in the laboratory is usually performed on scaled models (typically of decimetric size) made of synthetic materials (lucite, epoxy resins, silicone rubbers, etc.) immersed in a water tank. The emitters and the receivers are piezoelectric transducers at typically a few hundreds of kilohertz central frequency and of centimetric diameter (i.e., McDonald et al., 1983). Such experimental set-up exists in many laboratories in universities, research institutes, or oil companies, and seems to be quite satisfactory.

Nevertheless such physical models suffer from two drawbacks:

(1) Experiments must be performed in water which means that only marine seismics can conveniently be simulated. In fact, it is evidently possible to work without any water by directly contacting the emitter/receiver with the tested model. But problems, such as coupling reproducibility, induce serious pitfalls if one is interested not only on the time picking of the arrival times but also on more sophisticated analyses of the shapes of the waveforms.

(2) The second drawback is related to the scaling-down of the size of the emitter/receiver with respect to the wavelength. More precisely, in the laboratory typical wavelengths are millimetric to centimetric which is less or equal to the diameter of the emitter/receiver. Transposed to the field, with respect to the seismic wavelength (typically pluridecametric to hectometric), it is equivalent to having geophones of hectometric size.

The available techniques of generation and detection of ultrasound generally can be grouped into two types, namely the contact methods and the non-contact methods (i.e., Thompson and Thompson, 1985). The first type mainly includes the piezoelectric transducers, widely used in laboratories involved in Experimental Acoustics (i.e. Auld, 1973), Rock Physics (Bourbié et al., 1987) or Physical Modeling (McDonald et al., 1983). The inherent limitation of the contact transducers comes from the fact that they have to be coupled to the medium to be tested through a fluid or a solid bond. The presence of this bond leads to a limitation in non-destructive testing. For instance, the bond greatly reduces the speed of testing and the allowable temperature of inspection, and furthermore may potentially add sources of unreliability in the recorded signals. The non-contact probes include electrostatic, electromagnetic, and photoacoustic transducers. If one restricts the applications to seismic physical modeling, the only methods which can achieve both perfect coupling (because no physical contact) and "point source" with respect to the elastic wavelength are the photo-acoustic methods. These are the main reasons why we have adopted the methods based on photo-acoustic principles, which overcome some of the drawbacks of present "physical modeling" methods.

### Laser experimental set-up

As illustrated in Figure 1, for ultrasound generation we use a pulsed laser and for detection, a laser interferometer. These methods of emission and detection of ultrasonic waves using laser techniques are quite widely used in non-destructive evaluation where they are sometimes grouped under "photoacoustics" or more specifically "laser ultrasonics" [i.e., Montchalin (1986) and Hutchins and Tam (1986) for extensive reviews about these methods].

### Emitter

For the emission, we use a Nd:YAG laser operating in the infrared wavelength (1.06$\mu$m) with a maximum energy of 100mJ and a time duration of 20ns, which corresponds to a peak power of 5MW. A detailed description of the physical principle of ultrasound emission by laser technique would be far beyond the scope of this paper [see Hutchins and Tam (1986) for further details]. Nevertheless, refering to Figure 1 and without entering details, we can consider two types of mechanisms of ultrasound generation depending both on the optical power density and the surface condition. For the low power regime the ultrasonic wave generation occurs by thermoelastic expansion. The equivalent acoustic source is a double dipole without moment acting parallel to the surface of the sample. The experimental directivity diagrams corresponding to the *P*-wave and to the *S*-wave are plotted, respectively, on Figures 2a and 2b. For the high power regime, the considerable local increase of temperature of the tested surface results in substantial material ablation and plasma formation. As a consequence, due to momentum conservation, the material ablation causes a recoil force normal to the surface, which is acoustically equivalent to a point source acting normally to the surface. The experimental directivity diagrams plotted on Figures 2c and 2d correspond respectively to the *P*-wave and to the *S*-wave. We notice that the experimental data are in a good agreement with the theoretical directivity patterns computed following the elastic wave theory (White, 1965).

### Receiver

For the ultrasound detection we use a Mach-Zehnder heterodyne interferometer commercialized by B.M. Industries (France) according to Royer and Dieulesaint (1986) design. The physical principle of the detection system, sketched in Figure 3, is based on the measurement of the optical phase modulation induced by the moving surface to be inspected.

More precisely, a coherent light beam of wavelength $\lambda$ outcoming from a continuous wave (CW) laser is divided into two beams, namely a reference beam and a test beam. The reference beam is reflected by a fixed mirror. Thus its length $l_2$ is constant during the experiment. The test beam, of length $l_1$ at rest, is reflected by the tested surface in motion. Denoting $\delta(t)$ the variation with the time $t$ of the normal component of the surface displacement, the length of the test beam varies during the experiment and is equal to $l_1 - \delta(t)$. As a consequence the optical phase $\frac{4\pi}{\lambda} l_2$ of the reference light beam is constant whereas the phase $\frac{4\pi}{\lambda}$ $[l_1 - \delta(t)]$ of the test beam is modulated by the motion of the reflecting surface. The light energy $I$ of the interference beam is measured by a photodetector, which delivers a photo-current $i_p$ proportional to $I$. Under the hypothesis that the surface displacement $\delta(t)$ is much smaller than the optical wavelength $\lambda$, one can demonstrate (e.g., Pouet, 1991) that the AC component $i(t)$ of the photo-current $i_p$ is directly proportional to the surface displacement $\delta(t)$.

With this kind of detector it is possible to measure very small

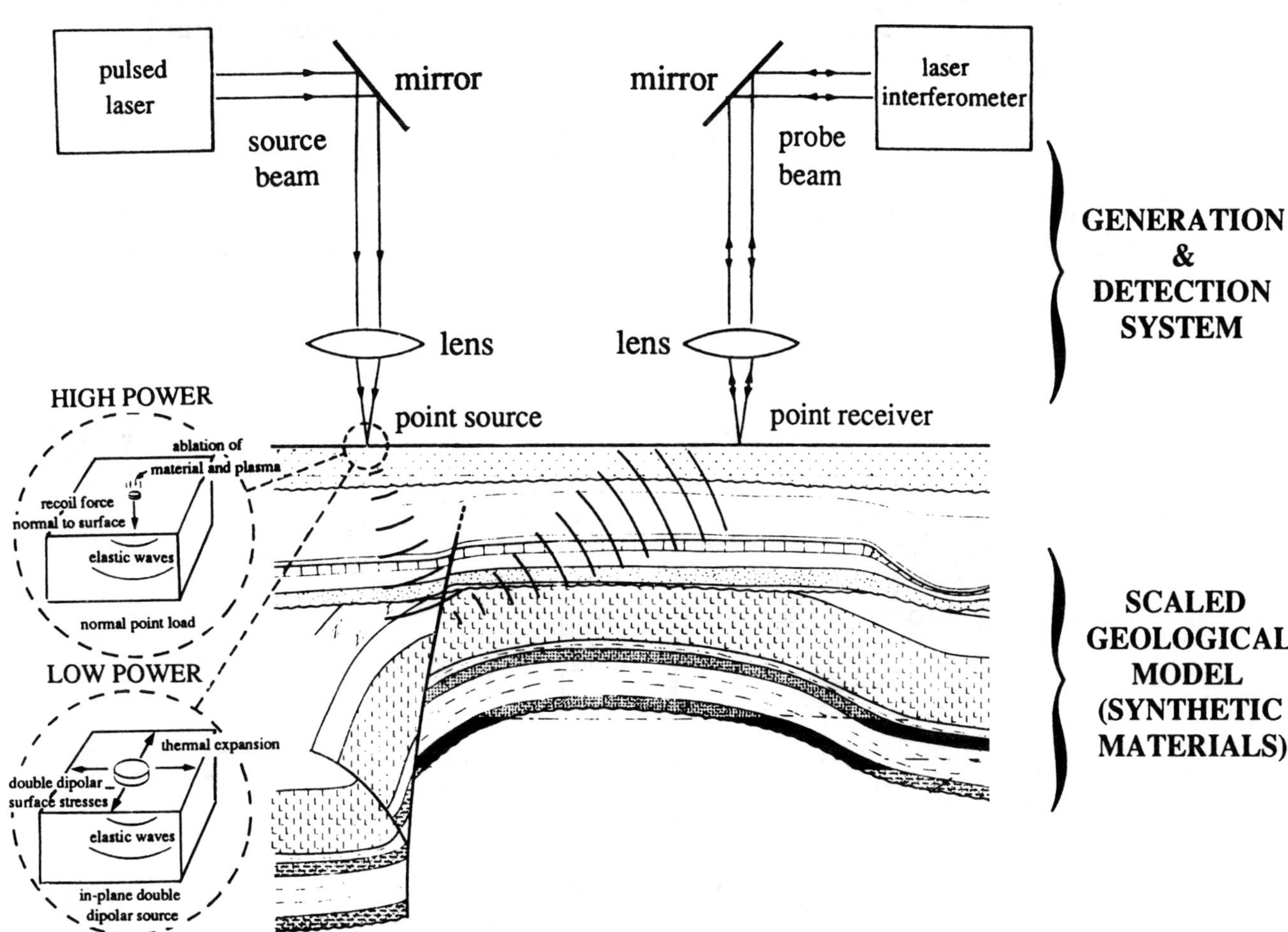


FIG. 1. Sketch of the laser ultrasonics experimental set-up adapted to seismic prospecting problems.

transient displacements (typically from .01 nm to 10 nm) of a small inspection area (limited only by the optical diffraction zone, namely a few $\mu$m) over a large frequency band (typically from 0.1 MHz to 50 MHz). But in our experiments, because of the high-frequency attenuation of the tested materials (resins, polymers, etc.) and the weak dynamic range of the interferometer (typically smaller than 60 dB), we achieve measurements over the 0.1 MHz-5 MHz frequency band [the sensitivity being in direct ratio to the square root of the frequency band (Montchalin, 1986)]. Nevertheless the frequency spectra of the recorded signals are much broader than those of signals provided by the standard piezoelectric transducers as illustrated in Figure 4. This makes a high resolution analysis of the signals easier and allows the study of ultrasonic properties of the tested materials over a broad frequency band (Pouet and Rasolofosaon 1989, Pouet 1991).

### Laser ultrasonics and conventional physical modeling techniques

In summary, compared to conventional techniques, "laser ultrasonics" has three main advantages:

(1) perfect coupling between the emitter/receiver and the model to be tested because there is no physical contact,

(2) high spatial resolution, thanks to the optical beam focusing by lenses that provides point emitters and receivers, and

(3) adaptability of the very broad frequency band (typically more than 6 octaves) to sophisticated analyses of the shapes of the waveforms.

We have to note that laser ultrasonics, in its present state of technological development, exhibits a major drawback. The poor sensitivity and the weak dynamic range of the laser interferometer limit the use of the complete laser system to favorable experimental conditions. These points are detailed in a later section.

The method is illustrated by two recent applications to seismic modeling reported in Pouet and Rasolofosaon (1990), Pouet (1991), and Martin et al. (1991, 1992).

## ASPECTS OF SEISMIC MODELING: LASER ULTRASONICS AND NUMERICAL MODELING

Now for a first illustration, namely a comparison between experimental data recorded with "laser ultrasonics" and numerical data computed by a finite difference method in the configuration of a cross-hole seismic experiment is given.

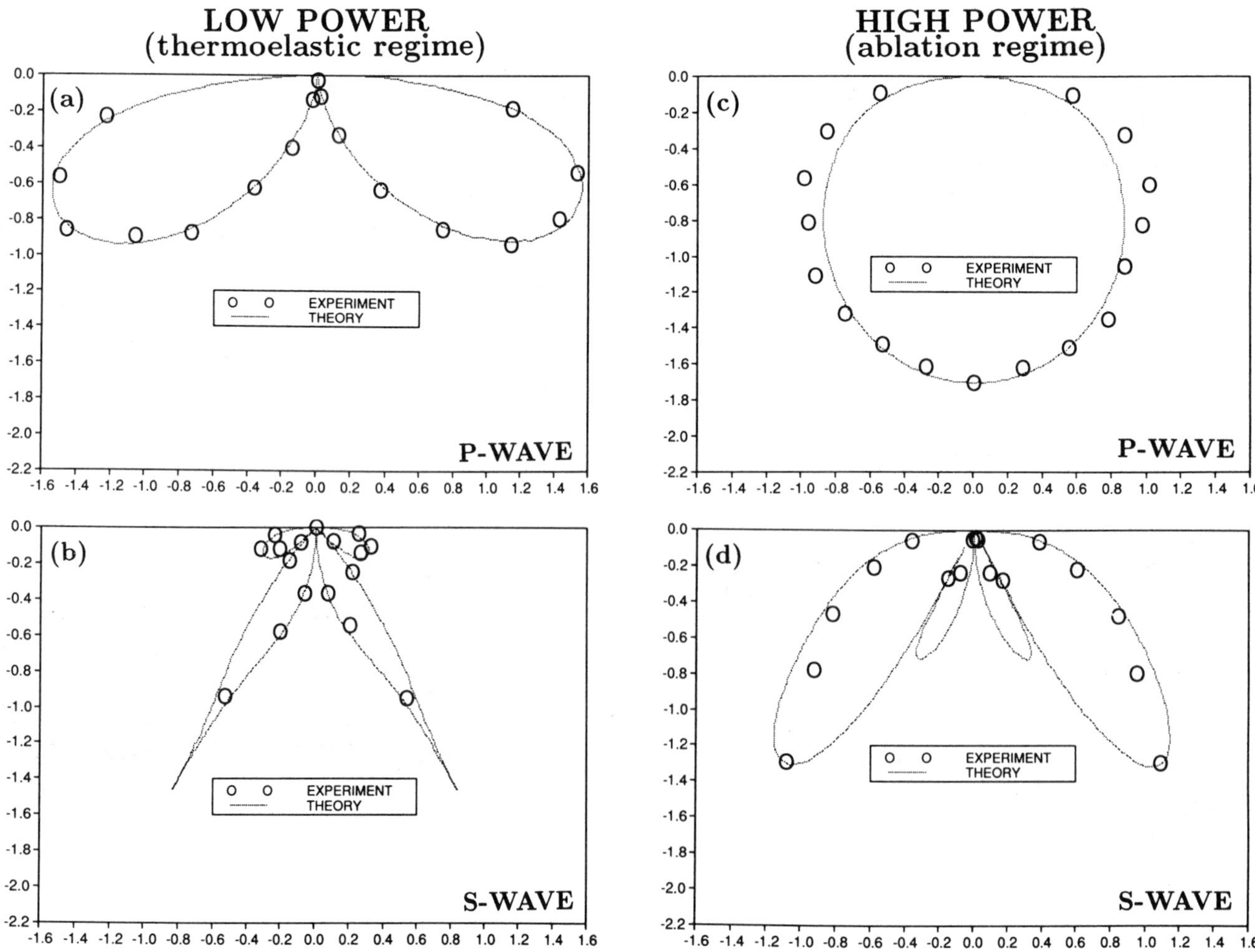


FIG. 2. Ultrasonic patterns of the laser source in aluminum (Poisson's ratio = 0.32)

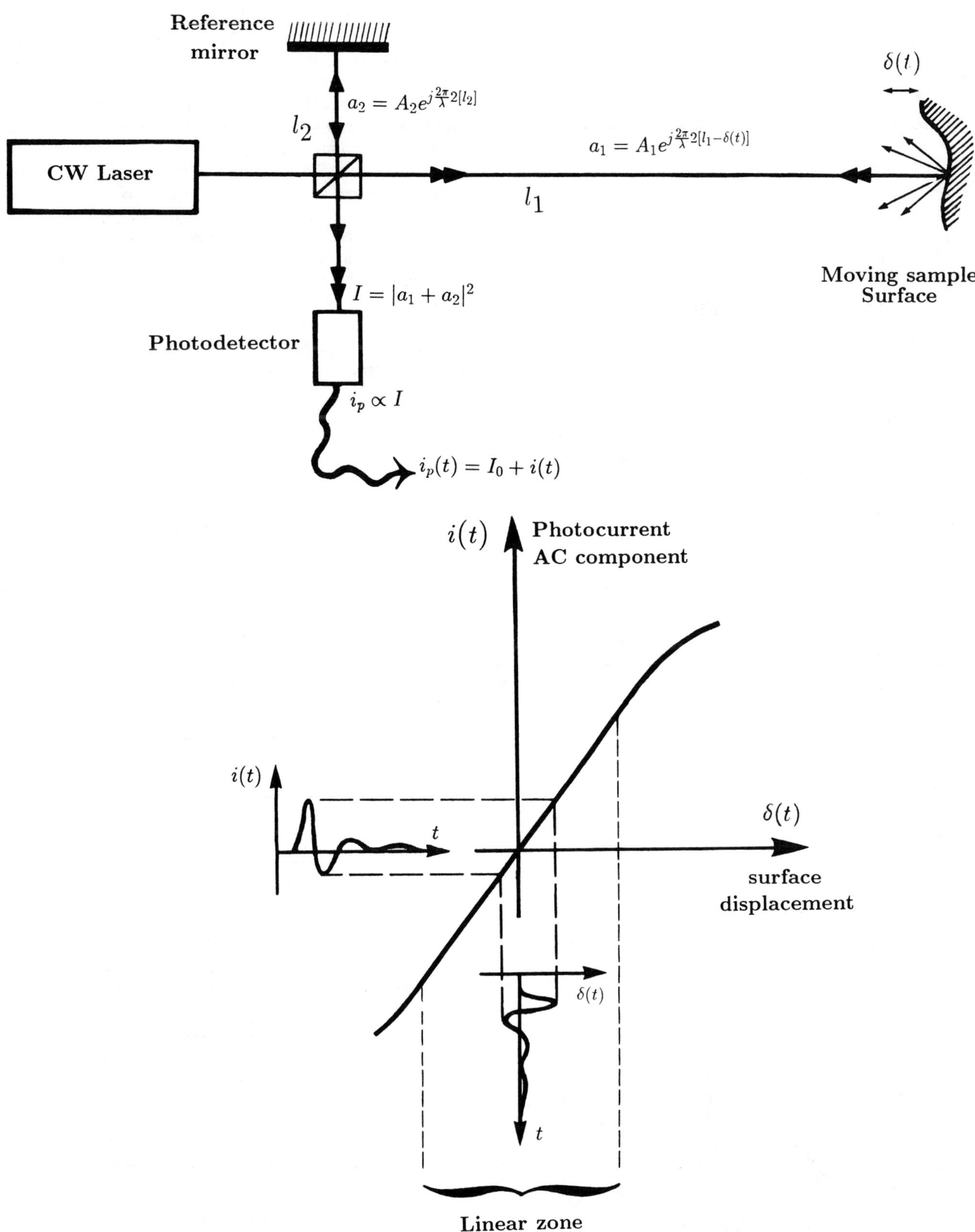


FIG. 3. Principle of the laser interferometer as an ultrasound detector.

## LASER ULTRASONICS

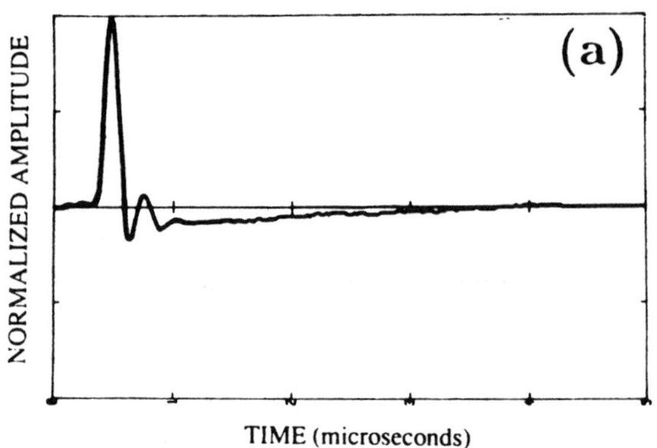


## PIEZOELECTRIC TRANSDUCERS

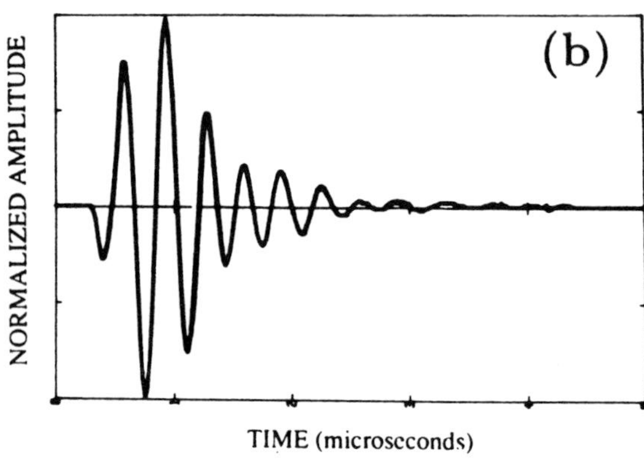


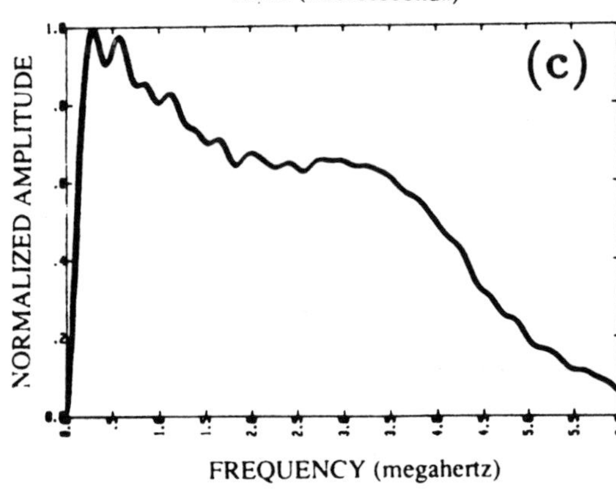


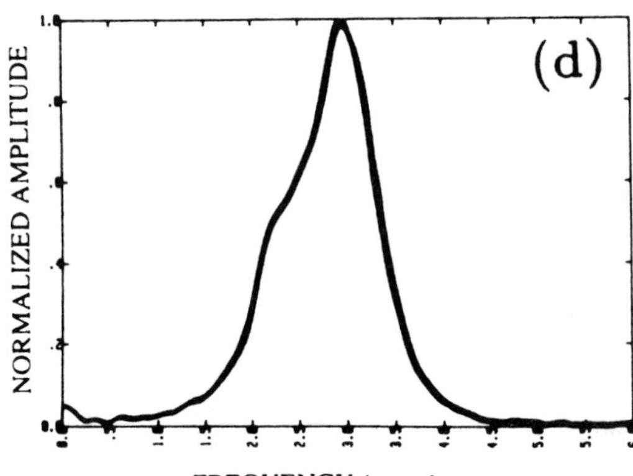


FIG. 4. Comparison between typical ultrasonic signals recorded with piezoelectric transducers (b) and (d) and signals recorded with the laser system (a) and (c) (time domain and frequency domain).

### Experimental set-up and procedures

In the experimental set-up, shown in Figure 5a, the model is made of lucite ($V_p$ = 2750 m/s; $V_s$ = 1350 m/s) and aluminum ($V_p$ = 6320 m/s; $V_s$ = 3250 m/s). The physical properties are not representative of field situations but have been chosen just to emphasize the rapidly increasing complexity of cross-hole data when both *P*- and *S*-waves are physically generated, even when the geometry of propagation is apparently simple. The point source is located at 2.5 mm from the interface on one side of the lucite surface. The point detection positions at the opposite surface are at constant offsets (0.5 mm). Contrary to what is sketched in Figure 5a the studied geometry is a single plane interface and not a "step" interface. Figure 5b shows the field equivalent. The typical scale factor from the field to the laboratory is around 1/100 000, both for time and length. More precisely the field source is located 250 m from the interface, the boreholes are spaced 2000 m apart, and the closest distance between two receivers is 50 m.

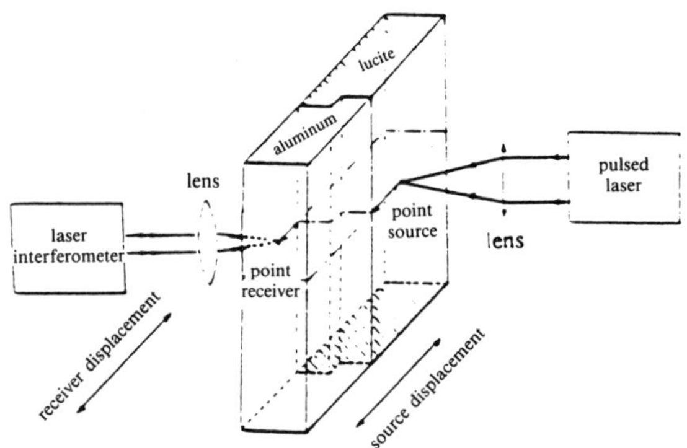


(a) Physical modeling experimental set-up

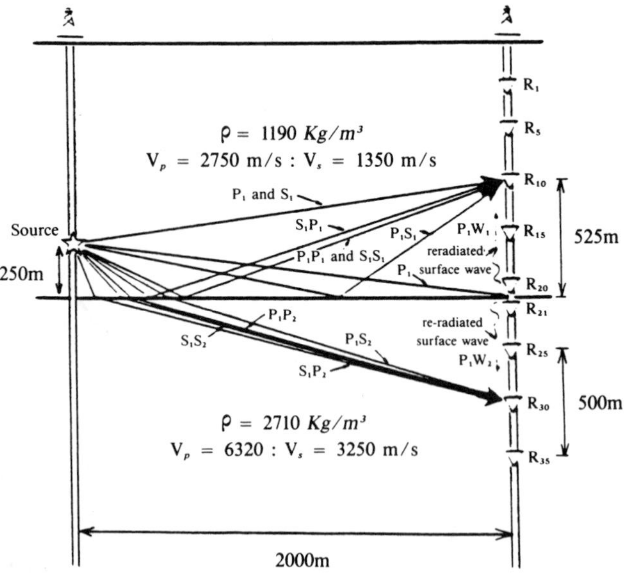


(b) Field equivalent

FIG. 5. Cross-hole seismic experiment. Sketch showing (a) the experimental set-up and (b) its field equivalent with the different types of arrivals. The approximate time and scale factor from the field to the laboratory is 1/100 000.

**Laser ultrasonics and finite-difference modeling**

The experimental data are compared to the numerical data computed using the finite difference method following a second order explicit scheme (both in space and in time) reported in Virieux (1986) and Collino (1989). Similar numerical results already have been published in Hu et al. (1988).

Figure 6a shows experimental data recorded with the laser setup sketch in Figure 5a. The corresponding numerical data are plotted in Figure 6b. The traces have all been equalized to improve the visualization of the arrivals.

The experimental data exhibit much more noise than the numerical data, especially in the aluminum part of the model and for times larger than 10μs. An explanation is that the presence of the large impedance contrast between lucite and aluminum implies that most of the energy is reflected. As a consequence, the energy transmitted in the aluminum is quite small and mixed with the experimental noise. Thus, in portions of the experimental seismograms where there are few arrivals, the noise is enhanced by the equalization process.

The numerical data which appears to be more reverberating than the experimental data, is probably induced by some minor-remaining numerical dispersion problems because the numerical parameters have not been totally optimized.

Even for a "single interface" model on both sets of data, a great variety of arrivals are observed. Three main types of waves can be distinguished:

(1) Bulk waves comprise the direct waves between emitter and receiver, the reflected waves in the first medium, and the transmitted waves in the second medium, both with or without conversion. The bulk waves are designated by $P_i$ and $S_i$ for the direct waves, and $P_iP_j$, $P_iS_j$, $S_iS_j$, $S_iP_j$ for the reflected and transmitted waves. The letters $P$ and $S$, respectively, stand for a $P$-wave and an $S$-wave, and the subscripts $i$ and $j$ correspond to the propagation media, 1 is for the lucite, and 2 for the aluminum.

Direct waves. — The large energy radiation of the $P_1$-wave and the progressive extinction of the $S_1$-wave when the emitter and the receiver are practically at symmetric positions are in good agreement with the directivities of the source and the receivers (see Figure 2). Most of the $P$-wave energy radiates globally in a direction parallel to the reflector whereas most of the $S$-wave energy radiates at about 30 degrees from this direction. Thus the $S_1$-wave amplitude increases (whereas the $P_1$-wave decreases) when the receiver is removed from the reflector. The numerical data show more clearly the $P_1$-wave decreasing.

Reflected waves. — The amplitude of the reflected wave $S_1S_1$ mainly depends on the emitter/receiver directivities because the variations of the reflection coefficient are negligible for all the possible angles of incidence of the experiment. Due to the $S_1$-wave radiation pattern, the amplitude of the reflected $S_1S_1$-wave increases when the receiver is removed from the reflector. The phase inversion of the $P_1P_1$-wave is clearly detected (compare with the phase of the $P_1$-wave). The phase inversion of the reflected $S_1P_1$-wave corresponds to the change of sign of its reflection coefficient.

Transmitted waves. — In the second medium, the first arrival corresponds to the $P_1P_2$-wave, that is to say an incident $P_1$-wave transmitted as a $P_2$-wave. The $S_1S_2$-wave is difficult to see because of geometric and acoustic characteristics of the model, and because of directivities of the emitter and the receiver. The $S_1P_2$-wave is lost in the noise, because it is weakly excited due to the radiation pattern of the $S_1$-wave.

(2) The surface waves are Rayleigh waves propagating on the boundary of the model and radiating bulk waves from the intersection of the borehole and the reflector. A first type of wave propagates just after emission and radiates on the end of the reflector (emitter side). A second type of wave is created when a bulk wave excites the other side of the reflector (receiver side). This result is very similar to results reported in White (1965), for instance, showing that each reflector intersecting the borehole or each sudden variation of the radius of the borehole along the well-bore act as secondary sources. The surface waves are designated by Ri. The $P_1$ bulk wave excites Rayleigh waves ($P_1R_1$ and

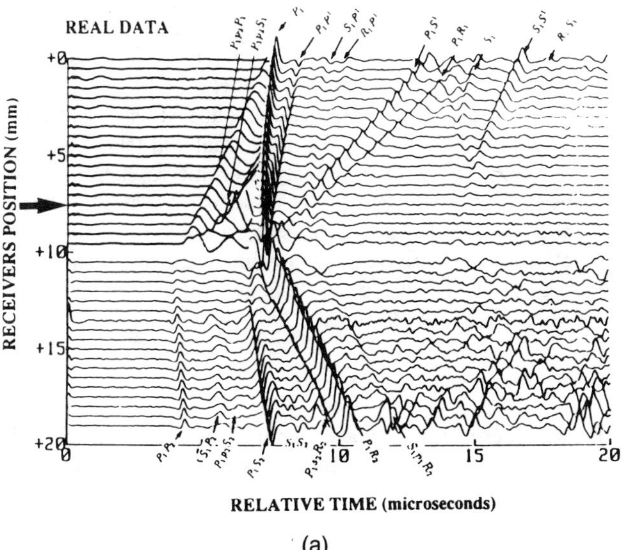


(a)

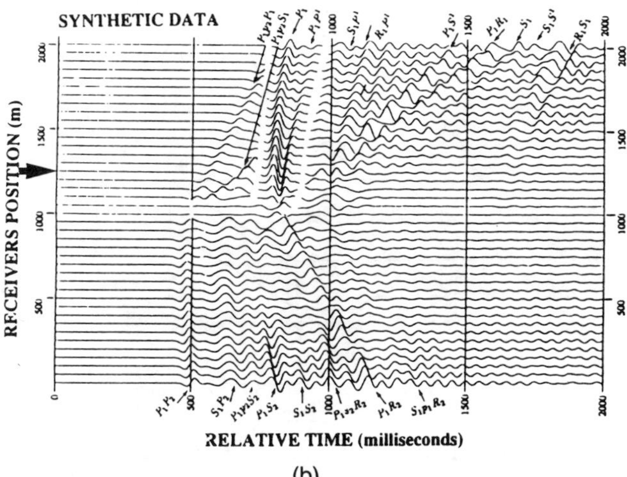


(b)

FIG. 6. Modeling of a cross-hole seismic experiment. Horizontal component of the displacement. (a) real data using the laser technique, and (b) synthetic data using a finite difference technique.

$P_1 R_2$ waves) propagating on both sides of the "receiver" surface. The observation of the seismograms leads to Rayleigh velocities of 1320 m/s in the first medium and of 3040 m/s in the second medium. The theoretical data practically gives the same velocities (1310 m/s in lucite and 3030 m/s in aluminum). These waves are clearly identifiable. As expected, we can notice that the $P_1$, $P_1 R_1$ and $P_1 R_2$ arrival times are identical when the receiver is at the level of the reflector. The $R_1 P_1$-wave and $R_1 S_1$-wave correspond to the excitation of a bulk wave $P_1$ and a bulk wave $S_1$ at the reflector (emitter side). The $R_1 S_2$-wave is not observable because it is certainly superimposed with other higher energy arrivals.

(3) The conical waves or head waves (Cerveny and Ravindra, 1971) are designated by $p_i$ *and* $s_i$. For instance a $P_i p_j S_k$-wave is an incident $P_i$-wave in the medium $i$, critically refracted as a $p_j$-wave in the medium $j$ and re-radiated as a $S_k$-wave in the medium $k$. In our favorable experimental configuraion, there are not less than 15 such waves (Pouet, 1991) because all the velocities in the first medium are smaller than all the velocities in the second medium. These are the $P_1 p_2 P_1$, $P_1 p_2 S_1$, $P_1 p_2 S_2$, $P_1 s_2 P_1$, $P_1 s_2 S_1$, $S_1 p_1 S_1$, $S_1 p_2 S_1$, $S_1 p_2 P_1$, $S_1 p_2 S_2$, $S_1 s_2 S_1$, $S_1 s_2 P_1$, $S_1 r_2 S_1$, $S_1 r_2 P_1$, $P_1 r_2 S_1$, and the $P_1 r_2 P_1$-wave. The notation $r_2$ corresponds to a pseudo-Rayleigh wave (e.g., Viktorov, 1967) of exponentially decreasing amplitude in the medium 2, that is to say in the aluminum, of which the velocity is close to the velocity of the Rayleigh wave in aluminum. Nevertheless only some of these waves are visible on the reported data. We can easily see the first $P_1 p_2 P_1$ and $P_1 p_2 S_1$ waves radiating in the first medium. In order to observe these arrivals we have magnified some parts of the seismograms (i.e., Martin et al., 1991, for further details). In the second medium, we observe one of the head waves, namely the $P_1 p_2 S_2$ wave. The $P_1 s_2 P_1$-wave is not identifiable. Some head waves can excite Rayleigh waves such as the $P_1 s_2 R_2$ and $S_1 p_1 R_2$ arrivals. We see also, the Rayleigh wave $S_1 p_1 R_2$ excited by the $S_1 P_1$-wave.

### Conclusions

Laser ultrasonics is an efficient method which will simulate in the laboratory seismic experiments in the elastic domain (with both $P$- and $S$-waves, eventually in 3-D configuration). We compared the experimental crosshole seismograms to the synthetic seismograms simulated by a finite-difference method. The agreement is quite satisfactory and constitutes a validation of the numerical program. Even in the case of simple geometry (plane single interface), the data exhibit a relative complexity because the propagation direction of the wave energy is globally parallel, and not perpendicular (as for conventional VSP or surface experiments), to the interfaces which generate numerous types of waves including direct, transmitted, reflected, converted, leaky, conical, and even guided waves (in the presence of waveguides). The identification of all the arrivals is not straightforward.

The previous results, and others reported in the literature, illustrate the challenge we face to correctly and completely interpret crosshole data. The challenge is worthwhile if one considers that geophysical tomography using borehole-to-borehole experiments is one way to remedy the low lateral resolution of surface-to-surface or borehole-to-surface methods in order to obtain the best image of the earth's interior.

## SOME ASPECTS OF SEISMIC MODELING AND IMAGING IN (AN) ISOTROPIC MEDIA USING A MIXED LASER/PIEZOELECTRIC SYSTEM

Next we analyze the effect of anisotropy on wave propagation and imaging, using the results of seismic prospecting experiments performed in the laboratory (Martin et al., 1992). We did not use the complete laser system but a mixed system, namely the pulsed laser as the source and a piezoelectric hydrophone as the receiver. The reason for the mixed system is that when working on solids in the reflection configuration, it was impossible to record simultaneously the high-energy arrivals (such as the Rayleigh waves) and the low-energy arrivals (such as the reflected arrivals) in order to separate them by further signal processing. Therefore, we had to experiment in a water tank and to use the mixed system of ultrasound emission and reception in order to generate no surface wave and to take advantage of the high sensitivity and large dynamic range of the hydrophone.

We focus our studies on reflections on a modified version of the well-known ridge and fault 2-D model initially proposed in French (1974). The physical models are made either of isotropic material (aluminum) or of anisotropic material (slate) of two different orientations (Transversely isotropic case: horizontal schistosity, and tilted case: planes of schistosity at 22.5° from vertical direction). We now comment on the main features of diffraction and reflection on the raw time sections, and analyze the processed data, "isotropicaly" migrated using a 2-D prestack program developed in Ehinger (1990).

### Experimental set-up and procedure

Figure 7 shows the experimental set-up. The source is the pulsed laser with the source beam focused in a water tank. This set-up

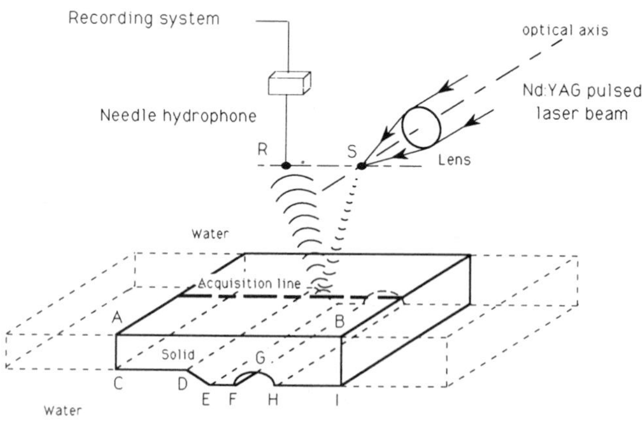


| | CD | DE | EF | FH | HI | AC | BI |
|---|---|---|---|---|---|---|---|
| | (mm) | | | | | | |
| Aluminium | 11,1 | 3,4 | 5 | 11,2 | 11,1 | 15 | 17 |
| Angers Slate | 11,5 | 3,4 | 5 | 11,2 | 11,5 | 28 | 30 |

Table 1

FIG. 7. Sketch showing the experimental set-up adapted to marine reflection seismics and the modified version of the 2-D ridge and fault French model. The dimensions of the models are listed in Table 1.

induces the formation of an explosive bubbling of the liquid and a formation of a rapidly expanding vapor cavity with the propagation of a compressional wave, as a conventional vapor-choc source. This compressional wave has an approximately spherical shape with a propagation velocity close to the sound velocity (i.e., Lyamshev and Naugol'nykh, 1980). For the reception, we use a needle hydrophone (i.e., Rasolofosaon, 1988). With respect to the wavelength in water, the source and the detector are approximately points and separated by a distance of 2 mm. Thus the source and receiver will be considered as coincident in further discussions. The 2-D seismic data consist in the recording of a single coverage reflection data at fixed distance to the model. In other words, the data are similar to a so-called zero-offset section in the field. A single acquisition line with a step of 0.3 mm is recorded for each model (Figure 7). For one position of the whole system (model, emitter, and receiver), an acquisition consists in the averaging of twenty laser shots.

### Geometry and properties of the studied samples

The particular geometry of the models is inspired by the 2-D ridge and fault model of French (1974). But contrary to French who studied the reflections of a compressional wave in water and the geometric structures (ridge and fault) drawn on the top of a solid model, we are interested in the reflections within an (an)isotropic solid. For this purpose, the geometric structures are represented on the bottom of a solid model. The 2-D model is shown in Figure 7 and the dimensions are listed in Table 1.

One model is isotropic and made of aluminum. The two other models are made of Angers slate, a highly anisotropic rock which has been studied in Arts et al. (1991). Following these authors, the slate exhibits 47 percent deviation from isotropy. The slate can be considered as a finely layered medium presenting macroscopic cleavage planes parallel to the layering. The Angers slate samples will be roughly approximated as a transversely isotropic medium presenting macroscopic "fractures" oriented perpendicular to the axis of orthotropy. The first anisotropic model in which the planes of schistosity are horizontal is commonly called Transversely Isotropic (T.I.) TI is the commonest form of anisotropy present in the earth subsurface. In the second anisotropic model, the planes of schistosity are 22.5° from the vertical axis in such a way that the axis of orthotropy is in the plane of the profile.

### Unmigrated profiles

Now we describe the raw time sections recorded on the three models.

The raw data corresponding to the isotropic case are shown in Figure 8a. The signal which corresponds to the signature of the reflection on the top of the model (label AB) and the noisy background signal before the reflection on the structures are considerably subsided. It is relatively easy to read the relief. We distinguish the horizontal reflectors labeled AB, CD, EF, and HI. We can also see the effect of the ridge (label C) on the wave propagation. At the point H, the waves are reflected by the structures of the anticline and of the horizontal plane (labeled HI). The top of the ridge is perfectly located, see the label G. The arrivals labeled F correspond to the superposition of the waves relected by the ridge and by the horizontal plane (label EF) and of the wave diffracted by the border of the fault (label D). The large diffraction energy from the limit point of the fault (label D) is easily identifiable.

On the "T.I." sample, the profile is shown in Figure 9a. The data are much more noisy than in the previous case. The signal, after the top reflection, has been attenuated for a better visualization of the results. We can see approximately the same effects as in the previous model. Each plane reflector is clearly identifiable (labels AB, CD, EF, and HI). The reflection on the anticline relief is shown by labels H and F. The radius of curvature of the wavefront reflected by the rounded ridge appears larger than in the isotropic case. This can be explained by the large increase of the group velocity from the horizontal direction to the vertical direction.

The raw data corresponding to the tilted anisotropic case are represented in Figure 10a. The time section is dominated by strong arrivals different from those corresponding to the reflections on the structural reliefs at the bottom of the model. These arrivals can be explained by the presence of macroscopic cleavage planes in the slate sample. We can distinguish two types of arrivals. The first type corresponds to the waves diffracted by the intersection of the cleavage planes with the top plane reflector. These arrivals are labeled J and exhibit an apparent velocity close to the water velocity. The second type labeled K, corresponds to reflections, converted or not, inside the model on the cleavage planes. Studies of these types of reflections have been reported in Pyrak-Nolte et al. (1987), and Rasolofosaon and Zinszner, (1989).

**Isotropic model : Aluminium**

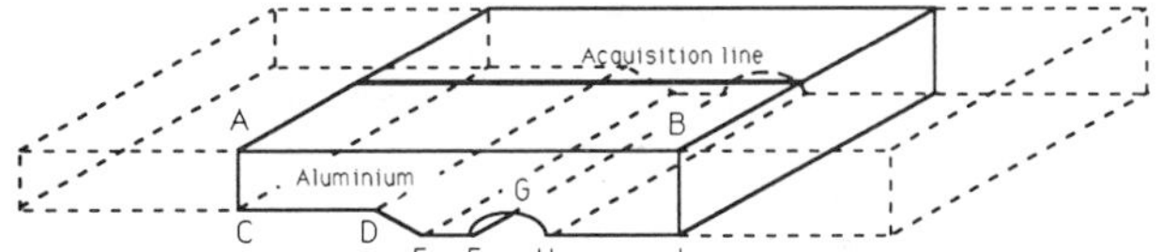


**a) Raw data**

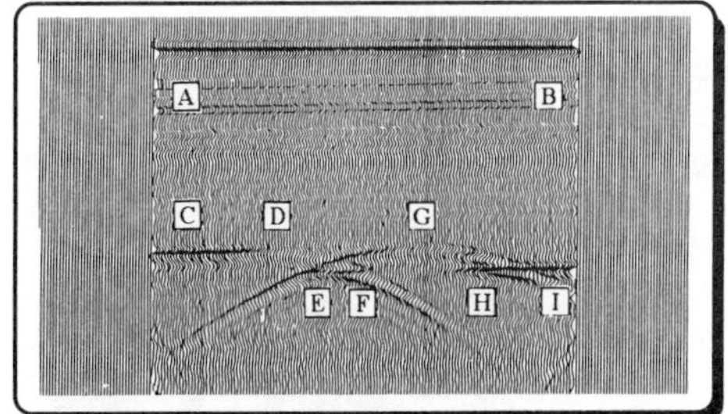


**b) Migrated section**

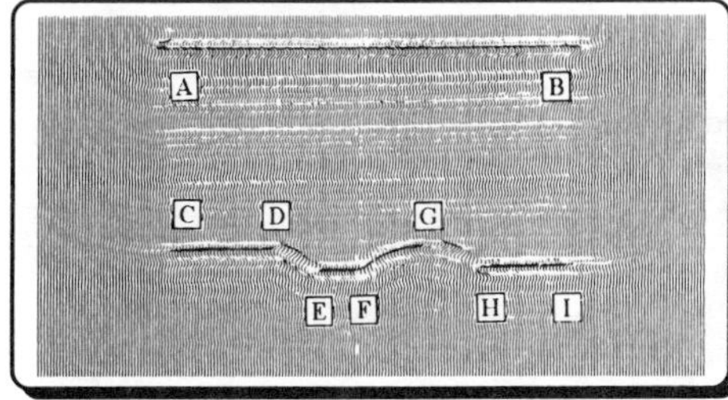


FIG. 8. (a) raw and (b) migrated data corresponding to the isotropic 2-D ridge and fault model (aluminum).

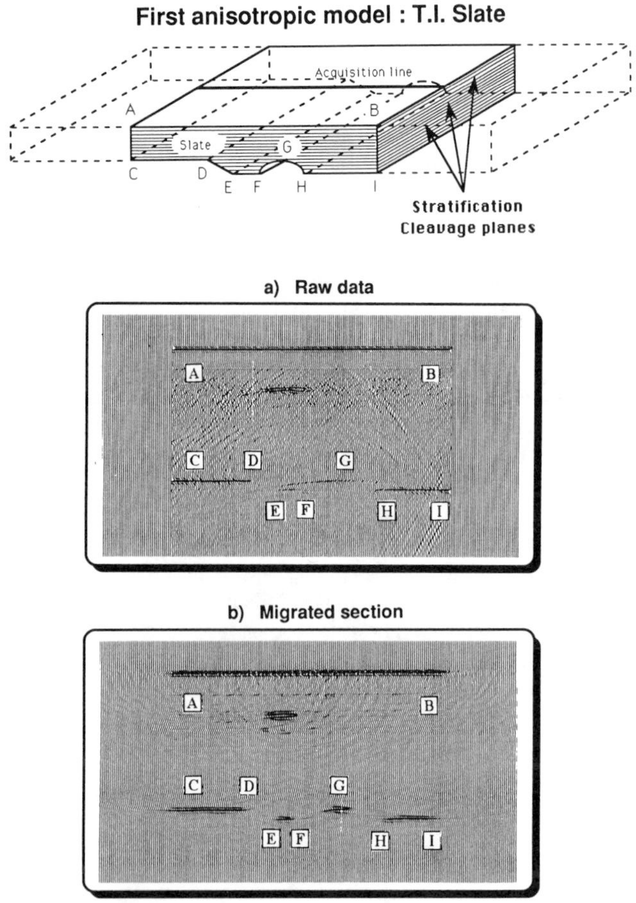

FIG. 9. (a) raw and (b) migrated data corresponding to the first anisotropic 2-D ridge and fault model (TI slate).

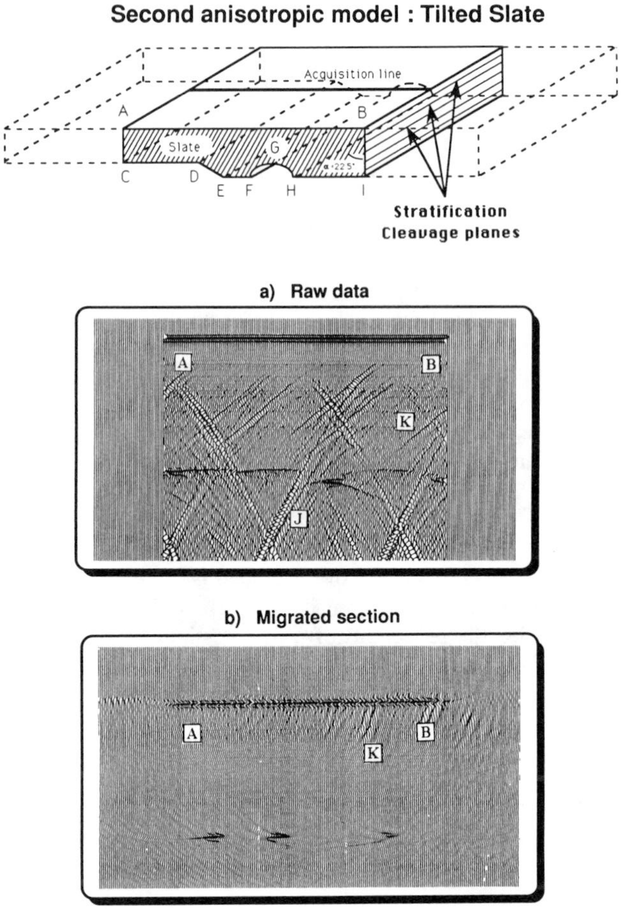

FIG. 10. (a) raw and (b) migrated data corresponding to the second anisotropic 2-D ridge and fault model (tilted slate).

Among these strong arrivals, we can guess the reflected arrivals on the structures. Compared to Figure 9a, the shape of the relief is much perturbed. However, knowing the geometry of the model, we distinguish the plane reflectors (labels AB, CD, EF, HI) and the rounded ridge (label G). The propagation of energy is favored in the direction of the laminations, which is sketched in by Figure 11. More precisely, a sub-normally incident wave vector in water is transmitted as a sub-normally wave vector in the slate, whereas the corresponding ray in the slate is no more sub-normal but orientated in a direction normal to the qP slowness curve (Auld, 1973), that is to say globally parallel to the stratification/cleavage planes. As a consequence, an "isotropic" interpretation of the phenomenon leads to an unavoidable mislocating of the reflectors, and explains the fact that the "apparent" locations of the reflectors are globally translated to the right.

## Migrated profiles

The above profiles are migrated using a 2-D acoustic and "isotropic" prestack migration program proposed in Ehinger (1990). The velocity model was made of two layers, a water layer and a solid layer. The chosen velocities were 1500 m/s and 6324 m/s for isotropic model and 1500 m/s and 5000 m/s for both of the anisotropic cases.

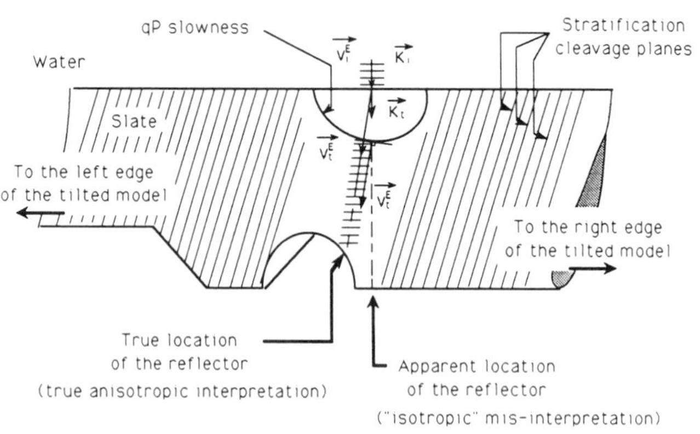


In the slate, the energy preferentially propagates globally parallel to the stratification/cleavage planes which leads to the mis-locating of the reflectors if an "isotropic" interpretation is made.

FIG. 11. Physical interpretation of the apparent mis-locating of the reflectors in the tilted anisotropic model. $K_I$ and $K_T$ designate the incident and the transmitted wave vectors, and $V_I^E$ and $V_T^E$ denote the incident and the transmitted group/energy velocity vectors.

The reconstructed image of the aluminum model gives a perfect representation of the relief (Figure 8b). However, one can see a decrease in the amplitudes on the sides of the rounded ridge and on the fault structures due to a decrease in the illuminating energy on these parts of the structure. This well-known result indicates that migration does not reconstitute true amplitudes.

The migration of the T.I. data (Figure 9b) still shows a good definition of the horizontal plane reflectors but a blurring image of the nonhorizontal structures (sides of the ridge and fault). This phenomenon is comparable to the previous situation but is more marked in this case by the fact that the medium is anisotropic. When the rays differ from the vertical direction, the energy propagates rapidly in the lateral direction. This physical phenomenon, associated with the use of an "isotropic" migration program, explains the bad resolution of the ridge and fault structures.

The third image (Figure 10b) shows the migrated profile of the Angers slate sample with a titled axis of symmetry. In a few words, the structural map is unrecognizable, the reflectors are mislocated, and the geometries of the rounded ridge and of the fault have disappeared. These results are in a global agreement with the predictions of previous theoretical works in the particular case of elliptically anisotropic media (Levin, 1978; Helbig, 1983; Blair and Korringa, 1987; Dellinger and Muir, 1988). The previously mentioned anisotropic effects are so much amplified that, without knowing the geometry of the model, it is very difficult to achieve even an approximate guess, at least if we refer to the results of numerous blind tests that we have done with experienced senior geophysicists.

From the migrated images, such as those shown in Figure 10b, we measure the difficulties we have to face to correctly migrate and/or interpret complex geologic structures in anisotropic media.

### Conclusions

We have learned from these experiments, about the laser experimental set-up and about modeling and imaging techniques.

Concerning the experimental set-up, the sensitivity and the dynamic range of the laser receiver are too weak to be used in many experimental configurations. For instance, in the reflection configuration it was impossible to work on solids, therefore we worked by immersion and used a combined system, namely a laser source and a piezoelectric hydrophone.

In addition, we analyzed some aspects of wave propagation and migration in anisotropic media. We studied a modified version of the well-known 2-D ridge and fault model of French (1974). The obtained data validated the migration program proposed in Ehinger (1990) in isotropic media. This validation on experimental data is complementary to a validation on synthetic data. We have shown the degradation of the images induced by anisotropy. These can be summarized as a defocusing effect (wrong migration velocity) and a deformation of the image (mispositioning of the reflectors).

Although the anisotropy of the studied material (slate) is too strong to be representative of common geologic media in seismic prospecting, we think this type of modeling can be of great help in testing imaging algorithms. The validation checked in extreme cases could, in principle, guarantee the validity in more realistic cases.

## CRITICAL EVALUATION OF THE LASER TECHNIQUE AND PROSPECTIVE VIEWS

The two previous examples, chosen from recent applications of "laser ultrasonics" to seismic physical modeling, have illustrated the great promise of the technique. The critical evaluation of this latter, developed in the following lines, is the result of some years of experimentation with different laser systems at the Rock Physics and Physical Modeling Laboratory of IFP. Therefore, in the light of this evaluation, we propose some prospective views about "laser ultrasonics" in the context of seismic physical modeling for tomorrow.

### Critical evaluation of "laser ultrasonics" applied to seismic modeling

Now that the numerous advantages of the technique have been thoroughly developed we shall now analyze the limitations of the laser system in its present design.

The present laser system technology exhibits actual limitations, namely:

(1) implementing the laser system is difficult compared to conventional techniques (e.g., piezoelectric transducers). An actual know-how both in optics and in acoustics is required.

(2) receiver sensitivity, which is quite low, critically depends on the roughness and the optical reflectivity of the surface to be tested. For instance, from a practical point of view, in order to improve the signal-to-noise-ratio (S/N) the surfaces to be tested were carefully polished and metal-plated (a thin film, a few nanometers, is deposited by vacuum evaporation) to achieve good mirror surfaces in order to increase the global optical efficiency.

(3) weak dynamic range restricts the use of the complete laser system to transmission experiments, such as in the cross-hole experiments previously reported. That is why, in previously reported reflection experiments, we worked by immersion in a water tank and combined the laser source with piezoelectric hydrophones.

(4) system adjustment, especially the laser interferometer, is rather tricky and requires special care. This is certainly the part of the system which is the most difficult to automate. Use of optical fibers would greatly facilitate positioning of both source and receiver, but their use creates serious technological problems of adaptability. More precisely, concerning the source, first the power provided by the pulsed laser is too severe to be safely injected into optical fibers, and second, serious problems of reproducibility of the source due to the ionization of the air arise, necessitating the use of a vacuum cell (Richou and Richou, 1992 and A. Lebrun, personal communication). And concerning the receiver, its coupling with an optical fiber would critically reduce its already low sensitivity.

In summary, the laser system is an outstanding research tool. From a physical point of view the system is nearly beyond reproach. The system provides more reliable simulations of seismic experiments and has overcome some drawbacks of cenventional techniques of physical modeling. But, due to current technological development, the system is not well suited to immediate large scale or industrial applications without previous additional and sizeable investments both, in research to overcome the aforementioned limitations, and in development to change from the present research prototype to an industrial prototype.

### Prospective views

As 3-D seismic is progressively becoming "a cost-effective tool for mapping subsurface structures" such as hydrocarbon reservoirs, at least if one refers to recent publications (i.e., Nestvold, 1992), the interest for 3-D seismic modeling, processing, and interpretation will doubtless continue to increase. Improvements in subsurface image resolution will certainly result from advancements in the technique.

Concerning specifically seismic modeling, in the present state of the technology "physical modeling" in the laboratory is the only method which can deal, at least in principle, with the 3-D problem in its whole generality. As a consequence "physical modeling" should be one of the main means that we have to try to understand the 3-D problems of wave propagation in complex media in order to improve our present techniques of acquisition, processing, and interpretation of 3-D seismic data.

Furthermore, improvement in physical modeling techniques is necessary in order to provide as reliable as possible test data on controlled media of propagation (geometry, rheology, etc.) in order, for instance, to validate numerical modeling or imaging programs or to allow "blind" interpretation tests. In this sense the photo-acoustical methods have recently contributed much to the discipline. Different research teams are still trying to go much deeper into various areas of investigation in domains including the improvement of the sensitivity and of the dynamic range of the systems, the multiplexing of the recording, or the coupling of optical fibers to the system. Such important efforts must be maintained.

Among these contributions, one recent breakthrough in laser ultrasonics, reported in Monchalin et al. (1989), opens the possibility of recording, at exactly the same point, by speckle interferometry the 3 components (3C) of vibration of a solid excited by an elastic wave. Testing this type of tool has begun and some investigations are in progress. The first results are quite satisfactory although the aforementioned limitations still remain. Thus, a new generation of 3C/3-D research tool could be expected which would really bring new dimensions in physical modeling. The main technological challenges are to increase the sensitivity and the dynamic range of the receiver and to succeed in completing the automation of the whole system. Overcoming these problems will certainly open the door to the industrial phase of laser ultrasonics and will revolutionize the domain of physical modeling and non-destructive evaluation.

## CONCLUSIONS

A general presentation and a critical evaluation of a technique called laser ultrasonics, widely known in non-destructive evaluation and recently applied to seismic physical modeling has been presented. This technique overcomes some drawbacks of conventional physical modeling techniques, namely coupling problems for full-elastic modeling and scaling the size of the emitter/receiver with respect to the wavelength.

To illustrate the great promise of the method we have reported two examples from the recent applications to seismic prospecting, one related to cross-well seismic modeling and the other related to reflection seismic modeling and imaging in isotropic and anisotropic media.

Up to now, the poor sensitivity and the weak dynamic range of the systems are the main reasons that limit the practical evolution of this technology from research to a full industrial application. But considering the continual evolution of the technology mainly in optics and electronics, laser ultrasonics is doubtless on the way to revolutionizing the domain of seismic physical modeling and non-destructive evaluation.

## ACKNOWLEDGMENTS

Research was done at the Rock Physics and Physical Modeling Laboratory of the Institut Français du Pétrole with partial financial support from the Commission of the European Communites (DGXII) in the framework of the Joule programme, Subprogram Energy from Fossile Sources, Hydrocarbons. We gratefully acknowledge M. Masson for his experimental help and Dr. A. Lebrun, from Electricité de France (D.E.R.), Drs. C. Jacquin and B. Zinszner for fruitful discussions. The imaging part of this paper benefited from illuminating discussions with E. de Bazelaire, from Elf-Aquitaine Co., and L. Nicoletis, P. Layotte, and C. Naville, from IFP, whom we acknowledge for their contribution to the global entropy.

## REFERENCES

Arts, R. J., Helbig, K., and Rasolofosaon, N. J. P., 1991. General anisotropic elastic tensor in rocks-Approximation, invariants and particular directions. 61st Ann. Internat. Mtg., Soc. Expl. Geophys., Abstracts, 1534-1537.

Auld, B. A., 1973, Acoustic fields and waves in solids: Wiley Intersc., New York.

Blair, J. M., and Korringa, J., 1987, Aberration-free image for SH reflection in transversely isotropic media: Geophysics, **52**, 1563-1565.

Bourbié, T., Coussy, O., and Zinszner, B. E., 1987, Acoustics of porous media: Gulf Publishing Co.

Cerveny, V., and Ravindra, R., 1971, Theory of seismic head waves: University of Toronto Press.

Collino, F., 1989, Computations on modeling, migration and inversion of seismic data: IFP report No. 37491 (in french).

Dellinger, J., and Muir, F., 1988, Imaging reflections in elliptically anisotropic media: Geophysics, **53**, 1616-1618.

Ehinger, A., 1990, FARS-A 2-D prestack shot record migration program: Prestack Structural Inversion (PSI) IFP research consortium annual report.

French, W. S., 1974, 2-D and 3-D migration of model-experiment reflection profiles: Geophysics, **39**, 265-277.

Helbig, K., 1983, Elliptical anisotropy - its significance and meaning: Geophysics, **48**, 825-832.

Hu, L. Z., McMechan, G. A., and Harris, J. M., 1988, Elastic finite-difference modeling of cross-hole seismic data: Bull. Seism. Soc. Am., **75**, 5, 1796-1806.

Hutchins, D. A., and Tam, A. C., 1986, Pulsed photoacoustic materials characterization: IEEE Trans. Ultras. Fer. Freq. Contr., **UFFC-33**, 5, 429.

Levin, F. K., 1978, The reflection, refraction, and diffraction of waves in media with an elliptical velocity dependence: Geophysics, **43**, 528-537.

Lyamshev, L. M., and Naugol'nykh, K. A., 1981, Optical generation of sound in a liquid: Thermal mechanism (review): Soviet Physics Acous., **27**, 5.

McDonald, J., Gardner, G. H. F., and Hilterman, F. J., 1983, Seismic studies in physical modeling: IHRDC Publ.

Martin, D., Etienne, G., and Rasolofosaon, N. J. P., 1991, Cross-well seismic modeling - Laser ultrasonics experiments versus finite difference simulations: 61st Ann. Internat. Mtg., Soc. Expl. Geophys., Abstracts, 683-686.

Martin, D., Ehinger, A., and Rasolofosaon, N. J. P., 1992, Some aspects of seismic modeling and imaging in anisotropic media using laser ultrasonics: Presented at the 62nd Ann. Internat. Mtg., Soc. Expl. Geophys.

Montchalin, J. P., 1986, Optical detection of ultrasound, IEEE Trans. Ultras. Fer. Freq. Contr.,. **UFFC-33**, 5, 485.

Montchalin, J. P., Aussel, J. D., Heon, R., Jen, C. K., Boudreault, A., and Bernier, R., 1989, Measurement of in-plane and out-of-plane ultrasonic displacements by optical heterodyne interferometry: J. Nondestructive Ev., Special issue on Optical methods, Ed. R. B. Thompson, Plenum Publ. Corp.

Nestvold, E. O., 1992, 3-D seismic: Is the promise fulfilled?. The Leading Edge, **11**, no. 6, 12-9.

Pouet, B., 1991, Physical modeling using laser ultrasonics, PhD thesis University of Paris 7 (in french).

Pouet, B., and Rasolofosaon, N. J. P., 1989, Ultrasonic intrinsic attenuation measurement using laser techniques: Proc. IEEE Ultrasonics Symposium, Montreal, Canada, 545-549.

Pouet, B., and Rasolofosaon, N. J. P., 1990, Seismic physical modeling using laser ultrasonics: 60th Ann. Internat. Mtg., Soc. Expl. Geophys., Expanded Abstracts, 841-844.

Pyrak-Nolte, L. J., Cook, N. G. W., and Myer, L. R., 1987, Seismic visibility of fractures: 28th US Symposium on Rock Mechanics, Tucson, 47-56.

Rasolofosaon, N. J. P., 1988, Influence of permeability on the attenuation of bulk and interface waves-Experiment versus theory: 58th Ann. Internat. Mtg., Soc. Expl. Geophys., Expanded Abstracts, 925-927.

Rasolofosaon, N. J. P., and Zinszner, B. E., 1989, Reflections on reflectors: 59th Ann. Internat. Mtg., Soc. Expl. Geophys., Expanded Abstracts, 560-564.

Richou, B., and Richou, J., 1992, Etude du couplage d'un laser Nd:YAG pulsé et d'une fibre optique: Opto 92, 12e journées professionnelles, Paris, France, CSI Publications, 512-514.

Royer, D., and Dieulesaint, E., 1989, Optical detection of sub-angstrom transient mechanical displacement: Proc. IEEE Ultrasonics Symposium.

Thompson, R. B., and Thompson, D. O., 1985, Ultrasonics in nondestructive evaluation: Proc. of the IEEE, **73**, 12, 1716-1755.

Viktorov, I. A., 1967, Rayleigh and Lamb waves: physical theory and applications: Plenum Press.

Virieux, J., 1986, P-SV wave propagation in heterogeneous media: Velocity-stress finite difference method: Geophysics, **51**, 4.

White, J. E., 1965, Seismic waves: radiation, transmission and attenuation: McGraw-Hill.

# Annotated Bibliography

A reprint book is constrained by publication costs to include only a few of the papers that are worthy of critical reading. This defect can be partially redressed by providing the interested reader with a guide to the larger literature. Even so, this reference list cannot pretend to be exhaustive. Absent from this list are the many fine theses, dissertations, internal company reports, and university consortium reports which we felt would be too difficult for the average reader to obtain.


Anderson, D. V., Northwood, T. D., and Barnes, C., 1952, Reflection of a pulse by a spherical surface: J. Acoust. Soc. Am., **24,** 276-283.
*Semiquantitative comparison of Kirchhoff theory predictions to experimental data, simulating a whole-earth internal reflection. Compare to Gregson (1967).*

Ass'ad, J. M., Kusky, T. M., McDonald, J. A., and Tatham, R. K., 1992, Implications of scale model seismology in the detection of natural fractures and microcracks: J. Seis. Expl., **1,** 61-76.
*Compares shear-wave splitting for the two different cases of an inclusion softer than the background medium (i.e., a microcrack), and an inclusion harder than the background medium (i.e., a filled microcrack).*

Ass'ad, J. M., Tatham, R. H., McDonald, J. A., Kusky, T. M., and Jech, J., 1993, A physical model study of scattering of waves by aligned cracks: Comparison between experiment and theory: Geophys. Prosp., **41,** 323-341.
*Hudson theory is shown to agree with the shear-wave splitting results for crack densities less than 7 percent. The range of agreement for the* P-*waves is smaller, from 0 to 3 percent crack density.*

Behrens, J. C., and Waniek, L., 1972, Modellseismik: Zeits. Geophys., **38,** 1-44.
*A comprehensive overview (as of 1972). In German.*

Bennett, A. D., 1962, Study of multiple reflections using a 1-D seismic model: Geophysics, **27,** 61-72.
*One-dimensional (rod) physical model simulating acoustic impedance contrasts with abrupt changes in rod diameter. Compare to Woods (1956).*

Berckhemer, H., and Waniek, L., 1970, Bibliography on Seismic Modelling, Parts 1, 2, and 3: Prague: European Seismological Commission.
*This lengthy (50 pages total) bibliography is a great guide to the more obscure or exotic physical modeling papers published before 1970. Arranged chronologically, this bibliography includes many works not in English. The only drawback to the use of the guide is its relative unavailability.*

Berckhemer, H., and Ansorge, J., 1963, Wavefront investigations in model seismology: Geophys. Prosp., **11,** 459-470.
*Two-dimensional physical model study of* P-*wave amplitudes and wave shapes primarily investigating the direct waves, the head waves, and the source radiation pattern.*

Berryman, L. H., Goupillaud, P. L., and Waters, K. H., 1958, Reflections from multiple transition layers, Part II, experimental investigation: Geophysics, **23,** 244-252.
*One-dimensional (rod) physical modeling to study effects of continuous velocity gradients. Velocity gradients were obtained by heating nylon, which has elastic constants that are strongly temperature dependent.*

Bullitt, J. T., and Toksöz, M. N., 1985, Three-dimensional ultrasonic modeling of Rayleigh wave propagation: Bull. Seis. Soc. Am., **75,** 1087-1104.
*Wedge mounts are used to maximize the Rayleigh wave response of the piezoelectric transducers. For a step discontinuity, at some incidence angles as much as 90 percent of the Rayleigh wave is converted to body waves. A simple approximate relation to describe the relative effect of incidence angle on Rayleigh wave transmission is given.*

Busby, J., and Richardson, E. G., 1957, The absorption of sound in sediments: Geophysics, **22,** 821-828.
*Compares attenuation in packs of natural sand to attenuation in packs of glass spheres.*

Carabelli, E., and Folicaldi, R., 1957, Seismic model experiment on thin layers, Geophys. Prosp., **5,** 317-327.
*Two-dimensional physical model study (including both* P- *and* S- *waves) investigating this layer reflection response.*

Cheadle, S. P., Brown, R. J., and Lawton, D. C., 1991, Orthorhombic anisotropy — A physical modeling study: Geophysics, **56,** 1603-1613.
*Experimental determination of the nine independent elastic constants for an orthorhombic medium. Also includes an excellent tutorial on the Kelvin-Christoffel equations.*

Chen, H.-w., and McMechan, G. A., 1993, 3-D physical modeling and pseudospectral simulation of common-source data volumes: Geophysics, **58,** 121-133.
*Physical model data are used as a test of a numerical forward-modeling algorithm. Includes a good discussion of why discrepancies between numerical and physical model data sets occur. Compare to Holt et al. (1990).*

Chen, S. T., 1982, The full acoustic wavetrain in a

laboratory model of a borehole: Geophysics, **47,** 1512-1520.
*Uses a 40:1 scale factor for borehole physical modeling. Shows the general nature of the full wavetrain and suggests the use of Stoneley-wave interpretation to infer shear-wave velocities.*

Chen, S. T., 1988, Shear-wave logging with dipole sources: Geophysics, **53,** 659-667.
*Uses 25:1 and 40:1 scale factors for borehole physical modeling. Shows that a dipole tool can allow shear-wave logging in formations whose shear-wave velocity is less than that of water.*

Chen, S. T., 1989, Shear-wave logging with quadrupole sources: Geophysics, **54,** 590-597.
*Uses a 6:1 scale factor for borehole physical modeling. Shows that a quadrupole tool can also allow shear-wave logging in soft formations, and at higher frequencies than a dipole tool.*

Chowdbury, D. K., and Dehlinger, P., 1963, Elastic wave propagation along layers in two-dimensional models: Bull. Seis. Soc. Am., **53,** 593-618.
*Interesting 2-D model study of waveguide and layered medium effects, especially long-wavelength effects. Compare to Melia and Carlson (1984, also reprinted in this volume).*

Coroneau, J., 1965, Etude du "Point Brilliant" sur modeles sismiques: Geophys. Prosp., **13,** 405-432.
*Models made of materials similar to Lavergne (1961) are used to investigate the interference zone between PP and PPP waves resulting in a "bright spot." A critical result is the difference between amplitudes calculated by plane-wave theory and amplitudes from spherical-wave theory. In French.*

Dampney, C. N. G., Mohanty, B. B., and West, G. F., 1972, A calibrated model seismic system: Geophysics, **37,** 445-455.
*Calibration was confirmed by comparing system output against Lamb's problem (i.e., the amplitude and waveform of the vertical component of a Rayleigh wave produced by an impulsive vertical force).*

Daniel, I. M., and Marino, R. L., 1971, Wave propagation in a layered model due to point source loading in a high-impedance medium: Geophysics, **36,** 517-532.
*Photoelastic technique captures snapshots of elastic waves caused by detonation of pentaerythritol tetranitrate (PETN). Compare to Riley and Dally (1966).*

Datta, S., and Bhowmick, A. N., 1969, Headwaves in two-dimensional seismic models: Geophys. Prosp., **17,** 419-432.
*Interference of the direct wave and the headwave is a function of dip angle and the sign of the dip, as well as the velocity contrast. Headwave amplitude decay is more regular in the down-dip profile than in the up-dip profile.*

de Bremaecker, J. C., 1958, Transmission and reflection of Rayleigh waves at corners: Geophysics, **23,** 253-266.
*Two-dimensional (plate) model using a polystyrene sheet. Detailed work with a number of interesting observations.*

Dix, C. H., 1958, The numerical computation of Cagniard's integrals: Geophysics, **23,** 198-222.
*Outlines a method for using an analog electrical computer to calculate exact synthetic seismograms. Interesting example of how the integration in the complex plane is accomplished by an actual physical experiment.*

Donato, R. J., 1960, Experimental investigation on the properties of Stoneley wave: Geophys. J., **3,** 441-443.
*Shows that the amplitude of a Stoneley (or rather, Scholte) wave decays exponentially away from the interface of a liquid over a solid.*

Donato, R. J., 1963, The S headwave at a liquid-solid boundary: Geophys. J., **8,** 17-25.
*Demonstrates that the amplitudes of the S headwave are in agreement with predictions from analytic theory. The theory is concisely but clearly developed.*

Donato, R. J., 1965, Measurements on the arrival refracted from a thin high-speed layer: Geophys. Prosp., **13,** 387-404.
*Shows that for refractor thicknesses up to 1/2 of a wavelength, the attenuation per unit distance decreases monotonically with the increase of the refractor thickness.*

Dubendorff, B., and Menke, W., 1986, Time-domain apparent-attenuation operators for compressional- and shear-waves: Experiment versus single-scattering theory: J. Geophys. Res., **91,** 14023-14032.
*Scattering from parallel cylindrical voids in a homogeneous block suggested that Kramers-Kronig theory underestimates the amount of scattering-related attenuation.*

Dubendorff, B., and Menke, W., 1990, Physical modeling observations of wave velocity and apparent attenuation in isotropic and anisotropic two-phase media: PAGEOPH, **132,** 363-400.
*Microcrack plate modeling. The attenuation of elastic waves along rays parallel to aligned microcracks is less than for the rays perpendicular to the microcracks. Comparison of experimental azimuthally dependent velocities to theories of Hudson, O'Connell and Budiansky, and Backus.*

Ebrom, D. A., Tatham, R. H., Sekharan, K. K., McDonald, D. A., and Gardner, G. H. F., 1990, Hyperbolic traveltime analysis of first arrivals in an azimuthally anisotropic medium: A physical modeling study: Geophysics, **55,** 185-191.
*Model study of a fractured rock analog (plexiglas sheets separated by films of water). Compares birefringence measured along a single ray to anisotropy deduced from moveout velocities. Compare to Sondergeld and Rai (1992) or Tatham et al. (1992).*

Evans, J. F., Hadley, C. F., Eisler, J. D., and Silverman, D., 1954, A three-dimensional seismic wave model with both electrical and visual observation of waves: Geophysics, **19,** 220-236.
*Piezoelectric transducers are the "electrical" observation tool, and Schlieren photography the "visual" observation tool in this early paper. Measurements of symmetric and antisymmetric modes excited in a plate were qualitatively compared to analytic solutions.*

French, W. S., 1975, Computer migration of oblique seismic reflection profiles: Geophysics, **40,** 961-980.
*Illustrates the limitations of 2-D seismic data when examining 3-D subsurface structures.*

Fuchs, K., 1965, Investigation on the wave propagation in wedge-shaped media: Z. Geophys., **31,** 51-89.
*Body-wave propagation in a wedge compared against theoretical calculations based on the method of images.*

Gangi, A. F., 1967, Experimental determination of *P*-wave/Rayleigh wave conversion coefficient at a stress-free wedge: J. Geophys. Res., **72,** 5685-5692.
*The experimental data are compared to theory and agree to within 10 percent. A minimum in the conversion coefficient occurs when a* P*-wave grazes a wedge. Conversely, the maximum conversion occurs when the* P*-wave encounters a wedge at normal incidence.*

Gangi, A. F., and Thomson, K. C., 1968, Wide-band transducer for seismic modeling: J. Geophys. Res., **73,** 4735-4739.

*Alternative design for a physical modeling source/receiver involves a six-step log-periodic design. The authors demonstrate improvements in bandwidth over more common piezoelectric transducers.*

Gangi, A. F., and Wesson, R. L., 1978, *P*-wave to Rayleigh-wave conversion coefficients for wedge corners: Model experiments: J. Comp. Phys., **29**, 370-388.
*Measurements are done for several wedge angles (10 degrees, 30 degrees, 60 degrees, 90 degrees, and 120 degrees). The 10-degree wedge data are compared to the analytic solution for a half-space.*

Gardner, G. H. F., French, W. S., and Matzuk, T., 1974, Elements of migration and velocity analysis: Geophysics, **39**, 811-825.
*Migration can serve as a form of velocity analysis: over-migration indicating too high a velocity field and under-migration indicating too low a velocity field. Physical model data illustrate these points.*

Gilbert, F., and Laster, S. J., 1962a, Excitation and propagation of pulses on an interface: Bull. Seis. Soc. Am., **52**, 299-319.
*The existence of an interface wave with prograde particle motion and a velocity intermediate between the* P *and* S *pulses is shown. Compare to Roever, Vining, and Strick (1959, also reprinted in this volume).*

Gilbert, F., Laster, S. J., Backus, M. M., and Schell, R., 1962b, Observations of pulses on an interface: Bull. Seis. Soc. Am., **52**, 847-868.
*Even when the conditions for existence of the Stoneley wave are not satisfied, a radiating pulse may still be generated. Compare to Gilbert and Laster (1962a).*

Goodman, R. E., and Appuhn, R. A., 1966, Model experiments on the earthquake response of soil-filled basins: Bull. Geol. Soc. Am., **77**, 1315-1326.
*These experiments involve strong motions and are thus more applicable to earthquake studies (or the very near field of a seismic source). The authors show the interesting result that a small amount of added water softens soil and hence increases surface motion of ground roll, but a large amount of added water diminishes the surface motion.*

Gregson, V., 1967, A model study of elastic waves in a layered sphere: Bull. Seis. Soc. Am., **57**, 959-981.
*An unusual 3-D model that simulates the seismic response of a layered spherical earth to a shallow earthquake. Compares experimentally derived dispersion curves to theory.*

Guha, S., 1965, Model seismic investigations on refracted waves: Geophys. Prosp., **13**, 659-664.
*Plate-model investigation relating inaccuracies in refraction time measurements to inferences of the velocity structure.*

Gupta, I. N., and Kisslinger, C., 1964, Model study of explosion-generated Rayleigh waves in a half-space: Bull. Seis. Soc. Am., **54**, 475-484.
*A shot on the free surface acts as a vertical impulse, while a buried shot acts as a compressional source. A change of polarity of the radiated field occurs when the source is dropped from the surface to the interior of the medium.*

Gupta, I. N., and Kisslinger, C., 1966, Radiation of body waves from near-surface explosive sources: Geophysics, **31**, 1057-1065.
*A near-surface source may be modeled as the sum of a vertical impulse plus a horizontal dipole without moment. The dipole component increases with source depth.*

Gupta, S. K., 1978, Reflections and refractions from curved interfaces — Model study: Geophys. Prosp., **26**, 82-96.
*Curved interfaces have the interesting property that PS- and SP-mode conversions are separated in traveltime. Compare to Ebrom et al. (1992, reprinted in this volume).*

Gutdeutsch, R., and Koenig, M., 1966, Component registration in two-dimensional model seismology: Studia Geophys. Geod., **10**, 314-22.
*Discusses theoretical and practical aspects of determining particle motion in plate models.*

Gutdeutsch, R., 1969, On Rayleigh waves in a wedge with free boundaries: Bull. Seis. Soc. Am., **59**, 1645-52.
*In 2-D model experiments, Rayleigh waves incident on a narrow (5-degree) wedge decompose into two waves which correspond to the symmetric and anti-symmetric modes of the plate.*

Hall, S. H., 1956, Scale model seismic experiments: Geophys. Prosp., **4**, 348-364.
*Early work oriented mostly towards identifying artifacts in water tank data collection.*

Harper, D. R., 1965, Observed reflection and diffraction wavelet complexes in two-dimensional seismic model studies of simple faults: Geophysics, **30**, 72-86.
*Interesting results from 2-D modeling. Compare to Trorey's article on numerical modeling of diffractions in* Geophysics, **35**, 762-784.

Helbig, K., 1958, Elastiche Wellen in anisotropen Medien, Teil 2: Gerlands Beitr. Geophys., **67**, 256-288.
*Excellent Schlieren-type photographs captured wave motion in a fluid surrounding an anisotropic half-cylinder. The figures in this paper can be appreciated even by those lacking a working knowledge of German.*

Henzi, A. N., and Dally, J. W., 1971, A photoelastic study of stress wave propagation in a quarter-plane: Geophysics, **36**, 296-310.
*As in Riley and Dally (1966), snapshots are made of progressing wavefronts. The observed reflection and transmission coefficients for a Rayleigh wave encountering a right angle corner are compared to the previous experimental work of deBremaecker (1958), Knopoff and Gangi (1960), and Pilant et al. (1964).*

Holt, R. M., Mittet, R., and Tuset, E. D., 1990, Laboratory verification of a 3-D fast finite-difference algorithm for seismic wave propagation: presented at EAEG/SEG Research Workshop on the Estimation and Practical Use of Seismic Velocities.
*Comparison of finite-difference elastic wave modeling to a physical model experiment. Compare to Chen and McMechan (1993).*

Hospers, J., and Rathore, J. S., 1991, Processing and interpretation of sideswipe and other external reflections from salt plugs: Geoexploration, **27**, 257-295.
*"Foreswipe" and "backswipe" (salt flank reflections) are events that are routinely discriminated against in processing, but might be useful in delineating the edge of the salt body. Includes both physical model and actual field data.*

Hoover, G. M., 1972, Acoustical holography using digital processing: Geophysics, **37**, 1-19.
*Plate model data are used to test an alternative imaging method (acoustical holography).*

Howell, B. F., and Baybrook, T. G., 1967, Scale-model study of refraction along an irregular interface: Bull. Seis. Soc. Am., **57**, 443-446.
*Short, but interesting, paper showing that refractions are attenuated if the refractor is rough on a scale*

*length similar to the wavelength of the refracted pulse.*

Hsu, C.-j., and Schoenberg, M., 1993, Elastic waves through a simulated fractured medium: Geophysics, **58,** 964-977.
*This is the full article associated with the expanded abstract of Hsu and Schoenberg in Chapter 3 of this reprint volume.*

Ivakin, B. N., 1960, Methods for controlling the density and elasticity of medium during the two-dimensional modeling of seismic waves: Bull. Acad. Sci. USSR, Geophys. Ser. (English transl.), No. 8, 761-771.
*Density and elastic constants in 2-D models can be reduced in a controlled manner by perforating the plate with a regular pattern. Wavelengths must be 8 to 10 times greater than the perforation spacing in order for the method to be successful.*

Ivakin, B. N., and Vasilev, Y. V., 1963, The wave properties of perforated plates for seismic modeling: Bull. Acad. Sci. USSR, Geophys. Ser. (English transl.), No. 2, 149-156.
*A triangular network of perforations in a plate model minimizes velocity anisotropy relative to a square network of perforations. Compare to Ivakin (1960).*

Jordan, N. F., 1966, Attenuation and dispersion of shear waves in Plexiglas: Geophysics, **31,** 622-624.
*A clear, concise experiment that experimentally demonstrates that wavelet matching for propagation through an attenuating medium requires not only a frequency-dependent reduction in amplitude but also a frequency-dependent phase shift (dispersion).*

Karrenbach, M., 1990, Three-dimensional time-slice migration: Geophysics, **55,** 10-19.
*Karrenbach used a physical model data set to test his one-pass 3-D time migration.*

Kasahara, K., 1953, Experimental studies on the mechanism of generation of elastic waves III: Bull. Earthquake Res. Institute: **31,** 235-243.
*A 3-D physical model excited by a vibrator to investigate Rayleigh wave generation. Interesting hodograms cruder than, but comparable to, Sorge's (1965) 2-D modeling.*

Kim, W. H., and Kisslinger, C., 1967, Model investigation of explosions in prestressed media: Geophysics, **32,** 633-651.
*Explosive sources placed on 2-D plates under tension showed anisotropy effects in the radiation patterns and the shot-hole cracking directions.*

Kim, J. Y., and Behrens, J., 1986, Experimental evidence of S*-wave: Geophys. Prosp., **34,** 100-108.
*Experimental confirmation of the non-geometrical S*-wave, which is generated at the surface (above a buried P-source) as an ordinary shear-wave.*

Klimentos, T., and McCann, C., 1988, Why is the Biot slow compressional wave not observed in real rocks?: Geophysics, **53,** 1605-1609.
*Compare to Plona's paper (1980) reprinted in this volume. The authors show experimentally that the addition of clay particles to porous rocks eliminates the Biot slow wave by restricting fluid movements through pore throats.*

Knopoff, L., 1955, Small three-dimensional seismic models: Trans. Am. Geophys. Union, **36,** 1029-1034.
*Early work emphasizing experimental technique.*

Knopoff, L., 1958, Surface motions of a thick plate: J. Appl. Phys., **29,** 661-70.
*Elastic physical modeling using Solenhofen limestone. Intriguing capture of many different conversion events. Includes a comparison of observed amplitudes to theory.*

Knopoff, L., and Gangi, A. F., 1959, Seismic reciprocity: Geophysics, **24,** 681-691.
*Compare to White (1960) in* Geophysics, *also reprinted in this volume.*

Knopoff, L., and Gangi, A. F., 1960, Transmission and reflection of Rayleigh waves by wedges. Geophysics, **25,** 1203-1214.
*The wavelet shape of Rayleigh waves reflected from a wedge can be interpreted as the sum of the incident wave and a line-source re-radiator placed at the vertex of the wedge.*

Knopoff, L., Gilbert, F., and Pilant, W. L., 1960, Wave propagation in a medium with a single layer: J. Geophys. Res., **65,** 265-278.
*Two-dimensional modeling of Rayleigh-wave dispersion and other attributes of P-SV motion in a layered medium.*

Koefoed, O., Van Ewyk, J. G., and Bakker, W. T., 1958, Seismic model experiments concerning reflected refractions: Geophys. Prosp., **6,** 382-393.
*Demonstrates that refractions may be reflected from oblique-cutting faults if the unit in which the refracted wave propagates is thin compared to the dominant wavelength.*

Kotcher, J. S., Gardner, G. H. F., and McDonald, J. A., 1984, Modeling the effect of static errors in areal seismic data caused by glacial erosion over carbonate reefs: Geophysics, **49,** 1-16.
*Compare to its companion paper, McDonald et al. (1981, reprinted in this volume). Color time slices demonstrate the ability of uncorrected static errors to undermine an interpretation.*

Krollpfeiffer, D., Dresen, L., Hsieh, C. H., and Chern, C. C., 1988, Detection and resolution of thin layers — A model seismic study: Geophys. Prosp., **36,** 244-264.
*Plate-model study of thinning layers. Compare to Tatalovic (1988).*

Kuo, J. T., and Thompson, G. A., 1963, Model studies on the effect of a sloping interface on Rayleigh waves: J. Geophys. Res., **68,** 6187-6197.
*Experiment primarily relevant to whole-earth geophysicists. The geometry is equivalent to a thinning crustal section. An intuitively appealing result of the experiment is that for a weakly sloping wedge (circa 2.5 degrees), the local phase velocity of the Rayleigh wave is about the same as for a flat-layered earth with the same (local) depth as the sloping model.*

Kuster, G. T., and Toksöz, M. N., 1974, Velocity and attenuation of seismic waves in two-phase media — Part II Experimental Results: Geophysics, **39,** 607-618.
*Experimental velocity data for suspensions are compared against six theories of velocity prediction. Attenuation data are also presented, and the relative contributions of several different mechanisms are evaluated.*

Lavergne, M., 1961, Etude sur modele ultrasonique de probleme des couches minces en sismique refraction: Geophys. Prosp., **9,** 60-73.
*Models made of Plexiglas, Duraluminum, and Laiton are used to simulate problems of refracted waves propagating along thin layers. Results show that velocity and depth determinations may be erroneous in the presence of such thin layers, and attenuation with offset is large. In French.*

Lavergne, M., 1966, Refraction le long des bancs minces rapides et effet d'ecran pour les marqueurs profonds: Geophys. Prosp., **14,** 504-527.
*Models made of the same materials as in Lavergne's 1961 paper were used to study the variation of amplitude with offset for refracted waves. For layers with thicknesses greater than a sixth of a wavelength,*

*refractions along high velocity basement were observed at significant distances. In French.*

Leggett, M., Goulty, N. R., and Kragh, J. E., 1993, Study of traveltime and amplitude time-lapse tomography using physical model data: Geophys. Prosp., **41,** 599-619.
*Suggests the use of seismic absorption as an additional tool for tomographically imaging the progress of fluids injected into reservoir rocks.*

Leslie, H. D., and Randall, C. J., 1990, Eccentric dipole sources in fluid-filled boreholes — Numerical and experimental results: J. Acoust. Soc. Am., **87,** 2405-2421.
*The variation of flexural-wave slowness for a formation is shown to be a weak function (circa 5 percent maximum difference) of tool eccentricity in the borehole. Waveshapes and amplitudes are compared to theory.*

Leslie, H. D., and Randall, C. J., 1992, Multipole sources in boreholes penetrating anisotropic formation — Numerical and experimental results: J. Acoust. Soc. Am., **91,** 12-27.
*These 20:1 scale factor borehole experiments show that, in anisotropic formations, there are two distinct Stoneley wave polarizations. A comparison with finite-difference modeling is made.*

Levin, F. K., and Hibbard, H. C., 1955, Three-dimensional seismic model studies: Geophysics, **20,** 19-32.
*Levin and Hibbard performed both reflection and transmission experiments using a two-layer system of cement and marble. Compare with results in Clay and McNeil, published in the same year.*

Lewis, D., and Dally, J. W., 1970, Photoelastic analysis of Rayleigh wave propagation in wedges: J. Geophys. Res., **75,** 3387-98.
*Quantitative measurements of transmission and reflection coefficients of Rayleigh waves as function of wedge angle. Includes comparison to theory and other experimenters' results.*

Lo, T.-w., Duckworth, G. L., and Toksöz, M. N., 1990, Minimum cross-entropy seismic diffraction tomography: J. Acoust. Soc. Am., **87,** 748-756.
*Experimental test for a tomographic inversion scheme to partially alleviate the angle-of-investigation limitations inherent in crosshole experiments.*

Mal, A. K., and Knopoff, L., 1966, Transmission of Rayleigh waves at a corner: Bull. Seis. Soc. Am., **56,** 455-66.
*Although primarily concerned with analytic methods for Rayleigh wave prediction, there is also a comparison of theory to experimental data.*

Matsunami, K., 1990, Laboratory measurements of spatial fluctuation and attenuation of elastic waves due to random heterogeneities: PAGEOPH, **132,** 197-220.
*Plate-model study involving waves propagating through a perforated aluminum sheet (the perforations are the heterogeneities of the article's title). The observed scattering attenuation has a maximum for "ka" in the neighborhood of 3 to 5 ("k" being wavenumber, and "a" being the correlation distance for the heterogeneities). In other words, attenuation is at a maximum when wavelength is about equal to the heterogeneity spacing. Compare to Levin and Robinson (1969, reprinted in this volume).*

McDonald, J. A., Gardner, G. H. F., and Hilterman, F. J., 1983, Seismic studies in physical modeling: Internat. Human Res. Dev. Corp.
*Presents several case histories of physical modeling applied to specific exploration problems. Introductory chapter gives excellent summary of the specific modeling techniques used at the Allied Geophysical Laboratories of the University of Houston.*

Menke, W., and Dubendorff, B., 1989, Seismology: Physical Model Studies *in* D. E. James, Ed., Encyclopedia of Geophysics, Van Nostrand and Reinhold, 1202-1211.
*Brief overview article. Especially good discussion of the factors influencing repeatable amplitude measurements.*

Nagy, P. B., 1993, Slow wave propagation in air-filled permeable solids: J. Acoust. Soc. Am., **93,** 3224-3234.
*Detailed study of the Biot slow* P*-wave. Compare to Plona (1980, reprinted in this volume) or Klimentos and McCann (1988). Nagy finds that Biot theory well predicts the phase velocity at all frequencies tested, but the attenuation at high frequencies is underpredicted by theory.*

Nakamura, Y., 1964, Model experiments on refraction arrivals from a linear transition layer: Bull. Seis. Soc. Am., **54,** 1-8.
*Studies the effects on headwave amplitudes (as a function of frequency) when a linear velocity transition layer is placed on top of the refractor. Results are interpreted so as to shed light on Moho refraction behavior.*

Nakamura, Y., 1966, Multi-reflected head waves in a single-layered medium: Geophysics, **31,** 927-939.
*Two-dimensional plate modeling demonstrates the complexities of refracted reflections, reflected refractions, and other multiple-interaction events.*

Northwood, T. D., and Anderson, D. V., 1953, Model seismology: Bull. Seis. Soc. Am., **43,** 239-245.
*Early work verifying that the Rayleigh wave excited by a surface source is much higher in amplitude than the associated body waves.*

O'Brien, P. N. S., 1963, A note on the reflection of seismic pulses with application to second-event refraction shooting: Geophys. Prosp., **11,** 59-72.
*Experimental demonstration that plane-wave theory does not well predict amplitude measurements made in the near-field. Specifically, there is a diminution of the measured reflection coefficient at critical angle compared to the plane-wave prediction.*

Pant, D. R., Greenhalgh, S. A., and Watson, S., 1988, Seismic reflection scale model facility: Expl. Geophys., **19,** 499-512.
*Probably the best introduction available to 2-D modeling as it is currently done.*

Pant, D. R., and Greenhalgh, S. A., 1989a, Multicomponent seismic reflection profiling over an orebody structure — A scale model investigation: Geophys. Res. Lett.: **16,** 1089-1092.
*Ore bodies, although generally too small to provide specular reflections, may be detectable by interpreting* P- *and* S- *diffracted arrivals.*

Pant, D. R., and Greenhalgh, S. A., 1989b, Blocking surface waves made by a cut — Physical seismic model results: Geophys. Prosp., **37,** 589-606.
*Compare to Tsuboi (1928) reprinted in this volume.*

Pant, D. R., and Greenhalgh, S. A., 1989c, A multicomponent offset VSP scale model investigation: Geoexploration, **26,** 191-212.
*Two-dimensional model with 2-component receiver for* P-SV *experiment.* P- and S- *wavefields (up-going and down-going) are separated in processing on the basis of intercept time, slowness, and polarization. Compare to Balch et al. (1991) reprinted in this volume.*

Pant, D. R., and Greenhalgh, S. A., 1989d, Lateral resolution in seismic reflection — A physical model study: J. Geophys., **97,** 187-198.

*Several different attributes of narrow, terminating reflectors are investigated, including detectability, horizontal extent, and dip.*

Pant, D. R., Greenhalgh, S. A., and Zhou, B., 1992, Physical and numerical model study of diffraction effects on seismic profiles over simple structures: Geophys. J. Int., **108,** 906-916.
*Resolution of a rectangular mound (e.g., a reef, or other laterally terminating structure) is investigated for mound heights ranging from 1/8 of a seismic wavelength up to 2 seismic wavelengths.*

Pilant, W. L., Knopoff, L., and Schwab, F., 1964, Transmission and reflection of surface waves at a corner, 3: Rayleigh waves (experimental): J. Geophys. Res., **69,** 291-297.
*Reflection and transmission coefficients of Rayleigh waves are measured for wedge angles from 30 degrees to 180 degrees.*

Poley, J., and Nooteboom, J. J., 1966, Seismic refraction and screening by thin high-velocity layers (a scale-model study): Geophys. Prosp., **14,** 184-203.
*Shows that derived refraction velocities are misleading when a thin, high-velocity layer is present above the inferred refractor. The effect becomes pronounced when the thin, high-velocity layer has a thickness in excess of 0.1 wavelength.*

Pouet, B. F., and Rasolofosaon, N. J. P., 1993, Measurement of broad-band intrinsic ultrasonic attenuation and dispersion in solids with laser techniques: J. Acoust. Soc. Am., **93,** 1286-1292.
*Using lasers both as source and receiver, the usual problems of narrow bandwidth in a physical model experiment are avoided. (The authors claim nearly 6 octaves of bandwidth. Piezoelectric devices are doing well when they achieve 2 octaves.) The wide bandwidth allows an experimental test of the Kramers-Kronig relations. Compare to Wuenschel (1965) or Martin et al. (1993, this volume).*

Pratt, R. G., and Worthington, M. H., 1988, The application of diffraction tomography to crosshole seismic data: Geophysics, **53,** 1284-1294.
*Physical modeling test of diffraction tomography using a thinning wedge velocity anomaly. Suggests that the Rytov approximation is better for determining gross velocity structure, but that the Born approximation is better for edge detection (i.e., imaging). Compare to East et al. (1988, also reprinted in this volume).*

Pratt, R. G., and Goulty, N. R., 1991, Combining wave-equation imaging with traveltime tomography to form high-resolution images from crosshole data: Geophysics, **56,** 208-224.
*A method is proposed whereby a velocity field determined from tomographic analysis is then used to back-propagate crosshole seismic data using a finite-difference wave equation implementation. The method is tested against physical model data in which the sedimentary layers have some added complications (thickening, faults, etc.).*

Pratt, R. G., Quan, L., Dyer, B. C., Goulty, N. R., and Worthington, M. H., 1991, Algorithms for EOR imaging using crosshole seismic data: An experiment with scale model data: Geoexploration, **28,** 193-220.
*Very interesting paper that argues in favor of tomographic inversions that include the full waveform (as opposed to just first-arrival traveltimes). For the test case (a model of a steam flood), the paper shows that differencing the two data sets collected before and after the flood also gives good results.*

Press, F., Oliver, J., and Ewing, M., 1954, Seismic model study of refractions from a layer of finite thickness: Geophysics, **19,** 388-401.
*Shows that the quality of velocity models derived from refraction picking depends upon the ratio of refractor thickness to seismic wavelength. Also gives examples of refraction-event shingling and layer masking.*

Press, F., 1957, A seismic model study of the phase-velocity method of exploration: Geophysics, **22,** 275-285.
*Suggests the use of phase velocity effects to infer the presence of thickness changes, lithology changes, and faults.*

Purnell, G. W., 1986, Observations of wave velocity and attenuation in two-phase media: Geophysics, **51,** 2193-2199.
*Tests several formalisms for predicting the* P*-wave velocity in a sediment suspension. Also shows observed attenuations, which are compared to expected losses as a result of only Rayleigh scattering. Compare to Kuster and Toksöz (1974).*

Pyrak-Nolte, L. J., Myer, L. R., and Cook, N. G. W., 1990, Anisotropy in seismic velocities and amplitudes from multiple parallel fractures: J. Geophys. Res., **95,** 11345-11358.
*Simulated fractured system consisting of steel plates which are placed under biaxial stress.* P*-wave and shear-wave transmission studies are analyzed in terms of Schoenberg's linear slip model. Compare to Hsu and Schoenberg (1990) reprinted in this volume.*

Rieber, F., 1937, Complex reflection patterns and their geological sources: Geophysics, **2,** 132-160.
*Companion paper to Rieber (1936, reprinted in this volume). Schlieren-type photographs are used to analyze reflections from both curved beds and faulted beds. The accompanying text serves us as a humbling reminder of an era when geophysicists directly interpreted structures from unprocessed shot records. Ironically, with increases in computer power we may see abandonment of CDP sections in the near future, and a return to the interpretation of (highly processed) shot records.*

Riley, W. F., and Dally, J. W., 1966, A photoelastic analysis of stress-wave propagation in a layered model: Geophysics, **31,** 881-899.
*Using an explosive (lead azide) as the source, snapshots are photographed of the wavefront in flight. Some of the events are interpreted in light of Cagniard theory, but the paper's thrust is observational.*

Riznichenko, Yu. V., 1957, The development of ultrasonic methods in seismology: Bull. Acad. Sci. USSR, Geophys. Ser. (English transl.), no. 11, 31-37.
*Review of physical modeling in the Soviet Union. Many references are cited, the majority of which are in Russian.*

Riznichenko, Yu. V., and Shamina, O. G., 1957, Elastic waves in a laminated solid medium, as investigated on two-dimensional models, Bull. Acad. Sci. USSR, Geophys. Ser. (English Transl.), no. 7, 17-37.
*Experimental study of wave propagation (primarily refractions) in a thin high-velocity layer.*

Riznichenko, Yu. V., and Shamina, O. G., 1959, Elastic waves in layers of finite thickness, Bull. Acad. Sci. USSR, Geophys. Ser. (English Transl.), 231-243.
*Thin refracting layers (relative to a wavelength) produce head waves much weaker than thick refractors. When the refractor was less than 1/10 of a wavelength thick, no refraction could be detected. Compare to Donato (1965).*

Riznichenko, Yu. V., Shamina, O. G., and Khanutina, R. V., 1961, Elastic waves with generalized velocity in two-dimensional bimorphic models, Bull. Acad. Sci. USSR, Geophys. Ser. (English Transl.), 321-334.
*Deals primarily with the technique of fused two-*

*layer (bimorphic) plate modeling. Shows that a succesful (i.e., non-dispersive) bimorphic model must have a thickness less than 1/10 of the wavelength of interest. Also shows that attenuation in a thin bimorphic plate is comparable to attenuation in a thin single-layer plate.*

Riznichenko, Yu. V., and Shamina, O. G., 1963, Modeling of longitudinal waves in the Earth's upper mantle, Bull. Acad. Sci. USSR, Geophys. Ser. (English Transl.), 134-148.
*Bimorphic 2-D models investigating the responses of different possible velocity/depth profiles of the crust and upper mantle.*

Rosenbaum, J. H., 1965, Refraction arrivals through thin high-velocity layers: Geophysics, **30,** 204-12.
*Develops a theory for refractions in thin high-velocity elastic layers embedded in either fluids or solids. A test is performed by observing the decay of amplitudes of refractions against horizontal distance for a lucite plate immersed in water.*

Rutty, M. J., and Greenhalgh, S. A., 1993, The correlation of seismic events on multicomponent data in the presence of coherent noise: Geophys. J. Int., **113,** 343-358.
*Plate-model data are used to confirm an algorithm for separating interfering coherent events.*

Rykunov, L. N., Khorosheva, V. V., and Sedov, V. V., 1960, A two-dimensional model of a seismic waveguide with soft boundaries: Bull. Acad. Sci. USSR, Geophys. Ser. (English Transl.), 1069-1071.
*A velocity gradient is induced by differential heating. Compare to Berryman et al. (1958), in which 1-D modeling of velocity gradients is studied.*

Sarrafian, G. P., 1956, A marine seismic model: Geophysics, **21,** 320-336.
*Early study of wave propagation in a shallow water layer. Considers multiple reflections, tilted water bottoms, and thin high-impedance layers.*

Schmidt, O., 1939, Uber Kopfwellen in der Seismik: Z. Geophys., **15,** 141-8.
*Schlieren-type photographs of wavefronts in a tank of liquid in which a refracting layer has been placed. Compare to Rieber (1936, 1937) or Helbig (1958). In German.*

Schwab, F., 1967, Accuracy in model seismology: Geophysics, **32,** 819-826.
*Suggests the use of multiple source positions and simultaneous recording at multiple receiver positions (with invocation of an appropriate symmetry condition) so as to correct amplitudes for inconsistent coupling.*

Schwab, F., and Burridge, R., 1968, The interface problem in model seismology: Geophysics, **33,** 473-480.
*The difficulty in obtaining a true welded contact interface is examined experimentally for 2-D physical models. A numerical model is developed to describe the observed results. Compare to Toksöz and Schwab (1964).*

Sherwood, J. W. C., 1962, The Seismoline, an analog computer of theoretical seismograms: Geophysics, **27,** 19-34.
*One-dimensional (lumped parameter electrical transmission line) analog computer using abrupt circuit impedance changes to model bed interfaces (abrupt acoustic impedance changes). Principal virtues lay in its flexibility and accuracy.*

Siskind, D. E., and Howell, B. F., 1967, Scale-model study of refraction arrivals in a three-layered structure: Bull. Seis. Soc. Am., **57,** 437-442.
*Study of wave propagation in a discretely layered medium of increasing velocity with depth. Results compared to expectations from true velocity gradient. Compare to Nakamura (1964).*

Slack, R. D., Ebrom, D. A., Tatham, R. H., and McDonald, J. A., 1993, Thin layers and shear-wave splitting: Geophysics, **58,** 1468-1480.
*A thin anisotropic layer must produce shear-wave splitting in excess of 1/8 of the dominant period of the wavelet to be clearly interpreted on hodograms.*

Sobotova, C., and Vanek, J., 1966, Velocity of longitudinal waves in multicomponent gels: Studia Geophys. Geod., **10,** 281-90.
*One difficulty in 3-D physical modeling is the simulation of a velocity gradient. One approach is detailed in this paper: the use of gels with variable composition, and hence variable velocity.*

Sondergeld, C. H., and Rai, C. S., 1992, Laboratory observations of shear-wave propagation in anisotropic media: The Leading Edge, **11,** no. 2, 38-43.
*Includes amplitude measurements of the different shear-wave polarizations in a simulated fractured medium. Compare to Ebrom et al. (1990) or Tatham et al. (1992).*

Tang, X. M., Toksöz, M. N., and Cheng, C. H., 1990, Elastic wave radiation and diffraction of a piston source: J. Acoust. Soc. Am., **87,** 1894-1902.
*Compares transducer measurements against elastic theory: a necessary exercise for multicomponent physical modeling.*

Tang, X. M., Cheng, C. H., and Toksöz, M. N., 1991, Stoneley-wave propagation in a fluid-filled borehole with a vertical fracture: Geophysics, **56,** 447-460.
*Compares experimentally derived dispersion and attenuation curves for borehole Stoneley waves (in the presence of a fracture) with numerically derived dispersion and attenuation curves.*

Tatalovic, R., McDonald, J. A., and Gardner, G. H. F., 1988, Modeling study of tuning effects for the calculation of the thickness and areal extent of thin beds: First Break, **6,** 395-404.
*Gives a nice example of how horizontal resolution is improved after application of migration. Also shows the factors limiting interpretation of thickness from amplitude measurements. Compare to Krollpfeiffer et al. (1988).*

Tatham, R. H., Matthews, M. D., Sekharan, K. K., Wade, C. J., and Liro, L. M., 1992, A physical model study of shear-wave splitting and fracture intensity: Geophysics, **57,** 647-652.
*Demonstrates a quasi-linear relationship between fracture density and shear-wave birefringence in a simulated fractured medium. Compare to Ebrom et al. (1990) and Sondergeld and Rai (1992).*

Teng, T. L., and Wu, F. T., 1968, A two-dimensional ultrasonic model study of compressional and shear-wave diffraction patterns produced by a circular cavity: Bull. Seis. Soc. Am., **58,** 171-178.
*Plate-model diffraction experiment for both* P- *and* S-*waves, emphasizing the analysis of the shadow zone boundary. The* P-*wave data is compared to a numerical model with good agreement.*

Toksöz, M. N., and Anderson, D. L., 1962, Generalized two-dimensional model seismology with application to anisotropic earth models: J. Geophys. Res., **68,** 1121-1130.
*An unusual experimental apparatus involving the simulation of anisotropy in a plate model. The relative effect of the anisotropy on the surface waves was greater than on the body waves.*

Toksöz, M. N., and Schwab, F., 1964, Bonding of layers in two-dimensional seismic modeling: Geophysics, **29,** 405-413.

*Interesting paper that shows that reflection coefficients at bonded interfaces in 2-D models are larger than would be expected of true welded contacts. Suggests that the cause of the large reflection coefficients is the presence of small zones of imperfect bonding.*

Upper Mantle Committee, 1966, Symposium on seismic models: Studia Geophys. Geod., **10,** 239-400.
*Many interesting papers, mostly oriented towards deep earth soundings.*

Uren, N. F., Gardner, G. H. F., and McDonald, J. A., 1990, The migrator's equation for anisotropic media: Geophysics, **55,** 1429-1434.
*A relation equivalent to the well-known migrator's equation is generalized to anisotropic media. As a test, data are used which were collected through phenolite, an anisotropic material.*

Vernik, L., and Nur, A., 1992, Ultrasonic velocity and anisotropy of hydrocarbon source rocks: Geophysics, **57,** 727-735.
*Interesting application of physical modeling to a problem on the border between exploration seismics and rock physics. Shows that source rocks (shales) laminated with kerogen demonstrate elevated anisotropy compared to shales lacking kerogen.*

White, J. E., 1965, Seismic waves-radiation, transmission, and attenuation: McGraw-Hill.
*Chapter 6 gives an introduction to physical modeling, with historical context and examples of data. Excellent set of references at the end of the chapter.*

White, R. M., 1958, Elastic-wave scattering at a cylindrical discontinuity in a solid: J. Acoust. Soc. Am., **30,** 771-785.
*Elegant experimental work quantitatively relating the observed and theoretical responses of gas-filled and liquid-filled boreholes in an elastic medium.*

Woods, J. P., 1956, The composition of reflections: Geophysics, **21,** 261-276.
*One-dimensional (pipe) acoustic physical modeling illustrating the convolutional nature of seismic reflections.*

Woods, J. P., 1975, A seismic model using sound waves in air: Geophysics, **40,** 593-607.
*Three-dimensional acoustic physical modeling using air rather than a water tank. Amplitude measurements and reflection character are discussed in light of diffraction and resolution issues. Compare to Hilterman (1970).*

Wuenschel, P. C., 1965, Dispersive body waves — An experimental study: Geophysics, **30,** 539-551.
*Provides experimental evidence for the linkage between attenuation and dispersion for waves propagating in linear elastic media. Compare to Pouet and Rasolofosaon (1993).*

Zlatev, P., Poeter, E., and Higgins, J., 1988, Physical modeling of the full acoustic waveform in a fractured fluid-filled borehole: Geophysics, **53,** 1219-1224.
*Uses a 12:1 scale factor for borehole physical modeling. The aperture of a fracture intersecting a borehole, as well as the angle with which the fracture intersects the borehole, affect the amplitude of a Stoneley wave propagating along the borehole.*